T0297500

Advances in Intelligent Systems and Computing

Volume 404

About this Series

The series "Advances in Intelligent Systems and Computing" contains publications on theory, applications, and design methods of Intelligent Systems and Intelligent Computing. Virtually all disciplines such as engineering, natural sciences, computer and information science, ICT, economics, business, e-commerce, environment, healthcare, life science are covered. The list of topics spans all the areas of modern intelligent systems and computing.

The publications within "Advances in Intelligent Systems and Computing" are primarily textbooks and proceedings of important conferences, symposia and congresses. They cover significant recent developments in the field, both of a foundational and applicable character. An important characteristic feature of the series is the short publication time and world-wide distribution. This permits a rapid and broad dissemination of research results.

More information about this series at http://www.springer.com/series/11156

Swagatam Das · Tandra Pal
Samarjit Kar · Suresh Chandra Satapathy
Jyotsna Kumar Mandal
Editors

Proceedings of the 4th International Conference on Frontiers in Intelligent Computing: Theory and Applications (FICTA) 2015

Editors
Swagatam Das
Indian Statistical Institute
Kolkata, West Bengal
India

Tandra Pal
National Institute of Technology
Durgapur, West Bengal
India

Samarjit Kar
National Institute of Technology
Durgapur, West Bengal
India

Suresh Chandra Satapathy
Anil Neerukonda Institute of Technology and Sciences
Visakhapatnam, Andhra Pradesh
India

Jyotsna Kumar Mandal
Kalyani University
Kalyani, West Bengal
India

ISSN 2194-5357 ISSN 2194-5365 (electronic)
Advances in Intelligent Systems and Computing
ISBN 978-81-322-2693-2 ISBN 978-81-322-2695-6 (eBook)
DOI 10.1007/978-81-322-2695-6

Library of Congress Control Number: 2015953238

Springer New Delhi Heidelberg New York Dordrecht London

Printed on acid-free paper

Springer (India) Pvt. Ltd. is part of Springer Science+Business Media (www.springer.com)

Preface

The FICTA (*Frontiers in Intelligent Computing: Theory and Applications*) is a series of conferences started in the year 2012. Since then it has become an international forum for exchanging ideas between researchers involved in intelligent computing, including both theory and applications. The fourth event in the series, *Fourth International Conference on Frontiers in Intelligent Computing: Theory and Applications (FICTA 2015)* was held in Durgapur, India, during November 16–18, 2015. The conference was organized by National Institute of Technology (NIT), Durgapur and was technically co-sponsored by the Kolkata Chapter of IEEE Computational Intelligence Society.

FICTA 2015 received an overwhelming submissions covering different areas related to intelligent computing and its applications. With the help of our program committee and reviewers, these submissions went through an extensive peer-review process. Finally, we accepted only 31 % of the submissions. This volume comprises 61 of the accepted papers, which were presented in the conference. These papers provided a comprehensive overview of the current research and direction for the future works in intelligent computing. There were a total of 17 sessions, including a special session, in the event to cover these papers.

In addition to the contributed papers, the conference featured three keynotes delivered by three eminent speakers: Prof. Hisao Ishibuchi, Graduate School of Engineering, Osaka Prefecture University; Prof. Yew-Soon Ong, Center for Computational Intelligence, Nanyang Technological University, Singapore, and Prof. Sanghamitra Bandyopadhyay, Indian Statistical Institute, Kolkata, India. It also featured invited talks by three renowned speakers: Prof. Dominik Slezak, Institute of Mathematics, University of Warsaw, Poland; Prof. Pawan Lingras, Professor and Director, Saint Mary's University, Halifax, Canada, and Dr. Debanik Roy, Board of Research in Nuclear Sciences (BRNS), Department of Atomic Energy (DAE), Government of India.

No such event can be successfully organized without active help and support from many. First and foremost, we would like to thank all participants for their contributions to the conference. We would like to thank all keynote speakers, invited speakers, session chairs, and authors for their excellent support to make

FICTA 2015 a grand success. We are indebted to the program committee members and the reviewers for providing timely and quality reviews. We take this opportunity to thank National Institute of Technology, Durgapur for providing financial support and other resources to organize the conference. We are also thankful to the Department of Science and Technology (DST), Govt. of India for providing financial support. Finally, we would like to thank all the volunteers for their tireless efforts for a long time in arranging every detail to run the conference smoothly.

November 2015

Swagatam Das
Tandra Pal
Samarjit Kar
Suresh Chandra Satapathy
Jyotsna Kumar Mandal

Proceedings

Patron:

Tarkeshwar Kumar

Hon. General Co-Chairs:

Jyotsna Kumar Mandal
Ponnuthurai Nagaratnam Suganthan
Suresh Chandra Satapathy

General Chair:

Tandra Pal

Steering Committee:

Bijay Ketan Panigrahi
Bhabendra Narayan Biswal
Siba Kumar Udgata

Program Chair:

Swagatam Das

Convener:

Samarjit Kar

Publicity Chair:

Amit Konar

Finance Co-Chairs:

Sajal Mukhopadhaya
Uday Mukherjee

Web Chair:

Subhamoy Changder

Student Activity Chair:

Goutam Sanyal

Advisory Committee:

Andrzej Skowron (Poland)
Ashok Deshpande (India)
Baoding Liu (China)
Basabi Chakraborty (Japan)
Dipti Prasad Mukherjee (India)
Kalyanmay Deb (USA)
Manoranjan Maiti (India)
Nikhil R Pal (India)
Nirmal Kumar Roy (India)
P N Suganthan (Singapore)
Partha Pratim Gupta (India)
Sanghamitra Bandyopadhya (India)
Swapan Bhattacharya (India)
Xiang Li (China)
Ying Tan (China)

Program Committee:

Amar Kishor, Indian Statistical Institute, Kolkata, India
Andrzej Skowron, Warsaw University, Poland
Ashok Deshpande, Pune University, India
Baoding Liu, Tsinghua University, Beijing, China
Bhabatosh Chanda, Indian Statistical Institute, Kolkata, India
Bibhash Mitra, Indian Institute of Technology Kharagpur, India

V. Suma, Dayananda Sagar Institutions, Bangalore, India
Vikrant Bhateja, Shri Ramswaroop Memorial Group of Professional Colleges, Lucknow, India
Xiaowei Chen, Nankai University, China

Organizing Committee:

Brig. A.S. Nijjar, National Institute of Technology Durgapur, India
Parthapratim Gupta, National Institute of Technology Durgapur, India
Nirmal Kumar Roy, National Institute of Technology Durgapur, India
Jyotiprakash Sarkar, National Institute of Technology Durgapur, India
Dipak Kumar Mondal, National Institute of Technology Durgapur, India
Suchismita Roy, National Institute of Technology Durgapur, India
Bibhash Sen, National Institute of Technology Durgapur, India
Partha Sarathee Bhowmik, National Institute of Technology Durgapur, India
Sujit Kumar Mondal, National Institute of Technology Durgapur, India
Kaustuv Nag, Jadavpur University, Kolkata, India
Prasenjit Dey, National Institute of Technology Durgapur, India
Siba Prasada Tripathy, National Institute of Technology Durgapur, India
Sujit Das, Dr. B.C. Roy Engineering College, Durgapur, India
Krishnendu Adhikary, National Institute of Technology Durgapur, India
Aloke Chattopadhyay, National Institute of Technology Durgapur, India
Ajit Kumar Bhagat, National Institute of Technology Durgapur, India
Vikash Kumar Singh, National Institute of Technology Durgapur, India
Dhirendra Mishra, National Institute of Technology Durgapur, India
Amit Bal, Central Mechanical Engineering Research Institute, India

Additional Reviewers of FICTA 2015

1. Abhijit Sharma
2. Amartya Sen
3. Amit Mandal
4. Amrita Ghosal
5. Anirban Mukhopadhyay
6. Anirban Sarkar
7. Arkajyoti Saha
8. Arnab Ghosh
9. Arundhati Tarafdar
10. Ashok Pradhan
11. Aurpan Majumder
12. Basant Subba
13. Chiranjoy Chattopadhyay
14. Debajyoti Sinha

15. Dinabandhu Bhandari
16. Dipankar Kundu
17. Dipti Prasad Mukherjee
18. Durbadal Mandal
19. Garisha Chowdhary
20. Himanshu Mittal
21. Indrajit Bhattacharya
22. Indrani Ray
23. Jija Dasgupta
24. Jyotsna Kumar Mandal
25. Kaustuv Nag
26. Kaushik Sarkar
27. Kavita Sharma
28. Mamata Dalui
29. Moumita Samanta
30. Mrinal Kanti Sarkar
31. Paramartha Dutta
32. Partha Pratim Mohanta
33. Prantik Chatterjee
34. Prasenjit Dey
35. Prashant Singh Rana
36. Rajani K. Mudi
37. Raju Pal
38. Rudra Mazumdar
39. Rupan Panja
40. Said Broumi
41. Sanchayan Santra
42. Sanghita Banerjee
43. Sanghita Bhattacharjee
44. Sanghita Bhattacharya
45. Sankha Mullick
46. Saptaditya Maiti
47. Sarbani Palit
48. Shimpi Singh Jadon
49. Shounak Datta
50. Somnath Mukhopadhyay
51. Soumitra Samanta
52. Soumya Sen
53. Subhasis Dasgupta
54. Subhendu Barat
55. Subhrabrata Choudhury
56. Subir Halder
57. Sudhindu Bikash Mandal
58. Sudipta Mondal
59. Sujit Das

60. Suman Sau
61. Swapna Agarwal
62. Swarup Chattopadhyay
63. Tamal Deb
64. Tapas Bhadra
65. Tapas Kumar Saha
66. Tirtharaj Dash
67. Ujjwal Bhattacharya

Contents

About the Editors

Swagatam Das received the B.E. Tel. E., M.E. Tel. E (Control Engineering specialization) and Ph.D. degrees, all from Jadavpur University, India, in 2003, 2005, and 2009, respectively. Currently, he is serving as an Assistant Professor at the Electronics and Communication Sciences Unit of Indian Statistical Institute, Kolkata. His research interests include evolutionary computing, pattern recognition, multiagent systems, and wireless communication. Dr. Das has published one research monograph, one edited volume, and more than 150 research articles in peer-reviewed journals and international conferences. He is the founding co-editor-in-chief of "Swarm and Evolutionary Computation", an international journal from Elsevier. He serves as an associate editor of the IEEE Trans. on Systems, Man, and Cybernetics: Systems and Information Sciences (Elsevier). He is an editorial board member of Progress in Artificial Intelligence (Springer), Mathematical Problems in Engineering, International Journal of Artificial Intelligence and Soft Computing, and International Journal of Adaptive and Autonomous Communication Systems. He is the recipient of the 2012 Young Engineer Award from the Indian National Academy of Engineering (INAE).

Tandra Pal received the B.Sc. degree (honors) in Physics, the B.Tech. degree in Computer Science from Calcutta University, Kolkata, India, the M.E. degree in Computer Science, and the Ph.D. degree in Engineering from Jadavpur University, Kolkata, India. Currently, she is an Associate Professor in Computer Science and Engineering Department of the National Institute of Technology, Durgapur, India. She is working here since 1994. She has co-authored more than 30 technical articles in international journals and conference proceedings. Her research interest includes fuzzy sets theory, fuzzy control, fuzzy decision making, artificial neural networks, evolutionary computing, and multiobjective genetic algorithms.

Samarjit Kar completed his Ph.D. in Mathematics from Vidyasagar University, West Bengal. He is presently an Associate Professor at the Department of Mathematics, National Institute of Technology, Durgapur, India. With over 15 years of experience in teaching, Dr. Kar is also a visiting professor at the Department of Mathematical Sciences, Tsinghua University, China. He has visited several universities and institutes in India and abroad. He has co-authored more than 120 technical articles in international journals, contributed volumes, and conference proceedings. He has published two textbooks and edited five contributed books with Springer, American Institute of Physics, Narosa, and U.N. Dhur & Sons Publications. Dr. Kar has supervised 10 Ph.D. students. He is the Associate Editor of Journal of Uncertainty Analysis and Application (Springer) and is currently associated with an ongoing project, Hybrid modeling of uncertainty analysis in environmental risk assessments under BRNS, Department of Atomic Energy (DAE), Government of India. His research interests include operations research and optimization, soft computing, uncertainty theory, and financial modeling. .

Dr. Suresh Chandra Satapathy is currently working as Professor and Head, Department of CSE at Anil Neerukonda Institute of Technology and Sciences (ANITS), Andhra Pradesh, India. He obtained his Ph.D. in Computer Science and Engineering from JNTU Hyderabad and M.Tech. in CSE from NIT, Rourkela, Odisha, India. He has 26 years of teaching experience. His research interests are data mining, machine intelligence, and swarm intelligence. He has acted as program chair of many international conferences and edited six volumes of proceedings from Springer LNCS and AISC series. He is currently guiding eight scholars for PhDs. Dr. Satapathy is also a Senior Member of IEEE.

Prof. Jyotsna Kumar Mandal has completed M.Sc. in Physics from Jadavpur University in 1986 and M.Tech. in Computer Science from University of Calcutta. He has been awarded Ph.D. in Computer Science and Engineering by Jadavpur University in 2000. Presently, he is working as Professor of Computer Science and Engineering and former Dean, Faculty of Engineering, Technology and Management, Kalyani University, for two consecutive terms. He started his career as a Lecturer at NERIST, Arunachal Pradesh in 1988. He has 28 years of experience in teaching and research. His area of research includes coding theory, data and network security, remote sensing and GIS based applications, data compression, error correction, visual cryptography, steganography, security in MANET, wireless networks, and unify computing. He has supervised 13 Ph.D. students. One student has submitted (2015) the thesis and eight are ongoing. He has supervised three M. Phil and more than 30 M.Tech. projects. He is a life member of Computer Society of India since 1992, CRSI since 2009, ACM since 2012, IEEE since 2013, Fellow member of IETE since 2012, and Vice Chairman of CSI Kolkata Chapter. He has conducted special sessions and chaired sessions in various International Conferences and delivered invited lectures. He is a reviewer of various International Journals and Conferences. He has organized more than ten international conferences of Science Direct, Springer, IEEE, and edited number of proceedings as Volume Editor. He has more than 350 articles, published in various journals and conferences. He has also published five books.

Part I
Special Session: Advances in Nature Inspired Algorithms for Engineering Optimization Problems

An Efficient Dynamic Scheduling Algorithm for Soft Real-Time Tasks in Multiprocessor System Using Hybrid Quantum-Inspired Genetic Algorithm

Debanjan Konar, Kalpana Sharma, Sri Raj Pradhan and Sital Sharma

Abstract This paper proposes a hybrid approach for dynamic scheduling of soft real-time tasks in multiprocessor environment using hybrid quantum-inspired genetic algorithm (HQIGA) combined with well-known heuristic earlier-deadline-first (EDF) algorithm. HQIGA exploits the power of quantum computation which relies on the concepts and principles of quantum mechanics. The HQIGA comprises variable size chromosomes represented as qubits for exploration in the Hilbert space 0–1 using the updating operator rotation gate. Earlier-deadline-first algorithm is employed in the proposed work for finding fitness values. In order to establish the comparison with the classical genetic algorithm-based approach, this paper demonstrates the usage of various numbers of processors and tasks along with arbitrary processing time. Simulation results show that quantum-inspired genetic algorithm-based approach outperforms the classical counterpart in finding better fitness values using same number of generations.

Keywords Multiprocessor system · Soft real-time task · Dynamic task scheduling · Quantum-inspired genetic algorithm · Earlier-deadline-first · NP complete

---

D. Konar (✉) · K. Sharma · S.R. Pradhan · S. Sharma
Department of Computer Science and Engineering, Sikkim Manipal Institute of Technology, Sikkim, India
e-mail: konar.debanjan@gmail.com

K. Sharma
e-mail: kalpanaiitkgp@yahoo.com

S.R. Pradhan
e-mail: srirajpradhan@gmail.com

S. Sharma
e-mail: sitaldahal.in@gmail.com

S. Das et al. (eds.), *Proceedings of the 4th International Conference on Frontiers in Intelligent Computing: Theory and Applications (FICTA) 2015*, Advances in Intelligent Systems and Computing 404, DOI 10.1007/978-81-322-2695-6_1

1 Introduction

Real-time systems are a quantitative notion of time in which the correctness of the system relies on the time at which the result is produced rather than its logical result of computation [1]. The timing constraint of the real-time tasks are associated with the deadlines [2]. Relying on the consequence of the task missing its deadline, real-time tasks are classified into hard and soft real-time tasks. A soft real-time task associates with timing constraints, although missing some deadlines are tolerable. Scheduling plays a very important role in addressing the aspect of timing constraints for hard and soft real-time tasks.

Scheduling dynamic real-time tasks on a multiprocessor [3, 4] system is a challenging proposition for researchers in the field of computer science and engineering. Dynamic task scheduling algorithms obviate the pre-allocation of tasks to the processor and the tasks are assigned to processors as and when they arrive [1]. The main problem of dynamic scheduling of real-time tasks in a multiprocessor system is higher turnaround time [2–4].

Several research initiatives based on different task scheduling algorithms have been entrusted on this topic over the decades such as heuristic approaches [5, 6], evolutionary algorithms [7, 8], and hybrid techniques [9–11]. Various evolutionary algorithms viz. classical genetic algorithms (GA)-based approaches [12–14] are routinely applied to exploit the suboptimal solutions of the problem which is typically an NP complete problem [15]. Genetic algorithm improves the average fitness of a population by exhibiting new population. The improvement in the average fitness value is guided by the selection of fitness function, population size, crossover probability, and mutation probability.

Hou et al. [14] proposed a genetic algorithm-based technique which solves multiprocessor scheduling for a directed task graph and reports a performance improvement by 10 % when compared with the rest. Wu et al. [7] suggested a novel genetic algorithm-based approach which used an incremental fitness function to obtain an optimal solution. The classical genetic algorithm employed in most of the scheduling technique tends to generate some illegal schedules which lead to high running time. Correa et al. [6] modified genetic algorithm by introducing heuristics in the crossover and mutation, which dramatically improves the quality of the solution although time taken by this heuristic approach is higher than classical counterparts. Bonyadi et al. [8] introduced a novel bipartite genetic algorithm which is centered on optimizing maximum finish time for scheduling task in multiprocessor environment. Recently, Cheng et al. [16] presented a dynamic real-time scheduling for multiprocessor tasks using genetic algorithm with a feasible energy function which yields good quality schedules. However, the classical genetic algorithm still suffers from the lack of generalization in generating valid schedules due its crossover and mutation operators and also time-consuming genetic operations used in most of these approaches for updating the schedules fail to meet the deadline constraint of the real-time tasks.

Quantum-inspired genetic algorithm (QIGA) [17, 18] is a promising field in research of artificial intelligence community in which the combination of evolutionary computing and genetic algorithms is employed.

In this paper, a hybrid QIGA (HQIGA) is proposed for dynamic soft real-time task scheduling in multiprocessor system guided by quantum-inspired genetic algorithm and earlier-deadline-first (EDF) heuristic. The proposed HQIGA algorithm adopts qubits probabilistic representation of variable length chromosomes which have better population diversity in schedules due to the probabilistic nature of the linear superposition of all possible states. It obviates use of crossover and mutation operation for generating new population of scheduled real-time task by employing quantum rotation gate operator which drives the individual schedules toward better solution. In this proposed work, fitness value for each individual chromosome is guided by EDF algorithm.

2 Multiprocessor Real-Time Task Scheduling Problem

The principle objective of the real-time task scheduling in a multiprocessor environment is to dynamically schedule as many tasks as possible such that each task meets its execution deadline while optimizing the total delay time of all of the tasks [1]. In this proposed work, a multiprocessor soft real-time task scheduling problem with m soft real-time tasks and n identical processors is commonly characterized as follows: each of m soft real-time tasks $T_i (1 \leqslant i \leqslant m)$ is to be processed on the processors $P_j (1 \leqslant j \leqslant n)$. It is assumed that each processor can process on at most one task at a time and all tasks are assumed to be non-preemptive [3] and having equal priority.

3 Real-Time Task Model

Consider a multiprocessor real-time system comprising m soft real-time tasks $T = \{T_1, T_2, T_3, \ldots, T_m\}$ and n identical processors $P = \{P_1, P_2, P_3, \ldots, P_n\}$. Each task $T_i \in T$ is characterized by the following: A_i: arrival time of the task T_i ; R_i: ready time of the task T_i; C_i : worst-case computation time of the task T_i; D_i : deadline of the task T_i; S_i : and start time of the task T_i; F_i : finish time of the task T_i. The task T_i is said to meet its deadlines if $R_i \leqslant S_i \leqslant D_i - C_i$ and $R_i + C_i \leqslant F_i \leqslant D_i$ [19].

3.1 Objective Function

In this paper, a suitable fitness function is employed to evaluate the quality of each individual schedule which is optimized using EDF algorithm [1], and is decided by the number of real-time tasks which are meeting their deadlines.

4 Quantum-Inspired Genetic Algorithm (QIGA)

To speed up the genetic operations like selection, crossover, and mutation, QIGA exploits the power of quantum computation which is inspired by the concept and principles of quantum mechanics. In recent years, Han et al. [18, 20] suggested QIGAs which employs the *qubits* (Q bit) representation for the chromosome individual. The novelty of QIGA lies in the representation of quantum chromosome in terms of qubits, which has better characteristic of population diversity than classical counterpart. Before explaining the QIGA, few necessary basic concepts of quantum computation are briefly addressed as follows.

A quantum bit or *qubit* [21] is defined as the smallest unit of information processing in quantum computing. In quantum computer, the qubit is linear superposition of two eigenstates $|0\rangle$ and $|1\rangle$. It can be designated as

$$|\phi\rangle = \alpha|0\rangle + \beta|1\rangle \tag{1}$$

where α and β are the complex numbers and specify the probability amplitudes associated with $|0\rangle$ and $|1\rangle$ *eigenstates* subjected to normalization criteria

$$|\alpha|^2 + |\beta|^2 = 1 \tag{2}$$

$|\alpha|^2$ and $|\beta|^2$ give the probability for finding the *qubit* in the $|0\rangle$ and $|1\rangle$ state, respectively.

A rotation gate $R(\theta)$ is applied to update a qubit individual as follows:

$$\begin{bmatrix} \alpha' \\ \beta' \end{bmatrix} = \begin{bmatrix} \cos\theta & -\sin\theta \\ \sin\theta & \cos\theta \end{bmatrix} \times \begin{bmatrix} \alpha \\ \beta \end{bmatrix} \tag{3}$$

where (α', β') is the modified qubit on updating the qubit (α, β) using a rotation angle θ.

In QIGA, genes of a chromosome are modeled relying on the concept of qubits for the probabilistic representation which establishes randomness and new dimension into the proposed algorithm.

Each chromosome individual Q can be represented as a string of n qubits which is encoded as

$$Q = \left[\begin{array}{c|c|c|c|c} \alpha_1 & \alpha_2 & \alpha_3 & \ldots & \alpha_n \\ \beta_1 & \beta_2 & \beta_3 & \ldots & \beta_n \end{array}\right] \tag{4}$$

where

$$|\alpha_i|^2 + |\beta_i|^2 = 1, i = 1, 2, 3, \ldots, n. \tag{5}$$

Figure 1 shows the structure of quantum chromosome which comprises with quantum genes.

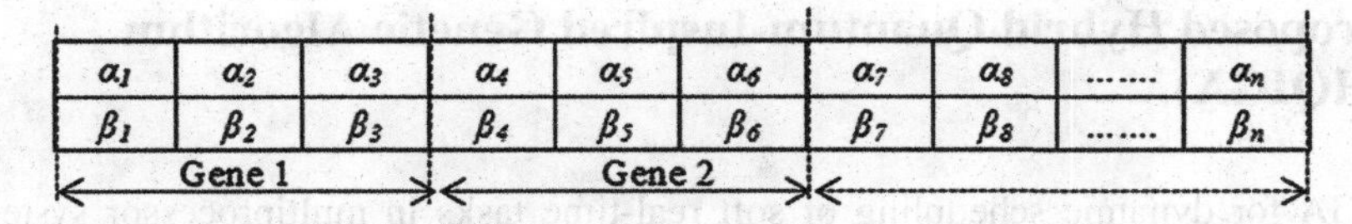

Fig. 1 Structure of quantum chromosome

The procedure of QIGA [18, 20] can be presented as follows:

Procedure QIGA

1 Begin

2 $t = 0$

3 Initially, randomly generate an initial population $P(t) = \{Q_1(t), Q_2(t), Q_3(t), Q_4(t), \ldots, Q_m(t)\}$ where Q_i(t) denotes the ith individual chromosome in the tth generation designated with qubits as

$$Q_i(t) = \left[\begin{array}{c|c|c|c|c} \alpha_1(t) & \alpha_2(t) & \alpha_3(t) & \ldots & \alpha_n(t) \\ \beta_1(t) & \beta_2(t) & \beta_3(t) & \ldots & \beta_n(t) \end{array}\right]$$

4 Convert $P(t)$ into solutions $S(t)$ using quantum measurement of $P(t)$ and evaluate the solutions obtained from $P(t)$

5 Accumulate the best solution among S(t) into the storage $B(t)$

6 If termination condition is met, then store the best solution in $B(t)$ else

7 Do

8 $t = t + 1$

9 Applying quantum gate (rotation gate), $R(\theta)$ update $P(t)$ and obtain $P(t + 1)$ as

$$P(t) = \begin{bmatrix} \cos\theta - \sin\theta \\ \sin\theta \ \cos\theta \end{bmatrix} P(t-1)$$

10 Convert $P(t + 1)$ into solutions $S(t + 1)$ using quantum measurement of $P(t + 1)$ and evaluate the solutions obtained from $P(t + 1)$

11 Accumulate the best solution among $S(t + 1)$ and $B(t)$ into the storage $B(t + 1)$ and store the best solution among $B(t + 1)$

12 Until termination condition is false

13 End

5 Proposed Hybrid Quantum-Inspired Genetic Algorithm (HQIGA)

HQIGA for dynamic scheduling of soft real-time tasks in multiprocessor system comprises basic structure of QIGA .The procedure of HQIGA is summarized as follows:

```
Procedure HQIGA
1 Begin
2 Generate task list of size N
3 While(size of the execution queue ! = N) do
4 Build the ready queue from the task list which has
arrived for execution
5 If(size of ready queue ⩾ 1) then
6 Set chromosome size ( CS) on basis of the ready queue
7 Generate random chromosome for population of size M
8 Assign processor to each gene of the valid population
by quantum measurement
9 Initialize Gen, Maxgen
10 Evaluate the valid population using the proposed
fitness function
11 Sort the population in descending order of their
fitness value
12 Choose the best chromosome among the population
13 While(the best chromosome not satisfying the
scheduling criteria and i! = Maxgen) do
14 Apply quantum rotation gates which rotate qubits of
the quantum chromosome with the best chromosome and
generate new population
15 Determine the fitness value of each chromosome in
the new population
16 Sort the chromosomes within the new population in
descending order of their fitness value
17 Choose the best chromosome among the population
18 i = i + 1
19 End while
20 Schedule the task in the corresponding processor
according to the best chromosome
21 End while
22 End
```

The fitness value of the quantum chromosome cannot be evaluated directly due to its qubits representation. A quantum measurement [20] is carried out on the qubits chromosome individuals to obtain binary strings. In the proposed algorithm, the angle of rotation in quantum rotation gate is adjusted using a lookup table.

6 Experimental Results and Discussions

In this work, test problem comprises a queue of 100, 200 and 300 soft real-time tasks which are randomly generated and these tasks are assigned to 10 and 20 processors (variable chromosome size). The tasks are ordered in descending order of their deadlines and the initial fitness value is evaluated by scheduling the tasks using earlier-deadline-first (EDF) algorithm [22]. In the experiments, GA and HQIGA have been implemented in MATLAB platform taking maximum 100 generations. GA algorithm has been tested with same number of tasks and processors as in HQIGA. In this implementation, the crossover probability is set to 1.0 and mutation probability ranges from 0.1 to 0.2 with population size of 30. The fitness values obtained from both algorithms are shown in Fig. 2 with variable chromosome size.

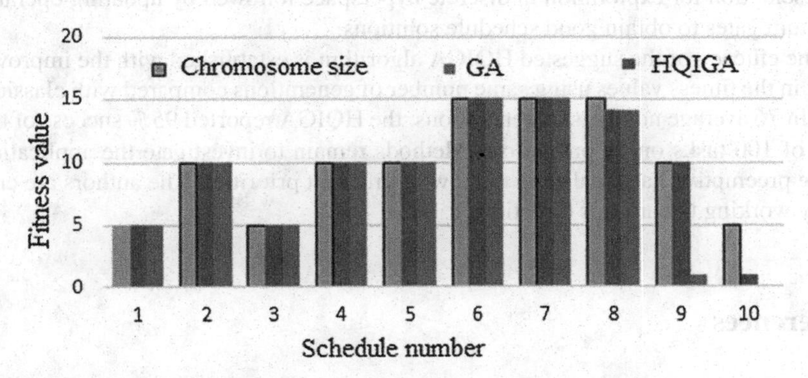

Fig. 2 Fitness value comparison of tasks in HQIGA and GA

Table 1 Percentage of success obtained for number of processors = 10

Number of tasks	HQIGA		GA	
	Percentage of success (%)	Average number of generations	Percentage of success (%)	Average number of generations
100	95	76	85	100
200	78	94	76	100
300	59	100	62	100

Table 2 Percentage of success obtained for number of processors = 20

Number of tasks	HQIGA		GA	
	Percentage of success (%)	Average number of generations	Percentage of success (%)	Average number of generations
100	100	01	100	01
200	100	25	100	36
300	100	54	99	70

The efficiency of the proposed algorithm is presented in Tables 1 and 2 by calculating the success ratio as follows:

$$Percentage\ of\ success = \frac{total\ number\ of\ scheduled\ tasks}{total\ number\ of\ tasks} * 100\,\% \qquad (6)$$

7 Conclusions

The proposed work presented a HQIGA incorporating traditional heuristic earlier-deadline-first (EDF) algorithm, suitable for dynamic scheduling of soft real-time task in multiprocessor system. The algorithm is governed by QIGA, relying on qubits representation for exploration in discrete hyperspace followed by updating operator quantum gates to obtain good schedule solutions.

The efficacy of the suggested HQIGA algorithm is established with the improvement in the fitness values using same number of generations compared with classical GA. In 76 average numbers of generations, the HQIGA reported 95 % success of the runs of 100 tasks on 10 processors. Methods remain to investigate the application of the preemptive hard real-time tasks with different priorities. The authors are currently working toward this direction.

References

1. Mall, R.: Real-Time Systems: Theory and Practice. Pearson Education, India (2007)
2. Ramamritham, K., Stankovic, J.A.: Scheduling algorithms and operating systems support for real-time systems. Proc. IEEE **82**(1), 55–67 (1994)
3. Manimaran, G., Siva Ram Murthy, C.: An efficient dynamic scheduling algorithm for multiprocessor real-time systems. IEEE Trans. Parallel Distrib. Syst. **9**(3), 312–319 (1998)
4. Srinivasan, A., Anderson, A.: Fair scheduling of dynamic task systems on multiprocessors. J. Syst. Softw. **77**, 67–80 (2005)
5. Kwok, Y.K., Ahmad, I.: Dynamic critical path scheduling: an effective technique for allocating task graphs to multiprocessors. IEEE Trans. Parallel Distrib. Syst. **7**(5), 506–521 (1996)
6. Correa, R.C., Ferreira, A., Rebreyend, P.: Schuduling multiprocessor tasks with genetic algorithms. IEEE Trans. Parallel Distrib. Syst. **10**(8), 825–837 (1999)
7. Wu, A.S., Yu, H.: An incremental genetic algorithm approach to multiprocessor scheduling. IEEE Trans. Parallel Distrib. Syst. **15**(9), 824–834 (2004). September
8. Bonyadi, M.R., Moghaddam, M.E.: A bipartite genetic algorithm for multi-processor task scheduling. Int. J. Parallel Program. **37**(5), 462–487 (2009)
9. Sivanandam, S.N., Visalakshi, P., Bhuvaneswari, A.: Multiprocessor scheduling using hybrid particle swarm optimization with dynamically varying inertia. Int. J. Comput. Sci. Appl. **4**(3), 95–106 (2007)
10. Chen, H., Cheng, A.K.: Applying ant colony optimization to the partitioned scheduling problem for heterogeneous multiprocessors. Special Issue IEEE RTAS 2005 Work-in-Progress **2**(2), 11–14 (2005)
11. Ercan, M.F.: A hybrid particle swarm optimization approach for scheduling flow-shops with multiprocessor tasks. In: Proceedings of the International Conference on Information Science and Security, pp. 13–16 (2008)

12. Dhingra, S., Gupta, S., Biswas, R.: Genetic algorithm parameters optimization for bi-criteria multiprocessor task scheduling using design of experiments. World Academy of Science, Engineering and Technology, International Journal of Computer, Control, Quantum and Information Engineering, vol. 8, no. 4 (2014)
13. Bohler, M., Moore, F. Pan, Y.: Improved multiprocessor task scheduling using genetic algorithms. In: FLAIRS Conference (1999)
14. Hou, E.S.H., Ansari, N., Hong, R.: A genetic algorithm for multiprocessor scheduling. IEEE Trans. Parallel Distrib. Syst. **10**(8), 113–120 (1999)
15. Gary, M.R., Johnson, D.S.: Computers and Imractibility: A Guide to the Theory of NP Completeness. W.H Freeman and Company, New York (1979)
16. Cheng, S.C., Huang, Y.M.: Dynamic Real-Time Scheduling for Multi-Processor Tasks using Genetic Algorithm. Comput. Softw. Appli. Conf. COMPSAC **2004**, 154–161 (2004)
17. Narayan, A., Moore, M.: Quantum-Inspired Genetic Algorithms. Proc. IEEE Evol. Comput. **1**, 61–66 (1996)
18. Han, K.H., Kim, J.H.: Genetic quantum algorithm and its application to combinatorial optimization problem. Proc. Congr. Evol. Comput. **1**, 1354–1360 (2000)
19. Eggers, E.: Dynamic Scheduling Algorithms in Real-time, Multiprocessor Systems. Term paper, EECS Department, Milwaukee School of Engineering, North Broadway, Milwaukee, WI, USA (1998–99)
20. Han, K.H., Kim, J.H.: Quantum-Inspired evolutionary algorithm for a class of combinatorial optimization. IEEE Trans. Evol. Comput. **6**(6), 580–593 (2002)
21. Mcmohan, D.: Quantum Computing Explained. Wiley, New Jersey (2008)
22. Dahal, K., Hossain, A., Varghese, B., Abraham, A. Xhafa, F., Daradoumis A.: Scheduling in multiprocessor system using genetic algorithms. In: Proceedings of the 7th IEEE Computer Information System and Industrial Mangement Applications, pp. 281–286 (2008)

Equitable Machine Learning Algorithms to Probe Over P2P Botnets

Pavani Bharathula and N. Mridula Menon

Abstract Cyber security has become very significant research area in line due to the increase in the number of malicious attacks by both state and nonstate actors. Ideally, one would like to properly secure the machines from being infected by viruses of any form. Nowadays, botnets have become an integral part of the Internet and the main drive for creating them is for financial gain. A bot conceals itself using a secret canal to communicate with its governing command-and-control server. Botnets are well-ordered from end to end using protocols such as IRC, HTTP, and P2P. Of all HTTP-based and IRC-based, P2P botnet detection became a challenging task because of its decentralized nature. The paper focuses on the techniques that are predominantly used in botnet detection and we formulate a method for detecting the P2P botnets using supervised machine learning algorithms such as random forest (RF), multilayer perceptron (MLP), and K-nearest neighbor classifier (KNN). We analyze the performance of selected algorithms there by revealing the best classification algorithm for detecting P2P botnets.

Keywords Botnet · Machine learning · Peer-to-peer (P2P)

1 Introduction

Even without our knowledge botnets have become a part and parcel in our Internet life, with an optimistic faith that we are not one amongst them right now. Traditionally, a botnet is described as a group of compromised computers that are controlled remotely using malicious software known as bots by the bot master [1, 15]. There are several reasons for the creation of bots namely financial gain, rivalry among

P. Bharathula (✉) · N. Mridula Menon
TIFAC-CORE in Cyber Security, Amrita Vishwa Vidyapeetham, Coimbatore, Tamil Nadu, India
e-mail: bharathula.pavani@gmail.com

N. Mridula Menon
e-mail: mridula.mmn@gmail.com

S. Das et al. (eds.), *Proceedings of the 4th International Conference on Frontiers in Intelligent Computing: Theory and Applications (FICTA) 2015*, Advances in Intelligent Systems and Computing 404, DOI 10.1007/978-81-322-2695-6_2

competing organizations, and many other unethical motives. Botnets are mostly used for profit: Identity theft, email spamming, software piracy [16]. The main aspect lies in the communication architecture of botnet. Most of the communication schemes among the traditional botnets include command-and-control server (C&C) [6].

Bot gets infected with a type of malware known as Trojans and reports to the server through Internet relay chat (IRC). Once infected, bot locates and connects to the server [16]. This connection session is used by the bot master for communicating and controlling the bots. These bots will get new commands from bot master to update its pattern [1]. Depending upon its structure and purpose, C&C server issues commands to perform spamming or to launch distributed denial-of-service (DDoS) attack on a particular target. Thus, the information related to botnet can be used to prevent activities like DDoS and phishing attacks.

Many existing techniques for detecting botnets are based on attack signatures [1]. Even though the signature-based system detects well-known botnets, they suffer from a major drawback that they are unable to detect the variants of the attack signatures and also show high false positive rate. The objective of this paper is to unveil the malicious activities of these botnets and to give the best classification algorithm to detect botnets.

We choose supervised machine learning algorithms because we work on labeled datasets alone. We prefer to use RF, KNN, and MLP algorithms for detecting P2P botnets due to the following advantages, i.e., robust to noisy training data, effective even if training data is large, able to extract patterns even from complicated data, and can handle any number of input variables.

Our contribution starts using the enormous collection of datasets to address the issues like how to identify the malicious contents that originated from botnets and which is the best classification algorithm for detecting P2P botnets among the three chosen algorithms. These questions are addressed in Sects. 3 and 4. Finally, we conclude the paper and outline our future enhancement in Sect. 5.

2 Related Works

Botnet detection involves analyzing capabilities of a botnet and also its network behavior. To detect botnet, there are two approaches, one is static analysis and other is dynamic analysis. Botnet has evolved from single host to distributed in network. P2P botnets were seen for the first time in the mid of 2000. Various types of botnets have been evolving every year that becomes a challenge to all the current detection mechanisms. Botnet named Zotob performed DDOS attack on online banking sites in US in 2005. Another species of Botnet named Kraken which was the worlds largest as of 2008 April infected so many fortune 500 companies and expanded to over 400,000 Bots.

The dangerous botnet Zeus seen in October 2012 was used to steal credit card information of banking customers. Whenever a new bot client is created, hacker injects malicious scripts to get the details of the computer not limited to memory, network information, and processor speed [5].

Pijush et al. [1] proposed the rule induction algorithm to detect P2P botnets which made the following assumptions: A) For every session of P2P botnets, flow of data occurs in two-way directions. B) Each botnet will have its own set of commands to interact with the bots in C&C. They generated rules based on decision tree classifier. The author measured their algorithm performance with three machine learning classification algorithms such as decision tree, linear support vector machines, and bayesian network. The comparison is done with the help of performance metrics such as accuracy, recall, f-measure, and precision.

Jignesh et al. [13] analyzed several botnet detection techniques and types of botnets. The techniques mentioned are honey pots and honey net, botnet detection using signatures, and host-based detection technique, detection using data mining technique, classification algorithms, clustering, and association rules. Their analysis revealed that data mining algorithms are convenient and effective in detecting botnets based on the characteristic features.

Fariba et al. [4] described botnet analysis system and it is implemented using naive Bayes and C4.5 algorithms for detecting HTTP-based botnet activity. They are detecting botnet behavior by aggregating the network flows and also using domain fluxing techniques.

According to David et al. [16], a theory based on traffic behavior analysis was implemented to detect botnet activity using machine learning. They classified network traffic behavior based on time intervals and did not depend on packet payload. They showed through experimentation that it is feasible to detect the unknown form of botnet activity but their method suffers from false positiveness.

Junjie et al. [15] suggested the technique for detecting the compromised machines using flow-clustering-based analysis. They proposed a method which detects stealthy P2P botnets even if illegitimate traffic is overlapped with legitimate traffic.

3 Methodology

Proposed system follows a hierarchy of three phases which have unique functionalities and are processed in a sequential order as shown in Fig. 1.

3.1 Data Acquisition

On the Internet, the information we communicate across the globe is in the form of packets which are transmitted to and fro the networks and they follow some standard protocols [3]. Network that ships data all over the place in small packets are called packet-switched networks especially in intranet. We capture these packets using Wireshark, which is a well-known network protocol analyzer.

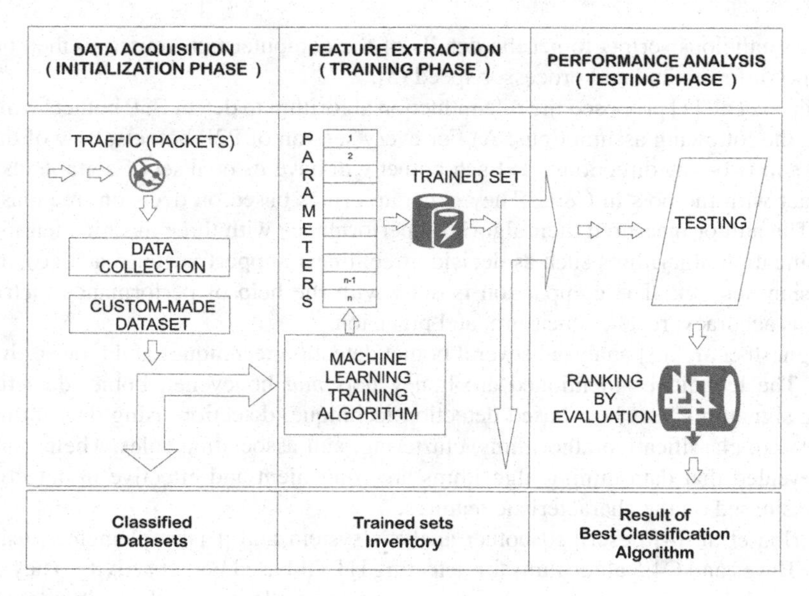

Fig. 1 Architecture

In our Intranet P2P botnet setup, we captured packets using Wireshark. In addition to this, we also collected standard datasets from malware capture facility projects and Information Security Centre of excellence UNB [9, 14]. The two common considerations while capturing packets manually are dealing with "Generic Receive Offload" (GRO) and its workmate "Large Receive Offload" (LRO). It provides a feature for the network card which accomplishes packet reassembling in advance before handled by the kernel. It may also cause issues with stream target-based rebuilding of packets by network interface. It is recommended to knock out these features while accumulating the datasets. The dataset collected contains pcap files for the botnet capture and it includes real botnet traffic. The dataset is mixed with real botnet and normal traffic. Each scenario has been captured in a pcap file. The total number of packets, the number of malicious packets, total number of features, and the size of each pcap file are shown in Table 1. This is directly fed into the next phase for the training of datasets. We train our collected data and then repeat the training process on data for tenfolds, i.e., tenfold cross validation for testing purpose.

3.2 Feature Extraction

Feature extraction is a vital pre-processing phase for analysis of packets. It is regularly disintegrated into feature building and feature assortment [8]. Feature extraction commonly known as flow monitoring, an indispensable component in

Table 1 Datasets information

Number of datasets	24
Number of features	90
Total number of packets	An avg. of 1,20,000
Number of malicious packets	72,000
Number of normal packets	48,000
Size of dataset	Each dataset is an avg. of 68 MB
Proportion of data	70 % from real sources and 30 % from simulated sources

irregularity detection, sums up network behavior from the evaluation of a series of packet stream [11]. Network feature extraction comprises organizing all the network packet symptoms required for analysis. It is mainly used to determine several characteristics of network behavior like entire traffic analysis and mediocre connection parameters [7, 12]. We extract all the essential and appropriate features from the datasets collected to process effectively with machine learning algorithms and train these datasets using supervised machine learning classification algorithms such as MLP (Multilayer Perceptron), KNN (K-Nearest Neighbor classifier), and RF (Random Forest).

We consider the following parameters: source and destination hosts IP addresses, ports, packets information like first and last packet, total packets, resets sent between two hosts, acknowledgement packets sent between two hosts, pure acknowledgement packets sent and received, unique bytes sent and received, actual data packets between two hosts, pushed data packets sent and received, syn and fin packets sent and received, window scale, urgent data packets between two hosts, segmentation size between two hosts, missed data and truncated data information between two hosts, idle time, and throughput. Of all the features, we consider optimized features using supervised attribute selection method for detecting botnets. We consider only bi-directional flow packets. If any single-directional flow is observed in packets, they are filtered out and the remaining are used for training the algorithm. We train the algorithms first by considering all the features and also after optimizing features.

3.3 Testing Phase

We feed these classified and filtered datasets to each algorithm that we have considered and analyze the performance on each algorithm with the datasets based on the parameters like accuracy, recall, precision, and time taken. A number of malicious packets , normal packets, and datasets information are already mentioned in Table 1. For testing purpose, we use tenfold cross validation and the results are tabulated in Table 3. The performance analysis reveals which one is the best classification algorithm to detect P2P botnets. In the test environment, there is a bot master who recognizes that a particular IP is bot-ridden, launches a connection to that IP address, and uses a dedicated bot control protocol to communicate with the septic computers. The majority of swarms which our approach detects are the infected computers that make prolonged or multiple short-term connections to a command-and-control server located somewhere remotely in the cyber space. The C&C server replies to these connections with a set of instructions to perform attacks. This approach is able to detect bots in network traffic devoid of deep packet examination, while still accomplishing greater detection rates with negligible exceptions.

4 Implementation

For testing, we use three supervised classification algorithms for detecting P2P botnets. First, we present the brief introduction to the algorithms chosen and then specify the analysis of results that are obtained from the algorithms.

4.1 Random Forest (RF)

According to Brieman, random forest algorithm is a large collection of decorrelated decision trees, where each tree is grown with respect to a random parameter. This randomization scheme is blended with bagging [2]. Bagging is to average the noisy and unbiased models to create a model with lower variance in terms of classification. Performing aggregation over ensemble will result in the final prediction. It uses the information of number of training sets and number of classifiers to determine the decision at each node while constructing tree. It uses the bootstrapping to estimate the error of the tree by predicting their classes. The construction of every single tree happens by randomly selecting features from the training set thereby evaluating the best split. This procedure is repeated on all the trees in the ensemble and the average voting of all these trees is considered as random forest prediction.

4.2 K-Nearest Neighbor (KNN)

KNN is a very simple and straightforward algorithm where it takes the input as a set of attributes and output depends upon whether the algorithm is used for classification purpose or regression purpose. It is classified based on the similarity measure. It is also called lazy learner as it does not require building a mode before its actual use. Simply to put, if we are using KNN classifier on set of n attributes, then it is classified based on the majority vote of k-nearest training records on all n attributes. Here, k indicates some user-specified constant and n indicates number of attributes.

4.3 Multilayer Perceptron (MLP)

MLP is a network of neurons called perceptron. It is a neural network with bunch of hidden layers between input and output layers. To train this type of network, it uses back propagation algorithm. Each node in the hidden layer is a function of nodes in the previous layer and the output node is nothing but the function of the node in the hidden layer [10]. With the help of back propagation technique, the input data is fed to the neural network repeatedly. In each iteration, it computes the error by comparing the output results with the desired one. Once the error is computed, it is back propagated to the network for adjusting weights thereby decreasing the error value in each iteration for convergence to the desired one.

We provide our analysis results with the help of three supervised machine learning algorithms namely random forest, KNN, and MLP. We use WEKA, a data mining environment to perform classification. Weka yields a set of machine learning (ML) algorithm which provides visualization tools for analyzing data and also for predictive modeling. Our results show very high true positive (TP) rate and very low false positive (FP) rate.

High true positive rate means that classifiers worked very well in detecting bot flows. Low false positive rate means that only a very small amount of normal web flows was misinterpreted as bot flows. We use the following performance metrics to compare our classification models:

$$Accuracy = TP + TN/TP + TN + FP + FN \quad (1)$$

$$Sensitivity\ or\ Recall\ or\ TP\ Rate = TP/TP + FN \quad (2)$$

$$F - measure = 2 * precision * recall/precision + recall \quad (3)$$

$$FP\ Rate = FP/FP + TN \quad (4)$$

$$Precision = TP/TP + FP \quad (5)$$

Here, sensitivity means true positive rate, that is, correctly detected bot flows. Precision is the proportion of correctly detected bot flows out of total number of flows classified as bot by our classifier. F-measure is test's accuracy measure.

Table 2 shows the effectiveness of different classification approaches. We applied k-cross validation where our observation revealed that k = 10 gives the best accuracy for most of the algorithms. We present the weighted average results obtained for three classification algorithms on our trained data with respect to TP rate, FP rate, and time taken to build the model.

Table 2 Average results obtained for three classification algorithms

Algorithm	FP rate	TP rate	Time taken (s) for classification phase
KNN	0.158	0.942	0.05
RF	0.257	0.937	3.04
MLP	0.083	0.906	7.71

Table 3 Performance results obtained for three classification algorithms

Algorithm	Accuracy	Precision	F-Measure	Recall
KNN	0.891	0.94	0.94	0.94
RF	0.756	0.937	0.93	0.957
MLP	0.911	0.915	0.906	0.906

In Table 3, we discuss the average results of the performance analysis of three classification algorithms on our datasets. Of all the three algorithms, MLP is showing promising results in predicting suspicious bot flows as per TP rate and FP rate.

5 Conclusion

In this paper, we provide the best classification algorithm among KNN, RF, and MLP for detecting suspicious P2P bots using machine learning technique. We perform our experimentation on real botnet datasets and also on our simulated datasets to classify P2P bots using network flow-based features and port-based analysis. Our observation results show that multilayer perceptron classifier gives very good results with an accuracy of 0.911 in detecting botnets. We propose certain features to be considered while detecting suspicious activity in the network.

References

1. Barthakur, P., Dahal, M., Ghose, M.K.: An efficient machine learning based classification scheme for detecting distributed command & control traffic of P2P botnets. p. 9 (2013)
2. Biau, G.: Analysis of a random forests model. JMLR. org, 1063–1095 (2012)
3. Gandotra, E., Bansal, D., Sofat, S.: Malware Analysis and Classification: A survey. Scientific Research Publishing (2014)
4. Haddadi, F., Morgan, J., et al.: Botnet behaviour analysis using ip flows: with http filters using classifiers. In: 28th International Conference on Advanced Information Networking and Applications Workshops (WAINA), pp. 7–12 (2014)
5. Li, L., Mathur, S., Coskun, B.: Gangs of the internet: towards automatic discovery of peer-to-peer communities. In: IEEE Conference on Communications and Network Security (CNS), pp. 64–72 (2013)
6. Lu, C., Brooks, R.: Botnet traffic detection using hidden markov models. In: Proceedings of the Seventh Annual Workshop on Cyber Security and Information Intelligence Research, p. 31 (2011)
7. Perényi, M., Dang, T.D., Gefferth, A., Molnár, S.: Identification and analysis of peer-to-peer traffic, pp. 36–46 (2006)
8. Rahbarinia, B., Perdisci, R., Lanzi, A., Li, K.: Peerrush: Mining for unwanted P2P traffic, pp. 194–208, Elsevier (2014)
9. Sebastian Garcia, V.U.: Malware capture facility project. http://mcfp.weebly.com/
10. Singh, K., Agrawal, S.: Comparative analysis of five machine learning algorithms for IP traffic classification. In: International Conference on Emerging Trends in Networks and Computer Communications (ETNCC), pp. 33–38 (2011)

11. Stevanovic, M., Pedersen, J.M.: Machine learning for identifying botnet network traffic (2013)
12. Strayer, W.T., Lapsely, D., Walsh, R., Livadas, C.: Botnet detection based on network behavior. Botnet Detection, pp. 1–24. Springer, New York (2008)
13. Vania, J., Meniya, A., Jethva, H.: A review on botnet and detection technique, pp. 23–29 (2013)
14. Victoria, U.: Isot research lab datasets. http://www.uvic.ca/engineering/ece/isot/datasets/
15. Zhang, J., Perdisci, R., Lee, W., Luo, X., Sarfraz, U.: Building a scalable system for stealthy P2P-botnet detection. IEEE, pp. 27–38 (2014)
16. Zhao, D., Traore, I., Sayed, B., Lu, W., Saad, S., Ghorbani, A., Garant, D.: Botnet detection based on traffic behavior analysis and flow intervals. Elsevier, pp. 2–16 (2013)

Directed Search-based PSO Algorithm and Its Application to Scheduling Independent Task in Multiprocessor Environment

Sneha Shriya, R.S. Sharma, Saurav Sumit and Sonu Choudhary

Abstract Particle swarm optimization (PSO) algorithm has proved to be a promising meta-heuristic algorithm to solve broad class of optimization problems which requires global search. Many variants of basic PSO have been proposed. To enhance the exploration capacity of basic PSO algorithm, a new technique called as *directed phase* is introduced in PSO. The proposed new phase is based on directed search optimization (DSO) which has capability of exploration and diversification which can accelerate the particles in the late iterations of PSO algorithm. Further, proposed algorithm along with PSO and DSO is implemented to solve task scheduling problem on homogeneous multiprocessor system and the results obtained are compared. Experimental results demonstrate that proposed work performs better and has the ability to be an adequate alternative to solve the optimization problem.

Keywords Multiprocessor scheduling · Particle swarm optimization · Directed search optimization

1 Introduction

Nowadays, researchers are more concerned about multiprocessor scheduling because of day-to-day increasing speed in the execution of a workload. Multiprocessor systems are emerging day-to-day because they offer enormous capacity to solve

S. Shriya (✉) · R.S. Sharma · S. Choudhary
Rajasthan Technical University, Kota, India
e-mail: sneha.shriya22@gmail.com

R.S. Sharma
e-mail: rssharma.kota@gmail.com

S. Choudhary
e-mail: sonuuce15@gmail.com

S. Sumit
National Institute of Technology, Hamirpur, India
e-mail: saurav.sumit.ss@gmail.com

S. Das et al. (eds.), *Proceedings of the 4th International Conference on Frontiers in Intelligent Computing: Theory and Applications (FICTA) 2015*, Advances in Intelligent Systems and Computing 404, DOI 10.1007/978-81-322-2695-6_3

applications which require parallel and complex computations. In order to explore the full potential of such computing systems, job scheduling decisions play a great role. Multiprocessor scheduling is an NP-hard problem [1–3]. The problem of mapping the tasks to multiple processors can be differentiated in terms of static mapping, dynamic mapping, independent task set, dependent task set, flow-shop task set, homogeneous processors, heterogeneous processors, or various qualitative parameters used for performance measure. This paper deals with off-line scheduling where schedule is computed off-line before the system begins to execute. Because the computation of the schedule is done off-line, the complexity of the algorithm used is not important [4]. An algorithm's only objective is to find good off-line schedule that can exploit full potential of the resources.

Traditional methods such as branch and bound, divide and conquer, and dynamic programming used in multiprocessor scheduling optimization tend to get stuck on local optimum point [5]. To compensate the drawback of traditional approach, a number of heuristic methods have been proposed. List scheduling approach has a problem of exponential increase in time complexity along with increase in problem size [6]. A meta-heuristic is a heuristic method for general-purpose optimization problem. Many used meta-heuristics are simulated annealing algorithms (SA) [7], genetic algorithms (GA) [5], tabu search (TS), neural networks [3], ant colony optimization (ACO) [2], and PSO [1]. In this paper, a new improved variant of PSO algorithm is proposed which can be implemented very efficiently. Then the proposed algorithm is used to solve the problem of assigning the tasks to multiprocessors.

2 Problem Definition

2.1 System Model

The system consists of a set of m processors having same memory and processing elements. A task will get the resources only from the processor allocated to it. There is a set $T(T_1, T_2, \ldots, T_n)$ of n independent and simultaneously available jobs and set $P(P_1, P_2, \ldots, P_m)$ of m identical processors. Processing time of task $T_i (i = 1, 2, \ldots, n)$ is denoted by C_i.

2.2 Assumptions

Following assumptions are made while modeling the problem:

- Processors are homogeneous and endlessly available from start-to-end time.
- At one time only one task can be handled by each processor.
- All tasks are mutually independent and each task can execute on only one processor and cannot be preempted.

- All set-up times are included in the execution time and are independent from the logical order of tasks.

2.3 Objectives

There are various qualitative parameters available in order to judge the performance of scheduling algorithms. Here, maximum span among all processors is taken to evaluate the performance of scheduling algorithm. Span for processor i named as Span_i could be defined as

$$\mathrm{Span}_{i\in\{1,2,3\ldots m\}} = \sum_{j=1}^{n} C_j.M_{ij}, \tag{1}$$

where $M_{ij} = 1$ if task T_j is assigned to processor P_i, otherwise $M_{ij} = 0$.

The makespan is stated as the maximum completion time for all processors. Maximum span can be defined as

$$\mathrm{Makespan} = \mathrm{Maximum}\left[\mathrm{Span}_{i\in\{1,2,3\ldots n\}}\right] \tag{2}$$

Besides makespan, in order to evaluate the efficiency of the obtained schedule, average utilization of the schedule is calculated. Average Utilization can be computed as

$$\text{Average Utilization} = \frac{\left(\sum_{i=1}^{m} \frac{\mathrm{Span}_i}{\mathrm{Makespan}}\right)}{m} \tag{3}$$

The objective of our problem is to map the task set T to the set of processors P while minimizing the makespan having considered the constraints.

3 Related Work

Mapping of tasks on multiprocessor and scheduling of mapped tasks on processors have been widely studied. Various researches have been done using heuristics for mapping the tasks to multiprocessor or grid computing environment. The main objective of this paper is to find a scheduling algorithm that can map tasks in a multiprocessor environment and also find a schedule with optimal makespan for a given number of processors.

Braun T presented a comparison of eleven heuristics for mapping meta-tasks without communication delays onto heterogeneous processors. Results show that GA gave the best results among the compared heuristics as in [8]. Chen and Cheng in [1] used ACO algorithm and proved that ACO gives better feasible solutions as compared to GA.

Abdelhalim in [1] applied PSO and also proposed a modified PSO that was based on the idea that as the PSO proceeds, the effect of the inertia factor decreases. Modified algorithm runs final results (particles positions) of PSO again with re-initializing inertia factor to one and the velocity v with new random values. Proposed algorithm produced better results than GA, ACO, and basic PSO.

Dexuan Zou in [9] has proposed DSO algorithm for constrained optimization problems which are used by Binodini Tripathy in [3] for work flow scheduling in grid environment and result showed that it outperformed GA and PSO in given environment.

4 Proposed Work

PSO [10] is a promising meta-heuristic for global search. Many variants of PSO have been proposed. In PSO with time-varying inertia, exploration can be done in initial cycles and exploitation is done in late cycles [1]. The main reason behind this is in the late cycles, particles are greatly effected by their personal best position and global best position. Also, in later cycles velocities of the particle become too small to approach at the nearest solution. Therefore, information flow among the particles becomes similar that can cause similar type of particles in the population, i.e., can result in loss of diversity among particles. Thus if there are many local optimum points then there is a possibility for the particles to stagnate at local optima. For resolution of the problem stated above, modification of the heuristic such that it can keep updating particle toward new best position even at the later iterations where PSO possess very small velocity is required. This paper proposes a new variant of PSO which prevents stagnation at local optimum and escaping from local optima is done by DSO [9].

The proposed algorithm has been designed with the following objectives:

- Identify good particles
- Update the position of these particles using DSO steps that can contribute to exploration and diversification in good particles.
- Then operate PSO to update velocity as well as position of particles.

The superior objective behind using DSO is that it can contribute in updating position of good particles even when the velocity becomes too low. DSO works in two regions. In forward region, it keep searching in forward direction at the same time in backward region, it avoids any premature convergence. Furthermore, DSO is able to create diversity among good particles through its mutation operator. The proposal has novelty that in each iteration DSO updates only those particles who possess good fitness probability. This approach take us out off unnecessary calculations because there is no need to update bad particles.

```
procedure DSO
    for each dimension j of particle i in iteration t do
        if rand() < p_α then
            x_v = X_i^j(t − 1) + (1 + α).(G_best^j − X_i^j(t − 1))
            if x_v is greater than x_upperbound then initialize x_v with X_upperbound
            if x_v is smaller than x_lowerbound then initialize x_v with X_lowerbound
            X_i^j(t) = X_i^j(t − 1) + rand().(x_v − X_i^j(t − 1))
        else
            x_s = X_i^j(t − 1) − β.(G_best^j − X_i^j(t − 1))
            if x_s is greater than X_upperbound then initialize x_s with X_upperbound
            if x_s is smaller than X_lowerbound then initialize x_s with X_lowerbound
            X_i^j(t) = X_i^j(t − 1) + rand().(x_s − X_i^j(t − 1))
        end if
        if rand() < p_m then
            X_i^j(t) = X_lowerbound + rand().(X_upperbound − X_lowerbound)
        end if
    end for
end procedure
```

4.1 Proposed Algorithm

Algorithm 1 DirPSO algorithm

```
Initialize parameters: Max_Iteration, ω_max, ω_min, C_1, C_2, α, β, P_α, P_m
Initialize population having solution and initial velocity randomly
while iteration t = Max_Iteration do
    Step 1: Calculate fitness value
    for each particle do
        Calculate fitness value
        Evaluate and update Personal best P_best for particle
        Update Global best G_best for particles
    end for
    Step 2: Directed phase( Update good particles )
    for each particle i do
        probability_(i) = ((0.9 * fitness_(i))/maxfitness) + 0.1
        if probability_(i) > rand() then
            Update position of particle using procedureDSO
        end if
    end for
    Step 3: PSO updation on velocity and position
    V_t^i = ω.V_{t−1}^i + c_1.r_1 (P_best(i) − X_{t−1}^i) + c_2.r_2 (G_best − X_{t−1}^i)
    X_t^i = X_{t−1}^i + V_t^i
end while
```

Here, V_t^i and X_t^i are velocity and position, respectively, of particle i at t iteration. c_1 and c_2 are the constants, and r_1 and r_2 are the random number in range [0,1]. α, β, P_α, and P_m are forward coefficient, backward coefficient, forward probability, and genetic mutation probability, respectively. ω is the inertia weight. In iteration, t value of ω is taken as

$$\omega_t = \omega_{\max} - \frac{\omega_{\max} - \omega_{\min}}{Max\_Iteration}.t \quad (4)$$

4.2 Problem Formulation

Multiprocessor scheduling is a problem having a discontinuous environment, but the PSO algorithm has been designed for continuous environments [11]. Therefore, it is a great challenge to model a scheduling problem according to PSO. Framing of task scheduling as an optimization problem is done as follows:

- Interval [0,1] is divided into m equal subintervals where m is total number of processor. Each subinterval is assigned to only one processor.
- We have a matrix $X[p, n]$ where p is the number of particles in initial population and n is the total number of tasks.
- X is initialized randomly with entries value between 0 and 1.
- For each particle, schedule is obtained according to entry value of X. Each row i of X corresponds to particle i. For particle i, the subinterval in which value X_{ij} value lie, task j is assigned to corresponding processor.

Example We have one particle of 4 dimension, i.e., 4 tasks, and have two processors. In such case, there will be two subinterval $(0, 0.5]$ and $(0.5, 1)$ which correspond to processor 1 and processor 2, respectively. After randomly initializing the particle$_i$, we get

$$X_i = [0.0123\ 0.7984\ 0.3783\ 0.4924]$$

Dimensions(tasks) having values in range $(0, 0.5]$ will be assigned to processor 1 and those whose values in $(0.5, 1]$ will be assigned to processor 2. Thus the obtained solution from corresponding particle position will be

$$\text{Solution}_i = [\text{proc}_1\ \text{proc}_2\ \text{proc}_1\ \text{proc}_1]$$

- Velocity of particles is presented by matrix $V[p, n]$ where V_{ij} is the velocity value of jth dimension of particle i.
- Personal best position of particles is presented by matrix $P_{\text{best}}[p, n]$ where each row presents the personal best position of corresponding particle.
- Global best position is presented by vector $G_{\text{best}}[n]$.
- Fitness is calculated using Eq. 2.

5 Simulation Results

To analyze the performance of the DirPSO algorithm (proposed), a multiprocessor scheduling problem is solved. For comparing the results obtained from implementing the proposed algorithm, the same multiprocessor scheduling problem is solved using PSO and DSO. For better analysis of the results, same task set, initial population, and other parameters are passed to all these algorithms. Makespan is taken as the criteria to measure the performance. Besides makespan, average utilization and standard deviation are also evaluated.

This paper focuses on mapping tasks on a multiprocessor system with three processors in order to evaluate and compare the efficiency of various methods. Multiprocessor scheduling problems with 10, 15, 20, 30, and 40 tasks where completion time for each task is taken randomly in the range [10,100] are simulated. For each result, experiment is performed 30 times and after taking average of the results obtained in every experiment it is compared. The parameters taken in simulations are same as in [3].

Experiment 1: In this experiment, algorithms have been implemented on different task sets for evaluation of makespan, average utilization, and standard deviation. Table 1 provides a report about the makespan. Figure 1 presents the results of average utilization and Fig. 2 of standard deviation. Results in Table 1, Figs. 1 and 2 reflect that all the time DirPSO outperforms in terms of optimization, efficiency, and stability in results as compared to basic PSO and DSO.

Table 1 Makespan for different task sets

No. of tasks	PSO	DSO	DirPSO
10	242.8	243.2	241
15	299.4	300	298.1
20	384.1	383.7	383
30	541.4	541.4	541
40	709.1	708.3	708

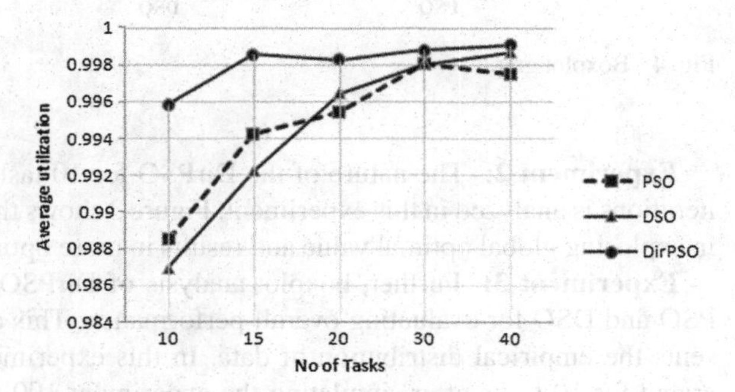

Fig. 1 Comparison of average utilization

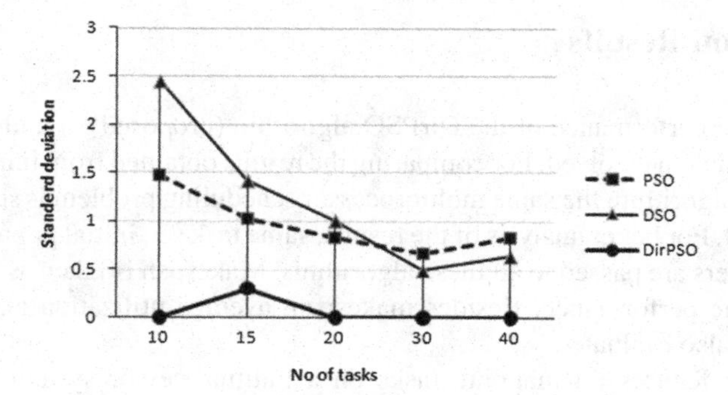

Fig. 2 Comparison of standard deviation

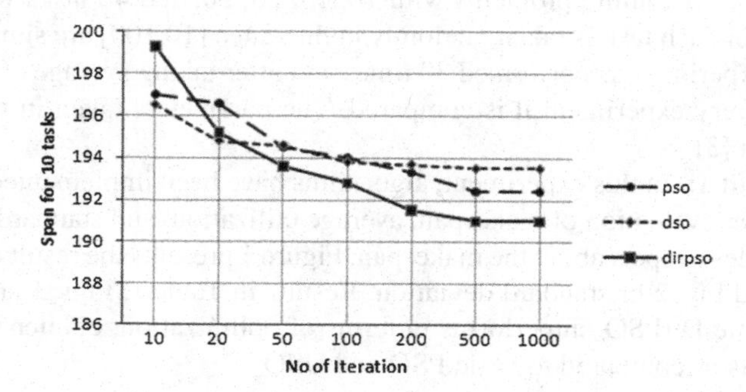

Fig. 3 Nature of algorithms at a different number of iteration

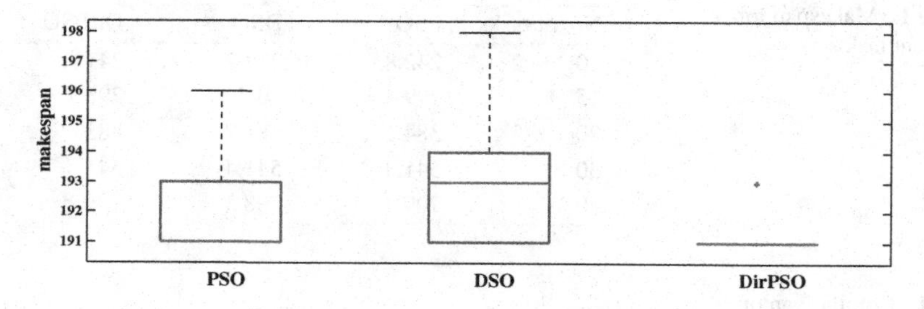

Fig. 4 Boxplot graph

Experiment 2: The nature of the DirPSO for 10 tasks at a different number of iterations is analyzed in this experiment. Figure 3 shows that DirPSO performs better in exploring global optimal value and results in more optimized value for makespan.

Experiment 3: Further, boxplot analysis of DirPSO is taken to compare with PSO and DSO for evaluating overall performance. This analysis graphically represents the empirical distribution of data. In this experiment, boxplot graph is generated for 10 tasks after simulating the experiment 100 times. Figure 4 shows that interquartile range and medians of DirPSO are comparatively low which concludes that proposed algorithm performs better to achieve its goals.

6 Conclusion

In this paper, a new DirPSO algorithm is proposed to avert stagnation at local optima. In proposed algorithm, a new phase is introduced in PSO termed as directed phase that only concerned about good particle. Directed phase has ability to direct good particles toward new promising position even when particle possess very small velocity. At the same time, it contributes to create diversity in position of good particles through its mutation operator. This property helps PSO algorithm to avoid stagnation at local optima. It is observed from simulations. DirPSO algorithm has exhibit performance superior to the PSO and DSO. Furthermore, the result obtained from proposed algorithm is more stable in nature. Proposed algorithm could be a good alternative to resolve global search optimization problems for which there presents many near optimum points. Since this paper is limited to scheduling of independent tasks on homogeneous processor, the effect of the algorithm considering task dependency and heterogeneity among processors and comparison of proposed algorithm to other variants of PSO may appear in some of future works.

References

1. Abdelhalim, M.: Task assignment for heterogeneous multiprocessors using reexcited particle swarm optimization. Int. Conf. Comput. Electr. Eng. IEEE 23–27 (2008)
2. Laalaoui, Y., Drias, H.: Aco approach with learning for preemptive scheduling of real-time tasks. Int. J. Bio-Inspired Comput. 383–394 (2010)
3. Tripathy, B., Dash, S., Padhy, S.K.: Dynamic task scheduling using a directed neural network. J. Parallel Distrib. Comput. **75**, 101–106 (2015)
4. Liu, J.W.S.: Real-time systems (2000)
5. Visalakshi, P., Sivanandam, S.N.: Dynamic task scheduling with load balancing using hybrid particle swarm optimization. Int. J. Open Probl. Comput. Math. **2**(3), 475–488 (2009)
6. Thanushkodi, K., Deeba, K.: A new improved particle swarm optimization algorithm for multiprocessor job scheduling. Int. J. Comput. Sci. Issues **8**(4), (2011)
7. Sivanandam, S.N., Visalakshi, P., Bhuvaneswari, A.: Multiprocessor scheduling using hybrid particle swarm optimization with dynamically varying inertia. IJCSA **4**(3), 95–106 (2007)
8. Braun, T., Siegel, H.J., Beck, N., Boni, L.L., Maheswaran, M., Reuther, A.I., Robertson, J., Theys, M.D., Yao, B., Hensgen, D.: A comparison of eleven static heuristics for mapping a class of independent tasks onto heterogeneous distributed computing systems. J. Parallel Distrib. Comput. **61**(6), 810–837 (2001)
9. Zou, D., Liu, H., Gao, L., Li, S.: Directed searching optimization algorithm for constrained optimization problems. Expert Syst. Appl. **38**(7), 8716–8723 (2011)
10. Kennedy, J., Mendes, M.: Population structure and particle swarm performance (2002)
11. HaghNazar, R., Rahmani, AM.: Prune pso: a new task scheduling algorithm in multiprocessors systems. Int. Conf. Networking Inf. Technol. IEEE 161–165 (2010)

Part II
Bioinformatics and Computational Biology

JUPred_MLP: Prediction of Phosphorylation Sites Using a Consensus of MLP Classifiers

Sagnik Banerjee, Debjyoti Ghosh, Subhadip Basu and Mita Nasipuri

Abstract Post-translational modification is the attachment of biochemical functional groups after translation from mRNA. Among the different post translational modifications, phosphorylation happens to be one of the most important types which is responsible for important cellular operations. In this research work, we have used multilayer perceptron (MLP) to predict protein residues which are phosphorylated. As features, we have used position-specific scoring matrices (PSSM) generated by PSI-BLAST algorithm for each protein sequence after three runs against 90 % redundancy reduced Uniprot database. For an independent set of 141 proteins, our system was able to provide the best AUC score for 36 proteins, highest for any other predictor. Our system achieved an AUC score of 0.7239 for all the protein sequences combined, which is comparable to the state-of-the art predictors.

Keywords Post-translational modification · Multilayer perceptron (MLP) · Position-specific scoring matrices (PSSM) · Shannon's entropy · Window consensus · Star consensus

S. Banerjee (✉)
Department of Electronics and Communications Engineering,
Institute of Engineering and Management, Kolkata, India
e-mail: sagnikbanerjee15@gmail.com

D. Ghosh · S. Basu · M. Nasipuri
Department of Computer Science and Engineering, Jadavpur University, Kolkata, India
e-mail: debjyoti.88@gmail.com

S. Basu
e-mail: subhadip@cse.jdvu.ac.in

M. Nasipuri
e-mail: mnasipuri@cse.jdvu.ac.in

S. Das et al. (eds.), *Proceedings of the 4th International Conference on Frontiers in Intelligent Computing: Theory and Applications (FICTA) 2015*, Advances in Intelligent Systems and Computing 404, DOI 10.1007/978-81-322-2695-6_4

1 Introduction

The biochemical modification of a protein by attachment of a functional group, after the process of translation from mRNA is called post-translational modification. A protein could undergo different types of structural or chemical alterations or modification could arise out of attachment of other biochemical functional groups such as acetate, phosphate, various lipids, and carbohydrates. Phosphorylation, one of the most important post-translational modifications, is basically the addition of a phosphate group (PO_4^{3-}) to a certain protein residue. For proper functioning of the cell phosphorylation is essential since it renders many proteins functional. High throughput experiments can lead to correct detection of sites which are phosphorylated but the process is extremely time-consuming and expensive. In order to ensure faster annotation computational approaches are often adopted. Most of such methods are comparatively faster than wet lab experiments and they also give appreciable results [1]. In this work we have proposed a window-based strategy for prediction using 23 features to represent an amino acid. It is important to conduct studies on kinases and protein substrates because it helps us to understand the biochemistry of diseases like cancer better. Serine (S), threonine (T), tyrosine (Y), and histidine (H) are the residues which undergo phosphorylation in eukaryotic proteins.

A lot of machine learning approaches have been proposed since 1999 [1]. Some tools predict phosphorylation along with the probable kinase that phosphorylates the residue. Most notable among them are AMS 3.0 [2] and a 2012 update AMS 4.0 [3]. In AMS 3.0, prediction was based on primary protein sequence and they used artificial neural networks. For features they chose ten physicochemical properties of amino acids. Most of the prediction systems extracted data from Swiss-Prot and PhosphoELM [4] dataset. In AMS 4.0, the researchers combined high-quality indices to improve their results. Apart from the auto-motif server (AMS) tools, there has been a range of several other predictors. For example SVM was used by PhosphoSVM [5] and PredPhospho [6], neural networks were used by NetPhos [7] and NetPhosK [8].

Even after almost a decade of research, perfect prediction of protein post-translational modification has not yet been achieved. The present research works attempts to provide best prediction for most of the protein sequences.

2 Materials and Methods

One of the most important aspects of machine learning is to find patterns within data. Classification is one major task in pattern recognition which categorizes a given sample to one of the provided classes. The multilayer perceptron (MLP) needs to be provided with a set of features for proper classification. In this work, we have used evolutionary information extracted from position-specific scoring matrices (PSSM) to serve as features during training and testing.

2.1 Features for MLP

Position-Specific Scoring Matrix

To represent motifs in biological sequences, using PSSMs is a widely used practice. PSSMs are used as features to encode sequences for prediction by aligning multiple sequences which is called multiple sequence alignment (MSA). Position-Specific Iterated BLAST or PSI-BLAST constructs PSSMs corresponding to protein sequences and operates in rounds. Positions that are highly conserved are rewarded with a high score and those that are weakly conserved are penalized with low scores (near zero).

For this work of the generation of PSSMs, number of iterations was limited to three. We downloaded 90 % redundancy-reduced protein sequences from UniProt-Swiss-Prot and PSSMs corresponding to these sequences were generated.

Shannon's Entropy

Shannon's entropy (SE) is used to figure out how much a sequence is conserved. It is computed from the set of 20 values that correspond to the weighted observed percentages in the PSSM. Hence, this quantitative feature, computed from the weighted observed percentages (WOP), is popularly used as a sequence conservation score in a classification task. For a given residue in a sequence, the Shannon entropy is calculated as follows [5]:

$$\mathrm{SE} = -\sum_{i=1}^{20} p_i \log p_i \tag{1}$$

Here, p_i is the value in the ith column of the WOP vector.

2.2 Multilayer Perceptron

Multilayer perceptrons (MLP) are made up of simple neurons called perceptrons. These neurons are interconnected in the form of a network to yield a MLP. Perceptrons map multiple inputs (real valued) to a single output. The weights that are fed in a perceptron are combined linearly and the mapping is done by a transformation or activation function which may be nonlinear in nature. First introduced by Rosenblatt in 1958, this concept can be mathematically formulated as follows:

$$y = \varphi i = (\sum_{i=1}^{n} w_i x_i + b) = \varphi(w^T x + b) \tag{2}$$

Fig. 1 Positive segment and negative segment

Here, y is the output, $\mathbf{x}$ is the input vector, $\mathbf{w}$ is the weight vector, φ is the activation function, and b is the *bias*.

2.3 *Star Consensus*

We have used the n-star consensus scheme to combine results from all the window sizes used for the experiment. In an n-star consensus scheme a particular site is declared as positive if at least n of the participating N predictors/classifiers vote the site as positive. For this work, we have considered only odd window sizes from 9 to 25 and the final result was obtained by performing a 9-star consensus.

First UniRef90 was downloaded from www.uniprot.org and then PSSMs were generated against this using the PSI-BLAST program for three iterations. For preparing training and test datasets, we downloaded reviewed data from UniProt resulting in 33,541 sequences which was reduced to 7242 sequences by removing redundancies using CD-HIT [9]. After eliminating sequences with 'X' we reached a final count of 7188 proteins. Of these, 141 sequences were chosen randomly, for testing, maintaining the organism-specific proportion in the original data and the remaining 7047 sequences were used for training.

Annotated residues from these sequences were considered as positive sites and non-annotated S, T, Y residues were considered as negative sites. Feature vectors were formed from the PSSMs by generating odd-sized windows (from 9 to 25) separately for positive and negative sites. These windows centered on an S/T/Y residue and incorporated 23 features per residue to form the feature vectors. Negative data exceeding positive data by a huge margin was clustered using Mini Batch K-Means so as cut down negative data count to approximately equal the positive data count [10]. Nine separate MLPs (for the 9 window sizes) were designed to predict phosphorylation sites for each one of the three residues (S/T/Y). The performances of these MLPs were evaluated for the independent test dataset and the results from the individual classifiers were combined using a 9-star consensus scheme (Fig. 1).

3 Results and Discussions

Under this section we have presented the results of the different experiments conducted on 141 proteins. Our system was trained with 7047 protein sequences. We trained three MLPs, one for each residue S, T, and Y and for each odd-sized

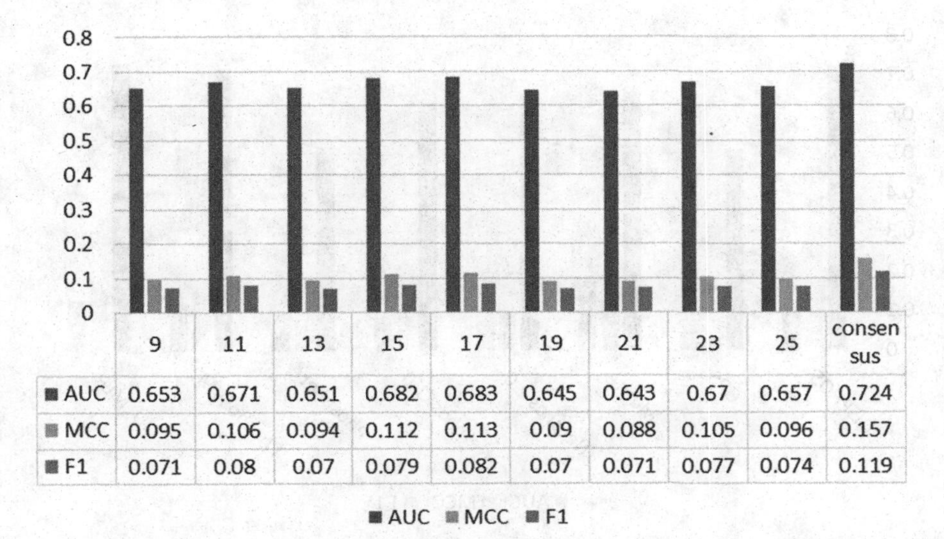

Fig. 2 Improvement in AUC, MCC, and F1 after application of star consensus scheme

window between 9 and 25. Star consensus scheme was applied to combine the results obtained from each one of these machines into a final prediction. This consensus scheme helped us improve our results considerably.

The contribution of star consensus can be easily seen in Fig. 2 and Table 1. The algorithm has improved all the metrics considerably. We then compared the performance of our predictor with other seven predictors namely, DISPHOS [11], GPS [12], Musite [13], NetPhos2.0 [7], NetphosK1.0 [7], Phosfer [14], PhosphoSVM [5]. A threshold of 0.5 was considered for DISPHOS, Netphos2.0, Phosfer, and PhosphoSVM. A specificity level of 0.95 was set for Musite. On an average the performance of our predictor was comparable to all the state-of-the-art predictors (Fig. 3).

Table 1 Improvement brought about by JUPred_MLP for a group of proteins

Protein name	AUC improvement	MCC improvement	F1 improvement
DOP1_DROME	0.0951	0.0566	0.0258
AT7L3_MOUSE	0.1	0.1059	0.0238
SC23A_BOVIN	0.0413	0.2413	0.2667
ORN_BOVIN	0.05	0.5257	0.6
STE2_YEAST	0.0763	0.1685	0.1414
GTPB1_HUMAN	0.0388	0.1676	0.1882
STF1_YEAST	0.2143	0.622	0.6

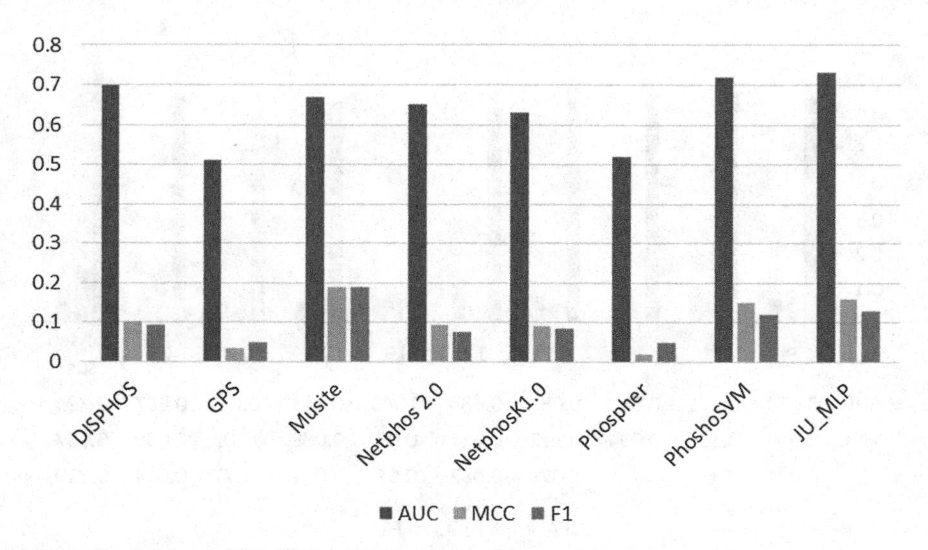

Fig. 3 Comparison of JU_MLP with other state-of-the-art predictors

Finally, we performed an experiment to find out in which of the proteins our predictor gave the best AUC score in which JUPred_MLP was the clear winner (Fig. 4). Among 141 proteins, our predictor scored the highest in 36 proteins in terms of AUC, bringing about a very high improvement in some of the proteins.

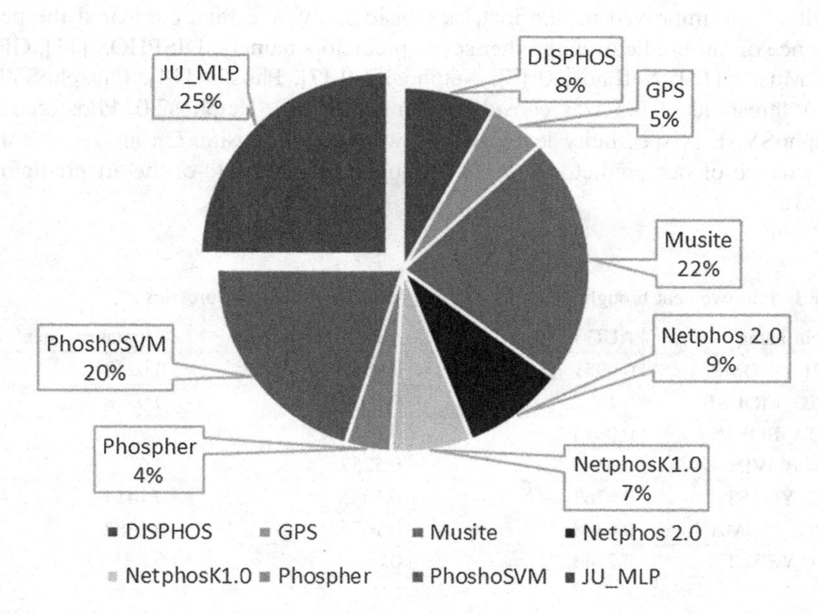

Fig. 4 Pie chart depicting the percentage of proteins for which each of the predictors gives the best AUC score

The proteins for which our predictor performed best have been summarized in Table 1. Our predictor has achieved the highest among the three metrics for a handful of proteins which have very important functions. The GEMI2_HUMAN protein is a catalyst which assists in the assembly of small nuclear ribonucleoproteins (snRNPs), which in turn are the building blocks of the spliceosome. STF1_YEAST is vital for stabilizing the ATPase–inhibitor complex, which normally disconnects when ATP is consumed externally. DOP1_DROME is involved in protein traffic between late Golgi and early endosomes. AT7L3_MOUSE forms a major component in regulating histone acylation complex SAGA. ORN_BOVIN is a protein which is in charge of cellular nucleotide recycling. SC23A_BOVIN forms a component of the COPII coat, which helps the transport of secretory, plasma membrane, and vacuolar proteins from the endoplasmic reticulum to the Golgi complex. Additionally for two proteins STF1_YEAST and ORN_BOVIN our predictor attained a perfect AUC, MCC, and F1 score of 1. AUC indicates the robustness of our predictor. Since JuPred_MLP has the highest AUC, it indicates that it has the greatest robustness among all other predictors. F1 is a score which combines both precision and recall. We chose F1 score to evaluate our predictor because a high value of F1 denotes a high precision as well as a high recall value. MCC is a correlation coefficient which works well even when the size of the classes is different. The value of MCC lies between −1 and +1. A coefficient of +1 denotes absolutely flawless prediction, a score of 0 denotes average prediction and a score of −1 denotes completely inverse prediction.

4 Conclusion

Almost all aspects of cell life are regulated by phosphorylation. There are several diseases which are attributed to abnormality in phosphorylation. In this work, we have proposed a non-kinase-specific protein phosphorylation prediction method. This predictor extracts information from position-specific scoring matrices in order to predict whether a residue is phosphorylated or not. In terms of AUC, our predictor provided the best results for 36 proteins brining about huge improvements in proteins which play a nontrivial roles in regulating cell activities. In this work, we had to use Mini Batch K-Means to reduce the huge amount of data due to a dearth of adequate computational resources. A further improvement can be brought about by performing the entire experiment in a distributed environment using Big Data technology [15].

Acknowledgments Authors are indebted to CMATER, department of computer science and Engineering, Jadavpur University for providing the necessary support for carrying out this experiment.

References

1. Trost, B., Kusalik, A.: Computational prediction of eukaryotic phosphorylation sites (2011)
2. Basu, S., Plewczynski, D.: AMS 3.0: prediction of post-translational modifications. BMC Bioinform. **11**, 210 (2010)
3. Plewczynski, D., Basu, S., Saha, I.: AMS 4.0: consensus prediction of post-translational modifications in protein sequences. Amino Acids **43**, 573–582 (2012)
4. Puntervoll, P., Linding, R., Gemünd, C., Chabanis-Davidson, S., Mattingsdal, M., Cameron, S., Martin, D.M.A., Ausiello, G., Brannetti, B., Costantini, A., Ferrè, F., Maselli, V., Via, A., Cesareni, G., Diella, F., Superti-Furga, G., Wyrwicz, L., Ramu, C., McGuigan, C., Gudavalli, R., Letunic, I., Bork, P., Rychlewski, L., Küster, B., Helmer-Citterich, M., Hunter, W.N., Aasland, R., Gibson, T.J.: ELM server: a new resource for investigating short functional sites in modular eukaryotic proteins. Nucleic Acids Res. **31**, 3625–3630 (2003)
5. Dou, Y., Yao, B., Zhang, C.: PhosphoSVM: prediction of phosphorylation sites by integrating various protein sequence attributes with a support vector machine. Amino Acids **46**, 1459–1469 (2014)
6. Kim, J.H., Lee, J., Oh, B., Kimm, K., Koh, I.: Prediction of phosphorylation sites using SVMs. Bioinformatics **20**, 3179–3184 (2004)
7. Blom, N., Gammeltoft, S., Brunak, S.: Sequence and structure-based prediction of eukaryotic protein phosphorylation sites. J. Mol. Biol. **294**, 1351–1362 (1999)
8. Hjerrild, M., Stensballe, A., Rasmussen, T.E., Kofoed, C.B., Blom, N., Sicheritz-Ponten, T., Larsen, M.R., Brunak, S., Jensen, O.N., Ganuneltoft, S.: Identification of phosphorylation sites in protein kinase A substrates using artificial neural networks and mass spectrometry. J. Proteome Res. **3**, 426–433 (2004)
9. Huang, Y., Niu, B., Gao, Y., Fu, L., Li, W.: CD-HIT suite: a web server for clustering and comparing biological sequences. Bioinformatics **26**, 680–682 (2010)
10. Biswas, A.K., Noman, N., Sikder, A.R.: Machine learning approach to predict protein phosphorylation sites by incorporating evolutionary information. BMC Bioinform. **11**, 273 (2010)
11. Xue, Y., Gao, X., Cao, J., Liu, Z., Jin, C., Wen, L., Yao, X., Ren, J.: A summary of computational resources for protein phosphorylation. Curr. Protein Pept. Sci. **11**, 485–496 (2010)
12. Xue, Y., Ren, J., Gao, X., Jin, C., Wen, L., Yao, X.: GPS 2.0, a tool to predict kinase-specific phosphorylation sites in hierarchy. Mol. Cell. Proteomics **7**, 1598–1608 (2008)
13. Gao, J., Thelen, J.J., Dunker, A.K., Xu, D.: Musite, a tool for global prediction of general and kinase-specific phosphorylation sites. Mol. Cell. Proteomics **9**, 2586–2600 (2010)
14. Trost, B., Kusalik, A.: Computational phosphorylation site prediction in plants using random forests and organism-specific instance weights. Bioinformatics btt031 (2013)
15. Banerjee, S., Basu, S., Nasipuri, M.: Big data analytics and its prospects in computational proteomics. In: Information Systems Design and Intelligent Applications. pp. 591–598. Springer, Berlin (2015)

Molecular Computing and Residual Binding Mode in ERα and bZIP Proteins from *Homo Sapiens*: An Insight into the Signal Transduction in Breast Cancer Metastasis

Arundhati Banerjee and Sujay Ray

Abstract The most provoking reason for death in breast cancer patients is the metastasis of breast cancer. Accumulating documentation states that signal transduction in human breast cancers initiate in estrogen-dependent manner with the signaling of estrogen receptor α-subunit (ERα) and XBP-1 (bZIP-domain) proteins. So, molecular level insight into the signaling mechanism is indispensable for future pathological and therapeutic developments. Thus, this current study discloses the stable residual participation of the two crucial human proteins for enhancing the signaling mechanism in breast tumor malignancies. For this purpose, 3D homology models of the respective proteins were prepared after the satisfaction of their stereo-chemical features. The protein–protein interaction was studied and protein complex was energy optimized. Revelation from the stability calculating parameters, solvent accessibility areas and interaction probes led to the inference of the most stable optimized complex and its residual participation (exceptional contribution of polar charged residues) for metastasis progression in breast cancer cells.

Keywords Homology modeling · Cell signal transduction · Docked protein–protein interaction · Energy optimization and breast cancer metastasis

A. Banerjee
Department of Biotechnology, National Institute of Technology, Durgapur, West Bengal, India
e-mail: arundhati.92star@gmail.com

S. Ray
Department of Biochemistry and Biophysics, University of Kalyani, Kalyani, West Bengal, India

S. Ray (✉)
Department of Biotechnology, Bengal College of Engineering and Technology, Durgapur, West Bengal, India
e-mail: raysujay@gmail.com

S. Das et al. (eds.), *Proceedings of the 4th International Conference on Frontiers in Intelligent Computing: Theory and Applications (FICTA) 2015*, Advances in Intelligent Systems and Computing 404, DOI 10.1007/978-81-322-2695-6_5

1 Introduction

One of the life-threatening phases of cancer is occupied by the metastatic breast cancer and has a lethal impact for the patients fighting against breast cancer. In the development and advancement of breast cancer, estrogen cell signal transduction accompanied by the estrogen receptor (ER) is highly associated [1]. In mammary epithelium, a major ER subunit, ERα participates in a pivotal role in breast cancer advancement [2, 3]. Estrogen binds to ERα, which is followed by the translocation of the ligand-activated ERα to the nucleus [1]. Now, post binding to the promoter of the target gene, transcription of the gene gets stimulated by genomic or nuclear signaling is performed by ERα [4, 5]. Another important human protein, human X box-binding protein 1 (XBP-1) is coupled with ERα activity in vitro and in vivo for breast tumors [6, 7]. It is known to elevate the transcriptional activity of ERα, in a ligand-free mode [6].

Human XBP-1, comprises a unique domain known as basic region leucine zipper (bZIP) domain which is efficiently responsible for important interactions including the one with ERα [8]. With the aid of SAGE (serial analysis of the expression of gene), bZIP-domain from XBP-1 has been properly documented to be expressed at elevated levels in ERα-positive malignant tumors in breast [6]. Therefore, in the progression of breast cancer advancement and succession, cell signal transduction by ERα protein holds a prior significance. This essential investigation in the efficient role of XBP-1 (bZIP-domain) in the transcriptional activity of ERα, at its molecular and computational level remains yet unexplored.

Therefore, in this present probe, the two vital proteins were modeled by homology or comparative modeling techniques after their sequence analysis. Each of the models was prepared finally after satisfying their varied stereo-chemical properties. Overall energy minimization of the 3D-modeled tertiary structures was executed for achieving a stable protein conformation. The protein monomers underwent protein–protein docking phenomena to interact with each other. Further energy optimization was performed for the docked complex. The stability and the strengthening impact on their binding were observed by the calculation of several stability determining parameters. Electrostatic surface potential calculations also aided to investigate the stable interactive complex. Residual participation from the respective proteins to cooperate among one another firmly was discerned and investigated. Earlier investigations [9, 10] documents the molecular level studies for many such diseases but breast cancer metastasis, a deadly life-threatening disease, was dealt nowhere.

In a digest, this extant computational exploration aimed into the root-source for the important responsibilities of the two indispensably important proteins in the signaling phenomena for the proliferation of malignant estrogen-controlled tissues. The examination of residual dependencies of the respective proteins from this study would be further beneficial for future therapeutic investigation. It might endow with an upcoming scope for the clinical research for the investigation of any small modulators for ERα or/and any specific discovery of drug to motivate the protein to reduce its tendency for causing the proliferation of the malignant breast cancer cells.

2 Materials and Methods

2.1 Analysis of Sequences and Template Search for Comparative Modeling

The initial step for homology modeling lies in analyzing the amino acid sequences for the proteins. So, the amino acid sequences of XBP-1 protein and ERα proteins from *Homo sapiens* were obtained, individually from NCBI (GI: 47678753, Accession No.: CAG30497.1 for XBP-1 and GI: 11907837, Accession No.: AAG41359.1 for ERα). The sequence analysis was verified by using Uniprot KB. For the purpose of the study, the only functionally most interactive and conserved region (domain) [11, 12], bZIP-domain from XBP-1 protein was extracted from EMBOSS from EMBL-EBI software packages [13], after identifying and analyzing the occupying region of the domain in the parent protein from pfam [14]. Results from BLASTP [15] against PDB [16] helped to deduce the templates for building homology models. HH-Pred, a very sensitive algorithm that uses homologous relationships between distantly related proteins [17], was also utilized for template search. Templates for bZIP-domain and ERα were obtained from the X-ray crystal structure of *Schizosaccharomyces pombe* and *Homo sapiens,* respectively, (PDB code: 1GD2_E with 99.46 % probability sharing 40 % sequence identity for bZIP-domain from HH-Pred search and 2OCF_A with 55 % sequence coverage and sharing 99 % sequence identity for ERα).

2.2 Comparative Modeling of bZIP-Domain and ERα Monomers

Homology modeling technique, also often familiar as comparative modeling technique was utilized with the aid of MODELLER9.14 software tool [18], for building up the individual 3D tertiary structures of the proteins. Back bone superimposition on each of the modeled protein's respective crystal templates (i.e., 1GD2_E for bZIP-domain and 2OCF_A for ERα), by using PyMOL [19], yielded the appropriateness for the modelled structures. For instance, the root mean squared deviation (RMSD) of ERα protein was perceived to be 0.291Å. So, Fig. 1 shows the ribbon-like representation for the superimposition of the modeled ERα protein (pink) on its crystal template (blue).

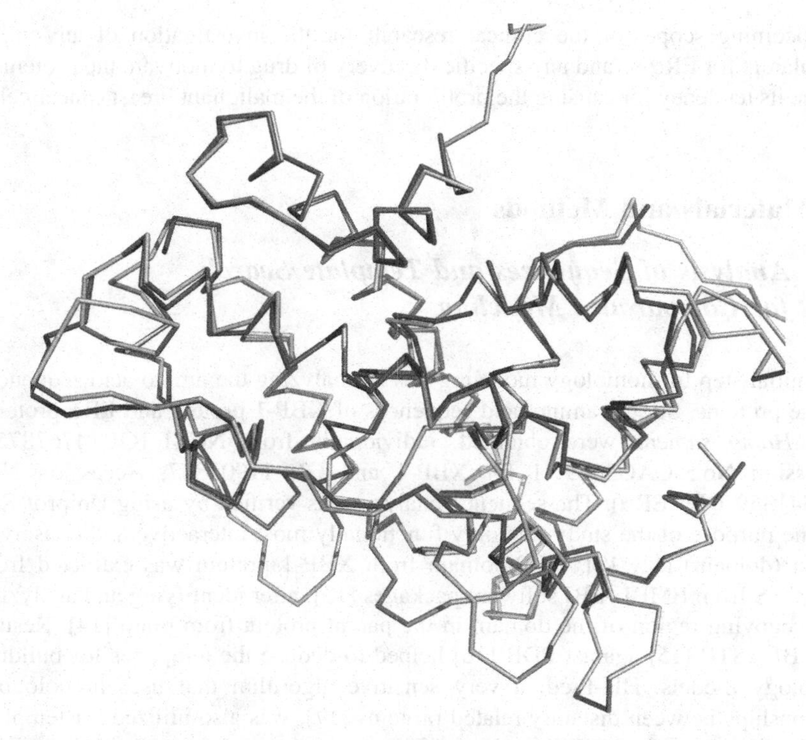

Fig. 1 Superimposition of the ERα protein on its X-ray crystal template (PDB ID: 2OCF_A)

2.3 *Optimization and Conformational Stability of Loop Regions Using ModLoop*

The homology modeled proteins were subjected further to ModLoop [20] for loop optimization. ModLoop [20] reduces and corrects the inaccuracies due to conformational discrepancy in the loop regions of the proteins. So, each of the structures were remodeled and optimized using ModLoop for proper conformation of ψ–φ angles.

2.4 *Energy Optimization of the Loop-Optimized Modeled Protein Structures*

Clearance the hostile geometries in the 3D models followed by protein refinement of the individual monomers was performed by ModRefiner tool [21]. The entire conformational search is done by physics based and knowledge based force field together. Thus native states of proteins are achieved from initial models (where, maximum stability among residues exists) with respect to positioning of backbone and spatial restraints [21]. It also thereby helps to minimize the overall energy of the protein structures, leading to a firmer and steady conformation [21].

2.5 *Stereo-Chemical Validation of the Modeled Proteins*

Verification of the stereo-chemical summary and overall quality of the modeled structures was performed with the assistance of PROCHECK [22] and ERRAT [23]. As per the predictions, quantitatively satisfying protein models were observed. There were no such amino acid residues that were perceived in the unfavored regions of the Ramachandran Plots for each of the modeled protein monomers [24].

2.6 *Protein–Protein Docking Simulations*

For the protein–protein interaction study, docking of the monomers were performed employing Cluspro2.0 [25] docking server. The ERα protein, having more amino acid residues, was uploaded as receptor and on the other hand, bZIP protein having comparatively lesser amino acid residues, was uploaded as ligand. Operating the advanced choice of ClusPro2.0 for modification of protein structures, the unstructured residues were reduced. A total of 10 docked bZIP-ERα complexes were presented by Cluspro2.0. The best cluster size among all the complexes was opted for further analysis. GRAMM-X [26] and ZDOCK [27] was also used to perform the protein–protein interactions. They offered with an inclusive outcome for the purpose.

2.7 *Molecular Dynamics Simulation of the Modeled Complex*

Improvement of high-resolution protein structures is increasingly essential to achieve a stable conformation of the protein with a diminished overall energy. For the purpose, FG-MD (Fragment-Guided Molecular Dynamics) was operated [28]. Besides implementing high resolution, it carries out atomic-level refinements for the protein structure using knowledge-based template information and physics-based MD simulations simultaneously [28]. To re-prepare the energy path of the MD simulation, special criteria was applied from the divided templates. Improvement of the local geometry of the refined protein structure was performed by eliminating steric clashes and upgrading torsion angle. Thus, it draws the structure near to its native state with higher level of accuracy [28].

2.8 *Stability Examination of the Energy Minimized and Simulated 'bZIP-ERα' Complex*

Overall optimization of energy leads to a better pattern of interaction and increased structural stability of the protein complex. So, the investigation of the stability of the modelled complex was scrutinized using FastContact server [29, 30]. Additionally, to infer a comprehensive result regarding a stable complex (that is, pre- and post-energy minimization complex protein structure), the net solvent accessibility area was estimated.

2.9 *Calculation of Interaction Patterns and Binding Modes in the Complex*

To explore the residues dependable for the protein–protein complexes from their respective positions on the protein structure, protein interaction calculator (P.I.C) web server [31] was utilized. The results were supported from the findings of the same by Discovery Studio software packages from Accelyrs and PyMOL [19]. It helped to analyze the net hydrogen bonding interactions, ionic interactions, aromatic interactions, and so on.

2.10 *Surface Electrostatic Potential Assessment*

Before and after energy optimization, the electrostatic potential on the surface for the 'bZIP-ERα' complex was computed and compared. The surface electrostatic potential was generated in vacuum electrostatics with the aid of PyMOL [19]. It was performed by mapping onto the protein surface using the units; kT/e.

3 Results

3.1 *Model's Structural Description for bZIP Monomer*

The functionally active human bZIP protein domain of XBP-1 protein had a Pfam accession number of PF07716.10. It occupied residue range of 69–120 amino acid residues in the parent XBP-1 protein. After homology modeling, the prepared model was observed to be analogous to its crystal template belonging from *Schizosaccharomyces pombe* (PDB Code: 1GD2; E chain). The 52 amino acid long protein begins with five residues in coil region, followed by α-helices (amino acid

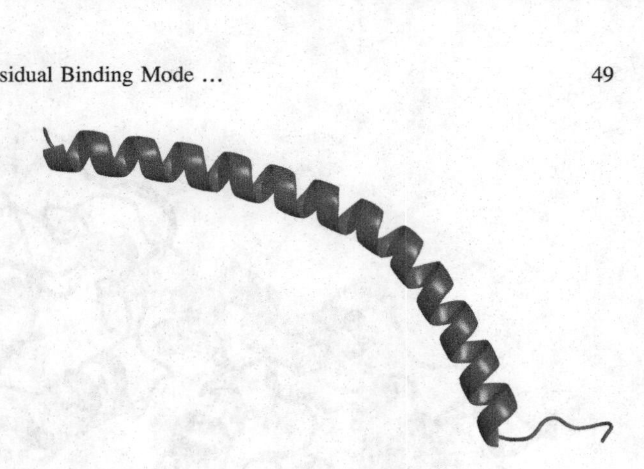

Fig. 2 The bZIP protein showing α-helices (*blue*) with tiny coil (*red*) regions at the N- and C-terminals of the protein

residues: 6–51) and ends with a single residue in coil region again. The structure is well depicted in Fig. 2 with α-helices in blue shades interspersed with red shaded coils.

3.2 Model's Structural Description for ERα Monomer

The functionally active human ERα protein from *Homo sapiens* after homology modeling was observed to be analogous to its X-ray crystal template (PDB Code: 2OCF; A chain). The 288 amino acid residue long protein monomer begins with methionine forming coil region and then followed by a helical region (amino acid residues: 2–5). Rest of the protein mainly forms several α-helical regions (amino acids: 18–23, 56–76, 91–93, 111–118, 120–126, 151–156, 160–177, 189–194, 212–232, 236–270 and 276–286) interspersed with coil regions and a set of antiparallel β-sheets. Six residues form antiparallel β-sheets (residues: 140–142 and 148–150). The structure again winds up with two residues in coil regions. The structure is presented in Fig. 3 with cyan and red shades illustrating α-helices and β-sheets, respectively, with interspersing magenta shaded coils.

3.3 Deduction of the Stability of bZIP-ERα Protein Complex

The variation in net interaction energy values of the bZIP-ERα protein complex (before energy optimization) and the energy minimized complex with MD simulations are well depicted in Table 1. It is lucid from the table that at the protein interfaces, the net interaction energy became stronger after the overall energy optimization of the modeled bZIP-ERα complex. Furthermore, from Table 1, the

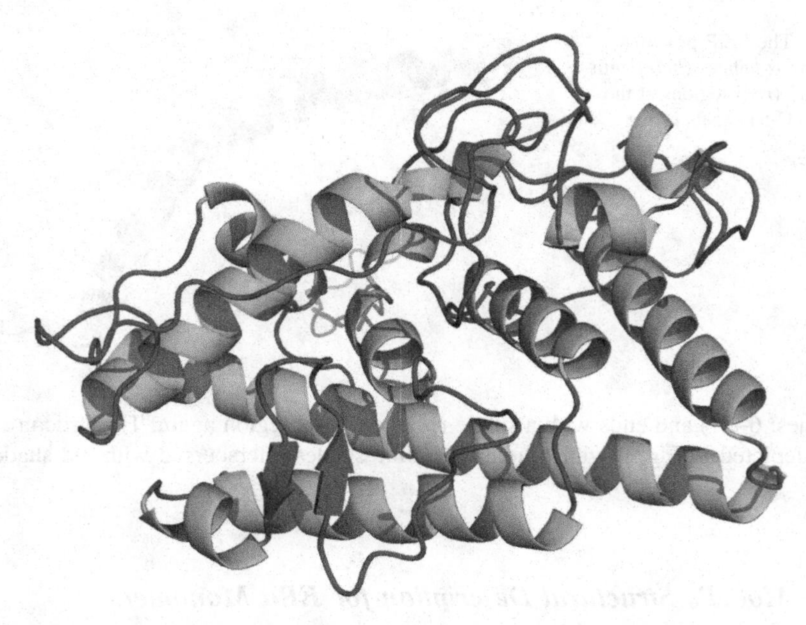

Fig. 3 ERα protein comprising tiny antiparallel β-sheets (*red*) and α-helices (*cyan*) with interspersing magenta coils

Table 1 Stability calculation of the optimized and simulated bZIP-ERα complex

Parameters/bZIP-ERα complex	Total interaction energy (kcal/mol)	Net solvent accessibility (Å^2)	Stabilizing/destabilizing
Before MD simulation	(–) 1615.04	18237.08 Å^2	Stable
MD simulated complex	(–) 1734.06	17572.69 Å^2	More stabilizing

reduction in the net solvent accessibility value for the complex implies that the energy optimized complex structure appeared to be a more interactive one.

3.4 *Protein–Protein Interactions in bZIP-ERα Complex*

The homology modeled complex structure of bZIP-ERα protein is well illustrated in Fig. 4. The examination using P. I. C web server [31] shows bZIP and ERα cooperate strongly with one another, not only through H-bonding but preponderantly by ionic–ionic interactions. Ionic–Ionic interactions that lead to a stronger and most interactive complex [32] were found to be increased in number after optimization and simulation. Tables 2 and 3 represent the ionic–ionic interactions

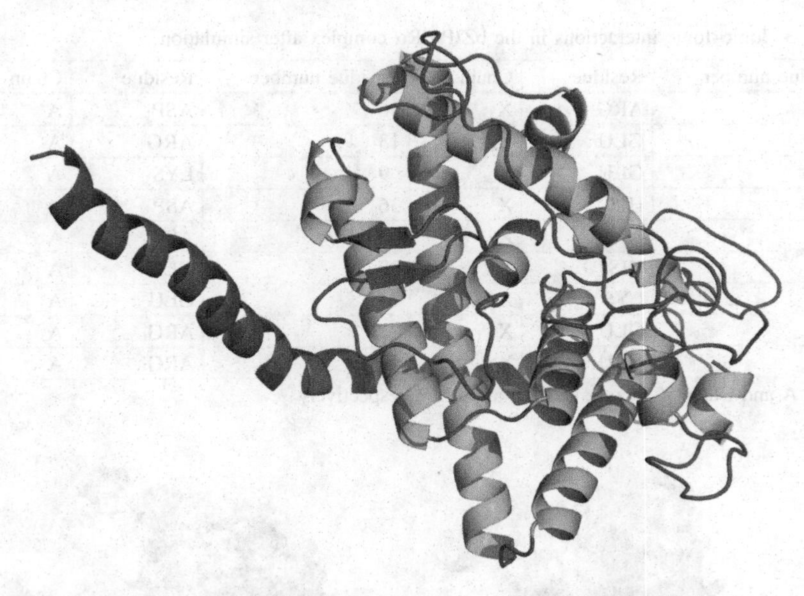

Fig. 4 The bZIP-ERα docked complex firmly interacting with bZIP showing α-helices (*blue*) and beginning with a tiny coil region (*red*) and ERα comprising tiny antiparallel β-sheets (*red*) and α-helices (*cyan*) with interspersing magenta coils

Table 2 Ionic–Ionic interactions in the bZIP-ERα complex before simulation

Residue number	Residue	Chain	Residue number	Residue	Chain
151	ARG	A	36	ASP	X
155	LYS	A	36	ASP	X
155	LYS	A	39	GLU	X
155	LYS	A	40	GLU	X
173	ARG	A	21	ASP	X
219	ASP	A	9	LYS	X
252	HIS	A	21	ASP	X

Chain A and Chain X represents ERα and bZIP, respectively

accomplished by the bZIP-ERα complexes before and after optimization of overall energy (followed by the MD Simulation).

3.5 Surface Electrostatic Potential Estimation

Fascinatingly, the vacuum electrostatic potential calculation also infers the energy minimized complex bZIP-ERα structure to be a more stable and highly interactive one. Figure 5 portrays the pictorial view for the comparable study of the

Table 3 Ionic–Ionic interactions in the bZIP-ERα complex after simulation

Residue number	Residue	Chain	Residue number	Residue	Chain
173	ARG	X	36	ASP	A
210	GLU	X	13	ARG	A
210	GLU	X	9	LYS	A
252	HIS	X	36	ASP	A
254	ARG	X	21	ASP	A
255	HIS	X	29	GLU	A
259	LYS	X	29	GLU	A
262	GLU	X	22	ARG	A
281	GLU	X	7	ARG	A

Chain A and Chain X represents ERα and bZIP, respectively

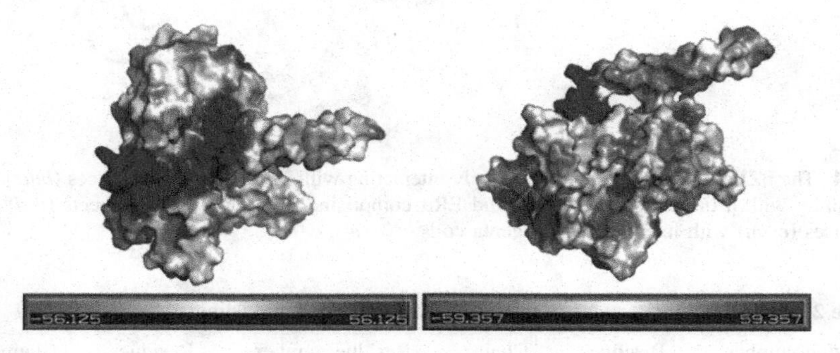

Fig. 5 Comparable view of the surface electrostatic potential change on the surfaces before (*left*) and after (*right*) energy minimization and simulation of bZIP-ERα protein complex

electrostatic potentials for the complex structures before and after energy minimization and corresponding MD simulation. The electrostatically positive zones are depicted by blue areas whereas electrostatically negative ones are depicted in red shades, in either of the two cases.

4 Discussion

In the contemporaneous work, the functional tertiary modeled protein structures of the bZIP-domain and ERα were built and analyzed. From the human XBP-1 protein, bZIP-domain is the only domain and the most important interactive zone for the detection of the proliferation of breast cancer cells [6, 8]. The varied interaction pattern in bZIP-ERα complex were analyzed, calculated, and illustrated. The net interaction energies were observed to get a turn-down from −1615.04 kcal/mol (prior-to optimization and simulation) to −1734.06 kcal/mol, after optimization of energy and simulation. Fascinatingly, in addition to that a descent was also observed in the value of net accessibility area for solvent. The reduction was rapidly from 18237.08 Å^2 (before optimization and simulation) to 17572.69 Å^2 after minimizing and simulating the complex structure. The electrostatic potential values also further, ensure that the optimized and simulated complex structure possessed an exceedingly compact, steady, and a firm interaction between one another. From the strengthening ionic–ionic interactions amongst the complexes, the final simulated complex was perceived to be more firmly interacting with greater number of ionic bonds, which is an increment to nine bonds from seven bonds. Solely, among total nine bonds, five positively charged residues from bZIP protein interacted with five negatively charged residues from ERα protein. Mainly, the positively charged arginine and negatively charged glutamine residues dominated the strengthening of the ionic bonds. From, bZIP protein, Glu210 alone, forms two ionic bonds with Arg13 and Lys9 from ERα. Again, Glu262 and Glu281 of bZIP protein were observed to interact with Arg22 and Arg7 from ERα. Negatively charged aspartic acid was found to interact only from the ERα protein. Asp36, therefore, formed two bonds with Arg173 and His252 of bZIP protein. Asp21 accompanied the interactions by binding Arg254. From ERα, Glu29 forms only two bonds with His255 and Lys259 from bZIP protein. So, in a digest, for the most stable complex, all the five types of polar charged residues (three positively charged-Lys, Arg, His and two negatively charged Asp and Glu) were observed to be satisfactorily indulged to fortify the interaction by the formation of the cavity to accommodate the bZIP protein.

Consequently, this contemporary study presents an acquaintance in the interaction between ERα and bZIP proteins from *Homo sapiens*. This residual level computational study to scrutinize the basis of the interaction is one of the most essential zones to be explored into. Previously, several molecular level studies [9, 10] were documented for other diseases but none dealt with the cell signal transduction in the enhancement of metastasis of breast tumors. This in silico discern therefore, unveils the residual participation, binding demonstration and analysis of the most stable complex (i.e., the energy optimized simulated complex) of ERα-bZIP protein for the signaling mechanism in the breast cancer malignancies. It endows with an avenue for the future therapeutic research in a lucid mode.

5 Conclusion and Future Prospect

The cooperative participation of the residues from the two essential human proteins (ERα and bZIP) for metastasis of breast tumors was the prime focus of this present investigation. Signals are triggered by the respective breast tumor cells at the time of progression toward invasion from tumorigenesis. This further triggers the extranuclear pathways for the cell signal transduction of ERα. As a result, it further provides an increment in the migratory functions of the responsible cells, followed by metastasis. This ERα protein is also associated with the bZIP protein for the increased signal transduction and metastasis. This duo protein interaction enhances during the advanced breast cancers. Thus, this study poses a cogent framework for the extranuclear cell signal transduction involving the metastatic control of ERα-positive tumors in collaboration with bZIP protein.

So, the present structural and computational molecular contribution of ERα–bZIP interactions was essential to be elucidated not only for the clinical progress in novel therapeutics for breast cancers but also for the improvement of the future production of refined modulators for ERα. Future scope lies in the in silico investigation of any mutation in either ERα or bZIP protein which might shed an impact on the cancer progression. It will further pave an outlook in the clinical and pharmaceutical research for the investigation of any small modulators for ERα or any certain drug discovery to mold the protein to lower its tendency for causing the efficient progression of metastasis in the breast cancer cells.

Acknowledgement Authors are deeply indebted for the immense help, paramount suggestions, and continuous encouragement rendered by Dr. Angshuman Bagchi, Assistant Professor, Department of Biochemistry and Biophysics, University of Kalyani, Kalyani, Nadia, India. Authors also render gratefulness to the Department of Biotechnology, National Institute of Technology, Durgapur as well as to the Department of Biotechnology, Bengal College of Engineering and Technology for their support and cooperation.

References

1. Sudipa, S.R., Ratna, K.V.: Role of estrogen receptor signaling in breast cancer metastasis. International J. Breast Cancer **2012**, 8 (2012). http://dx.doi.org/10.1155/2012/654698. Article ID 654698
2. Warner, M., Nilsson, S., Gustafsson, J.Å.: The estrogen receptor family. Curr. Opin. Obstet. Gynecol. **11**(3), 249–254 (1999)
3. Hewitt, S.C., Couse, J.F., Korach, K.S.: Estrogen receptor knockout mice: what their phenotypes reveal about mechanisms of estrogen action. Breast Cancer Res. **2**(5), 345–352 (2000)
4. McKenna, N.J., Lanz, R.B., O'Malley, B.W.: Nuclear receptor coregulators: cellular and molecular biology. Endocr. Rev. **20**(3), 321–344 (1999)
5. McDonnell, D.P., Norris, J.D.: Connection and regulation of the human estrogen receptor. Science **296**(5573), 1642–1644 (2002)
6. Ding, L., Yan, J., Zhu, J., Zhong, H., Lu, Q., Wang, Z., Huang, C., Ye, Q.: Ligand-independent activation of estrogen receptor alpha by XBP-1. Nucleic Acids Res. **31** (18), 5266–5274 (2003)
7. Sengupta, S., Sharma, C.G.N., Jordan, V.C.: Estrogen regulation of X-box binding protein-1 and its role in estrogen induced growth of breast and endometrial cancer cells. Horm. Mol. Biol. Clin. Investig. **2**(2), 235–243 (2010). doi:10.1515/HMBCI.2010.025
8. Liou, H.C., Boothby, M.R., Finn, P.W., Davidon, R., Nabavi, N., Zeleznik-Le, N.J., Ting, J.P., Glimcher, L.H.: A new member of the leucine zipper class of proteins that binds to the HLA DR alpha promoter. Science **247**(4950), 1581–1584 (1990). doi:10.1126/science.2321018.PMID2321018
9. Simanti, B., Amit, D., Semanti, G., Rakhi, D., Angshuman, B.: Hypoglycosylation of dystroglycan due to T192M mutation: a molecular insight behind the fact. Gene **537**, 108–114 (2014)
10. Angshuman, B.: Structural characterizations of metal ion binding transcriptional regulator CueR from opportunistic pathogen *Pseudomonasaeruginosa* to identify its possible

involvements in virulence. Appl. Biochem. Biotechnol. (2014). doi:10.1007/s12010-014-1304-5
11. Jones, S., Stewart, M., Michie, A., Swindells, M.B., Orengo, C., Thornton, J.M.: Domain assignment for protein structures using a consensus approach: characterization and analysis. Protein Sci. **7**(2), 233–42 (1998). doi:10.1002/pro.5560070202. PMC 2143930. PMID 9521098
12. George, R.A., Heringa, J.: An analysis of protein domain linkers: their classification and role in protein folding. Protein Eng. **15**(11), 871–879 (2002). doi:10.1093/protein/15.11.871. PMID 12538906
13. McWilliam, H., Li, W., Uludag, M., Squizzato, S., Park, Y.M., Buso, N., Cowley, A. P., Lopez, R.: Analysis tool web services from the EMBL-EBI. Nucleic Acids Res. **41**(Web Server issue), W597-600 (2013). doi:10.1093/nar/gkt376. PMID:(23671338)
14. Punta, M., Coggill, P.C., Eberhardt, R.Y., Mistry, J., Tate, J., Boursnell, C., Pang, N.: Forslun: the Pfam protein families database. Nucleic Acids Res. **40**(D1), D290–D301 (2011). http://dx.doi.org/10.1093/nar/gkr1065. PMC 3245129. PMID 22127870
15. Altschul, S.F., et al.: Basic local alignment search tool. J. Mol. Biol. **25**, 403–410 (1990)
16. Berman, M.H., et al.: The protein data bank. Nucleic Acids Res. **28**, 235–242 (2000). doi:10.1093/nar/28.1.235
17. Johannes, S., Andreas, B., Andrei N.L..: The HHpred interactive server for protein homology detection and structure prediction. Nucleic Acids Res. **33**, W244–W248 (2005). Web Server issue. doi:10.1093/nar/gki408
18. Sali, A., Blundell, T.L.: Comparative protein modelling by satisfaction of spatial restraints. J. Mol. Biol. **234**, 779–815 (1993)
19. DeLano, W.L.: The PyMOL molecular graphics system DeLano scientific, San Carlos (2002). doi:10.1093/nar/gki408
20. Fiser, A., Sali, A.: ModLoop: automated modeling of loops in protein structures. Bioinformatics **19**(18), 2500–2501 (2003)
21. Xu, D., Zhang, Y.: Improving the physical realism and structural accuracy of protein models by a two-step atomic-level energy minimization. Biophys. J. **101**, 2525–2534 (2001). doi:10.1016/j.bpj.2011.10.024
22. Laskowski, R.A., et al.: PROCHECK: a program to check the stereochemistry of protein structures. J. Appl. Crystallogr. **26**, 283–291 (1993)
23. Colovos, C., Yeates, T.O.: Verification of protein structures: patterns of nonbonded atomic interactions. Protein Sci. **2**, 1511–1519 (1993)
24. Ramachandran, G.N., Sashisekharan, V.: Conformation of polypeptides and proteins. Adv. Protein Chem. **23**, 283–438 (1968)
25. Comeau, S.R., et al.: ClusPro: an automated docking and discrimination method for the prediction of protein complexes. Bioinformatics **20**, 45–50 (2004)
26. Vakser, I.A.: Protein docking for low-resolution structures. Protein Eng. **8**, 371–377 (1995)
27. Chen, R., et al.: ZDOCK: an initial-stage protein docking algorithms. Proteins. **51**, 82–87 (2003)
28. Zhang, J., Liang, Y., Zhang, Y.: Atomic-level protein structure refinement using fragment-guided molecular dynamics conformation sampling. Structure **19**, 1784–1795 (2011)
29. Mina, M., Gokul, V., Luis, R.: The role of electrostatic energy in prediction of obligate protein-protein interactions. Proteome Sci. **11**, S11 (2013). doi:10.1186/1477-5956-11-S1-S11
30. Camacho, C.J., Zhang, C.: FastContact: rapid estimate of contact and binding free energies. Bioinformatics **21**(10), 2534–2536 (2005)
31. Tina, K.G., Bhadra, R., Srinivasan, N.: PIC: protein interactions calculator. Nucleic Acids Res. **35**, W473–W476 (2007)
32. Baldwin, R.L.: How Hofmeister ion interactions affect protein stability. Biophys. J. **71**(4), 2056–2063 (1996)

SVM-Based Pre-microRNA Classifier Using Sequence, Structural, and Thermodynamic Parameters

K.A. Sumaira, A. Salim and S.S. Vinod Chandra

Abstract microRNAs are single-stranded noncoding RNA sequences of 18–24 nucleotide length. They play important role in post-transcriptional regulation of gene expression. Last decade witnessed immense research in microRNA identification, prediction, target identification, and disease associations. They are linked with up/down regulation of many diseases including cancer. The accurate identification of microRNAs is still complex and time-consuming process. Due to the unique structural and sequence similarities of microRNAs, many computational algorithms have been developed for prediction of microRNAs. According to the current status, 28645 microRNAs have computationally discovered from the genome sequences, and have reported 1961 human microRNAs (miRBase version 21, released on June 2014). There are several computational tools available for predicting the microRNA from the genome sequences. We have developed a support vector machine-based classifier for microRNA prediction. Top ranked 19 sequence, structural, and thermodynamic characteristics of validated microRNA sequence databases are employed for building the classifier. It shows an accuracy of 98.4 % which is higher than that of existing SVM-based classifiers such as Triplet-SVM, MiRFinder, and MiRPara.

Keywords microRNA · Svm classifier · Computational prediction · Structural parameters · Thermodynamic characteristics

1 Introduction

RNAs are single-stranded long sequences that are formed from the DNA sequences through transcription process. With the help of hydrogen bonding between the bases, a nucleotide sequence of RNA could form a nonlinear structure, called secondary structure [15, 16]. The components of a secondary structure can be classified as stem loop (hairpin loop), bulge loops, interior loops, and junctions (Multi-loops) [8].

K.A. Sumaira · A. Salim (✉) · S.S. Vinod Chandra
College of Engineering Trivandrum, Kerala, India
e-mail: salim.mangad@gmail.com

S. Das et al. (eds.), *Proceedings of the 4th International Conference on Frontiers in Intelligent Computing: Theory and Applications (FICTA) 2015*, Advances in Intelligent Systems and Computing 404, DOI 10.1007/978-81-322-2695-6_6

Functionally, RNAs are responsible for protein synthesis and RNAs such as messenger RNA (mRNA), ribosomal RNA (rRNA), and transfer RNA (tRNA) have its own roles in this process [5]. A family of noncoding RNA, around 22 nt long, found in many eukaryotes including humans is called microRNA. The process of formation of microRNAs has many stages, initially longer primary transcript (pre-microRNA) is formed, which in turn converted into a pre-microRNA, and processes mandate presence of ribo-nucleolus Drosha, Exportion-5 [4, 9]. The pre-microRNAs are characterized by a hairpin-like structure. microRNAs play different roles in gene regulation by binding to specific sites in mRNA and causes translational repression or cleavage [22]. Due to the change in gene expression, microRNAs role as suppressor /oncogenes in different cancers such as colon, gastric, breast, and lung cancers are proved [3]. microRNA also helps for the proper functioning of brain and nervous system, and have regulatory roles in several other diseases like deafness, Alzheimers disease, Parkinson disease, Down's syndrome, and Rheumatoid arthritis [1, 12]. microRNA-based cancer detection and therapy is underway [18]. As the in vivo identification of microRNAs is time consuming and complex, many computational tools had been developed to predict most provable microRNA sequences. The methods employed for computational prediction of microRNAs vary from search in conserved genomic regions, measuring structure, sequence, thermodynamic characteristics of RNA secondary structures, to properties of reads of next-generation sequencing data, together with advances in machine learning techniques [19].

Comparing DNA sequences of related species for conserved noncoding regions having regulatory functions were the initial approach employed for microRNA prediction. miRScan [11] and miRFinder [20] are examples of such tools. The sequence characteristics, especially the properties of blocks of three of consecutive nucleotides, namely triplet structure along with other parameters are used in Triplet-SVM [6], MiPred [17], and MiRank [25]. MiRank, developed by Yunpen et.al, works with a ranking algorithm based on random walks and reported prediction accuracy is 95 %. Peng et.al developed MiPred which classifies real and pseudo-microRNA precursors using random forest prediction model. MiPred has reported 88.21 % of total accuracy, and while combining the P-value randomization, the accuracy of prediction increased to 93.35 %. Mpred [18, 21] is a tool which uses artificial neural network for pre-microRNA validation and microRNA prediction by hidden Markov model. MiRPara [23], Triplet-SVM, and MiRFinder are the SVM-based classifier where reported accuracy of MiRPara is 80 % and that of Triplet-SVM is 90 %. MiR-Para divides the input sequences into number of fragments of length around 60 nucleotides, filter out the fragments having an hairpin structure, extracts 77 different parameters from the sequence, and fed to SVM classifier. Triplet-SVM classifies the real and pseudo-microRNA precursor using structure and triplet sequence features. The positive training dataset collected miRNA registry database and the pseudo-miRNA datasets from the protein coding regions. MiRFinder tried to distinguish between microRNA and nonmicroRNA sequences using different representations of the sequence states such as paired, unpaired, insertion, deletion, and bulge with different symbols. They constructed the positive training data with the pre-miRNA sequences of human, mouse, pig, cattle, dog, and sheep collected from miRBase, and constructed the negative dataset with the sequences extracted from the UCSC genome pairwise alignments. MiRFinder used RNAfold [7] to predict the secondary structure of the sequences.

The tools discussed above uses different subsets of structural, sequence, and thermodynamic properties of secondary structure of microRNA sequence. Still there is relevance for a better tool with reduced feature set and higher level of accuracy. The main motivation of this work is to develop a classifier with high sensitivity (True Positive Rate), high specificity (True Negative Rate), low false positive rate, and an accuracy greater than 95 %. We have developed an SVM-based classifier and trained by the properties extracted from the experimentally validated database of human microRNAs.

2 SVM-Based Classifier Model

Figure 1 shows the system model. A trained and tested classifier could be able to predict whether a given input sequence is a probable microRNA or not. Figure 2 shows the preprocessing steps required for microRNA identification from an input gene sequence. The length of gene sequence vary from few hundreds to several thousands of nucleotides. A moving window divides input sequence into subsequence of length 100 with step size of 30. The candidate sequences with a lesser base pairing value than a threshold value can be discarded in the initial screening. The known microRNA sequences have at least 17 base pairs, and hence sequences having 17 or more pairs are only passed to feature extraction phase.

2.1 Training Data Preparation

Sufficient quantities of positive and negative samples of data are required to train and test a classifier. The quality of the training dataset determines the accuracy of the

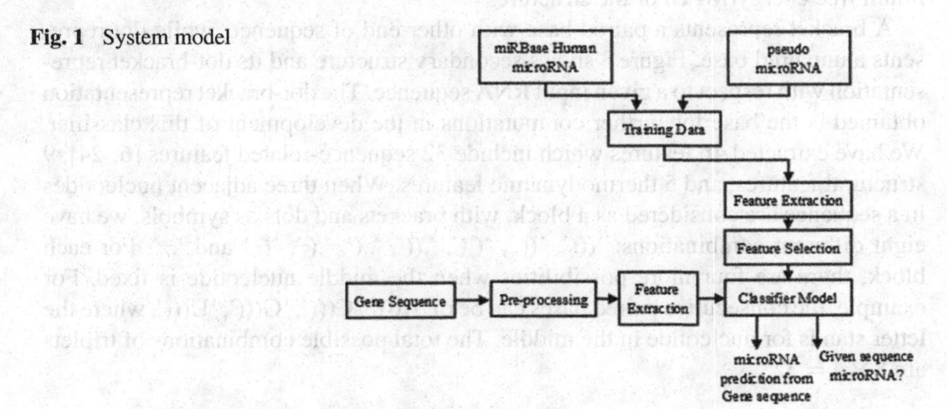

Fig. 1 System model

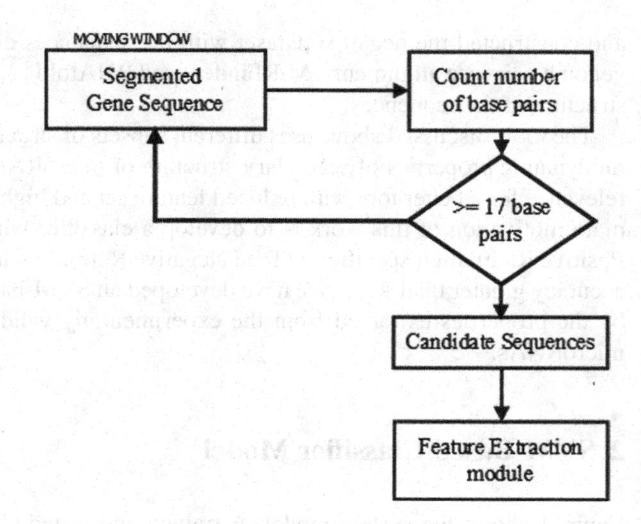

Fig. 2 Gene sequence preprocessing and feature extraction

classifier. miRBase is a primary microRNA sequence repository keeping identified pre-microRNA sequences, mature sequences, and gene coordinate information [10]. Presently, the database contains sequences from 223 species. 500 human microRNA sequences downloaded from miRBase database are used as positive dataset. The negative training dataset is prepared from the coding region of RNA, by filtering out sequences that contain a hairpin-like structure. The reason behind this selection is that the real microRNAs are characterized by their hairpin loop along with other properties. 500 sequences are selected for the negative dataset also.

Feature Extraction and Selection A major discriminating property of RNA secondary structure is free energy due to the hydrogen bonding between its bases, called minimum free energy (MFE). Several computational algorithms based on dynamic programming have been developed to find MFE. RNAfold [7] is one such algorithm. RNAfold generates the secondary structure in dot-bracket notation and predicts minimum free energy(MFE) of the structure.

A bracket represents a paired base with other end of sequence, while dot represents a unpaired base. Figure 3 shows secondary structure and its dot-bracket representation with respect to a given input RNA sequence. The dot-bracket representation obtained is the base for further computations in the development of this classifier. We have extracted 46 features which include 32 sequence-related features [6, 24], 9 structural features, and 5 thermodynamic features. When three adjacent nucleotides in a sequence are considered as a block, with brackets and dots as symbols, we have eight different combinations: '(((', '((.', '(.(', '.((', '..(', '.(.', '(..' and '...'. For each block, there are four more possibilities when the middle nucleotide is fixed. For example, the consecutive paired bases can be of 'A(((' 'C(((', 'G(((', 'U(((', where the letter stands for nucleotide in the middle. The total possible combinations of triplets are $8 \times 4 = 32$.

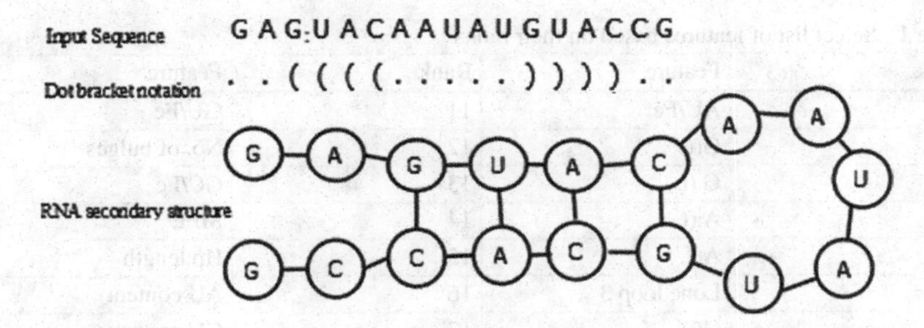

Fig. 3 Secondary structure and *dot-bracket* representation corresponding to a typical RNA sequence

The following structural features were selected from the secondary structure of the sequence.

1. Base Count: Total number of base pairs.
2. Base Content: The ratio of total number of base pairs to the total number of nucleotides in that sequence.
3. Lone loop 3: The count of lone loops that have 3 nts. (a lone loop is the one with first and last nucleotides of the loop as Watson Crick or wobble base pair).
4. Lone loop 5: The count of lone loops that have 5 nts.
5. AU content: The ratio of number of AU base pairs to the total number of base pairs.
6. GC content: The ratio of number of GC base pairs to the total number of base pairs.
7. GU content: The ratio of number of GU base pairs to the total number of base pairs.
8. Hairpin length: The number of nucleotides in the hairpin loop.
9. Number of Bulges: Total number of bulges.

The features related with the structural stability in terms of energy value due to the bonding of bases are known as thermodynamic features [14].

1. Minimum Free Energy: Minimum free energy of the structure.
2. MFE content: The ratio of MFE to the number of nucleotides in the sequence [24].
3. GC/Fe: The ratio of number of GC pairs to the MFE.
4. AU/Fe: The ratio of number of AU pairs to the MFE.
5. GU/Fe: The ratio of number of GU pairs to the MFE.

This is quite large number of parameters, and dimensionality reduction is applied based on principle component analysis (PCA) [2]. PCA is a mathematical method for dimensionality reduction. This can be viewed as rotation of axes of original variable coordinate system to new orthogonal axes called principal axes, which coincide with the direction maximum variation of original observations. Thus, principal components represent a reduced set of uncorrelated variables corresponding to the original set of correlated variables. We used WEKA [13] to build the classifier. Based on the

Table 1 Select list of features based on their ranks

Rank	Feature	Rank	Feature
1	AU/Fe	11	GU/Fe
2	G((.	12	No. of bulges
3	G.((	13	GC/Fe
4	A.((	14	MFE
5	A((.	15	Hp length
6	Lone loop 3	16	AU content
7	G(((	17	GU content
8	C(((	18	GC content
9	Bspair count	19	A(((
10	Lone loop 5		

value of variance specified, WEKA chooses sufficient number of Eigen vectors to account original data. Ranking of attribute can be performed with WEKA by selecting an option to transform back to original space. The top ranked 19 features are only used for final classification, as there is very little improvement in accuracy when others are also considered. The selected features and their rank are shown in Table 1. This includes seven features from sequence-related features such as 'A(((', 'C(((', 'G(((', 'A((.', 'G((.', 'A.((' and 'G.(('; and eight features from structure-related group; and four from thermodynamic group. Although many subsets of these features are used by other computational tools for microRNA prediction, we uniquely identified three new features. They are ratio of GC and free energy (GC/Fe), ratio of AU and free energy (AU/Fe), and ratio of GU and free energy (GU/Fe). It is evident that they have decisive role as they have ranked 1st, 11th, and 13th in the select list of attributes.

Machine Learning Support vector machines(SVM) are supervised learning model with associated learning algorithms [6, 23]. Given a set of training examples, each marked as belonging to one of two classes, an SVM training algorithm builds a model that assigns new examples into the appropriate class, making it a non-probabilistic binary classifier. SVMs effectively do this classification by a technique called kernel trick, implicitly mapping their inputs into high-dimensional feature space. A linear classifier is based on discriminant function of the form $f(x) = \omega^T \cdot x + b$, where ω is the weight vector, and b is the bias. The set of points $\omega^T \cdot x = 0$ define a *hyperplane*, and b translates *hyperplane* away from the origin. A nonlinear classifier is based on discriminant function of form $f(x) = \omega^T \phi(x) + b$, where ϕ is a nonlinear function. Performance of the SVM classifier with a linear kernel, and two nonlinear kernels, namely radial basis function kernel (RBF) and Pearson VII kernel (PUK), are analyzed. The RBF kernel is defined by

$$K(x, y) = e^{(-\gamma \, \|x - y\|^2)} \tag{1}$$

and Pearson VII kernel is defined by

$$K(x,y) = \frac{1}{\left(1 + \left(\frac{2\sqrt{\|x-y\|^2}\sqrt{2^{\left(\frac{1}{\omega}\right)}-1}}{\sigma}\right)^2\right)^{\omega}} \tag{2}$$

where ω and σ control half width and trailing factor of peak, respectively.

3 Performance Analysis of the Classifier

The performance of the classifier with linear and nonlinear kernel, with complete and reduced feature set, is evaluated. Table 2 shows the confusion matrix in SVM with PUK kernel function when 10-fold cross validation is employed. A classifier gives best result when it reaches high TP and TN rates. The efficiency and quality of a tool depend upon a number of factors such as sensitivity (TP rate), specificity (TN rate), and accuracy. The accuracy of the classifier can be calculated using the following equations [23, 24]:

$$\text{Sensitivity} = \frac{\text{TP} * 100}{\text{TP} + \text{FN}} \tag{3}$$

$$\text{Specificity} = \frac{\text{TN} * 100}{\text{TN} + \text{FP}} \tag{4}$$

$$\text{Accuracy} = \frac{(\text{TN} + \text{TP}) * 100}{\text{TP} + \text{FP} + \text{TN} + \text{FN}} \tag{5}$$

When all the 46 features are used with PUK kernel and 10-fold cross validation, the sensitivity, specificity, and accuracy reached 98.6 %, and this recorded as the best result from the classifier. However, if top ranked 19 features are used, the classifier provides sensitivity, specificity, and accuracy as 98.4 % (same value for all the parameters). When compared with the performance with whole feature set, variation is insignificant, but computational cost will be definitely higher in the former case. The classifier performance under different conditions is shown in Table 3. It is also evident from the data in the table, when RBF kernel is used, that change in value of parameter γ from 0.01 to 1 makes considerable increase in accuracy.

Table 2 Confusion matrix in SVM

a	**b**	
491	9	**a** = Yes
7	493	**b** = No

Table 3 Performance of the classifier under cross validation and separate test set in SVM

Validation method	No. of features	Kernel	TP rate	FP rate	Precision	Recall	F-Measure	ROC area
10 fold CV	19	Linear	0.974	0.026	0.974	0.974	0.974	0.974
10 fold CV	19	RBF, $\gamma = 0.01$	0.964	0.036	0.964	0.964	0.964	0.964
10 fold CV	19	RBF, $\gamma = 1$	0.982	0.018	0.982	0.982	0.982	0.982
10 fold CV	**19**	**PUK**	**0.984**	**0.016**	**0.984**	**0.984**	**0.984**	**0.984**
10 fold CV	**46**	**PUK**	**0.986**	**0.014**	**0.986**	**0.986**	**0.986**	**0.986**
10 fold CV	46	RBF, $\gamma = 1$	0.985	0.015	0.985	0.985	0.985	0.985
Test set (74/26)	19	PUK	0.985	0.017	0.985	0.985	0.985	0.984
Test set 66/34)	19	PUK	0.985	0.015	0.985	0.985	0.985	0.985

Linear and different nonlinear kernel functions are examined. Best result (accuracy 98.6) is with nonlinear kernel—Pearson VII kernel function

ROC is plot of fraction of true positives out of the total actual positives (TPR = true positive rate) versus the fraction of false positives out of the total actual negatives (FPR = false positive rate), at various threshold settings. TPR is also known as sensitivity or recall in machine learning. The FPR is also known as the fall-out and can be calculated as one minus specificity. The ROC curve is then the sensitivity as a function of fall-out. In general, if both of the probability distributions for detection and false alarm are known, the ROC curve can be generated by plotting the cumulative distribution function (area under the probability distribution from −inf to +inf) of the detection probability in the y-axis versus the cumulative distribution function of the false alarm probability in x-axis. Figure 4 A shows the ROC curve of the classifier, with RBF and PUF kernel functions. Area under the ROC curve should be high for an excellent classifier. In our classifier, the area under the ROC curve is 0.984, which indicates TP rate attains its highest values when FP rate is as low as 0.016.

Figure 4 A shows the ROC curve of the classifier with RBF, and

MiRPara, Triplet-SVM, MiRFinder, etc. are the main examples of SVM-based classifiers. We tried to compare performance of our SVM classifier with that of above three tools. Figure 5 B shows the comparison of their accuracy.

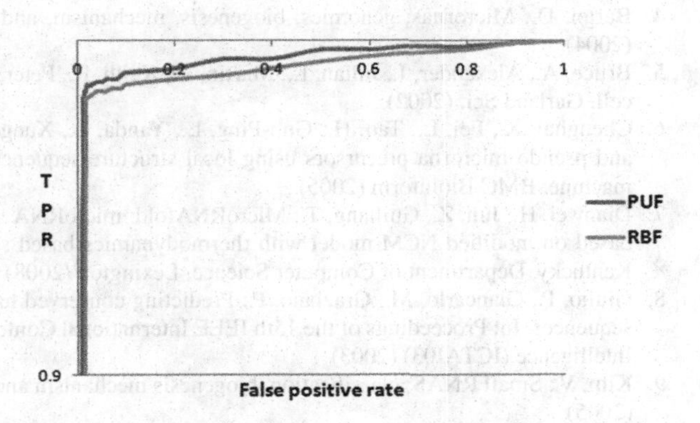

Fig. 4 ROC *curves* of SVM with RBF kernel and PUK kernel

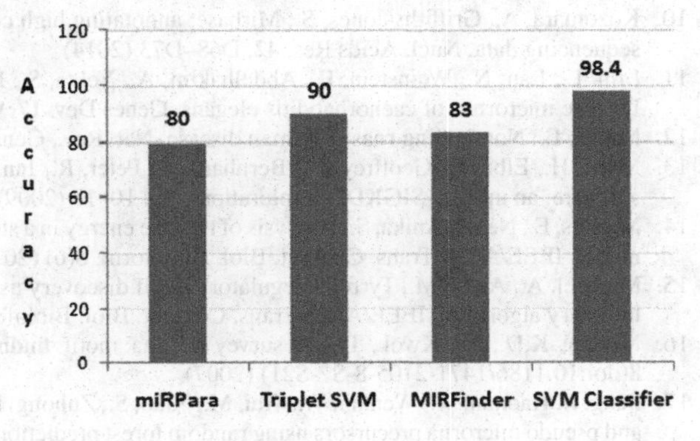

Fig. 5 Comparison of accuracy of different microRNA prediction tools

4 Conclusion

The classifier that we developed distinguishes microRNAs and nonmicroRNAs very accurately. When compared with other tools that employ SVM as the classifier, our results sense better possibly due to the use of hybrid future set, precise feature selection, and selection of best classifier algorithm. The accuracy of our tool is 98.4 % which is higher than that of existing SVM-based classifier such as MiRFinder, Triplet-SVM, and MirPara. The classifier sensitivity is 98.4 % and specificity is 98.4 % which is also higher than that of existing classifiers.

References

1. Anastasis, O., Martin, R., Panayiota, P.: Image encryption based on chaotic modulation of wavelet coefficients. IEEE Trans. Inf. Technol. Biomed. **13**(1), (2009)
2. Arnaz, M., Robert, X.G.: Pca-based feature selection scheme for machine defect classification. IEEE Trans. Instrum. Measur. **53**(6), (2004)
3. Aurora, E.K., Frank, J.S.: Oncomirs micrornas with a role in cancer. Nat. Rev. Cancer **6**, 259–270 (2006)
4. Bartel, D.: Micrornas: genomics, biogenesis, mechanism, and function. Cell **116**, 281–297 (2004)
5. Bruce, A., Alexander, J., Julian, L., Martin, R., Keith, R., Peter, W.: Molecular biology of the cell. Garland Sci. (2002)
6. Chenghai, X., Fei, L., Tao, H., Guo-Ping, L., Yanda, L., Xuegong, Z.: Classification of real and pseudo microrna precursors using local structure-sequence features and support vector machine. BMC Bioinform (2005)
7. Dianwei, H., Jun, Z., Guiliang, T.: MicroRNAfold: microRNA secondary structure prediction based on modified NCM model with thermodynamics-based scoring strategy. University of Kentucky, Department of Computer Science, Lexington (2008)
8. Giulio, P., Giancarlo, M., Graziano, P.: Predicting conserved hairpin motifs in unaligned rna sequences. In: Proceedings of the 15th IEEE International Conference on Tools with Artificial Intelligence (ICTAI03) (2003)
9. Kim, V.: Small RNAS: classification, biogenesis mechanism and function. Mol. Cell **19**, 1–15 (2005)
10. Kozomara, A., Griffiths-Jones, S.: Mirbase: annotating high confidence micrornas using deep sequencing data. Nucl. Acids Res. **42**, D68–D73 (2014)
11. Lim, L., Lau, N., Weinstein, E., Abdelhakim, A., Yekta, S., Rhoades, M., Burge, C., Bartel, D.: The micrornas of caenorhabditis elegans. Genes Dev. **17**, 991–1008 (2003)
12. Manel, E.: Non-coding rnas in human disease. Nat. Rev., Genet. (2011)
13. Mark, H., Eibe, F., Geoffrey, H., Bernhard, P., Peter, R., Ian H., W.: The weka data mining software: an update. SIGKDD Explorations, pp. 10–18 (2009)
14. Markus, E., Nebel, Anika, S.: Analysis of the free energy in a stochastic rna secondary structure model. IEEE/ACM Trans. Comput. Biol. Bioinform. **8**(6) (2011)
15. Michael, A., Andy, M., Tyrrell: Regulatory motif discovery using a population clustering evolutionary algorithm. IEEE/ACM Trans. Comput. Biol. Bioinform. **4**(3) (2007)
16. Modan, K.D., Ho-Kwok, D.: A survey of dna motif finding algorithms. BMC Bioinfor. **8**(doi:10.1186/1471-2105-8-S7-S21) (2007)
17. Peng, J., Haonan, W., Wenkai, W., Wei, M., Xiao, S., Zuhong, L.: Mipred: classification of real and pseudo microrna precursors using random forest prediction model with combined features. Nucleic Acids Res. **35**(1) (2007)

18. Reshmi, G.: Vinod Chandra, S., Janki, M., Saneesh, B., Santhi, W., Surya, R., Lakshmi, S., Achuthsankar, S.N., Radhakrishna, P.: Identification and analysis of novel micrornas from fragile sites of human cervical cancer: computational and experimental approach. Genomics **97**(6), 333–340 (2011)
19. Salim, A., Vinod Chandra, S.: Computational prediction of micrornas and their targets. J. Proteomics Bioinform. 7:7, 193–202 (2014)
20. Ting-Hua, H., Bin, F., Max, F.R., Zhi-Liang, H., Kui, L., Shu-Hong, Z.: Mirfinder: an improved approach and software implementation for genome-wide fast microrna precursor scans. BMC Bioinform. (2007)
21. Vinod Chandra, S., Reshmi, G.: A pre-microrna classifier by structural and thermodynamic motifs (2009)
22. Vinod Chandra, S., Reshmi, G., Achuthsankar, S.N., Sreenathan, S., Radhakrishna, P.: MTAR: a computational microrna target prediction architecture for human transcriptome. BMC Bioinform. **10**(S1), 1–9 (2010)
23. Wu, Y., Wei, B., Liu, H., Li, T., Rayner, S.: Mirpara: a SVM-based software tool for prediction of most probable microrna coding regions in genome scale sequences. BMC Bioinform. **12**(107) (2011)
24. Ying-Jie, Z., Qing-Shan, N., Zheng-Zhi, W.: Identification of microrna precursors with new sequence-structure features. J. Biomed. Sci. Eng. **2**, 626–631 (2009)
25. Yunpen, X., Xuefeng, Z., Weixiong, Z.: Microrna prediction with a novel ranking algorithm based on random walks. Bioinformatics **24** (2008)

DCoSpect: A Novel Differentially Coexpressed Gene Module Detection Algorithm Using Spectral Clustering

Sumanta Ray, Sinchani Chakraborty and Anirban Mukhopadhyay

Abstract Microarray-based gene coexpression analysis is widely used to investigate the regulation pattern of a group (or cluster) of genes in a specific phenotype condition. Recent approaches look for differential coexpression patterns, where there exists a significant change in coexpression pattern between two phenotype conditions. These changes happen due to the alternation in regulatory mechanism across different phenotype conditions. Here, we develop a novel algorithm DCoSpect to identify differentially coexpressed modules across two phenotype conditions. DCoSpect uses spectral clustering algorithm to cluster the differential coexpression network. The proposed method is assessed by comparing with state-of-the-art techniques. We show that DCoSpect outperforms the state of the art in terms of significance and interpretability of detected modules. The biological significance of the discovered modules is also investigated using GO and pathway enrichment analysis.

Keywords Microarray gene expression · Differential coexpression · Spectral clustering · Correlation

1 Introduction

The DNA microarray is constructed from a number of DNA samples involving an mRNA molecule hybridization to the DNA template. These microarrays are used in a wide range of applications like gene expression profiling, disease detection,

---

S. Ray (✉) · S. Chakraborty
Department of Computer Science and Engineering, Aliah University,
Kolkata 700156, West Bengal, India
e-mail: sumantababai86@gmail.com

A. Mukhopadhyay
Department of Computer Science and Engineering, University of Kalyani,
Kalyani 741235, India

S. Das et al. (eds.), *Proceedings of the 4th International Conference on Frontiers in Intelligent Computing: Theory and Applications (FICTA) 2015*, Advances in Intelligent Systems and Computing 404, DOI 10.1007/978-81-322-2695-6_7

comparative genomic hybridization, SNPs, and in many field of bioinformatics [9]. Among these, gene expression profiling provides vast scope for the study of gene functional coherence which is useful for finding biological pathways and regulatory networks along with mining disease-related genes like tumors [10].

As gene differential expression does not include pairwise (or group-wise) analysis, hence various biological processes such as function of protein products and alteration in tumor cell growth process could not be entirely studied under it [2, 3, 5, 6]. Therefore, differential coexpression analysis has been developed as a complementary method. Differentially coexpressed genes exhibit high degree of coherence in one class while low degree of coherence in the other, which means they have substantially different levels of coherence in their expression profile. In mathematical terms, differential gene expression aims at changes in first-order moments (means), while differential gene coexpression opts for changes in second-order moments (covariance). Moreover, gene expression analysis can be carried out gene by gene, whereas differential coexpression analysis is generally found among pairs or clusters [3]. Differential coexpression may indicate the disruption of a regulatory mechanism possibly caused by disregulation of pathways or mutations of transcription factors.

Several studies exist for identifying differentially coexpressed modules in two different phenotype conditions [2, 4, 12, 13]. For example, CoXpress [13] is an R package that identifies differentially expressed genes from a dataset by employing hierarchical clustering to figure out relationships among genes. Differentially coexpressed pathways have also been determined by a technique called dCoxS [2]. A recently proposed method called DICER(Differential Correlation in Expression for meta-module Recovery) [1] aims to identify two groups of clusters; one is differentially coexpressed and the other is meta-module. CLICK [11] is an algorithm that clusters multi-condition gene expression patterns. It utilizes graph-theoretic and statistical techniques to identify tight groups of highly similar dements(kernels) that are likely to belong to the same true cluster. Another approach called Diffcoex [12] identifies differentially coexpressed modules across two conditions using a statistical framework. It avails a prominent and powerful tool for coexpression analysis, i.e., WGCNA [4] followed by hierarchical clustering to group differentially coexpressed genes pairwise.

In this article we have proposed a novel framework DCoSpect based on spectral clustering [7, 8] to detect differentially coexpressed modules across two conditions in a dataset. DCoSpect is easy to implement and it applies the knowledge of spectral clustering. It utilizes top eigenvectors and eigenvalues through the eigenvalue gap of a matrix for clustering. As Fig. 1 shows, the correlation matrices for two conditions are calculated and then a differentially coexpressed matrix is compiled by pairwise correlation analysis of genes from the two correlation matrices. Finally, spectral clustering is applied in order to get groups of differentially coexpressed genes.

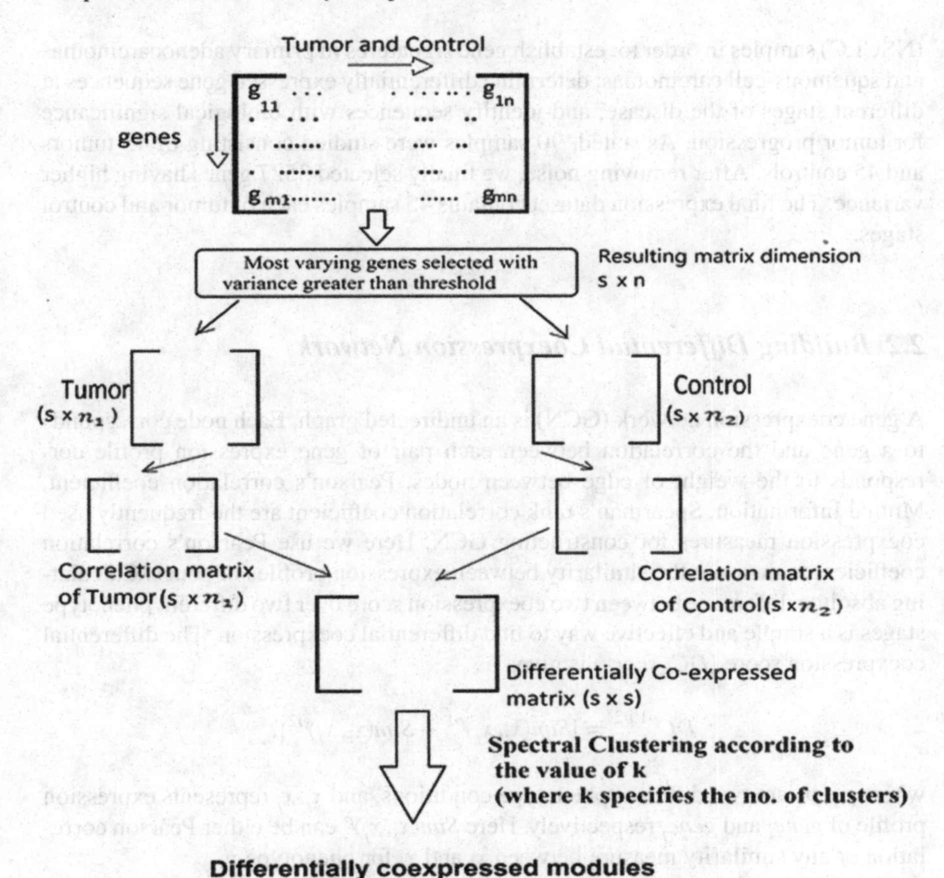

Fig. 1 Diagrammatic representation of DCoSpect

2 Method

In this section we describe our proposed method. This section consists of three subsections involving : dataset preparation, building differential coexpression network, and detection of module using spectral clustering.

2.1 Dataset Preparation

We downloaded the series GSE18842 dataset from the GEO database (http://www.ncbi.nlm.nih.gov/geo/), which contains tumor versus control genomic study. In this study, microarray analysis was performed in a set of 90 non-small cell lung cancer

(NSCLC) samples in order to: establish gene signatures in primary adenocarcinomas and squamous-cell carcinomas; determine differentially expressed gene sequences at different stages of the disease; and identify sequences with biological significance for tumor progression. As stated, 90 samples were studied consisting of 45 tumors and 45 controls. After removing noise, we finally selected 3527 genes having higher variance. The final expression dataset contains 45 samples each of tumor and control stages.

2.2 Building Differential Coexpression Network

A gene coexpression network (GCN) is an undirected graph. Each node corresponds to a gene and the correlation between each pair of gene expression profile corresponds to the weight of edge between nodes. Pearson's correlation coefficient, Mutual Information, Spearman's rank correlation coefficient are the frequently used coexpression measures for constructing GCN. Here we use Pearson's correlation coefficient to measure the similarity between expression profiles of genes. Calculating absolute difference between two coexpression score over two different phenotype stages is a simple and effective way to find differential coexpression. The differential coexpression score (*DC_score*) is given as

$$DC_{i,j}^{p1,p2} = |Sim(x_i, x_j)^{p1} - Sim(x_i, x_j)^{p2}|,$$

where p_1, p_2 are two different phenotype conditions, and x_i, x_j represents expression profile of $gene_i$ and $gene_j$ respectively. Here $Sim(x_i, x_j)^p$ can be either Pearson correlation or any similarity measure between x_i and x_j for phenotype p.

Here two coexpression networks are calculated employing the Pearson's correlation coefficient method. These two coexpression networks are built only for those genes that we have considered after extracting it by setting a cutoff. From these two networks pairwise genes are examined and a differential coexpression network is built using the absolute difference similarity method.

2.3 Detecting Modules Using DCoSpect

DCoSpect is based on spectral clustering which uses top eigenvectors and eigenvalues of matrices derived from data. Here we describe some basic concept of similarity graphs, graph Laplacians, and graph cut problem used in spectral clustering algorithm.

Similarity Graphs Given some datasets $x_1 \ldots x_n$ and some measure for similarity $s_{i,j}$, a similarity graph $G = (V, E)$ is constructed consisting of the data points x_i as its vertices v_i. Any two data points x_i and x_j are only connected when their edge weight has value equal to the similarity measure given, i.e., $w_{ij} = s_{ij}$. Here we used the DC_score as a similarity measure to compile the differential coexpression graph.

Graph Laplacians Spectral graph theory consists of the description of graph Laplacians. There are two types of graph Laplacians : normalized and unnormalized. All of these have various properties which lead to several types of spectral clustering. Assuming G to be an undirected weighted graph and W as its weight matrix where $w_{ij} = w_{ji} \geq 0$. The unnormalized graph Laplacian is given as: $L = D - W$, where D is the degree matrix (which is a diagonal matrix) given by $d_i = \sum_{j=1}^{n} W_{ij}$. The normalized graph Laplacian leads to two different spectral clustering matrices corresponding to the definition of two graph Laplacians. The first one is given as similarity $L_{sym} = D^{-1/2}LD^{-1/2}$ and the second is given as random walk $L_{rw} = D^{-1}W$. In DCoSpect we can choose any one of the above graph Laplacians according to the type of spectral clustering.

Graph Cuts The similarity graphs and the graph Laplacians defined above are the basis required to do clustering. The graph cuts generally involve partitioning the graph formed in such a way that data points within a group have higher weights while data points from two different groups have lower weights. There are three types of the graph cut problem: The mincut problem, the RatioCut problem, and the Ncut problem. The mincut problem does not gives the result that we actually want in clustering since it often just separates a single vertex from the rest of the graph. The

---

Algorithm 1 Algorithm DCoSpect

Input: Gene expression dataset of tumor and control stage of non-small cell lung cancer (NSCLC) samples.

Output:Differential coexpression modules.

Step 1. Preparing the dataset

Let $Q \in R^{m\times n}$ be the matrix where $Q_1, Q_2, \ldots Q_n$ are the columns containing tumor and control states and the rows contain the gene samples. Here $m \times n$ is equal to 54676×90.

Select most varying genes by calculating the variance of each expression profile and filter out those that have variance less than a certain threshold.

Separate the control state from the tumor state. Hence two matrices are derived from Q. One is $T \in R^{(s\times n_1)}$ (s=3527, n_1=45) where $T_1 \ldots T_{n_1}$ are the columns representing only tumor state. Another matrix is $C \in R^{(s\times n_2)}$ (s=3527, n_2=45) where $C_1 \ldots C_{n_2}$ are the columns representing only control state..

Step 2. Building the differential coexpression network

Calculate the correlation matrices T_cormat and C_cormat for both matrices T and C as follows: $T\_cormat(i,j) = corr(x_i^{tumor}, x_j^{tumor})$, and $C\_cormat(i,j) = corr(x_i^{control}, x^{control_j})$, where x^p represents gene expression profile of stage $p = tumor, control$, .

Build differential coexpression matrix $D \in R^{(s\times s)}$(where s=3527) as follows: $D_{i,j} = |T\_cormat(i,j) - C\_cormat(i,j)|$

Step 3. Performing clustering

Calculate the normalized graph Laplacian of matrix D. Let the resulting matrix be L.

Compute the first k eigenvectors $u_1 \ldots u_k$ of the normalized graph Laplacian L_{rw}

Let $U \in R^{n\times k}$ be the matrix containing the vectors $u_1 \ldots u_k$ as columns.

Cluster the points $(y_i)_{i=1}^{n}$ into clusters $C_1 \ldots C_k$ using the k-means algorithm.

Retrieve clusters $A_1 \ldots A_k$ by $A_i = j|y_j \in C_i$.We use $k = 30$

RatioCut problem or the Ncut problem gives better results but can be shown to be NP-hard problems. Considering $A_i, \ldots A_k$ to be the number of connected components in the graphs L_{sym} and L_{rw}, where k represents a number of smallest eigenvectors, then the RatioCut problem performs partitioning by taking into consideration the number of vertices in each group while Ncut problem does this using edge weights.

The DcoSpect algorithm is shown in Algorithm 1

3 Results

In this section we describe the performance of our proposed method with that of state-of-the-art techniques such as DiffCoEx, CoXpress, and CLICK. All these methods determine differential coexpressed modules across two different conditions by utilizing a scoring technique to represent differential coexpression of a gene pair and group them to specific module by some clustering algorithm.

3.1 Comparison with Some State-of-the-art

We applied CLICK, CoXpress, DiffCoex, and proposed method DCoSpect on the tumor and control stage of gene expression dataset and found a set of differentially coexpressed modules in each case. We inspect absolute change in the correlation of each detected module. For this purpose we calculate the average value of absolute change in correlation for each module detected by all methods and plot these values to show the capability of capturing the differential coexpression patterns. Figure 2 shows the extent of differential coexpression patterns detected by the four methods. The X-axis represents absolute change in correlation pattern across acute and chronic

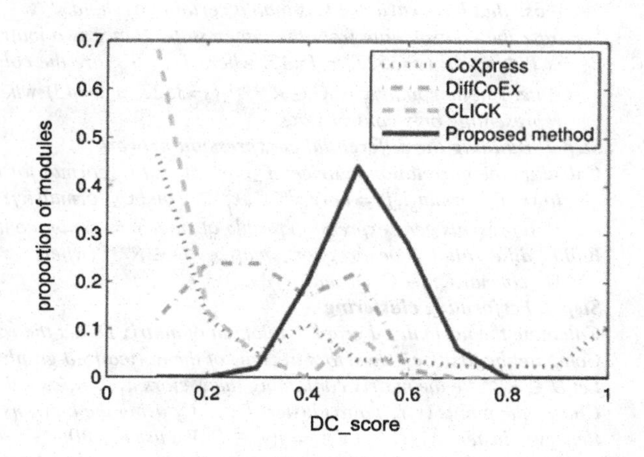

Fig. 2 The distribution of within module absolute change in correlation for DiffCoex, CoXpress, CLICK, and the proposed DCoSpect algorithm

Table 1 GO-terms and KEGG pathway of some identified differentially coexpressed modules

Module (Sl No.)	*DC_score*	GO term(bp)	KEGG pathway
1	0.8216	Positive regulation of molecular function (GO:0044093) (1.7E-5)	Colorectal cancer (1.0E-3)
2	0.8168	Generation of precursor metabolites and energy (GO:0006091)(1.3E-5)	Pancreatic cancer (1.1E-5)
3	0.8131	Generation of precursor metabolites and energy (GO:0006091)(9.1E-6)	Parkinson's disease (3.4E-3)
4	0.8070	Positive regulation of ubiquitin-protein ligase activity during mitotic cell cycle(GO:0051437)(8.1E-5)	Melanoma (6.6E-6)
5	0.8093	Lymphocyte differentiation (GO:0030098) (1.5E-2)	Thyroid cancer (1.6E-2)
6	0.7302	Anaphase-promoting complex-dependent proteasomal ubiquitin-dependent protein catabolic process(GO:0031145)(5.4E-8)	Parkinson's disease(4.5E-6)
7	0.7260	DNA unwinding during replication(GO:0006268) (1.7E-4)	Intestinal immune network for IgA production (1.0E-3)
8	0.7100	mRNA metabolic process (GO:0016071) (2.2E-5)	Systemic lupus erythematosus (4.2E-2)
9	0.7003	Anaphase-promoting complex-dependent proteasomal ubiquitin-dependent protein catabolic process (GO:0031145) (4.7E-6)	Proteasome (1.4E-5)

stages, while the Y-axis stands for the proportion of differentially coexpressed modules that achieve the corresponding value. From Fig. 2 it is clear that our proposed method is superior in detecting differential coexpression changes within modules compared to other methods. By inspecting the distribution of correlation changes between each pair of modules of all the methods, it is evident that our proposed method is significantly shifted toward higher values compared to others. This signifies the superiority of our method in capturing stronger signals with respect to other methods.

3.2 Biological Validation of Modules

In this section we biologically validate the differential coexpresed modules identified by the proposed DCoSpect algorithm. For this purpose we have performed Gene Ontology and pathway-based analysis of the identified modules.

GO and Pathway Enrichment We have investigated to what extent Gene Ontology terms and pathways are associated with the identified differential coexpression modules. For this purpose we collected the GO terms from the GO database and associated these terms with the identified modules. From the KEGG database we also identified significant pathways that are involved in different differential coexpression modules. Table 1 shows significant GO terms and pathways discovered from the identified modules. A careful observation of Table 1 reveals that some of the identified GO terms are common among the modules. This is because of the overlap in the identified modules. In Table 1 we consider only the relevant GO terms and pathways of those modules that have *DC_score* greater than 0.7. From Table 1 it is noticeable that several cancer pathways are often associated with different differential coexpression modules.

4 Conclusion

In this study we have developed a spectral clustering-based framework to identify differential coexpression modules from two microarray datasets corresponding to two different phenotypes. First, a differential coexpression network is built from the microarray dataset and then our proposed algorithm DCoSpect is applied to this network. The resulting modules have significant differential coexpression (measured in *DC_Score*) between two phenotypes.

We compared DCoSpect with three state-of-the-art algorithms: CoXpress, DiffCoEx, and Click with respect to the ability for capturing differential coexpression patterns. As DCoSpect utilizes spectral clustering-based framework, so it is flexible enough to capture the modularization property of graph structure. DCoSpect performs significantly better than other methods for capturing the differential coexpression patterns. The detected modules are also enriched with significant GO terms and KEGG pathway.

References

1. Amar, D., Safer, H., Shamir, R.: Dissection of regulatory networks that are altered in disease via differential co-expression. Plos Comput. Biol. (2013). doi:10.1371/journal.pcbi.1002955
2. Cho, S., Kim, J., Kim, J.: Identifying set-wise differential co-expression in gene expression microarray data. BMC Bioinform. **10**(109) (2009)
3. Kostka D, S.R.: Finding disease specific alterations in the co-expression of genes. Bioinformatics **20**(Sup 1), i194–i199 (2005)
4. Langfelder, P., Horvath, S.: Wgcna: an r package for weighted correlation network analysis. bmc bioinformatics. BMC Bioinform. **9**(559) (2008)
5. Lee, H., Hsu, A., Sajdak, J., Qin, J., Pavlidis, P.: Coexpression analysis of human genes across many microarray data sets. Genome Res. **14**(6), 1085–1094 (2004)
6. Li, K.C.: Genome-wide co-expression dynamics: theory and application. Proc. Natl. Acad. Sci. USA **99**(26), 16875–16880 (2002)

7. Luxburg, U.: A tutorial on spectral clustering. Stat. Comput. **17**(4), 395–416 (2007). Springer Link
8. Ng, A., Jordan, M., Weiss, Y.: On spectral clustering: analysis and an algorithm. NIPS, 849–856 (2001)
9. Pollack, J., Srlie, T., Perou, C., Rees, C.A., Jeffrey, S., Lonning, P., Tibshirani, R., Botstein, D., Dale, A., Brown, P.: Microarray analysis reveals a major direct role of dna copy number alteration in the transcriptional program of human breast tumors. PNAS **99**(20), 12963–12968 (2002)
10. Quackenbush, J.: Microarray analysis and tumor classification. N. Engl. J. Med. **354**(23), 2463–2472 (2006)
11. Sharan, R., Maron-Katz, A., Shamir, R.: Click and expander: a system for clustering and visualizing gene expression data. Bioinformatics **19**(14), 1787–1799 (2003)
12. Tesson, B., Breitling, R., Jansen, R.: Diffcoex: a simple and sensitive method to find differentially coexpressed gene modules. BMC Bioinform. **11**(497) (2010)
13. Watson, M.: Coxpress: differential co-expression in gene expression data. BMC Bioinform. **7**(509) (2006)

Intelligent Topological Differential Gene Networks

Mrityunjay Sarkar and Aurpan Majumder

Abstract Microarray gene expression profiles are frequently explored to understand the causal factors associated with some disease. To date, most of the research being conducted is restricted upon comparison of expression values across more than one condition or the discovery of genes having altered interaction levels with neighbours across conditions. Therefore, differential expression (DE), gene correlation and co-expression have been intensively studied using microarray gene expression profiles. However, in the recent past the focus has been shifted towards conglomeration of differential expression and differential connectivity properties to gain a better insight of the problem, such as investigating the topological overlap (TO) of the network formed by DE genes using the generalized topological overlap measure (GTOM). In this work, we explore through the unweighted–TO networks which requires selection of a smart threshold to transform the GTOM structure into a differential network. The essence of our work lies in the generation of a series of GTOM threshold pairs across different conditions from which the best threshold pair for a network (across different conditions) is selected by comparing the cumulative effect of TO and *p*-value obtained from the series of threshold pairs.

Keywords Peripheral blood · Influenza · Topological overlap (TO) · Generalized topological overlap measure (GTOM) · Unweighted TO · cTOP

M. Sarkar (✉)
Department of ECE, D.I.A.T.M Durgapur, Durgapur, India
e-mail: mrityu1488@gmail.com

A. Majumder
Department of ECE, N.I.T Durgapur, Durgapur, India
e-mail: aurpan.nitd@gmail.com

S. Das et al. (eds.), *Proceedings of the 4th International Conference on Frontiers in Intelligent Computing: Theory and Applications (FICTA) 2015*, Advances in Intelligent Systems and Computing 404, DOI 10.1007/978-81-322-2695-6_8

1 Introduction

Over the last few decades, analyzing gene expression data using microarrays has been used as a common tool for disease research [1]. In general, the healthy (control) samples are compared with the diseased ones (mainly across different stages) to identify tissues having different expression levels across conditions. However, this approach ignores the fact that most of the biological functions require the orchestrated action of many genes. Thus, over and above traditional differential expression (henceforth will be represented by DE) of each gene, the differential connectivity (henceforth will be represented by DC) studies have been done based on gene correlation and co-expression [2]. For many years, DE and DC analysis have been carried out separately, until the recent past where efforts have been put through to adjoin the concepts of DE and DC to have a thorough understanding of the involved network.

Before we proceed further, it is worth to mention that the study of differential correlation or co-expression aims at a deeper understanding of the abnormalities compared to the simple differential expression analysis [3]. As given in [4], change in the coding region and the posttranslational modifications (e.g. phosphorylation, acylation, methylation, etc.) can modify the protein activity without any change in the gene expression level, but does alter the interaction pattern with other genes. We can also consider the example of transcription factors (TF) where a number of down-stream targets could be regulated by a master gene. Unfortunately, in diseased tissues where the regulatory mechanism is dysfunctional, the expression of the gene module will be unordered or random. Gene modules showing this sort of altered association can be detected only by DC analysis, but might be overlooked by DE analysis [3]. In order to amalgamate DE and DC concepts, a simple concept [5, 6] has been applied in recent studies, which is to perform DC analysis on those networks formed by the DE genes only.

Ray and Zhang [7] first developed the differential connectivity measure using a topological overlap (TO) based approach. The concept was to select those DE genes having disjoint set of neighbours across conditions. Based upon the concept, an earlier algorithm have been developed [5] considering all the gene–gene interaction patterns instead of taking only a few interaction patterns, which was the main drawback of [7].

In our previous work [6], we have again extended the concept given in [5], by incorporating two kinds of TO concepts to explore the dynamic interactions. The uniqueness lies in the implementation of generalized topological overlap measure (GTOM) which we have used for the computation of TO instead of simple correlation values and developed a concept called TOP comprising of the TO values as well as gene wise significance values (p-value).

Now, as explored in our previous work [6], the unweighted TOP analysis gives us fruitful results compared to weighted TOP analysis. Accordingly, in this work we have restricted our analysis over unweighted TOP analysis only. Transformation of a fully connected GTOM-based weighted structure into a network requires fixing a threshold [3, 5, 6]. In our current piece of work, we have generated a multitude of GTOM thresholds to cover the entire range of GTOM value in the data. This enables us to detect different kinds of topological overlap changes at the same significance level; strong changes for a few genes, but moderate changes for an appreciable number of genes.

In this work, we deal with two states. In this connection, we develop a sorted set of GTOM threshold values across each condition. We have designed our algorithm in such a way that it would choose only those threshold pairs for which the TOP value is minimum. In other words, we can claim on implementing a global network topology by considering a threshold common for every gene in a network in any particular condition/state instead of considering gene-specific threshold.

We have tested our proposed algorithm with a peripheral blood dataset having (control) healthy samples as well as samples having initial (day 0) and severe (day 6) stages of influenza [8]. As noted in our previous study, we have followed the same strides towards construction of three different networks. The first network comprises of states control and day 0, in the second network states control and day 6, and finally a common network based on the first and second network. We have performed our analysis separately for these three networks. In the results, we have shown that the best threshold pair happens to be an intermediate value in the sorted series of threshold levels and definitely not the extreme levels.

The rest of the paper is as follows. In the next section, we have discussed about the methodology followed by the thorough analysis of the results obtained.

2 Methodology

Algorithm 1. The Intelligent Threshold Selection for First/Second Network

1. Find DE genes between the conditions *control* and *day* 0/*day* 6 of influenza.
2. Evaluation of GTOM measure corresponding to DE genes in each condition using Eq. 2.
3. Go for the Fisher transformation of GTOM matrix and find the maximum value.
4. Choose the condition specific threshold value Δth and set maximum threshold $\text{th}_{max} = |\text{Fisher transformed GTOM values}|_{max}$
5. $n = 1$.
6. while($n \times \Delta\text{th} \leq \text{th}_{max}$)
   begin
   - Find the specific interactions for each and every threshold Δth in each condition.
   - Evaluation of TO by Eq. 3.
   - Go for significance testing and evaluation of TOP by Eq. 4.
   - Calculation of cumulative TOP or cTOP scores for each threshold.
   - $n = n + 1$.

   End

7. Selection of best threshold pairs through comparison of cTOP values for all thresholds.

In a broad sense, the total algorithm can be divided into two segments. The first segment contains the procedure to generate multiple thresholds, and the computation of TOP values using these thresholds. In the second segment, we have discussed about selection of best threshold among them, simply by comparing the different TOP results. The different steps for the execution of the algorithm are discussed below.

2.1 Network Construction

In our work, we have constructed three differential networks, because of three different conditions. Each differential network will consist of two different conditions, having same set of genes with altered expression levels across conditions. Conditions of the first network are control (not affected by influenza) and day 0 (initial observation of peripheral blood of patients having symptoms of influenza). For the second network, the conditions are control and day 6 (seventh day observation of peripheral blood of patients having severe symptoms of influenza).

Next, we find the differentially expressed (DE) genes individually for the first (DE genes between control and day 0, henceforth represented by DE1) and second (DE genes between control and day 6, henceforth represented by DE2) network. The third network is constructed simply by taking the common genes between DE1 and DE2 (hereafter represented as cDE) similar to [6], and carried out the same analysis stated above on these cDE genes.

Algorithm 2. The Intelligent Threshold Selection for Common Network

1. Find the common DE genes between network 1 and network 2.
2. Rest follows step 2 through 7 under Algorithm 1.

With these DE genes (DE1, DE2 and cDE) in hand, we proceed to construct a fully connected network following the algorithm proposed by Ruan and Zhang [9], which have also been implemented in [6]. In a simple sense, it will lead us to a graph where the genes will be represented by nodes and their connection strength by edges. At this stage, we assume a weighted network structure where the edges of the graph are Pearson correlation coefficient (PCC) value between two genes. In this way, we yield six network graphs conveying two differential networks for each of the first, second and common networks.

2.2 GTOM Calculation

As per the algorithm, we compute the generalized topological overlap measure for each DE gene, be it a ϵ DE1 or ϵ DE2 or ϵ cDE for all the six differential networks using the PCC value of the concerned gene with others. Now, as discussed in [7], the traditional GTOM value between gene i and j can be written as per [10].

$$t_{ij} = \begin{cases} \dfrac{l_{ij} + a_{ij}}{\min\{k_i, k_j\} + 1 - a_{ij}} \ldots\ldots\ldots i \neq j \\ 1 \ldots\ldots\ldots\ldots\ldots\ldots\ldots\ldots\ldots i = j \end{cases} \quad (1)$$

where $l_{ij} = \sum_u a_{iu} a_{uj}$ (a_{iu} is the PCC value between gene i and u, similarly, a_{ju} is the PCC value between gene j and u) $k_i = \sum_u a_{iu}$ where the index u runs over all nodes of the network except node i. Equation (1) can be rewritten as

$$t_{ij} = \begin{cases} \dfrac{|N_m(i) \cap N_m(j)| + a_{ij}}{\min\{|N_m(i) \cap N_m(j)|\} + 1 - a_{ij}} \ldots\ldots i \neq j \\ 1 \ldots\ldots\ldots\ldots\ldots\ldots\ldots\ldots\ldots\ldots\ldots\ldots\ldots\ldots i = j \end{cases} \quad (2)$$

where $N_m(i)$ represents the neighbours of node i excluding i itself, and $N_m(i) \cap N_m(j)$ means the set of common neighbours shared by gene i and j, for an m order GTOM measure (suggesting that the common node is reachable from node i within m steps and vice versa). As there are two conditions, thus we calculate the GTOM measure between every pair of genes in both the conditions. Here, we have chosen the order $m = 1$.

The above set of equations highlights upon the fact that GTOM value depends on a connectivity measure between every pair of genes directly as well as via all other genes. Accordingly, it would give us an idea about gene pairs having strong connectivity not only between themselves but via indirect gene regulations too. In other words, genes having highly connected neighbourhood across both conditions can be studied further.

2.3 Construction of Multiple Unweighted Networks Using Multiple Thresholds

Here, we elaborate steps 3–6 of our proposed Algorithm 1. As prior investigation [6] clarifies about unweighted TO outperforming the weighted TO measure, thus in this work we proceed through the unweighted approach only. In this context, to convert the weighted GTOM measure to unweighted matrices we proceed as follows.

2.3.1 First of all, we have computed the Fisher transformed GTOM values using the underlying formula [3] $z_{ij} = 0.5 \times \frac{1 + \text{GTOM}_{ij}^{q}}{1 - \text{GTOM}_{ij}^{q}}$, where $q = 1$.

2.3.2 Unweighted GTOM networks are constructed from a comprehensive series of z-score thresholds. Let the maximum value of $z = z_{\text{max}}$ (selected from the entire z matrix). In this work, we have generated 100 different thresholds, and the best threshold is chosen amongst them. In order to attain the same, a set of thresholds T is chosen from the series of 100 equidistant values between 0 to z_{max}. Thus, the minimum threshold (i.e. the difference between two thresholds) can be written as $\Delta \text{th} = \frac{z_{\text{max}}}{100}$

So, if we denote the series of thresholds as $T = \{T_1, T_2, T_3, \ldots, T_{100}\}$, then we can write that $T_1 = \Delta\text{th}$, $T_2 = T_1 + \Delta\text{th}, \ldots, T_{100} = T_1 + 99\Delta\text{th}$

2.3.3 Finally, we construct 100 sub-networks (in each condition) by considering each threshold individually, as given in Algorithm 1. It is noteworthy that these thresholds are not gene-specific (local), but they are network-specific (global).

Let A be a matrix whose entries give us the significant interactions and for a threshold T_m, we have found that the strength of interaction (Fisher transformed value) between gene i and j is greater that T_m (means the ith row and the jth column of A is significant). Hence, we put $A_{ij} = 1$, else $= 0$.

2.4 Unweighted TO Analysis

On having the matrix A, of significant interactions, we perform our computation of unweighted topological overlap TO_i for each gene i as given in Algorithm 1. For a particular threshold T_m, if a gene i in condition1 has X_i significant interactions and in condition2 it is Y_i, then as per [6], the TO of gene i between two networks can be defined as

$$TO_i = (X_i \cap Y_i)/\max(X_i, Y_i) \tag{3}$$

2.5 Calculation of TOP Value

Following Algorithm 1, we measure the p-value of each gene based upon the TO value of the corresponding gene. For this, we performed random shuffling of the sample labels between control and disease states. Next, based upon the shuffled states, we have evaluated the Fisher transformed GTOM matrices to compute the TO of each gene.

Then, we have evaluated the average of TO measure and *p*-value individually for every gene and termed that TOPavg. An equation for the same is given below

$$\mathrm{TOPavg}_i = (\mathrm{TO}_i + p\mathrm{value}_i)/2 \quad (4)$$

Now, let us assume that we are having *N* genes. In this connection, we obtain the TOP score individually for all the *N* genes. Adding up the individual TOP scores, we obtain a cumulative TOP score, cTOP. Here, we denote the cumulative TOP score for *i*th threshold as cTOP_i.

2.6 Selection of Best Threshold (Intelligent Threshold Selection)

As we have taken 100 different thresholds, thus there are 100 cTOP values. We put an end to our algorithm in step 7, where we have compared all the cTOP values only to find the minimum cTOP and the corresponding best threshold. It should be noted that as each network consists of two conditions, this algorithm yields up a pair of thresholds.

2.7 Preprocessing

Here, we have followed the steps as discussed in [6]. We had to reduce our dataset obtained from [8], with accession no. GSE 27131, because of its large size (33,297 genes with 21 time instants). Thus, to continue with our algorithm using this entire data would lead us to space and time complexity issues. The steps of preprocessing are stated below.

- In order to normalize the skewness of the data set for different entries and to make the distribution uniform first we have gone for 'log-normal' distribution.
- Next, we have evaluated the standard deviation of all genes over different expression levels. Then, a threshold value is chosen (mean of the standard deviations) and those genes which posses a standard deviation greater than the threshold are finally chosen to start off with.

3 Results and Discussion

The data set is the temporal analysis of (peripheral blood) patients having severe pdm (H1N1) influenza. The data consists of the expression profiles of seven control (healthy) and 14 patient samples. Patient samples are again equally distributed across two kinds of conditions, condition1 consists of first day (i.e. day 0, initial stage of influenza) and condition2 consists of seventh day (i.e.day 6, severe stage of influenza) peripheral blood expression profiles. Thus, on the whole we have (3*7) or 21 time instants with 33,297 genes. On conducting the necessary preprocessing, the data is reduced to have 10,000 most variant genes.

Our main intention is to form the differential networks using DE genes. First network comprises of DE genes (DE1) between control and day 0 (initial stage of influenza), whereas the second network comprises of DE genes (DE2) between control and day 6 (severe stage of influenza). We have done this by DEGseq [11] (It is an R package to find the differentially expressed genes. Here, depending upon expression values of genes at different time instants, specifically between conditions and by a particular threshold p-value/z-score/q-value DE genes are computed). Next we have searched for common DE genes (cDE) between DE1 and DE2. Completion of this task has given us 73, 82 and 42 DE1, DE2 and cDE genes, respectively. We have carried out our intelligent threshold selection in unweighted TOP analysis individually for all the three kinds of DE networks. The corresponding steps of validation/significance are given below.

Following the steps of our algorithm at first, we have distributed the Fisher transformed [3] GTOM values over 100 equidistant interaction levels, and then taking each sample as a threshold we have computed the cumulative TOPavg value. Finally, comparing the TOPavg values, we compute the best threshold pair applicable for a network (first, second or common) across two conditions.

In Fig. 1, we have given the distribution of the TOPavg values over the 100 threshold values for all the three networks. It is clear from the figure that the best threshold values do not reside on the extreme left (minimum) or right (maximum) side of the distribution, but it tends to be an intermediate value a bit inclined toward the right hand side of the distributions. For the first network, the threshold values are distributed within 0.002–0.0236 and within 0.003–0.0273 in the two conditions, respectively. Our results for intelligent threshold selection have given us 0.0170 and 0.0197 to be the best for first network across control and day 0. Similarly, for the second network the threshold values are distributed from 0.002 to 0.0215 and from 0.002 to 0.0248 across control and day 6, where the application of our algorithm yields up 0.0131 and 0.0151 to be the best threshold pair for this network. For the common network, our algorithm finds 0.0145 to be the best for both the conditions where the thresholds are distributed from 0.002 to 0.0204 in both day 0 and day 6.

Now, our search area specially focuses on the gene wise TOPavg analysis for the selected threshold pairs across individual networks. Results obtained using the best threshold pairs are given in Fig. 2. From Fig. 2, it is evident that the number of significant genes in common network is much higher than first network, which is again greater than the second network. In order to validate our findings, next we have gone for a physical network wise validation using BioLayout [12]. In Figs. 3, 4 and 5, we have given the networks formed by the 73, 82 and 42 DE genes in first, second and common network across conditions. From the plots, it is quite apparent that there are major differences in connectivity of genes between conditions. The genes are having a low connectivity with neighbours across control state, compared

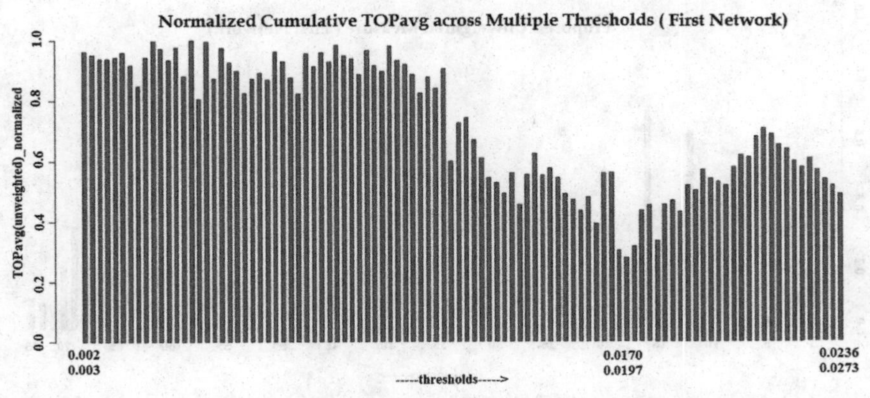

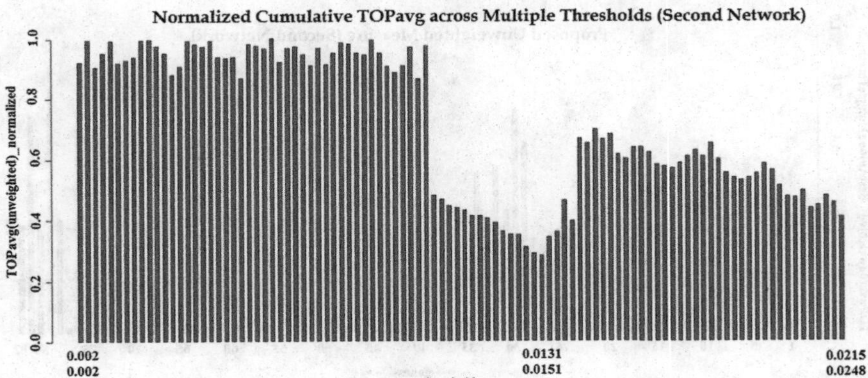

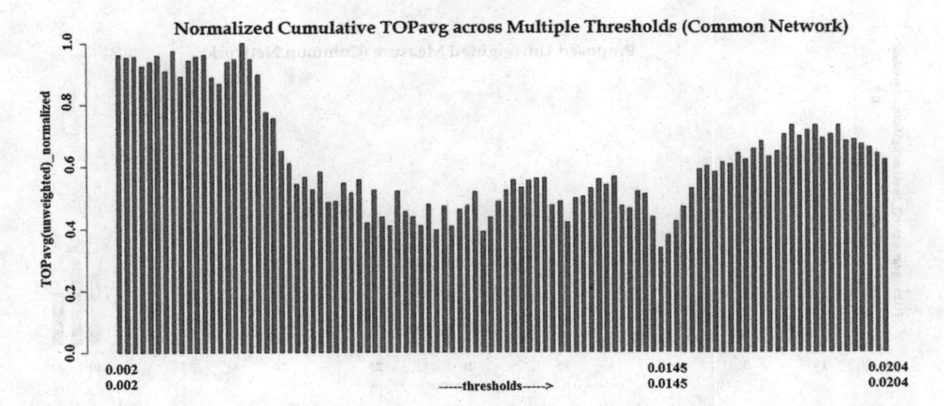

Fig. 1 Plot of normalized *cTOP* value against each threshold pair by proposed methodology across first (*top panel*), second (*middle panel*) and common (*bottom panel*) network

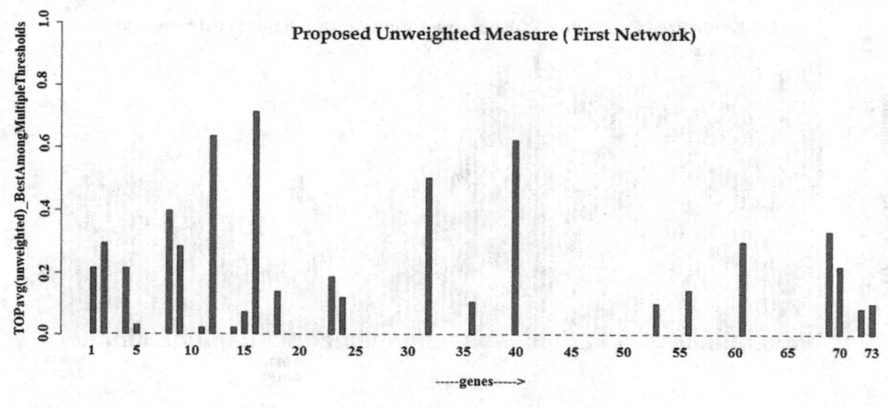

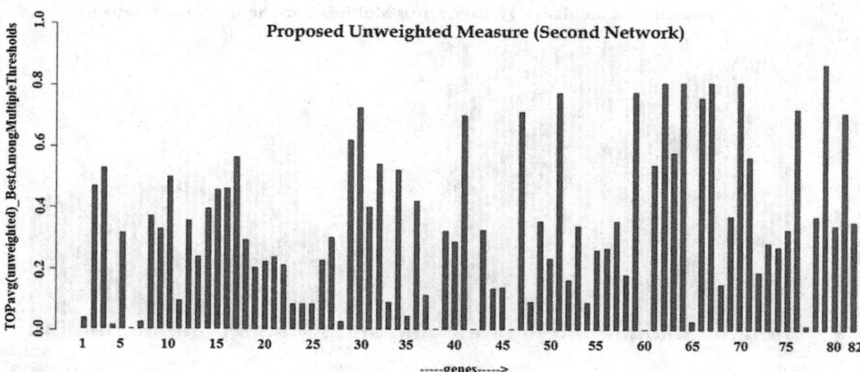

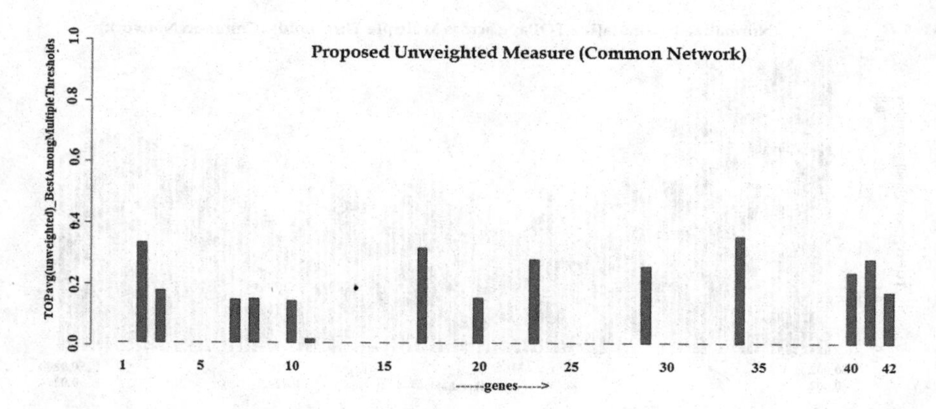

Fig. 2 Plot of *TOPavg* against each gene by proposed methodology using the best threshold pairs across first (*top panel*), second (*middle panel*) and common (*bottom panel*) network

to diseased state. Now, if this is true then it means the concerned genes must possess a low TO value. In order to mathematically validate our assumption, we have searched for the TO values (using best thresholds) of the genes. As expected, genes showing altered connectivity across conditions are showing very low TO values. As an example, an exhaustive search of genes in first network shows us ELL2, IGLV6-57, IGKC, HIST1H3J, DTL and Chr2:89629867-89630178 genes having different connectivity properties possessing TO values 0, 0, 0, 0, 0 and 0.05, respectively. In the second network, such genes are CEACAM6, SUCNR1, MPO, GGH, MS4A3, TDRD9, CLEC5A, LCORL and ANLN acquiring strong connectivity in control network, but weak connectivity in diseased network. The corresponding TO values are 0.051, 0.293, 0.056, 0.316, 0.267, 0.131, 0.043, 0 and 0.17, respectively. In common network, the best genes are C19orf59, HIST1H2AB, BUB1, HIST1H1B and CD177, where all the genes possess 0 TO value. In order to gain more impact of our algorithm, we have also searched for TO values of those genes which do not posses significant change in connectivity. As an example, let us take HBZ, RAB13 and ANKRD22 from the second network. As all of them are having a weak connectivity in both conditions, they also possess high TO values such as 0.423, 0.56 and 0.47.

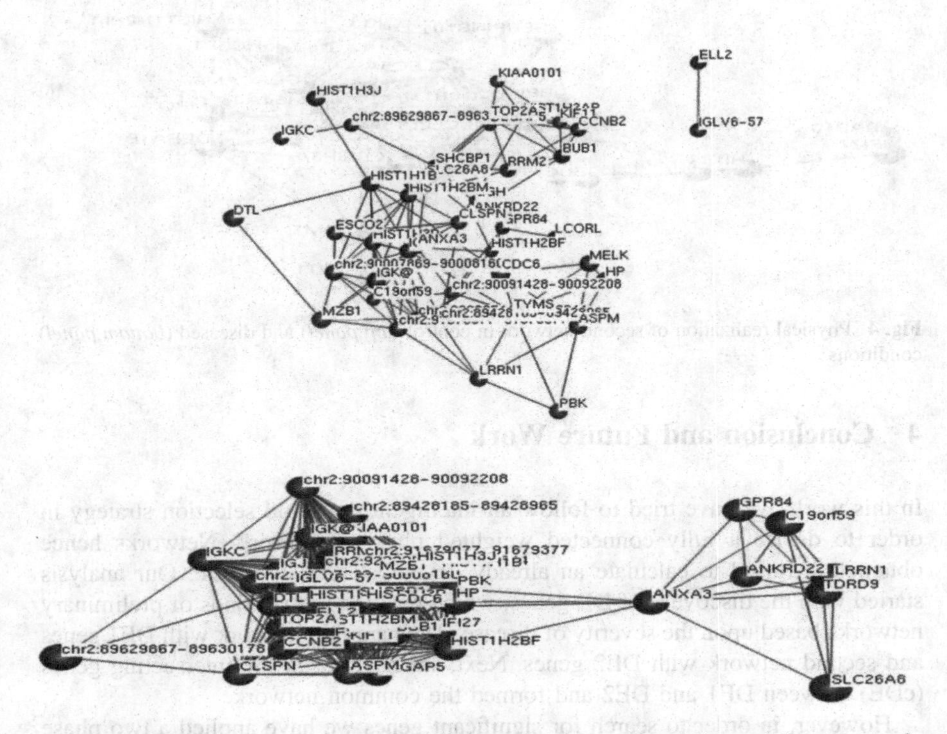

Fig. 3 Physical realization of first network in control (*top panel*) and diseased (*bottom panel*) conditions

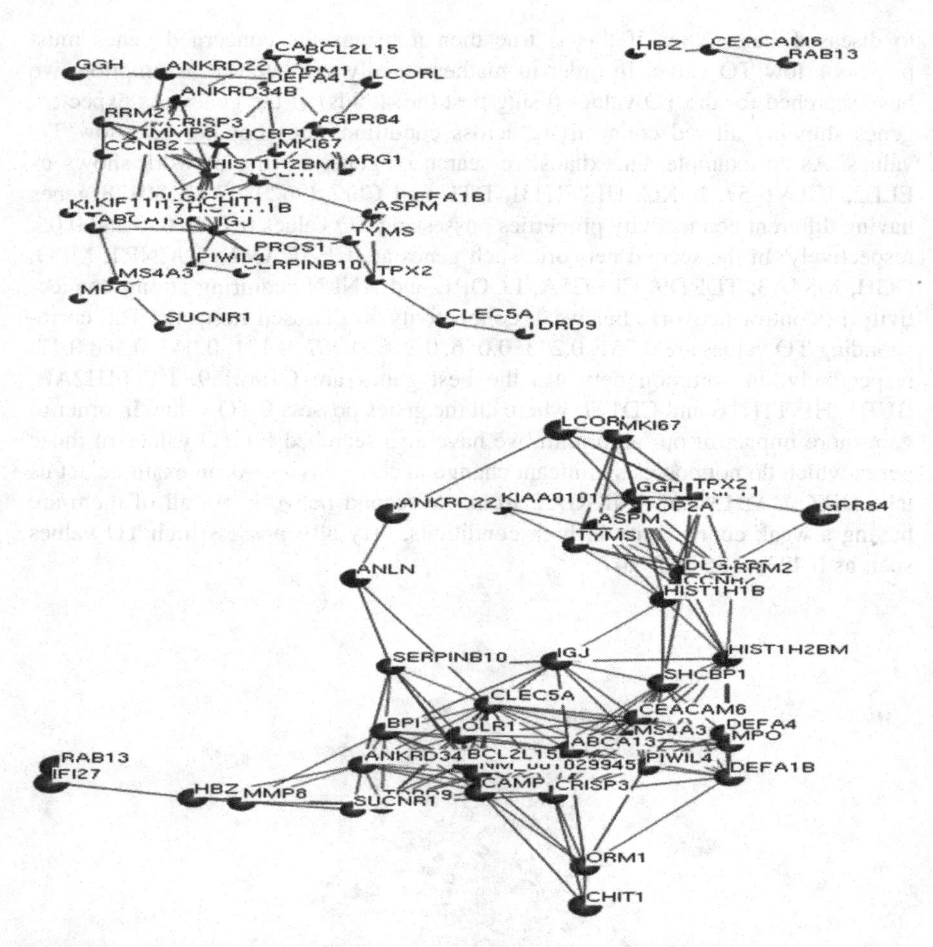

Fig. 4 Physical realization of second network in control (*top panel*) and diseased (*bottom panel*) conditions

4 Conclusion and Future Work

In this work, we have tried to follow an intelligent threshold selection strategy in order to devise a fully connected weighted physical network. Networks hence obtained are used to calculate an already developed measure TOP. Our analysis started with the discovery of DE genes, which resulted in two kinds of preliminary networks based upon the severity of disease. They are first network with DE1 genes and second network with DE2 genes. Next, we have taken the intersecting genes (cDE) between DE1 and DE2 and formed the common network.

However, in order to search for significant genes we have applied a two phase filtering strategy. It is the first phase of filtering exerted on the Fisher transformed

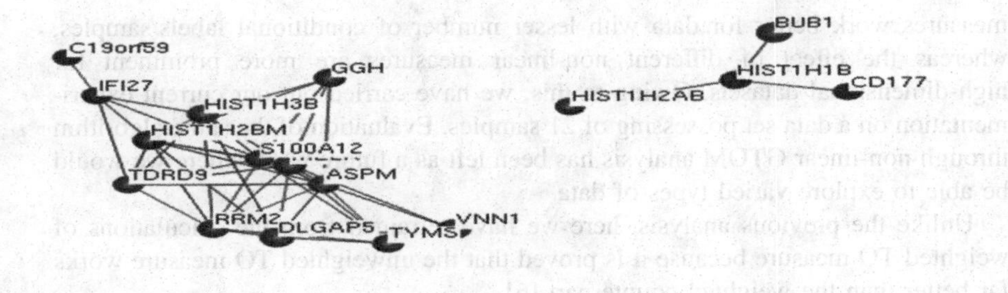

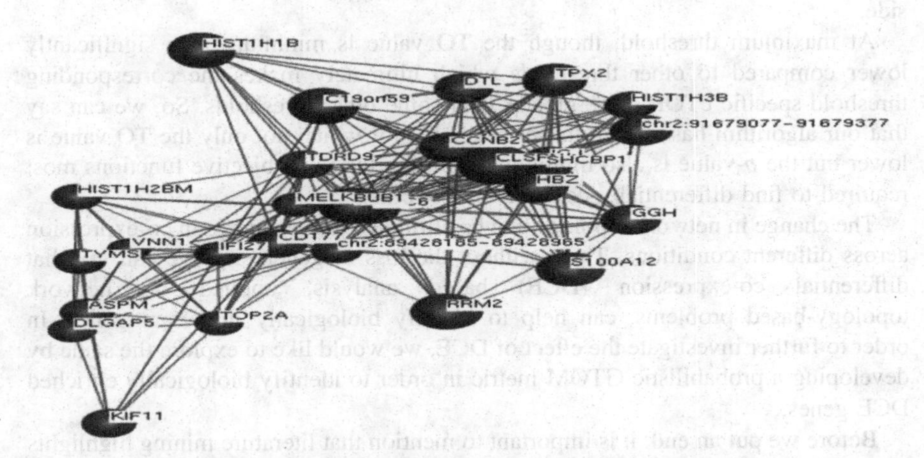

Fig. 5 Physical realization of common network in control (*top panel*) and diseased (*bottom panel*) conditions

GTOM measures considering a series of equidistant thresholds. Here, we have taken into consideration 100 equidistant thresholds which yields the best threshold value (as shown in Fig. 1) to be of the order 10^{-4}. Hence, further significant improvement in the best threshold value by considering greater than 100 thresholds is less likely to occur and inevitably will increase the time and space complexity of our algorithm. Accordingly, we have restricted ourselves to 100 equidistant thresholds only. After that for each threshold value, we have computed the topological overlap (TO) measure to apply the second phase of filtering. Based upon this TO result, we have gone for significance testing, and finally by taking an average of the TO and *p*-value, i.e. TOP is calculated for each gene. We bring an end computing the cumulative TOP, i.e. cTOP for each threshold. Best threshold pairs are then *calculated* simply comparing the cTOP values.

Our present work evaluates the TO score based upon GTOM matrix, which is calculated via Pearson correlation measure. Now, as given in [13] and experimentally established in [14], in gene association networks, linear correlative measures work better for data with lesser number of conditional labels/samples, whereas the effect of different non-linear measures are more prominent for high-dimensional datasets. Owing to this, we have carried out our current experimentation on a data set possessing of 21 samples. Evaluation of the same algorithm through non-linear GTOM analysis has been left as a future work where we would be able to explore varied types of data.

Unlike the previous analysis, here we have refrained from the calculations of weighted TO measure because it is proved that the unweighted TO measure works far better than the weighted counterpart [6].

From Fig. 2, we observe that for each network the best threshold value is not the highest one (extreme right), but an intermediate value, inclined towards the right side.

At maximum threshold, though the TO value is minimum it is significantly lower compared to other thresholds which ultimately makes the corresponding threshold-specific cTOP score greater than some other thresholds. So, we can say that our algorithm has chosen those thresholds for which not only the TO value is lower but the *p*-value is also minimum, fulfilling the two objective functions most required to find differentially connective networks.

The change in network topology can be attributed to the change in co-expression across different conditions. Thus, without the loss of generosity, we can say that differential co-expression (DCE) based analysis, applied on network topology-based problems, can help to identify biologically important genes. In order to further investigate the effect of DCE, we would like to explore the same by developing a probabilistic GTOM metric in order to identify biologically enriched DCE genes.

Before we put an end, it is important to mention that literature mining highlights on a number of algorithms to find genes having differential activity across conditions. In order to find the best problem-specific algorithm among them, it is good to take the help of some statistical properties like specificity and reproducibility [3] which could be handful while comparing many of the algorithms.

References

1. Allison, D.B., Cui, X., Page, G.P., Sabripour, M.: Microarray data analysis: from disarray to consolidation and consensus. Nat. Rev. Genet. **7**, 55–65 (2006)
2. Lai, Y., Wu, B., Chen, L., Zhao, H.: A statistical method for identifying differential gene-gene co-expression patterns. Bioinformatics **20**(17), 3146–3155 (2004)
3. Bockmayr, M., Klauschen, F., Györffy, B., Denkert, C., Budczies, J.: New network topology approaches reveal differential correlation patterns in breast cancer. BMC Syst. Biol. **7**, 78 (2013)
4. de la Fuente, A.: From 'differential expression' to 'differential networking' identification of dysfunctional regulatory networks in diseases. Trends Genet. **26**, 326–333 (2010)
5. Majumder, A., Sarkar, M.: Exploring different stages of Alzheimer's disease through topological analysis of differentially expressed genetic networks. Int. J. Comput. Theory Eng. **6**(5), 386–391 (2014)

6. Sarkar, M., Majumder, A.: TOP: an algorithm in search of biologically enriched differentially connective gene networks. In: Proceedings of 5th Annual International Conference on Advances in Biotechnology (BIOTECH 2015), pp. 124–133. GSTF, Singapore (2015)
7. Ray, M., Zhang, W.X.: Analysis of Alzheimer's disease severity across brain regions by topological analysis of gene co-expression networks. BMC Syst. Biol. **4**, 136 (2010)
8. Berdal, J.E., et al.: Excessive innate immune response and mutant D222G/N in severe A (H1N1) pandemic influenza. J. Infect. **63**(4), 308–316 (2011)
9. Ruan, J., Dean, A.K., Zhang, W.: A general co-expression network-based approach to gene expression analysis: comparison and applications. BMC Syst. Biol. **4**, 8 (2010)
10. Yip, A.M., Horvath, S.: Gene network interconnectedness and the generalized topological overlap measure. BMC Bioinform. **8**, 22 (2007)
11. Wang, L., Feng, Z., Wang, X., Wang, X., Zhang, X.: DEGseq: an R package for identifying differentially expressed genes from RNA-seq data. Bioinformatics **26**(1), 136–138 (2010)
12. Theocharidis, A., Dongen, S. v., Enright, A. J., and Freeman, T.C.: Network visualization and analysis of gene expression data using BioLayout *Express*^{3D}. Nat. Protocols. **4**, 1535–1550 (2009)
13. Chatterjee, S., and Hadi, A.S.: Regression analysis by example. 4th edn. Wiley, ch 2, pp. 21–45 (2006)
14. Majumder, A., Sarkar, M.: Paired transcriptional regulatory system for differentially expressed genes. Lecture Notes Inf. Theory **2**(3), 266–272 (2014)

Part III
Signal Processing

Impact of Threshold to Identify Vocal Tract

Soumen Kanrar

Abstract Speaker identification process is to identify a particular vocal cord from a set of existing speakers. In the speaker identification processes, the unknown speaker voice sample targets each of the existing speakers in the system and gives a predication. The predication is more than one existing known speaker voice and is very close to the unknown speaker voice. It is a one to many mapping. The mapping function gives a set of predicated values associated with the order pair of speakers. In the order pair, the first coordinate is the unknown speaker, and the second coordinates is the existing known speaker from the speaker recognition system. The set of predicated values helps to identify the unknown speaker. The identification process makes a comparison of the unknown speaker model with the models of the existing voice in the system. In this paper, the model is a Gaussian mixture model built by the extraction of the acoustic feature vectors. This paper presents the impact of the decision threshold based on false accepts and false reject for an unknown number of speaker conversion in the speaker identification result. In the simulation, the considered known speaker voices are collected through different channels. In the testing, the GMM voice models of the known speakers are distributed among the numbers of clusters in the test dataset.

Keywords Speaker identification · Gaussian mixture mode · Acoustic feature vectors · Decision threshold · False accept · False reject

1 Introduction

The acoustic signal corresponding to articulation is independent of language. Text-independent and language-independent voices are identified by the tracking of the vocal tract. The human utterances described in terms of a sequence of segments,

S. Kanrar (✉)
Vehere Interactive Pvt Ltd, Calcutta 53, West Bengal, India
e-mail: Soumen.kanrar@veheretech.com

S. Das et al. (eds.), *Proceedings of the 4th International Conference on Frontiers in Intelligent Computing: Theory and Applications (FICTA) 2015*, Advances in Intelligent Systems and Computing 404, DOI 10.1007/978-81-322-2695-6_9

and on the further crucial assumption that each segment can be characterized by an articulator target [1]. 'Articulation' is the activity of the vocal organs in making a speech sounds. The aforesaid biometrics offers greater potentiality over the traditional methods in person recognition by GMM-based speaker identification [2–5]. In particular, voice recognition technology produces relatively low to medium error rate, and it has a high public acceptance rate due to not noticeable nature of voice sample. In general, the voice is the acoustic signal characteristic of a person's individual articulation. The articulation is the aspect of pronunciation involving the articulator organs. So the identification of the unknown speaker is to identify the vocal track of the speaker from the number of existing speaker model present in the system [1]. Vocal cords produce acoustic energy by vibrating as air passes between then. If the claim speaker voice is very near to an existing model in the system or numbers of models, the initial stage of testing basically drills with the one to many matching. The numerical score gives the best probable prediction about the speaker with the list of the voice model present in the voice recognition system. We observe a large amount overlap between the impostors and the true speaker model in the speaker recognition system. It increases the equal error rate higher. The score normalization reduces the equal error rate by the z-norm and T-norm [6]. The decision threshold is very crucial to maintain the false accept and false reject [7]. The decision threshold truly depends on the environment from where the voice is collected, i.e., the collected voice sample is fully contaminated or less contaminated or pure and clean with white noise [8]. The paper organized as section one gives brief description of the problem. Section 2 presents the architecture of the identification procedure. The model description is presented in the Sect. 3. Sections 4 and 5 are describes about the simulation and conclusion remarks.

2 System Architecture

The procedure for the initial stage of checking is based on the Gaussian mixture model. The Gaussian mixture model is created through a number of steps. The acoustic feature is extracted from the MFCC (Mel Frequency Cepstral Coefficient) [9] according to Fig. 1. Melody's frequency Cepstral coefficients (MFCC) are collective built up by the individual Melody frequency Cepstral (MFC). MFC is a physical representation of the short-term power spectrum of an acoustic signal in a particular frequency band on a linear cosine transform of the log power spectrum [10]. The extracted acoustic feature from the voice signal after normalization produces various acoustic classes. These acoustic classes belong to an individual speaker voice or a set of speakers. The GMM is the soft representation of the various acoustic classes of an individual person's voice or a set of speakers. The probability of a feature vector of being in the acoustic classes is represented by the mixture of different Gaussian probability distribution functions.

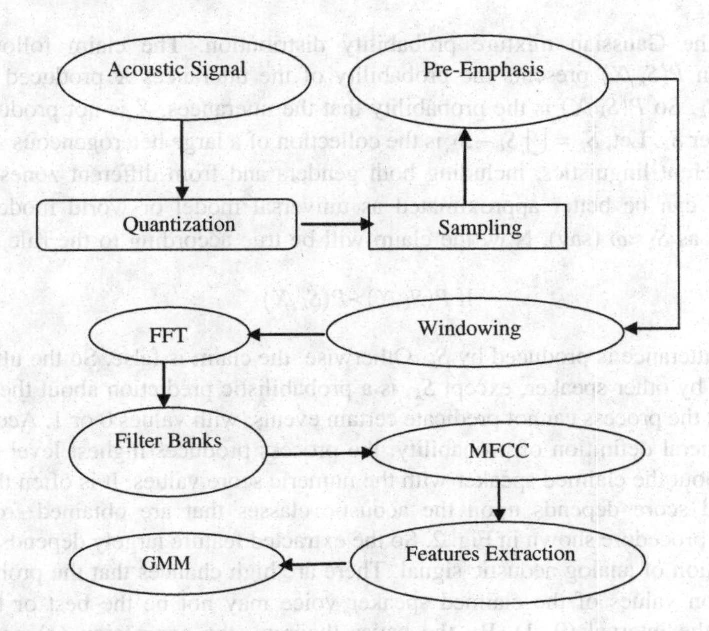

Fig. 1 GMM creation flow diagram

3 Model Development

Let us consider X as a random vector, i.e., $X = \{x_1, x_2, x_3, \ldots\ldots, x_L\}$ as a set of L vectors, each x_i is a k-dimensional feature vectors belong to the one particular acoustic class. L is the number of acoustic classes and the vectors x_i are statistically independent. So the probability of the set X for the λ speaker model can be expressed as $\log P(X \mid \lambda) = \sum_{i=1}^{L} \log P(x_i \mid \lambda)$. The distribution of vector x_i with the k-dimensional components are unknown. It is approximately modeled by a mixture of Gaussian densities, which is a weighted sum of $l \leq k$ component's densities, which can be expressed as $P(x_s|\lambda) = \sum_{i=1}^{l} w_i N(x_s, \mu_i, \sum_i)$, here w_i is the mixture weight, where $1 \leq i \leq l$ and $\sum_{i=1}^{l} w_i = 1$. Each $N(x_s, \mu_i, \sum_i)$ is a k variate Gaussian component density presents as

$$N(x_s, \mu_i, \sum_i) = \frac{e^{-\{0.5(x_s - \mu_i)' \sum_i^{-1} (x_s - \mu_i)\}}}{(2\Pi)^{k/2} |\sqrt{\sum_i}|},$$

μ_i is the mean vector and $\sum_i$ is the covariance matrix. $(x_s - \mu_i)'$ is the transpose of $(x_s - \mu_i)$. In the speaker identification from the set of speakers, $\{S_i\}$ where i is countable finite and X is given utterances, if we claim that the utterance produce by the speaker S_k from the set of speakers $\{S_i\}$. So the basic goal is how to valid claim that the speaker S_k makes the utterance X. The utterance X is a random variate that follows the Gaussian mixture probability distribution. The claim follows the expression $P(S_k/X)$ presents the probability of the utterances X produced by the speaker S_k. So $P(\bar{S}_k/X)$ is the probability that the utterances, X is not produced by the speaker S_k. Let, $\bar{S}_k = \bigcup_i S_i - S_k$ is the collection of a large heterogeneous speaker from different linguistics, including both genders and from different zones of the globe. $\bar{S}_k$ can be better approximated as universal model or world model. It is presented as $\bar{S}_k \approx \omega$ (say). Now the claim will be true according to the rule,

$$\text{If } P(S_k/X) \succ P(\bar{S}_k/X) \tag{1}$$

then the utterance is produced by S_k. Otherwise, the claim is false. So the utterance produced by other speaker, except S_k, is a probabilistic prediction about the claim. However, the process cannot predicate certain events, with values 0 or 1. According to the general definition of probability, the process produces highest level of prediction about the claimed speaker with the numeric score values. It is often that this predicated score depends upon the acoustic classes that are obtained from the long-step procedure shown in Fig. 2. So the extracted feature largely depends on the digitalization of analog acoustic signal. There are high chances that the probability comparison values of the claimed speaker voice may not be the best or highest value in the interval (0, 1). By the bayes theorem, the expression (1) produced $\frac{P(X/S_k)P(S_k)}{P(X)} > \frac{P(X/\omega)P(\omega)}{P(X)}$, since we assume that X is not silent, then clearly $P(X) \neq 0$. We get,

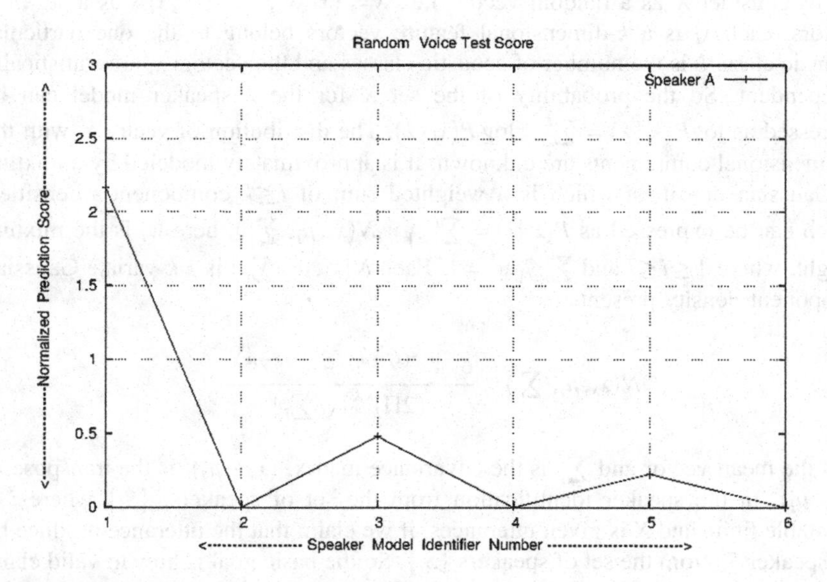

Fig. 2 Test Score for Monologue of Person A

$$\frac{P(X/S_k)}{P(X/\omega)} > \frac{P(\omega)}{P(S_k)} = \lambda_k. \tag{2}$$

λ is a preassumed threshold. To compact the all possible predictions, we consider the log on the both side.

$\log \frac{P(X/S_k)}{P(X/\omega)} > \log \lambda_k = \lambda$, the predicted values indicate how closer the claimed speaker is to the existing speaker's voices after comparison. The predicated values are Gaussian in nature so further compactness can be done on the predicted values by the static $\frac{\frac{P(X/S_k)}{P(X/\omega)} - \mu}{\sigma} > \lambda$. μ is the mean and σ is the covariance of the predicated score values of the known speakers voice models.

4 Simulation Result and Discussion

The speaker identification is a compression of the prediction of a said speaker with the number of existing speaker models present in the voice identification system. We make 100 numbers of speaker models with 10 clusters. Each cluster contains 10 speaker's models. We consider three speakers A, B, and C for testing purpose. We use initial known voice models of the speakers A, B and C to the three different clusters. The known voice model for the Speaker 'A' is placed into the first cluster with speaker identifier number 1. The known voice model for the Speaker 'B' is placed into the second cluster with speaker identifier number 2. Voice model for a known speaker is placed into the first cluster with speaker identifier number 3 as an impostor speaker of 'C'. The known voice model for the speaker 'C' is placed into the third cluster with speaker identifier number 4. The other 96 impostor speaker's models are placed along the 10 cluster make the each cluster size 10. At the first stage of identification, we consider two-sample voice of the speaker 'A ' and 'B.' These are pure voices of the respective speakers. Figure 2 presents the predicted score values of the simulation. All figures are presented in the first 6 speaker's identifier matching prediction.

Figure 2 presents the predicted matching value in the list of target models. The new voice sample of the Speaker 'A' match with the model identifier 1 with normalized value 2.2, the predicted normalize score value with the models identifier number 2, 3, 4, 5, and 6 are 0.0,0.45, 0.0, 0.34, and 0.0, respectively. These models belong to different clusters. Figure 3 presents the predicted normalized score value for the new voice sample of the speaker 'B.' The simulated result show the prediction about the new voice sample match with the model identifier number 1, 2, 3,4, 5, and 6 are (0.0, 3.0, 0.0, 0.0, 0.5, and 0.0), respectively.

The new voice sample matches with the model identifier number 2 with normalized score value 3.0, and also with a model identifier number of 5 with normalized score value 0.5. The above simulation result gives a physical significance of the environment to select the decision threshold of the normalized score values.

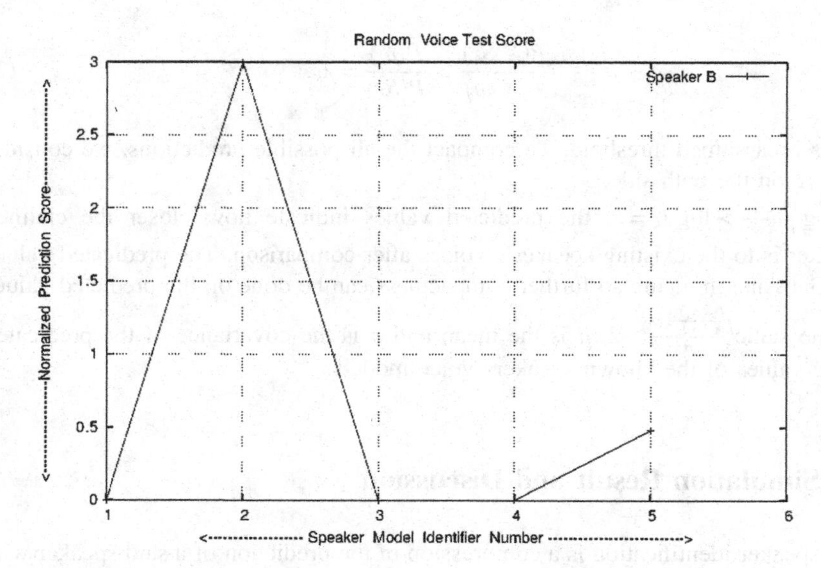

Fig. 3 Test Score for Monologue Person B

According to Figs. 2 and 3, we consider (1.0) as a decision threshold for level of acceptance and rejection to maintain the equal error rate at minimum level. According to this decision threshold, Fig. 2 shows that the new voice is the voice of the speaker A, and Fig. 3 shows the new voice is Speaker's B voice. Figures 4, 5, and 6 present the predicted score for the conversion voice sample AB, AC, and BC. In the target list, the speaker voice models 'A,' 'B,' and 'C' are present. Through the simulation, we would like to identify in which conversion speaker 'C' is present? Figure 4 presents normalized predicted score value for the conversion voice sample of the speaker A and C. The predicted normalized score is 2.8 for the model identifier 4.

The model identifier 4 is the speaker 'C'. According to the GMM-based hypothetical testing prediction about the speaker 'C,' is present in this conversion voice sample. The simulation result does not contribute any prediction about the speaker 'A,' but truly speaker 'A' is present in the conversion. So it is a false rejection for speaker 'A.' Figure 5 presents the predicated normalized score for the conversion voice sample for speaker 'B' and 'C.' The simulation result shows that the model identifier 2 is present in the conversion with the predicted score value 0.7, and the model identifier 4 is present in the conversion with the normalized score value 2.6. Since we have considered the accepted level of threshold value as (1.0), hence we conclude that model identifier 4, i.e., speaker 'C' is present in this conversion strongly. The model identifier 2, i.e., speaker 'B' present in the conversion is very less, and according to the level of acceptance, we ignore the presence, but truly the speaker 'B' is present in the conversion. It is the false rejection. The simulation result for the conversion voice sample of the speakers 'A' and 'B' are presented in Fig. 6. The predicted normalized score values shown that

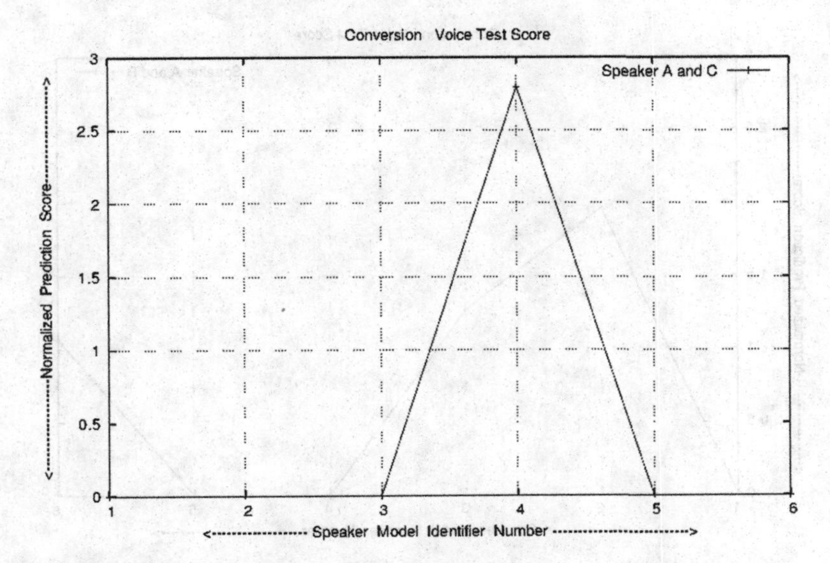

Fig. 4 Test Score for Conversion Voice of Speakers A and C

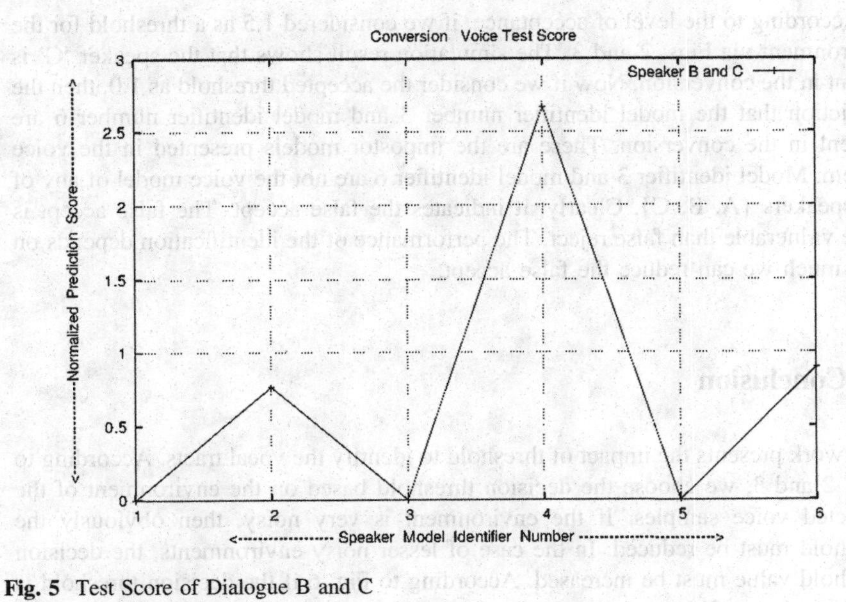

Fig. 5 Test Score of Dialogue B and C

the model identifier 2, i.e., speaker 'B' is present in the conversion with the value 2.0. The model identifier 3, i.e., another speaker but not 'A,' 'B,' and 'C' exists in the conversion with the predicted normalized score 1.3, and the model identifier 6, i.e., a speaker except {A, B, C} matches with the score value (1.2).

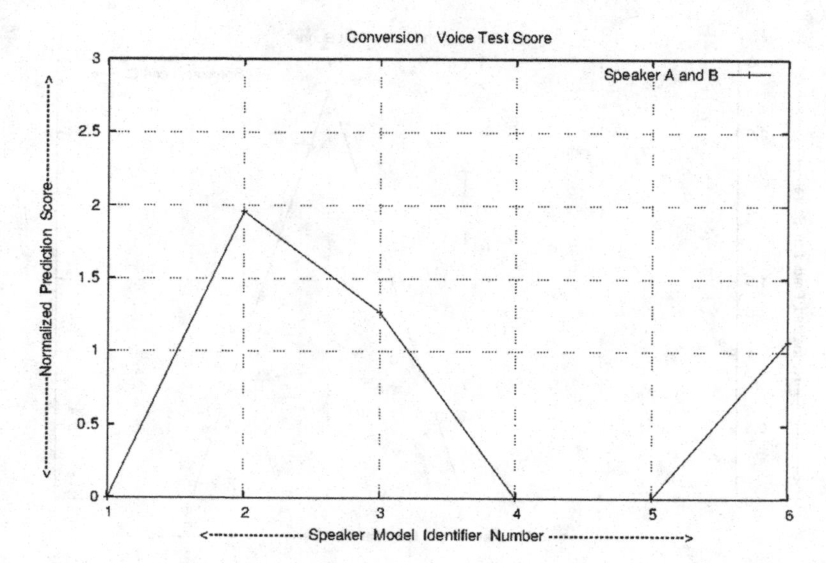

Fig. 6 Test Score for Conversion A and B

According to the level of acceptance, if we considered 1.5 as a threshold for the environment via Figs. 2 and 3. The simulation result shows that the speaker 'C' is absent in the conversion. Now if we consider the accepted threshold as 1.0, then the prediction that the model identifier number 3 and model identifier number 6 are present in the conversion. These are the impostor models presented in the voice system. Model identifier 3 and model identifier 6 are not the voice model of any of the speakers {A, B, C}. Clearly, it indicates the false accept. The false accept is more vulnerable than false reject. The performance of the identification depends on how much we can reduce the false accept.

5 Conclusion

This work presents the impact of threshold to identify the vocal tracts. According to Figs. 2 and 3, we choose the decision threshold based on the environment of the collected voice samples. If the environment is very noisy, then obviously the threshold must be reduced. In the case of lesser noisy environments, the decision threshold value must be increased. According to Fig. 6 if the decision threshold is considered as 1.0 then obviously the model identifier number 3, i.e., speaker 'C' must be identified as present in the conversion but the speaker 'C' is truly not present in the conversion. It will be a case of false accepts, which will brings the worst impact on the performance of the identification procedure. The main conclusion drawn from the simulation result is to first select the decision thrashed from the known speakers according to the environment. The specified threshold applies to identify the unknown speaker through the process. The unknown speaker voice should be collected from the same environment as well as for the GMM model creation voice also. The model voice and test voice should be collected through the same channel to achieve the better system performance.

References

1. Bimbot, F., et al.: A tutorial on text-independent speaker verification. EURASIP J. Appl. Signal Process. **4**, 430–451 (2004)
2. Reynolds, D., Quatieri, T., Dunn, R.: Speaker verification using adapted Gaussian mixture models. Digit. Signal Proc. **10**, 19–41 (2000)
3. Xiang, B., et al.: Short-time gaussianization for robust speaker verification. Proc. ICASSP **1**, 681–684 (2002)
4. Reynolds, D.: Automatic speaker recognition using Gaussian mixture speaker models. Linciln Lab. J. **8**(2), 173–191 (1995)
5. Reynolds, D.: Speaker identification and verification using Gaussian mixture speaker models. Speech Commun. **17**, 91–108 (1995)
6. Auckenthaler, R., Carey, M., Lloyd-Thomas, H.: Score normalizing for text-independent speaker verification system. Digit. Signal Process. **10**, 42–52 (2000)
7. Mirghafori, N., Heck, L.: An adaptive speaker verification system with speaker dependent a priori decision thresholds. In: Proceedings of the ICSLP, Denver Colorado (2002)
8. Morrison, G.S.: Measuring the validity and reliability of forensic likelihood-ratio System. Sci Justice **5**, 91–98 (2011). doi:10.1016/j.scijus.2011.03.002
9. Apsingeker, V.R., DeLeon, P.: Speaker model clustering for efficient speaker identification in large population applications. IEEE Trans. Audio Speech Language Process. **17**(4), 848–853 (2009)
10. Kanrar, S., et al.: Detect mimicry by enhancing the speaker recognition system. Adv. Intell. Syst. Comput. **339**, 21–31 (2015). doi:10.1007/978-81-322-2250-7_3

Application of Compressed Sensing in Cognitive Radio

Naveen Kumar and Neetu Sood

Abstract In the last few years, compressed sensing (CS) has been well used in the area of signal processing and image compression. Recently, CS has been earning a great interest in the area of wireless communication systems. CS exploits the sparsity of the signal processed for digital acquisition to reduce the number of measurement, which leads to reductions in the size, power consumption, processing time, and processing cost. This paper presents application of CS in cognitive radio (CR) networks for spectrum sensing and channel estimation. The effectiveness of the proposed CS-based scheme is demonstrated through comparisons with the existing conventional spectrum sensing and channel estimation methods.

Keywords Compressed sampling • Cognitive radio • Spectrum sensing • Channel estimation • Noncontiguous orthogonal frequency division multiplexing

1 Introduction

According to Nyquist's sampling theorem, a continuous-time band-limited signal $x(t)$ with bandwidth $B > 0$ can be exactly recovered from twice as many samples per second as the highest frequency present in the signal, i.e., $2B$ also known as the Nyquist rate [1]. However, around 2004, Donoho [2] proved that at given knowledge about sparsity of a signal, the signal may be reconstructed back with even fewer samples than required by Nyquist's sampling theorem, also known as compressed sensing (CS).

N. Kumar (✉) · N. Sood
Department of Electronics & Communication, Dr. B. R. Ambedkar NIT Jalandhar, Jalandhar, India
e-mail: naaveen@live.com

N. Sood
e-mail: soodn@nitj.ac.in

S. Das et al. (eds.), *Proceedings of the 4th International Conference on Frontiers in Intelligent Computing: Theory and Applications (FICTA) 2015*, Advances in Intelligent Systems and Computing 404, DOI 10.1007/978-81-322-2695-6_10

To implement wideband spectrum sensing, cognitive radio (CR) needs fast analog-to-digital converter (ADC) but the achievable sampling rate of ADC is only 3.6 Gsps. Capitalizing on the wideband signal spectrum sparseness, CS technique can be employed in spectrum sensing in CR network [3]. Tian and Giannakis [4] firstly applied CS theory to wideband CR networks for acquiring spectrum at sub-Nyquist sampling rates. In most scenarios in CR networks, the number of used channels is comparatively much lesser than total channels; those are vacant at a particular time and space. Therefore, when dealing with channel estimation problem in CR system where the channel band is really wide and dynamic and occupation information of the channels is compressible, CS can be exploited, since CS does not require any knowledge of the underlying multipath channel, based on the fact that a sparse structure is exhibited by the physical multipath channels in angle delay Doppler spreading, especially at large signal space dimensions, it is advantageous to utilize sparse channel estimation method based on convex/linear programming, which can be proved to outperform the existing least square-based methods [5]. Jia et al. [6] presented channel estimation algorithm for OFDM-CR, based on OMP and applied sparsity adaptive matching pursuit (SAMP) algorithm for the first time for channel estimation in NC-OFDM systems. Moreover, for the reconstruction time-consuming of SAMP algorithm was too large, modified adaptive matching pursuit (MAMP) algorithm was introduced as an improved SAMP algorithm. Qi et al. [7] introduced sparse channel estimation (SCE) scheme in OFDM-CR, where pilot design was formulated as an optimal column selection problem and constrained cross entropy optimization-based scheme was proposed to obtain an optimized pilot pattern.

The remainder of this paper is structured as follows Sect. 2 presents CR system model for spectrum sensing and channel estimation. In Sect. 3, CS-based spectrum sensing and channel estimation scheme is proposed. Section 4 demonstrates and summarizes the performance advantages of proposed CS-based scheme over traditional energy detection spectrum sensing and maximum likelihood ratio-based channel estimation techniques. Section 5 concludes the paper.

2 System Model and Problem Statement

Suppose that CR system aims to find spectral holes (SHs) in the frequency range of 0 to W Hertz, as shown in Fig. 1. During the spectrum sensing interval, all CR nodes keep quiet. Thus, the continuous signal received at the receiver of CR network, i.e., $x(t)$, is composed of primary users' (PUs) signals and additive white Gaussian noise (AWGN).

Mathematically, using sub-Nyquist sampling rate f_S ($f_S < 2W$), the compressed samples y ($y \in C^{M \times 1}$, $M = \tau f_S \ll N$) can be written as

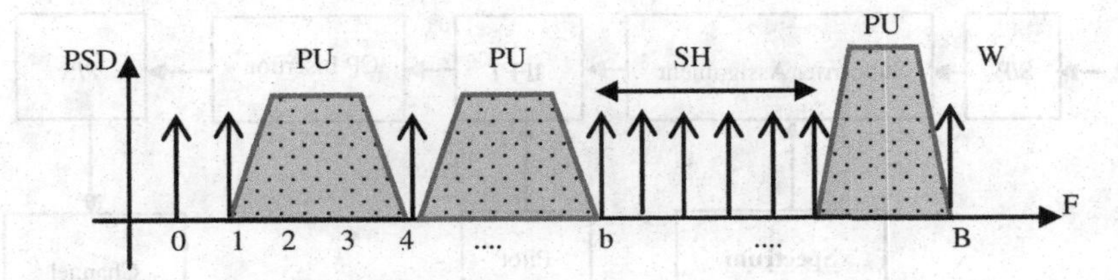

Fig. 1 Frequency frame of wideband cognitive radio

$$y = \varphi x = \varphi \psi_S \quad (1)$$

where φ is an $m \times N$ sensing matrix, ψ_S is sparsifying matrix and y is the measurement vector of m measurements. Let $\mu(\varphi, \psi)$ be the coherence between φ and ψ, $S = k/N$ be the sparsity measure, then we can choose minimum number of measurements required for reconstruction of x from y

$$m \geq C_0 \cdot \mu^2(\varphi, \psi) \cdot S \cdot \log N \quad (2)$$

for a Gaussian measurement matrix m, where C_0 is a constant. An estimate can be obtained by solving the CS reconstruction problem [8]:

$$\hat{x} = \arg\min||x||_1 \quad \text{s.t. } y = \varphi \psi x \quad (3)$$

After spectrum reconstruction secondary users (SUs) sense the recovered channel spectrum in order to identify frequency holes. If K symbol periods are allocated for channel sensing, the problem can be described as the hypothesis testing problem, mathematically:

$$\mathcal{H}_0\colon z_i = n_i \quad i = 1, 2, \ldots, K \quad (4)$$

$$\mathcal{H}_1\colon z_i = x_i + n_i, \quad i = 1, 2, \ldots, K \quad (5)$$

After spectrum sensing, CR adopts noncontiguous orthogonal frequency division multiplexing (NC-OFDM) technique that decomposes wideband into orthogonal sub-channels. The sub-channels are activated when the spectrum is idle and when it is not available corresponding sub-channels are deactivated.

For S number of nonzero elements, the vector is S-sparse, discrete Fourier transform (DFT) size is N, active sub-carriers are M and pilot sub-carriers (c_p) are K ($K \leq M$). The CP length is greater than the maximum possible path delay (Fig. 2).

OFDM symbol data $X(n)$ contains mapping signals and pilot signals. After removing CP, discrete Fourier transform (DFT) is applied to the received time-domain signal y_n for $n \in [0, N-1]$ to obtain $k \in [0, N-1]$. The discrete-time channel model is:

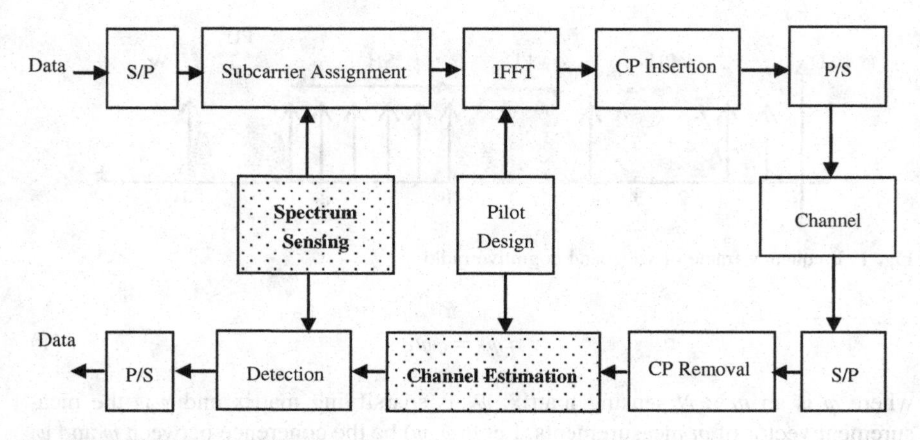

Fig. 2 NC-OFDM-based CR system

$$h(n) = \sum_{l=0}^{l-1} h_l \delta(n-l) \tag{6}$$

where the impulse response vector of the channel $h = [h_0, h_1, \ldots, h_{l-1}]^{\mathrm{T}}$ remains unchanged in multiple OFDM symbol period of time reflects the slow time variation of the channel. The relation between the transmitted pilots and received pilots can be written as:

$$\begin{bmatrix} y(c_{p_1}) \\ y(c_{p_2}) \\ \vdots \\ y(C_{pk}) \end{bmatrix} = \begin{bmatrix} x(c_{p_1}) & 0 & \cdots & 0 \\ 0 & \ddots & & 0 \\ \vdots & & \ddots & \vdots \\ 0 & 0 & \ldots & x(c_{pk}) \end{bmatrix} \cdot F_{K\times L} \cdot \begin{bmatrix} h(1) \\ h(2) \\ \vdots \\ h(L) \end{bmatrix} + \begin{bmatrix} Z(1) \\ Z(2) \\ \vdots \\ Z(K) \end{bmatrix} \tag{7}$$

where Z is additive white Gaussian noise (AWGN) and is $F_{K\times L}$ is a DFT sub matrix given by:

$$F_{K\times L} = \frac{1}{\sqrt{N}} \begin{bmatrix} 1 & w^{c_{p1}} & \ldots & w^{c_{p1}\cdot(L-1)} \\ 1 & \ddots & & w^{c_{p2}\cdot(L-1)} \\ \vdots & & \ddots & \vdots \\ 1 & w^{c_{pK}} & \ldots & w^{c_{pK}(L-1)} \end{bmatrix} \tag{8}$$

where $w^{nl} = \mathrm{e}^{-j\frac{2\pi nl}{N}}$. Let $A = XF_{K\times L}$, then (7) can be written as:

$$y = Ah + Z \tag{9}$$

Since the channel delay spread is much larger than sampling period [9], particularly for OFDM systems with over sampling, most components of h are either zero or nearly zero, which implies that h is sparse. With this a priori condition, CS theory can be applied to estimate h.

3 Compressed Spectrum Sensing Algorithm

Consider a discrete-time signal $x \in \mathcal{P}^N$, which can be expressed as $X(n)$ where $n = 1, 2, \ldots, N$. The claim of compressive sensing is that from m ($m \ll n$) measurements, we can reconstruct the original signal x with nonadaptive linear measurements. This does not violate the Shannon–Nyquist sampling theorem as reconstruction of only sparse signals is possible. According to Eq. (1):

$$y = \varphi\hat{x} = \varphi\psi_s = \varphi\psi_s F^{-1}\hat{X} \quad (10)$$

where $F^{-1}\hat{X}$ is the inverse Fourier transform of $\hat{x}$. Using Eq. (10) problem reconstruction of $\hat{x}$ can be converted into the problem of reconstruction of $\hat{X}$:

$$\hat{x} = \arg_x \min \|\hat{X}\|_0 \quad \text{s.t. } y = \left(\varphi F^{-1}\right)\hat{X} \quad (11)$$

Basis pursuit (BP) [10] can be used for signal reconstruction, which transforms the sparseness constraint on into a convex optimization problem solvable by linear programming:

$$\hat{x} = \arg_x \min \|\hat{X}\|_1 \quad \text{s.t. } y = \left(\varphi F^{-1}\right)\hat{X} \quad (12)$$

To deal with the signals with noise components, some variants of LASSO algorithm can be developed by minimizing the usual sum of squared errors:

$$\hat{x} = \arg\min \|\hat{X}\|_0 \quad \text{s.t. } \|\varphi F^{-1}\hat{X} - y\|_2 \le \varepsilon \quad (13)$$

where ε is recovery error threshold. The problem can be solved with a two-step scheme: first, use compressed measurements y to estimate the sparse sequence and second, reconstruct signal $\hat{x}$ according to ψ_s.

Orthogonal matching pursuit (OMP) algorithm [11–13] suggests the reconstruction under the conditions of a given iteration number, as the iterative process is forced to stop, OMP algorithm needs a lot of linear measurement to ensure accurate reconstruction. The basic idea of the OMP algorithm is to select the columns of measurement matrix with greedy iterative algorithm, make sure the correlative value between the columns selected in each iteration and the current redundant vector is maximum, and then subtract the correlative value from the sampling vector and repeat iteration until the number of iterations achieves the sparse degree *s*. OMP algorithm selects an atom in each iteration to update the atom collection, which will certainly pay a large time for reconstruction. The number of iterations is closely related to sparse degree *S* and the number of samples *m*, with their increase, time consumption will also increase significantly.

Problem with OMP algorithm is that it is not adaptive, pre-estimate of the sparse degree of the sparse signal is needed, and the reconstruction accuracy is not satisfactory. In reality, the sparse degree of the sparse channel is usually unknown. Sahoo et al. [14] proposed extended OMP-CS algorithm in order to improve the accuracy of reconstruction and make the algorithm adaptive. In the ExtOMP-CS algorithm, one key issue is how to choose the step size. Unlike the OMP-CS algorithm, the iteration times of ExtOMP-CS algorithm is not certain and is related to step size, and computational complexity and computational time are higher in the ExtOMP-CS algorithm than OMP-CS algorithm.

An extension to OMP algorithms is the compressed sampling matching pursuit (CoSaMP) algorithm [15]. The basis of the algorithm is OMP but CoSaMP can be shown to have tighter bounds on its convergence and performance [16].

Algorithm 1 Compressed Sampling Matching Pursuit Algorithm

1. Input: S, y, ϕ
2. $x(0) \leftarrow 0$
3. $v \leftarrow y$
4. $k \leftarrow 0$
5. while Halting condition false do
6. $k \leftarrow k + 1$
7. $z \leftarrow \phi^T v$:signal proxy
8. $\Omega \leftarrow z^{2S}$:find the largest 2S components of the signal proxy *(Identification)*
9. $\Gamma \leftarrow \Omega \ \mathrm{supp}\ (x^{(k-1)})$:merge the support of the signal proxy with the support of the solution from the previous iteration *(Support Merge)*
10. $\bar{x} \leftarrow \arg\min ||(\phi x - y)||_2$:estimate a solution via least squares with the constraint that the solution lies on a particular support *(Estimation)*
11. $x^k \leftarrow x^s$:takes the solution estimate and compresses it to the required support *(Pruning)*
12. $v \leftarrow y - \phi x$: update the sample, namely the residual in F-space *(Sample Update)*
13. end while
14. $\hat{x} \leftarrow x(k)$
15. Output $\hat{x}$:such that it is S-sparse and $y = \phi x$

For any 2S-sparse channel vector x, CoSaMP algorithm produces the channel estimator $\hat{x}$ that satisfies

$$||x-\hat{x}||_2 \leq C \ \max\{\varepsilon, 1/\sqrt{s}||x-\hat{x}_{2S}||_1 + ||z||_2\} \tag{14}$$

for a given parameter ε, and x_{2S} is a best 2S-sparse approximation to x.

Having estimated $\hat{x}$, SU finds the presence of PUs in a certain sub-band using energy detector. Let the energy $E_{\mathrm{p}} = \sum_{i\,\varepsilon\,\mathrm{sub-channel}\,m} \hat{x}_i$ received in sub-channel M. The spectrum availability is decided by:

$$\mathcal{H}(n) = \begin{cases} \mathcal{H}_0 & Z_i \geq \lambda \\ \mathcal{H}_1 & Z_i \leq \lambda \end{cases} \tag{15}$$

The threshold λ is a decision threshold and is a design parameter for the CR receiver system. P_{d} probability of detection and P_{f} probability of false alarm are two probabilities used for performance evaluation of the scheme.

$$P_{\mathrm{f}} = P_{\mathrm{r}}\{Z \geq \lambda | H_0\} \tag{16}$$

$$P_{\mathrm{d}} = P_{\mathrm{r}}\{Z < \lambda | H_1\} \tag{17}$$

After deactivating, the active sub carriers random pilots are assigned. As wireless channels are rapidly decaying, the channel response h is highly sparse because of the small number of significant multipath components. A sparse high-resolution signal h can be recovered with high probability with a constraint from the measurements y. The corresponding model can be written as

$$\min_{h\,\in\, c^N} ||h||_0 \quad \text{s.t. } y = Ah \tag{18}$$

Problem (18) is NP-hard problem and even for moderate N, it is not possible to solve. ℓ_1 relaxation model with the same constraints can be used as an alternative.

$$\min_{h\,\in\, c^N} ||h||_1 \quad \text{s.t. } y = Ah \tag{19}$$

The reconstruction can never be exact due to the noise present in the measurements. Using a final de-noising step based on least square problem, noise can be eliminated.

4 Performance Analysis and Simulation Result

In NC-OFDM, the power spectral density (PSD) of P_{th} sub-carrier signal is characterized of the form

$$\Gamma_k(f) = K \cdot \mathrm{sinc}^2((f - f_{\mathrm{P}})T_{\mathrm{S}}) \tag{20}$$

where K is the signal level, fk is the sub-carrier center frequency, T_{S} is the OFDM symbol duration and TG is guard interval. Assuming independent symbols in different sub-carriers, the PSD of an NC-OFDM signal is obtained as

$$\Gamma(f) = \sum_{P} \Gamma_P(f) \tag{21}$$

where index P is the number of active subcarriers. In this paper, a wideband spectrum of 0–100 MHz is considered with total six sub-bands (B_1–B_6). Among these sub-bands, B_1, B_3, and B_5 located at 1–10 MHz, 20–35 MHz, 70–75 MHz, have relatively high PSD in the range of 0.0277–0.1126, as shown in Fig. 3 by level 16–25. For CS, the compression ratio is set to 75 % and the noise level is 8 dB.

4.1 Recovered PSD

Using algorithm 1 based the CS scheme, wideband spectrum can be successfully reconstructed back at sub-Nyquist rate, as shown in Fig. 4.

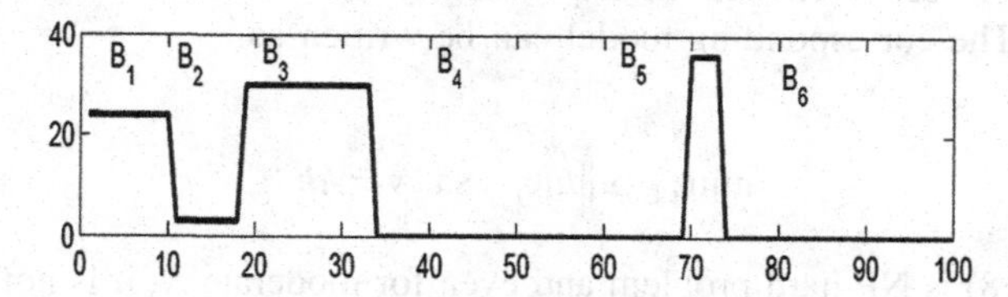

Fig. 3 Wideband spectrum $\hat{X}$

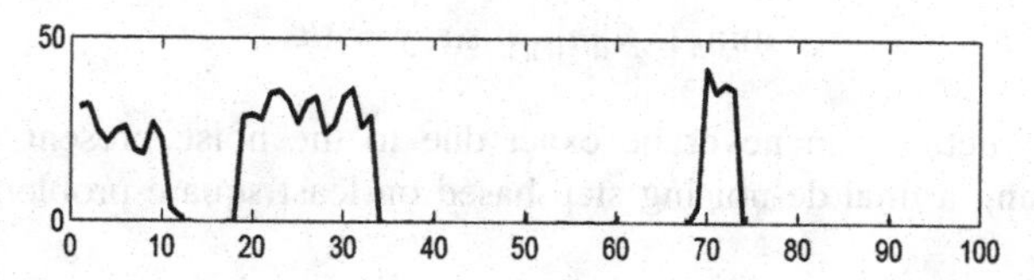

Fig. 4 Reconstructed spectrum

4.2 Probability of Detection Performance

The performance of the conventional generalized likelihood ratio test scheme and CS-based spectrum sensing scheme is evaluated and compared via the probability of detection P_d for a constant false alarm rate of $P_f = 0.08$ (Fig. 5).

4.3 BER Performance

An OFDM-based CR system is considered with $M = 1024$ subcarriers, after spectrum sensing without any false alarm or missing detection and deactivating

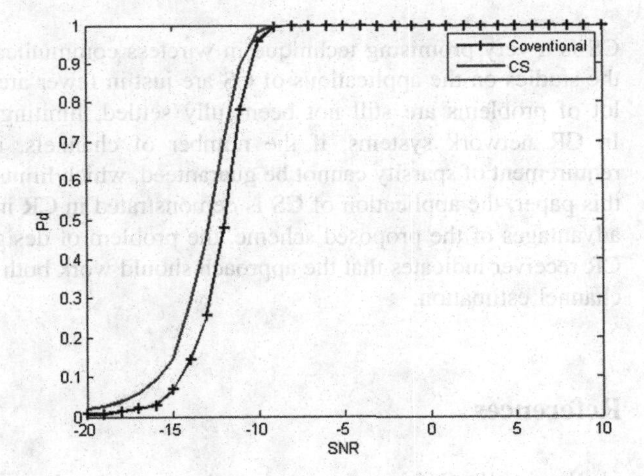

Fig. 5 Comparison of probability of detection performances

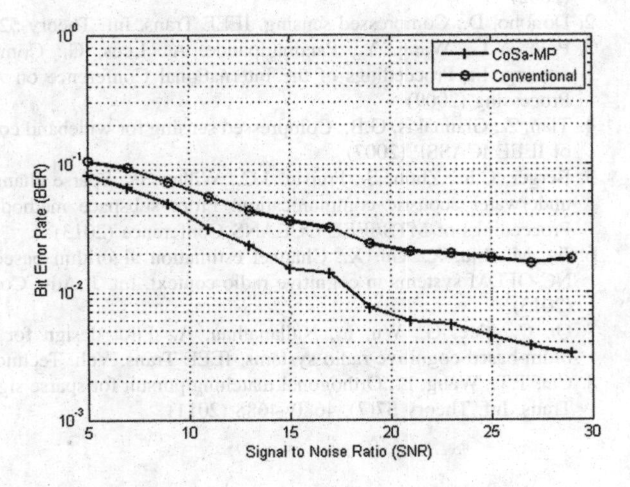

Fig. 6 Comparison of BER performances

those subcarriers occupied by PUs, there are 512 remaining OFDM subcarriers for SUs, including three noncontiguous subcarrier blocks, i.e., {1, 2, …, 256}, {513, 514, …, 640} and {897, 898, …, 1024}, with the number of subcarriers in each block being 256, 128, and 128, respectively. A sparse multipath channel h is considered with $L = 60$ taps where 5 nonzero taps are placed randomly. The channel estimation performance is evaluated now using the designed pilot patterns.

Figure 6 shows BER performance of two schemes—proposed CS channel estimation and channel estimation scheme based on LS. Improved BER performance of the proposed CS-based scheme can be seen from the above figure, over the conventional LS-based scheme.

5 Conclusion

CS is a very promising technique in wireless communication networks. However, the studies on the applications of CS are just in fewer areas. Even in these areas, a lot of problems are still not been fully settled, limiting the performance of CS. In CR network systems, if the number of channels, is not large enough, the requirement of sparsity cannot be guaranteed, which limits the advantages of CS. In this paper, the application of CS is demonstrated in CR networks and based on the advantages of the proposed scheme, the problem of designing a high-performance CR receiver indicates that the approach should work both for spectrum sensing and channel estimation.

References

1. Nyquist, H.: Certain topics in telegraph transmission theory. Proc. IEEE **90**(2), 280–305 (2002)
2. Donoho, D.: Compressed sensing. IEEE Trans. Inf. Theory **52**(4), 4036–4048 (2006)
3. Polo, Y.L., Wang, Y., Pandharipande, A., Leus, G.: Compressive wide-band spectrum sensing. In: Proceedings of the International Conference on Acoustics, Speech, and Signal Processing (2009)
4. Tian, Z., Giannakis, G.B.: Compressed sensing for wideband cognitive radios. In: Proceedings of IEEE ICASSP (2007)
5. Berger, C.R., Zhou, S., Preisig, J.C., Willett, P.: Sparse channel estimation for multicarrier underwater acoustic communication: From subspace methods to compressed sensing. In: Proceedings of MTS/IEEE OCEANS Conference (2013)
6. Jia, M., Liu, X., Gu, X.: Channel estimation algorithm based on compressive sensing for NC-OFDM systems in cognitive radio context. Int. J. Adv. Comput. Technol. **5**(1), 343–351 (2013)
7. Qi, C., Yue, G., Wu, L., Nallanathan, A.: Pilot design for sparse channel estimation in ofdm-based cognitive radio systems. IEEE Trans. Veh. Technol. **63**(2), 982–987 (2014)
8. Cai, T.T., Wang, L.: Orthogonal matching pursuit for sparse signal recovery with noise. IEEE Trans. Inf. Theory **57**(7), 4680–4688 (2011)

9. Qi, C., Wu, L.: Application of compressed sensing to DRM channel estimation. In: Proceedings of 73rd IEEE VTC-Spring, Budapest, Hungary, pp. 1–5 (2011)
10. Baraniuk, R.: A lecture on compressive sensing. IEEE Signal Process. Mag. **24**(4), 118–121 (2007)
11. Tropp, J.A., Gilbert, A.C.: Signal recovery from partial information by orthogonal matching pursuit. IEEE Trans. Inf. Theory **53**, 4655–4666 (2007)
12. Jiang, X., Jeng, W.J., Cheng, E.: A fast algorithm for sparse channel estimation via orthogonal matching pursuit. In: IEEE 73rd Vehicular Technology Conference, Budapest, pp. 1–5 (2011)
13. Cai, T.T., Wang, L.: Orthogonal matching pursuit for sparse signal recovery with noise. IEEE Trans. Inf. Theory **57**(7), 4680–4688 (2011)
14. Sahoo, S.K., Makur, A.: Signal recovery from random measurements via extended orthogonal matching parsuit. IEEE Trans. Signal Recovery **63**(10), 2572–2581 (2015)
15. Needell, D., Tropp, J.A.: CoSaMP: iterative signal recovery from incomplete and inaccurate samples. Commun. ACM (2008)
16. Shen., Y., Zhang, H., Liu, G., Liu, H., Wu, H., Xia, W.: A novel method on compressive sampling matching pursuit. In: 11th World Congress on WCICA, pp. 5567–5572 (2014)

Compressed Sensing-Based NBI Mitigation in Ultra-WideBand Energy Detector

Priyanka G. Patil and Gajanan K. Birajdar

Abstract Recently, ultra-wideband (UWB) energy detectors are developed enormously with compressed sensing (CS) theory in multipath fading environment. As wideband communication is sensitive to narrowband interference (NBI), it is necessary for efficient UWB energy detector to mitigate NBI-affected measurements without harming samples containing important information. According to the traditional sampling theorem, UWB requires huge bandwidth for short range communication with little utilization. To avoid this wastage of frequency band, CS process uses sub-Nyquist rate and provides compressed version of received signal. In this paper, reconstruction-based energy detector which is robust to NBI is presented. To mitigate the NBI-affected measurements, notch out method is employed at the detector in this article. Energy detection of the UWB detector before adding NBI and after mitigating NBI is compared. Experimental results show that the presented energy detector is robust to NBI due to superior performance of the notch out method.

Keywords Compressed sensing · Energy detector · Narrowband interference · Ultra-wideband

1 Introduction

Communication field is developing with wireless nature in many applications, which requires signal of higher bandwidth like ultra-wideband signal. Especially, in short range communication, ultra-wideband (UWB) signals have great importance because of its unique features like high data rate capability for communication, multipath immunity, fine time resolution and higher spatial capacity [5–20]. The UWB signal

P.G. Patil (✉) · G.K. Birajdar
Department of Electronics & Telecommunication, Pillai HOC College of Engineering & Technology, Raigad 410206, Maharashtra, India
e-mail: priyankagpatil91@gmail.com

G.K. Birajdar
e-mail: gajanan123@gmail.com

S. Das et al. (eds.), *Proceedings of the 4th International Conference on Frontiers in Intelligent Computing: Theory and Applications (FICTA) 2015*, Advances in Intelligent Systems and Computing 404, DOI 10.1007/978-81-322-2695-6_11

can be defined as the signal having fractional bandwidth greater than 20 % of the central frequency and absolute bandwidth greater than 0.5 GHz [6]. In the UWB architecture, analog-to-digital converter (ADC) consumes most of the power due to higher sampling rate.

According to the Shannon–Nyquist–Whittaker–Kotelnikov sampling theorem [9–15], a band-limited signal can be recovered completely at the receiver, when it is sampled at twice the maximum frequency of the signal. But most of the large frequency signals are sparse in nature, so these signals require sampling depend on amount of information [15]. This can be achieved by compressive sampling (CS) [1–3], which states that the sparse signal can be recovered easily from lower sampling rate than the traditional sampling rate. In the process of CS, measurement matrix and reconstruction algorithm plays crucial role in compressing and recovering the signal, respectively.

The measurement matrix has to satisfy restricted isometry property (RIP) to provide recoverable compressed version of the received signal. But the challenging task is to recover the original signal from less number of samples. The recovery process is based on the solution of the linear system through optimization [1]. The effectiveness of CS theory is decided by speed and accuracy of the reconstruction algorithm. Reconstruction algorithms can be classified into three fundamental types [9] as shown in Fig. 1. Each algorithm of Fig. 1 has its own pros and cons in specific reconstruction problem. So selection of the algorithm is dependent on the type of problem. Also, there are many solvers of linear optimization problem for reconstructing the signal.

In this paper, CS-based noncoherent detector which is highly robust to narrowband interference (NBI) is presented. The two types of CS-based energy detectors of UWB pulse position modulation (PPM) are proposed in [7], which are employed with the help of generalized maximum likelihood (GML) principle. In the first type of detector, energy detection takes place after reconstruction of the compressed signal and in second, compressed signal is directly applied to energy detection. These two detectors are analyzed using theoretical expression of bit error probability (BEP)

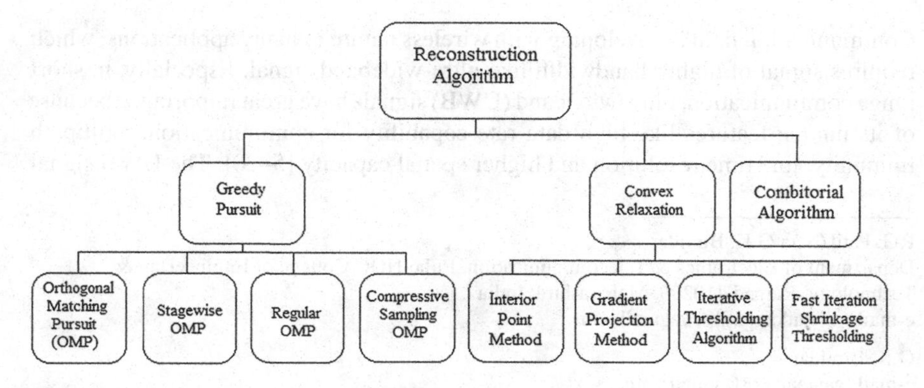

Fig. 1 Classification of reconstruction algorithms in CS

in [7]. It is observed that the direct compressed detector performs well only under exact synchronization compared to reconstruction-based detector. In [7], measurement process has been assumed identical. If this process changed due to some perturbation, direct compressed detector would deteriorate whereas, reconstruction-based detector is robust to changing measurement process. Therefore, reconstruction-based detector is more preferable than the earlier.

1.1 Contribution

Due to wideband, CS-based UWB energy detectors are very sensitive to NBI, though its performance is good in multipath fading environment. The NBI problem in this detector is handled by the notch out method implemented in [10]. In this paper, random measurement matrix is used to compress the received signal. Then, we choose appropriate ensembles which are coherent with NBI to ensure that only few CS measurements are affected by NBI and then suppress those samples. The remaining measurements are used to recover the original signal using approximated massage passing (AMP) algorithm. This recovered signal is provided to the energy detector.

The remaining paper is organized as: Section 2 reviews prior work related to this article. Section 3 provides the proposed system model. Experimental results are presented in Sect. 4. Finally, Sect. 5 concludes the paper.

2 Related Work

In wireless communication, tremendous development has been occurred in the field of compressed sensing-based UWB communication. In [11], the novel receiver is proposed for IR-UWB communication using CS, which is characterized by bursty traffic and severe power constraints. The receiver does not need high rate ADC and wideband analog delay lines. The receiver can acquire and track the channel response irrespective of the environmental condition and operate in severe intersymbol interference and its performance is close to maximum likelihood principle. The noncoherent receivers with PPM are less complex and require low power consumption [2–13, 15]. In [13], it is shown that the information rate can be achieved close to channel capacity for moderate signal-to-noise ratio (SNR) with PPM and UWB bandwidth. The receiver proposed in [2] can operate without knowledge of channel response. In [8], the channel estimation method for both narrowband and wideband has been proposed with lower sampling rate than Nyquist rate using finite rate of innovation theory. The CS theory not only reconstructs the sparse signal, but also provides the generalized likelihood radio test (GLRT) detector for impulse UWB [16]. The GLRT detector of [16] is further extended with matching pursuit (MP) algorithm for pilot-assisted IR-UWB detection in [18]. In [12–14, 16–21], methods for channel estimation are provided for CS-based UWB communication, while in [4], time delay estimation is provided. The undersampling method using CS and analog-to-information converter (AIC) is proposed in [19]. The original signal can be theoretically subsampled by projection matrix according to CS theory, but the multiplication of matrix and signal needs already sampled received signal. The random matrix is not realizable using hardware and undersampling is uncontrollable. These problems can be solved by replacing random matrix with AIC in CS measuring projection stage, as described in [19]. But this method does not guarantee the precise reconstruction of sampled signal.

The NBI mitigation methods are proposed in [10, 17]. In [17], the NBI subspace is estimated when UWB signal is absent. While detecting the UWB symbol, a measurement matrix is designed, which incorporates the UWB subspace and null subspace of the NBI. But this method needs only low pulsing rate and random CS measurement matrix to identify NBI sparsity subspace by taking a discrete cosine transform. These problems can be overcome using the NBI mitigation method proposed in [10]. Here, pulsing rate is very high and CS ensembles inherently employ Fourier analysis, which makes NBI subspace apparent from CS measurements. Also, this method is flexible for imperfect timing of signals. Therefore, it is preferable to implement 'notch out' method described in [10].

3 System Model

The kth message symbol is transmitted with the $X_k(t)$ UWB information signal of length T, containing F_N frames of length F_T, so that $T = F_N * F_T$. The signal is transmitted with m-ary PPM by delaying the transmitted pulses by $T_m \overset{\Delta}{=} \frac{F_T}{m}$. Consider that the transmitted kth symbol, $b_k \in (0, 1, \ldots, m-1)$ can be represented as, $X_k(t) = \sum_{i=0}^{F_N-1} d(t - (i + kF_N)F_T - b_kT_m)$, where $d(t)$ is second derivative of Gaussian pulse of duration $T_d << T_m$. If $h(t)$ is represented as impulse response of Gaussian communication channel, then the received signal as shown in the block diagram of the proposed detector in Fig. 2. is,

$$R(t) = X_k(t) \star h(t) + W_k(t) + I_k(t) \quad (1)$$

where $W_k(t)$ and $I_k(t)$ are the additive noise and interference symbol of bandwidth B_i, corresponding to kth information symbol, respectively, and $X_k(t) \star h(t) = g_k(t)$ is the received pulse waveform of bandwidth B_u. For Nyquist rate sampling of the symbol, we take N samples per frame, whereas N/m samples for each slot. Then, the ith sampled frame of kth symbol for $j = 0, 1, \ldots, N-1$, is represented as,

$$R^i_{k,j} = R(iF_T + \frac{jF_T}{N}) = g^i_{k,j} + W^i_{k,j} + I^i_{k,j} \quad (2)$$

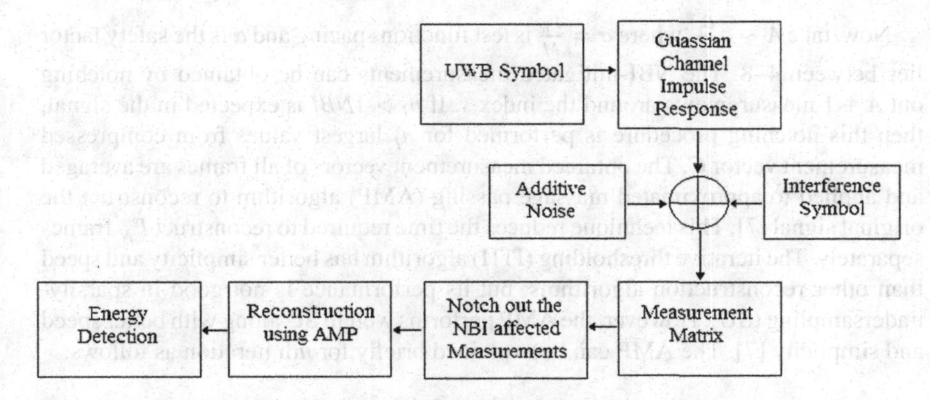

Fig. 2 Block diagram of NBI-mitigated reconstruction-based UWB energy detector

But the received pulse is holding sparsity, hence most of the sampled values do not contain any information, which are eliminated from signal. This practice can be replaced by CS method, which consumes less power.

In CS, measurement matrix, ϕ transforms the signal as $\mathbb{R}^N \rightarrow \mathbb{R}^M$, where $M << N$ and ϕ contains M number of linear functions in its rows. The ratio of M to N is called as undersampling ratio. The measurement matrix have crucial role in compressing the signal. For this, it should satisfy restricted isometry property (RIP) [1]. This property is satisfied by the random matrices like Gaussian, Bernoulli, and also, structured matrix like Fourier [7]. For reconstruction-based detectors, ϕ contains rows which are approximately orthogonal to each other [7]. Now, the received signal is applied to $M \times N$ matrix ϕ to get compressed version of signal. For ith frame, applying CS to (2), we get,

$$C_k^i = \phi R_k^i = \phi(g_{k'j}^i + W_{k'j}^i + I_{k'j}^i) = G_k^i + V_k^i + Z_k^i, \quad (3)$$

where C_k^i is the $M \times 1$ measurement vector. Similarly, V_k^i and Z_k^i are compressed versions of noise and NBI symbol, respectively. The UWB measurements are deteriorated due to addition of NBI symbol and UWB symbol. As UWB symbols are sensitive to NBI, it is difficult to handle this problem. In this paper, we have applied the notch out method [10] to suppress the NBI-affected measurements from vector. In this method, first step is to choose Fourier ensemble of magnitude $1/\sqrt{N}$ and its frequency is selected from $(F_c + \frac{\omega}{2})$ to $(F_c - \frac{\omega}{2})$, which are decoherent with UWB signal and coherent with NBI, to ensure that only few measurements are affected by NBI. Then, we implement notch to mitigate NBI-affected measurements from compressed vector C. For at most $1NBI$ we find,

$$s = \max_{s \in (0,1,\ldots,m-1)} |C_s| \quad (4)$$

Now, take $A \sim \alpha \frac{B_i}{\sigma}$, where $\sigma = \frac{B_u}{M}$ is test function spacing and α is the safety factor lies between 4–8. The NBI-mitigated measurements can be obtained by notching out $A + 1$ measurements around the index s. If $n_I > 1NBI$ is expected in the signal, then this notching procedure is performed for n_I largest values from compressed measurement vector C. The obtained measurement vectors of all frames are averaged and applied to approximated massage passing (AMP) algorithm to reconstruct the original signal [7]. This technique reduces the time required to reconstruct F_N frames separately. The iterative thresholding (ITH) algorithm has better simplicity and speed than other reconstruction algorithms, but its performance is not good in sparsity-undersampling (SU). However, the AMP performs well in SU along with better speed and simplicity [7]. The AMP can be explained briefly for nth iteration as follows:

$$x_k^{[n+1]} = S(x_k^{[n]} + \phi^T y_k^{[n]}, \beta^{[n]}), \tag{5}$$

where

$$y_k^{[n]} = C_k - \phi x_k^{[n]} + \frac{1}{\mu} y_k^{[n-1]} \langle S'(x_k^{[n]} + \phi^T y_k^{[n]}, \beta^{[n]}) \rangle \tag{6}$$

Here, β is iteratively updating threshold and $\langle S'(\cdot) \rangle$ is the average of derivative of soft thresholding over N samples. The value provided by x_k^n is the reconstructed vector of signal. Now, these samples are provided to reconstruction-based energy detector which is proposed in [7]. For N_m nonzero samples, detection is done as follows:

$$\hat{b}_k^{(R-ED)} = \max_{b_k} \sum_{l=0}^{N_m-1} \left[\frac{1}{F_N} \sum_{i=0}^{F_N-1} [\check{d}_k]_{iN+b_k N_m+l} \right]^2 \tag{7}$$

The Nyquist rate energy detector is obtained by replacing reconstructed samples with Nyquist rate samples in (7). For analysis of energy detection, the bit error probability (BEP) of both Nyquist rate energy detector and reconstruction-based energy detector is given in same [7].

$$P_e^{(R-BEP)} = 1 - \frac{2\Gamma(\frac{N}{2})}{\frac{N}{4}[\Gamma(\frac{N}{4})]^2} \left[\frac{\sigma_R \sigma_w}{\sigma_R^2 + \sigma_w^2} \right]$$

$$\times 2F_1\left(1, \frac{N}{2}; \frac{N}{4} + 1; \frac{\sigma_R^2}{\sigma_R^2 + \sigma_w^2}\right), \tag{8}$$

where $2F_1(\cdot, \cdot; \cdot; \cdot)$ is the Gaussian hypergeometric function.

$$P_e^{(N-BEP)} = 1 - \frac{2\Gamma(\frac{N}{2})}{\frac{N}{4}[\Gamma(\frac{N}{4})]^2} \left[\frac{\sigma_{a0} \sigma_{F0}}{\sigma_{a0}^2 + \sigma_{F0}^2} \right]$$

$$\times {}_2F_1\left(1, \frac{N}{2}; \frac{N}{4}+1; \frac{\sigma_{a0}^2}{\sigma_{a0}^2+\sigma_{F0}^2}\right), \quad (9)$$

where $\sigma_{a0}^2 \stackrel{\Delta}{=} (1+\frac{\sigma^2}{F_N})$ and $\sigma_{F0}^2 \stackrel{\Delta}{=} (\frac{\sigma^2}{F_N})$

4 Experimental Results

This section presents the simulation results for the detector presented in previous section. We consider that the UWB signal is transmitted along Gaussian distributed physical channel with zero mean and unit variance. The received signal is compressed by applying it to the random measurement matrix. We consider that the number of interference, $n_I = 1$ for experiment and we can simulate this detector for multiple numbers of interferences. In AMP algorithm, the threshold policy is in the form of $\beta^{[n]} = \tau\sigma_w^{[n]}$, which is infeasible in practice. So, threshold can be updated as suggested in [7],

$$\beta^{[n]} = \beta + \frac{1}{\mu}\beta^{[n-1]}\langle S'(x_k^{[n-1]} + \phi^T y_k^{[n-1]}, \beta^{[n-1]})\rangle, \quad (10)$$

where β is a constant.

Figure 3 shows the reconstruction-based energy detection (R-ED) for frame number $F_N = 30, 20, 10$. The signal-to-noise ratio (SNR) is varying and compression ratio has kept constant, $\mu = 0.5$. It is observed that as the number of frames increases, energy detection also increases. In Fig. 3, energy detection of $F_N = 30$ is much greater than energy detection of $F_N = 10$.

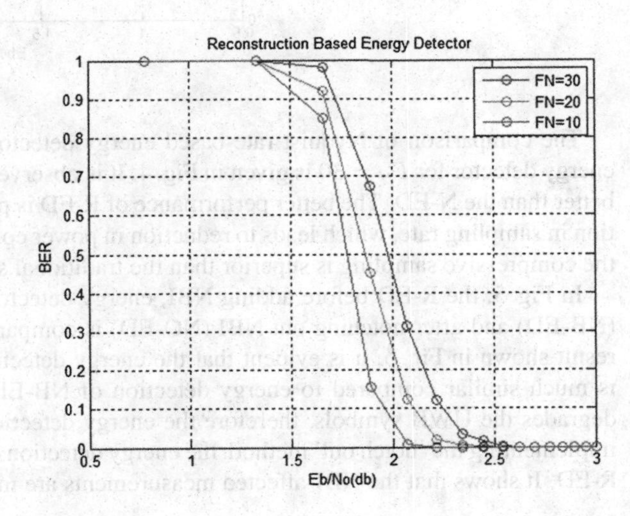

Fig. 3 Reconstruction-based energy detection for different number of frames in a symbol

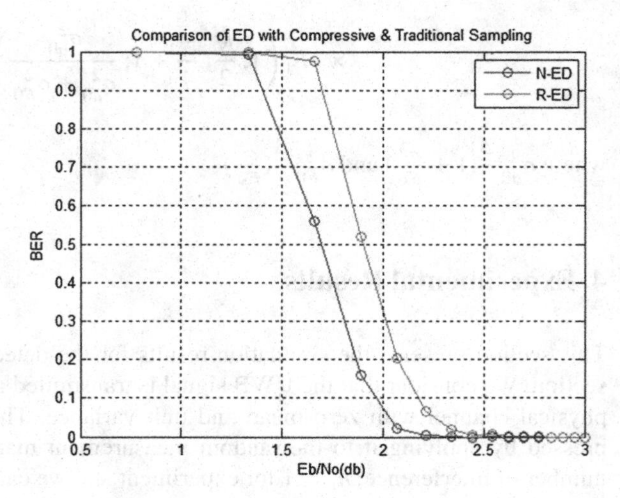

Fig. 4 Comparison between CS energy detection and traditional sampling energy detection

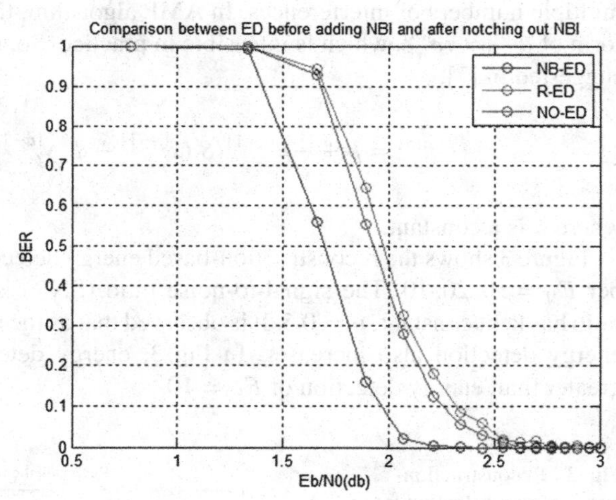

Fig. 5 Comparison between energy detectors before adding NBI, after adding NBI, and after notching it out

The comparison of Nyquist rate-based energy detector (N-ED) with CS-based energy detector for $F_N = 30$ is given in Fig. 4. It is observed that the R-ED performs better than the N-ED. The better performance of R-ED is possible because of reduction in sampling rate, which leads to reduction in power consumption at ADC. Thus, the compressive sampling is superior than the traditional sampling.

In Fig. 5, the R-ED before adding NBI, energy detector in the presence of NBI (NB-ED) and after notching out NBI (NO-ED) is compared. From the experiment result shown in Fig. 5, it is evident that the energy detection of R-ED and NO-ED is much similar compared to energy detection of NB-ED. It shows that the NBI degrades the UWB symbols, therefore the energy detection of signal is low. After implementing the 'notch out' method, the energy detection of NO-ED increases up to R-ED. It shows that the NBI-affected measurements are mitigated without harming much of the compressed measurements of the UWB signal. Thus, the presented energy detector is robust to NBI.

5 Conclusion

In this paper, we have developed compressed sensing-based UWB energy detector which is robust to narrowband interference. This detector consumes less power due to reduction in sampling rate than the traditional sampling rate. The wideband communication is sensitive to NBI; therefore we have implemented the method which mitigates the NBI-affected measurements from compressed version of the received signal without harming much of the other measurements. From the simulated results, it is observed that CS-based energy detection is better than the Nyquist rate energy detection.

References

1. Candes, E.J., Romberg, J., Tao, T.: Robust uncertainty principles: Exact signal reconstruction from highly incomplete frequency information. IEEE Trans. Inf. Theory **52**(2), 489–509 (2006)
2. Carbonelli, C., Mengali, U.: M-ppm noncoherent receivers for uwb applications. IEEE Trans. Wirel. Commun. **5**(8), 2285–2294 (2006)
3. Donoho, D.L.: Compressed sensing. IEEE Trans. Inf. Theory **52**(4), 1289–1306 (2006)
4. Gedalyahu, K., Eldar, Y.C.: Time-delay estimation from low-rate samples: a union of subspaces approach. IEEE Trans. Signal Process. **58**(6), 3017–3031 (2010)
5. Ghavami, M., Michael, L.B., Kohno, R.: Ultra Wideband Signals and Systems in Communication Engineering, second edn. Wiley, New York (2007)
6. Gishkori, S.: Compressive sampling for wireless communications. Ph.D. thesis, TU Delft, Fac. EEMCS (2014). ISBN 9789461863294
7. Gishkori, S., Leus, G.: Compressive sampling based energy detection of ultra-wideband pulse position modulation. IEEE Trans. Signal Process. **61**(15), 3866–3879 (2013)
8. Kusuma, J., Maravic, I., Vetterli, M.: Sampling with finite rate of innovation:channel and timing estimation for UWB and GPS. In: New Frontiers in Telecommunications, vol. 5 (2003)
9. Needell, D., Tropp, J.A.: Cosamp: iterative signal recovery from incomplete and inaccurate samples. ACM Trans. Commun. **53**(12), 93–100 (2010)
10. Oka, A., Lampe, L.H.J.: Compressed sensing reception of bursty uwb impulse radio is robust to narrow-band interference. In: IEEE GLOBECOM, pp. 1–7 (2009)
11. Oka, A., Lampe, L.: A compressed sensing receiver for uwb impulse radio in bursty applications like wireless sensor networks. Phys. Commun. **2**(4), 248–264 (2009)
12. Paredes, J.L., Arce, G.R., Wang, Z.: Ultra-wideband compressed sensing: channel estimation. IEEE J. Signal Process. **1**(3), 383–395 (2007)
13. Souilmi, Y., Knopp, R.: On the achievable rates of ultra-wideband ppm with non-coherent detection in multipath environments. IEEE Int. Conf. Commun. **5**, 3530–3534 (2003)
14. Unser, M.: Sampling–50 years after shannon. Proc. IEEE **88**, 569–587 (2000)
15. Vetterli, M., Marziliano, P., Blu, T.: Sampling signals with finite rate of innovation. IEEE Trans. Signal Process. **50**(6), 1417–1428 (2002)
16. Wang, Z., Arce, G.R., Paredes, J.L., Sadler, B.M.: Compressed detection for ultra-wideband impulse radio. In: IEEE Signal Processing Advances in Wireless Communications (SPAWC) 2007, pp. 1–5 June 2007

17. Wang, Z., Arce, G., Sadler, B., Paredes, J., Hoyos, S., Yu, Z.: Compressed uwb signal detection with narrowband interference mitigation. In: IEEE International Conference on Ultra-Wideband, 2008. ICUWB 2008, vol. 2, pp. 157–160. IEEE (2008)
18. Wang, Z., Arce, G.R., Sadler, B.M., Paredes, J.L., Ma, X.: Compressed detection for pilot assisted ultra-wideband impulse radio. In: IEEE International Conference on Ultra-Wideband, 2007. ICUWB 2007, pp. 393–398. IEEE (2007)
19. Wang, W., Wang, S., Yang, J., Liu, H.: Under-sampling of ppm-uwb communication signals based on cs and aic. Springer publication on Circuits Systems and Signal Processing (March 2015)
20. Win, M.Z., Scholtz, R.A.: Impulse radio: how it works. IEEE Commun. Lett. **2**(2), 36–38 (1998)
21. Zhang, P., Hu, Z., Qiu, R.C., Sadler, B.M.: A compressed sensing based ultra-wideband communication system. In: Proceedings of the IEEE International Conference on Communications, pp. 4239–4243. IEEE Press, Piscataway, NJ, USA (2009)

Part IV
Data Mining

A Novel Data Mining Scheme for Smartphone Activity Recognition by Accelerometer Sensor

Yajnaseni Dash, Sanjay Kumar and V.K. Patle

Abstract The prime objective of activity recognition is to recognize the actions performed by a person with the surrounding environment and forming different observation sets. It is necessary to choose the appropriate classifier for the data collected through accelerometer sensors incorporated in mobile phones, which have limited resources such as energy and computing power. In this paper, standard classification techniques of data mining like random forest (RF), multilayer perceptron (MLP), logistic regression, classification via regression, and J48 and RepTree have been implemented to compare the performance and accuracy of different classifiers by reducing the computational cost. In this experiment, it was found that RF required quite short time than MLP (0.64 vs. 270.07 s, respectively) to build the model and gives the better accuracy (92.6 % vs. 92.1 %, respectively). This study has concluded that RF has better performance score than other classification techniques applied in this study.

1 Introduction

The state of the user and its environment can be captured by computation of the recognized activity through heterogeneous sensors to facilitate adaptation to the external computing resources. Attachment of these sensors to the user's body allows continual monitoring of various physical activities in the form of signals. Activity recognition is one of the synergistic contexts between human and

---

Y. Dash (✉) · S. Kumar · V.K. Patle
School of Studies in Computer Science & IT, Pt. Ravishankar Shukla University, Raipur 492010, Chhattisgarh, India
e-mail: yajnasenidash@gmail.com

S. Kumar
e-mail: sanraipur@rediffmail.com

V.K. Patle
e-mail: patlevinod@gmail.com

S. Das et al. (eds.), *Proceedings of the 4th International Conference on Frontiers in Intelligent Computing: Theory and Applications (FICTA) 2015*, Advances in Intelligent Systems and Computing 404, DOI 10.1007/978-81-322-2695-6_12

computer. Activity means any physical activity of persons such as walking, jogging, seating, and standing. Activity recognition is a recent research field, where sensor containing mobile phones, i.e., smartphones can be used to record the physical activities. By applying classification techniques of data mining, we can find that a particular person is doing which activity more or less than a limit and accordingly the smartphone can form a signal. For example, if a person is performing less physical activity then he may become obese, thus an alarm can be send from smartphone to the user. Similarly it can send calls directly to voice mails when the user is exercising. Many more applications of activity recognition are there. Accelerometer senses the acceleration whenever there occurs a change in speed of any body part movement. A variety of researches on context-aware services have been focusing upon recognizing the individual activity based on minute wearable sensors because of the successful development of micro-electro-mechanical systems (MEMS) technology [1]. A context-aware system uses location as the main outline of context. However, addition of low-priced sensors (e.g., measurement of acceleration, audio, and light) to mobile and pervasive platforms, in combination with progresses in machine learning and data mining, facilitates systems for building a more affluent model of the user's context. For instance, body-mounted accelerometers can able to recognize various human activities, i.e., from common activities (walking or sitting) to higher-level activities (car driving or bus riding) [2]. Particularly, by combining numerous sensors for physical movements and bio-signals could spectacularly amplify the recognition accuracy through capturing details of users' current states [3]. At a particular instance of time, a person is busy in which particular activity that will be predicted by this research work. In this study, different data mining techniques were applied and a comparison of each algorithm was made to predict performance of the activities from the sensory input.

2 Related Work

Activity recognition (AR) is a fastest growing field, but AR with smartphone sensors is a difficult task owing to intrinsic noisy character of the input data and limited resource of the target platform. The rich features of smartphones like computing power, multi-tasking ability, and sensory inputs like tilt and acceleration of smartphones lends itself for studies of AR with imperative objective of identifying a number of activities, such as walking, jogging, sitting, and standing from the sensory inputs of the device.

Currently, activity recognition has been growing as a novel research area because of the increasing accessibility of accelerometers in mobile phones, and other prospective applications. Researchers have employed a combination of accelerometers and other sensors to accomplish activity recognition. Information can be extracted from variety of sources like environment [4] and body worn sensors [5, 6] to achieve the recognition process. Human AR has been studied by researchers since the last decade. Vision sensor, inertial sensor, and the mixture of both are used in the existing approaches. Machine learning-based algorithms are often applied for classification of the activities because of its reliable and correct outcomes. The earliest studies in accelerometer-based AR focused on the utilization of multiple accelerometers placed on different parts of the user's body. Bao and Intille [7] collected data from 20 users by 5 biaxial accelerometers and used different data mining techniques for classification. Authors [8] tested subjects by carrying the phone in the suitable location, i.e., in their pants pockets. Krishnan et al. [9] assembled data from 3 users using 2 accelerometers for recognizing five different activities. Authors claimed that data from a thigh accelerometer was not sufficient to classify various activities and hence several accelerometers were needed. The body activities were again monitored by 3 accelerometers and data gathered from 10 subjects [10]. Tapia et al. [11] implemented a real-time system to identify 30 gymnasium activities by gathering data from 5 accelerometers located on several body parts of 21 users. Performance was increased slightly by incorporating data from a heart monitor as well as the accelerometer data. Mannini and Sabitini [12] employed 5 triaxial accelerometers and recognized 20 activities from13 users. Foerster and Fahrenberg [13] collected data of 31 male subjects from 5 accelerometers. They built hierarchical classification model for differentiation of various postures. Parkka et al. [14] developed a system using 20 diverse types of sensors for recognition of activities. Another system was formed by Lee and Mase [15] for recognizing location of user and their activities by a wearable sensor module. Nishkam et al. [16] used sensors on wrist, chest, waist, and thighs to achieve better classification performance.

Some researchers used wearable sensors on different body parts as discussed above. However, the problem arises for common users to bear the uncomfortable situation as the sensor repositioning is required after dressing. Novel opportunities of AR research are coherent by the use smartphones where the user is a rich source of context information, and the phone is the sensing tool with embedded built in sensors such as dual cameras, microphones, gyroscopes, and accelerometers. Authors [17] presented an approach to utilize an Android smartphone for human AR employing its embedded triaxial accelerometers. Generally, researchers apply supervised classification algorithms for activity recognition. The algorithms are trained with labeled samples to generate classification model for the input data. As supervised classification algorithms require accurate computations for producing models from training data, so the implementations are being done in servers. Some implementations in smartphones were presented in [7, 18–22].

3 Classification of Wireless Sensor Data

Classification is one of the major tasks in data mining. It is the separation of one class of elements from other class of elements using different classifiers. We have employed three different classifier namely functions, meta and decision tree classifier. Logistic regression (LR), multilayer perceptron (MLP) known as function-based classifier, whereas classification via regression (CVR) comes under meta classifier. Decision trees such as J48, RepTree (RT), and random forest (RF) were also applied to evaluate the performance by classifying the sensor data. All these algorithms were explored in order to find the more appropriate algorithm based on its accuracy on the selected dataset.

A multimodal LR model was used for classification purpose with a ridge estimator. If there are k classes for n instances with m attributes, then the parameter matrix A to be calculated will be $m*(k-1)$ matrix. MLP is a classifier that uses back-propagation to classify instances. The network can be monitored and customized during training time. The nodes in this network are all sigmoid. CVR perform classification using regression methods with binary value and one regression model is built for each class value. J48 used to generate pruned or unpruned C4.5 decision tree. RT is fast decision tree learner that can builds a decision or regression tree using information gain or variance and prunes it using reduced error pruning (with backfitting). It can only one time sorts the values for numeric attributes, and the missing values are dealt with by splitting the consequent instances into pieces (i.e., as in C4.5). RF is used for constructing a forest of random trees.

4 Experimental Results

This section outlines our experimental analysis followed by presentation and discussion of our results for the activity recognition task. The dataset was collected from wireless sensor data mining (WISDM) Lab [23]. This WISDM Lab is concerned with collecting the sensor data from smartphones and other modern mobile devices (e.g., tablet computers, music players, etc.). Mining techniques can be applied to these sensor data for imperative knowledge discovery. After collection of data, we preprocessed and applied several classification techniques on it to predict the user activities using sensors. We have employed tenfold cross validation for execution of all the experiments in WEKA [24]. Here the total numbers of instances are 5418. The detailed result which is showing the confusion matrices associated with each of the six learning algorithms are presented in Tables 1, 2, 3, 4 and 5.

5 Performance Evaluation of Each Algorithm

The performance of various algorithms were evaluated and presented in Table 6 and Fig. 1. In Table 7, the training and simulation errors were represented. Percentage of correctly predicted records and accuracy of activity recognition of each class is summarized in Table 8. In Fig. 2a–c, comparisons between various parameters including percentage of instances, error score, and time required to execute different algorithms were shown.

Table 1 Confusion matrix for LR

Actual classes	Predicted classes					
	Walk	Jog	Up	Down	Sit	Stand
Walk	1980	9	57	34	0	1
Jog	18	1603	1	2	0	1
Up	177	6	317	128	4	0
Down	129	2	203	190	3	1
Sit	0	0	5	5	288	8
Stand	4	0	6	0	11	225

Table 2 Confusion matrix for CVR

Actual classes	Predicted classes					
	Walk	Jog	Up	Down	Sit	Stand
Walk	2009	11	33	28	0	0
Jog	15	1590	12	6	1	1
Up	81	20	432	95	2	2
Down	86	10	102	326	1	3
Sit	0	0	3	1	288	14
Stand	2	0	2	5	5	232

Table 3 Confusion matrix for J48 decision tree

Actual classes	Predicted classes					
	Walk	Jog	Up	Down	Sit	Stand
Walk	1988	19	37	34	2	1
Jog	1563	1563	31	13	0	1
Up	59	37	427	106	1	2
Down	53	14	126	334	1	0
Sit	3	1	2	1	295	4
Stand	2	3	1	0	0	240

Table 4 Confusion matrix for RT decision tree

Actual classes	Predicted classes					
	Walk	Jog	Up	Down	Sit	Stand
Walk	2053	9	12	6	1	0
Jog	38	1565	15	7	0	1
Up	9	11	479	133	0	0
Down	10	6	155	357	0	0
Sit	2	1	8	4	286	5
Stand	5	2	9	6	7	217

Table 5 Confusion matrix for RF decision tree

Actual classes	Predicted classes					
	Walk	Jog	Up	Down	Sit	Stand
Walk	2037	7	16	20	1	0
Jog	13	1596	8	7	0	1
Up	30	19	490	89	2	2
Down	39	8	118	360	2	1
Sit	0	1	1	1	300	3
Stand	2	0	4	3	3	234

Table 6 Simulation result of each algorithm

Algorithms used (total instances, 5418)		Correctly classified instances (% value)	Incorrectly classified instances (% value)	Time taken (in s)	Kappa statistics
Functions	LR	84.9575 (4603)	15.0425 (815)	133.68	0.7918
	MLP	92.1189 (4991)	7.8811 (427)	270.07	0.8926
Meta	CVR	90.0148 (4877)	9.9852 (541)	13.85	0.8629
Decision trees	J48	89.4611 (4847)	10.5389 (571)	0.89	0.8559
	RT	91.4913 (4957)	8.5087 (461)	0.6	0.8839
	RF	92.5987 (5017)	7.4013 (401)	0.64	0.8989

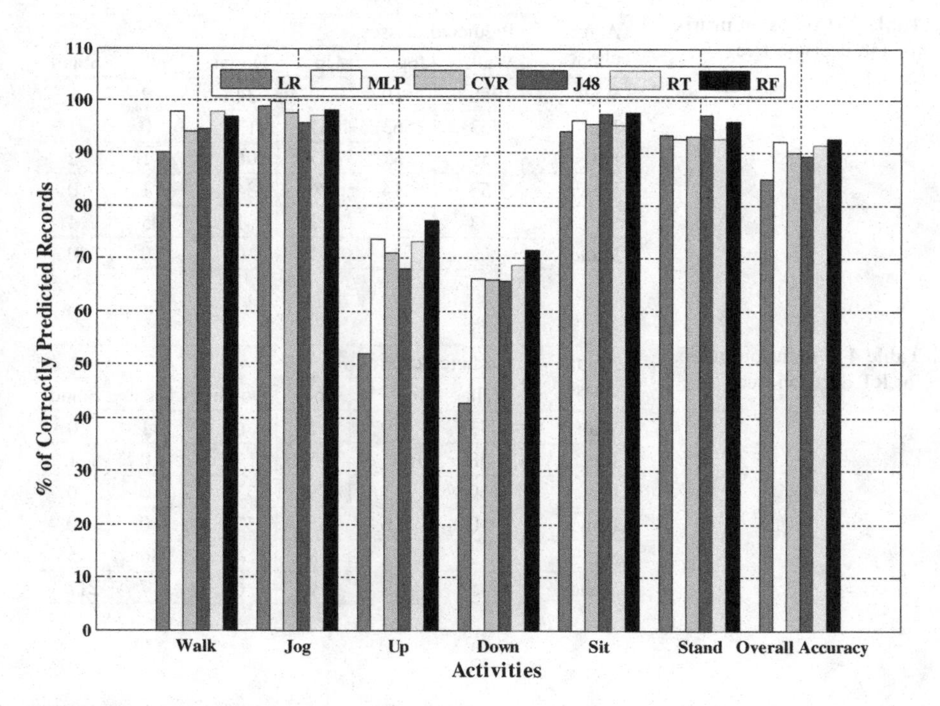

Fig. 1 Percentage of correctly classified records

Table 7 Training and simulation errors

Algorithms used (total instances, 5418)		Mean absolute error (MAE)	Root mean squared error (RMSE)
Functions	LR	0.0666	0.1893
	MLP	0.0285	0.1481
Meta	CVR	0.0563	0.1581
Decision trees	J48	0.0383	0.1741
	RT	0.0358	0.149
	RF	0.0505	0.1394

Table 8 Accuracy of activity prediction of each class

Activities	Percentage of correctly predicted records					
	LR	CVR	MLP	J48	RT	RF
Walk	90.2	94.0	**97.8**	94.6	**97.8**	**97**
Jog	**98.8**	**97.7**	**99.8**	95.8	97.2	**98**
Up	51.9	71.1	73.7	68.0	73.1	77.2
Down	42.8	65.9	66.1	65.7	68.6	71.4
Sit	94.1	95.5	96.3	**97.5**	95.3	97.7
Stand	93.4	93.2	92.6	97.2	92.7	96.1
Overall accuracy	84.9	90.0	**92.1**	89.5	91.4	**92.6**

6 Discussion

The abstract results for the AR experiments are represented in Tables 7 and 8. The predictive accuracy related to each of the activities specified in these tables for each of the six learning algorithms. Table 8 reveals that in most of cases higher levels of accuracy (above 90 %) can be achieved for the two common activities, walking, and jogging. Since jogging involves more intense alteration in acceleration, so it is easier to recognize than walking. It is very difficult to identify the two stair climbing activities as these two identical activities are often confused with each other. Sitting and standing activities were easily detected from the accelerometer data and can be well identified, as these two activities cause the device to modify its orientation.

Experimental results indicate that among the six learning algorithms RF (92.6 %) and MLP (92.1 %) are consistently perform best, but the MLP does not perform best overall as its time complexity is high. The performance of an algorithm is evaluated according to its time and space complexity. Hence, the total time taken to build the model is also a very influential parameter in comparing the classification algorithm.

In this experiment, the time required was shortest in case of RT (0.6 s) as compared to others, but the overall accuracy is 91.4 %. The next one after RT is RF which has taken 0.64 s to build the model and gives the highest accuracy of 92.6 %,

Fig. 2 **a** Comparision between algorithms on basis of % of instances, **b** comparison between algorithms on basis of error score, **c** comparison between algorithms on basis of time taken (s)

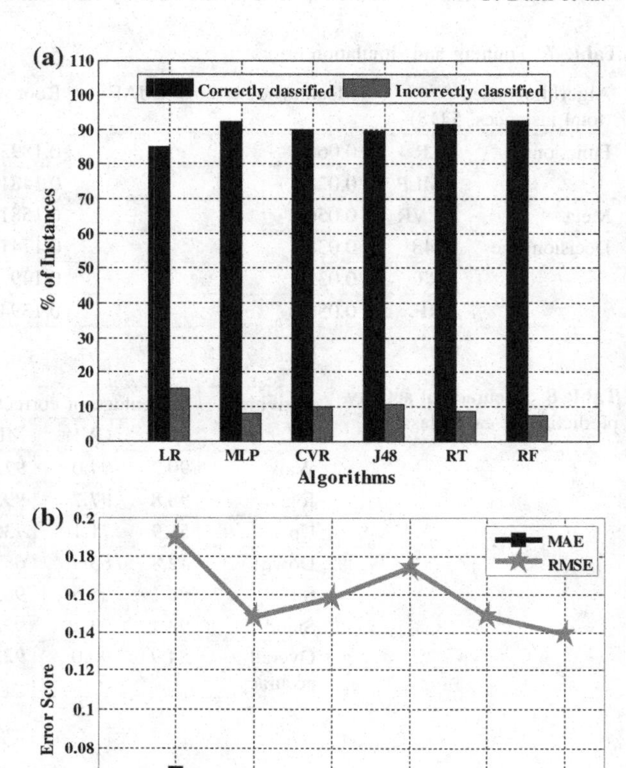

whereas MLP has taken 270.07 s to build the model with an accuracy of 92.1 %. In a previous experiment, authors [17] have shown the overall accuracy of MLP as 91.7 % but in our study it was 92.1 %. Authors [17] have applied J48, logistic regression, MLP, and straw man techniques for classification purpose. In the current study, we have applied other techniques too for better comparison purpose.

Kappa statistic is used to assess the accuracy of any particular measuring cases, it is usual to distinguish between the reliability of the data collected and their validity based on the kappa statistic criteria, the accuracy of this classification purposes is substantial [25]. The average kappa score of the RF algorithm is around 0.8989. The algorithm having a lower rate of error will be preferred as it has more powerful classification capability. We can study the errors resultant from the training of the six selected algorithms from Table 8. In this experiment, very common indicators for measuring error are employed namely mean absolute errors

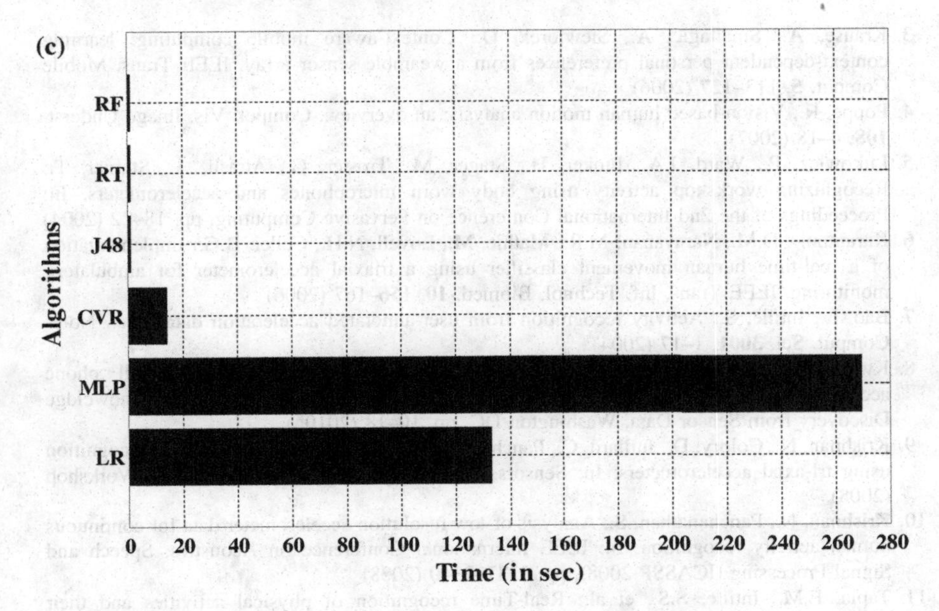

Fig. 2 (continued)

(MAE) and root mean squared errors (RMSE). In MLP, MAE is less as compared to RF, whereas in RF, RMSE is lowest as compared to MLP. So RF is a better option than MLP.

7 Conclusion

In this paper, a novel ensemble scheme was addressed to choose the appropriate classifier for the smartphone-based activity recognition system with wearable sensors. We have tested the activity recognition dataset using various classifiers. Thus, by observing accuracy, time complexity, kappa score, and error rate, we conclude that random forest decision tree results in better outcomes than other algorithms like MLP, RT, and CVR. Further investigation will be carried out with these issues as our future work.

References

1. Pantelopoulos, A., Bourbakis, N.G.: A survey on wearable sensor-based systems for health monitoring and prognosis. IEEE Trans. Syst. Man Cyber.-Part C: Appl. Rev. **40**, 1–12 (2010)
2. Davies, N., Siewiorek, D.P., Sukthankar, R.: Activity-based computing. IEEE Pervasive Comput. **7**, 20–21 (2008)

3. Krause, A., Smailagic, A., Siewiorek, D.: Context-aware mobile computing: learning context-dependent personal preferences from a wearable sensor array. IEEE Trans. Mobile Comput. **5**, 113–127 (2006)
4. Poppe, R.: Vision-based human motion analysis: an overview. Comput. Vis. Image Underst. **108**, 4–18 (2007)
5. Lukowicz, P., Ward, J.A., Junker, H., Stager, M., Troster, G., Atrash, A., Starner, T.: Recognizing workshop activity using body worn microphones and accelerometers. In: Proceedings of the 2nd International Conference on Pervasive Computing, pp. 18–22 (2004)
6. Karantonis, D.M., Narayanan, M.R., Mathie, M., Lovell, N.H., Celler, B.G.: Implementation of a real-time human movement classifier using a triaxial accelerometer for ambulatory monitoring. IEEE Trans. Inf. Technol. Biomed. **10**, 156–167 (2006)
7. Bao, L., Intille, S.: Activity recognition from user-annotated acceleration data. Lect. Notes Comput. Sci. **3001**, 1–17 (2004)
8. Kwapisz, J.R., Weiss, G.M., Moore, S.A.: Activity recognition using cell phone accelerometers. In: Proceedings of the 4th International Workshop on Knoweldge Discovery from Sensor Data, Washington DC, pp. 10–18 (2010b)
9. Krishnan, N., Colbry, D., Juillard, C., Panchanathan, S.: Real time human activity recognition using tri-axial accelerometers. In: Sensors, Signals and Information Processing Workshop (2008)
10. Krishnan, N., Panchanathan, S.: Analysis of low resolution accelerometer data for continuous human activity recognition. In: IEEE International Conference on Acoustics, Speech and Signal Processing (ICASSP 2008), pp. 3337–3340 (2008)
11. Tapia, E.M., Intille, S.S., et al.: Real-Time recognition of physical activities and their intensities using wireless accelerometers and a heart rate monitor. In: Proceedings of the 11th IEEE International Symposium on Wearable Computers, pp. 1–4 (2007)
12. Mannini, A., Sabatini, A.M.: Machine learning methods for classifying human physical activity from on body accelerometers. Sensors **10**, 1154–1175 (2010)
13. Foerster, F., Fahrenberg, J.: Motion pattern and posture: correctly assessed by calibrated accelerometers. Behav. Res. Methods Instrum. Comput. **32** (2000)
14. Parkka, J., Ermes, M., Korpipaa, P., Mantyjarvi, J., Peltola, J., Korhonen, I.: Activity classification using realistic data from wearable sensors. IEEE Trans. Inf. Technol. Biomed. **10**, 119–128 (2006)
15. Lee, S.-W., Mase, K.: Activity and location recognition using wearable sensors. IEEE Pervasive Comput. **1**, 24–32 (2002)
16. Nishkam, R., Nikhil, D., Preetham, M., Littman, M.L.: Activity recognition from accelerometer data. In: Proceedings of the Seventeenth Conference on Innovative Applications of Artificial Intelligence, pp. 1541–1546 (2005)
17. Kwapisz, J.R., Weiss, G.M., Moore, S.A.: Activity recognition using cell phone accelerometers. SIGKDD Explor. Newslett. **12**, 74–82 (2011)
18. Cheverst, K., Davies, N., Mitchell, K., Friday, A., Efstratiou, C.: Developing a context-aware electronic tourist guide: some issues and experiences. **2**, 17–24 (2000)
19. Thurau, C., Hlavac, V.: Pose primitive based human action recognition in videos or still images. In: Proceedings of the 2008 IEEE Conference on Computer Vision and Pattern Recognition, pp. 1–8 (2008)
20. Wu, O.W.H., Bui, A.T., Batalin, M., Au, L.K., Binney, J.D., Kaiser, W.J.: MEDIC: medical embedded device for individualized care. Artif. Intell. Med. **42**, 137–152 (2008)
21. Fahriddin, M., Song, M.G., Kim, J.Y., Na, S.Y.: Human Activity Recognition Using New Multi-Sensor Module in Mobile Environment (2011)
22. Deng, W.-Y., Zheng, Q.-H., Wang, Z.-M.: Cross-person activity recognition using reduced kernel extreme learning machine. Neural Netw. (2014)
23. http://www.cis.fordham.edu/wisdm
24. WEKA at http://www.cs.waikato.ac.nz/~ml/weka
25. Kappa: http://www.dmi.columbia.edu/homepages/chuangj/kappa

Mining and Ranking Association Rules in Support, Confidence, Correlation, and Dissociation Framework

Subrata Datta and Subrata Bose

Abstract Existing methods in association rule mining based on traditional support-confidence framework generates huge number of frequent patterns and association rules often ignoring the dissociation among items. Moreover these procedures are unable to order the rules by comparing them to find which one is better than whom. We have introduced a new algorithm for mining frequent patterns based on *support* and *dissociation* and thereafter generating rules based on *confidence* and *correlation*. The association rules have been ranked based on a composite index computed from the four measures. The experimental results obtained after implementation of the proposed algorithm justify our approach.

Keywords Association rule mining • Dissociation • Correlation • Ranking rules

1 Introduction

Frequent itemset generation and mining of association rules [1–6] are two important and challenging tasks in the field of data mining and knowledge discovery. Fundamentally association rules express hidden relationships among frequent patterns of items present in a database. Opinion differs in measurement of association. Well-known methods follow support for frequent itemset generation and confidence for measurement of association rules [1, 2]. This is popularly termed as support-confidence framework for association rule mining in the literature [7–10].

S. Datta (✉) · S. Bose
Neotia Institute of Technology, Management and Science, Kolkata, India
e-mail: subrataju2008@gmail.com

S. Bose
e-mail: subratabose@yahoo.com

S. Das et al. (eds.), *Proceedings of the 4th International Conference on Frontiers in Intelligent Computing: Theory and Applications (FICTA) 2015*, Advances in Intelligent Systems and Computing 404, DOI 10.1007/978-81-322-2695-6_13

Let DB be a database of transactions of a set of items I. Each transaction contains a subset of the items. Support is a parameter which indicates the frequency of appearance of a set of items or an itemset in DB. An itemset is said to have a support s if s % of the transactions in DB contain the itemset. Table 1 represents a sample database of 4 items—A, B, C and D. Support of A in the sample database is 0.5. An itemset is called frequent if its support is greater than or equal to user-defined minimum support. An association rule is an expression of the form $A \rightarrow B$ where itemsets A, B $\subset$ I, and $A \cap B = \phi$. Itemset A is the antecedent and B is the consequent of the rule. The confidence of the association rule $A \rightarrow B$ is the probability of occurrence of B given that A has occurred or the conditional probability P(B/A) = P(AB)/P(A) = support(AUB)/support(A). A rule is considered to be a valid association rule if its support and confidence are greater than and equal to user-specified minimum support (minsupp) and minimum confidence (minconf), respectively. Note that users have no easy way to determine minsupp and minconf to generate desired number of rules. So depending on the values of these minimums, the existing algorithms can perform extremely slow and generate huge number of results or generate none or too few results, omitting valuable information [11]. As much low the minimum thresholds are number of frequent patterns and association rules are high. Again huge number of rules without ordering or ranking is too many information or equivalently no useful information. Ranking would help the user to give preference of one rule over the others. The problem is important because in practice users have limited resources for analyzing the results. In this work, we look for a mechanism which restricts generation of too many rules and attempts to bring out the best.

1.1 Dissociation

Association measured through support is a binary association where the joint occurrence of all items of an itemset is only considered. On the contrary, dissociation represents those transactions where one itemset is transacted but the other is not. In Table 1 association of {A, B}, {C, D}, and {B, D} are 0.5 each but their respective dissociations are 0.1, 0.4, and 0.6. Table 4 summarizes this result. Therefore, {A, B} is more associative than {C, D} and similarly of {C, D} is more associative than {B, D}. High association should imply low dissociation or vice versa. The concept of measuring association through dissociation was first introduced in the literature by Pal et al. [12]. They considered togetherness[1] as the measure for frequent itemset generation in lieu of support to get the effect of dissociation in association.

[1]Togetherness is the ratio of association to association plus dissociation. Togetherness is same as Jaccard Similarity coefficient.

Table 1 Synthetic database

TID	Items
1	A, B, D
2	A, B, C, D
3	D
4	C, D
5	A, B, D
6	C, D
7	A, B
8	C, D
9	B, D
10	A, B, C, D

Table 2 Contingency table for X → Y

	Y	$\overline{Y}$	$\sum$ row
X	f_{11}	f_{10}	f_{11}
$\overline{X}$	f_{01}	f_{00}	f_{11}
$\sum$ col.	f_{+1}	f_{+0}	N

Table 2 is a 2 × 2 contingency table for a given rule X → Y. The cells of this table represent the possible combinations of values of X and Y and store the frequency associated with each combination. N is the size of the dataset D on which these frequencies have been computed. In terms of Table 2,

$$dissociation\,(X, Y) = (f_{10} + f_{01})/N \tag{1}$$

and

$$togetherness\,(X, Y) = f_{11}/(f_{11} + f_{10} + f_{01}) \tag{2}$$

Using Eqs. (1) and (2) from Table 3 we get, dissociation(A, B) = 0.1 and its togetherness = 0.5/(0.5 + 0.1) = 0.83. Table 4 displays association, dissociation, and togetherness of few selected rules based on Table 1 data. It can be easily seen that togetherness ignores the effect of null transactions. This has been overcome by Bose et al. [13]. The authors suggested weighted support[2] (ws) to replace togetherness. ws is based on support (s), Jaccard similarity coefficient (j) [14], and null

[2] *ws* is defined as $\Phi s + (1 - \Phi)j$ where $\Phi = f_{00}/N$ i.e., percentage of null transactions in the database for the rule.

Table 3 Contingency table for A → B

	B	$\overline{B}$	$\sum$ row
A	5	0	5
$\overline{A}$	1	4	5
$\sum$ col.	6	4	10

Table 4 Measurement of dissociation of some patterns

Patterns	Association (a)	Dissociation (d)	Togetherness a/(a + d)
{A, B}	0.5	0.1	0.83
{A, D}	0.4	0.6	0.4
{B, D}	0.5	0.5	0.5
{C, D}	0.5	0.4	0.56
{A, B, D}	0.4	0.6	0.4

transaction impact factor (ϕ). ws aimed at measuring association through dissociation and provided a balanced view of support and Jaccard similarity coefficient which is same as togetherness.

1.2 Observations on Support in Frequent Pattern Generation

Most common measure for frequent set generation is support. Well-known algorithms such as Apriori [1, 2], FP-growth [3] all use support as the basic measure for frequent set generation. We have the following observations on support.

(1) Specifying minimum support threshold (minsupp) by the user requires profound knowledge of data. If not several dummy runs may be required to strike a suitable minsupp.
(2) For some real-world datasets, when minsupp is low, there are too many frequent itemsets to be useful [11]. Number of frequent patterns generated is inversely proportional to minsupp.
(3) A pattern which is frequent w.r.t. low minsupp may be infrequent w.r.t comparatively high minsupp.
(4) Two different frequent patterns having same support might differ in dissociation. It is natural to treat the one with low dissociation superior than another with high dissociation.

1.3 Effect of Dissociation in Rule Generation

Dissociation plays a role in frequent pattern generation [12, 13] and association rule formation. By generating frequent sets only on the basis of support ignoring their dissociation is likely to generate rules which may have strong dissociation. So dissociation is an important issue in frequent pattern generation and hence association rule mining. From Eq. (1), in percentage term dissociation of a rule $d = (f_{10} + f_{01})/N = 1 - f_{11}/N - f_{00}/N = 1 - s - \varphi$ where s is the support and φ is the null transaction impact factor [14]. Dissociation does not make sense for 1-itemsets. So dissociation is not defined for them. In this work, we have proposed to use support-dissociation framework for frequent set generation.

1.4 Our Contribution

In this paper, we have worked on two issues of AR mining—(a) find association rules from frequent itemsets which meet user-defined maxdisso and mincorr in addition to traditional minsupp and minconf, i.e., to generate rules within limits of defined correlation and dissociation and (b) order the rules by an index computed from support, confidence, correlation, and dissociation measures of the rules.

1.5 Related Work

Researchers are trying to find out appropriate frequent patterns and association rules since last three decades. Some well-known works are [3–5]. Scholars like Antonie et al. [6], Wu et al. [7] have separately mined 11 transactions as positive association rules and 10, 01, and 00 transactions as negative association rules. Sun et al. [8] mined weighted association rules based on ranking of transactions, but not the rules. Webb et al. [15] suggested for K-optimal rule discovery method which is limited to the value of K. Vo. et al. [16], Das et al. [17], Tzanis et al. [18] worked for association rule mining but the authors have not focused on superiority measurement of generated association rules. Li et al. [19], Tan et al. [20] have tried to analyze the performance of an association rule. Fournier-Viger et al. [21, 22] have proposed methods for mining Top-K association rules where K is user-defined input. Mallik et al. [23] proposed a ranking algorithm which is basically weighted updated form of Apriori and related to specially microarray/bead chip data. Wang et al. [24] proposed a classifier HARMONY which mines the best rules for classification. The method by the authors is an instance-centric rule generation approach and works for each and

Table 5 Frequent patterns in support versus support-dissociation framework

Frequent patterns in support framework	{A}	{B}	{C}	{D}	{A, B}	{A, D}	{B, D}	{C, D}	{A, B, D}
Support	0.5	0.6	0.5	0.9	0.5	0.4	0.5	0.5	0.4
Dissociation	NA	NA	NA	NA	0.1	0.6	0.5	0.4	0.6
Frequent patterns in support-dissociation framework	{A}	{B}	{C}	{D}	{A, B}	Infrequent	Infrequent	{C, D}	Infrequent

every instance until it finds one of the highest confidence rules which increases its time complexity. We have adopted a straightforward approach for ranking rules based on four parameters as stated in Sect. 1.4. A very few authors have worked on dissociation present in association rules in case of frequent pattern mining and ranking of rules. Morzy [25] has mined dissociation rules. But his definition of dissociation is sophisticated negative rules which is quite different from the concept of dissociation [12, 13, 26] adopted in this work.

2 Proposed Method

2.1 *Frequent Pattern Generation in Support-Dissociation Framework*

Our proposed method uses support-dissociation framework for frequent pattern generation. In addition to minsupp, we have proposed user-defined maximum dissociation *maxdisso*. It signifies that any pattern should possess dissociation not more than *maxdisso* to become frequent. Table 5 shows a comparative study in between support framework and support-dissociation framework based on *minsupp* = *0.4* and *maxdisso* = *0.4*. We have not subjected the 1-itemset to dissociation as dissociation makes sense only for 2-itemsets and higher.

2.2 *Association Rule Mining in Support-Confidence-Correlation-Dissociation Framework*

We propose to generate rules from the frequent patterns in support-confidence-correlation-dissociation framework instead of traditional support-confidence.

Table 6 Association rules and their rank using Rank Index

No.	Rule	Support	Confidence	Correlation	Dissociation	Rank Index	Rank
1	{A} ⇒ {B}	0.5	1.0	0.82	0.1	3.22	1
2	{B} ⇒ {A}	0.5	0.83	0.82	0.1	3.05	2
3	{C} ⇒ {D}	0.5	1.0	0.33	0.4	2.43	3
4	{D} ⇒ {C}	0.5	0.56	0.33	0.4	1.99	4

Correlation coefficient[3] ρ measures the strength of the linear relationship between a pair of two variables [6]. In this work, we have applied minimum correlation threshold *mincorr*. Applying minsupp = 0.4, minconf = 0.5, mincorr = 0.1, and maxdisso = 0.4, the association rules obtained from Table 1 dataset are shown under Column 2 of Table 6.

2.3 Ranking of Association Rules

A global ordering of rules has been proposed in [19] in confidence × support × number of attributes in LHS of the rules. We propose to order association rules w.r.t. *Rank Index* which is computed as

$$\textit{Rank Index} = s + c + \rho + (1 - d) \quad (3)$$

the sum of support, confidence, correlation, and complement of dissociation. This is a heuristic measure based on the idea of maximizing support, confidence, correlation, and minimizing dissociation. Its value range is (minsupp + minconf + 1 − maxdisso + mincorr, 4.0). It is evident that higher the value of rank index stronger is the rule and so the rules are ordered in descending order of rank. If two or more rules have same rank index they can be further ordered in confidence, support, correlation, and dissociation sequence within themselves. Rank index and rank are shown under Columns 7 and 8 of Table 6, respectively.

[3]If the two variables are independent then ρ equals 0. $\rho = +1$ signifies positive correlation and $\rho = -1$ signifies negative correlation.

2.4 Algorithm

```
Inputs: Transaction DB D, minsupp, maxdisso, min-
conf, mincorr.
Output: Lk : Frequent sets, AR: Association rules
ordered in Ranks
(1)   Scan D and find the set of all frequent 1-
   itemset L1 applying minsupp
(2)   L= L1 /* L is the set of all frequent itemsets */
(3)   for (k=2; Lk-1≠0; k++) {
(4)   Ck ←Lk-1 ⋈ L1 /* Ck is candidate set*/
(5)   ∀ i ∈ Ck {
(6)   s← support (D, i)  /* compute support of itemset i */
(7)   d← dissociation (D, i) /* compute its dissociation */
(8)   if s≥ minsupp AND d≤ maxdisso then
(9)   Lk ← Lk ∪ {i}    /* itemset i added to Lk */
(10)  end if
(11)  }
(12)  L = L ∪ Lk
(13)  }
(14)  AR← Ø   /* AR is the set of association rules*/
(15)  for each i in L {
(16)  ∀ A, B (i = A∪ B) {
(17)  c← confidence (A, B)     /* c is the confidence*/
(18)  ρ← correlation (A, B)    /* ρ is the correlation*/
(19)  if c≥ minconf  AND ρ ≥ mincorr then
(20)  AR← AR ∪ {A → B}; AR.Count++
(21)  AR.RI ← s + c + (1-d) + ρ /* RI is the Rank Index */
(22)  end if
(23)  }
(24)  }
(25)  Sort AR in RI × c× s × ρ × (1-d) order
(26)  return AR
(27)  }
```

3 Experimental Results

We have performed our experiments on Extended Bakery Datasets,[4] a collection of transaction data of a chain of bakery stores of size 1, 5, 20, and 75 K. Table 7 is the summary of our experimental results on the 20 K dataset. Table 8 shows Top 4

[4]These datasets can be downloaded from http://wiki.csc.calpoly.edu/datasets/wiki/.

Table 7 Summary of experimental results on datasets of size 20 K

Run #	minsupp	minconf	Max disso	mincorr	# of frequent itemsets	# of AR	Max RI	Min RI
1	0.001	0.25	–	–	1703	1023	2.730	1.077
2	0.001	0.25	–	0.1	1703	354	2.730	1.372
3	0.001	0.25	0.25	0.1	1702	353	2.730	1.372
4	0.001	0.25	0.20	0.1	1548	338	2.730	1.372
5	0.001	0.25	0.15	0.1	936	192	2.475	1.437
6	0.001	0.25	0.10	0.1	159	68	2.067	1.579

Table 8 Sample association rules in Rank Index order on datasets of size 20 K

No.	Rule	Support	Confidence	Dissociation	Correlation	RI
1	{40, 41} → {23, 24}	0.025	0.948	0.157	0.919	2.730
2	{24, 40} → {23, 41}	0.025	0.917	0.157	0.915	2.695
3	{23, 41} → {24, 40}	0.025	0.917	0.157	0.915	2.695
4	{23, 24, 43} → {40, 41}	0.020	0.993	0.197	0.865	2.682

Table 9 Run to run comparison of association rules of datasets of size 20 K

Rule (generated in RUN #)	Rule measures	Remarks
{3, 18, 33} → {35} [1]	s = 0.001, c = 0.88, d = 0.239, ρ = 0.096 and RI = 1.757	Will not appear in Run #2 due to low correlation
{37} → {7} [1]	s = 0.036, c = 0.50, d = 0.11, ρ = 0.347 and RI = 1.752	Will appear in all runs except Run #6 when it violates max dissociation
{2, 16, 45} → {32} [2]	s = 0.001, c = 0.91, d = 0.254, ρ = 0.103 and RI = 1.783	Will not appear in Run #3 as it will violate max dissociation
{3, 33, 35} → {18} [3]	s = 0.001, c = 0.957, d = 0.239, ρ = 0.102 and RI = 1.839	Will not appear in Run #4 as it will violate max dissociation
{23, 24, 43} → {40, 41} [4]	s = 0.02, c = 0.993, d = 0.197, ρ = 0.865 and RI = 2.682	Will not appear in Run #5 as it will violate max dissociation
{7, 37} → {11} [5]	s = 0.034, c = 0.942, d = 0.148, ρ = 0.642 and RI = 2.457	Will not appear in Run #6 as it will violate max dissociation

Association Rules (AR) ordered in Rank Index (RI) of RUN #3 of 20 K dataset. Table 9 shows six sample Association Rules which were dropped from RUN #1 to #5 successively. Graphical view of the effect of dissociation in AR mining is shown in Fig. 1.[5]

[5]In the figures X-Axis represents max dissociation, Primary Y-Axis represents frequent itemsets & Rules and Secondary Y-Axis represents Max and Min Rank Index.

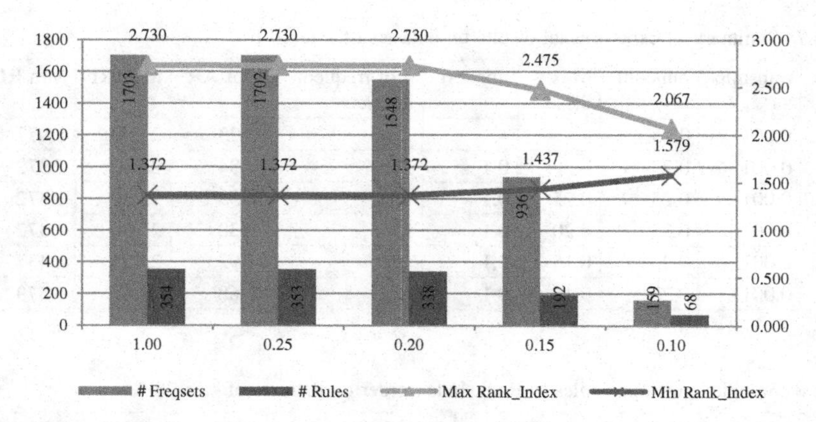

Fig. 1 Effect of dissociation in AR mining on extended bakery dataset of 20,000 transactions for minsupp = 0.001, minconf = 0.25 and mincorr = 0.1

4 Conclusion

Traditional support framework alone cannot distinguish among patterns having equal support. Moreover, sometimes it happens that generated patterns are being treated as frequent patterns though they possess high dissociation. Dissociation is the opposite of association. We have introduced a support-dissociation framework for frequent pattern generation where we have made a scope for the user to define minimum support and maximum allowable dissociation. For rule generation, we have introduced support-confidence-correlation-dissociation framework instead of traditional support-confidence framework. Strength of relationship and strength of separation between antecedent and consequent part of a rule are expressed by correlation and dissociation, respectively. Finally, we have proposed a heuristic measure called Rank Index for ordering association rules based on the idea of maximizing support, confidence, correlation, and minimizing dissociation. Our experimental study shows the effectiveness of the proposed method. The Rank Index proposed in this work assumes equal importance of each of its factor. But depending on the nature of the dataset being mined, these factors may have to be dealt with different priorities or weights. For example, a sparse dataset will generally have low support and high dissociation. Future focus lies in developing computationally efficient algorithms for robust association rule mining using support-confidence-correlation-dissociation framework, introduction of weight-based rank index, and finding specific areas of application.

References

1. Agarwal, R., Imielinski, T., Swami, A.: Mining association rules between sets of items in large datasets. In: Proceedings of the ACM SIGMOD'93, pp. 207–216 (1993)
2. Agarwal, R., Srikant, R.: Fast algorithms for mining association rules. In: 20th VLDB conference, pp. 487–499. Santiago, Chile (1994)
3. Pei, J., Han, J.: Mining frequent patterns by patterns-growth: methodology and implications. In: ACM SIGKDD, Dec 2000
4. Han, J., Pei, J., Yin, Y.: Mining Frequent patterns without candidate generation. In: Proceedings of the ACM SIGMOD (2000)
5. Craus, M., Archip, A.: A generalized parallel algorithm for frequent itemset mining. In: 12th WSEAS International Conference on Computers, Heraklion, Greece, 23–25 July 2008
6. Antonie M.L., Zaiane, O.R.: Mining positive and negative association rules: an approach for confined rules. In: Proceedings of the International Conference on Principles and Practice of Knowledge Discovery in Databases, pp. 27–38 (2004)
7. Wu, X., Zhang, C., Zhang, S.: Efficient mining both positive and negative association rules. ACM Trans. Inf. Syst. **22**(3), 381–405 (2004)
8. Sun, K., Bai, F.: Mining weighted association rules without pre-assigned weight. IEEE Trans. Knowl. Data Eng. **20**(4), 489–495 (2008)
9. Ramaraj, E., Rameshkumar, K.: Ranking mined association rule: a new measure. JTES **1**(1), 57–61 (2009)
10. Lavrac, N., Flach, P., Zupan, B.: Rule Evaluation Measure: A Unifying View. ILP-99, LNAI vpl, vol. 1634, pp. 174–185, Springer, Heidelberg (1999)
11. Zheng, Z., Kohavi, R., Mason, L.: Real world performance of association rule algorithms. In: Proceedings of the Seventh ACM-SIGKDD International Conference on Knowledge Discovery and Data Mining, ACM, New York, NY (2001)
12. Pal, S., Bagchi, A.: Association against dissociation: some pragmatic consideration for frequent itemset generation under fixed and variable thresholds. ACM SIGKDD Explor. **7**(2), 151–159 (2005)
13. Bose, S., Datta, S.: Frequent pattern generation in association rule mining using weighted support. In: IEEE C3IT, pp. 1–5 (2015)
14. Jaccard Index from Wikipedia http://en.wikipedia.org/wiki/Jaccard_index
15. Webb, G.I., Zhang, S.: K-optimal rule discovery. Data Min. Knowl. Discov. **10**(1), 39–79 (2005)
16. Vo, B., Le, B.: Fast algorithm for mining generalized association rules. Int. J. Database Theory Appl. **2**(3), 1–10 (2009)
17. Das, A., Ng, W.K., Woon, Y.K.: Rapid association rule mining. In: ACM CIKM, Atlanta, GA, USA, 5–10 Nov 2001
18. Tzanis, G., Berberidis, C., Vlahavas, I.: On the discovery of mutually exclusive items in a market basket database. In: Proceedings of the 2nd ADBIS Workshop on Data Mining and Knowledge Discovery (ADMKD 2006), Thessaloniki, Greece, 06 Sept 2006
19. Li, W., Han, J., Pei, J.: CMAR: accurate and efficient classification based on multiple class-association rules. In: Proceedings of the 2001 IEEE International Conference on Data Mining (Proceeding ICDM'01), pp. 369–376 (2001)
20. Tan, P.N., Kumar, V., Srivastava, J.: Selecting the right interestingness measure for association patterns. In: ACM SIGKDD'02, Edmonton, Alberta, Canada (2002)
21. Fournier-Viger, P., Wu, C.W., Tseng, V.S.: Mining Top-K association rules. In: Proceedings of the 25th Canadian Conference on Advances in Artificial Intelligence (Proceeding Canadian AI'12), pp. 61–73. Springer, Berlin, Heidelberg, ©2012
22. Fournier-Viger, P., Tseng, V.S.: Mining Top-K non-redundant association rules. In: ISMIS 2012, LNCS, vol. 7661, pp. 31–40. Springer, Heidelberg (2012)

23. Mallik, S., Mukhopadhyay, A., Maulik, U.: RANWAR: rank-based weighted association rule mining from gene expression and methylation data. IEEE Trans. Nanobiosci. **14**(1), 59–66 (2014)
24. Wang, J., Karypis, G.: HARMONY: efficiently mining the best rules for classification. In: Proceedings of the 2005 SIAM Conference on Data Mining, pp. 205–216 (2005)
25. Morzy, M.: Efficient mining of dissociation rules. In: DaWaK 2006, LNCS, vol. 4081, pp. 228–237. Springer, Heidelberg (2006)
26. Datta, S., Bose, S.: Discovering association rules partially devoid of dissociation by weighted confidence. In: IEEE 2nd International Conference on Recent Trends in Information Systems, Jadavpur University, India, 9–11 July, 2015 (in press)

Mining Closed Interesting Subspaces to Discover Conducive Living Environment of Migratory Animals

G.N.V.G. Sirisha and M. Shashi

Abstract This paper presents the suitability of subspace clustering techniques to identify the conducive living environment of migratory animals given the geographical and weather conditions prevailing at various locations where the animals thrive. The set of collaborative weather and geographical conditions prevailing at different locations where animals move define the conducive living environment/conditions of animals and hence their accessibility in turn influence the migration behavior of animals. The concept of closed interesting subspaces in density divergence context for multidimensional data is proposed by the authors to model the conducive living conditions of migratory animals. A grid-based subspace mining algorithm namely SCHISM which is originally meant for extracting the maximal interesting subspaces was adapted for finding closed interesting subspaces. Migratory Burchell's Zebra movement data collected from MoveBank was used for this analysis purpose.

Keywords Data mining · Subspace clustering · Animal migrations · Closed interesting subspaces · Ecology · Sequence mining

1 Introduction

Animal migration is usually a seasonal movement of animals in search of food, suitable breeding sites, or to escape bad weather or other environmental conditions [1]. Migratory animals help us in protecting our food, seed dispersal, pollination,

G.N.V.G. Sirisha (✉)
Department of CSE, S.R.K.R. Engineering College, Bhimavaram, India
e-mail: sirishagadiraju@gmail.com

M. Shashi
Department of CS & SE, Andhra University College of Engineering,
Andhra University, Visakhapatnam, India
e-mail: smogalla2000@yahoo.com

S. Das et al. (eds.), *Proceedings of the 4th International Conference on Frontiers in Intelligent Computing: Theory and Applications (FICTA) 2015*, Advances in Intelligent Systems and Computing 404, DOI 10.1007/978-81-322-2695-6_14

and fertilizing the plants [2]. They help in balancing the ecosystem. We need to safeguard them to preserve our ecosystem. Human activities like urbanization, deforestation, overgrazing, agricultural land conversion etc. are leading to habitat loss or habitat fragmentation of migratory animals. This leads to extinction of the migratory animals.

If we are able to identify the favorable environmental conditions and vegetation type for migratory animals, we can reduce habitat loss and habitat fragmentation due to human intervention. It also helps in creating artificial habitats for migratory animals in case of habitat loss to prevent their extinction. Availability of long-term data on migrations annotated with atmospheric observations and landscape variations helps us to analyze the favorable environmental conditions of migratory animals. The ability to locate animals at high spatial and temporal granularities using the GPS and sensor technologies and access to environmental data from different online sources provide rich data which can be analyzed using different statistical techniques, data mining algorithms etc.

Using such data, the authors proposed to apply closed interesting subspace mining techniques to extract patterns that reflect conducive living environment or habitats of migratory animals.

1.1 Subspace Clustering

Clustering is the process of grouping objects such that all the objects in each group are similar to one another. In traditional clustering, the similarity between objects is often determined using distance between objects over all the dimensions. Because of the existence of irrelevant and redundant attributes, the objects that are homogenous in only a subset of attributes cannot be detected when seen in full dimensional space [3–5].

Subspace clustering addresses this problem and aims at exploring various subspaces that define clusters specific to such subspaces. The clusters that are existing in subspaces of the multidimensional data space are called subspace clusters. A subspace constitutes the subset of relevant attributes shared by the members of a cluster. The subset of relevant attributes shared by members of one cluster may be different from the subset of relevant attributes shared by members of another cluster.

Let $\mathbf{D} = \mathbf{O} \times \mathbf{A}$ be a dataset represented in the form matrix, where $\mathbf{O}$ is the set of objects and $\mathbf{A}$ is the set of attributes. A subspace cluster $\mathbf{C}$ is a submatrix $\mathcal{O} \times \mathcal{A}$, where the set of objects $\mathcal{O} \subseteq \mathbf{O}$ is homogeneous in the set of attributes defined by the subspace $\mathcal{A} \subseteq \mathbf{A}$ [6].

Subspace clustering algorithms discover clusters that exist in multiple, possibly overlapping subspaces, thus allowing an object to be a member of multiple clusters. Subspace clustering techniques can be classified into different types based on the type of data they handle, dimensionality of cluster solutions, and the approaches used for clustering data.

Different subspace clustering algorithms are devised for handling continuous valued data, categorical data, sequence data, stream data, and to provide either 2D or 3D cluster solutions. A 2D cluster solution defines each cluster in two dimensions, with the first dimension representing the objects of the cluster while the second dimension representing the set of attributes shared by the members of a cluster. In other words, each cluster defines a set of objects that are homogenous in a subspace defined by the set of attributes.

It may be noted that the cluster solution given by traditional clustering algorithms is one-dimensional as it groups the objects into clusters in a predefined problem space specified in terms of relevant set of attributes. In 3D data cluster solution, first dimension represents objects, second dimension represents attributes describing the objects, and the third dimension represents an attribute that has to be handled differently like time or location.

Based on the approach taken, subspace clustering algorithms can be classified as grid-based, density-based, and window-based algorithms.

1.2 Animal Migrations

A lot of research is being carried out to find the factors that influence migration and how migrants respond to changes in climate. In [7], the authors have studied the reasons for adult albatrosses to make long-range trips to preferred, productive areas and how wind assistance facilitates their return flights and how their outbound flights are hampered by head winds.

Altitudinal migration of Galapagos tortoises was studied by Stephen Blake et al. in [8]. The authors studied the roles of environmental variation and individual traits in partial migration of tortoises. A study on migratory behavior of woodland caribou showed that the variability in the movement behavior of woodland caribou is attributed to the local environmental conditions [9] All of the above methods used various statistical techniques to analyze animal movement data.

Li et al. has applied data mining techniques on spatiotemporal data to find periodic movements of animals [10].

Hattie et al. studied the effect of changes in environment on speed and onset of migration of zebras [11]. Hattie et al. have proposed an array of increasingly complex models representing alternative hypotheses regarding the environmental cues and controls for movement. These hypotheses were tested on the movement data of zebras in Botswana. The best models that explained the influence of environmental conditions on departure date and movement speed of migrating zebras were identified. Their analysis has shown that the movement speed of zebras was influenced by Normalized Difference Vegetation Index (NDVI) and rainfall. The date at which zebras started their migration was influenced by cumulative precipitation.

1.3 Overview of Proposed Methodology

This research has taken "Migratory Burchell's zebra in northern Botswana" migratory data [11, 12] annotated with environmental parameters and vegetation indices from MoveBank as a case study to show that closed interesting subspace mining can be used in identifying the conducive living conditions of migratory animals.

Migratory animals move in search of conducive living environment. The climate and vegetation types that are favorable to the development of migratory animals is called conducive living environment. Conducive living conditions detection requires the application of subspace clustering. This is because animal movement is influenced by a subset of atmospheric and landscape variations. We need to identify clusters each of which is a combination of such attributes value pairs (where zebras thrive or move in search of) that defines the conducive living conditions preferred by many zebras. Multiple clusters exist as the atmospheric, vegetative, and geographical conditions defining the conducive living conditions differ with seasons, locations, animal life cycle stage, and their combinations. This requires a clustering algorithm which can identify a cluster that appears significantly and is present in a subset of attributes.

This paper proposes a 2D grid and density-based closed interesting subspace mining algorithm for identifying the conducive living conditions of migratory animals.

SCHSIM [4] a grid- and density-based interesting subspace mining algorithm which mines maximal interesting subspaces was adapted for finding the closed interesting subspaces. A maximal interesting subspace represents the maximal set of collaborative environmental and geographical conditions that appear frequently in the locations where zebras move. A subset of these environmental and geographical conditions may influence the movement of zebras more than the other environmental and geographical conditions defining the maximal interesting subspace. The identification of such strongly influencing factors will be missed if we mine only the maximal interesting subspaces.

So, in multidimensional spaces, the concept of closed interesting subspaces in density divergence context is proposed by the authors. The proposed approach can be used to find conducive living environment of any migratory animal given the atmospheric observations and landscape variations along the path in which migratory animals moved. Though closed interesting subspaces are used to model the conducive living conditions of migratory animals in this paper, the concept of closed interesting subspaces is applicable for identifying interesting subspaces in any multivariate time series.

2 Literature Review

There are a number of recent studies for mining interesting subspaces from multidimensional datasets. All of the existing approaches either mine all the interesting subspaces or only the maximal interesting subspaces. Grid-based subspace clustering partitions the data space into grids. Grid cells which have high density are used to form subspace clusters. CLIQUE [13] is a pioneering algorithm in this category. By discretizing the continuous valued attributes, CLIQUE has proposed to apply frequent itemset mining techniques for identifying subspaces clusters. It first identifies all interesting subspaces at all dimensionalities. A K-dimensional subspace corresponds to k-itemset. The number of objects (support in frequent item mining) that lie in a subspace corresponds to its density. Only those subspaces whose density exceeds a threshold τ are mined. Clusters are then discovered by grouping the connected dense subspaces at each dimensionality. ENCLUS [14] uses subspace entropy for selecting interesting subspaces. These interesting subspaces are then used for discovering the clusters. The same clustering model as CLIQUE is used for cluster discovery. MAFIA [15] is a major extension of CLIQUE which uses adaptive grids for cluster discovery.

All the above algorithms use a global density threshold to identify dense units. Because of "density divergence problem," these algorithms fail to discover high-quality clusters in all subspaces at all dimensionalities. Density divergence problem [4, 16] refers to the phenomenon that cluster densities vary in different subspace cardinalities. As the number of dimensions increase, the data points (database objects) are more spread out in higher dimensional space and they are sparsely distributed. Thus as subspace dimensionality increases, it is constrained by additional attributes or dimensions. Hence, as the dimensionality of subspace increases the number of points in the subspace decreases. If we use the same global density threshold to identify the dense subspaces at all dimensionalities, we may not find high-quality subspaces at all dimensionalities. To overcome this problem, the density threshold value should decrease with increase in subspace cardinality (dimensionality).

SUBCLU [17] uses DBSCAN [18] cluster model of density connected objects to discover the subspace clusters. A cluster is defined as a maximal set of density connected points. It takes two input parameters ε and m. SUBCLU uses the same parameter values at all subspace dimensionalities, so SUBCLU may not identify high-quality clusters at all subspace dimensionalities.

FIRES [19] a filter–refinement based subspace clustering algorithm is proposed to overcome the scalability and density divergence problems faced by most density-based subspace clustering algorithms. As it mines only the maximal subspace clusters, some clusters that appear significantly at lower dimensionalities will not be detected.

INSCY [20] and Scalable Density-Based Subspace Clustering [21] are two other efficient density-based subspace clustering algorithms. Both INSCY and Scalable Density-Based Subspace Clustering do not deal with the density divergence problem. DUSC [22] is a density-based subspace clustering algorithm which uses

DBSCAN model for cluster discovery. It uses a density measure that is adaptive to the dimensionality.

DENCOS [16] is a subspace clustering algorithm proposed to solve the density divergence problem. It uses density threshold that adapts to subspace cardinality.

SCHISM [4] is a grid-based interesting subspace mining algorithm that overcomes the density divergence problem. It is a highly scalable algorithm taking an order of minutes to mine subspaces from datasets which are as large as 3 million. Though DENCOS also overcomes the density divergence problem, it uses a non-parametric method in calculating the density thresholds at different subspace cardinalities. SCHISM uses a parametric method to find the density thresholds at different subspace cardinalities using Chernoff-Hoeffding's inequality. It mines maximal subspaces by constraining a subspace with additional dimensions while ensuring the density thresholds are met.

Due to density divergence, the a priori property is violated by those lower dimensional subspaces failing to satisfy their density thresholds even if they are part of a dense subspace of higher dimensionality. Hence, unlike the traditional maximal frequent patterns representing all their subpatterns, a dense maximal subspace may not imply all the subspaces that are subsets of it to be dense. Hence we proposed the concept of closed interesting subspaces in the context of density divergence. Density divergence is inevitable in subspace clustering. This research adapts SCHISM to identify closed interesting subspaces which define the conducive living environment of migratory animals.

3 Discovery of Preferred Living Conditions of Migratory Animals Using Density-Based Closed Interesting Subspace Mining

3.1 Problem Definition

Movement trajectories data when annotated with surrounding atmospheric observations and underlying landscape information forms a rich dataset which can be analyzed to find the conducive living conditions of animals.

Many of the previous works in movement ecology have used different statistical models to find effect of changing environment on animal movement. By taking the long-term movement data annotated with atmospheric observations and landscape information, the authors proposed to apply data mining techniques to automatically extract patterns that explain the conducive living conditions of animals.

Migratory Burchell's zebra in northern Botswana data from MoveBank is taken as a case study to show that closed interesting subspace mining can be used in identifying the conducive living conditions of migratory animals. This research aims to find the set of collaborative environmental conditions and vegetation types that attract zebras.

3.2 Application of Closed Interesting Subspace Mining on Zebra Data

In multidimensional spaces, the concept of closed interesting subspaces in density divergence context is proposed in this research and SCHISM algorithm was adapted to find the closed interesting subspaces. The overview of the algorithm is given below. It consists of three steps.

Algorithm: Mine Closed Interesting Subspaces from Multivariate Datasets

Input: Multivariate Dataset
Output: Closed Interesting Subspaces
Method:
Step 1: Data Preparation
Step 2: Discretize the database to convert multivariate continuous valued data to discrete valued data
Step 2: Convert the dataset from horizontal to vertical format.
Step 3: Mine closed interesting subspaces.

Subspace Definition: A subspace is defined an axis -aligned hyper-rectangle $[l_1, h_1] \times [l_2, h_2] \times \cdots \times [l_d, h_d]$ where $l_i = (aD_i)/\xi$, and $h_i = (bD_i)/\xi$, a, b are positive integers and $a < b < \xi$, ξ is the number of divisions of an axis, d is the number of dimensions. If $h_i - l_i = D_i$, the subspace is unconstrained in dimension i whose range is given as D_i. An m-subspace is a subspace constrained in m dimensions, denoted as S_m [4].

Data Preparation.
The dataset used for this study consisted of 7 adult migratory zebras data. The data is recorded from October 2007 to June 2009. Zebra's location is given in terms of latitude and longitude. Location data is annotated with Weekly-Precipitation, Moderate-resolution Imaging Spectroradiometer (MODIS) Land Terra GPP 1 km 8d GPP, MODIS Land Aqua GPP 1 km 8d GPP, MODIS Land Terra GPP 1 km 8d PsnNet, MODIS Land Aqua GPP 1 km 8d PsnNet variables constituting records with 7 attributes. There are 53,793 such records. Except Weekly_Precipitation, the other variables data is obtained from MoveBank [12]. The original dataset contained many other variables like timestamp, individual zebras identifier, tag type etc. which were removed because they are irrelevant to influence migration behavior.

Weekly-Precipitation gives the weekly cumulative precipitation in the Tropical Rainfall Measuring Mission (TRMM) grid where the location defined by latitude and longitude is present. The rest of the variables indicate the vegetation productivity. For the calculation of Weekly-precipitation timestamp variable is also considered. The trajectory in which zebras moved is covered by 24 TRMM grids.

For all these 24 TRMM grids, daily precipitation is collected from January 1, 2007 to January 1, 2010. Daily precipitation for all these grids in the specified time period is obtained from IRI data library from Columbia University [23]. These daily precipitation values are then used for finding the weekly precipitation. For each TRMM grid, the weekly precipitation is obtained by adding the daily precipitations of the preceding 7 days of the week that ends with the given day. For every location where a zebra has moved, the weekly precipitation is taken as weekly precipitation of the grid at the timestamp in which zebra is located there.

Zebras followed a highly directed movement during migration which can be captured by applying principal component analysis to latitude and longitude. Latitude and longitude attributes are combined to form a new attribute using principal component analysis. This attribute is named as lat-long. Lat-long is the first principal component obtained by applying PCA to latitude and longitude. The first principal component explained 98.3 % variance in the dataset containing latitude and longitude. Prcomp function from R Language is used for finding the principal components. Latitude and longitude are scaled and centered which normalizes them using Z-score normalization before applying PCA. Weights of the variables longitude and latitude in the computation of the first principal component are 0.7071068 and −0.7071068, respectively.

The range of values of the following 6 variables Lat-Long, Weekly-Prec, MODIS Land Terra GPP 1 km 8d GPP, MODIS Land Aqua GPP 1 km 8d GPP, MODIS Land Terra GPP 1 km 8d PsnNet, MODIS Land Aqua GPP 1 km 8d PsnNet are [−2.21778, 1.49096], [0, 110.1495], [0, 0.03395], [3.99e–009, 0.022086], [−0.01248, 0.027102], [−0.00887, 0.014134], respectively. The values of Weekly_Precipitation dimension follow a skewed distribution. So, the values are transformed by applying $\log_2$ transformation except 0 value. Most of the values of the attribute are 0 and it is given a separate interval. First interval corresponds to 0 precipitation. The rest of the transformed values are then discretized into 9 intervals using equiwidth binning. The ranges of values of the rest of the variables are discretized into 10 intervals each using equiwidth binning.

Data Discretization.
The intervals that are generated after discretization are analogous to items in frequent itemset mining. Each interval corresponds to an attribute-value pair. The problem of interesting subspace mining is thus analogous to frequent itemset mining. Each interval is a *1-D* subspace. A *k-D* subspace is a combination of *k* intervals from *k* dimensions. Each database record (object) corresponds to a transaction and the intervals correspond to items. *K*-subspace corresponds to *k*-itemset. The number of objects (support in frequent item set mining) that lie in a subspace corresponds to its density. Only those subspaces whose density exceeds the density threshold are mined. These subspaces are called dense subspaces or interesting subspaces

Data Transformation.
After discretization the dataset is converted to vertical format, i.e., for each *1-D* subspace we store the set of data records (ridset) where the *1-D* subspace occurred.

Mining Closed Interesting Subspaces.
SCHISM algorithm is adapted to find the closed interesting subspaces. The proposed algorithm uses a depth first search with backtracking. The algorithm first mines *1-D* and *2-D* subspaces. *p-D* subspaces are used in mining the *(p + 1)-D* subspaces. This is done by taking a *p-D* subspace as input and the set of interesting *1-D* subspaces that can be used to constrain the *p-D* subspace in an interval of another dimension. Each such *1-D* subspace can be used to constrain a *p-D* subspace to produce a *(p + 1)-D* subspace, only if the number of points (objects) that exists in this new *(p + 1)-D* subspace satisfy the density threshold at *(p + 1)* dimensionality [4]. This process continues until the algorithm can no longer extend the given subspace. At this stage, the algorithm backtracks to produce new interesting subspaces by extending other dense subspaces. When the algorithm backtracks, SCHISM algorithm is adapted in this research to check if any of the subsets of the dense subspaces are closed. All the closed subspaces are saved.

Handling Density Divergence.
In the problem of mining interesting subspaces as the dimensionality of subspace increases it is constrained by additional attributes or dimensions. Hence, a higher dimensional subspace excludes some of the data points covered by lower dimensional subspace. Volume increases with dimensionality and consequently as the dimensionality of the subspace increases the density of subspace decreases which is referred to as density divergence problem. So in order to find all interesting subspaces of different dimensionalities we have to use high-density threshold at low subspace dimensionalities and low-density threshold in high subspace dimensionalities. Therefore, the candidate elimination process based on antimonontonicity property used in frequent itemset mining algorithms can no longer be used in mining the interesting subspaces.

To overcome the density divergence problem, SCHISM sets different density thresholds for different subspace dimensionalities. All the subspaces with same dimensionality will have same density threshold. Chernoff-Hoeffding bound is used in calculating the density threshold at a given dimensionality.

Closed Interesting Subspace.
As a natural consequence of density divergence, the density of a higher dimensional subspace is expected to reduce proportionately as a ratio of density thresholds of corresponding subspaces. The concept of closed pattern/subspace is framed by the authors as follows. A dense subspace of *p*-dimensions is closed if all of its extensions have their density reduced by more than the expectation in accordance with the ratio of density thresholds in corresponding subspaces.

Let S_p denote a dense subspace constrained in p dimensions.

S_p is closed if $\forall q > p, \frac{density(S_q)}{density(S_p)} < \frac{density\_Threshold(q)}{density\_Threshold(p)}$ where S_q an extension of S_p is a dense subspace constrained in q dimensions.

Though this research adapted SCHISM to find dense/closed interesting subspaces, the proposed adaptation is equally applicable irrespective of the way density thresholds are fixed.

3.3 Density Threshold as a Function of No. of Constrained Dimensions

Density thresholds calculation in the proposed algorithm is done in the same way as SCHISM [4]. In SCHISM density thresholds are fixed in the following way. A p-subspace is said to be interesting if the ratio of number of points in the subspace to the total number of points in database is greater than the right hand term in (1)

$$\frac{n_p}{n} \geq \frac{E[X_p]}{n} + \sqrt{\frac{1}{2n}\ln\left(\frac{1}{\tau}\right)} \quad (1)$$

With $E[X_p] = n\left(\frac{1}{\xi}\right)^p$ where ξ is the number of intervals into which each dimension is discretized.

The interestingness measure $thresh_{SCHISM}(p) = \frac{E[X_p]}{n} + \sqrt{\frac{1}{2n}\ln\left(\frac{1}{\tau}\right)}$ is a nonlinear monontonically decreasing function in the number of dimensions p, in which the subspace S_p is constrained. To increase the efficiency and effectiveness of finding interesting subspaces the formula for the density threshold calculation is further optimized by Sequeira et al. [4].

4 Result Analysis

The dataset consists of 53,793 records described using 6 attributes. The attributes are discretized into intervals as discussed in Sect. 3.2. Each such interval corresponds to an attribute-value pair. A subspace is a combination of one or more such intervals that appear significantly in the dataset. The significance of subspace is given by its density. All the subspaces whose density exceeds the corresponding density thresholds are called interesting subspaces.

Algorithms that handle density divergence set high-density thresholds at low subspace dimensionalities and low-density thresholds at high subspace dimensionalities. Hence, unlike the traditional maximal frequent patterns representing all their subpatterns, a dense maximal subspace may not imply all the subspaces that are subsets of it to be dense. SCHISM being a maximal interesting subspace miner will miss some of the interesting subspaces at low subspace dimensionalities. Hence we proposed the concept of closed interesting subspaces in the context of density divergence. Density divergence is inevitable in subspace clustering.

To handle the density divergence problem, nonlinear monotonically decreasing threshold is used as the dimensionality of subspace increases. The density threshold is calculated as shown in Eq. 1. Its value depends on ξ and τ for a given dataset. The performance of SCHISM is good for ξ values ranging from 5 to 15 [4]. So for mining the closed interesting subspaces, ξ is taken as 10 in the experiments. The performance of the proposed algorithm is tested for τ values ranging from $\frac{10000}{n}, \frac{1000}{n}, \frac{100}{n}, \frac{10}{n}, \frac{1}{n}, \frac{1}{10n}, \frac{1}{100n}$ and so on to $\frac{1}{1000000n}$ where n is the number of records. Figure 1 shows $\log_{10}\tau$ versus number of patterns. Figure 2 shows $\log_{10}\tau$ versus time in sec.

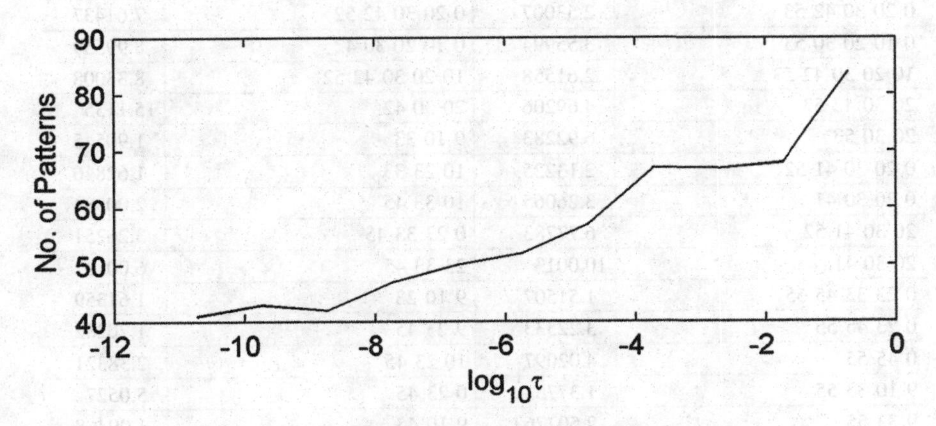

Fig. 1 Number of closed interesting subspaces found for different values of $\log_{10}\tau$

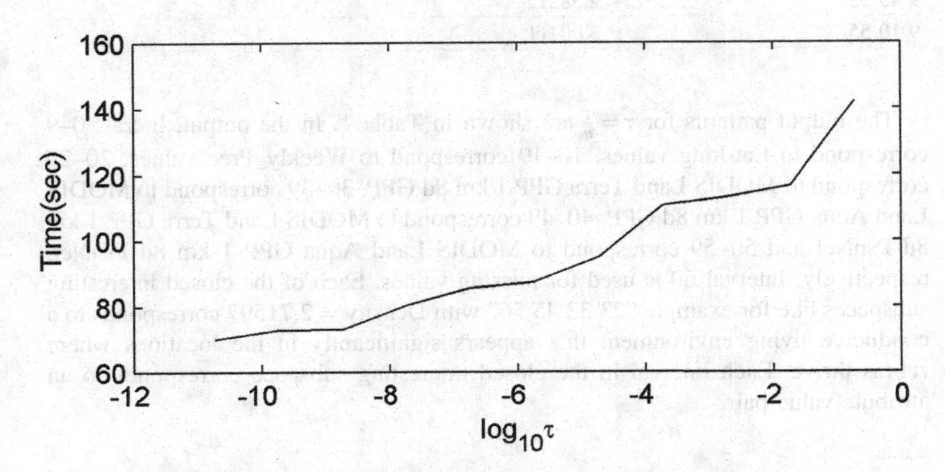

Fig. 2 Time taken by closed interesting subspace miner for different values of $\log_{10}\tau$

Table 1 Closed interesting subspaces mined for $\tau = \frac{1}{n}$

Closed interesting subspace	Density	Closed interesting subspace	Density
8 30	2.80892	9 55	5.54347
8 10	6.47482	10 23 45 55	1.51135
17 20 30	2.59699	10 23 55	2.02814
9 17	7.237	10 33 55	2.56725
23 33 45 56	2.71597	10 45 55	3.32757
10 56	3.55622	10 55	6.22386
18 20 30 41 52	2.32186	23 33 45 55	3.19558
18 20 30	3.92058	23 45 55	6.55662
9 18	11.8752	33 45 55	4.48385
0 10 20 30 42 53	2.36834	0 10 20 30 42 52	6.6254
0 20 30 42 53	2.53007	0 20 30 42 52	7.61437
0 10 20 30 53	3.55994	0 10 20 30 42	8.99559
10 20 30 42 53	2.61558	10 20 30 42 52	8.33008
20 30 42 53	4.69206	20 30 42	15.1135
20 30 53	6.92283	9 10 33	1.90545
0 20 30 41 52	2.13225	10 23 33	1.62846
0 20 30 41	3.26065	10 33 45	2.00584
20 30 41 52	6.77783	0 23 33 45	3.26251
20 30 41	10.0013	23 33 45	6.00822
0 23 33 45 55	1.51507	9 10 23	1.61359
0 23 45 55	3.22533	9 23 45	1.70282
0 45 55	4.02097	10 23 45	2.38321
9 10 33 55	1.3775	0 23 45	5.0527
9 33 55	2.50776	9 10 45	1.99468
9 10 45 55	1.31244	9 45	4.90398
9 45 55	2.58212		
9 10 55	3.00411		

The output patterns for $\tau = \frac{1}{n}$ are shown in Table 1. In the output, literals 0–9 correspond to Lat-long values, 10–19 correspond to Weekly_Prec values, 20–29 correspond to MODIS Land Terra GPP 1 km 8d GPP, 30–39 correspond to MODIS Land Aqua GPP 1 km 8d GPP, 40–49 correspond to MODIS Land Terra GPP 1 km 8d PsnNet and 50–59 correspond to MODIS Land Aqua GPP 1 km 8d PsnNet, respectively. Interval 60 is used for missing values. Each of the closed interesting subspaces like for example "23 33 45 56" with Density = 2.71597 corresponds to a conducive living environment that appears significantly in the locations where zebras thrive. Each interval in the closed interesting subspace corresponds to an attribute-value pair.

5 Conclusions

This research identifies the conducive living conditions prevailing at locations where zebras thrive and hence migrate in search of such conditions. Though many researchers from movement ecology field have done research on effect of environmental variation on animal migration, to our knowledge this is the first work where subspace clustering is applied to the field of movement ecology. The authors integrated the idea of closed patterns with density divergence of subspaces to propose the concept of closed interesting subspaces. SCHISM the grid-based interesting subspace mining algorithm which overcomes the density divergence problem is adapted to mine the closed interesting subspaces. The closed interesting subspaces are used to model the conducive living conditions of migratory animals. The results of this research are used for identifying migratory patterns of zebras as a separate project by the authors, which involve the application of sequential pattern mining algorithms.

Acknowledgements Our sincere thanks to Hattie L.A. Bartlam-Brooks for providing access to "Migratory Burchell's zebra in northern Botswana" data in MoveBank.

References

1. BBC: Nature migration videos, news and facts. http://www.bbc.co.uk/nature/adaptations/Animal_migration
2. Blake, S., Wikelski, M., Cabrera, F., Guezou, A., Silva, M., Sadeghayobi, E., Yackulic, C.B., Jaramillo, P.: Seed dispersal by Galapagos tortoises. J. Biogeogr. **39**, 1961–1972 (2012)
3. Beyer K., Goldstein J., Ramakrishnan R., Shaft U.: When is nearest neighbors meaningful?. In: ICDT, pp. 217-235 (1999)
4. Sequeira, K., Zaki, M.: SCHISM: a new approach to interesting subspace mining. J. Bus. Intell. Data Min. **1**, 137–160 (2005)
5. Kriegel, H.-P., Kroger, P., Zimek, A.: Clustering high-dimensional data: a survey on subspace clustering, pattern-based clustering and correlation clustering. ACM Trans. Knowl. Discov. Data **3**, 1 (2009)
6. Sim, K., Gopalkrishnan, V., Zimek, A., Cong, G.: A survey on enhanced subspace clustering. J. Data Min. Knowl. Discov. **26**, 332–397 (2013)
7. Dodge S., Bohrer G., Weinzierl R., Davidson S.C., Kays R., Douglas D., Cruz S., Han J., Brandes D., Wikelski M.: The environmental-data automated track annotation (Env-DATA) system: linking animal tracks with environmental data. J. Mov. Ecol. **1**, 3 (2013)
8. Blake, S., Yackulic, C.B., Cabrera, F., Tapia, W., Gibbs, J.P., Kümmeth, F., Wikelski, M.: Vegetation dynamics drive segregation by body size in Galapagos tortoises migrating across altitudinal gradients. J. Anim. Ecol. **82**, 310–321 (2012)
9. Avgar, T., Mosser, A., Brown, G.S., Fryxell, J.M.: Environment and individual drivers of animal movement patterns across a wide geographical gradient. J. Anim. Ecol. **82**, 96–106 (2013)
10. Li, Z., Han, J., Ding, B., Kays, R.: Mining periodic behaviors of object movements for animal and biological sustainability studies. J. Data Min. Knowl. Discov. **24**, 355–386 (2012)

11. Bartlam-Brooks, H.L.A., Beck, P.S.A., Bohrer, G., Harris, S.: In search of greener pastures—using satellite images to predict the effects of environmental change on zebra migration. J. Geophys. Res.: Biogeosci. **188**, 1–11 (2013)
12. Movebank Data Repository: http://www.datarepository.movebank.org/handle/10255/move.343, doi:10.5441/001/1.f3550b4f
13. Agrawal R., Gehrke J., Gunopulos D., Raghavan P.: Automatic subspace clustering of high dimensional data for data mining applications. In: ACM SIGMOD International Conference on Management of Data, pp. 94–105 (1998)
14. Chun-Hung. ENCLUS: entropy-based subspace clustering for mining numerical data. In: ACM SIGKDD International Conference on Knowledge Discovery and Data Mining, pp. 84–93. ACM, New York (1999)
15. Goil S., Nagesh H., Choudhary A.: MAFIA: efficient and scalable subspace clustering for very large data sets. Technical Report 9906-010, Northwestern University, 1999
16. Chu, Y.-H., Huang, J.-W., Chuang, K.-T., Yang, D.-N., Chen, M.-S.: Density conscious subspace clustering for high-dimensional data. IEEE Trans. Knowl. Data Eng. **22**, 16–30 (2010)
17. Kailing K., Kriegel H-P., Kröger P.: Density-connected subspace clustering for high dimensional data. In:.4th SIAM International Conference on Data Mining, pp. 246–256 (2004)
18. Ester M., Kriegel H-P., Sander J., Xu X.: A density-based algorithm for discovering clusters in large spatial databases with noise. In: 2nd International Conference on Knowledge Discovery and Data Mining, pp. 226–231. Portland (1996)
19. Kriegel H-P., Kroger P., Renz M., Wurst S.: A generic framework for efficient subspace clustering of high dimensional data. In: 5th International Conference on Data Mining, pp. 250–257. Houston, TX (2005)
20. Assent I., Krieger R., Müller E., Seidl T.: INSCY: indexing subspace clusters with in process removal of redundancy, In: Eighth IEEE International Conference on Data Mining, pp. 414–425 (2008)
21. Muller, E., Assesnt, I., Gunnemann, S., Seidl, T.: Scalable density based subspace clustering. In: 20th ACM Conference on Information and Knowledge Management, pp. 1076–1086 (2011)
22. Assent I., Krieger R., Muller E., Seidl T.: DUSC: dimensionality unbiased subspace clustering. In: ICDM, pp. 409–414. Omaha, Nebraska (2007)
23. NASA GES-DAAC TRMM_L3 TRMM_3B42 v7 daily Surface Rain from all Satellite and Surface data tables. http://iridl.ldeo.columbia.edu/SOURCES/.NASA/.GES-DAAC/.TRMM_L3/.TRMM_3B42/.v7/.daily/precipitation/datatables.html

Frequent Patterns Mining from Data Cube Using Aggregation and Directed Graph

Kuldeep Singh, Harish Kumar Shakya and Bhaskar Biswas

Abstract An algorithm has been proposed for mining frequent maximal itemsets from data cube. Discovering frequent itemsets has been a key process in association rule mining. One of the major drawbacks of traditional algorithms is that lot of time is taken to find candidate itemsets. Proposed algorithm discovers frequent itemsets using aggregation function and directed graph. It uses directed graph for candidate itemsets generation and aggregation for dimension reduction. Experimental results show that the proposed algorithm can quickly discover maximal frequent itemsets and effectively mine potential association rules.

Keywords Association rule · Data cube · Frequent itemsets · Support count · Directed graph · Maximal itemsets

1 Introduction

Frequent pattern mining has been a major task in data mining. Frequent pattern mining finds interesting association or correlation among a large number of itemsets. Finding frequent patterns play an important role in association rule mining, correlation. Frequent pattern are the patterns that appear repeatedly in the dataset.

K. Singh (✉) · H.K. Shakya · B. Biswas
Department of Computer Science and Engineering, IIT-BHU, Varanasi,
Uttar Pradesh, India
e-mail: kuldeep.rs.cse13@iitbhu.ac.in

H.K. Shakya
e-mail: hkshakya.rs.cse@iitbhu.ac.in

B. Biswas
e-mail: bhaskar.cse@iitbhu.ac.in

S. Das et al. (eds.), *Proceedings of the 4th International Conference on Frontiers in Intelligent Computing: Theory and Applications (FICTA) 2015*, Advances in Intelligent Systems and Computing 404, DOI 10.1007/978-81-322-2695-6_15

It was first proposed by Agrawal [1] in 1994. In this paper, a frequent pattern mining technique is proposed that use directed graph together with aggregation to discover knowledge from data cube. Aggregation transforms 3D model views to 2D (tabular) view of dataset using sum-based measure. Directed graph provides fast and easy construction of candidate generation. Candidate generation using directed graph saves time and memory consumption because, most candidate itemsets are frequent. Frequent 2-itemsets are required for the directed graph generation. First item of a frequent 2-itemset works as origin and second item is like destination for an edge in the directed graph.

Many algorithms have been developed for searching association rules. The main challenge in mining association rules is developing fast and efficient algorithms that can handle large volume of data, minimum time scans database and find associated rule quickly. Most of the proposed apriori-like algorithms for mining association rules are wasting lots of time to generate candidate itemsets. FP-Growth algorithm is also very useful for finding association rule. The FP-Growth algorithm does not generate candidate itemsets and so takes less time to find frequent itemsets [2]. But it also has limitations with respect to space and time. The proposed algorithms overcome these limitations. Frequent itemsets can be of two types, closed and maximal. A frequent itemset is called maximal if it has no superset that is frequent [3]. An itemset is closed if it is frequent but none of its superset has the same support count [4].

The proposed algorithm uses directed graph for candidate itemsets generation. It saves time and memory consumption because it generates minimum number of candidate itemsets, those are likely to be frequent. Frequent 2-itemsets are required for the directed graph generation. First item of a frequent 2-itemset works as origin and second item like destination for an edge in the directed graph [5]. Directed graph gives us minimum candidate itemsets that are likely to be frequent itemsets.

Proposed algorithms generate frequent patterns using data cube reduction. Data reduction technique used to reduce the dataset that is much smaller in volume. Reduced dataset should maintain the integrity of original dataset. Mining on the reduced dataset produce exactly the same or almost the same results. If reconstruction of original dataset from reduced dataset is without any loss of information, then the reduced dataset is called lossless. Otherwise, we loss the information and produce approximation of original dataset, then the reduced dataset is called loosy. There are many methods of data reduction. Dimensionality reduction and numerosity reduction techniques are the main forms of data reduction. Dimensionality reduction is the process of reducing the number of attributes or variables. There are many dimensionality reduction methods which transform the original dataset into a smaller spaced dataset, e.g., wavelet transformation, principal component analysis, and attribute subset selection [6]. Numerosity reduction techniques replace the original dataset by smaller form of data representation. Numerosity reduction may be parametric of nonparametric. Parametric methods are used to estimate the data, only the data parameters needs to be stored, instead of the original dataset, e.g., regression and log-linear model. Nonparametric methods are used to reduce representations of the dataset, e.g., histograms, clustering, sampling, and data cube aggregation. A data cube allows data to be viewed in multiple dimensions. The data cube contains three dimensional item, time, and location. It can return the total sales for any combination of the three dimensions. Apex cuboid refers to the case where the group-by is empty. It contains the total sum of all sales. The base cuboid is the least generalized of the cuboids. The apex cuboid is the most generalized of the cuboid. If we start at the base cuboid and explore upward, this is similar to roll-up operation.

2 Frequent Patterns Discovery Using Directed Graph

In this section, we present the proposed algorithm. The proposed algorithm is shown in Fig. 1.

```
Input: Data-Cube, min_sup;
Output: Set of Maximal frequent itemset.
1. Condense one predicate of Data-Cube using aggregation
2. Find total_item_count of all the items
3.        If total_item_count < min_sup than
4.                Remove those items
5.        Otherwise add 1-itemsets into L1
6. Sort the items in ascending order according to their total_item_count
7. Generate candidate 2-itemsets using lexicographical order with L1
8. Calculate total_item_count of candidate 2-itemsets
9.        If total_item_count < min_sup than
10.               Remove those itemsets
11.       Otherwise add 2-itemsets into L2
12. Generate directed graph using L2
13.       If (Ii,Ij) in L2 than
14.               Draw a directed edge from Ii to Ij in directed graph
15. Traverse the directed graph and generate candidate k-itemsets
16. Calculate total_item_count of k-itemsets
17.       If total_item_count < min_sup than
18.               Remove those itemsets
19.       Otherwise add k-itemsets into Lk
20. Discover Maximal Frequent itemset from Lk
21. Return Maximal Frequent itemset
```

Fig. 1 The proposed algorithm

2.1 Algorithm Details

The proposed algorithm can be divide into fives main steps on the basis of broad tasks. These main steps are explained below.

Perform Aggregation on Data Cube.
The data cube condenses using aggregation. We simply computed by aggregation the counts from cells contained in the one predicate. The resulting dataset is smaller in volume without loss of information necessary for frequent pattern mining [7–9]. We did this aggregation because memory consumption is reduced. Aggregate function and the group-by operator produce 1D aggregate or more dimensional aggregates. Aggregation functions return a single value.

Generate the set of frequent 1-itemsets L_1.
Calculate total_item_count of items by counting item_freq of items. If total_item_count of I_j item is smaller than min_sup, itemset I_j in not a frequent 1-itemset so removed. Otherwise, itemset I_j is frequent and is added to the set of frequent 1-itemset L_1. Sorting helps us to generate minimum candidate itemsets as in [10]. Sorting has been done of all frequent 1-itemsets by total_item_count in ascending order, these are likely to be frequent.

Generate the set of frequent 2-itemsets.
In this step, we generate candidate 2-itemsets in lexicographical order using L_1. Count the total_item_count of the candidate 2-itemset $\{I_i, I_j\}$. If the total_item_-count of candidate 2-itemset $\{I_i, I_j\}$ is greater than min_sup, then itemset $\{I_i, I_j\}$ frequent 2-itemset and added into L_2, otherwise removed.

Construction of Directed Graph using frequent 2-itemsets.
We construct the directed graph by using frequent 2-itemsets L_2. If itemset $\{I_i, I_j\}$ frequent 2-itemset, then draw a directed edge from I_i to I_j in directed graph [11, 12]. First item of 2-itemset is the origin and second item is the destination. We draw all directed edge by using itemset of L_2. After drowning all the edges, we present directed graph.

Generate all the Candidate k-itemsets using directed graph.
This proposed algorithm traverses once to the directed graph and generate candidate itemsets by using the directed neighbor nodes of itemset. Start from I_i and traverse all the possible reachable nodes. After completed traversing it leads to formation of simple paths; these simple paths are our candidate itemsets. Nodes of possible simple paths are the elements of these itemsets.

Find Frequent itemsets.
In the final step, we collect only frequent itemsets. For finding frequent k-itemsets, count the total_item_count of the candidate k-itemset $\{I_i, I_j, \ldots I_k\}$. If the total_item_count of candidate k-itemset $\{I_i, I_j, \ldots I_k\}$ is greater than min_sup, then $\{I_i, I_j, \ldots I_k\}$ frequent k-itemset and added into L_k. Otherwise, remove infrequent

itemsets. Finally, we got the final set of frequent k-itemsets L_k. Now, we verify whether frequent k-itemsets L_k are maximal or not. The itemsets that do not fulfill the property of maximal itemsets are removed; otherwise added into final result set.

2.2 *Example*

We present a simple example to help illustrate working of the proposed algorithm. The data cube showed in Fig. 2. The user has given minimum support threshold as 5. We can find out the new minimum support threshold by multiplying the number of aggregated cell with old minimum support threshold. So the new minimum support threshold is (min_sup) is (20 = 5*4) 20.

Perform Aggregation on Data Cube.
The data cube is aggregated into a tabular dataset as shown in Fig. 2b. The data cube can be aggregated so that the resulting data summarize the total sales in all locations instead of sale per city, as shown in Fig. 2. The attributes of dimension location are grouped. The data cube consist the sales per year, for the year 2011–2014. We are interested in the total sale per year, rather than the total obtained per city. Aggregation returns the total sales for combination of the dimension location. The dimension attributes are grouped, where locations are grouped. Therefore, we perform aggregation summarization of location predicates. Figure 2a shows the condensed dataset.

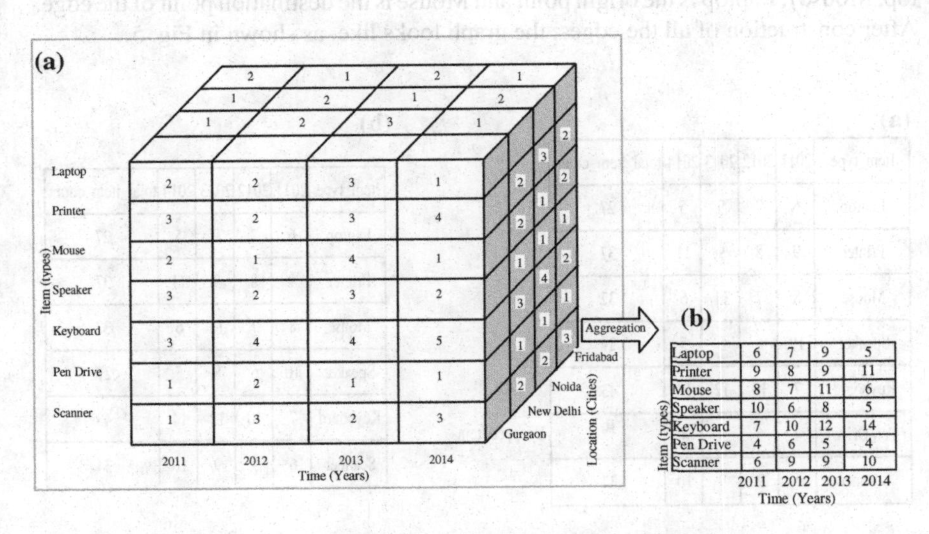

Fig. 2 Sales data cube of computer devices for the year 2011–2014. On the *right*, data are aggregated to provide the total sale of all locations

Generate the set of Frequent 1-itemsets L_1.
Scan the condensed dataset and calculate total_item_count of items by counting values as shown in Fig. 3a. Those items that have total_item_count less than min_sup are removed. Item pen drive is removed because it has less total total_item_count to the min_sup. After removal of item pen drive, the resulted set is shown as Fig. 3b. Then, sort all frequent 1-itemsets by total_item_count in ascending order. Now, the set of sorted itemsets are L_1 = {Laptop, Speaker, Mouse, Scanner, Printer, Keyboard}.

Generate Candidate 2-itemsets:-.
Generate candidate 2-itemsets using sorted 1-itemsets L_1. Count the total_item_count of the candidate 2-itemset {I_i, I_j} as shown in Fig. 4a. If the total_item_count of candidate 2-itemset is {I_i, I_j} greater than or equal to min_sup, then {I_i, I_j} frequent 2-itemset and added into L_2. Otherwise, remove 2-itemset {I_i, I_j}. Itemset {Laptop, Speaker} and {Speaker, Scanner} are infrequent because total_item_count of these itemsets less than min_sup. Now, the frequent 2-itemsets are shown in Fig. 4b.

Construction of Directed Graph using Frequent 2-itemsets.
Construct the directed graph using L_2. If {I_i, I_j} in L_2, then draw a directed edge from I_i to I_j in directed graph. A graph consists of two things node (V) and direct edged {Mouse, Scanner, Keyboard} or {L_1} and E = {{Laptop, Mouse}, {Laptop, Scanner}, {Laptop, Printer}, {Laptop, Keyboard}, {Speaker, Mouse}, {Speaker, Printer}, {Speaker, Keyboard}, {Mouse, Scanner}, {Mouse, Printer}, {Mouse, Keyboard}, {Scanner, Printer}, {Scanner, Keyboard}, {Printer, Keyboard}}. For itemset {Laptop, Mouse}, Laptop is the origin point and Mouse is the destination point of the edge. After construction of all the edges, the graph looks like, as shown in Fig. 5.

(a)

Item_type	2011	2012	2013	2014	total_item_count
Laptop	6	7	9	5	27
Printer	9	8	9	11	37
Mouse	8	7	11	6	32
Speaker	10	6	8	5	29
Keyboard	7	10	12	14	43
Pen Drive	4	6	5	4	19
Scanner	6	9	9	10	34

Dropped item

(b)

Item_type	2011	2012	2013	2014	total_item_count
Laptop	6	7	9	5	27
Printer	9	8	9	11	37
Mouse	8	7	11	6	32
Speaker	10	6	8	5	29
Keyboard	7	10	12	14	43
Scanner	6	9	9	10	34

Fig. 3 The generation of frequent 1-itemsets. **a** Candidate 1-itemsets. **b** Frequent 1-itemsets

(a)

Candidate 2-itemsets	2011	2012	2013	2014	total_item_count
{Laptop, Speaker}	6	6	6	5	23
{Laptop, Mouse}	6	7	9	5	27
{Laptop, Scanner}	6	7	9	5	27
{Laptop, Printer}	6	7	9	5	27
{Laptop, Keyboard}	6	7	9	5	27
{Speaker, Mouse}	8	6	6	5	25
{Speaker, Scanner}	6	6	6	5	23
{Speaker, Printer}	9	6	6	5	26
{Speaker, Keyboard}	7	6	6	5	24
{Mouse, Scanner}	6	7	9	6	28
{Mouse, Printer}	8	7	9	6	30
{Mouse, Keyboard}	7	7	11	6	31
{Scanner, Printer}	6	8	9	10	33
{Scanner, Keyboard}	6	9	9	10	34
{Printer, Keyboard}	7	8	9	11	35

Dropped itemsets

(b)

Frequent 2-itemsets	2011	2012	2013	2014
{Laptop, Mouse}	6	7	9	5
{Laptop, Scanner}	6	7	9	5
{Laptop, Printer}	6	7	9	5
{Laptop, Keyboard}	6	7	9	5
{Speaker, Mouse}	8	6	6	5
{Speaker, Printer}	9	6	6	5
{Speaker, Keyboard}	7	6	6	5
{Mouse, Scanner}	6	7	9	6
{Mouse, Printer}	8	7	9	6
{Mouse, Keyboard}	7	7	11	6
{Scanner, Printer}	6	8	9	10
{Scanner, Keyboard}	6	9	9	10
{Printer, Keyboard}	7	8	9	11

Fig. 4 The generation of frequent 2-itemsets. **a** Candidate 2-itemsets with total_item_count. **b** The frequent 2-itemsets

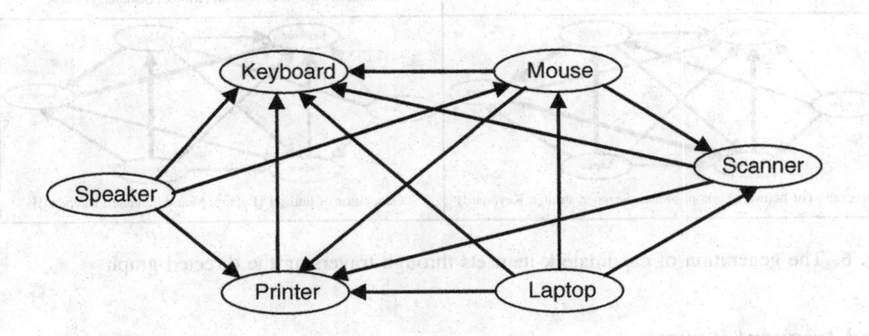

Fig. 5 The directed graph

Generate candidate k-itemsets using directed graph.
Generate candidate k-itemsets by traversal of directed graph. Start from I_i and traverse all the reachable nodes. An I_j is reachable from Ii if there is a simple path from I_i to I_j. After completion of traversing, we found that there are many simple paths; these simple paths are our candidate itemsets. The candidate k-itemsets generated are shown in Table 1. Figure 6 shows the process of candidate itemset generation through traversing the directed graph. Node Keyboard is not having any candidate itemset because it did not have any outgoing edge or node Keyboard; not a source node for any other node.

Table 1 The candidate k-itemsets with source node

Source node	k-candidate itemsets
Laptop	{Laptop, Mouse, Scanner}, {Laptop, Mouse, Scanner, Printer}, {Laptop, Mouse, Scanner, Printer, Keyboard}, {Laptop, Mouse, Scanner, Keyboard}, {Laptop, Mouse, Printer}, {Laptop, Mouse, Printer, Keyboard}, {Laptop, Mouse, Keyboard}, {Laptop, Scanner, Printer}, {Laptop, Scanner, Printer, Keyboard}, {Laptop, Scanner, Printer}, {Laptop, Printer, Keyboard}
Speaker	{Speaker, Mouse, Scanner}, {Speaker, Mouse, Scanner, Printer}, {Speaker, Mouse, Scanner, Printer, Keyboard}, {Speaker, Mouse, Scanner, Keyboard}, {Speaker, Mouse, Printer}, {Speaker, Mouse, Printer, Keyboard}, {Speaker, Mouse, Keyboard}, {Speaker, Printer, Keyboard}
Mouse	{Mouse, Scanner, Printer}, {Mouse, Scanner, Printer, Keyboard}, {Mouse, Scanner, Keyboard}, {Mouse, Printer, Keyboard}
Scanner	{Scanner, Printer, Keyboard}
Printer	{ }

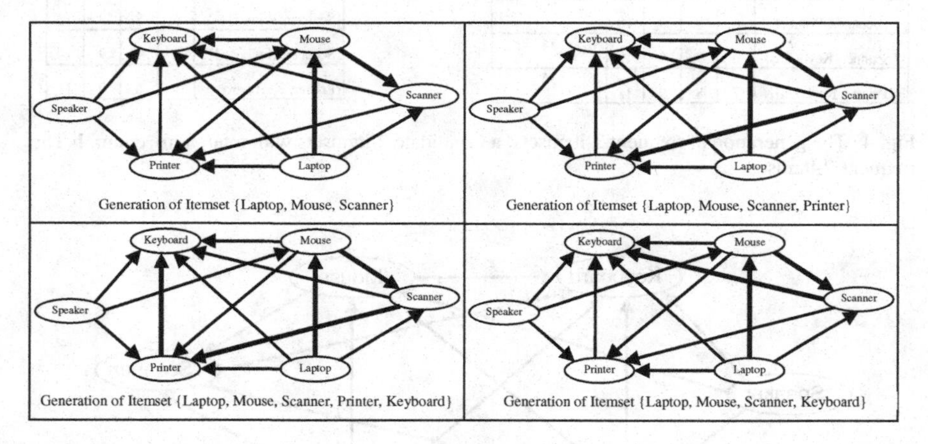

Fig. 6 The generation of candidate k-itemsets through traversing the directed graph

Find frequent itemsets.
At last, we check whether each candidate k-itemsets is frequent or infrequent itemsets. Count the total_item_count of the candidate k-itemset $\{I_i, I_j, \ldots I_k\}$. The total_item_count of itemsets {Speaker, Mouse, Scanner}, {Speaker, Mouse, Scanner, Printer}, and {Speaker, Mouse, Scanner, Printer, Keyboard} and {Speaker, Mouse, Scanner, Keyboard} are less than Sup_count, so remove these itemsets. All other itemsets are having greater total_item_count than min_sup, so these are added into frequent k-itemset L_k.

In the end, we find maximal frequent itemsets from L_k. All frequent k-itemsets are the subsets of the maximal frequent itemset {Laptop, Mouse, Scanner, Printer, Keyboard} and {Speaker, Mouse, Printer, Keyboard}. In maximal itemset we did not include frequent subsets. So, remove the subsets of the maximal itemsets.

3 Experiment

To find experimental results, we have used mushroom and accidents dataset obtained from UCI [13] and synthetic dataset which are created. The algorithms were implemented in Java and tested on a windows platform. Mushroom dataset have total number of instances: 8124, number of attributes: 23, number of items: 119. Dimensions come from the former five attribute of the mushroom dataset. We analyses that as support count increases, the execution time goes down as shown in Fig. 7.

Another dataset used for experiment result is accidents [14]. Accidents dataset have total number of Instances: 3196, number of attributes: 37, number of items: 75. Dimensions come from the former five attributes of the accidents dataset. The result shows that lesser the minimum support threshold takes more time to execute as shown in Fig. 8.

The proposed algorithm takes lesser time when minimum support count threshold is higher. The frequent pattern generated by Apriori and proposed algorithm do not have much difference. Candidate Generation: To better understand the execution time behavior of the proposed algorithm, we explicitly evaluated the number of candidate itemsets generated by the proposed algorithm. The interesting observation here is that the number of candidate itemsets counted by proposed algorithm is less than other traditional algorithms. Mostly, the candidates generated by proposed algorithm are frequent so we can say that candidate itemsets generated by proposed algorithm are likely to be frequent.

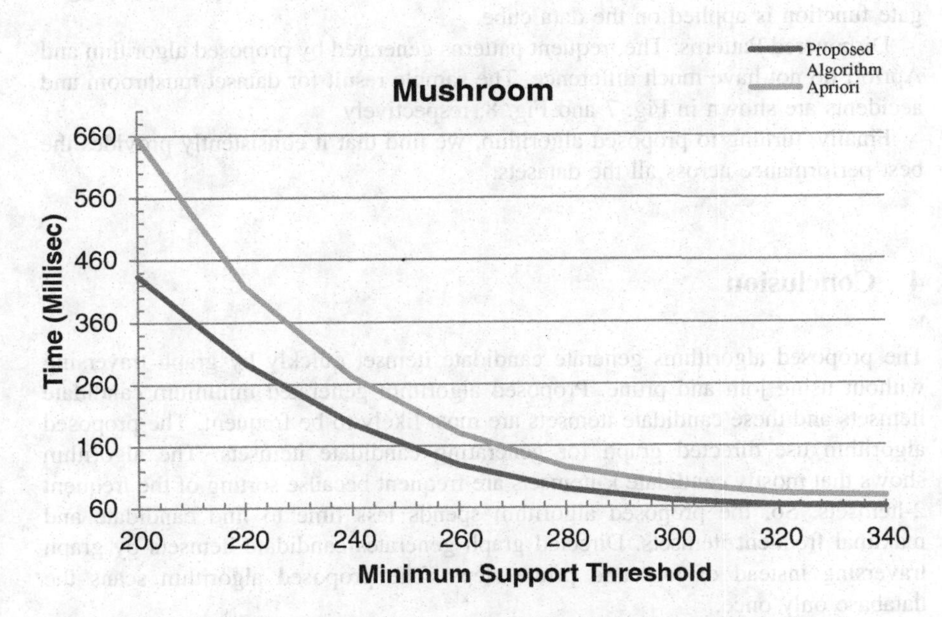

Fig. 7 Execution time with different minimum support thresholds for mushroom dataset

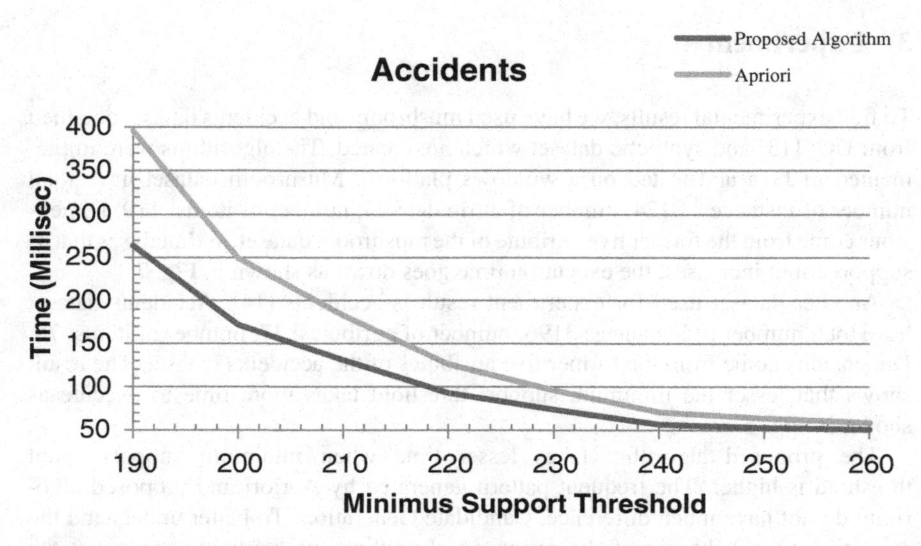

Fig. 8 Execution time with different minimum support thresholds for accidents dataset

Memory consumption: We also monitored the memory consumption overhead of the various algorithms. We found that the memory occupancy of the proposed algorithm is comparable with other algorithms. The consumption was related less in condensed data cube. Condensed Data cube requires less memory because aggregate function is applied on the data cube.

Discovered Patterns: The frequent patterns generated by proposed algorithm and Apriori do not have much difference. The sample result for dataset mushroom and accidents are shown in Fig. 7 and Fig. 8, respectively.

Finally, turning to proposed algorithm, we find that it consistently provides the best performance across all the datasets.

4 Conclusion

The proposed algorithms generate candidate itemset quickly by graph traversing without using join and prune. Proposed algorithm generated minimum candidate itemsets and these candidate itemsets are most likely to be frequent. The proposed algorithm use directed graph for generating candidate itemsets. The algorithm shows that mostly candidate k-itemsets are frequent because sorting of the frequent 2-itemsets. So, the proposed algorithm spends less time to find candidate and maximal frequent itemsets. Directed graph generates candidate itemsets by graph traversing instead of join and prune steps. The proposed algorithm scans the database only once.

5 Future Work

The above described work can be enhanced for directed graph generation. This directed graph can be generated by using 1-frequent itemset that can enhance the performance of the proposed algorithm.

References

1. Agrawal, R., Srikant, R.: Fast algorithms for mining association rules in large databases. In: Proceedings of the 20th International Conference on Very Large Databases, pp. 487–499. Santiago, Chile (1994)
2. Han, J., Pei, J., Yin, Y.: Mining frequent patterns without candidate generation. In: Proceedings of the ACM-SIGMOD Conference Management of Data, pp. 1–12 (2000)
3. Amir, A., Anumann, y., Feldman, R., Fresko, M.: Maximal association rules: a tool for mining associations in text. J. Intell. Inf. Syst. **25**(3), 333–345 (2005)
4. Zaki, M.K.: Closed itemset mining and non-redundant association rule mining. In: Encyclopedia of Database Systems, Springer (2009). doi: 10.1007/978-0-387-39940-9_66
5. Rosen, K.H.: Discrete Mathematics and Its Application, 7th edn. McGraw-Hill Publication, Columbus (2012)
6. Han, J., Kamber, M.: Data Mining Concepts and Techniques, 3rd edn. Morgan Kaufmann Publishers, San Francisco (2012)
7. Harsola, S.K., Deshpande, P.M., Haritsa, J.R.: IceCube: efficient targeted mining in data cubes. In: ICDM, pp. 894–899 (2012)
8. Gray, J., Chaudhuri, S., Bosworth, A., Layman, D., Reichart, D., Venkatrao, M., Pellow, F., Pirahesh, H.: Data cube: a relational aggregation operator generalizing group-by, cross-tab, and sub-totals. Data Min. Knowl. Discov. Arch. **1**(1), 29–53 (1997)
9. Messaoud, R.B., Rabaseda, S.L.: Enhanced mining of association rules from data cubes. In: Proceedings of the 9th ACM International Workshop on Data Warehousing and OLAP, pp. 11–18 (2006)
10. Liu, H., Wang, B.: An association rule mining algorithm based on a Boolean matrix. Data Sci. J. **6**, s559–s565 (2007)
11. Liu, N., Ma, L.: Discovering frequent itemsets an improved algorithm of directed graph and array. In: 4th IEEE International Conference on Software Engineering and Service Science (ICSESS), pp. 1017–1020 (2013)
12. Xu, W.X., Wang, R.J.: A fast algorithm of mining multidimensional association rules frequently. In: International Conference on Machine Learning and Cybernetics, pp. 1199–1203. China (2006)
13. Schlimmer, J.: https://archive.ics.uci.edu/ml/machine-learning-databases/mushroom/agaricus-lepiota.data
14. Shapiro, A.: https://archive.ics.uci.edu/ml/datasets/Accident (KRKPA7)

Part V
Document Image Analysis

A Modified Parallel Thinning Method for Handwritten Oriya Character Images

Soumen Bag and Glory Chawpatnaik

Abstract Binary image thinning has wide applications in image processing, machine vision, and pattern recognition. Thinning is a preprocessing step to obtain single-pixel-thin skeleton for document imaging and pattern analysis. Indian languages are complex in character shape than Latin, Chinese, Japanese, and Korean languages. The performance of existing thinning methods cannot reach the state of the art in the field of document image processing for Indian languages. Our objective is to improve the performance of complex-shaped character skeletonization, particularly for Oriya script. This paper presents an iterative parallel thinning method which is well suited to handwritten Oriya characters. Some fundamental requirements of thinning like preservation of stroke connectivity, removal of spurious strokes have been ensured in this method. Experimental results show its efficacy comparing with other existing thinning methods.

Keywords Elimination rule · Handwritten Oriya script · OCR · Parallel thinning · Template · Skeleton

1 Introduction

Thinning is one of the primary phases in digital image analysis. It is most often, although not exclusively, used for images of printed or handwritten characters so we describe it here in this context. Thinning and skeletonization are synonymous on many articles. Usually, we consider that the skeleton is the result of any thinning process. Thinning is an important preprocessing step in various applications of image analysis and pattern recognition.

S. Bag (✉)
Department of Computer Science and Engineering, ISM Dhanbad, Dhanbad, India
e-mail: bagsoumen@gmail.com

G. Chawpatnaik
Department of Computer Science and Engineering, IIIT Bhubaneswar, Bhubaneswar, India
e-mail: glory.cse@gmail.com

S. Das et al. (eds.), *Proceedings of the 4th International Conference on Frontiers in Intelligent Computing: Theory and Applications (FICTA) 2015*, Advances in Intelligent Systems and Computing 404, DOI 10.1007/978-81-322-2695-6_16

Datta and Parui [4] have presented a parallel thinning method that maintain proper pixel connectivity and generates single-pixel-thick images. Each iteration contains four sub-iterations. The sub-iterations apply two 1×3 and two 3×1 templates for the removal of right, top, left, and bottom boundary pixels. The method uses a 3×3 window to prevent pixel disconnectivity and recession of stroke length. To improve the thinning performance, Han et al. [5] have presented a fully parallel algorithm which uses information in a 5×5 template. But its computational cost is higher than the Dutta and Parui method. Also these algorithms are inefficient in some cases like information loss of the pattern. To solve this problem, a template-based contour following thinning method is proposed by Leung et al. [7]. The proposed method uses a lookup table with 256 entries based on the different possibilities of pixel connectivity. To improve the thinning performance, Huang et al. [6] have introduced a template-based parallel thinning method. This method uses a set of pixel elimination rules based on the total number of object pixels in 8-neighborhood. They have also defined an information loss measurement technique which may occur if too many pixels are deleted from the image. In that case, the information loss is rewarded by the replacement of the thinned image with its corresponding contour image. Later, Zhu and Zhang [9] have proposed a pixel substitution-based post-processing technique to retain the true shape of resultant skeleton. The proposed method is used on different languages to improve the stroke connectivity after skeletonization.

It is noticed that describing shape of an object is often necessary in image analysis. Thinning simplifies the shape of an object and helps to store the essential structural information of the given object in less memory space. These are the motivating factors of our research work. We observe that no thinning methods exist in the literature for Oriya language. But Oriya is one of the twenty-two official Indian languages recognized by the Indian constitution [2]. Oriya, an Indo-Aryan language is spoken by about 31 million people mainly in Odisha, and also in West Bengal and Jharkhand and globally over 45 million people speak Oriya. So, Oriya has a big importance as language and script both. We notice that the existing thinning methods on different languages have some limitations for this Oriya language. Figure 1 shows different types of shape distortions present in thinned Oriya character images. These shape distortions make the character recognition problem much more complex in nature. Sometimes the resultant skeleton may cause discontinuities if it thins further. These

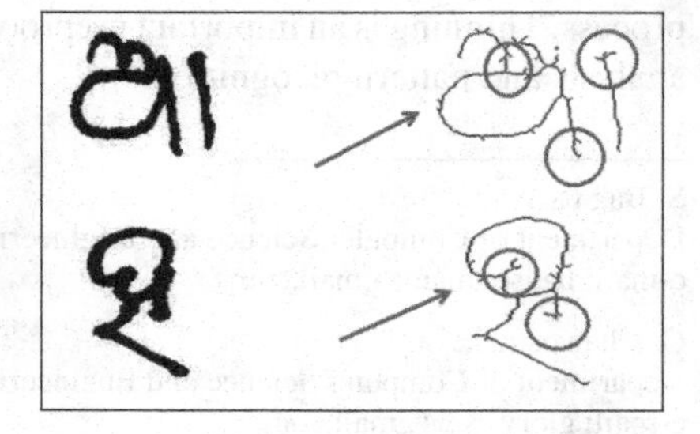

Fig. 1 Shape distortion of thinned character images. *Left Column* Input image; *Right Column* Thinned image with spurious stroke and junction or end point distortion as indicated by *red arrow* and *circle*, respectively

limitations motivated us for the further investigation and to find the possible solution to overcome the above limitations.

The remaining paper is arranged as follows. Section 2 presents the proposed thinning methodology for Oriya character images. Experimental results and comparative analysis are reported in Sect. 3. The concluding remarks are given in Sect. 4.

2 Proposed Method

This section describes the design of our proposed parallel thinning method for Oriya character images based on predefined templates. The proposed method consists of three stages. The first one is the binarization of the scanned input images, second one is the skeletonization of character images using different pixel elimination rules, and finally we get the output thinned images by preserving the pixel connectivity using a set of templates used in [6]. The detailed methodology is given below.

We convert the input RGB image into gray scale image. Then we perform image binarization using an adaptive form of Otsu's method [8]. This adaptive form works efficiently in multiscale framework to handle noises in different scales. This method has the potentiality to preserve stroke connectivity for degraded and noisy images. The detailed methodology is reported in [3].

After image binarization, we label black/foreground/object pixels as '1' and white/background pixels as '0'. We see that in parallel thinning methods, object pixels are tested for deletion based on the structural results of only the previous iteration. Our proposed method uses the information of 3×3 windows as templates to eliminate boundary pixels. But to maintain proper stroke connectivity, we use a set of masks with window size 3×4, 4×3, and 4×4. The pixel elimination rules (as shown in Fig. 2) are categorized based on the total number of 8-connected neighbors of an object pixel. The first and second column denote the total number of 8-connected neighbors and the elimination rules, respectively. Then according to the amount value, the corresponding elimination rules are applied to the center pixel of the 3×3 template. All the elimination rules are applied concurrently to each object pixel p. Now, if p matches with any defined elimination rules with a consideration that p is the center pixel of each template, then we mark that pixel p for deletion. One major drawback of these elimination rules is that they remove two-pixel-width rectangular patterns, resulting in loss of information or pattern connectivity. To avoid this problem, the pixel p is preserved (unmarked) if it matches any of the templates defined in Fig. 3. Thus Fig. 2 is used to eliminate the background pixels to get the thinned output and Fig. 3 is used to preserve the pixel connectivity in the resultant thinned pattern. At the end of an iteration, if there is no such marked pixel for deletion then the algorithm stops; otherwise the marked pixels are removed and the next pass starts. The algorithmic steps of the proposed method are given below.

1. The handwritten data is converted into digital form by scanning.
2. The input RGB image is converted into gray scale image.

3. The gray scale image is binarized using an adaptive binarization method [3].
4. Every possible combination of 3×3 templates corresponding to the index number (amount value) starting from 3 to 7 (excluding 6) is generated. These set of templates are shown in Fig. 2.
5. Count the amount value for every object pixel. The pixel removal is determined based upon the amount value and template sets.
6. If an object pixel p is flagged for elimination then it is again tested with a set of templates as shown in Fig. 3. If it matches with any of the templates then the pixel p is preserved (unmarked).
7. Repeat Steps 4–6 until no pixel can be removed to get the resultant thinned image.

Fig. 2 Set of elimination rules. Here blank denotes *white pixel* and '1' denotes *black (object) pixel*

Amount	Elimination Rules
0	Never
1	Never
2	Never
3	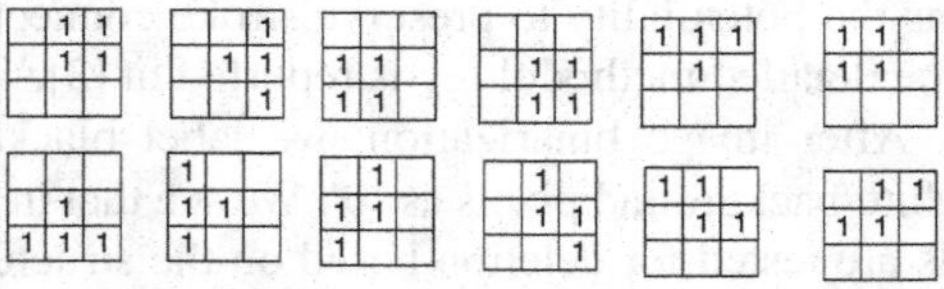 -11/-11/--- ; --1/-11/--1 ; ---/11-/11- ; ---/-11/-11 ; 111/-1-/--- ; 11-/11-/--- ; ---/-1-/111 ; 1--/11-/1-- ; -1-/11-/1-- ; -1-/-11/--1 ; 11-/-11/--- ; -11/11-/---
4	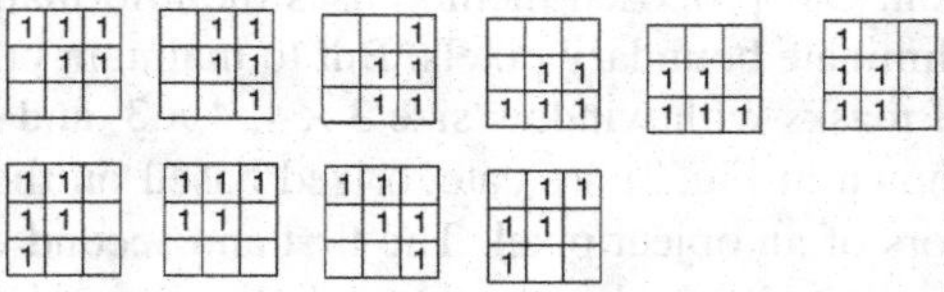 111/-11/--- ; -11/-11/--1 ; --1/-11/-11 ; ---/-11/111 ; ---/11-/111 ; 1--/11-/11- ; 11-/11-/1-- ; 111/11-/--- ; 11-/-11/--1 ; -11/11-/1--
5	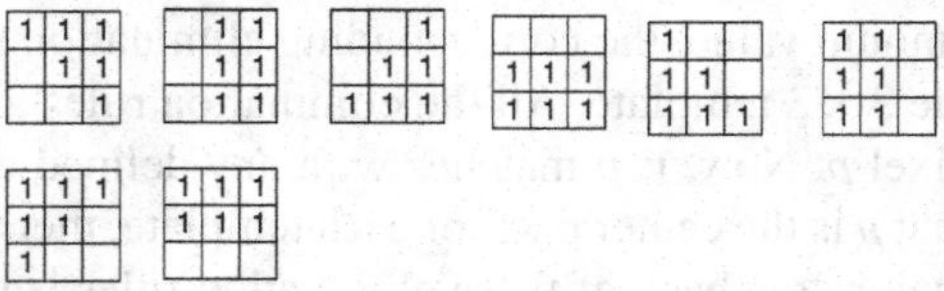 111/-11/--1 ; -11/-11/-11 ; --1/-11/111 ; ---/111/111 ; 1--/11-/111 ; 11-/11-/11- ; 111/11-/1-- ; 111/111/---
6	Never
7	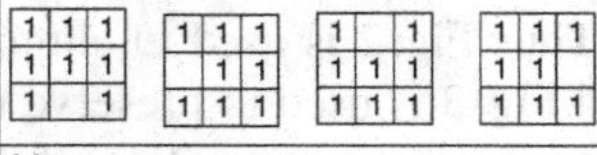 111/111/1-1 ; 111/-11/111 ; 1-1/111/111 ; 111/11-/111
8	Never

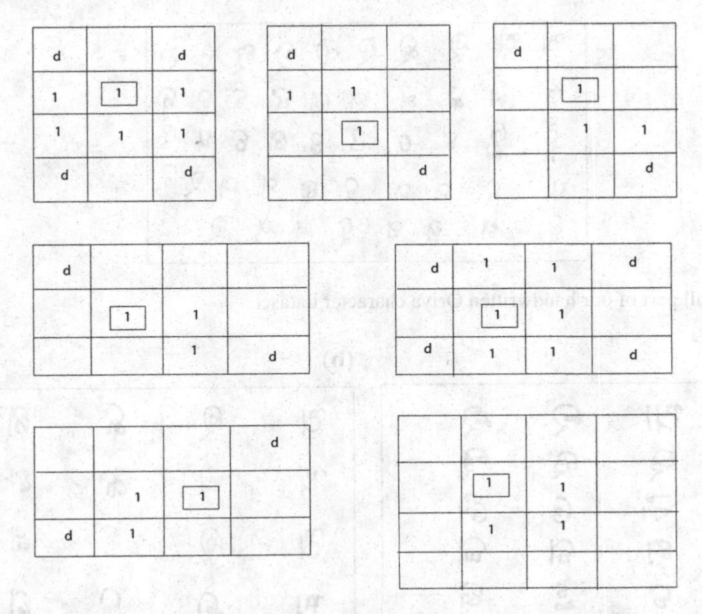

Fig. 3 Set of templates to preserve pixel connectivity. Here, blank and 'd' denote '0' and don't care, respectively

3 Experimental Results and Discussion

This section presents the experimental results and related discussion of our proposed method.

3.1 Dataset

We have collected Oriya characters from several handwritings of different persons written by black ink pen. Our main consideration is on only isolated character images. The overall dataset size is 15,268. Figure 4 shows a small part of our handwritten Oriya character dataset. The program implementation was done by MATLAB (R2010a).

3.2 Test Results

At the time of experimental analysis, we consider the following criteria for assessing the performance of thinning method.

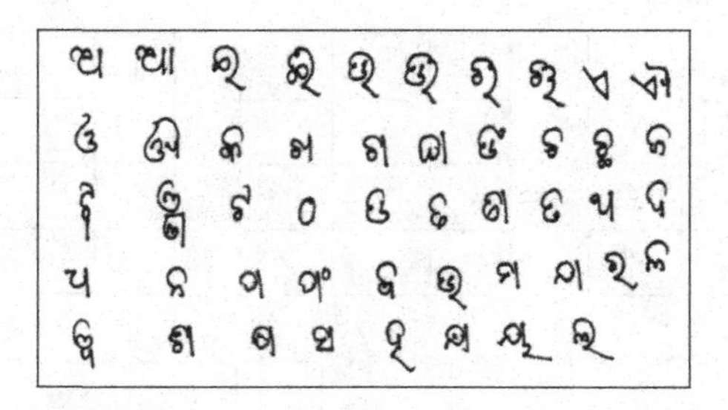

Fig. 4 Small part of our handwritten Oriya character dataset

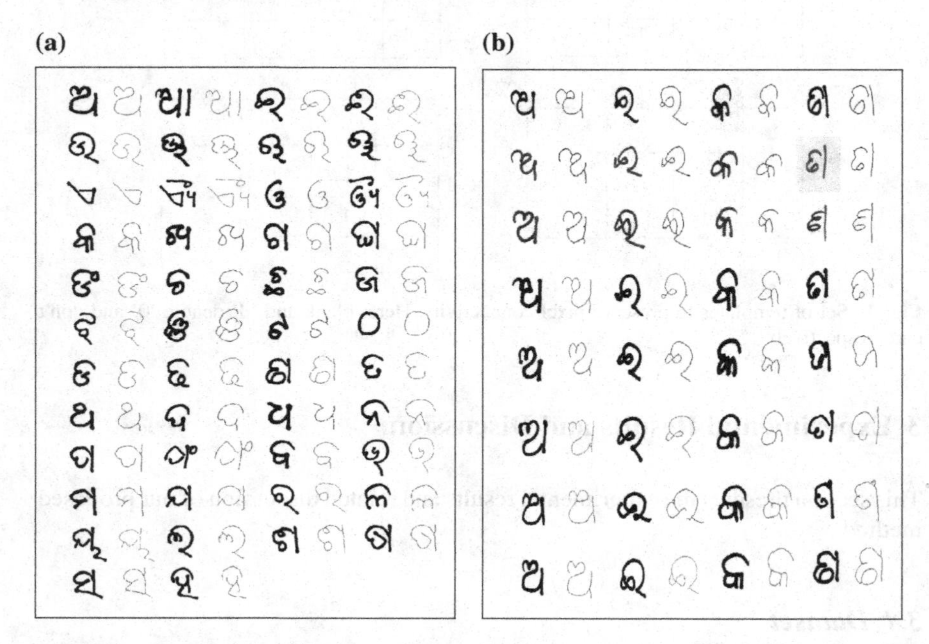

Fig. 5 Test results of the proposed thinning method. **a** *CASE I* Characters written by same person; **b** *CASE II* Characters written by different persons. In these figures, alternative columns represent input images and their corresponding thinned shapes, respectively

- Preservation of stroke connectivity.
- Preservation of true shape.
- Single-pixel-width resultant skeleton.
- No recession of skeleton length.
- Minimum deviation in between thinned skeleton and its actual medial axis.

We have considered two different cases for our experimental analysis. In CASE I, we have considered that the whole dataset is written by a single person. Here we have mainly tested the thinning efficiency of our proposed method on Oriya characters. In CASE II, we have tested the efficiency of the proposed method to handle shape variation of Oriya characters. The detailed explanation is given below.

CASE I (Characters written by same person): Figure 5a shows a collection of different handwritten Oriya characters of same handwriting. We obtain good results by testing the scanned Oriya character set on our proposed method.

CASE II (Characters written by different persons): We have observed that our proposed method is efficient to handle shape variation of a same character written by different persons. Figure 5b shows the thinned output of different Oriya characters written by different individuals. In this figure, the first, third, fifth, and seventh columns contain the Oriya characters 'a', 'i', 'ka', and 'na', respectively, and those are written by different individuals. We have observed that the resulting outputs are handwriting independent. Also some of the basic characteristics of thinning are also maintained such as it is clearly shown from the output that it preserves pixel connectivity and produces less spurious strokes. The experimental results indicate that the proposed method is efficient and the thinned image is closed to the true shape of the input image.

3.3 Comparison Results

We have considered Datta and Parui [4] and Huang et al. [6] as two well-known existing methods for comparative analysis. We have applied these methods on handwritten Oriya character images as shown in Fig. 6 (1st column) and compared the results with our proposed method. We have seen that both the methods fail to maintain the

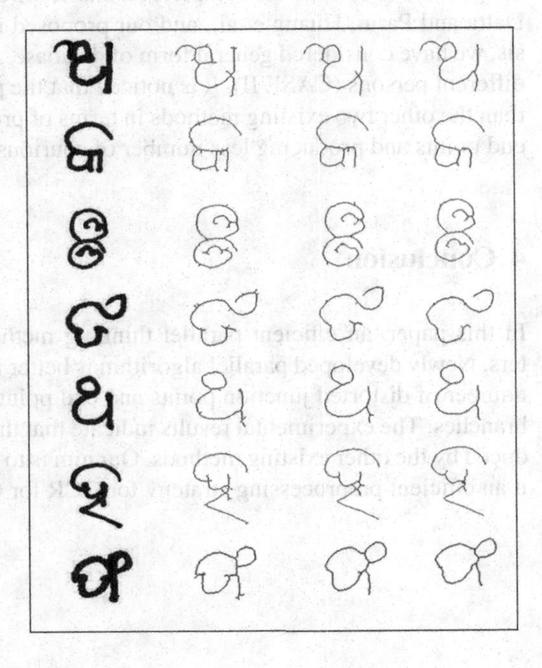

Fig. 6 Comparison results with other existing methods. 1st *Column* Handwritten Oriya character images; 2nd *Column* Datta's method [4]; 3rd *Column* Huang's method [6]; 4th *Column* Proposed method

Table 1 Quantitative analysis among Datta and Parui, Huang et al., and our proposed method

Language	Shape distortion	Datta and Parui	Huang et al.	Proposed method
Oriya	Junction point shape distortion	63.98	39.16	24.68
	End point shape distortion	45.67	29.87	18.26
	Spurious strokes on curvature region	43.34	27.60	16.20

actual shape of resultant images at their junction or end points. They also produce spurious strokes which make them improper for any further processing of optical character recognition (OCR) (Fig. 6, 2nd and 3rd columns). Finally, after comparing the test results of our proposed method (Fig. 6, 4th column) with all other test results, we have seen that our method generates better results for Oriya character images by preserving all the basic structural properties of skeletonization.

We have also done a quantitative analysis among Datta and Parui, Huang et al., and ourselves. We have considered different measurement techniques used in [1] to quantify different types of shape distortions as reported earlier. We measure the shape distortion at junction or end points as a ratio of the number of distorted junction (end) points and total number of junction (end) points in whole dataset. The ratio of the number of spurious strokes on curvature regions and the total number of strokes is used as a measurement of spurious stroke distortion. The overall quantitative measurement is done by manually counting the number of distorted shapes in thinned images. Table 1 shows the complete results of different shape distortion errors among Datta and Parui, Huang et al., and our proposed method. In our quantitative analysis, we have considered general form of database, i.e., different characters written by different persons (CASE II). It is noticed that the proposed method is much efficient than the other two existing methods in terms of preserving true shapes at junction or end points and producing less number of spurious strokes in curvature region.

4 Conclusion

In this paper, an efficient parallel thinning method is proposed for Oriya characters. Newly developed parallel algorithm is better in terms of pixel connectivity, less number of distorted junction points and end points, and less generation of spurious branches. The experimental results indicate that this method is better than those produced by the other existing methods. Our aim is to extend this work in future to make it an efficient preprocessing strategy for OCR for Oriya script.

References

1. Bag, S., Harit, G.: Skeletonizing character images using a modified medial axix-based strategy. Int. J. Pattern Recognit. Artif. Intell. **25**, 1035–1054 (2011)
2. Bag, S., Harit, G.: A survey on optical character recognition for bangla and devanagari scripts. Sadhana **38**, 133–168 (2013)
3. Bag, S., Bhowmick, P., Behera, P., Harit, G.: Robust binarization of degraded documents using adaptive-cum-interpolative thresholding in a multi-scale framework. In: International Conference on Image Information Processing, pp. 1–6. IEEE Press, New York (2011)
4. Datta, A., Parui, S.K.: A robust parallel thinning algorithm for binary images. Pattern Recognit. **27**, 1181–1192 (1994)
5. Han, N.H., La, C.W., Rhee, P.K.: An efficient fully parallel thinning algorithm. In: International Conference on Document Analysis and Recognition, pp. 137–141. IEEE Press, New York (1997)
6. Huang, L., Wan, G., Liu, C.: An improved parallel thinning algorithm. In: International Conference on Document Analysis and Recognition, pp. 780–783. IEEE Press, New York (2003)
7. Leung, W., Ng, C.M., Yu, P.C.: Contour Following Parallel Thinning for Simple Binary Images. In: International Conference Systems Man, and Cybernetics, pp. 1650–1655. IEEE Press, New York (2000)
8. Otsu, N.: A threshold selection method from gray-level histogram. IEEE Trans. Syst. Man Cybern. **9**, 62–66 (1979)
9. Zhu, X., Zhang, S.: A shape-adaptive thinning method for binary images. In: International Conference on Cyber worlds, pp. 721–724. IEEE Press, New York (2008)

Offline Signature Verification Using Artificial Neural Network

Chandra Subhash, Maheshkar Sushila and Srivastava Kislay

Abstract In this paper, we have used a neural networks (NN)-based approach to train the offline signature classifier and use it to classify and recognize samples based on some predecided feature sets. We have taken five parameters as a part of the feature set area, mean, standard deviation, centroid value, and the number of even-positioned black pixels. We believe that these features are enough to differentiate between signature images and our experiments have provided us with encouraging results. The NN classifier so implemented provides an efficiency of around 95 % for our sample data. The efficiency has been compared with and shown to be much better than the contemporary signature verification algorithms.

Keywords Biometric · Artificial neural network · Activation function · Standard deviation · Centroid

1 Introduction

Biometric is a measure of identification or verification that is unique for each person. The fact that biometric features cannot be misused by third-party miscreants makes it an ideal authorization verifier tool. Handwritten signature is one of the oldest biometrics. From times immemorial signatures served as an authenticity establishing biometric [1]. Documents were authenticated in the Roman Empire (AD 439) by affixing handwritten signatures to the documents. The importance of signature in establishing authenticity in real-life scenarios cannot be denied and thus the efficient

C. Subhash (✉) · M. Sushila · S. Kislay
Department of Computer Science and Engineering, Indian School of Mines,
Dhanbad 826004, Jharkhand, India
e-mail: subhash08mit@gmail.com

M. Sushila
e-mail: sushila_maheshkar@yahoo.com

S. Kislay
e-mail: kk.ism070@gmail.com

S. Das et al. (eds.), *Proceedings of the 4th International Conference on Frontiers in Intelligent Computing: Theory and Applications (FICTA) 2015*, Advances in Intelligent Systems and Computing 404, DOI 10.1007/978-81-322-2695-6_17

automation of the signature verification system is the need of the hour. Based on the signature acquisition method followed, signature verification can be classified into two types: online and offline verification. Online verification system records the motion, location, velocity, acceleration, and pressure of the pen when signature is produced [9], whereas offline signature verification uses 2-D image of the signature during data acquisition [11]. In this paper, we are primarily concerned with offline signature recognition & verification using NN [15]. The relative apathy shown toward the research of offline signature verification system when compared to online signature verification can be attributed to many factors [12]. One of these is the employment of simple and unreliable pattern matching algorithms. Another problem faced often is that some of the procedures might deliver unexpected results in the long run in online signature verification. Hence, in this paper offline signature verification is proposed.

2 Preliminaries

Artificial neural networks (ANNs): ANN is a computational model based on the structure and functions of biological neural networks. Information that flows through the network affects the structure of the ANN because a neural network changes or learns, in a sense based on that input and output [8].

Mathematically, a neuron's network function y(k) is represented as

$$y(k) = F(\sum_{i} w_i(k) * x_i(k) + b) \quad (1)$$

where x_i is the input value in discrete time k, w_i is the weight value in discrete time k, b is the bias, F is the transfer function, and $y_i(k)$ is the output value in discrete time k.

3 Proposed Technique

In this paper, the verification and recognition of offline signature samples using ANNs is presented. We have employed ANNs primarily because they follow a paradigm which models human learning patterns. Figure 1 illustrates our approach toward the problem statement. The proposed technique is explained in detail in the next section:

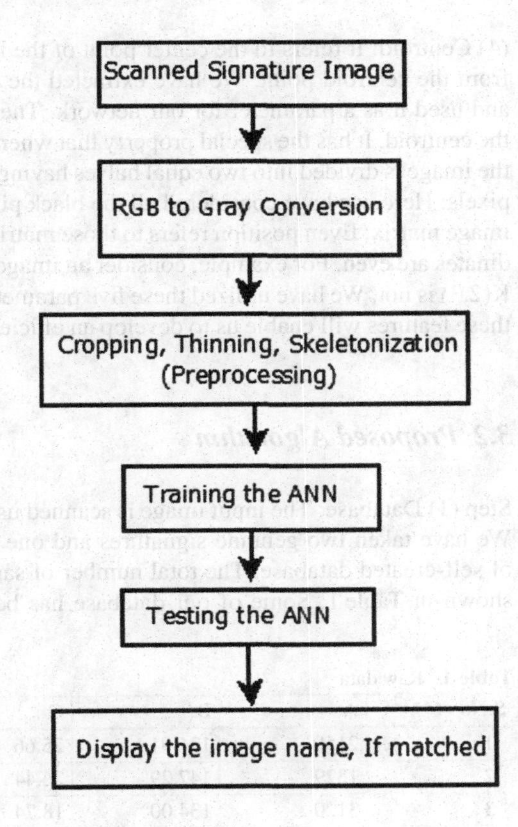

Fig. 1 Proposed technique

3.1 Features Used

In order to train the NN, we have used the following features that serve as the input matrix to the said network [14]. The input matrix X is organized as a set of five features. In other words, (X→[A, B, C, D, E]), where A, B, C, D, and E are as follows: (1) Area: It refers to the total number of black pixels present in the binary image which represent the signature strokes. Basically, area is the count of black pixels in the signature image. The given binary image has 978 black pixels. It thus has an area of 978 units. (2) Mean: An image is stored as a matrix having a finite number of pixel values. The range of pixel values lies between 0(black) and 255(white). Mean is the average of the pixel values of the entire image. The above image has an average pixel value of 87. It thus has a mean of 87 units. (3) Standard deviation(SD): It measures the amount of variation on a set of data values. If the data points tend to very close to the mean data value, then variation is less otherwise, spread over a wider range of values.

$$SD = \sqrt{\frac{X - \bar{X}}{X}} \tag{2}$$

(4) Centroid: It refers to the center point of the image. All the edges are equidistant from the centroid point. We have extracted the pixel value at the centroid position and used it as a parameter for our network. The center point of the image is called the centroid. It has the special property that when a topdown line is drawn through it, the image is divided into two equal halves having the same intensity values. (5) Even pixels: Here, we have considered all the black pixels present at even positions in the image matrix. Even position refers to those matrix positions for which both the coordinates are even. For example, consider an image K, K(2,2) is even positioned while K(2,3) is not. We have utilized these five parameters to train the NN. We believe that these features will enable us to develop an efficient NN-based pattern classifier [10].

3.2 Proposed Algorithm

Step (1) Database: The input image is scanned using a high-/medium-grade scanner. We have taken two genuine signatures and one forged signature for seven persons of self-created database. The total number of samples in our database is thus 21 as shown in Table 1. Some of our database has been depicted in Fig. 4. In addition,

Table 1 Raw data

S.No	A	B	C	D	E
1	2150	136.91	25.66	45	5057
2	1329	147.29	26.44	158	5281
3	3120	134.00	18.74	146	4811
4	678	169.33	26.29	157	5428
5	1507	147.13	29.38	102	5231
6	1117	139.40	21.53	135	5317
7	1006	146.64	17.46	155	5348
8	419	150.13	15.31	159	5492
9	616	181.73	24.89	148	5454
10	906	242.60	35.29	280	5385
11	1447	139.21	23.19	152	5228
12	1103	146.66	20.82	153	5327
13	1387	139.13	22.74	125	5249
14	790	151.16	19.31	157	5406
15	880	144.52	20.85	153	5390
16	1243	157.46	24.48	95	5282
17	837	148.10	19.22	181	5391
18	1218	146.53	18.66	150	5288
19	961	145.76	17.53	158	5372
20	1795	135.53	28.54	142	5159
21	1645	137.89	27.43	144	5191

Table 2 Standard data

S.No	A	B	C	D	E
1	263	248.35	33.39	255	5431
2	210	251.27	20.37	254	5547
3	79	252.65	14.09	252	5580
4	311	247.12	37.24	255	5413
5	261	247.91	31.57	255	5436
6	933	244.74	38.66	255	5374
7	607	240.61	52.31	255	5351
8	208	249.67	21.72	252	5545
9	606	240.56	46.55	255	5351
10	688	237.48	56.97	253	5322
11	1131	242.17	46.59	21	5321
12	1903	233.77	63.19	162	5115
13	519	247.14	29.41	254	5474
14	606	2239.87	52.13	255	5333
15	599	246.41	32.20	137	5447
16	1120	243.21	42.49	253	5327
17	1903	233.76	63.19	162	5115
18	1074	243.40	41.49	255	5325
19	1740	235.52	60.53	250	5178
20	925	244.50	41.89	250	5372
21	443	243.54	46.29	255	5388
22	346	246.43	36.86	255	5410
23	310	247.27	34.23	255	5427
24	530	248.52	31.64	255	5473
25	1222	241.22	51.58	253	5296
26	887	243.99	41.75	208	5378
27	527	247.66	29.37	255	5474
28	426	247.68	27.08	249	5502
29	705	245.88	33.50	255	5437
30	1108	242.75	41.81	192	5335
31	390	249.65	24.85	255	5512
32	847	245.78	37.08	253	5385
33	542	248.57	29.86	237	5472
34	1066	243.95	41.51	255	5335
35	972	244.47	40.39	187	5357
36	385	249.57	25.30	255	5509
37	629	248.23	31.65	255	5444
38	363	250.47	22.80	255	5502
39	454	243.49	46.54	255	5387
40	429	244.27	44.68	255	5397

(continued)

Table 2 (continued)

S.No	A	B	C	D	E
41	671	246.29	33.45	251	5433
42	745	236.27	58.41	255	5313
43	1436	238.54	55.20	255	5593
44	457	248.58	27.83	255	5482
45	1575	237.12	57.35	86	5208
46	379	250.32	24.05	232	5410
47	382	245.24	43.68	255	5598
48	307	250.87	21.42	255	5519
49	547	248.92	31.45	255	5470
50	314	246.65	39.72	255	5419

we have tested the results on the standard GPDS database [4] of 10 person each having five signatures shown in Table 2. Step (2) Image preprocessing: The input image undergoes some preprocessing to enable the future computations to proceed smoothly. The preprocessing steps include conversion of RGB image into binary form, thinning of image, skeletonization as shown in Figs. 2, 3, and 5, respectively. Step (3) Training the NN: The input properties are extracted and organized as an input array to the back propagation network. The selected feature vectors are directed as inputs to the NN. The network is trained using several input patterns. Step (4) Testing and Verifying: The trained NN is used to classify the signature images as either genuine or forged. If the image is a match, then it displays the image name, otherwise declare the image as a forgery.

Fig. 2 Binarized image

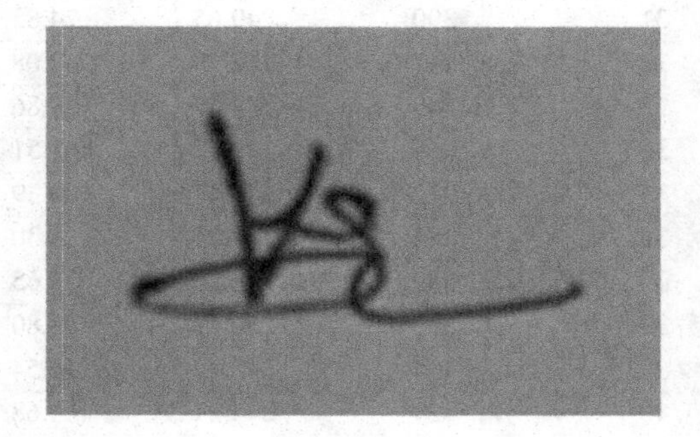

Fig. 3 Given RGB image

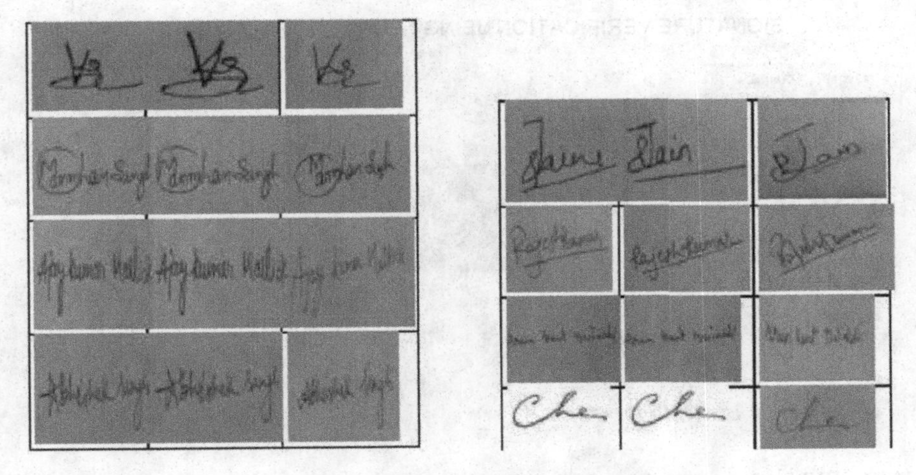

Fig. 4 The image database used to generate the input set during NN training

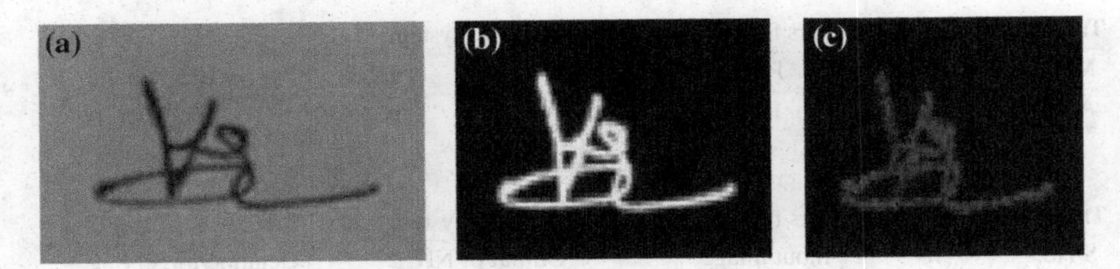

Fig. 5 **a** Original scanned image **b** Inverted binary image **c** Thinned image

4 Experimental Results and Discussions:

A user can select signature database through GUI interface as shown in Fig. 6. The neural network is trained with signature database content. Then simulation is done and features are extracted [7].

The results of the signature classifier are depicted below. The result is demonstrated on a database of five image samples. For our classification purpose, we have taken a threshold value of 93 on the output obtained by simulating the test image against the network. The database of 21 signatures, having two genuine and one forged signature of each signee, was tested. Table 3 shows the following results. Thus, the total efficiency obtained using the above approach comes out to be above 95 % as shown in Table 4.

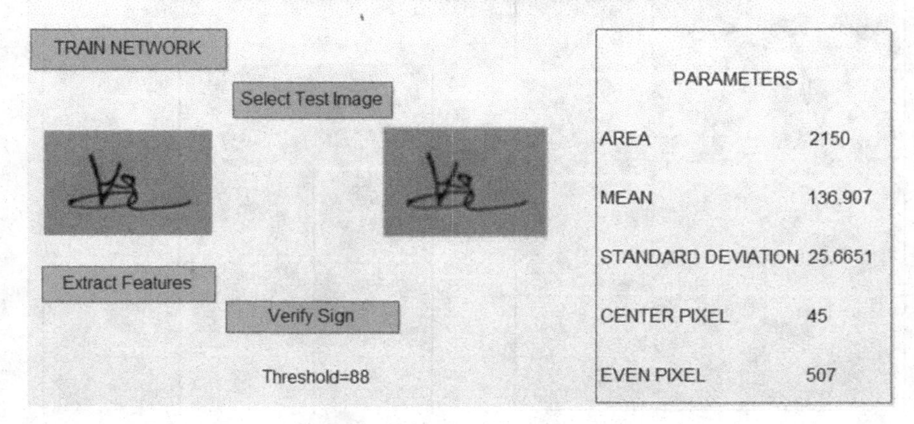

Fig. 6 GUI for the created signature classifier

Table 3 Various parameters for our proposed verification system

Method	Samples	FAR	FRR	TAR	TRR	EFF
ANN based SV	24	1.3105	7.3202	98.6895	92.8629	95.7762

Table 4 Various parameters for our proposed verification system

S No.	Input image	Output of NN	Genuine/forgery
1		98.23	Genuine
2		97.16	Genuine
3		34.67	Forgery
4		20.992	Forgery

The FAR/FRR histogram given in Fig. 7 demonstrates the performance of our network. Here, a value of 1 represents that a false result is obtained over every iteration of the network, while a value of 0.5 implies that the probability of a false result occurring over these images is near zero. The parameter readings for the several surveyed systems in various cases are depicted in Table 5.

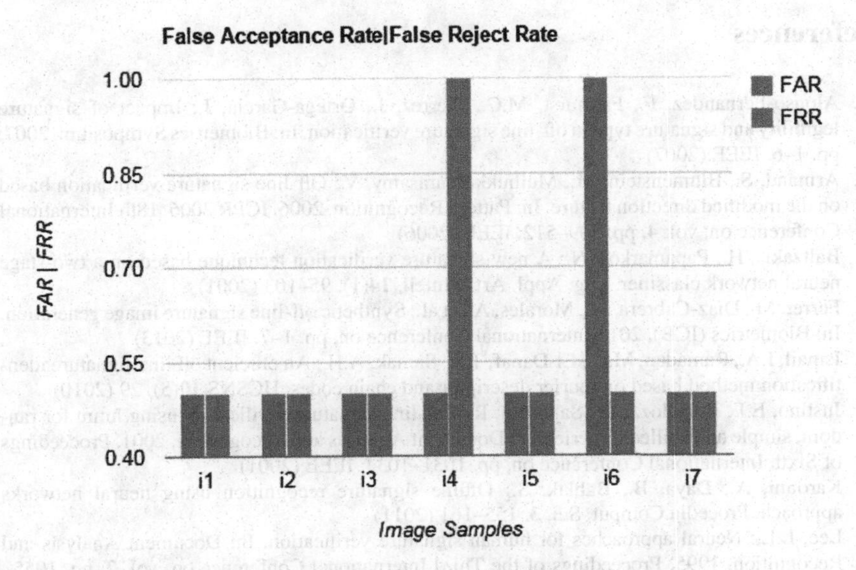

Fig. 7 The Performance of the system as a function of its FAR, FRR

Table 5 Comparison of the several established techniques of offline signature verification

Method	Samples	FAR	FRR	TAR	TRR	EFF
Kshitij Isoidea et al. [13]	240	4.16	7.29	92.70	95.83	94.22
H. Baltzakisa et al. [3]	2000	9.81	3.00	97.00	90.01	94.12
Ismail et al. [5]	2400	1.71	2.88	–	–	96.20
Justino et al. [6]	1600	22.00	10.04	–	–	91.12
Armand et al. [2]	2106	8.20	6.30	–	–	91.80

5 Conclusion

Overall error rate (OER) of a system depends heavily on the false acceptance rate (FAR) and false rejection rate (FRR). These two values in turn depend on the threshold variance value taken for the system. We have taken a threshold value of 93 % for our NN classifier. We have obtained an FAR of 2.8 % and an FRR of around 2 % using the above-mentioned threshold value. During testing, the network was found to classify signatures (either genuine or forged) efficiently and the overall correctness of the system was found to be over 95 %. Classification ratio is more than 93 %. The proposed algorithm effectively classifies signature as exact or forged using simple geometric features. It is robust and effectively differentiates between semi-skilled, random, and simple forgeries. We can improve the efficiency of the ANN classifier by several ways viz. changing the training approach, altering the number of training parameters, and using a larger training dataset. We can employ different training approaches, each having their own merits and demerits.

References

1. Alonso-Fernandez, F., Fairhurst, M.C., Fierrez, J., Ortega-Garcia, J.: Impact of signature legibility and signature type in off-line signature verification. In: Biometrics Symposium, 2007, pp. 1–6. IEEE (2007)
2. Armand, S., Blumenstein, M., Muthukkumarasamy, V.: Off-line signature verification based on the modified direction feature. In: Pattern Recognition, 2006. ICPR 2006. 18th International Conference on, vol. 4, pp. 509–512. IEEE (2006)
3. Baltzakis, H., Papamarkos, N.: A new signature verification technique based on a two-stage neural network classifier. Eng. Appl. Artif. Intell. **14**(1), 95–103 (2001)
4. Ferrer, M., Díaz-Cabrera, M., Morales, A., et al.: Synthetic off-line signature image generation. In: Biometrics (ICB), 2013 International Conference on, pp. 1–7. IEEE (2013)
5. Ismail, I.A., Ramadan, M.A., El-Danaf, T.S., Samak, A.H.: An efficient off line signature identification method based on fourier descriptor and chain codes. IJCSNS **10**(5), 29 (2010)
6. Justino, E.J., Bortolozzi, F., Sabourin, R.: Off-line signature verification using hmm for random, simple and skilled forgeries. In: Document Analysis and Recognition, 2001. Proceedings of Sixth International Conference on, pp. 1031–1034. IEEE (2001)
7. Karouni, A., Daya, B., Bahlak, S.: Offline signature recognition using neural networks approach. Procedia Comput. Sci. **3**, 155–161 (2011)
8. Lee, L.L.: Neural approaches for human signature verification. In: Document Analysis and Recognition, 1995, Proceedings of the Third International Conference on, vol. 2, pp. 1055–1058. IEEE (1995)
9. Lejtman, D.Z., George, S.E.: On-line handwritten signature verification using wavelets and back-propagation neural networks. In: Document Analysis and Recognition, 2001. Proceedings of Sixth International Conference on, pp. 992–996. IEEE (2001)
10. Mahar, J.A., Mahar, M.H., Khan, M.K.: Comparative study of feature extraction methods with k-nn for off-line signature verification. In: Emerging Technologies, 2006. ICET'06. International Conference on, pp. 115–120. IEEE (2006)
11. McCabe, A., Trevathan, J., Read, W.: Neural network-based handwritten signature verification. J. Comput. **3**(8), 9–22 (2008)
12. Oliveira, L.S., Justino, E., Sabourin, R.: Off-line signature verification using writer-independent approach. In: Neural Networks, 2007. IJCNN 2007. International Joint Conference on, pp. 2539–2544. IEEE (2007)
13. Sisodia, K., Anand, S.M.: Off-line handwritten signature verification using artificial neural network classifier. Int. J. Recent Trends Eng. **2**(2), 205–207 (2009)
14. Trier, Ø.D., Jain, A.K., Taxt, T.: Feature extraction methods for character recognition-a survey. Pattern Recognit. **29**(4), 641–662 (1996)
15. Zhang, B.l., Fu, M.Y., Yan, H.: Handwritten signature verification based on neuralgas based vector quantization. In: Pattern Recognition, 1998. Proceedings. Fourteenth International Conference on, vol. 2, pp. 1862–1864. IEEE (1998)

A Simple and Effective Technique for Online Handwritten Bangla Character Recognition

Shibaprasad Sen, Ram Sarkar and Kaushik Roy

Abstract In this paper, we have proposed a simple but effective feature extraction technique following the distance-based features to recognize online handwritten isolated Bangla basic characters. In this approach, a character is divided into N number of segments and then distances are calculated among each other. These distance values are then used as features for recognition purpose. On evaluation of this feature set on 10,000 Bangla character samples (50-class character set) by various classifiers, the method yields reasonably good result with 98.20 % success rate.

Keywords Online character recognition · Handwritten Bangla script · Stroke classification · Distance-based feature

1 Introduction

Nowadays, a large section of the society is very much accustomed to enter data through keyboard for the different activities in their daily life. Though this makes life very convenient, there are some inherent difficulties using keyboard such as: (a) there may be a chance of mistyping (b) the typing speed of different individuals are different, and (c) as the size of the devices is becoming smaller, sometimes it is quite tricky to handle a keyboard. Rather, if people are asked to write the same thing freely by pen, perhaps they can write faster with less probability of misspelling. Moreover, it would save some amount of time also. Devices like A4

S. Sen (✉)
Future Institute of Engineering and Management, Kolkata, India
e-mail: shibubiet@gmail.com

R. Sarkar
Jadavpur University, Kolkata, India
e-mail: raamsarkar@gmail.com

K. Roy
West Bengal State University, Barasat, India
e-mail: kaushik.mrg@gmail.com

S. Das et al. (eds.), *Proceedings of the 4th International Conference on Frontiers in Intelligent Computing: Theory and Applications (FICTA) 2015*, Advances in Intelligent Systems and Computing 404, DOI 10.1007/978-81-322-2695-6_18

Take Note, Smartphone with Android, iPad, etc., are used for this purpose where people can write freely and the written data are saved in the form of online information. Regarding the research on online handwritten character recognition for Indian regional languages, many effective works [1–4] are available for Devanagari script only. But, researchers are yet to pay substantial attention in online Bangla handwriting recognition domain. Though, right now, there are almost 220 millions of people who speak Bangla all over the world. Among them, few works in the literature of online Bangla character recognition; in [5], authors have proposed a technique toward unsupervised feature generation that is based on dissimilarity space, embedding the local neighborhoods around the pen points along the trajectory. Authors showed its benefit for classification due to its high capability of discriminative representation. In [6], authors have concentrated on segmentation and recognition of online Bangla cursive texts. Here, authors have segmented the text into primitives that may be a basic character, a compound character, or may represent the part of a basic character or compound character. Then, they have tried to recognize these primitives. A technique by combining online and offline approach to segment handwritten online Bangla text and then is reported in [7]. Here, authors have used directional features to recognize these strokes. In another work mentioned in [8], authors have extracted constituent strokes from characters and then stroke level sequential and dynamic information are obtained from pen movements on writing pads, which are used as features for experiment. Authors, in [9], are working toward annotation of unconstrained Bangla handwriting samples and have tried to construct a big volume of annotated database of online handwriting samples. A GUI-based semi-automatic approach has been developed for annotation at character boundary levels and a scheme is built to represent these annotated data in XML representation. According to authors in [10], *difficulties* to design a unconstrained Bangla handwriting recognition system is due to the reason that Bangla has a large volume of alphabet set that consists almost 300 shapes, many of which are complex enough and also there exists a lot of similar shape characters. In the paper [10], authors have described preliminary results on limited vocabulary of Bangla cursive handwriting, based on a combination of Multi-Layer Perceptron (MLP) and Support Vector Machine (SVM). A new approach has been proposed by the authors in [11] to recognize cursively written Bangla words. This method has described the feature in sub stroke level for the script and a writing model is proposed based on Hidden Markov Model (HMM). In [12], authors have applied two known feature extraction techniques, called point-float and direction code histogram on Bangla script. Effectiveness of the system is tested by Nearest Neighbor, MLP, and HMM. In [13], sequential and dynamic information obtained from pen movements on the writing pads act as features for online Bangla handwriting recognition. These features are then fed to quadratic classifier for recognition purpose. In spite of the efforts of the research community, still online handwritten Bangla character recognition is in infant stage. In this paper, we have proposed a technique to recognize handwritten online Bangla characters in which we have customized the distance-based feature to suitably fit into the current problem.

অ	আ	ই	ঈ	উ	ঊ	এ	ঐ	ও	ঔ
ঋ	ক	খ	গ	ঘ	ঙ	চ	ছ	জ	ঝ
ঞ	ট	ঠ	ড	ঢ	ণ	ত	থ	দ	ধ
ন	প	ফ	ব	ভ	ম	য	র	ল	শ
ষ	স	হ	ড়	ঢ়	য়	ং	ঃ	ঁ	ৎ

Fig. 1 The shapes of Bangla basic characters

2 About Bangla Script

In India, Bangla is familiar as second most popular language. It is the official language as well as script in Bangladesh also. It is a Brahmin script although its exact derivation is disputed. Writing style shares some similarities with the Dravidian-language scripts, particularly in the shapes of some vowel letters, but it is generally more similar to the Aryan-language scripts, in particular Devanagari. Bangla is an Abugida—i.e., every consonant letter represents a syllable containing an inherent vowel—written from left to right. Basic character set of Bangla script consists of 39 consonants and 11 vowels. Figure 1 shows the basic Bangla characters set.

3 Challenges

The main challenge to work with online handwriting recognition domain arises mainly because writing style of different individuals is different. Due to this fact, number and order of strokes for different samples of a particular character vary significantly. In the literature, stroke is defined as the collection of pen points between one pen down and pen up [8]. Figure 2 describes such scenarios very clearly when আ is written in four different ways by varying number and order of the strokes. In the first case, three strokes are used to represent the shape while only two strokes are sufficient to describe the same shape in the second case. In the third and fourth ways, same four strokes are used to represent আ where the only difference is in the order of strokes.

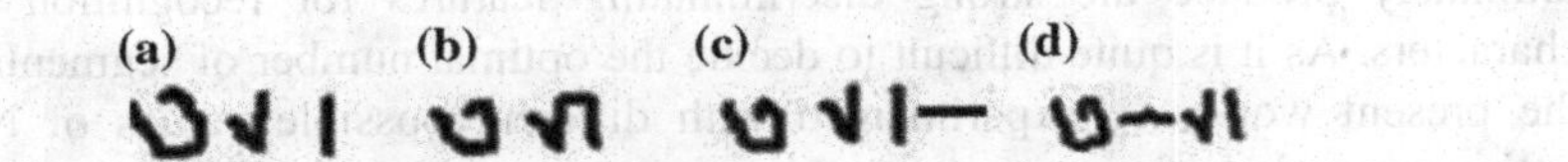

Fig. 2 a–d Several ways to write আ by varying number and order of strokes

4 Database Preparation

We have collected 200 samples for each character sample from 100 different individuals. There are 50 different characters in Bangla script, as described earlier, which makes the total data samples 10,000. During data collection, we have not imposed any restriction to the writers while writing except, we just told that for writing constituent strokes they must follow the basic stroke database of Bangla script [8]. A4 Take note was used for data collection purpose. Take note outcome is represented by a set of pen points p_j, for all j = 1, …., M. Where p_j is the pen position with x coordinate x_j and y coordinate y_j along with pen up/pen down status. M is the total number of pen points used to write a character sample. In the preprocessing step, at first we have removed duplicate points because these make no sense. If p_j and p_k are two consecutive pen points, then jth point p_j will be retained with respect to kth point p_k if and only if Eq. (1) is satisfied.

$$x^2 + y^2 > m^2 \tag{1}$$

Where $x = x_j - x_k$ and $y = y_j - y_k$. Here, the value of m is considered as 0 to remove all duplicate points. Then, we have normalized the character sample into 64 points by keeping the number of strokes and structure intact. Normalized points are then scaled to a fixed window of size 512 × 512.

5 Feature Extraction Using Distance-Based Feature

A different approach toward distance-based feature for Uyghur character is presented in [14]. For Bangla script, due to its cursive nature, we assumed that with suitable modification, distance-based feature would fit into this pattern classification problem. In this feature extraction technique, characters are considered as a single entity irrespective of the number of constituent strokes and divided into N hypothetical segments. As to represent a single segment, two points are required; we have divided the character shape into N + 1 sample pen points p_i, where i varies from 1 to N + 1. We have computed the distance from each point to every other point and these distance values are considered as the features for the mentioned experiment. Closer observation to the Bangla character set reveals that there are some characters which are structurally almost similar. To cope up with this, we need some local information in certain regions of the character image which would ultimately produce the strong discriminating features for recognition of such characters. As it is quite difficult to decide the optimal number of segmentation for the present work, we experimented with different possible values of N to get optimum number of segments. For this, we have considered the values of N as 6, 8, 10, 16, 32, 48, and 55, respectively. Higher value of N indicates the more the

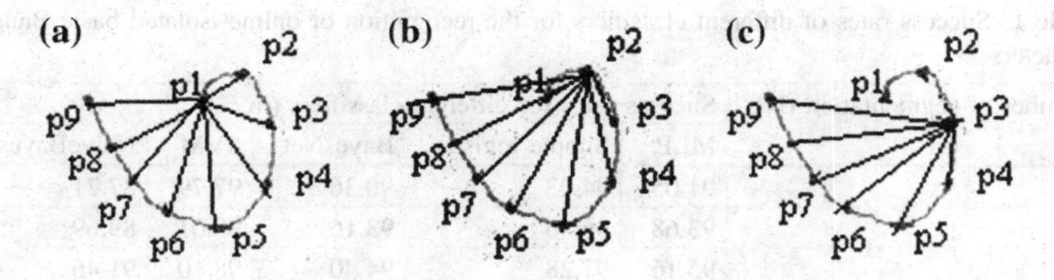

Fig. 3 **a–c** Distance calculation from one pen point to the rest

number of segments, i.e., it would provide more detailed/close view about the character sample. Algorithm for distance-based feature is shown in Algo 1.

STEP 1: START
STEP 2: i = 1;
STEP 3: j = i+1;
STEP 4: compute the distance d between p_i and p_j;
STEP 5: j = j+1;
STEP 6: if (j <=N + 1) then goto STEP 4, otherwise goto STEP 7;
STEP 7: i = i+1;
STEP 8: if (i < N+1)goto STEP 3, otherwise goto STEP 9;
STEP 9: END

Algorithm 1. Algorithm used to compute distance-based features

Here, if we consider 8 (i.e., N = 8) segments, then total number of pen points is 9, from p_1 to p_9, which is shown in Fig. 3a–c. From the Algorithm in Algorithm 1, it is clear that there will be a total of eight iterations. In the first iteration, distances are computed from pen point p_1 to p_2, p_3, p_4, p_5, p_6, p_7, p_8, and p_9, respectively. Therefore, eight feature values are generated in this iteration. In the second iteration, distances are calculated from pen point p_2 to all other pen points except p_1, because it is already computed in the first iteration. Thus, it produces seven feature values in the second iteration. Following the similar approach, in the last iteration only one distance value is computed from pen point p_8 to p_9. Thus, a total of 36 (i.e., 8 + 7 + 6 + 5 + 4 + 3 + 2 + 1) feature values are generated for an eight segmented Bangla character sample. In general, for an N segmented character shape, N (N + 1)/2 number of feature values will be produced by the distance-based feature extraction procedure. In Fig. 3a–c, red points refer to pen points and black lines denote the distances between pen points.

6 Result and Discussion

For the evaluation of the present technique, online handwritten Bangla basic character set which consists of 50 different shapes is considered. Each class contains 200 different samples, thereby we have used a total of 10,000 character

Table 1 Success rates of different classifiers for the recognition of online isolated basic Bangla characters

Number of segmentation (N)	Success rates for different classifiers (in %)				
	MLP	Simple logistic	BayesNet	SVM	NaiveBayes
6	91.05	94.03	90.36	97.79	87.71
8	93.68	96.46	93.16	98.01	89.69
10	95.16	97.28	94.40	98.10	91.46
16	95.79	97.57	95.50	98.20	92.75
32	96.36	97.79	93.53	95.42	92.45
40	96.77	97.80	93.28	90.64	91.68
48	96.67	97.79	93.16	83.65	91.28
52	96.58	97.69	92.88	79.70	91.14
55	96.22	97.58	92.51	75.41	91.04

samples in the experiment. Some standard classifiers like MLP, Simple Logistic, BayesNet, NaiveBayes, and SVM are used for recognition of the characters. We have applied fivefold cross-validation on total dataset and Table 1 reflects the corresponding success rates of different classifiers. As we continue to divide the character sample by more number of segments, up to a certain level recognition rate increases gradually but after that it starts falling downwards. This is because, as we are increasing the number of segments, length of the structural units of the character sample gets decreased, indicating the closed/detailed view of the character sample. Extracted features from such structural units then plays better performance toward recognition of the character sample. It can be observed easily from Table 1—initially success rate is increasing steadily with increasing number of segments. Table 1 reveals the maximum recognition rate when we have segmented the character sample into 40 segments for MLP and Simple Logistic classifiers whereas all the other three classifiers such as BayesNet, SVM and NaiveBayes show maximum success rates for 16 segments. But when we break the character sample beyond these levels, then the size of those structural units become less informative, so extracted features from such small units fail to describe the character sample properly. Hence, recognition rate starts to go down slowly. Gray cells in the Table 1 represent maximum success rates for a particular classifier.

Figure 4 graphically describes the behavior of different classifiers for different number of segmentations. Black line represents recognition result of MLP, whereas pink, green, violet, and blue curves describe the recognition rates of Simple Logistic, SVM, NaiveBayes, and BayesNet, respectively. From the figure, it is observed that SVM recognizes the Bangla character samples with highest recognition accuracy of 98.20 % when N = 16, It is worth mentioning that SVM produces better result because it has certain advantages over other classifiers such as: first, it has a regularization parameter that helps to avoid the over-fitting. Second, as it uses kernel trick, anyone can build in expert knowledge about the problem via engineering the kernel. For conducting the present experiment, we have chosen C-SVC type SVM from LibSVM library, whose kernel type is using radial basis

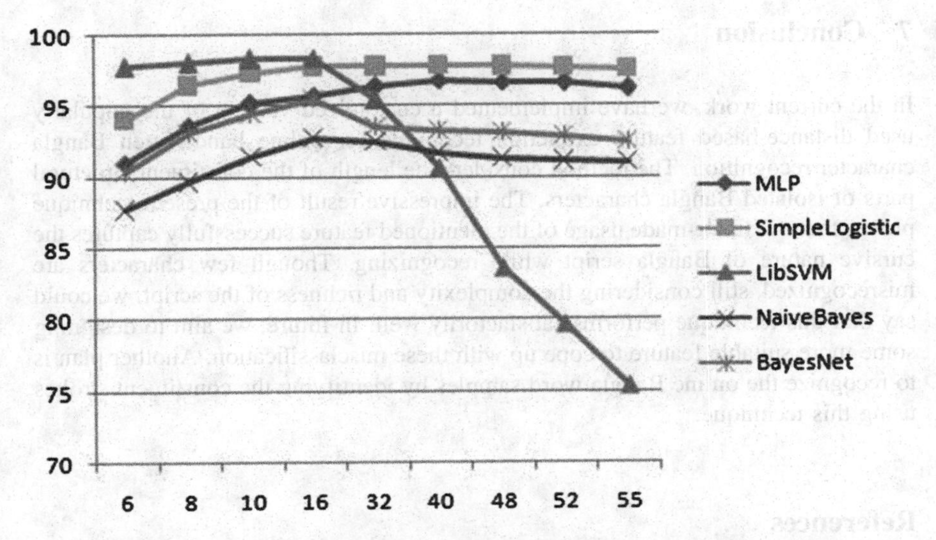

Fig. 4 Graphical description of performance of the different classifiers while recognizing the online handwritten Bangla basic characters with varying number of segments of characters considering during feature extraction

Table 2 Some misclassified characters (experimented by SVM with N = 16)

Original character with number of samples	Mostly misclassified as
ঢ (200)	ট (6)
ঈ (200)	ই (3)
স (200)	য (3)

function: exp (-gamma*|u-v|^2) with eps = 0.001, gamma = 0.0. From the experimental result, we can observe that sometimes ঢ is misclassified as ট, ঈ is misclassified as ই, স is misclassified as য , etc. It can also be noticed that 17 other character samples are wrongly classified as ব, whereas 15 other Bangla character samples are misclassified as র. Close comparison of the misclassified characters revels that strong structural similarity is the sole reason for this misclassification. In Table 2, some of the misrecognized Bangla characters along with the actual characters are shown.

In [8], we have used point-based and structural features for online character recognition with 92.9 % accuracy whereas in [15], an approach has been developed by us combining offline and online features to recognize online isolated Bangla characters with an accuracy of 83.92 %. As the present technique is evaluated on different datasets, so it is difficult to compare this technique with the earlier ones. But in order to have some idea about how good the current technique is, we have mentioned the recognition accuracies of the past works. Looking into results it can be claimed that the present technique performs much better.

7 Conclusion

In the current work, we have implemented a customized version of the popularly used distance-based feature extraction technique for online handwritten Bangla character recognition. The method considers the length of the constituent structural parts of isolated Bangla characters. The impressive result of the present technique proves that our tailor-made usage of the mentioned feature successfully captures the cursive nature of Bangla script while recognizing. Though few characters are misrecognized, still considering the complexity and richness of the script, we could say that this technique performs satisfactorily well. In future, we aim to designing some more suitable feature to cope up with these misclassification. Another plan is to recognize the online Bangla word samples by identifying the constituent strokes using this technique.

References

1. Kubatur, S., Sid-Ahmed, M., Ahmadi, M.: A neural network approach to online Devanagari handwritten character recognition. In: International Conference on High Performance Computing and Simulation (2012) doi: 10.1109/HPCSim.2012.6266913
2. Swethalakshmi, H., Sekhar, C.C., Chakravarthy, V.S.: Spatiostructural features for recognition of online handwritten characters in Devanagari and Tamil scripts. In: Proceedings of International Conference on Artificial Neural Networks, vol. 2, pp. 230–239 (2007)
3. Swethalakshmi, H., Jayaraman, A., Chakravarthy, V.S., Sekhar, C.C.: On-line handwritten character recognition for Devanagari and Telugu scripts using support vector machines. In: Proceedings of International Workshop on Frontiers in Handwriting Recognition, pp. 367–372 (2006)
4. Joshi, N., Sita, G., Ramakrishnan, A.G., Deepu, V.: Machine recognition of online handwritten Devanagari characters. In: Procedings of International Conference on Document Analysis and Recognition, pp. 1156–1160 (2005)
5. Frinken, V., Bhattacharya, N., Pal, U.: Design of unsupervised feature extraction system for on-Line handwriting recognition. In: 11th IAPR International Workshop on Document Analysis and Systems, pp. 355–359 (2014)
6. Bhattacharya, N., Pal, U., Kimura, F.: A system for Bangla online handwritten text. In: International Conference on Document Analysis and Recognition, pp. 1335–1339 (2013)
7. Bhattacharya, N., Pal, U.: Strokes segmentation and recognition from Bangla online handwritten text. In: Frontiers in Handwriting Recognition, pp. 740–745 (2012)
8. Roy, K., Bandhopadhyay, A., Mondal, R.: Stroke-Database design for online handwriting recognition in Bangla. Int. J. Modern Eng. Res. pp. 2534–2540 (2012)
9. Bhattacharya, U., Banerjee, R., Baral, S., Dey, R., Parui, S.K.: A semi automatic annotation scheme for Bangla online mixed cursive handwriting samples. In: International Conference on Frontiers in Handwriting Recognition, pp. 680–685 (2012)
10. Mohiuddin, S.k., Bhattacharya, U., Parui, S.K.: Unconstrained Bangla online handwriting recognition based on MLP and SVM. In: Proceedings of Joint Workshop on Multilingual OCR and Analytics for Noisy Unstructured Text Data (2011) doi:10.1145/2034617.2034635
11. Fink, G.A., Vajda, S., Bhattacharya, U., Parui, S., Chaudhuri, B.B.: Online Bangla word recognition using sub-stroke level features and hidden Markov models. In: International Conference on Frontiers in Handwriting Recognition, pp. 393–398 (2010)

12. Mondal, T., Bhattacharya, U., Parui, S.K., Das, K., Mandalapu, D.: On-line handwriting recognition of indian scripts—the first benchmark. In: 12th International Conference on Frontiers in Handwriting Recognition, pp. 200–205 (2010)
13. Roy, K., Sharma, N., Pal, U.: Online Bangla handwriting recognition system. In: International Conference on Advances in Pattern Recognition, pp. 117–122 (2006)
14. Hamdulla, A., Simayi, W., Ibrayim, M., Tursun, D.: Research on on-line Uyghur handwritten character recognition technology based on modified center distance feature. In: Int. J. Signal Process. Image Process. Pattern Recogn. pp. 409–424 (2014)
15. Sen, S.P., Paul, S.S., Sarkar, R., Roy, K., Das, N.: Analysis of different classifiers for online bangla character recognition by combining both online and offline information. In: 2nd International Doctoral Symposium on Applied Computation and Security Systems (2015)

Visual Analytic-Based Technique for Handwritten Indic Script Identification—A Greedy Heuristic Feature Fusion Framework

Sk. Md. Obaidullah, Chayan Halder, Nibaran Das and Kaushik Roy

Abstract Script identification from multi-script handwritten document images has been a subject of considerable discussion in the literature. In this paper, a novel *feature fusion framework (FFF)* using *structural appearance (SA)* and *directional morphological filter (DMF)* is proposed based on the idea of *visual analytic (VA)*. A dataset of 181 handwritten document pages distributed over 2450 line and 20,260 word images of Bangla, Devanagari, Roman, Oriya, and Urdu scripts is built and considered for experimentation. Experimental result shows a significant improvement of the identification rate by the *VA-FFF* over the *SA* and *DMF* technique if they had been applied individually.

Keywords Handwritten script identification · Visual analytic · Greedy heuristic feature fusion · Structural appearance · Directional morphological features · MLP classifier

---

Sk. Md. Obaidullah (✉)
Department of Computer Science & Engineering, Aliah University, Kolkata, West Bengal, India
e-mail: sk.obaidullah@gmail.com

C. Halder · K. Roy
Department of Computer Science, West Bengal State University, Kolkata, West Bengal, India
e-mail: chayan.halderz@gmail.com

K. Roy
e-mail: kaushik.mrg@gmail.com

N. Das
Department of Computer Science & Engineering, Jadavpur University, Kolkata, West Bengal, India
e-mail: nibaran@gmail.com

S. Das et al. (eds.), *Proceedings of the 4th International Conference on Frontiers in Intelligent Computing: Theory and Applications (FICTA) 2015*, Advances in Intelligent Systems and Computing 404, DOI 10.1007/978-81-322-2695-6_19

1 Introduction

In the literature, many works are reported for efficient recognition of characters by OCR algorithms with very high accuracy and low computational complexity. However, the prerequisites for those algorithms are the knowledge of the particular script by which the language is written. By short definition, a script provides the set of graphemes for writing a particular language. Language-specific script is very common throughout the world, but situations are there when multiple languages are written using a single script. For example, Bangla, Assamese, and Manipuri languages are commonly written by Bangla script. Now if a country uses language/s which is written using a single script, the task of designing OCR for those languages is very simple. It can be considered as a problem of recognizing a set of specific script characters by the OCR. However, the situation is very complex for a multilingual/ multi-script country like India, where 22 official languages [1] and 12 scripts are present. Apart from this, Roman is also a very popular script in India which is used to write English and Santali languages. So here an official document can be written using multiple languages and/or multiple scripts. In our day-to-day life, we come across various multi-script documents like preprinted application forms, postal documents, etc. So for a country like India, designing of a general OCR system considering all official languages/scripts is very challenging. To overcome this, an automatic script identification system can be developed and the output of this system can be provided as input to the script-dependent OCR [2]. So a system to identify the script of the document images is of pressing need [3].

Many works are reported so far related to Indic scripts. Printed script identification has got more attention from the research community compared to handwritten one. The average identification rate is also very impressive for printed documents in comparison with handwritten types [4–9]. Because of versatility of writing style, variation in interline, inter- and intra-word spacing, character sizes, and printed script identification system generally does not yield expected results in handwritten cases. That is why script identification from handwritten document images is still an open challenge for document image processing researchers. Among the reported works in handwritten domain, Hochberg et al. [10] proposed a technique to identify Arabic, Chinese, Cyrillic, Devanagari, Japanese, and Latin scripts using some structure-based features like sphericity, aspect ratio, white holes, etc. L. Zhou et al. [11] proposed a scheme to identify Bangla and English scripts using connected component profile-based features. V. Singhal et al. [12] identified four popular scripts namely Roman, Devanagari, Bangla, and Telugu from line-level handwritten document images using Gabor filter and gray-level co-occurrence matrix-based rotation invariant texture features. This work is targeted to capture the texture variability among different scripts. In another work, Obaidullah et al. [1] identified six popular Indic scripts namely Bangla, Devanagari, Malayalam, Urdu, Oriya, and Roman. The work was carried out using some structural features at document level like fractal dimension, circularity, etc. In a recent work, M. Hangarge et al. [13] proposed a word-level directional DCT-based approach to identify six different Indic scripts. Recently, R. Pardeshi et al. [14] proposed a word-level approach to identify eleven handwritten Indic scripts using different transform-based features and KNN classifier.

The paper is organized as follows: Proposed methodology is discussed in Sect. 2. Experimental details are discussed in Sect. 3. Finally, conclusion and future scopes are given in Sect. 4, and references are available in the last section.

2 Proposed Methodology

2.1 Preprocessing

The collected documents are digitized using an HP flatbed scanner at 300 dpi. Initially, the digitized images are stored using 256 different intensity levels. An existing two-stage-based binarization algorithm [15] which is a combination of local and global approach is applied afterward to generate the two tone/binary images. The main motivation of using this two-stage approach is that the output binarized image will be at least as good as if only global technique would have been used. The following section discusses about the feature extraction process.

2.2 VA-FFF: Visual Analytic-Based Feature Fusion Framework

A greedy heuristic feature fusion framework based on visual analytic (VA) has been proposed for the present work. This VA is carried out observing the visual appearance of the graphemes of the scripts considered and feature properties were selected heuristically. Totally, 54 features are computed using this technique, out of which 42 are from structural appearance (SA) based and 12 are from directional morphological filter (DMF) based. The following section discusses about the motivation behind choosing this VA-FFF technique.

It has been observed that a prior knowledge about the structure of image components helps extracting several important features to categorize one script from another. But identification of such features is difficult when graphemes of two or more scripts have some similarity in their structure or shape. To overcome this limitation, directional morphological filters can help to categorize the variations within the scripts by identifying the presence of different directional strokes like horizontal, vertical, slanting lines, or some curve with arbitrary orientation. This visual observation motivated us to develop a novel feature fusion technique which is presented here. A brief description about different features used during the fusion *(VA-FFF)* process is provided below:

Structural Appearance of the Image Component Contour (SA).
One of the important types of feature used in the present work is based on the structure of the image component. Connected component analysis approach is followed here. All picture elements in a connected component share similar pixel intensity values and are in some way connected with each other. There are several methods like 4-connected, 8-connected, m-connected, etc. algorithms to compute connected component of an image [16]. Two pixels with coordinates (x′, y′) and (x″, y″) are called 4-connected and 8-connected, respectively if, and only if, following condition satisfies:

$$|x' - x''| + |y' - y''| = 1(4 - \textit{connected}) \text{ and}$$
$$\max(|x' - x''|, |y' - y''|) = 1(8 - \textit{connected})$$

Then inner or interior and outer or exterior contours are computed on these connected components. Following structural features are computed on both outer and inner contours. It is to be noted that while calculating the structural features, components of very small size or large size have been discarded as their feature values are not quite helpful in identifying script types.

Chain Code.
The presence of horizontal, vertical, or other type of lines in the scripts is important structural feature for identification [9]. For example, a polygon/object can be represented as a sequence of steps in one of the eight directions; each step is designated by an integer from 0 to 7. For our case, Bangla and Devanagari scripts contain horizontal lines on the upper part of the scripts which are called 'Matra' or 'Shirorekha' This is a unique distinguishing feature of these two scripts from the rest. Similarly, some slanting lines in some other scripts can also be determined.

Convex Hull.
The feature value obtained from surroundings created by the convex curve/hull can be used to characterize different script types [9]. The hull is computed for every selected component's inner and outer contours by the proposed method. From this hull, different important statistical information may be obtained and used as feature values.

Circularity.
The roundness of the graphemes in script like Oriya can be captured by this feature [9]. As mentioned earlier, this feature is carried out on both inner and outer contours following the same sequence of steps. The difference of the radius from that of the minimum enclosed circle (c_2) to the fitted circle (c_1) is then evaluated which gives us the characteristics of the circularity of those contours ($c = c_2 - c_1$). The value of c will be zero for complete circular components. The value of c is stored in the feature vector and the process is iterated for both inner and outer contours.

Bounding Box.
Using this feature, the up-right bounding rectangles of the outer contours and inner contours of selected components are computed [9]. The ratio of width and height determines whether the contour is square (ratio = 1), horizontal (ratio > 1), or vertical (ratio < 1). Here, a minimum of all the widths and heights of bounded boxes for the outer and inner contours are stored. Following figure shows computation of rectangularity on Devanagari script.

Directional Morphological Filter (DMF).
In present work, we have introduced the idea of directional morphological filters or DMF which is used to capture different directional strokes present in different scripts. Heuristically, four user-defined kernels namely H kernel, V kernel, LD kernel, and RD kernel are defined by visually analyzing the properties of graphemes of different scripts. They are 3 × 11, 11 × 3, 11 × 11, and 11 × 11 matrices correspondingly where horizontal, vertical, right diagonal, and left diagonal pixels are 1 and rest of them are 0.

Morphological Reconstruction using DMF.
The image dilation and erosion operations for the present work have been customized with user-defined directional kernel DMFs and can be mathematically represented by the following Eqs. (1) and (2):

$$\text{Custom\_dilate}(x, y) = \max_{(x', y' \varepsilon \text{DMF})} \text{src}\left(x + x', y + y'\right) \quad (1)$$

$$\text{Custom\_erode}(x, y) = \min_{(x', y' \varepsilon \text{DMF})} \text{src}\left(x + x', y + y'\right) \quad (2)$$

where 'DMF' is the directional morphological kernels and 'src' is the source image and DMF ε {H kernel, V kernel, LD kernel, RD kernel}.

In the proposed method, four kernels namely H kernel, V kernel, LD kernel, and RD kernel have been developed. Then the operations carried out on the binarized image are as follows: First, the binarized image is resized with an image size which is fixed empirically. Then the resized image is dilated using a default kernel. Afterward, the dilated image is then eroded four times using four different structural elements (H kernel, V kernel, LD kernel, and RD kernel). Finally, the ratio of those eroded images with the dilated one is obtained and calculation of the average and standard deviation of the height and width of the component of the eroded images is also computed. This process helps in indiscriminating individual script type using the different morphological structures present on those scripts. For example, Urdu scripts are more likely to be diagonally structured than horizontally structured, which completely differs from Bangla or Devanagari scripts having horizontal ('shirorekha') structures. Hence, using horizontal structuring element the difference between these script types can be solely evaluated. Based on the above-mentioned set of features, a visual analytic (VA) inspired greedy heuristic feature fusion algorithm that is proposed.

Algorithm I:
Algorithm VA-FFC (Fs, Fm: Set)
{F is the set of features, Fs: Structural, Fm: Directional Morphological, X: Selection Set}
Fc $\leftarrow \phi$ *{Fc is the set that contains final combined features}*
For every fsi of Fs and fmj of Fm, where i, j>=1
While (Fs & Fm != ϕ*) and !solution(Fc) do*
    X $\leftarrow$ *select(Fsi or Fmj based on Visual Analytic)*
    Fs $\leftarrow$ *Fs / {X} & Fm* $\leftarrow$ *Fm / {X}*
If feasible (Fc U {X}) then Fc $\leftarrow$ *Fc U {X}*
If ((solution (Fc) >= solution (Fc)) && (solution (Fc) >= solution (Fm))) then return Fc
Else return "The VA based combined features are not optimum and different VA combination needs to be taken"

In Algorithm I, the feasibility criteria are defined as follows: the combined set Fc must lie within the accuracy rate produced by Fs or Fm if applied separately and the absolute accuracy rate which is 100 %. The selection() function chooses the features based on VA techniques observing graphemes of different script characters. The objective() function of the above algorithm is to produce feature fusion identification rate, which is higher compared to if only single type of feature was applied. The said algorithm is implemented to generate the feature set VA-FFF from feature sets SA and DMF.

2.3 Classification

In any pattern, recognition work after feature extraction classification is to be carried out, which comprises stages namely training, testing, and validation. MLP neural network with back-propagation learning algorithm [1, 9, 15] is used for classification. In present work, the number of features is 54 and output is 5, so experimentally we have determined the number of neurons in the hidden layer as 17. Finally, a MLP with configuration of 54-17-5 has been designed.

3 Experimental Details

3.1 Dataset Development

One of the most imperative tasks is the data collection. Data are collected from different people with varying handwriting to incorporate versatility in the database. Totally, 181 handwritten document images with almost equal distribution are collected and used for experimentation. The images are digitized using HP flatbed

Table 1 Statistics of the dataset used for present experiment

Script name	No. of documents	No. of lines	No. of words	No. of writers
Bangla	40	480	4800	40
Devanagari	37	370	3700	37
Roman	40	400	3600	30
Oriya	24	480	3840	24
Urdu	40	720	4320	20
Total	**181**	**2450**	**20260**	**151**

scanner at 300 dpi and stored as gray-scale images. Table 1 provides an estimated statistical overview about the dataset used for present experiment.

3.2 *Experimental Protocol*

Training phase of classification initiates the learning process of distinguishable properties for each target class. During test phase, the dissimilarity measure of the classes is evaluated. Generation of training and test set data is very crucial decision for any classification scheme. For present work, whole data set is divided into training and test sets into 1:1 ratio. Following section describes about the outcome of the test phase.

3.3 *Results and Analysis*

The results generated by the classifier are provided in the following subsection. Table 2 shows the average accuracy rate obtained for different feature categories, namely structural (SA), morphological (DMF), and fusion (VA-FFF). As per our observation, VA-FFF performs significantly better over the other two.

Confusion matrix for each of the three features is shown in Table 3. In all the cases, Devanagari scripts are misclassified at a maximum rate. It is evident from SA and VA-FFF that Devanagari script has a maximum similarity with Bangla. The reason is due to the presence of 'Matra' or 'Shirorekha' feature in both the scripts and similarity of few graphemes in both the scripts.

The average Bi-script and Tri-script identification accuracy is shown in Table 4. Out of ten Bi-script combinations, 100 % accuracy rate has found in seven cases. For Bangla, Devanagari, and Oriya script combination, the identification rate has dropped notably. Same scenario may be observed in Tri-script combination also. Here, the identification rate for combination of above-mentioned three scripts (BDO) has dropped 4.8 % from the average Tri-script rate. It can be found from the alphabet set that some of the graphemes of these three scripts have some structural

Table 2 Average accuracy rate using three different categories of feature sets *SA*, *DMF* and *VA-FFF*

Sl. No.	Feature set considered	Average five-script accuracy rate (%)
1.	Structural appearence *(SA)*	93.4
2.	Directional morphological filter *(DMF)*	92.3
3.	Fusion approach *(VA-FFF)*	95.6

Table 3 Confusion matrix on test set for different categories of feature sets *SA*, *DMF*, *VA-FFF*

Structural Feature (*SF*)						Morphological feature (*DMF*)					Combination of SA and DMF				
Script	Classified as					Classified as					Classified as				
	B	*D*	*R*	*O*	*U*	*B*	*D*	*R*	*O*	*U*	*B*	*D*	*R*	*O*	*U*
B	19	0	0	0	0	19	0	0	0	0	19	0	0	0	0
D	2	14	1	1	0	0	14	2	1	1	3	14	0	1	0
R	0	0	24	0	0	0	1	23	0	0	0	0	24	0	0
O	2	0	0	9	0	0	0	0	11	0	0	0	0	11	0
U	0	0	0	0	18	0	0	0	0	16	0	0	0	0	18

Table 4 Bi-script and tri-script accuracy rates (%) using feature fusion approach *(VA-FFF)*

B D	*B R*	*B O*	*B U*	*D R*	*D O*	*D U*	*R O*	*R U*	*O U*	*μ*
Bi-script combination accuracy rate (%) using VA-FFF										
94.8	100	96.9	100	100	93.4	100	100	100	100	98.5
R	*BDO*	*BDU*	*BRO*	*BRU*	*BOU*	*DRO*	*DRU*	*DOU*	*ROU*	*μ*
Tri-script combination accuracy rate (%) using VA-FFF										
94.9	92	96.6	98.1	100	100	96	98.3	94	98.1	96.8

similarity. The issue needs to be taken care of by investigating some micro-level distinguishing features of these three scripts for highest dissimilarity measurement.

4 Conclusion and Future Scope

In the present work, structural appearance and directional morphological features based on fusion framework for script identification from offline handwritten documents of five Indic scripts have been proposed. Experimental result shows an overall recognition accuracy of 95.6 % for VA-FFF on test set using MLP classifier which is significantly better than SA and DMF techniques if they had been applied individually. The Tri-script and Bi-script identification rates for VA-FFF are found to be 96.8 and 98.5 %, respectively. Present result is comparable with existing methods available in the literature. Some misclassification issues of Bangla,

Roman, and Oriya scripts are identified, which will be addressed by the authors in future work. Scope can be further extended to develop benchmark handwritten document image database for all official Indic scripts which are not available till date.

References

1. Obaidullah, S.M., Das, S.K., Roy, K.: A system for handwritten script identification from indian document. J. Pattern Recognit. Res. **8**(1), 1–12 (2013)
2. Chaudhuri, B.B., Pal, U.: A complete printed Bangla OCR. Pattern Recognit. **31**, 531–549 (1998)
3. Ghosh, D., Dube, T., Shivprasad, S.P.: Script recognition—a review. IEEE Trans. Pattern Anal. Mach. Intell. **32**(12), 2142–2161 (2010)
4. Chaudhuri, B.B., Pal, U.: An OCR system to read two Indian language scripts: Bangla and Devanagari (Hindi). In: Proceedings of 4th International Conference on Document Analysis and Recognition, Uhn. pp. 18–20 (1997)
5. Hochberg, J., Kelly, P., Thomas, T., Kerns, L.: Automatic script identification from document images using cluster-based templates. In: IEEE Trans. Pattern Anal. Mach. Intell. **19**, 176–181 (1997)
6. Chaudhury, S., Harit, G., Madnani, S., Shet, R. B.: Identification of scripts of Indian languages by combining trainable classifiers. In: Proceedings of Indian Conference on Computer Vision, Graphics and Image Processing, Bangalore, India (2000)
7. Dhanya, D., Ramakrishnan, A.G., Pati, P.B.: Script identification in printed bilingual documents. Sadhana **27**(part-1), 73–82 (2002)
8. Pati, P.B., Ramakrishnan, A.G.: Word level multi-script identification. Pattern Recognit. Lett. **29**(9), 1218–1229 (2008)
9. Obaidullah, S.M., Mondal, A., Das, N., Roy, K.: Script Identification from printed Indian document images and performance evaluation using different classifiers. Appl. Comput. Intell. Soft Comput. **2014**(Article ID 896128), 12 (2014)
10. Hochberg, J., Bowers, Cannon, K.M., Kelly, P.: Script and language identification for handwritten document images. Int. J. Doc. Anal. Recognit. **2**(2–3), 45–52 (1999)
11. Zhou, L., Lu, Y., Tan, C.L.: Bangla/English Script identification based on analysis of connected component profiles. In: Lecture Notes in Computer Science, vol. 3872/2006, pp. 243–254 (2006)
12. Singhal, V., Navin, N., Ghosh, D.: Script-based classification of hand-written text document in a multilingual environment. In: Research Issues in Data Engineering, pp. 47–54 (2003)
13. Hangarge, M., Santosh, K.C., Pardeshi, R.: Directional discrete cosine transform for handwritten script identification. In: Proceedings of 12th International Conference on Document Analysis and Recognition, pp. 344–348 (2013)
14. Pardeshi, R., Chaudhury, B.B., Hangarge, M., Santosh, K.C.: Automatic handwritten indian scripts identification. In: Proceedings of 14th International Conference on Frontiers in Handwriting Recognition, pp. 375–380 (2014)
15. Roy, K., Banerjee, A., Pal, U.: A system for word-wise handwritten script identification for indian postal automation. In: Proceedings of IEEE India Annual Conference, pp. 266–271 (2004)
16. Bradski, G., Kaehler, A.: Learning OpenCV. O'Reilly Media, Sebastopol (2008)

Offline Writer Identification from Isolated Characters Using Textural Features

Chayan Halder, Sk. Md. Obaidullah and Kaushik Roy

Abstract Study on behavioural biometric has gained renewed interest from researchers in recent years. Writer identification and verification is one of the areas that has promising prospect in real-life applications like forensic, security, access control, HOCR (Handwritten Optical Character Recognizer), etc. We could not find any complete system for writer identification/verification on Indic scripts including Bangla. In this proposed method, we have modified and evaluated the performance of FFT (Fast Fourier Transform), GLCM (Gray-Level Co-occurrence Matrix), DCT (Discrete Cosine Transform) on our general unconstrained Bangla character database. The database is a collection of total 53250 Bangla characters (38250 alphabets + 7500 Bangla numerals + 7500 Bangla vowel modifiers) from 150 writers with 5 sets from each writer. Modification on FFT, GLCM and DCT to use as textural features and combination of those features produces promising results. The results show that our method is comparable with other available works and capable of handling large volume of data.

Keywords Individuality of handwriting · Writer identification · Bangla handwriting analysis · FFT · GLCM · DCT · Textural features · LIBLINEAR

1 Introduction

The biometric identification has gained renewed interest in research area due to the fact that it produces extremely accurate results and secured access to data if developed properly. It also has the advantage over password protection security system

C. Halder (✉) · K. Roy
Department of Computer Science, West Bengal State University, Kolkata 126, West Bengal, India
e-mail: chayan.halderz@gmail.com

K. Roy
e-mail: kaushik.mrg@gmail.com

Sk. Md. Obaidullah
Department of Computer Science & Engineering, Aliah University, Kolkata, West Bengal, India
e-mail: sk.obaidullah@gmail.com

S. Das et al. (eds.), *Proceedings of the 4th International Conference on Frontiers in Intelligent Computing: Theory and Applications (FICTA) 2015*, Advances in Intelligent Systems and Computing 404, DOI 10.1007/978-81-322-2695-6_20

where users do not have to remember, write down, worried about forgetting or losing the password as they themselves are the security key. Identification and verification of writer from handwriting is a technique belongs to behavioural biometric analysis as it is based on the pattern of writing that the writer has learnt over time. It has a wide variety of prospects in different fields like forensic, graphology, financial institutions, security, criminal justice, access control, archaeology (in case of identifying writer of ancient documents) and HOCR (Handwritten Optical Character Recognizer). Writer identification and verification is an area of research in document level image processing of computer vision and pattern recognition field. Researchers have been participating in this area for more than three decades [1–3], but in recent times researchers gained a renewed interest in this area [4–18]. There are various works available on literature mostly based on Roman and non-Indic scripts [4–14]. Some insights of these works can be found in [17].

In recent years, there is a trend to work on multiple scripts for writer identification and verification but most of them are on non-Indic scripts. This trend can be seen in [12–14]. In [12], Fiel and Sablatnig have worked on English and Dutch languages. They have used scale invariant feature transform (SIFT) to reduce the negative effects of binarization and K-means clustering for writer retrieval and K-NN classifier for writer identification and achieved highest accuracy of 98.9 % for both writer retrieval and writer identification on IAM database (1539 documents form 657 writers) and TrigraphSlant dataset (188 documents by 47 writers). Djeddi et al. [13] have worked on ICFHR 2012 Latin/Greek database (126 Greek writers; each contributing two in Greek and two in English). They have computed probability distributions of gray-level run-length, edge-hinge features, edge-direction, combination of codebook and visual features, extracted from chain code and polygonized representation of contours and autoregressive (AR) coefficients features. The K-NN and SVM classifiers are used in their work to get highest accuracy of 92.06 % for only Greek, 83.33 % for only English and 73.41 % for Greek + English. Jain and Doermann in [14] worked with the IAM, ICFHR 2012 Writer Id dataset, DARPA MADCAT databases to extract Connected Component, Vertical cut, Seam cut and Contour gradient features from handwritten samples of English, Greek and Arabic languages. They have used K-NN classifier to get highest accuracy of 96.5 %, 98 % for English on IAM and ICFHR datasets, respectively, and 97.5 and 87.5 % for Greek and Arabic, respectively. Although there is not much system on writer identification based on Indic scripts, few scattered works by various researchers are available in literature. Details on these can be found in [18]. In [15], Halder et al. have proposed writer identification on isolated Devanagari characters using a database of 250 handwritten documents of isolated characters from 50 writers. Using 64 dimensional chain code feature and LIBLINEAR classifier, they have achieved 99.12 % accuracy.

As far as our knowledge of literature, there is no complete system on writer identification and verification based on Bangla script. Being the second most popular language in India and sixth most popular language in the world in terms of population, the work on writer identification based on Bangla handwriting has not gained so much attention due to the fact that there is no such standard dataset of Bangla handwritten documents with author identity. Only some bits and pieces of research have been carried out on Bangla handwriting based on some locally collected very small dataset. The 400 dimensional gradient features with SVM classifier is used by Chanda et al. [16] to developed text-independent writer identification method on Bangla and achieve accuracy of 95.19 % for the same. Halder and Roy in [17] used only isolated Bangla numerals for writer identification from 450 documents of 90 writers using 400 and 64 dimensional features and LIBLINEAR and MLP classifiers. Numeral 5 proved to be most individual with accuracy of 43.41 % and highest writer identification accuracy of 97.07 % has been achieved. In [18], Halder and Roy worked on all isolated Bangla characters (alphabets + numerals + vowel modifiers) from same set of writers as the work of [17]. The 400 and 64 dimensional features have been used along with LIBLINEAR and MLP classifiers to get highest individuality of 55.85 % for the character GA ("গ") and got writer identification accuracy of 99.75 %.

The variation within a writers handwritings is known as intra-writer variation, and the variation between handwritings of two different writers is called inter-writer variation. So, during the analysis of handwriting for identification of writers, there are two major points need to be considered: identification of inter-writer variation and at the same time minimizing the effect of intra-writer variation. This will help the identification of writers more accurately. The intra- and inter-writer variation can be seen in Fig. 1a–c where single isolated characters from three different writers are shown in (a), the intra-writer variation of the three different writers is shown in (b), and in (c) the inter-writer variation is shown. In Fig. 1b five samples of a single character of same writer is superimposed to create a single character which shows us the visual intra-writer variation, whereas in Fig. 1c for a single character five samples each from ten different writers are superimposed to create a single character which gives us the inter-writer variation. The detail understanding of the individuality of Bangla characters can be found in [18]. In our current experiment, we emphasize more on this textural difference that can be visualized. The goal of this experiment is to find out the visual variations with precision using automated techniques for which we have opted for textural features. We are also planing to do similar experiments using these type of textural features on other Brahmi scripts like: Devanagari, Gujarati, Gurmukhi etc.

The rest of the paper are outlined as follows: Sect. 2 describes about the experimental dataset and preprocessing techniques. In Sects. 3 and 4, textural feature extraction techniques and the classification techniques with experimental design are described. Section 5 describes experimental results with analysis and at last the paper will be concluded in Sect. 6.

2 Dataset and Preprocessing

The current experiment is conducted using a dataset of 750 handwritten documents of isolated Bangla characters from 150 writers consisting of total 53250 Bangla characters (38250 alphabets, 7500 Bangla numerals and 7500 Bangla vowel modifiers).

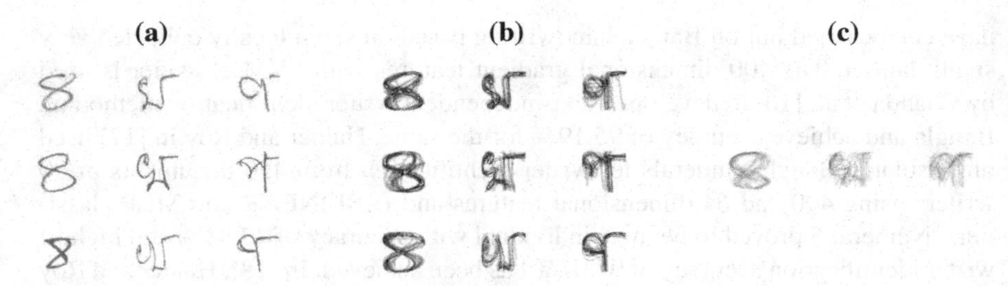

Fig. 1 **a** Example of three different isolated characters from three different writers, **b** Example of three superimposed characters of each writer from the same three writers, **c** Example of three superimposed characters from ten different writers including the three writer from (**a**) and (**b**)

The documents were scanned in 300 dpi and digitized in gray-scale mode. The details about the data collection, type of data and digitization of the raw collected handwritten data can be found in [18].

In our proposed work, the isolated images are not directly used for feature extraction. The characters of each writers are superimposed onto each other to create a single character image of each character category belong to that writer only. First, the bounding box of isolated gray character images are calculated. After applying global binarization, the images are normalized to fixed 128 × 128 pixels size. Next, the normalized 128 × 128 binary images are projected into a white 128 × 128 image. In the projection technique, for each object pixel of the original image, corresponding pixel of the white 128 × 128 image has been decremented by a fixed value that is calculated using the formula (1). Thus a single gray character image is created by capturing the writing variation. This procedure is repeated for each character category writer wise. Figure 1b shows some sample superimposed images that are used for feature extraction.

$$\left\lceil \frac{N}{s} \right\rceil \tag{1}$$

where N: Total number of gray levels.
Where s: Total number of image samples that are used to create a superimposed image.

3 Feature Extraction

In this proposed work, the textural features of the superimposed images are extracted for individuality of handwriting and writer identification by distinguishing between the intra-writer and inter-writer variations. Here the FFT (Fast Fourier Transform) and DCT (Discrete Cosine Transform) are modified along with GLCM (Gray-Level Co-occurrence Matrix) to extract the textural features. In general, the FFT and DCT are used to capture the varying frequency of an image but in this experiment, the varying gray-level intensity (gray level frequency) values i.e. the textural differences are calculated by modifying FFT, DCT and GLCM. MFFT (Modified Fast Fourier Transform), MGLCM (Modified Gray Level Co-occurrence Matrix) and MDCT (Modified Discrete Cosine Transform) are used to get textural features of the images. The MFFT is used to get the textural and structural features of the images by the means of phase spectrum. The MGLCM is used to get the local variation among gray-level pixel values, probability of occurrence, uniformity and closeness of the distribution of the gray-level pixel values. The MDCT is used to get the similar textural measures like MFFT but with less computational cost.

3.1 *MFFT (Modified Fast Fourier Transform)*

The Fourier transform has many different variations. Among those, the Discrete Fourier Transform (DFT) is very widely used. The Fast Fourier Transform (FFT) is a quicker version of DFT where the computational overhead is lower compared to DFT. Using FFT, pixel values of an image along a row or column can be transformed into a periodic sequence of complex number. The 2D FFT function computes transformation of a given 2D image of length *MXN* by:

$$F(x, y) = \sum_{m=0}^{M-1} \sum_{n=0}^{N-1} f(m, n) e^{-j2\pi(x\frac{m}{M}+y\frac{n}{N})} \quad (2)$$

In the current method, the modified FFT feature is calculated on the superimposed images using the following steps: First, the 128 × 128 dimensional feature vector has been computed using FFT algorithm then using Gaussian filter and equation (3) MFFT is calculated on the images to get 64 dimension features.

$$F(x) = \left\{ \frac{f(m)}{M}, F(x) \leq 1 \right\} \quad (3)$$

where $M = max(f(m)), f(m) = \sum_{n=1}^{N} m_n$ and N: Total number of feature dimension for a single column of the feature set

3.2 *MGLCM (Modified Gray-Level Co-occurrence Matrix)*

The GLCM (Gray-Level Co-occurrence Matrix) is a statistical calculation of how often different combination of gray-level pixels occurs in an image. It has been the workhorse for textural analysis of images since the inception of the technique. GLCM matrix describes the frequency of occurrence of one gray level with another gray

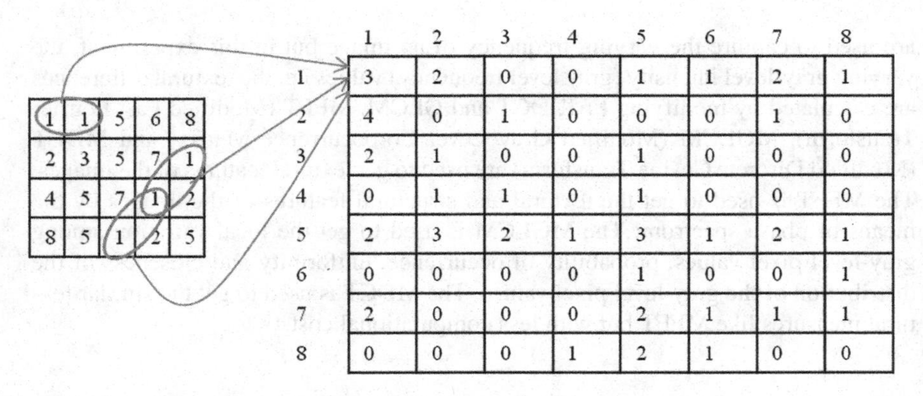

Fig. 2 GLCM calculation for both type of pixel pairs in all four directions

level in a linear relationship within a defined area. The co-occurrence matrix is computed based on two parameters, which are the relative distance between the pixel pair d measured in pixel number and their relative orientation φ. Normally, φ is quantized in four directions ($0^0, 45^0, 90^0$ and 135^0). A sample GLCM calculation is shown in Fig. 2.

In the current approach, the GLCM is calculated with two pixel pairs in all four directions considering both type of pairs like $P[i, j]$ and $P[j, i]$. Then using the equation (3), the MGLCM feature vectors are calculated to get 8 dimensional features.

3.3 MDCT (Modified Discrete Cosine Transform)

In terms of image analysis, Discrete Cosine Transform (DCT) is one of the widely used transformation technique. The approach involves taking the transformation of the image as a whole and separating the relevant coefficients. The DCT uses mainly three basic frequency components namely low, middle and high to get detail information from an image. The DCT is very similar to FFT. The main difference between a DCT and a DFT is that the DCT uses only cosine functions, while the DFT uses both sin and cosine. The 2D DCT function computes the transformation of a given *MXN* image by:

$$F(x,y) = \left(\frac{2}{M}\right)^{\frac{1}{2}} \left(\frac{2}{N}\right)^{\frac{1}{2}} \sum_{m=0}^{M-1} \sum_{n=0}^{N-1} \Lambda(m)\Lambda(n)$$

$$\cos\left[\frac{\pi x}{2M}(2m+1)\right] \cos\left[\frac{\pi y}{2N}(2n+1)\right] f(m,n) \tag{4}$$

where $f(m, n)$: The gray intensity value of the pixel (m, n).

To get the 64 dimensional MDCT features of the superimposed images, first, the 128×128 dimensional feature vector has been computed using DCT then Gaussian filter and equation (3) is used.

4 Classification Techniques and Experimental Design

In the current experiment, the main concentration is on the extraction and evaluation of textural features on Bangla isolated characters to analyse individuality of characters and writer identification. The LIbLINEAR classifier is used for classification. It has been observed from [16–18] that the classifiers like LIBSVM, LIBLINEAR and MLP are usually give better performance for handwritten Bangla script. The LIBLINEAR is chosen here over LIBSVM and MLP due to the fact that it has lower computational cost and gives better results in terms of classification.

4.1 LIBLINEAR

The support vector machine (SVM) is basically a linear classifier. The LIBSVM supports many kernel transformations to turn a nonlinear problem into linear problem. In LIBSVM, the SMO-type algorithm is used, which works both for Kernel and linear SVM, but the main drawback of this is that it has a complexity of $O(n^2)$ or $O(n^3)$. On the other hand, the LIBLINEAR is implemented using the SVM but with some modification like it does not support kernel transformation so the complexity become $O(n)$ (n is the number of samples in the training set). The LIBLINEAR is suitable for most cases when the amount of data with instances or features to be classified is large. The SVM Type parameter L2-Loss support vector machine (dual) is used for the current work. Details of LIBLINEAR can be found in [19].

For classification, out of 5 sets from each writer 4 sets are used for training and 1 for testing. But the training and testing sets are selected at random basis, i.e. random 5 fold cross validation is used so that the classification has no biased result. Out of total 53250 Bangla characters randomly selected 42600 characters are used for training and 10650 are used for testing during a single fold.

5 Experimental Results and Analysis

The proposed experiment has been carried out on total 53250 Bangla characters from 150 writers taking five samples of isolated characters from each writer. The textural features MFFT, MGLCM and MDCT are used for feature extraction, and LIBLINEAR has been used for classification in case of both individuality calculation and writer identification with five-fold cross validation. In this experiment, the

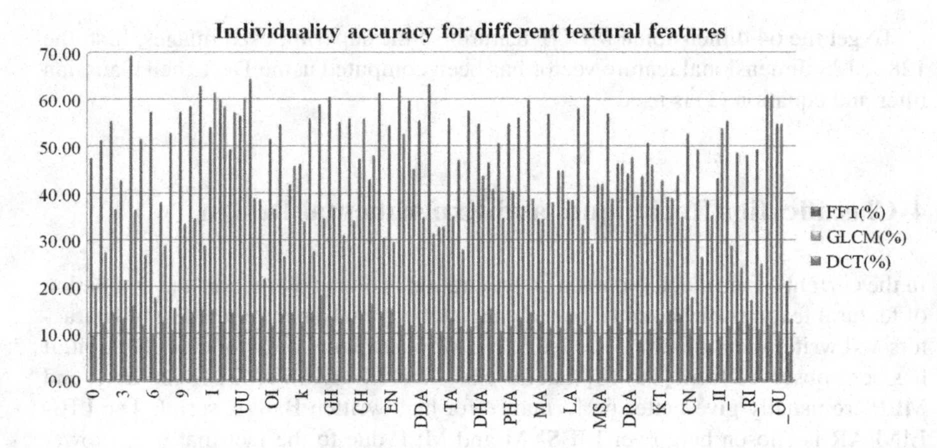

Fig. 3 Individuality of all characters for MFFT, MGLCM, MDCT

individuality is calculated for each feature and also for the combination of features. The writer identification is also calculated in the same manner. It has been found that the MFFT feature is the most effective in terms of accuracy of individuality and writer identification.

5.1 Individuality Accuracy

The individuality accuracy of each character has been measured by finding the writer identification accuracy of a single character. From the classification result of LIBLINEAR, it has been found that the MFFT feature has most impact on individuality among the three features where MGLCM has the least. The feature set of numeral 4 extracted by MFFT scored the highest individuality accuracy of 69.44 % among all characters. Figure 3 shows the individuality results of all characters based on three different features and LIBLINEAR classifier.

5.2 Writer Identification Accuracy

In this phase, the objective is to identify an unknown writer among the writers in the reference set. To evaluate the system out of 750 documents from 150 writers, five-fold cross validation technique has been used. It has also been consciously maintained that no two folds contain exactly the same set of writers. Every document in training and testing set contains 71 isolated characters each. During the training and testing procedure, the textural features are extracted and feature sets are created and then the LIBLINEAR classifier is used to achieve writer identification. The writer

Table 1 Writer identification results

Feature	Dimension	Accuracy (%)
Modified fast fourier transform (MFFT)	64	97.84
Modified gray-level co-occurrence matrix (MGLCM)	8	76.86
Modified discrete cosine transform (MDCT)	64	96.77
MFFT+MGLCM+MDCT	136	98.62

Table 2 Comparison of methods

Method	Writers (data details)	Samples/ writer	Feature(s)	Top individuality rate (%)	Top writer identification rate (%)
Halder and Roy	90 (Numerals)	5	400 and 64 dimensional	43.41	97.07
Halder and Roy	90 (Alpha-numeric)	5	400 and 64 dimensional	55.85	99.75
Current method	150 (Alpha-numeric)	5	MFFT, MGLCM and MDCT	69.44	98.62

identification accuracies of all the features and combination of features are produced in Table 1.

5.3 Comparison of Results

To the extent of our knowledge, though individuality calculation and writer identification on isolated Bangla characters is present in literature but these kind of textural features are never attempted for the writer identification on Bangla script. The existing works that can be compared regarding the database used include the works of Halder and Roy [17, 18]. Table 2 shows the comparison of different methods. Although the writer identification rate is 1.13 % less compared to the work of [18] but the number of writers are almost twice of that work.

6 Conclusion

The lack of standard database and inadequate writer identification methods based on Bangla script motivate us to work in this field. In this work, some modifications on the FFT, DCT and GLCM are carried out to produce textural features that reflect the individuality of handwriting and writer identification on isolated Bangla characters using LIBLINEAR classifier. We have achieved writer identification rate of 98.62 % for 150 writers. In comparison with other approaches, our method produces promising results. The number of writers that are used also indicates that our approach has practical feasibility in real-life applications and holds the capability of dealing with large handwritten datasets.

Future plan includes incrementing the size of the database both in terms of writer as well as the number of samples per writer for both Bangla isolated chars and unconstrained text which will produce a standard Bangla database for writer identification. Combination of features, classifiers and some script-dependent features will be considered for writer identification on isolated characters, and other textural features can also be applied in combination with the current features. We will perform this kind of experiment on unconstrained text for writer identification. It can be considered that this type of research can also be performed on other Brahmi scripts like: Devanagari, Gujarati, Gurmukhi etc., and we are looking forward to work with those scripts in future.

Acknowledgments One of the author would like to thank Department of Science and Technology (DST) for support in the form of INSPIRE fellowship.

References

1. Azari, B.: Automatic handwriting identification based on the external properties of the samples. IEEE Trans. Syst. Man Cybern. **13**(4), 635–642 (1983)
2. Zimmerman, K., Varady, M.: Handwriter identification from one-bit quantized pressure patterns. Pattern Recognit. **18**(1), 63–72 (1985)
3. Plamondon, R., Lorette, G.: Automatic signature verification and writer identification—the state of the art. Pattern Recognit. **22**(2), 107–131 (1989)
4. Said, H.E.S., Tan, T.N., Baker, K.D.: Personal identification based on handwriting. Pattern Recognit. **33**(1), 149–160 (2000)
5. Marti, U.V., Messerli, R., Bunke, H.: Writer identification using text line based features. In: 6th ICDAR, pp. 101–105 (2001)
6. Srihari, S.N., Cha, S.H., Arora, H., Lee, S.: Individuality of handwriting. J. Forensic Sci. **47**(4), 1–17 (2002)
7. Bensefia, A., Paquet, T., Heutte, L.: Information retrieval based writer identification. In: 7th ICDAR, pp. 946–950 (2003)
8. Bulacu, M., Schomaker, L.: Text-independent writer identification and verification using textural and allographic features. IEEE Trans. Pattern Anal. Mach. Intell. **29**(4), 701–717 (2007)
9. Bulacu, M., Schomaker, L., Brink, A.: Text-independent writer identification and verification on offline Arabic handwriting. In: 9th ICDAR, pp. 769–773 (2007)
10. Al-Maadeed, S., Mohammed, E., AlKassis, D., Al-Muslih, F.: Writer identification using edge-based directional probability distribution features for Arabic words. In: AICCSA, pp. 582–590 (2008)
11. He, Z., You, X., Tang, Y.Y.: Writer identification of Chinese handwriting documents using hidden Markov tree model. IEEE Trans. Pattern Recognit. **41**(4), 1295–1307 (2008)
12. Fiel, S., Sablatnig, R.: Writer retrieval and writer identification using local features. In: 10th IAPR International Workshop on DAS, pp. 145–149 (2012)
13. Djeddi, C., Siddiqi, I., Souici-Meslati, L., Ennaji, A.: Text-independent writer recognition using multi-script handwritten texts. Pattern Recognit. Lett. **34**, 1196–1202 (2013)

14. Jain, R., Doermann, D.: Writer Identification using an alphabet of contour gradient descriptors. In: 12th ICDAR, pp. 550–554 (2013)
15. Halder, C., Thakur, K., Phadikar, S., Roy, K.: Writer identification from handwritten Devanagari script. In: 2nd INDIA, pp. 497–505 (2015)
16. Chanda, S., Franke, K., Pal, U., Wakabayashi, T.: Text independent writer identification for Bengali script. In: 20th ICPR, pp. 2005–2008 (2010)
17. Halder, C., Roy, K.: Individuality of isolated Bangla numerals. J. Netw. Innov. Comput. **1**, 33–42 (2013)
18. Halder, C., Roy, K.: Individuality of isolated Bangla characters. In: ICDCCom, pp. 1–6 (2014)
19. Fan, R.E., Chang, K.W., Hsieh, C.J., Wang, X.R., Lin, C.J.: Liblinear: a library for large linear classification. J. Mach. Learn. Res. **9**, 1871–1874 (2008)

PWDB_13: A Corpus of Word-Level Printed Document Images from Thirteen Official Indic Scripts

Sk. Md. Obaidullah, Chayan Halder, Nibaran Das and Kaushik Roy

Abstract We present *PWDB_13*, a Word-level printed document image corpus from thirteen official Indic scripts, which consists of 26,000 words with equal distribution of each of the thirteen script types, collected by an automated process. A realistic classification framework based on four major regions of India has been proposed which represent the work as a unique one. Benchmarking is done with respect to *PSI* or printed script identification problem as it is very relevant in multi-script scenario. The result is said to be impressive observing the volume of the corpus and intrinsic complexities of Indic scripts. *PWDB_13* will bridge the gap of unavailability of a complete document image dataset on all official Indic scripts and freely available to the researchers for noncommercial use.

Keywords Document image corpus · Printed script identification · W–R hybrid transform · MLP classifier · Benchmark result

Sk. Md. Obaidullah (✉)
Department of Computer Science & Engineering, Aliah University,
Kolkata, West Bengal, India
e-mail: sk.obaidullah@gmail.com

C. Halder · K. Roy
Department of Computer Science, West Bengal State University,
Kolkata, West Bengal, India
e-mail: chayan.halderz@gmail.com

K. Roy
e-mail: kaushik.mrg@gmail.com

N. Das
Department of Computer Science & Engineering, Jadavpur University,
Kolkata, West Bengal, India
e-mail: nibaran@gmail.com

S. Das et al. (eds.), *Proceedings of the 4th International Conference on Frontiers in Intelligent Computing: Theory and Applications (FICTA) 2015*, Advances in Intelligent Systems and Computing 404, DOI 10.1007/978-81-322-2695-6_21

1 Introduction

One challenging field of research in document image processing is optical character recognition or in short OCR. OCR algorithms are used for conversion of image to textual information. Input document images may be generated from different sources like camera, scanner, etc. Application domain of OCR technology has a wide range starting from automatic processing, archiving, indexing, and retrieval of huge volume of data to many other real-life requirements which suit toward the development of a paperless world. In general, any OCR algorithm is designed for a particular script which means that it can read characters from the specific scripts only. So for countries where multiple scripts are used simultaneously, there exists script-specific OCR. An example of such country is our subcontinent where 23 different languages are used as communication medium and 13 different scripts are used to write them [1, 2]. Now, to model the solution to the problem of OCR development in such multi-script environment, one feasible approach is to develop a preprocessor which will identify the script a priori by which the document is written and then forward it to script-specific OCR. Another issue is when a collection of different documents written by different scripts exists in a consolidated manner then to sort them script wise a manual interventions is required. One example of such environment is postal document sorting in India. Here, an optical script identification system or OSI can be proposed to automate the entire process. In the light of above two discussions, it is indeed clear that there is a pressing need of development of an automatic OSI engine for country like India.

Development of state-of-the-art techniques for script identification demands availability of standard dataset of all the official Indic scripts to train, validate, and test the system. This is the another real challenge in this field of research. A survey on the dataset development for document image processing research can be found from the work of Obaidullah et al. [3]. It can be observed that till date no dataset has been developed which covers all the thirteen official Indic scripts. In this paper, we propose PWDB_13, which is a rich corpus having 26,000 printed words from each of the thirteen official Indic scripts. A new classification framework based on four major regions of India has also been proposed and benchmark result with respect to POSI or printed optical script identification problem [4–9] has been obtained.

The rest of the paper is organized as follows: Data acquisition and representation procedure are discussed in Sect. 2. Experimental details and benchmarking are discussed in Sect. 3. Finally, conclusion and future scopes are given in Sect. 4 and references are available in the last section.

2 Data Acquisition and Representation Procedure

This section describes about the sources of data collection, different preprocessing techniques used to normalize the data as per expectation and final representation of the data.

2.1 *Preprocessing*

Collected document images are preprocessed which includes segmentation from page/block level images to Word-level images and further converting them from gray-scale to binary version. Following section discusses about the preprocessing techniques.

Segmentation into Word-level Images.
An automated word segmentation technique has been employed to extract Word-level images from the digitized images. Inter-word/line spacing is very much regular and prominent in case of printed documents when compared to handwritten documents, which helps the segmentation process. Initially, a *LSE* (Line Structuring Element) has been designed whose dimension was fixed experimentally. Then morphological dilation operation is applied using *LSE* on the complemented version of the threshold image. It will create single block for each of the word image. Then component labeling is done and word blocks are extracted applying bounding box technique on the original image file. Figure 1 shows the word segmentation techniques followed.

Gray-scale to Binary Conversion.
A two-stage-based binarization method has been used for the present work [10]. The flow diagram of the technique has shown below which itself is a self-explanatory:

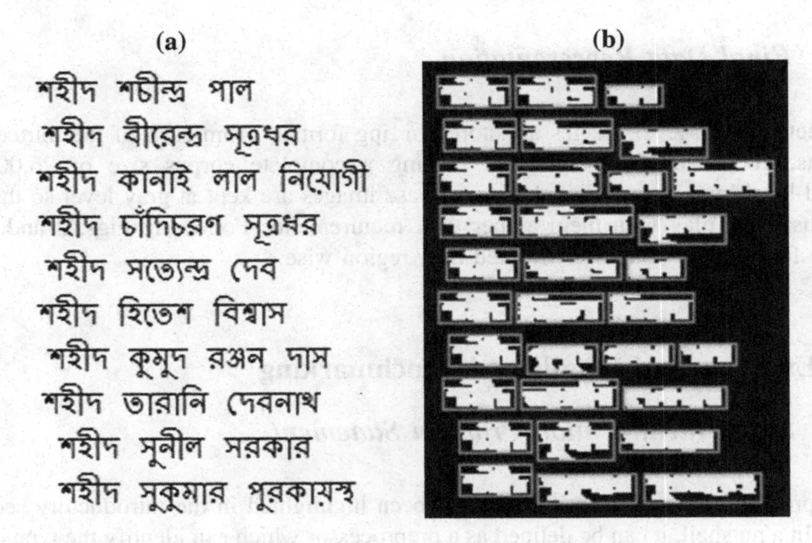

Fig. 1 **a** Original Bangla document image fragment, **b** segmented word blocks

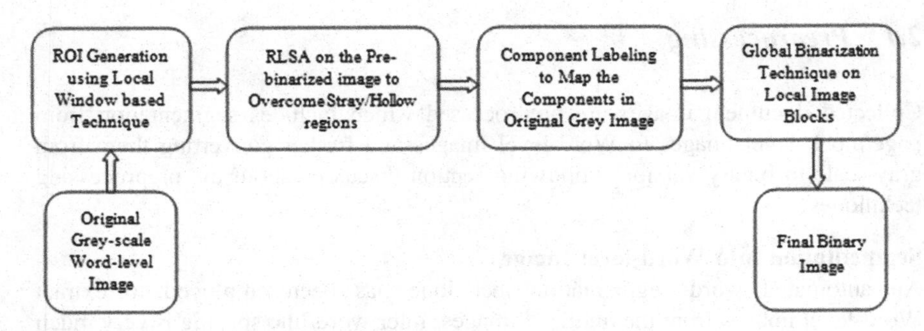

Fig. 2 Two-stage-based binarization technique

Eastern	North	South	West
পাতাবাহার	ब्राम्हीण	ಪ್ರತಿಯೊಬ್ಬ	आजकल
हालाँकि	महत्वपूर्ण	മേത്ത	વિસ્તારમાં
ସବଥାଇ	استعمال	அளவில்	ਸਮੱਸਿਆ
back	جومدری	వ్యాఖ్యల పైన	permitted

Fig. 3 Sample word-level images from different regions of India

2.2 *Final Data Representation*

Collected Word-level images are stored in .jpg format. From each of the thirteen scripts, 2000 words are collected making a complete corpus size of 26,000 Word-level images. In original corpus, these images are kept at gray level so that any user can play with them as per their requirements. Following Figs. 2 and 3 show few sample images distributed over region wise.

3 Experimental Details and Benchmarking

3.1 *Script Identification Problem Statement*

The problem of script identification has been highlighted in the introductory section. In a nutshell, it can be defined as a preprocessor which can identify the type of the script first by which the document is written before applying the script-specific *OCR*. Its relevancy and usefulness have already been discussed. All the script identification works can be classified into two broad categories, namely printed

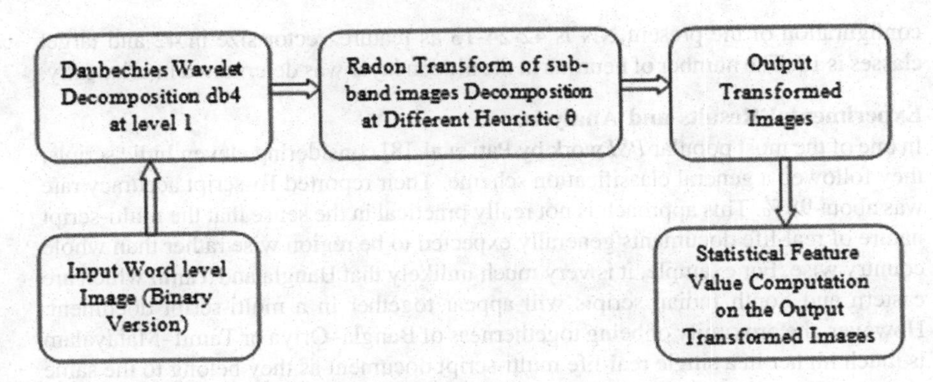

Fig. 4 Block diagram of the W–R hybrid transform feature computed

script identification of *PSI* and handwritten script identification or *HSI* based on the type of the input data. For present work, the benchmarking has been done with respect to *PSI* problem.

Computation of Feature Vector.
A multi-dimensional feature vector has been constructed by applying a novel *W–R hybrid transform*-based technique which is discussed below.

W–R Hybrid Transform.
Wavelet and Radon transforms [11–15] are two frequency-domain techniques which are hybridized for the present work. Wavelets are used successfully for multi-resolution analysis by which a signal can be decomposed into different frequencies with different resolutions. Initially, original images are transformed applying Daubechies wavelet *db4* which is capable enough to represent components of an image signal with linear and quadratic coefficients. Four sub-band images with coefficients, namely approximation coefficients (*cA*), horizontal coefficients (*cH*), vertical coefficients (*cV*), and diagonal coefficients (*cD*), have been produced at level 1 decomposition. Then radon transform is applied on each of these sub-band images at different angles $\theta = \{0°, 30°, 60°, 90°, 120°, 150°, \text{and } 180°\}$ to obtain *W–R Hybrid Transform* images. Afterward, different statistical measures, namely entropy, mean, and standard deviation, are computed on these hybrid transformed images to constitute the final feature vector. The following figure shows computation of *W–R Hybrid Transform* technique. Figure 4 presents the block diagram of the feature extraction process.

3.2 Classifier

Different well-known classifiers are tested and MLP neural network [9] has been chosen finally best on some statistical measures for the present work. The configuration of the present NN is 42-24-13 as feature vector size in 42 and target classes is 13. The number of neurons in the hidden layer was determined heuristically.

Experimental Results and Analysis.
In one of the most popular *PSI* work by Pati et al. [8] considering eleven Indic scripts, they followed a general classification scheme. Their reported Bi-script accuracy rate was about 98 %. This approach is not really practical in the sense that the multi-script nature of real-life documents generally expected to be region wise rather than whole country wise. For example, it is very much unlikely that Bangla and Tamil which are eastern and South Indian scripts will appear together in a multi-script document. However, the proximity of being togetherness of Bangla–Oriya or Tamil–Malayalam is much higher in a single real-life multi-script document as they belong to the same regions. Two scripts from different regions are structurally and visually very dissimilar, so any simple method will be able with classify them with 100 % accuracy. Our critical observation toward the work of Pati et al. is that their Bi-script result was over fitted as they had taken average of all the script combinations for eleven scripts together. That is why we have proposed a new classification framework based on four regions of India which is really a practical approach.

Following section shows performances of script identification based on different regions of India where four scripts namely Bangla, Devanagari, Oriya, and Roman are considered. Table 1 shows confusion matrix for all eastern Indian scripts and Bi-scripts identification rates have been shown in Table 2. Average Eastern Indian all-script identification rate and average Bi-script identification rate have been found to be 93.9 % and 98.32 %, respectively. In a similar pattern, North, South, and West Indian script identification performances have been shown in Tables 3, 4, 5, 6, 7 and 8, respectively.
Eastern Indic Script Group.
For North Indian region, five scripts namely Devanagari, Dogri, Kashmiri, Roman, and Urdu are considered. Average all-script and Bi-script identification rates have been found to be 88.8 and 97.57 %. Here, due to the presence of two very similar nature scripts namely Kashmiri and Urdu, the average rate has dropped notably. From Table 4, it can be found that while nine Bi-script combinations have identification rate of above 98 %, Kashmiri–Urdu combination is only 81.6 %.
North Indic Script Group.
The performance of South Indic region has been shown in Tables 5 and 6. Five scripts namely Kannada, Malayalam, Roman, Tamil, and Telugu, are considered here. Here, all-script average identification rate has been found to be 90.2 % and average Bi-script identification rate is reported as 98.23 %.
South Indic Script Group.
In West region, four scripts namely Devanagari, Gujarati, Gurumukhi, and Roman are considered. Detailed performances can be found from Tables 7 and 8. Average all-script and Bi-script identification rates have been found to be 89.4 and 97.22 %. Here, the performance of Devanagari–Gurumukhi script combination has been reduced significantly compared to others due to their structural similarity.
Western Indic Script Group.

Table 1 Confusion matrix – Eastern Indic Script Group

Script	BEN	DEV	ORY	ROM
BEN	626	46	15	9
DEV	27	646	7	22
ORY	5	0	628	4
ROM	1	24	6	654
Average accuracy rate (%): 93.9				

Table 2 Bi-script performance – Eastern Indic Script Group

Script group	AAR (%)
BEN, DEV	95.2
BEN, ORY	99.5
BEN, ROM	99
DEV, ORY	99.1
DEV, ROM	98.1
ORY, ROM	99
Average	**98.32**

Table 3 Confusion matrix – North Indic Script Group

Script	DEV	DOG	KAS	ROM	URD
DEV	667	9	4	14	13
DOG	2	676	0	0	0
KAS	7	0	509	6	119
ROM	32	6	5	620	6
URD	7	0	147	5	546
Average accuracy rate (%): 88.8					

Table 4 Bi-script performance – North Indic Script Group

Script group	AAR (%)
DEV, DOG	99.5
DEV, KAS	99.4
DEV, ROM	98.1
DEV, URD	99.1
DOG, KAS	100
DOG, ROM	99.5
DOG, URD	100
KAS, ROM	99.6
KAS, URD	81.6
ROM, URD	98.9
Average	**97.57**

Table 5 Confusion matrix – South Indic Script Group

Script	KAN	MAL	ROM	TAM	TLU
KAN	654	7	4	7	35
MAL	16	622	14	20	6
ROM	6	27	578	24	6
TAM	11	7	17	631	3
TLU	104	10	5	6	580
Average accuracy rate (%): 90.2					

Table 6 Bi-script performance – South Indic Script Group

Script group	AAR (%)
KAN, MAL	98.7
KAN, ROM	99.7
KAN, TAM	98.8
KAN, TLU	95.2
MAL, ROM	97.8
MAL, TAM	98.1
MAL, TLU	98.4
ROM, TAM	97.5
ROM, TLU	98.9
TAM, TLU	99.2
Average	**98.23**

Table 7 Confusion matrix – West Indic Script Group

Script	DEV	GUJ	GUR	ROM
DEV	590	13	82	11
GUJ	7	671	2	22
GUR	79	23	532	3
ROM	12	31	5	637
Average accuracy rate (%): 89.4				

Table 8 Bi-script performance – West Indic Script Group

Script group	AAR (%)
DEV, GUJ	98.7
DEV, GUR	91.2
DEV, ROM	98.1
GUJ, GUR	98.9
GUJ, ROM	98
GUR, ROM	98.4
Average	**97.22**

Table 9 Comparison with existing method

Method	Dataset size and No. of scripts	Feature used	Scripts	Accuracy	
Pati et al. [8]	20,000 and eleven	Gabor filter and DCT	Bangla, Roman, Devanagari, Gujarati, Kannada, Malayalam, Oriya, Gurumukhi, Tamil, Telugu, Urdu	Bi-script: 98 %	
				Eleven-script: 89 %	
Proposed method	26,000 and thirteen	W–R hybrid transform	Script region wise (total 13)	All-script	Bi-script
			Eastern	93.9	98.32
			North	88.8	97.57
			South	90.2	98.23
			West	89.4	97.22

3.3 *Comparative Study*

Table 9 shows the comparative study with existing methods. Our results are really impressive in terms of the number of scripts considered, the volume of the corpus, simplistic feature extraction technique, and overall performance rate. The present result can be considered as a benchmark one on the present dataset.

4 Conclusion and Future Scope

PWDB_13, a Word-level printed document image corpus of thirteen official Indic scripts, has been proposed here. A region-based realistic classification framework has been introduced here which represent this work as unique one. Benchmarking is done with respect to PSI problem which is very much relevant in multi-script scenario. The result is compared with existing similar kind of methodologies and shows its effectiveness in terms of performance and realistic nature. Future scope includes increasing the volume of the corpus and introduction of sophisticated techniques to handle few of the misclassification issues to some possible extent.

References

1. Obaidullah, S.M., Das, S.K., Roy, K.: A system for handwritten script identification from Indian document. J. Pattern Recognit. Res. **8**(1), 1–12 (2013)
2. Ghosh, D., Dube, T., Shivprasad, S.P.: Script recognition—a review. IEEE Trans. Pattern Anal. Mach. Intell. **32**(12), 2142–2161 (2010)

3. Obaidullah, S.M., Rahaman, Z., Das, N., Roy, K.: Development of document image database for offline handwritten Indic script identification—a state-of-the-art. Int. J. Appl. Eng. Res. **9**(20) special issue, 4625–4630, Research India Publication
4. Chaudhuri, B.B., Pal, U.: An OCR system to read two Indian language scripts: Bangla and Devanagari (Hindi). In: Proceedings of 4th International Conference on Document Analysis and Recognition, pp. 18–20. University Health Network (1997)
5. Hochberg, J., Kelly, P., Thomas, T., Kerns, L.: Automatic script identification from document images using cluster-based templates. IEEE Trans. Pattern Anal. Mach. Intell. **19**, 176–181 (1997)
6. Chaudhury, S., Harit, G., Madnani, S., Shet, R.B.: Identification of scripts of Indian languages by combining trainable classifiers. In: Proceedings of Indian Conference on Computer Vision, Graphics and Image Processing, Bangalore, India (2000)
7. Dhanya, D., Ramakrishnan, A.G., Pati, P.B.: Script identification in printed bilingual documents. Sadhana **27**(part-1), 73–82 (2002)
8. Pati, P.B., Ramakrishnan, A.G.: Word level multi-script identification. Pattern Recognit. Lett. **29**(9), 1218–1229 (2008)
9. Obaidullah, S.M., Mondal, A., Das, N., Roy, K.: Script identification from printed indian document images and performance evaluation using different classifiers. Appl. Comput. Intell. Soft Comput. **2014**(Article ID 896128), 12 (2014)
10. Roy, K., Banerjee, A., Pal, U.: A system for word-wise handwritten script identification for Indian postal automation. In: Proceedings of IEEE India Annual Conference, pp. 266–271 (2004)
11. Obaidullah, S.M., Halder, C., Das, N., Roy, K.: A corpus of word-level offline handwritten numeral images from official Indic scripts. In: International Conference on Computer and Communication Technologies. AISC Series, Springer, Hyderabad (2015)
12. Mandal, J.K., Sengupta, M.: Authentication/secret message transformation through wavelet transform based subband image coding (WTSIC). In: International Symposium on Electronic System Design 2010, pp. 225–229. Bhubaneswar, India (2010). ISBN: 978-0-7695-4294-2
13. Bhateja, V., Urooj, S., Mehrotra, R., Verma, R., Ekuakille, A.L., Verma, V.D.: A composite wavelets and morphology approach for ECG noise filtering. PReMI **2013**, 361–366 (2013)
14. Dey, N., Das, A., Chaudhuri, S.S.: Wavelet based normal and abnormal heart sound identification using spectrogram analysis. Int. J. Comput. Sci. Eng. Technol. (IJCSET) **3**(6) (2012)
15. Pardeshi, R., Chaudhury, B.B., Hangarge, M., Santosh, K.C.: Automatic handwritten Indian scripts identification. In: Proceedings of 14th International Conference on Frontiers in Handwriting Recognition, pp. 375–380 (2014)

Part VI
Applications of Metaheuristic Optimization

Application of Krill Herd Algorithm for Optimum Design of Load Frequency Controller for Multi-Area Power System Network with Generation Rate Constraint

Dipayan Guha, Provas Kumar Roy and Subrata Banerjee

Abstract In this paper, a novel biologically inspired algorithm, namely krill herd algorithm (KHA), is proposed for solving load frequency control (LFC) problem in power system. The KHA is based on the simulations of herding behavior of individual krill. Three unequal interconnected reheat thermal power plants equipped with different classical controllers are considered for simulation study and their optimum settings are determined using KHA employing integral square error-based fitness function. The appropriate value of generation rate constraint (GRC) of the steam turbine is included in the study to confirm the effectiveness of proposed method. Performances of several classical controllers are compared with their nominal results and some other recently published algorithms. Additionally, two-stage lag–lead compensator with superconducting magnetic coil is designed to improve the existing results in coordination with LFC. Finally, random pulse load perturbation is given to the system to identify the robustness of proposed controller.

1 Introduction

The main utility of a power system network is to provide continuous power supply to the customers with an acceptable quantity as well as maintained equality between total generation with load demand plus associated system losses. Mainly, two techniques are offered to reach equilibrium condition in the power system, namely (i) MW control, and (ii) MVAr control. MW power control is more significant subject to the power engineer since it is the basic governing elements of revenue and requirements.

D. Guha (✉)
Department of Electrical Engineering, Dr. B. C. Roy Engineering College, Durgapur, West Bengal, India
e-mail: guha.dipayan@yahoo.com

P.K. Roy
Department of Electrical Engineering, Jalpaiguri Government Engineering College, Jalpaiguri, West Bengal, India

S. Banerjee
Department of Electrical Engineering, NIT-Durgapur, Durgapur, West Bengal, India

S. Das et al. (eds.), *Proceedings of the 4th International Conference on Frontiers in Intelligent Computing: Theory and Applications (FICTA) 2015*, Advances in Intelligent Systems and Computing 404, DOI 10.1007/978-81-322-2695-6_22

Any sudden change in the load demand causes series deviation of area frequency and interchange tie-line power from their nominal values, thus resulting in instability to power system unit. With rapid progression of power system network, conventional control mechanism by fly-ball governor technique of alternator is no longer competent to compensate system oscillations caused by load perturbation; hence, a secondary control loop in the form of load frequency controller (LFC) is introduced in power system to damp out the system oscillations and retained system stability.

Literature review confirms that Chon [1] was initiated work on LFC for controlling bulk power transmission using tie-line bias control scheme. However, Elgerd and Fosha [2] was the pioneer of introducing optimal control theory in LFC area. Kothari et al. [3] studied two-area hydro-thermal power system using continuous–discrete controller. Different classical control strategies in LFC area are available in [4–8]. Besides classical controllers, different intelligent controllers like neural network [9, 10], genetic algorithm (GA) [11, 12], fuzzy logic [7, 8], ANFIS technique [13], etc. are also proposed to solve LFC problem. These types of intelligent controllers require deep insight knowledge of their structure, having complex dynamics, mathematically rigors which prove their insufficiency in practical implementation. On the other hand, the classical controllers require low user skills, simple structure, easily implemented, and a favorable ratio between cost and its performances that makes it for often used in industries.

Having knowledge of aforesaid discussion, this paper aims to study the dynamic behavior of three unequal reheat thermal power systems employing a novel, nature-inspired, population-based optimization technique called krill herd algorithm (KHA) considering GRC. Furthermore, the investigation is forwarded to observe the impact of different classical controllers on power system dynamics and a comparative study has been presented. To improve the dynamic stability of concerned power system, an optimal coordinated LFC–SMES controller is developed using KHA and incorporated in power system. Finally, robustness of proposed technique is established with random load perturbation.

The abbreviations used in this paper are discussed as follows: i: Subscript referred to area 'i' (1–3), P_{ri}: Rated power of area 'i' in MW, $T_{i,j}$: Synchronizing coefficient of transmission line, R_i: Governor speed regulation parameter in Hz/pu MW, T_{gi}: Steam turbine time constant of area 'i,' K_{ri}: Steam turbine reheat coefficient of area 'i,' T_{ri}: Steam turbine reheat time constant of area 'i,' T_{ti}: Steam turbine time constant of area 'i,' B_i: Frequency bias constant of area 'i,' T_{pi}: Power system time constant of area 'i,' K_{pi}: Power system gain of area 'i,' ΔP_{gi}: Incremental generation change in area 'i,' Δf_i: Incremental change in frequency of area 'i,' and $\Delta P_{tie,\,i-j}$: Incremental change in tie-line power connecting between area 'i' and area 'j.'

Rest of the paper is organized as follows: Sect. 2 describes the problem formulation followed by the selection of fitness function. Section 3 gives an outline of generation rate constraint. Section 4 offers an overview of SMES controller.

The proposed KHA is elaborated in Sect. 5. Section 6 highlights the algorithmic steps of KHA applied to LFC problem. Comparative transient responses and discussion are given in Sect. 7. Finally, Sect. 8 concludes the present study.

2 Problem Formulation

The LFC system investigated for the present study is composed of an interconnection of three unequal thermal systems (area1:2000 MW, area2:4000 MW, area3:8000 MW), which has been widely used in literature. Areas are equipped with single-stage reheat turbine. An appropriate value of GRC is also included in the study to investigate its impact on power system stability. Transfer function model of the test system is available in [14] and 1 % step load perturbation in area1 is considered for investigation. Nominal values of system parameters are taken from [14] and presented in Table 1. Different classical controllers like integral (I), proportional-integral (PI), proportional-integral-derivative (PID), integral double derivative (IDD), proportional-integral-double derivative (PIDD) controllers are optimally designed using KHA employing integral square error-based fitness function, defined in (1). The T.F models of IDD and PIDD controllers are defined by (2) and (3), respectively.

$$J_{ISE} = \int_{t=0}^{T_{Simulation}} (ACE_i)^2.dt = \int_{t=0}^{T_{Simulation}} \left(B_i \Delta f_i + \Delta P_{tie,i,j}\right)^2 dt \quad (1)$$

$$G_C(s) = \frac{K_i}{s} + K_{DD} s^2 \quad (2)$$

$$G_C(s) = K_p + \frac{K_i}{s} + K_{DD} s^2 \quad (3)$$

In (1), ACE stands for area control error which is treated as control output of LFC. ACE represents the deficiency or excess of power generation corresponding to the load demand; mathematically, it is defined in (4).

$$B_i^* \Delta(f_i) + \Delta(P_{tie,\,i,j}) \quad (4)$$

Table 1 Nominal values of system parameters

Parameter	Value	Parameter	Value	Parameter	Value	Parameter	Value
P_{r1}	2000 MW	P_{r2}	4000 MW	P_{r3}	8000 MW	T_{sg}	10 s
T_{sg}	0.08 s	$T_{i,j}$	0.0866 s	T_t	0.3 s	a_{12}	−0.5
T_{PS}	20 s	K_{PS}	120	K_r	0.5	a_{23}	−0.5
R_i	2.4 Hz/pu MW	B_i	0.425	f	60 Hz	a_{13}	−0.25

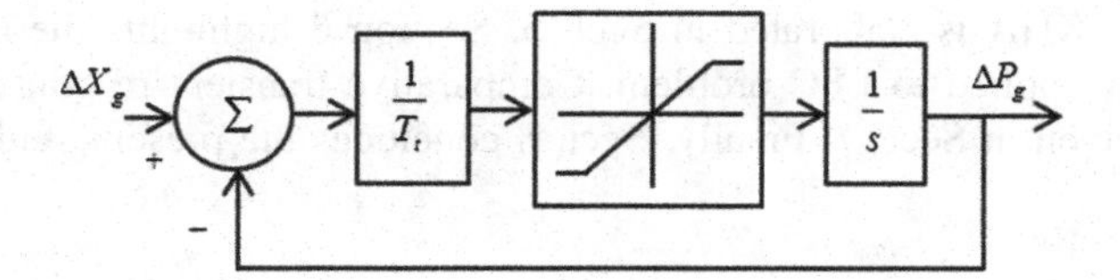

Fig. 1 Representation of GRC with steam turbine

3 Generation Rate Constraint (GRC)

GRC is mainly produced by the mechanical and thermodynamics constraint of steam turbine. It is represented by a limiter with steam turbine as shown in Fig. 1. It is quite realistic to consider GRC in LFC study since output of power generation is always changed at a maximum specified rate. If this constraint is not measured, then power system unit experiences undesirable large momentary disturbances, thus resulting wear and tear of controller. The GRC value considered for present study is 0.1 p.u./min [15].

4 Superconducting Magnetic Energy Storage (SMES)

The power circuit diagram of SMES is shown in Fig. 2 [16]. It mainly consists of (i) superconducting magnetic coil, and (ii) 12-pulse bridge converter connected with 3-phase power grid by Δ-Y/Y-Y step-down transformers. The flow of current through the coil is maintained unidirectional so that voltage appearing across the

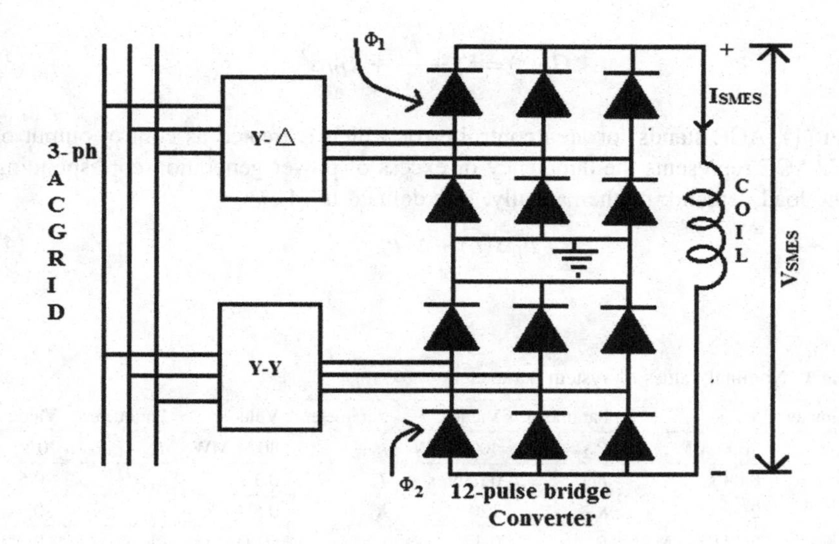

Fig. 2 Power circuit diagram of SMES unit

coil can be altered by controlling firing angle of power circuit converter. The DC voltage appears across the coil, current through the coil, and energy stored in the coil are defined in (5), (6), and (7), respectively.

$$V_{SMES} = V_{\max,0}(Cos\alpha_1 + Cos\alpha_2) \tag{5}$$

$$I_{SMES} = \frac{1}{L_{SMES}}\int_{t_0}^{t} V_{SMES}d\tau + I_{\max,0} \tag{6}$$

$$W_{SMES} = 0.5L_{SMES}(I_{SMES})^2 \tag{7}$$

Once the magnetic coil is charged to its rated value, it is ready for load frequency control. Change of frequency or ACE is sensed and used to control the voltage across the coil by controlling firing angles Φ_1 and Φ_2. For the present study, proposed coordinated LFC–SMES controller is placed near at area1 and area3, and frequency deviation of respective area is considered as an input to coordinated controller.

5 Krill Herd Algorithm (KHA)

In the recent time, different nature-inspired optimization algorithms are widely employed to solve different complex and nonlinear real-time problems. KHA is a newly developed optimization method based on life cycles and herding mechanism of krill's algorithm. KHA is developed by Gandomi et al. [17] in 2012 to solve some benchmark problem. The significant characteristic of krill is that they can form large swarms. When predators attack krill swarm, the krill density is drastically reduced. After attacking, increasing krill density and reaching the food location are defined as the initialization phase in KHA. The time-dependent position of individual krill in 2D solution space can be identified by forming the following three steps: (a) movement induced by other krill, (b) foraging motion, and (c) random diffusion. The generalized Lagrangian model in n-dimensional solution space is defined in (8):

$$\frac{dX_i}{dt} = N_i + F_i + D_i \tag{8}$$

where N_i is the motion induced by other krills, F_i is the foraging motion, and D_i is the physical diffusion. The movement of individual krill is obtained using (9):

$$N_i^{new} = N_i^{\max}\left(\alpha_i^{local} + \alpha_i^{t\arg et}\right) + w_n N_i^{old} \tag{9}$$

where $N_i^{\max}$ is the maximum induced speed, w_n is the inertia weight selected between [0, 1], and N_i^{old} is the last induced motion.

$$\alpha_i = \sum_{j=1}^{NN} \hat{K}_{i,j} \hat{X}_{i,j}, \quad \hat{X}_{i,j} = \frac{X_j - X_i}{\|X_j - X_i\| + \varepsilon}, \quad \hat{K}_{i,j} = \frac{K_i - K_j}{K^{worst} - K^{best}} \tag{10}$$

where K^{worst} and K^{best} are the worst and best fitness value obtained so far, respectively. K_i is the objective function value of ith − krill individual, K_j is the fitness of jth − neighbor, X is the related positions, and NN is the number of neighbor. The sensing distance $(d_{s,i})$ for individual krill is obtained using (11):

$$d_{s,i} = \frac{1}{5N} \sum_{j=1}^{N} \|X_i - X_j\| \tag{11}$$

where N is the number of individual krill. The global optimum solution is computed using (12):

$$\alpha_i^{target} = 2\left(rand + \frac{l}{l_{\max}}\right) K_{i,best} X_{i,best} \tag{12}$$

where l is the current iteration and $l_{\max}$ is the maximum iteration. The krill motion is consists of foraging motion, motion induced by the other krill, and physical diffusion. The formulation of foraging motion is based on two main parameters, i.e., food location and past experience about food location [18]. Foraging motion of ith − krill individual is obtained as follows:

$$F_i = 0.02\left(2(1 - l/-0ptl_{\max}) K_i^{food} X_i^{food} + \beta_i^{best}\right) + w_f F_i^{old} \tag{13}$$

where w_f is the inertia weight, F_i^{old} is the last foraging motion, and β_i^{best} is the effect of best krill found in the population so far. Physical diffusion of krill individual is a random process and can be generated using (15):

$$X^{food} = \sum_{i=1}^{N} \frac{X_i}{K_i} / -0pt \sum_{i=1}^{N} \frac{1}{K_i} \tag{14}$$

$$D_i = D^{\max} \delta \tag{15}$$

where $D^{\max}$ is the diffusion speed and δ is a random directional vector. The above-defined motions frequently change position of krill toward the best fitness solution. The position of individual krill in the time interval $[t, t + \Delta t]$ is defined as

$$X_i(t + \Delta t) = X_i(t) + \Delta t \frac{dX_i}{dt} \tag{16}$$

6 KHA Applied to LFC Problem

In the present study, the proposed KHA is successfully designed and implemented to find an effective solution of LFC problem. The algorithmic steps of KHA for solving LFC problem are enumerated as follows:

Step 1. Initialize the input parameters of KHA, i.e., $N_{\max} = 0.01, w_n = 0.8, w_f = 0.9$, $V_f = 0.02$, $D^{\max} = 0.06$, number of krill $(n_p) = 50$ (population size).
Step 2. Randomly generate initial population (krill position) within the defined search space. In LFC study, this step resembles initialization of control variables, i.e., controller gains and frequency stabilizer gains.
Step 3. Evaluate fitness value of individual krill using (1). Based on the fitness value, filter out some elite solutions and non-elite solutions are updated by the proposed method.
Step 4. Update the current krill's motion using (9), (13), and (15).
Step 5. Modify the position of individual krills according to (16) and use current population for next step of generation.
Step 6. Increment the generation count as $t = t + 1$ and check whether controlled variables lie within the defined limit or not. If present fitness value is greater than maximum value, then keep maximum value as present solution otherwise go with present value. Similarly, if present fitness value is less than minimum value, then keep minimum value and evaluate objective function using (1).
Step 7. Sort current solution from best to worst and replace worst values by the new modified set.
Step 8. Go to Step 2 until termination criterion is fulfilled.
Step 9. Use optimal setting of control variables evaluated in step 7 and plot transient responses to calculate typical system specifications

7 Results and Discussion

The main purpose to perform this simulation study is to test the effectiveness of proposed KHA in LFC area. A widely employed test system, namely three unequal-area thermal power plants with and without GRC, is considered for design and analysis. The program was written in R2009a MATLAB and executed on personal computer having 2.4 GHz core i3 processor with 2 GB RAM. 1 % step load perturbation (SLP) in area1 is considered to investigate the dynamic stability of concerned power system under normal and disturbed conditions. The optimum values of proposed controllers are determined using KHA employing ISE-based fitness function and at the end of optimization, the controller gains and minimum fitness values are given in Table 2. Transient responses of concerned power system without GRC are depicted in Fig. 3. Figure 4 demonstrates the frequency and

Table 2 Optimum values of different controllers

Types of controllers	Optimal gains of controllers (− *ve*)												ISE value ($\times 10^{-4}$)
	K_{i1}	K_{i2}	K_{i3}	K_{p1}	K_{p2}	K_{p3}	K_{d1}	K_{d2}	K_{d3}	K_{DD1}	K_{DD2}	K_{DD3}	
I	0.655	0.004	0.002	–	–	–	–	–	–	–	–	–	2.9519
PI	0.745	0.017	0.016	0.066	0.932	0.991	–	–	–	–	–	–	2.6408
PID	0.998	0.027	0.011	0.984	0.014	0.511	0.996	0.790	0.946	–	–	–	**0.5683**
IDD	0.788	0.852	0.014	–	–	–	–	–	–	0.173	0.004	0.004	2.3789
PIDD	0.868	0.611	0.967	0.362	0.004	0.002	–	–	–	0.999	0.761	0.979	1.2733

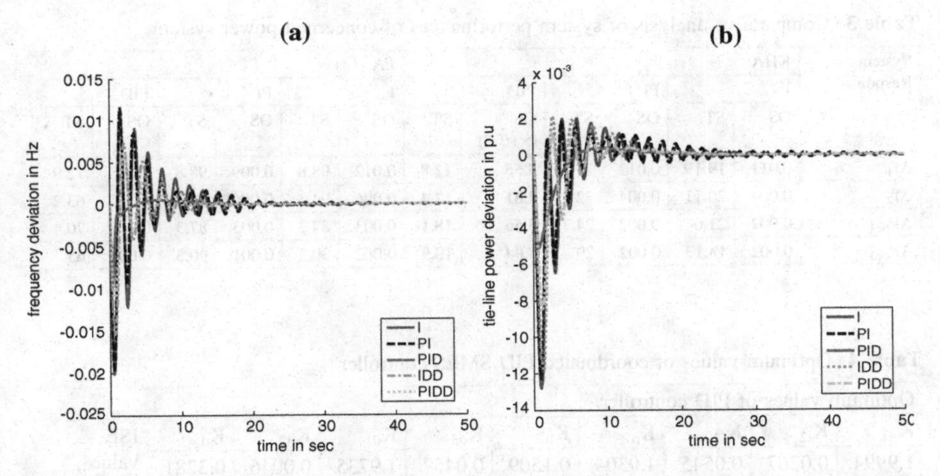

Fig. 3 Output responses of test system without GRC **a** changes in frequency, **b** changes in tie-line power

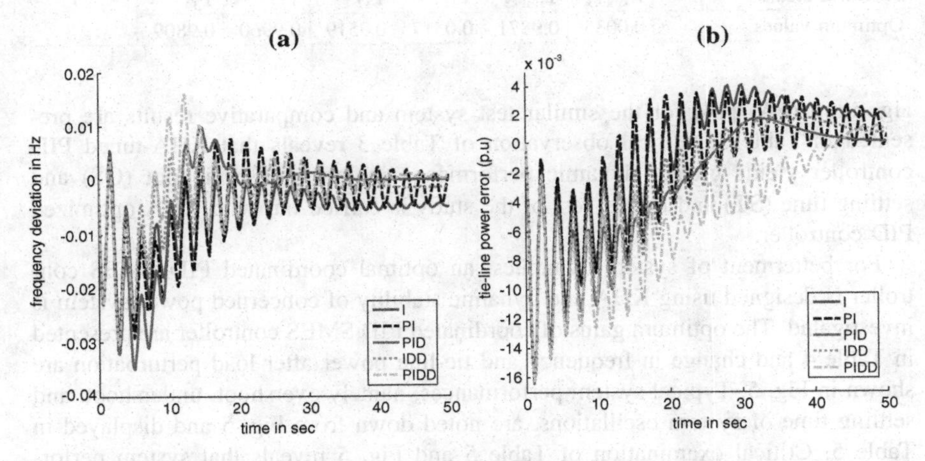

Fig. 4 Output response of test system with GRC **a** frequency error, **b** tie-line power error

tie-line power deviation caused by sudden load perturbation in the presence of GRC. It is seen from Table 2 that proposed KHA-based PID controller finds minimum fitness value **(ISE = 0.5683)** compared to other controller structure as listed in Table 2. Furthermore, it is noted from Figs. 3 and 4 that transient performance of concerned power system is improved remarkably with KHA-tuned PID controller compared to others. Hence, it can be concluded that KHA-tuned PID controller outperforms other controller as listed in Table 2.

To establish the superiority of proposed KHA, the numerical simulation results obtained with KHA are compared with those results obtained using firefly

Table 3 Comparative analysis of system performances of concerned power system

System Responses	KHA						FA [14]					
	I		PI		PID		I		PI		PID	
	OS	ST	OS	ST	OS ($\times 10^{-4}$)	ST	OS	ST	OS	ST	OS	ST
Δf_1	0.011	19.19	0.012	21	**5.55**	**12.8**	0.012	98.8	0.009	98.8	0.008	72.9
Δf_2	0.004	20.11	0.004	32	**4.81**	**17.4**	0.006	94.3	0.006	88.2	0.005	63.2
ΔP_{tie1}	0.002	21.6	0.002	24.7	**3.66**	**18.1**	0.003	87.3	0.003	87.3	0.002	70.4
ΔP_{tie3}	0.002	18.13	0.002	29	**8.04**	**18.9**	0.002	90.3	0.001	90.3	0.001	65

Table 4 Optimum values of coordinated PID SMES controller

Optimum values of PID controller									ISE value
K_{i1}	K_{i2}	K_{i3}	K_{p1}	K_{p2}	K_{p3}	K_{d1}	K_{d2}	K_{d3}	
1.9994	0.0267	0.0545	1.9304	0.1509	0.0457	1.9735	0.0046	0.3281	
SMES in area-1			K_{smes1}	T_{smes1}	T_{11}	T_{12}	T_{13}	T_{14}	4.9937×10^{-5}
Optimum values			0.9525	0.5082	0.9558	0.6906	0.9983	0.6132	
SMES in area-3			K_{smes3}	T_{smes3}	T_{31}	T_{32}	T_{33}	T_{34}	
Optimum values			0.0035	0.9871	0.0317	0.9519	0.9960	0.9809	

algorithm (FA) [14] for the similar test system and comparative results are presented in Table 3. Critical observation of Table 3 reveals that KHA-tuned PID controller yields greater dynamic performance in terms of overshoot (OS) and settling time (ST) and hence, rest of the study is carried out with KHA-optimized PID controller.

For betterment of system dynamics, an optimal coordinated PID SMES controller is designed using KHA and dynamic stability of concerned power system is investigated. The optimum gains of coordinated PID SMES controller are presented in Table 4 and change in frequency and tie-line power after load perturbation are shown in Fig. 5. Typical system performances, namely overshoot, undershoot, and settling time of system oscillations, are noted down from Fig. 5 and displayed in Table 5. Critical examination of Table 5 and Fig. 5 reveals that system performances are improved remarkable with proposed KHA-tuned coordinated PID SMES controller.

Figure 6 shows the convergence profile of KHA. To evaluate the effectiveness and robustness of proposed KHA, a random pulse load perturbation of width 40 s and amplitude of 0.01 p.u. is applied to area1 of concerned test system. Frequency and tie-line power deviation after load perturbation are shown in Fig. 7. It is clearly evident from Fig. 7 that the amplitude and settling time of frequency and tie-line power deviations with random pulse load perturbation are significantly improved with KHA-optimized PID SMES coordinated controller which shows the robustness of designed controller by KHA.

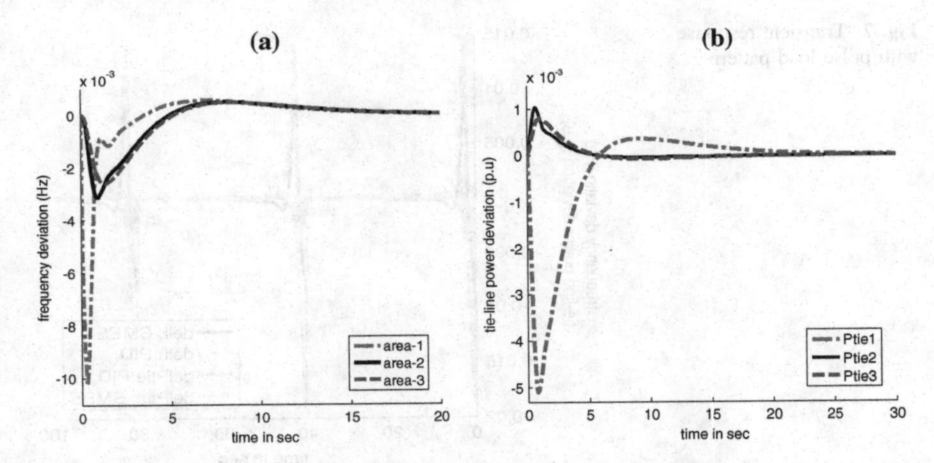

Fig. 5 System performances with coordinated PID SMES controller considering GRC **a** frequency error, **b** tie-line power error

Table 5 System performances with coordinated PID SMES controller after load perturbation

Output variable	Without GRC			With GRC		
	OS	US	ST	OS	US	ST
Δf_1	4.93×10^{-4}	−0.0060	12.03	0.0016	−0.0112	16.19
Δf_2	4.05×10^{-4}	−0.0017	15.35	0.0013	−0.0038	18.77
Δf_3	4.1×10^{-4}	−0.0016	15.49	0.0013	−0.0039	18.08
ΔP_{tie1}	2.75×10^{-4}	−0.0022	17.54	9.68×10^{-4}	−0.0055	22.79
ΔP_{tie2}	4.44×10^{-4}	-8.88×10^{-5}	17.2	0.0011	-1.96×10^{-4}	20.9
ΔP_{tie3}	4.17×10^{-4}	-4.18×10^{-5}	17.69	0.0011	-1.44×10^{-4}	23.3

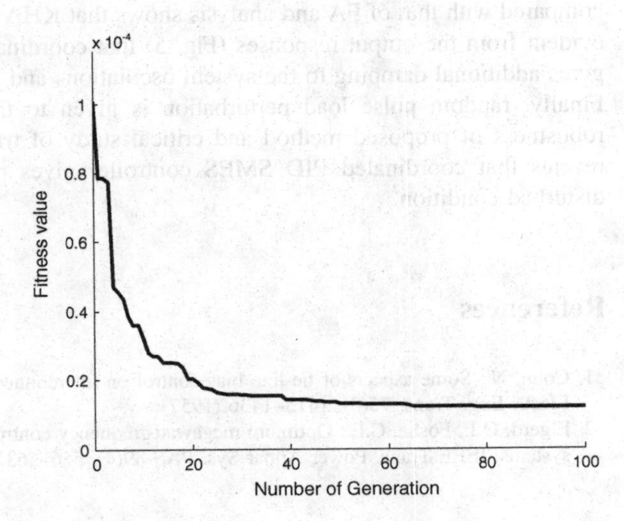

Fig. 6 Convergence profile of KHA

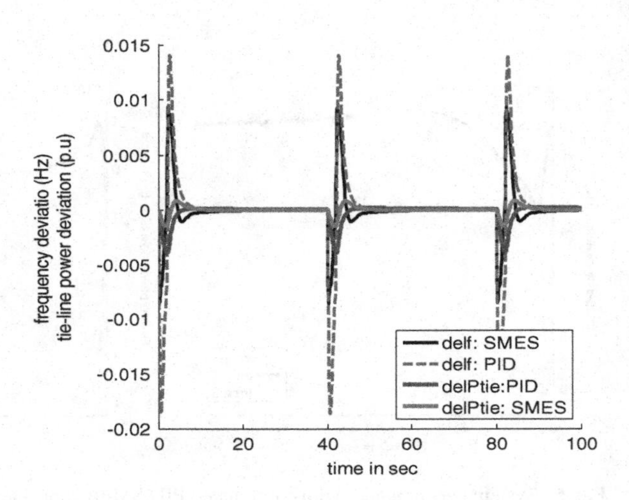

Fig. 7 Transient response with pulse load pattern

8 Conclusion

In this article, authors investigate the effectiveness of proposed KHA-based optimal controllers for an interconnected multi-area LFC system. The classical controller gains are simultaneously optimized by KHA employing ISE-based fitness function. The system is studied with turbine non-linearity (GRC) and output responses show (Fig. 4) that GRC has destabilizing effect to the system dynamics. The system response in terms of frequency and tie-line power deviations with PID controller was compared to that of other classical controllers and output results are presented. Summary of performances reveals that KHA-tuned PID controller gives better performance than other controllers. To demonstrate the superiority of KHA, the system performances in terms of overshoot, undershoot, and settling time are compared with that of FA and analysis shows that KHA outperforms FA. It is also evident from the output responses (Fig. 5) that coordinated PID SMES controller gives additional damping to the system oscillations and improves system stability. Finally, random pulse load perturbation is given to the system to identify the robustness of proposed method and critical study of transient responses (Fig. 7) reveals that coordinated PID SMES controller gives robust performance under disturbed condition.

References

1. Cohn, N.: Some aspects of tie-line bias control on interconnected power systems. Am. Inst. Electr. Eng. Trans. **75**(3), 1415–1436 (1957)
2. Elgerd, O.I., Fosha, C.E.: Optimum megawatt-frequency control of multi-area electric energy systems. IEEE Trans. Power Appar Sys. PAS-89(4), 556–563 (1970)

3. Kothariet, M.L., Satsangi, P.S., Nanda, J.: Sampled data AGC of interconnected reheat thermal stations consisting GRC. IEEE Trans. Power Apparatus Syst. PAS-100(5), 2334–2342 (1981)
4. Van Ness, J.E.: Root loci of LFC system. IEEE Trans. Power Apparatus Syst. PAS (82), 712–726 (1963)
5. Barcelo, W.R.: Effect of power plant response on optimal LFC system design. IEEE Trans. Power Apparatus Syst. PAS-92(1), 254–258 (1973)
6. Nanda, J., Kaul, B.L.: AGC of an interconnected power system. Proc. IEE. **125**(5), 385–390 (1978)
7. Ghoshal, S.P., Goswami, S.K.: Application of GA based optimal integral gains in Fuzzy based active power frequency control of non-reheat and reheat thermal generating systems. Int. J. Electr. Power Syst. Res. **67**(2), 79–88 (2003)
8. Ghoshal, S.P.: Optimizations of PID gains by partial swarm optimizations in Fuzzy based automatic generation control. Int. J. Electr. Power Syst. Res. **72**, 203–212 (2004)
9. Kalra, P.K., Srivastava, A., Chaturvedi, D.K.: Possible applications of neural networks to power system operation and control. Int. J. Electr. Power Syst. Res. **25**(2), 83–90 (1992)
10. Douglas, L.D., Green, T.A., Kramer, R.A.: New approaches to the AGC non-conforming load problem. IEEE Trans. Power Syst. **19**(2), 619–628 (1997)
11. Abdel-Magid, Y.L.: GA applications in load frequency control. In: Proceeding of IEE Conference Publication No. 414, Shetheld, U.K., pp. 207–212 (1995)
12. Aditya, S.K.: Design of load frequency controllers using GA for two-area interconnected hydro-power system. Electr. Power Compon. Syst. **31**, 81–94 (2003)
13. Khuntia, S.R., Panda, S.: Simulation study for AGC of a multi-area power system by ANFIS approach. Appl. Soft Comput. **12**, 333–341 (2012)
14. Debbarma, S., Saikia, L.C., Sinha, N.: Solution of automatic generation control problem using firefly algorithm optimized $I^{\lambda}D^{\mu}$ controller. ISA Trans. **53**(2), 358–366 (2014)
15. Lu, C.F., Liu, C.C., Wu, C.J.: Effect of battery energy storage system on load frequency control considering governor dead band and GRC. IEEE Trans. Energy Convers. **10**(3), 555–561 (1995)
16. Hemeida, A.M.: A fuzzy logic controlled superconducting magnetic energy storage SMES frequency stabilizer. Electr. Power Syst. Res. **80**, 651–656 (2010)
17. Gandomi, A.H., Alavi, A.H.: Krill herd: A new bio-inspired optimization algorithm. Commun. Nonlinear Sci. Numer. Simul. **17**, 4831–4845 (2012)
18. Bacanin,N., Pelevic, B., Tuba, M.: Krill herd algorithm applied for portfolio optimization. Math. Comput. Bus. Manuf. Tourism. 39–44 (2013)

Optimal Allocation of Distributed Generator Using Chemical Reaction Optimization

Sneha Sultana, Santanu Roy and Provas Kumar Roy

Abstract In this paper, an effort is taken to attain optimal DG placement using chemical reaction optimization (CRO) algorithm to minimize the power loss in radial distribution system. The suggested methodology is fruitfully applied on 33-bus and 69-bus radial distribution systems. Simulation results have been compared with other population-based optimization technique like genetic algorithm (GA), particle swarm optimization (PSO), hybrid GA and PSO (GA/PSO), oppositional cuckoo optimization algorithm (OCOA), and bacteria foraging optimization algorithm (BFOA). The simulation results imply that the suggested methodology offers reasonable efficiency and it outperforms the other artificial optimization techniques that are available in the recent literature.

1 Introduction

Presently, electrical power system is the integrated bulk system where generation plants connected to transmission network and in future it will be more distributed where small generating units will be directly connected to the distribution network or directly to the consumer site. This type of generating units is known as distributed generator (DG). Most of the distribution systems are radial in nature due to their simplicity with unidirectional power flow. Power loss in a radial distribution system is high due to large reactance-to-resistance ratio. The power loss shared by the distribution system is large contrasted to the generation and transmission system. DG resource has unique characteristics to reduce this power loss and also it has the capability to get better voltage profile, improvement of voltage deviation index,

S. Sultana (✉) · S. Roy
Dr. B. C. Roy Engineering College, Durgapur, West Bengal, India
e-mail: sneha.sultana@gmail.com

P.K. Roy
Jalpaiguri Government Engineering College, Jalpaiguri, West Bengal, India

S. Das et al. (eds.), *Proceedings of the 4th International Conference on Frontiers in Intelligent Computing: Theory and Applications (FICTA) 2015*, Advances in Intelligent Systems and Computing 404, DOI 10.1007/978-81-322-2695-6_23

continuity of power flow, improvement of voltage stability index, and reliability improvement for both the consumer as well as the electricity suppliers. Therefore, it is very important for the researchers to find the correct position of DG as well as to calculate the proper rating of DG in order to minimize the power loss in radial distribution system. Presently, there are many methods existing so as to find the ideal sizing and sitting of DG.

Acharya et al. proposed an analytical approach [1] to determine the ideal position of DG for the purpose of reducing the power losses in the radial distribution system. Kumar et al. implemented a comparison of novel, combined loss sensitivity, index vector, and voltage sensitivity index methods [2] for optimal location and sizing of DG in a distribution network. Hung et al. proposed an improved analytical expression [3] method that was limited to DG type, which was capable of delivering real power only. Mithulananthan et al. implemented an improved analytical approach [4] for the purpose of multiple DG placements to attain a high loss reduction. Muttaqi et al. implemented an analytical approach [5] derived from algebraic equations for equally distributed loads to decide the ideal operation, rating, and position of the DG so as to support the voltage profile in radial distribution system.

Besides the analytical approach, many researchers proposed classical-based optimization approach so as to locate the most favorable position of DG. Moreover, many researchers proposed population-based optimization technique so as to locate the most favorable position of DG in radial distribution system. Selvan et al. implemented a scheduling of DG units in the distribution system employing hierarchical agglomerative clustering algorithm (HACA) [6]. Nekooei et al. developed an enhanced multi-objective harmony search algorithm (HSA) [7] for the best possible position of DGs in radial distribution system. Rueda-Medina et al. developed a mixed-integer linear programming (MILP) [8] approach to solve the problem of most favorable category, rating, and actual position of DGs in radial distribution systems. Junjie et al. implemented immune algorithm (IA) [9] for the purpose of sizing and sitting of DG. Shukla et al. minimized active power loss by placing DG in radial distribution system. The problem was formulated as optimization problem and solution was obtained using genetic algorithm (GA) [10]. Moradi proposed a novel combined GA/PSO [11] for the best possible position and rating of DG on distribution systems. Safari et al. suggested a hybrid GA and PSO (HGAPSO) [12] aiming at optimal DG allocation in distribution network. Kim et al. proposed a hybridized method that combined the GA with fuzzy set theory [13] which determined the best possible placement of DG and their capacities in distribution networks, simultaneously. Mohammadi et al. implemented a GA-based tabu search (TS) algorithm [14] so as to locate the finest place of DG unit. Abu-Mouti et al. proposed a new meta-heuristic optimization approach, based on an ABC algorithm [15], to establish the best possible site, rating, and power factor of DGs in a distribution system. Moravej et al. proposed an optimization that is cuckoo search algorithm (CSA) [16] for the best possible DG allotment to reduce power loss in distribution network. Falaghi et al. developed an ant colony optimization (ACO) [17] method as an optimization tool for solving the DG rating and position problems. Nara et al. implemented a TS [18] method for solving the best possible DG size as well as to minimize the system losses. A novel approach based on bacterial foraging optimization algorithm (BFOA) [19] was used to reduce the system losses, operational cost, and to get better voltage stability. Garcia et al. developed a modified teaching learning-based optimization (MTLBO) [20] to establish the optimal DG location and sizing in distribution system.

The above-stated algorithms have been extensively used in DG allocation problem but they suffer from some disadvantages such as premature convergence, taking long computational time, and trapping into local optima during simulation. To overcome these limitations, the present authors established a novel, easy, and proficient population-based optimization technique to resolve DG allocation problem. In this paper, the authors implemented CRO algorithm to reduce power loss in radial distribution networks. The result shows that CRO algorithm is well-organized, fast-converging, and capable of handling complex power system network. The idea of CRO has been used by various researchers to enhance the usefulness of population-based algorithms in different areas [21–24].

The rest of this paper is organized as follows: Problem formulation is given in Sect. 2. The key points of the proposed CRO technique are described in Sect. 3. In Sect. 4, the suggested technique applied to DG allocation problem in radial distribution system is illustrated. Two cases based on large-scale power systems are studied and the simulation results are illustrated in Sect. 5. Section 6 summarized the conclusion.

2 Problem Formulation

2.1 Objective Function

In a radial distribution system due to large reactance-to-resistance ratio, power loss is high. So the goal of our objective function is to reduce the system power loss.

The real power loss in a radial distribution is

$$P_{RLOSS} = \sum_{i=2}^{n_N} \left(P_{gqi} - P_{dqi} - v_{pi} v_{qi} Y_{pqi} \cos(\partial_{pi} - \partial_{qi} + \theta_{pqi})\right) \quad (1)$$

Here, P_{RLoss} is the real power loss; P_{gqi} is real power generation at bus qi; P_{dqi} is active power demand at bus qi; v_{qi} is voltage magnitude at bus qi; Y_{pqi} is the magnitude of the bus admittance matrix between pi and qi; ∂_{pi} is voltage angle of bus pi; ∂_{qi} is voltage angle of bus qi; and θ_{pqi} is the phase angle between pi and qi of the bus admittance matrix.

2.2 Constraints

Load Balance Constraint

For each bus, the following balance constraint equation should be fulfilled:

$$P_{gqi} - P_{dqi} - v_{qi} \sum_{j=1}^{M} v_{qj} Y_{qj} \cos(\delta_{qi} - \delta_{qj} + \theta_{qj}) = 0 \quad (2)$$

$$Q_{gqi} - Q_{dqi} - v_{qi} \sum_{j=1}^{M} v_{qj} Y_{qj} \cos\left(\delta_{qi} - \delta_{qj} + \theta_{qj}\right) = 0 \quad (3)$$

Here, $qi = 1, 2 \ldots\ldots\ldots n_N$.

Voltage Limits

In a distribution system, the generator voltage is the sum between the load and bus voltage and the impedance of the line, and if the power flow in the distribution network is increased, then X/R ratio of the line will be increased because of the presence of resistive element of the line. So the voltage must be kept within the standard limits of each bus.

$$v_{qi}^{\min} < v_{qi} < v_{qi}^{\max} \quad (4)$$

DG Power Constraints

A DG power capacity depends on the energy resources of any given location; therefore, it is essential to keep the DG power capacity in its minimum and maximum levels:

$$P_{gqi}^{\min} \leq P_{gqi} \leq P_{gqi}^{\max} \quad (5)$$

Thermal Limits

In distribution lines, the thermal limit for the network must not be exceeded to the following values:

$$|S_{qi}| \leq |S_{qi}^{\max} \quad i = 1, 2, 3, \ldots\ldots\ldots\ldots\ldots, M \quad (6)$$

3 Chemical Reaction Optimization (CRO)

CRO is a population-based optimization technique which is first developed by Lam and Li [25] in 2010, a process that leads to the transformation of one set of chemical substances to another. Multiple steps in chemical reaction are very common occurrence. Chemical reactions are usually characterized by a chemical change. Different types of molecules are involved in chemical processes. A molecule possesses two kinds of energies: potential energy (PEN) and kinetic energy (KEN).

Reactions involved in CRO:

There are four types of reactions taken into considerations which are elaborated below.

3.1 On-Wall Ineffective Collision

A single molecule is involved in this reaction and it is allowed to collide on the wall of a container and the bounces back. Therefore, the present molecular structure (α) is changed and it becomes a new molecule (α'). Thus the KEN for the new molecule (α') is as follows:

$$KEN_{\alpha'} = (PEN_{\alpha} + KEN_{\alpha} - PEN_{\alpha'}) \times Q \tag{7}$$

Here, Q is a number, arbitrarily selected from the range of $[KEN_{lossrate}, 1]$. $KEN_{lossrate}$ is the loss rate of kinetic energy.

3.2 Decomposition

Decomposition means a molecule hits the barrier and then the molecule converts into two or more new molecules with different structures, i.e., $\alpha \rightarrow \alpha_1' + \alpha_2$.

In decomposition, two steps are considered: (i) the molecule has sufficient energy to complete the decomposition; and (ii) the molecule should get energy from the energy center.

Step 1: The KE of the resultant molecules is shown as follows:

$$KEN_{\alpha_1'} = (PEN_{\alpha} + KEN_{\alpha} - PEN_{\alpha_1'} - PEN_{\alpha_2}) \times Q \tag{8}$$

$$KEN_{\alpha_2} = (PEN_{\alpha} + KEN_{\alpha} - PEN_{\alpha_1'} - PEN_{\alpha_2}) \times (1 - Q) \tag{9}$$

Here, Q is a random number generated from the interval [0, 1] considering the constraint as follows:

$$PEN_{\alpha} + KEN_{\alpha} \geq PEN_{\alpha_1'} + PEN_{\alpha_2} \tag{10}$$

Step 2: Here, the KE of the resultant molecules is shown as follows:

$$KEN_{\alpha_1'} = (PEN_{\alpha} + KEN_{\alpha} - PEN_{\alpha_1'} - PEN_{\alpha_2} + buffer) \times n1 \times n2 \tag{11}$$

$$KEN_{\alpha_2} = (PEN_{\alpha} + KEN_{\alpha} - PEN_{\alpha_1'} - PEN_{\alpha_2} + buffer) \times n3 \times n4 \tag{12}$$

Here, $n1$, $n2$, $n3$, and $n4$ are random numbers uniformly generated from the interval [0,1] and the constraint as follows:

$$PEN_{\alpha} + KEN_{\alpha} + buffer \geq PEN_{\alpha_1'} + PEN_{\alpha_2} \tag{13}$$

3.3 Inter-Molecular Ineffective Collision

This is a reaction involving two molecule collisions. When the two molecules collide with each other, they are changed into two new molecules, if the following energy criteria satisfy:

$$PEN_{\alpha_1} + KEN_{\alpha_1} + PEN_{\alpha_2} + KEN_{\alpha_2} \geq PEN_{\alpha_1'} + PEN_{\alpha_2} \tag{14}$$

Then KE values of the two new molecules are as follows:

$$KEN_{\alpha_1'} = (PEN_{\alpha_1} + KEN_{\alpha_1} + PEN_{\alpha_2} + KEN_{\alpha_2} - PEN_{\alpha_1'} - PEN_{\alpha_2}) \times r \tag{15}$$

$$KEN_{\alpha_2} = (PEN_{\alpha_1} + KEN_{\alpha_1} + PEN_{\alpha_2} + KEN_{\alpha_2} - PEN_{\alpha_1'} - PEN_{\alpha_2}) \times (1 - r) \tag{16}$$

Here, r is a arbitrary number generated from the interval [0, 1].

3.4 Synthesis

Here, two or more molecules crash with each other, combined to generate a new molecule. The new molecular structure is entirely different from the original one.

Suppose that two molecular structure α_1 and α_2 collide to form a single molecule with structure α'. The condition is as follows:

$$KEN_{\alpha'} = PEN_{\alpha_1} + KEN_{\alpha_1} + PEN_{\alpha_2} + KEN_{\alpha_2} - PEN_{\alpha'} \tag{17}$$

4 CRO Algorithm Applied to DG Allocation Problem in Radial Distribution System

The following steps must be taken to apply the CRO.

Step1: The preliminary position of each habitat should be randomly selected while satisfying different inequality constraints of the control variables. In the proposed area, DG size and locations are considered as the control variables.

Step2: Power losses can be found by running the load flow problem. In this paper, a direct load flow algorithm based on the BIBC (Bus-Injection to Branch-Current) matrix and the (BCBV) (Branch-Current to Bus-Voltage) matrix [26] used on-wall ineffective collision using (7).

Step3: Calculate the fitness value for each habitat using (8) to (10).

Step4: Update DG size and its position using (13).

Step5: Check whether the independent variables (rating of DG) violate the operating limits or not. For this, conditions are followed using (14) to (16).

Step6: Go to Step 3 until the current iteration number reaches the pre-specified maximum iteration number.

5 Test Systems and Results

This segment exhibits the fruitfulness of the suggested chemical reaction optimization (CRO) algorithm. For establishment, this suggested algorithm is executed in 12.66 kV 33 and 69-bus IEEE radial distribution systems to calculate the actual position as well as the rating of various DGs to reduce the total power losses. This suggested methodology successfully executed on constant power (CP) load model during light load (0.5), full load (1.0), and heavy load (1.6) conditions. In this proposed methods, the coding and simulations are executed on a personal laptop having core i-5, 2.5 GHz processor and 4 GB of RAM with the help of MATLAB software.

5.1 Test Case 1 (for 33-Bus Radial Distribution Systems)

This test system consists of 33 buses and 32 branches along with a total real and reactive power loads of 3.715 MW and 2.3 MVAR, respectively. The line and load data are taken from [28]. The single-line diagram of 33-bus radial distribution system is shown in Fig. 1. The simulation results obtained by the proposed CRO and the other population-based optimization methods like OCOA [27] and BFOA [19] are taken from the literature and are listed in Table 1 and the actual position as

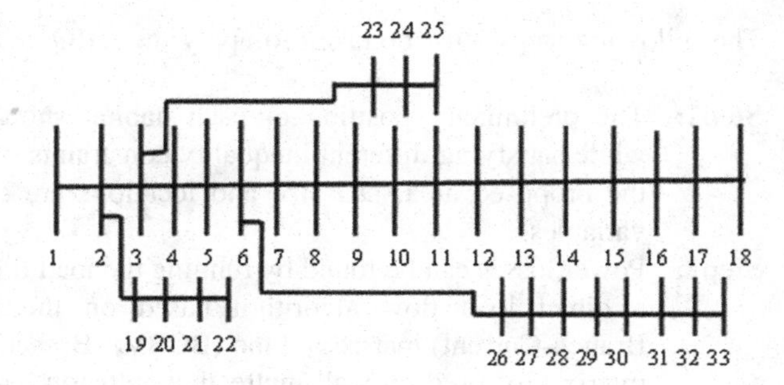

Fig. 1 Single-line diagram of 33-bus radial distribution system

Table 1 Performance analysis of the proposed methods on 33-bus system at constant power (CP) load model

Parameters			CRO			OCOA [27]			BFOA [19]		
CP Load model	Light load (0.5)	DG Size (MW)	0.38	0.54	0.52	0.39	0.53	0.51	0.34	0.07	0.49
		Location of DG	14	24	30	13	24	30	14	18	32
		Power loss (KW)	17.6344			17.6367			21.61		
		CT Time (s)	3.1			3.7			4.2		
	Full load (1.0)	DG Size (MW)	0.80	1.09	1.05	0.77	1.09	1.06	0.65	0.19	1.06
		Location of DG	13	24	30	14	24	30	14	18	32
		Power loss (KW)	69.4213			72.7899			89.90		
		CT Time (s)	3.7			4.0			4.9		
	Heavy load (1.6)	DG Size (MW)	105.31	3.67	1.79	1.30	1.77	1.71	0.96	0.44	1.77
		Location of DG	19	24	32	13	24	30	14	18	32
		Power loss (KW)	138.3589			193.381			243.63		
		CT Time (s)	3.9			4.1			5.3		

well as the ratings of various DGs for the objective function is also listed in that Table. From the table, it is shown that power loss is 17.6344 kW during light load, 69.4213 kW during full load, and 138.3589 kW during heavy load achieved by suggested CRO which is better than the OCOA and BFOA that are taken from the literature. So it may be concluded means here author describes that from the observation table it is clearly shown that the proposed CRO method attains power loss which is significantly less than the other population based optimization technique like PSO, GA, GA/PSO, OCOA and BFOA at unity as well as 0.866 power factor. From this table it is also clear that computational time for the proposed CRO method is less than OCOA and BFOA.

The simulated results of the suggested CRO are also compared with other population-based optimization methods like GA [11], PSO [11], GA/PSO [11], BFOA [19], and OCOA [27], at unity power factor and tabulated in Table 2. It is seen that the total power loss at unity and 0.866 power factor obtained by proposed

Table 2 Optimal location and settings of multiple DGs and corresponding loss using CRO for 33-bus system at unity and 0.866 power factor

Power factor (Unity)							Power factor (0.866)		
Method	GA [11]	PSO [11]	GA/PSO [11]	BFOA [19]	OCOA [27]	CRO	BFOA [19]	OCOA [27]	CRO
$P_{DG,TLoss}$	106.30	105.35	103.40	89.90	72.7906	72.7853	37.85	15.3489	15.3488
DG Location	11	8	11	14	14	13	14	13	13
	29	13	16	18	24	24	18	24	24
	30	32	32	32	30	30	32	30	30
DG size (MW)	1.5000	1.1768	0.9250	0.6521	0.7685	0.80214	0.6798	0.7575	0.75824
	0.4228	0.9816	0.8630	0.1984	1.0994	1.09144	0.1302	1.0266	1.02651
	1.0714	0.8297	1.2000	1.0672	1.0708	1.05244	1.1085	1.2146	1.21140
CT (s)	NA	NA	NA	NA	4.34	4.21	NA	4.35	4.1

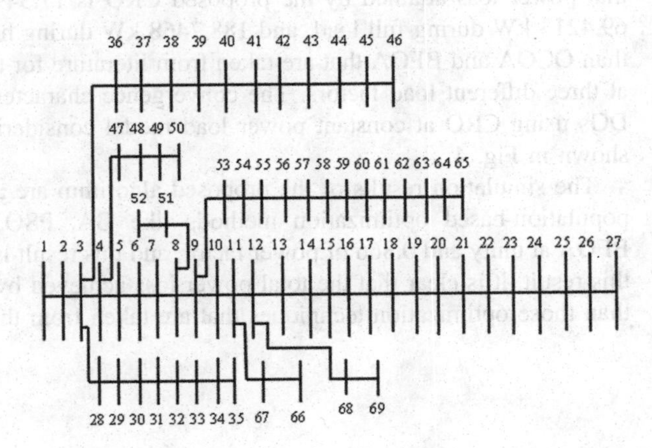

Fig. 2 Single-line diagram of 69-bus radial distribution system

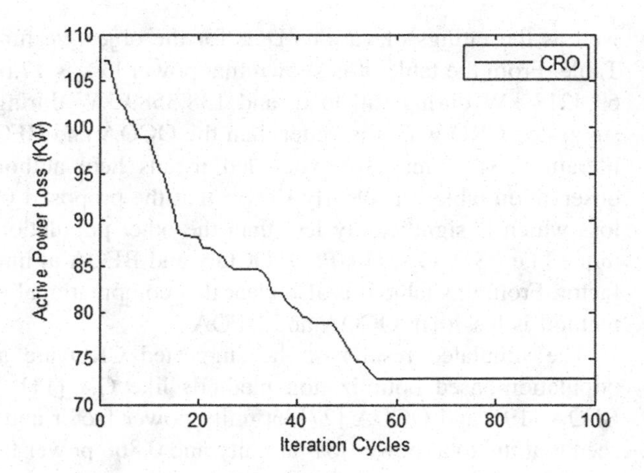

Fig. 3 Power loss convergence graphs using CRO of 33-bus system for multiple DGs at unity power factor

CRO method are significantly small in compared to other population-based optimization approaches. The convergence characteristic of loss for multiple DGs using CRO at unity power factor is shown in Fig. 3.

5.2 Test Case 2 (69- Bus Radial Distribution System)

It is a large-scale distribution network. It has 68 branches with the total real and reactive power loads of this system of 3.80 MW and 2.69 Mvar, respectively. The single-line diagram of this test system is shown in Fig. 2. The line and load data of this system are taken from [28]. To show the usefulness of the suggested algorithm, this large-scale distribution network of CP load model at three different load factors is considered. The simulation results are listed in Table 3. From the table, it is seen that power loss attained by the proposed CRO is 17.3493 kW during light load, 69.4213 kW during full load, and 188.7468 kW during heavy load which is better than OCOA and BFOA that are taken from literature for that particular load model at three different load factors. The convergence characteristic of loss for multiple DGs using CRO at constant power load model considering full load condition is shown in Fig. 4.

The simulation results of the proposed algorithm are also compared with other population-based optimization methods like GA, PSO, GA/PSO, OCOA, and BFOA at unity and 0.866 of power factor, and this result is shown in Table 4. From this result, it is clear that the total power loss achieved by proposed CRO is better than those optimization techniques that are taken from the literature.

Table 3 Performance analysis of the suggested methods on 69-bus system at constant power (CP) load model

Parameters			CRO			OCOA [27]			BFOA [19]		
CP Load Model	Light Load (0.5)	DG Size (MW)	0.31	0.33	0.86	0.22	0.25	0.76	0.19	0.65	0.24
		Location of DG	14	49	61	55	18	61	27	61	65
		Power loss (KW)	17.3493			17.4482			18.342		
		CT Time (s)	4.8			5.2			6.3		
	Full Load (1.0)	DG Size (MW)	0.48	0.40	1.71	0.53	0.39	1.71	0.30	1.35	0.45
		Location of DG	11	18	61	10	18	61	27	61	65
		Power loss (KW)	69.4213			69.515			75.238		
		CT Time (s)	5.3			5.5			6.8		
	Heavy Load (1.6)	DG Size (MW)	0.95	1.73	1.13	0.75	1.9479	0.8726	0.49	1.88	0.56
		Location of DG	17	61	63	20	61	63	27	61	65
		Power loss (KW)	188.7468			189.217			205.67		
		CT Time (s)	5.6			5.9			7.2		

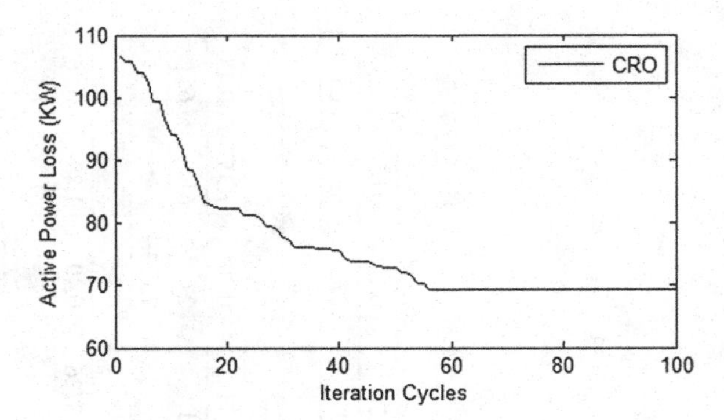

Fig. 4 Power loss convergence graphs using CRO of 69-bus system for multiple DGs at constant power, full load condition

Table 4 Ideal position and settings of multiple DGs and corresponding loss using CRO for 69-bus system at unity and 0.866 power factor

Power factor (Unity)							Power factor (0.866)		
Method	GA [11]	PSO [11]	GA/PSO [11]	BFOA [19]	OCOA [27]	CRO	BFOA [19]	OCOA [27]	CRO
$P_{DG,TLoss}$	89.0	83.21	81.1	75.23	69.5432	69.3987	12.90	6.2014	6.166
DG Location	21	61	63	27	10	11	27	11	12
	62	63	61	65	19	18	61	18	18
	64	17	21	61	61	61	65	61	61
DG size (MW)	0.9297	1.1998	0.8849	0.2954	0.5446	0.53177	0.3781	0.5656	0.4223
	1.0752	0.7956	1.1926	0.4476	0.3799	0.38099	1.3361	0.3937	0.3939
	0.9925	0.9925	0.9105	1.3451	1.7159	1.72183	0.3285	1.6891	1.7751
CT (s)	NA	NA	NA	NA	5.53	5.3	NA	5.54	4.8

6 Conclusion

In this paper, a new approach namely chemical reaction optimization (CRO) is proposed for the purpose of ideal sizing and sitting of DG in radial distribution system. This technique is tested on two systems namely an IEEE 33-bus and IEEE 69-bus radial distribution systems to calculate the ideal positions and ratings of DG. In order to judge the superiority of the suggested CRO approach, it is compared with other population-based optimization techniques including GA, PSO, GA/PSO, OCOA, and BFOA. The study suggests that the efficiency and effectiveness of the proposed method are better than the other population-based optimization technique. Therefore, the proposed CRO method can be a very favorable method for solving the placement of DG in radial distribution system.

References

1. Acharya, N., Mahat, P., Mithulananthan, N.: An analytical approach for DG allocation in primary distribution network. Int. J. Electr. Power Energy Syst. **28**(10), 669–678 (2006)
2. Murthy, V.V.S.N., Kumar, A.: Comparison of optimal DG allocation methods in radial distribution systems based on sensitivity approaches. Int. J. Electr. Power Energy Syst. **53**(10), 450–467 (2013)
3. Hung, D. Q., Mithulananthan, N.: Analytical expressions for DG allocation in primary distribution networks. IEEE Trans. Energy Convers. **25**(3) (2010)
4. Hung, D.Q., Mithulananthan, N.: Multiple distributed generator placement in primary distribution networks for loss reduction. IEEE Trans. Ind. Electric. **60**(4), 1700–1708 (2013)
5. Muttaqi, K.M., Le, D.T., Negnevitsk, M., Ledwich, G.: An algebraic approach for determination of DG parameters to support voltage profiles in radial distribution networks. IEEE Trans. Smart Grid **5**(3), 1351–1360 (2014)
6. Vinothkumar, K., Selvan, M.P.: Hierarchical agglomerative clustering algorithm method for distributed generation planning. Electr. Power Energy Syst. **56**, 259–269 (2014)
7. Nekooei, K., Farsangi, M.M., Hossein, N.P., Lee, K.Y.: An Improved multi-objective harmony search for optimal placement of DGs in distribution systems. IEEE Trans. Smart Grid **4**(1), 557–567 (2013)
8. Rueda-Medina, A.C., Franco, J.F., Rider, M.J., Feltrin, A.P., Romero, R.: A mixed-integer linear programming approach for optimal type, size and allocation of distributed generation in radial distribution systems. Electr. Power Syst. Res. **97**, 133–143 (2013)
9. Junjie, M.A., Yulong, W., Yang, L.: Size and location of distributed generation in distribution System based on immune algorithm systems engineering procedia **4**, 124–132 (2012)
10. Shukla, T.N., Singh, S.P., Naik, K.B.: Allocation of optimal DG using GA for minimum power losses in distribution systems. Int. J. Eng. Sci. Technol. **2**(3), 94–106 (2010)
11. Moradi, M.H., Abedini, M.: A combination of genetic algorithm and particle swarm optimization for optimal DG location and sizing in distribution systems. Int. J. Electr. Power Energy Syst. **34**(1), 66–74 (2012)
12. Safari, A., Jahani, R., Shayanfar, H.A., Olamaei, J.: Optimal DG allocation in distribution network. Int. J. Electr. Electron. Eng. **4**(8), 550–553 (2010)
13. Kim, K.H., Lee, Y.J., Rhee, S.B., Lee, S.K., You, S.K.: Dispersed generator placement using fuzzy-GA in distribution systems. In: Power Engineering Society Summer Meeting, IEEE, **3**, 1148–1153 (2002)
14. Mohammadi, M., Mehdi, N.: Optimal placement of multitypes DG as independent private sector under pool/hybrid power market using GA-based Tabu search method. Int. J. Electr. Power Energy Syst. **51**, 43–53 (2013)
15. Abu-Mouti, F.S., El-Hawary, M.E.: Optimal distributed generation allocation and sizing in distribution systems via artificial bee colony algorithm. IEEE Trans. Power Deliv. **26**(4), 2090–2101 (2011)
16. Moravej, Z., Akhalaghi, A.: A novel approach based on cuckoo search for DG allocation in distribution system. Int. J. Electr. Power Energy Syst. **44**(1), 672–679 (2013)
17. Falaghi, H., Haghifam, M.R.: ACO based algorithm for distributed generation sources allocation and sizing in distribution networks. IEEE Lausanne Power Tech. 555–560, (2008)
18. Nara, K., Hayashi, Y., Ikeda, K., Ashizawa, T.: Application of tabu search to optimal placement of distributed generators. In: IEEE Power Engineering Society Winter Meeting; (2001)
19. Mohamed, I.A., Kowsalya, M.: Optimal size and siting of multiple distributed generators in distribution system using bacterial foraging optimization. Swarm Evol. Comput. **15**, 58–65 (2014)
20. García, J.A.M., Mena, A.J.G.: Optimal distributed generation location and size using a modified teaching–learning based optimization algorithm. Int. J. Electr. Power Energy Syst. **50**, 65–75 (2014)

21. Yi, S., Lam, A.Y.S., Li, V.: Chemical reaction optimization for the optimal power flow problem. In: Conference on Evolutionary Computation Brishbane, QLD, pp. 1–8. 10–15 June 2012
22. Bhattacharjee, K., Bhattacharya, A., Haldar, S.: Chemical reaction optimisation for different economic dispatch problems. Gener. Transm. Distrib. IET. **8**(3), 530–541 (2014)
23. Bhattacharjee, K., Bhattacharya, A.: Chemical reaction optimization applied in economic dispatch problems. In: Conference on ACES 2014, pp. 1–6, 1–2 Feb 2014
24. Mukherjee, A., Roy, P.K.: Chemical reaction optimization for transient stability constrained optimal power flow. In: 1st International Conference on Non Conventional Energy (ICONCE 2014), IEEE Conference, pp. 316–321, Kalyani, Nodia, 16–17 Jan 2014
25. Lam, A., Li, V.: Chemical-reaction-inspired metaheuristic for optimization. IEEE Trans. Evol. Comput. **14**, 381–399 (2010)
26. Teng, J.H.: A direct approach for distribution system load flow solutions. IEEE Trans. Power Deliv. **18**(3), 882–887 (2003)
27. Roy, S., Sultana, S., Roy, P. K.: Oppositional cuckoo optimization algorithm (OCOA) to solve DG allocation problem in radial distribution system. In: 2015 1st IEEE International Conference onRecent Development in Control Automation and Power Engineering (RDCAPE), Amity University, Noida, India, 12–13 March 2015
28. Sahoo, N.C., Prasad, K.: A fuzzy genetic approach for network reconfiguration to enhance voltage stability in radial distribution systems. Int. J. Energy Convers. Manage. Energy Syst. **47**, 3288–3306 (2006)

A Novel Combined Approach of *k*-Means and Genetic Algorithm to Cluster Cultural Goods in Household Budget

Sara Sadat Babaie, Ebadati E. Omid Mahdi and Tohid Firoozan

Abstract Setting up household spending and leading to an efficient and optimal usage is one of the important issues that every family faces. In this paper, we consider a sample of 35,000 households, among them we evaluate features of households that allocated a larger share of the budget to cultural goods and cluster them to extract common social and economic characteristics using Combination of *k*-means and genetic algorithm. GA as a meta-heuristic optimization algorithm increases the speed of achieving optimal solutions in *k*-means algorithm. The families' priorities to spend their budget in rural and urban areas show that in most of the families with a high level portion of cultural goods, food and drinks, smokes, and education are three categories which have a higher priority than other groups. The results show the highest accuracy that there are three well separated and compact clusters, which for the fitness are accredited by Davies–Bouldin index by calculating inter and intra distances.

Keywords Machine learning · Genetic algorithm · *k*-means · Household budget · Cultural goods · Davies–Bouldin index

---

S.S. Babaie
Department of Knowledge Engineering and Decision Science, Kharazmi University, #242, Somayeh Street, Between Qarani and Vila, Tehran, Iran
e-mail: sara.babaiee@gmail.com

E.E. Omid Mahdi (✉)
Department of Mathematics and Computer Science, Kharazmi University, #242, Somayeh Street, Between Qarani and Vila, Tehran, Iran
e-mail: omidit@gmail.com

T. Firoozan
Faculty of Management, Kharazmi University, #242, Somayeh Street, Between Qarani and Vila, Tehran, Iran
e-mail: t_firoozan@yahoo.com

S. Das et al. (eds.), *Proceedings of the 4th International Conference on Frontiers in Intelligent Computing: Theory and Applications (FICTA) 2015*, Advances in Intelligent Systems and Computing 404, DOI 10.1007/978-81-322-2695-6_24

1 Introduction

In recent years, due to recognizing the importance of household budget planning in social, economic and culture, lots of research has been done in this case. How to set up household spending and leading to an efficient and optimal usage of that is one of the important issues that every family faces. As respects the economic and social factors have a large influence on the choice of lifestyle and manner of allocating budget to different parts of the household budget. In this paper, we consider a sample of 35,000 households; among them evaluate features of households that allocated a larger share of the budget to cultural goods and extract common social and economic characteristics of the households using data mining techniques. In recent years, machine learning and data mining techniques which aim to find patterns in the data with minimal user intervention, provide information for crucial decisions and in different fields such as forecasting and classification of data has many successes. The study attempts to use data mining techniques to analyze the behavior of households, life style and assess consumption cost based on cultural goods in their budget to compare them with other families, which pay less attention to this class of goods in their budget.

The *k*-means algorithm as a popular unsupervised learning algorithm is used for clustering datasets of household. Clusters with different cultural goods in the share of household expenditure in the various provinces of the country are obtained. To improve the performance of this algorithm and escape from local optimum, a meta-heuristic optimization algorithm called genetic algorithm is used to determine the optimal number of clusters. Finally, examine each of the clusters and consider the role of cultural goods on other household budget spending. Spending priorities for the rest of the household budget for families with different portion cost of cultural goods is calculated to extract differences of clusters, which have a higher share of their budget in the cost of cultural goods. It is performed by taking into account various parameters in the cost of funds for each of the sampled households in rural and urban areas. The clusters are considered based on portion cultural goods in household budgets (proportion of households with very high, high), other costs in each of the clusters are analyzed and spending priorities for each cluster is obtained. In each clusters other costs related to the household sample are prioritized. Comparisons can be useful to realize the social and economic characteristics of clusters. For example, we can consider families with a high proportion of cultural goods, how much money spends in the communications sector in comparison with other clusters.

2 Literature Review

Work with a large amount of data and gain information from them is one of the important concerns to find the best decision in governmental problem, organization, banking, business, etc. To gain information and find the easiest usage of that, machine learning techniques are very helpful. Clustering is widely used these days to partitioning dataset in some groups with the most similarities among them. *k*-means algorithm, one of the most popular clustering algorithms, is applied in this paper. The ultimate answer gain from *k*-means algorithm is not always optimal solution so to eliminate this problem and scape from the local optimum genetic algorithm is utilized. Introducing some of the works on these algorithms is considered in this part. In addition, the application of them in household budget is presented.

2.1 *Data Clustering*

Clustering can be considered as an important algorithm, which is used for unsupervised learning. Many data mining problems can be stated as the problem of clustering where intelligent or semi-intelligent agent without having any background information must be able to partition datasets by considering the similarity among them [1–4]. Clustering as a data driven algorithm is applied in many fields, like image segmentation [5–7], Clustering weblog data [8, 9], city-planning [10, 11], marketing [12, 13], banking system [14–16], etc. *k*-means as one of the reliable data clustering algorithm is presented in the next section.

2.2 k-*Means Method Clustering*

Despite the simplicity of *k*-means method clustering, it is a basic method for many other clustering methods (such as fuzzy clustering) [17]. The algorithm performs clustering based on the minimum distance of each data point from the center of a cluster [18, 19]. The standard *k*-means was first proposed by Liyod [20], which points are required to be selected randomly according to the number of clusters; then the squared distances of each point to centroids are calculated and each one belongs to nearest clusters [21]. The procedure continues until the termination condition appears [22]. The situation will be reached where the centroids do not move anymore [23]. Clustering problems, in order to minimize the sum of squared deviations, are nonlinear and non-convex and have a large number of local optimum [24, 25]. Owing to this problem, a combination of meta-heuristic algorithms and *k*-means is considered [26].

2.3 *Combination of* k-*Means Method and Genetic Algorithm*

Combining genetic algorithm and *k*-means to get out of the local optimum and to achieve an optimal solution is conducted by researchers [27–29] and also deterministic number of clusters and initialization of population [30] to decrease the runtime of performance are considered in this paper. In [31], for enhancing the finding of global answer exchange centroids which are neighbors. To calculate the fitness of chromosome and cluster separation, measure of the Davies–Bouldin index (DBi) [32] is an authentic metric of how well the clustering has been done.

2.4 *Household Budget Analysis*

Household budget planning in the family expenses is necessary and is essential for low-income families and poor families. In rich families, according to poor management of household expenses, need to allocate large amounts of revenue to expenses so as a result, further work is imposed on the family man. For low-income families, mismanagement costs sometimes ruin families. So estimating the expenditure of families on food [33], house [34], health care [35], education [36, 37], and cultural goods, managing how to expend money are best for finding the ways of household budget planning improvement [38–40].

3 Research Methodology

Combination of *k*-means and genetic algorithm for the cultural goods in household budget is provided here. The *k*-means algorithm is an iterative algorithm to obtain the optimal clusters of objects, but determination of the number of clusters is the subject for which GA can be useful. Before clustering all 36,880 records, data decile rating of the datasets is used. The first three out of ten of deciles among them are utilized as groups of high cultural goods consumption. The whole steps are explained below in Pseudocode in Fig. 1.

Step 1. The first step involves processes on datasets using Excel 2013. Data is collected from the Central Statistics of Iran in 2014. The dataset contains four characters: 1. Address: the address of the family consists of four categories 1.1. City or town 1.2. Provinces code 1.3. The number of primary sampling units 1.4. Household's series 2. Commodity code, which has 72 different goods 3. Shop way includes: 3.1. Buy 3.2. Homemade 3.3. For public services 3.4. For cooperative services 3.5. For private services 3.6. Agricultural businesses 3.7. Non-Agricultural businesses 3.8. Free 4. Cost of commodities. The small part of data is shown in Table 1.

Since every family may have more than one cultural good in their basket, so just for this stage the aggregation of each household should be considered, then according to the result deciles are formed. The processes in these steps are performed in Excel 2013.

Three first deciles of the dataset export to Matlab software. After this step the records decrease to 1,740 records in rural.

```
Read data records
Define the Problem:
Population Size (PopSize); Crossover Percentage (P_c); Number of Offspring (N_c);
    Mutation probability (P_m);
    Mutation probability (P_m); Expected cluster number (K_min, K_max); Maximum
    number of iterations (MaxIt);
SET the row data in the form which can be used in the process
FOR <X=1: all numbers of Population Size>
    Generate a number K, in the range K_min to K_max
    Choose K_i points randomly from the dataset
    Distribute these points randomly in the chromosome
    FOR <Y=1: number of clusters (K_i)>
        Perform clustering by assigning each point to closest center
        Compute DB index (Dbi)
        Set minimum of the DB index as Pop_best
    END FOR
END FOR
FOR <It=1: MaxIt>
    Select offspring by Roulette Wheel or Tournament or Random
    Apply Single point crossover with prob. P_c
    Perform Mutation with prob.P_m:
    Compute DB index (Db_i) for PopC, PopM
    Set Pop= [Pop; Pop_C; Pop_M]
    Set minimum of the DB index as Pop_best
END FOR
FOR< Z=1: all numbers of features>
    Implementation of k-means algorithm in Pop_best
END FOR
```

Fig. 1 Pseudocode of the proposed model

Table 1 Small part of the dataset

	Location	Provinces	Commodity	Shopping way	Cost
Household 1	Rural	Tehran	Notebook	Homemade	25,000
Household 2	Urban	Tehran	Gym	Buy	800,000
Household 3	Rural	Kermanshah	Cinema	Cooperative services	Free
Household 4	Urban	Booshehr	CD	Buy	25,000
Household 5	Urban	Zanjan	Fertilizer	Buy	35,000
Household 6	Rural	Fars	Book	Buy	750,000
Household 7	Urban	Markazi	Toy	Buy	500,000
.	.	.	.	.	.
.	.	.	.	.	.

Step 2. In this step the proper algorithm for clustering datasets are evaluated and selected. The appropriate algorithm aims to achieve optimum results with the lower run time, which is performed in this paper is a combination of *k*-means and one of the best meta-heuristic algorithm called genetic. Both of these algorithms can be performed to cluster dataset, but the hybrid of them helps to escape from the local optimum and decrease the run time.

Step 3. The implementation of hybrid genetic and *k*-means is done in this stage. The number of *k* is assumed between *Kmin* and *Kmax*, for each iteration of GA based on population size. k_i centers are randomly chosen, and these points in the chromosome are randomly distributed. Clustering each point, according closest center is applied. To calculate fitness of each chromosome, the Davies–Bouldin index (DBi) is performed. Determination of *DB* is explained as follows: at first the average dispersion of cluster_k and cluster_j called e_i, e_j and also D_{jk}, which is the Euclidean distance between cluster_k and cluster_j are calculated. R_k for each in cluster_k is obtained (Eq. 1).

$$R_k = \max_{j \neq k} R_{j,k} = \frac{e_j + e_k}{D_{j,k}} \tag{1}$$

The Davies–Bouldin index is defined as follows: (Eq. 2)

$$DB_k = \frac{1}{k} \sum_{k=1}^{k} R_k \tag{2}$$

Then the detected centers import to *k*-means algorithm for clustering datasets. The flowchart of the proposed model is shown in Fig. 2.

4 Results

Among all the households, cultural goods were used in 17,751 urban and 21,871 rural families. Some of the households have used more than one commodity; so after recognizing these families and obtaining the whole costs of each family, which are spent on cultural goods, costs are ranked to deciles. First three deciles are chosen and the datasets decrease to 1,740 records. Then among these families, the expenditure in each category in household budget is calculated. The results show that in the rural without considering food the first priority of these families is cloths and the rest of them are as follows: housing, health care, transport, miscellaneous commodity, entertainment, furniture, restaurant and junk food, communications, education, and finally drinks and smokes. The priority consumption costs in rural families are shown in Fig. 3.

The arrangement in the figure is: (1) Miscellaneous commodity (2) Drinks and smokes (3) Clothes (4) Housing (5) Furniture (6) Healthcare (7) Transports (8) Communication (9) Entertainment (10) Education (11) Junk food and restaurant.

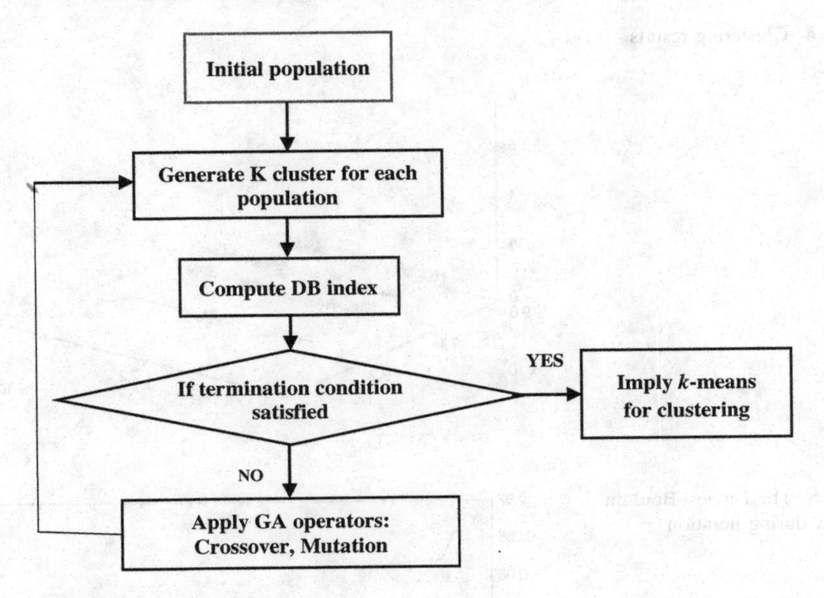

Fig. 2 Proposed model flowchart

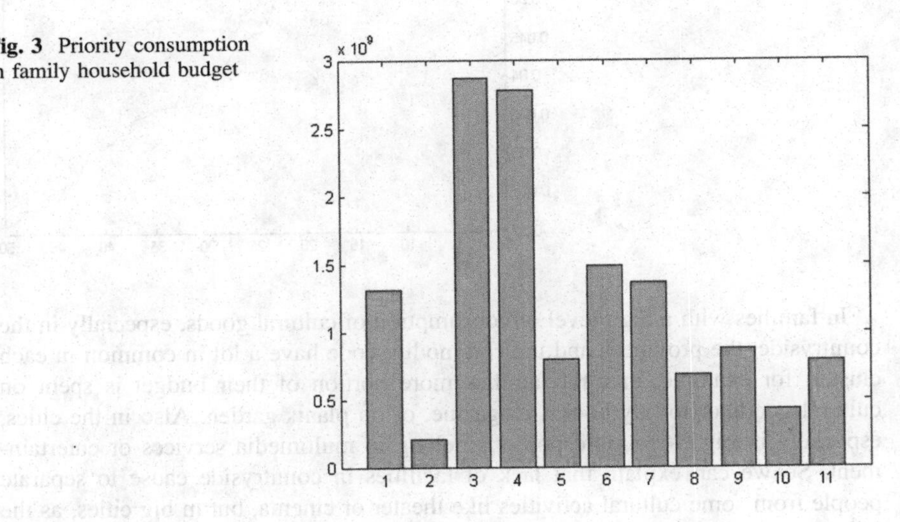

Fig. 3 Priority consumption in family household budget

The combination of *k*-means and GA algorithm is used in this stage to cluster datasets. To estimate the fitness of each chromosome, the Davies–Bouldin index is performed. The optimal number of clusters obtained from genetic algorithm is 5. The clustering results are shown in Fig. 4.

The Davies–Bouldin index, which is obtained from the best population in each iteration, is also shown in Fig. 5.

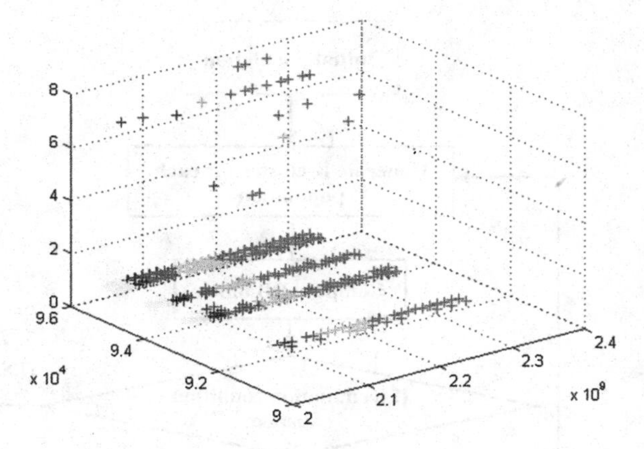

Fig. 4 Clustering results

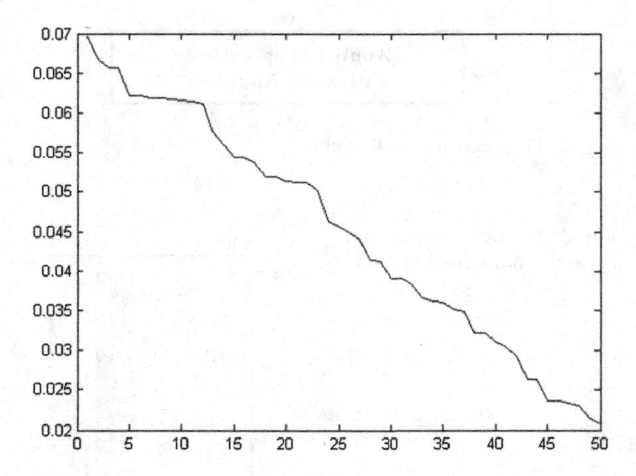

Fig. 5 The Davies–Bouldin index during iteration

In families with a high level of consumption of cultural goods, especially in the countryside, the provinces and the commodity code have a lot in common in each cluster, for example, in rural families more portion of their budget is spent on cultural goods or to buy books, magazine, or on plants garden. Also in the cities, especially in big cities more portions belong to multimedia services or entertainment. So we can explain that lack of facilities in countryside cause to separate people from some cultural activities like theater or cinema, but in big cities, as the type of living is variable, their activities are focused on technologies, reading books or surf the nature or growing plants. The whole of the datasets groups by calculating the intra- and inter-clusters results very well. The Davies–Bouldin index helps to find the distances with more inter and less intra distance among objects.

5 Conclusion and Future Works

The combination of *k*-means and genetic algorithms are considered in this paper for clustering household's cultural goods consumption. GA as a meta-heuristic optimization algorithm increases the speed of achieving optimal solutions. Also introducing the best initial points by GA helps *k*-means to escape from local optima. Households with high level of consumption of cultural goods have a lot in common in the clusters. According to the results, since some of cultural goods are not prevalent in some areas may be one of the main causes is lack of facilities. So, recognizing these shortages and demand of each part of cultural goods helps us to improve the facilities in the areas.

Demand forecasting methods can be divided into two categories:

1. Qualitative methods (subjective)
2. Quantitative methods (computational)

Using neural networks as a computational method can help us to predict the people demands.

References

1. Ebadati, E.O.M., Babaie, S.: Implementation of two stages *k*-means algorithm to apply a payment system provider framework in banking systems. In: Silhavy, R. et al. (eds.) Artificial Intelligence Perspectives and Applications, pp. 203–213. Springer International Publishing, New York (2015)
2. Shahabi, C., Banaei-Kashani, F.: A framework for efficient and anonymous web usage mining based on client-side tracking. In: WEBKDD 2001—Mining Web Log Data Across All Customers Touch Points, pp. 113–144. Springer, New York (2002)
3. Ebadati, E.O.M., et al.: Impact of genetic algorithm for meta-heuristic methods to solve multi depot vehicle routing problems with time windows. Cienc. Tec. Sci. Technol. **29**(7), 9 (2014)
4. Razavi, S., et al.: An efficient grouping genetic algorithm for data clustering and big data analysis. In: Acharjya, D.P., Dehuri, S., Sanyal, S. (eds.) Computational Intelligence for Big Data Analysis, pp. 119–142. Springer International Publishing, New York (2015)
5. Chuang, K.-S., et al.: Fuzzy c-means clustering with spatial information for image segmentation. Comput. Med. Imaging Graph. **30**(1), 9–15 (2006)
6. Wu, Z., Leahy, R.: An optimal graph theoretic approach to data clustering: theory and its application to image segmentation. Pattern Anal. Mach. Intell. IEEE Trans. **15**(11), 1101–1113 (1993)
7. Cai, W., Chen, S., Zhang, D.: Fast and robust fuzzy c-means clustering algorithms incorporating local information for image segmentation. Pattern Recogn. **40**(3), 825–838 (2007)
8. Xie, Y., Phoha, V.V.: Web user clustering from access log using belief function. In: Proceedings of the 1st International Conference on Knowledge Capture. ACM, (2001)
9. Oyanagi, S., Kubota, K., Nakase, A.: Application of matrix clustering to web log analysis and access prediction. In: Proceedings of the WEBKDD (2001)
10. Ip, A., Fong, S., Liu, E.: Optimization for allocating BEV recharging stations in urban areas by using hierarchical clustering. In: Advanced Information Management and Service (IMS), 2010 6th International Conference on 2010. IEEE

11. Baoxing, Q.: On the coordinated development of urbanization in China with a note on the cluster structure. City Plan. Rev. **6**, 000 (2003)
12. Punj, G., Stewart, D.W.: Cluster analysis in marketing research: review and suggestions for application. J. Mark. Res. **20**(2), 134–148 (1983)
13. Liu, H.-H., Ong, C.-S.: Variable selection in clustering for marketing segmentation using genetic algorithms. Expert Syst. Appl. **34**(1), 502–510 (2008)
14. Alam, P., et al.: The use of fuzzy clustering algorithm and self-organizing neural networks for identifying potentially failing banks: an experimental study. Expert Syst. Appl. **18**(3), 185–199 (2000)
15. Tabak, B.M., et al.: Directed clustering coefficient as a measure of systemic risk in complex banking networks. Physica A **394**, 211–216 (2014)
16. Kumar, M.V., Chaitanya, M.V., Madhavan, M.: Segmenting the banking market strategy by clustering. Int. J. Comput. Appl. **45**, 10–15 (2012)
17. Alpaydin, E.: Introduction to machine learning. MIT press, Cambridge (2014)
18. MacQueen, J.: Some methods for classification and analysis of multivariate observations. In: Proceedings of the 5th Berkeley symposium on mathematical statistics and probability (1967)
19. Jain, A.K.: Data clustering: 50 years beyond *k*-means. Pattern Recogn. Lett. **31**(8), 651–666 (2010)
20. Lloyd, S.: Least squares quantization in PCM. Inf. Theor. IEEE Trans. **28**(2), 129–137 (1982)
21. Kanungo, T., et al.: An efficient *k*-means clustering algorithm: analysis and implementation. Pattern Anal. Mach. Intell., IEEE Trans. **24**(7), 881–892 (2002)
22. Jagannathan, G., Wright, R.N.: Privacy-preserving distributed *k*-means clustering over arbitrarily partitioned data. In: Proceedings of the Eleventh ACM SIGKDD International Conference on Knowledge Discovery in Data Mining, ACM (2005)
23. Senthil Kumar, A., Mythili, S.: Parallel implementation of genetic algorithm using *k*-means clustering. Int. J. Adv. Netw. Appl. **3**(6) (2012)
24. Bradley, P.S., Fayyad, U.M.: Refining Initial Points for *k*-means Clustering. In: ICML, Citeseer (1998)
25. Khan, S.S., Ahmad, A.: Cluster center initialization algorithm for *k*-means clustering. Pattern Recogn. Lett. **25**(11), 1293–1302 (2004)
26. Ahmadyfard, A., Modares, H.: Combining PSO and *k*-means to enhance data clustering. In: Telecommunications, 2008. IST 2008. International Symposium on, 2008. IEEE
27. Wu, F.-X., Zhang, W.-J., Kusalik, A.J.: A genetic *k*-means clustering algorithm applied to gene expression data. In: Advances in Artificial Intelligence, pp. 520–526. Springer, New York (2003)
28. Chang, D.-X., Zhang, X.-D., Zheng, C.-W.: A genetic algorithm with gene rearrangement for *k*-means clustering. Pattern Recogn. **42**(7), 1210–1222 (2009)
29. Roy, D.K., Sharma, L.K.: Genetic *k*-means clustering algorithm for mixed numeric and categorical data sets. Int. J. Artif. Intell. Appl. **1**(2), 23–28 (2010)
30. Poikolainen, I., Neri, F., Caraffini, F.: Cluster-based population initialization for differential evolution frameworks. Inf. Sci. **297**, 216–235 (2015)
31. Laszlo, M., Mukherjee, S.: A genetic algorithm that exchanges neighboring centers for *k*-means clustering. Pattern Recogn. Lett. **28**(16), 2359–2366 (2007)
32. Davies, D.L., Bouldin, D.W.: A cluster separation measure. Pattern Anal. Mach. Intell., IEEE Trans. **2**, 224–227 (1979)
33. Verma, M., Hertel, T.W., Preckel, P.V.: Predicting within country household food expenditure variation using international cross-section estimates. Econ. Lett. **113**(3), 218–220 (2011)
34. Selim, H.: Determinants of house prices in Turkey: Hedonic regression versus artificial neural network. Expert Syst. Appl. **36**(2), 2843–2852 (2009)
35. Brown, S., Hole, A.R., Kilic, D.: Out-of-pocket health care expenditure in Turkey: analysis of the 2003–2008 household budget surveys. Econ. Model. **41**, 211–218 (2014)
36. Tang, H.-W.V., Yin, M.-S.: Forecasting performance of grey prediction for education expenditure and school enrollment. Econ. Educ. Rev. **31**(4), 452–462 (2012)

37. Miningou, É.W., Vierstraete, V.: Households' living situation and the efficient provision of primary education in Burkina Faso. Econ. Model. **35**, 910–917 (2013)
38. Chanda, A.: The rise in returns to education and the decline in household savings. J. Econ. Dyn. Control **32**(2), 436–469 (2008)
39. Wallenborn, G., Wilhite, H.: Rethinking embodied knowledge and household consumption (2014)
40. Ahmad, E., Arshad, M.: Household budget analysis for Pakistan under varying the parameter approach. East Asian Bureau of Economic Research (2007)

A PSO with Improved Initialization Operator for Solving Multiple Sequence Alignment Problems

Rohit Kumar Yadav and Haider Banka

Abstract In this paper, an improved particle swarm optimization (PSO) technique is explored for aligning multiple sequences. PSO has recently emerged as a new randomized heuristic method for both real-valued and discrete optimization problems. This is a nature-inspired algorithm based on the movement and intelligence of swarms. Here each solution is represented in encoded form as 'position' like 'chromosome' in genetic algorithm (GA). The fitness function is designed accordingly to optimize the objective functions, i.e., maximizing the matching components of the sequences and reducing the number of mismatched components in the sequences. The performance of the proposed method has been tested on publicly available benchmark datasets (i.e., Bali base) to establish the potential of PSO to solve alignment problem with better and/or competitive performance. The results are compared with some of the well known existing methods available in literature such as DIALIGN, HMMT, ML-PIMA PILEUP8, and RBT-GA. The experimental results showed that proposed method attained better solutions than the others for most of the cases.

Keywords Multiple sequence alignment (MSA) • Pairwise alignment • Particle swarm optimization (PSO) • Dynamic programming (DP)

1 Introduction

MSA is the most predominant and demanding assignment for solving computational biology problems. Biological sequence can be analyzed with the help of multiple sequence alignment. MSA is the combination of nucleotide or amino acid

R.K. Yadav (✉) · H. Banka
Department of Computer Science and Engineering, Indian School of Mines, Dhanbad 826004, India
e-mail: rohit.ism.123@gmail.com

H. Banka
e-mail: hbanka2002@yahoo.com

S. Das et al. (eds.), *Proceedings of the 4th International Conference on Frontiers in Intelligent Computing: Theory and Applications (FICTA) 2015*, Advances in Intelligent Systems and Computing 404, DOI 10.1007/978-81-322-2695-6_25

sequences. It is helpful to understand the primary and secondary structure of given sequence. Phylogenetic distance of given sequences can be also find with the help of MSA. Prediction of molecular structure [1–3] of given sequences is also a very important task in computational biology. It can also be found with the help of MSA. So we can handle different types of tasks in computational biology using MSA. This is the major reason researchers are interested to solve MSA problem in efficient manner. MSA problem is meaningful only when all sequences have a common origin. The goal of MSA is to compare row with the same non-idle character. Also to compare row with same non-idle character to different character as far as possible. Hence we can say that MSA is an important problem in bioinformatics to know genetic and phylogenetic relationship. In the past, there are many methods to solve this problem.

Dynamic programming method can be solved MSA problem in efficient manner. To upgrade the solution of MSA problem dynamic programming (DP) use large domain scoring function. Needleman and Wunsch [4] proposed a method to solve MSA problem using DP in 1970. But DP crash speedily if the length and number of sequence is increasing. In other word we can say that if the length and number of sequence is increase in MSA then the time complexity of DP is also increase exponentially. Then the MSA problem leads to NP-hard. Hence our aim is to solve MSA problem in efficient way. In other words, we have to maximize the similarity among the sequences (which is given in MSA problem) in less complexity.

The progressive alignment method is another technique to solve MSA problem. This technique is useful because using this MSA problem can be solved in less time complexity [5, 6]. According to this procedure we start with the most similar sequence and then move to dynamically more divergent sequence or group of sequences. The standard representative of progressive methods is CLUSTALW [7].

Iterative method is another option to solve MSA problem. The procedure of iterative method is it starts with initial alignment and frequently filters solutions until no more improvement possible. Hence in this case results are not depending on initial alignment. The main objective of the iterative approach for MSA is to globally improve the quality of a sequence alignment. Simulated annealing [8, 9] is a kind of iterative or stochastic method.

Evolutionary algorithms (EAs) are population-based global search algorithms. When using EAs for MSA, an initial generation is generated by random manner, and then the steps of an EA are applied to improve the similarities among the sequences. There are some other genetic algorithm (GA) based methods for MSA, such as SAGA [10], RBT-GA [11] and others. The authors [11] solved 34 problems from reference sets 2 and 3 of the benchmark Balibse 2.0 dataset. The drawback of these evolutionary methods is also local optima due to poor diversity of the solutions.

In proposed method, to solve problem of local optima we used DP in initialization operator. In PSO, there is an inbuilt elitism (best solution always keeping in next generation) operator. Because of this modification of initial operator and PSO elitism property proposed method can solve efficiently MSA problem.

This paper is organized as follows. After the introduction, a discussion about the particle swarm optimization is given in Sect. 2. Section 3 presents the steps of the proposed method, the experimental study of the proposed and other methods are discussed in Sect. 4. In Sect. 5, our conclusions are provided.

2 Basics of PSO

Particle swarm optimization (PSO) is an artificial intelligence technique. It is used to find the approximation solution in the large search space. It is derived by Dr. Eberhart and Dr. Kennedy in 1995 [12] inspired by bird flock or fish school. Equations 1 and 2 are shown below for updating the position and velocity.

$$V_{i,d}(t+1) = w * V_{i,d}(t) + c_1 * r_1 * (\text{pbest} - X_{i,d}) + c_2 * r_2 * \left(\text{pbest}_{\text{gbest},d} - X_{i,d}\right) \quad (1)$$

$$X_{i,d}(t+1) = X_{i,d}(t) + V_{i,d}(t+1) \quad (2)$$

where V is the particle velocity, X is the present position, C_1 and C_2 accelerating constants, r_1 and r_2 are the random numbers between 0 and 1, w is the inertial weight, and t is the iteration.

3 Proposed Method

The steps of our method using PSO are: initialize particle's position, update the position of particle using velocity and position equation which is given in (1) and (2). The steps of this method are explained below.

3.1 Initial Generation

The aim of this step is to generate initial solutions. In this step we generate initial solution by put the gap between residues. How we generate initial population is given below as an example. Initial MSA has been given in Fig. 1a. After that we find pairwise alignment each and every pair using Needleman and Wunsch algorithm [4]. First pair (1, 2) is shown in Fig. 1b. Now we find pairwise alignment of first pair (1, 2) which is shown in Fig. 1c. Since this is a problem of 4 sequences, we have 4 * (4 – 2)/2 = 6 pairs. Similarly we can find pairwise alignment of each pairs. Now According to our method we can generate random permutation between 1 to N. In this particular example we have 4 sequences. So we have to generate random permutation from 1 to 4. Suppose we generate random permutation (3, 4, 1, and 2). Then complete alignment of initial MSA is shown in Fig. 1d. Similarly we can generate K number of solutions using K times Random permutation generates between 1 to N. At this stage our initial population has created.

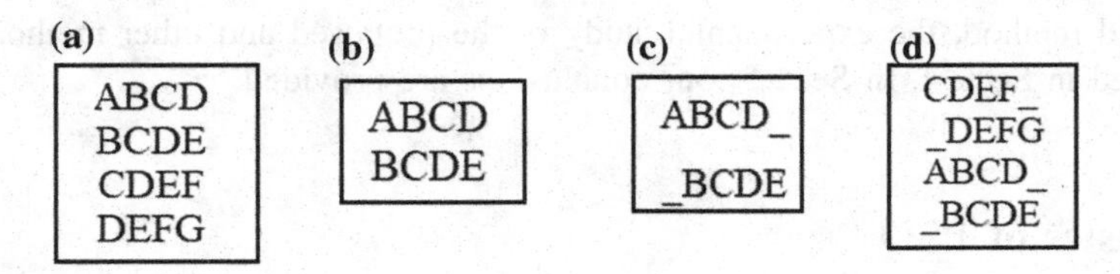

Fig. 1 A complete process to generate initial solution

3.2 *Particle Representation*

In PSO each solution is represented as particle.

$$X_i = (X_i^1, \ldots.., , X_i^d, \ldots\ldots, X_i^{n)} \text{ where } 1 \leq i \geq N \quad (3)$$

where N is the number of particles and n is the number of dimensions in a particle. We are taking 1 solution out of 200 solutions. Here we define how to encode MSA problem into particle. Initial solution has been given in Fig. 2.

Binary encoding Scheme- In the encoding scheme put 1 in position of gap and put 0 in the position of protein sequence. After encoding scheme it has been shown in Fig. 3. After that we are taking decimal value of this binary encoded value from bottom to top of each column. Hence our particle representation of this solution is X_1 = (1, 0, 0, 8, 2, and 4). Hence the number of columns in the MSA is the number of dimension in Particle. Now according to encoding scheme we can generate particle of all MSA which has come after initials generation.

3.3 *Fitness Function*

Multiple sequence alignment uses WSPM function for measure the fitness. Here each column score is equal to summing the product of score of each pair of symbol

C	G	A	-	G	T
A	T	G	T	C	-
T	G	T	T	-	T
-	C	C	A	T	C

Fig. 2 Initial solution

	0	0	0	1	0	0
	0	0	0	0	0	1
	0	0	0	0	1	0
	1	0	0	0	0	0
Particle →	1	0	0	8	2	4

Fig. 3 Encoding scheme

and their pair weight. The score of the entire alignment is then summed over all column scores by using (4)

$$S = \sum_{l=1}^{L} S_l, \text{ where } S_l = \sum_{i=1}^{N-1} \sum_{j=i+1}^{N} W_{ij} \text{Cost}(A_i, A_j) \quad (4)$$

Here, S is the cost of multiple sequence alignments. L is the number of columns in the entire alignment. S_l is the cost of the lth column of L length. N is the number of sequences, W_{ij} is the weight of sequence i and j. It defines diversity between two sequences.

W_{ij} = (Number of Mismatch character in the alignment)/(Total align length) Cost(A_i, A_j) is the alignment score between two aligned sequences A_i and A_j. When $A_i \neq$ "_" and $A_j \neq$ "_" then cost (A_i, A_j) is determined from the percentage of acceptable mutations matrix. Also when $A_i =$ "_" and $A_j =$ "_" then cost (A_i, A_j) = 0. Finally, the cost function cost (A_i, A_j) includes the sum of the substitution costs of the insertion or deletions when $A_i =$ "_" and $A_j \neq$ "_" or $A_i \neq$ "_" and $A_j =$ "_" then cost (A_i, A_j) = 1.

3.4 Illustration

Suppose in initialization we are taking 5 solutions. And initial velocity of each particle of each dimension is 0. After initialization state, 1 solution out of 100 has been shown in Fig. 4. Now according to our encoding scheme this solution becomes a particle (2, 8, 0, 0, 5). Similarly we can find particle of each solutions.

Hence, P_1 = (2, 8, 0, 0, 5) P_2 = (4, 4, 0, 1, 2) P_3 = (4, 2, 1, 0, 2) P_4 = (8, 2, 4, 0, 2) P_5 = (4, 1, 0, 2, 4)

$pbest_1$ = (2, 8, 0, 0, 5) $pbest_2$ = (4, 4, 0, 1, 2) $pbest_3$ = (4, 2, 1, 0, 2) $pbest_4$ = (8, 2, 4, 0, 2) $pbest_5$ = (4, 1, 0, 2, 4)

After that we can find the fitness value of each particle according to Eq. (4)

Fit (P_1) = 11 Fit (P_2) = 20 Fit (P_3) = 15 Fit (P_4) = 10 Fit (P_5) = 5

Fit ($pbest_1$) = 11 Fit ($pbest_2$) = 20 Fit ($pbest_3$) = 15 Fit ($pbest_4$) = 10 Fit ($pbest_5$) = 5

A	-	C	T	G
C	T	G	A	-
-	T	C	A	G
T	C	C	G	-

Fig. 4 Initial alignment

Now update the velocity and position using Eqs. (1) and (2)

$$V_{i,d}(t+1) = w^* V_{i,d}(t) + c_1{}^* r_1{}^* (\text{pbest} - X_{i,d}) + c_2{}^* r_2{}^* \left(\text{pbest}_{\text{gbest},d} - X_{i,d}\right)$$
$$V_{i,d}(t+1) = 0.729^*0 + 2.05^*0.29^*(2-2) + 2.05^*0.49^*(4-2)$$
$$= 0 + 0 + 2.009$$
$$= 2.009$$

Now update the position

$$X_{i,d}(t+1) = X_{i,d}(t) + V_{i,d}(t+1)$$
$$= 2 + 2.009 = 4.009$$
$$= 4$$

We are taking floor value of updated position because our encoding scheme is discrete valued function. Hence updated position is 4.

Similarly, we can update each particle of each dimension.

Hence, $P_1 = (4, 0, 0, 0, 11)$ $P_2 = (4, 2, 0, 1, 8)$ $P_3 = (2, 2, 0, 1, 1)$ $P_4 = (3, 4, 0, 1, 2)$ $P_5 = (1, 1, 0, 1, 2)$

We can find fitness value of new particles

Fit (P_1) = 40 Fit (P_2) = 30 Fit (P_3) = 10 Fit (P_4) = 5 Fit (P_5) = 12

After that, we can update the pbest value using algorithm which is given in Algorithm 1.

Algorithm 1. For updating pbest value

```
1 Begin
2   for i=1 to N
3     If(Fit (Pi)> Fit (pbest_i)
4          pbest_i = Pi
5     End If
6   End for
7 End
```

Since 40 > 11

Hence, $\text{pbest}_1 = (4, 0, 0, 0, 11)$ $\text{pbest}_2 = (2, 2, 0, 1, 1)$ $\text{pbest}_3 = (4, 2, 1, 0, 2)$ $\text{pbest}_4 = (8, 2, 4, 0, 2)$ $\text{pbest}_5 = (1, 1, 0, 1, 2)$

And gbest = 1

Hence our best particle is (4, 0, 0, 0, 11)

Now, we can put the gap in each column according to our decoding scheme which is shown in Fig. 5a. After that we can put protein sequences in vacant positions to make final MSA. Final MSA has been shown in Fig. 5b.

(a)

4	0	0	0	11
				-
-				
				-
				-

(b)

A	C	T	G	-
-	C	T	G	A
T	C	A	G	-
T	C	C	G	-

Fig. 5 Final solution after one generation. **a** Decoding Scheme. **b** Final MSA

4 Experimental Study

In experimental section, we study the production of propound method after that we compare our algorithm to other existence algorithm.

The implementation parameter of PSO is given in Table 1. It has taken from [13].

4.1 Effect of Initialization Operator

In this paper, we have used DP method to generate initial population and we claim that proposed method perform better. To prove superiority of our components, we have designed two sets of experiments. In these sets, PSO runs with simple initial operator and improved initial operator. We have taken five datasets randomly from Bali base database. Out of these five datasets, three from reference set 2 and two from reference set 3. After that we find the corresponding Bali score w.r.to best WSPM solution of each datasets. We saw that PSO with improved initial operator perform better than PSO with simple initial operator in all five datasets. This experiment shows that superiority of our proposed initial operator. The details of these experiments are reported in Table 2.

Table 1 Implementation parameter of PSO

N	200
V_{min}	0
V_{max}	$2^{\text{Number of sequences}}$
w	0.758
C_1	2.050
C_2	2.050
r_1	Random value between 0 and 1
r_2	Random value between 0 and 1

Table 2 Performance test PSO with simple and improved mutation operator

Datasets	PSO with improved initial operator	PSO with simple initial operator
1aboA	**0.825**	0.745
1tvxA	**0.911**	0.815
1ubi	**0.754**	0.695
Kinase	**0.839**	0.612
1ped	**0.804**	0.723
Avg score	**0.826**	0.718

4.2 Comparing Proposed Method with Other Existence Methods

In this paper, we have taken randomly 16 datasets from Bali base database. Out of these 16, 12 from reference set 2 and 4 from reference set 3. We have taken the results of other existence method for corresponding datasets from paper RBT-GA [11].

Performance of Proposed Method w.r.to Reference 2 Datasets

We have taken 12 datasets from this reference set. All are materially different in terms of length and number. We have compared proposed method with other well known existence methods such as DIALIGN, HMMT, ML-PIMA, PILEUP-8, and RBT-GA to examine the performance of our proposed method. Table 3 shows that out of 12 datasets, proposed method perform better than other existence methods in seven test cases. In five test cases, proposed method could not achieve best solution but it is always close to the best solution. The overall performance is that, from the average score in Table 3 proposed method performs better than other methods.

Table 3 Experimental results on reference 2

Datasets	DIALI	HMMT	ML-PIMA	PILEUP-8	RBT-GA	Proposed
1aboA	0.384	0.724	0.220	0.000	0.812	**0.838**
1idy	0.000	0.353	0.000	0.000	**0.997**	0.559
1csy	0.000	0.000	0.000	0.114	0.735	**0.797**
1r69	0.675	0.000	0.675	0.450	**0.900**	0.251
1tvxA	0.000	0.276	0.241	0.345	**0.891**	0.534
1tgxA	0.630	0.622	0.543	0.318	0.835	**0.886**
1ubi	0.000	0.053	0.129	0.000	0.795	**0.799**
1wit	0.724	0.641	0.463	0.476	**0.825**	0.757
2trx	0.734	0.739	0.702	0.87	**0.982**	0.787
1sbp	0.043	0.214	0.054	0.177	0.778	**0.816**
1havA	0.000	0.194	0.238	0.493	0.792	**0.817**
1uky	0.216	0.395	0.306	0.562	0.625	**0.719**
Avg score	0.283	0.350	0.297	0.317	0.830	**0.708**

Table 4 Experimental results on reference 3

Datasets	DIALI	HMMT	ML-PIMA	PILEUP-8	RBT-GA	Proposed
1idy	0.000	0.227	0.000	0.000	0.548	**0.623**
1r69	0.524	0.000	**0.905**	0.000	0.374	0.425
1ubi	0.000	0.366	0.000	0.268	0.310	**0.598**
1uky	0.139	0.037	0.148	0.083	0.350	**0.411**
Avg score	0.165	0.157	0.263	0.087	0.395	**0.514**

Performance of Proposed Method w.r.to Reference 3 Datasets

We have taken 4 datasets out of 12 from reference set 3. The experimental results illustrated in Table 4 show that in three test cases proposed method performs better, even in one test case where proposed method did not perform was close to the best. Table 4 also shows that average solution of proposed method outperformed the other existence methods.

5 Conclusion

This paper addresses a novel PSO (with improved initial operator) for alignment problem. We have tested a significant number of experiments for evaluating the performance of our proposed algorithm. PSO method was tested with simple initial operator as well as improved initial operator to show the dominance and efficacy of the proposed initial operator. The proposed method used weighted pair function for calculating fitness. We have taken a significant number of datasets from Bali base database version 2.0 which is given in RBT-GA to show the efficacy of our algorithm. After that, we have taken corresponding Bali score of the best solution to compare with other existence methods. We saw that proposed method performed better for most of the test cases. While proposed method for some of the test cases did not perform better but it was always close to the best solution. Improved initial operator is the reason behind the efficacy of proposed method. Hence we can judge that proposed method is better than other existence methods.

References

1. Gusfield, D.: Algorithms on Strings, Trees, and Sequences: Computer Science and Computational Biology. Cambridge University Press, New York (1997)
2. Sankoff, D.: Matching sequence under deletion-insertion constraints. Proc. Natl. Acad. Sci. USA. **69**, 4–6 (1972)
3. Gupta, S.K., Kececioglu, J.D., Schaffer, A.A.: Improving the practical space and time efficiency of the shortest-paths approach to sum-of-pairs multiple sequence alignment. J. Comput. Biol. **2**, 459–472 (1995)

4. Needleman, S.B., Wunsch, C.D.: A general method applicable to the search for similarities in the amino acid sequences of two proteins. J. Mol. Biol. **42**, 161–245 (1970)
5. Taylor, W.R.: A flexible method to align large numbers of biological sequences. J. Mol. Evol. **28**, 161–169 (1988)
6. Feng, D., Doolittle, R.F.: Progressive sequence alignment as a prerequisite to correct phylogenetic trees. J. Mol. Evol. **25**, 351–360 (1987)
7. Thompson, J.D., Higgins, D.G., Gibson, T.J.: CLUSTAL W: improving the sensitivity of progressive multiple sequence alignment through sequence weighting position-specific gap penalties and weight matrix choice. Nucleic Acids Res. **22**, 4673–4680 (1994)
8. Kim, J., Pramanik, S., Chung, M.J.: Multiple sequence alignment using simulated annealing. Comput. Appl. Biosci. **10**, 419–426 (1994)
9. Lukashin, A.V., Engelbrecht, J., Brunak, S.: Multiple alignment using simulated annealing: Branch point definition in human mRNA splicing. Nucleic Acids Res. **20**, 2511–2516 (1992)
10. Notredame, C., Higgins, D.G.: SAGA: sequence alignment by genetic algorithm. Nucleic Acids Res. **24**, 1515–1524 (1996)
11. Taheri, J., Zomaya, A.Y.: RBT-GA: a novel Meta heuristic for solving the multiple sequence alignment problem. BMC Genom. **10**, 1–11 (2009)
12. Kennedy, J., Eberhart, R.: Particle swarm optimization. In: Proceedings of the IEEE International Conference on Neural Networks. vol 4, pp. 1942–1948 (1995)
13. Daniel, B., Kennedy, J.: Defining a standard for particle swarm optimization. In: Proceedings of the 2007 IEEE Swarm Intelligence Symposium, pp. 120–127 (2007)

Profit Maximization of TSP with Uncertain Parameters Through a Hybrid Algorithm

Aditi Khanra, Manas Kumar Maiti and Manoranjan Maiti

Abstract Here, a model of travelling salesman problem (TSP) with uncertain parameters is formulated and solved using a hybrid algorithm. For the TSP, there are some fixed number of cities and the costs and time durations for travelling from one city to another are known. Here, a travelling salesman (TS) visits and spends some time in each city for selling the company's product. The return and expenditure at each city are dependent on the time spent by the TS at that city and these are given in functional forms of t. The total time limit for the entire tour is fixed and known. Now, the problem for the TS is to identify tour programme and also to determine the stay time at each city so that total profit out of the system is maximum. In reality, profit and cost for spending time in a city by the salesman are finite but fuzzy in nature. So fuzzy expenditure and fuzzy return are taken to make the problem realistic. Here, the model is solved by a hybrid method combining the particle swarm optimization (PSO) and ant colony optimization (ACO). The problem is divided into two subproblems where ACO and PSO are used successively iteratively in a generation using one's result for the implicitly other. Numerical experiments are performed to illustrate the models. Some behavioural studies of the models and convergency of the proposed hybrid algorithm with respect to iteration numbers and cost matrix sizes are presented.

Keywords Ant colony optimization · Particle swarm optimization · Travel cost · Travel time · Profit · Fuzzy returns · Fuzzy expenditure · Hybrid algorithm · Fuzzy credibility

A. Khanra(✉)
Bohichberia High School (H.S.), Purba Medinipur, West Bengal, India
e-mail: aditikhanra@gmail.com

M.K. Maiti
Department of Mathematics, Mahishadal Raj College, Mahishadal, Medinipur 721628, West Bengal, India
e-mail: manasmaiti@yahoo.co.in

M. Maiti
Department of Applied Mathematics, Vidyasagar University, Medinipur 721102, West Bengal, India
e-mail: mmaiti2005@yahoo.co.in

S. Das et al. (eds.), *Proceedings of the 4th International Conference on Frontiers in Intelligent Computing: Theory and Applications (FICTA) 2015*, Advances in Intelligent Systems and Computing 404, DOI 10.1007/978-81-322-2695-6_26

1 Introduction

The travelling salesman problem (TSP) [1] has a list of cities and the distances between each pair of cities. Now there is a salesman who wants to find a shortest route to visit each city exactly once and return to the original city. Different types of TSPs have been solved by the researchers during last two decades. Among these, TSP with time windows [2], stochastic TSP [3–6], asymmetric TSP [7], TSP with precedence constraints [8], probabilistic TSP (PTSP) [8], etc. are worth mentioning. In TSP with precedence constraints, there exists an order in which the cities should be visited. In asymmetric TSP, cost of travelling from vertex (node/city) v_i to v_j is not equal to the cost of travelling from vertex v_j to v_i. In stochastic TSP, each vertex is visited with a given probability and goal is to minimize the expected distance/cost of a priori tour. In TSP with time windows, each vertex is visited within a specified time windows. PTSP is a TSP problem where each customer has a given probability of requiring a visit.

During the last decades, several algorithms emerged to approximate the optimal solution of TSP such as tabu searching method [9], neural networks [10, 11], simulated annealing [12], genetic algorithms (GA) [4, 6, 13–16], ant colony optimization (ACO) [17–21] and particle swarm optimization (PSO) [22, 23].

ACO is an important soft computing technique for solving optimization problems. In ACO, the behaviour of real ants to find the shortest path between their nest and food source has been used. Several ACO algorithms are available to solve the well-known NP-Hard TSP. Dorigo and Gambardella (1997) described an artificial ant colony capable of solving the TSP. In 2007, Cheng and Mao presented a modified ant colony system for solving the TSP with time windows. Ibanez and Blum proposed a Beam-ACO which is a hybrid method combining ACO with beam search to solve TSP in 2010. In 2011, Ghafurian and Javadian proposed an ACO algorithm for solving fixed destination multi-depot multiple travelling salesmen problems. In 2013, Bai, Yang, Chen, Hu, and Pan proposed a model inducing max–min ant colony optimization for asymmetric TSP.

The particle swarm optimization (PSO) algorithm has been introduced by Kennedy and Eberhart [24, 25] and is inspired by the emergent motion of a flock of birds searching for food. Since its introduction, PSO has seen many improvements and applications. Among them, a brief discussion on its improvement in different directions and applications is given in [26]. Some researchers tried to improve the performance of PSO by variable parameters [27, 28]. An important improvement was adding the inertia weight into the updating rule of PSO [28]. Ratnaweera et al. [28] used time-varying strategies of inertia weight. Later, Clerc and Kennedy [17] presented a constriction factor to form a new inertia weight, which has proven effective and is the current state-of-the-art. Inspite of the above developments, there are some lacunas/gaps in forming the realistic TSPs, These are as follows:

(i) Although the visit of a salesman is organized in order to get a return for the company, till now, none has considered the returns at each city out of the tour. **(ii)** Most of the TSPs are concerned with the minimization of tour cost or travel time. However, for selling or canvassing a product, a salesman has to spend some time at each city and incurs some expenditure, which may vary, for this. This has not been overlooked by the TSP researchers. **(iii)** Normally, a salesman is asked by his/her company to finish the entire activities including the tour (travel and stay times) within a specified time limit. This constraint has been taken into consideration. **(iv)** Although ultimate goal of a company is to make the profit through the sales men/women, till now, no TSP has been formulated as a profit maximization problem considering fuzzy returns and fuzzy expenditures. **(v)** In reality, profit and cost for spending time in a city by the salesman are finite but fuzzy in nature. **(vi)** In order to bridge the above gaps, a new TSP has been formulated and solved.

The above assumptions are realistic and have applications in medicine-producing firms with respect to their medical representatives and salesmen. For a medical firm, normally a medical representative/salesman is asked to tour a number of towns/subtowns within a limited time. In this process, he/she fixes the paths and spends some time at the towns/stations in such a way that total outcome/benefit of the tour is maximum. He/she incurs some costs for travel, pays for stay at stations and earns indirectly through doctors/medical shops by canvassing and presentation. Here, returns from a city and cost for spending time in a city by the salesman are finite but fuzzy in nature. All these are done within the limited time fixed by the medical firm. The proposed TSP incorporating the above assumptions is the most appropriate for the real-life medical house problems and incorporated in the proposed model.

In this paper, for the proposed TSP, a salesman visits some fixed number of cities following the TSP rule and spends sometime at each visited city. He/she earns some returns which are uncertain and incurs some uncertain expenditure at each city and these are stay time dependent in some functional forms of time. Here, total allowable time for the entire tour including stay times is fixed. Now, the problem for the salesman is to fix the tour programme and the stay times at each city so that total profit from the tour is maximum. Thus a TSP is formulated as maximization problem with fuzzy returns and stay expenditures.

In comparison to the usual TSPs, here, proposed TSP consists of two sub-optimization models, (i) minimization of total travel costs between the cities and (ii) allocation of stay times at the cities, so that total outcomes due to stay are maximum . Then a trade-off between these two sub-models is made. For these twofold TSP optimization problems, a hybrid algorithm combining the algorithms of ACO and PSO is designed and applied successfully. Here, ACO and PSO are used successively and iteratively in a generation using one's result for the other. The proposed TSP is illustrated with numerical examples. Some interesting model behaviours are presented with different sizes of cost matrices. It is shown that more stay at stations does not fetch more profit. Some parametric studies of the proposed hybrid algorithm with respect to iteration number and cost matrix sizes are also presented.

Rest of the paper is organized as follows. In Sect. 2, some mathematical prerequisites required for mathematical formulation of the model are presented. Models are formulated in Sect. 3. In Sect. 4, hybrid ACO-PSO system is described. In Sect. 5, experimental results are presented. Finally, a brief discussion, models' behavioural and sensitivities studies and conclusion are, respectively, drawn in Sects. 6, 7 and 8.

2 Mathematical Prerequisites

2.1 Fuzzy Number [29]

A fuzzy number $\tilde{a}$ is a fuzzy set in $\Re$ (set of real number) characterized by a membership function, $\mu_{\tilde{a}}(x)$, which is both normal (i.e. $\exists$ at least one point $x \in \Re$, s.t. $\mu_{\tilde{a}}(x) = 1$) and convex.

2.2 Triangular Fuzzy Number (TFN):

A TFN $\tilde{a} = (a_1, a_2, a_3)$ has three parameters a_1, a_2, a_3, where $a_1 < a_2 < a_2$, whose membership function $\mu_{\tilde{a}}(x)$ is given by

$$\mu_{\tilde{a}}(x) = \begin{cases} \frac{x-a_1}{a_2-a_1} & \text{for } a_1 \leq x \leq a_2 \\ \frac{a_3-x}{a_3-a_2} & \text{for } a_2 \leq x \leq a_3 \\ 0 & \text{otherwise} \end{cases}$$

2.3 Possibility (Pos), Necessity (Nes) Measure:

Let $\tilde{a}$ and $\tilde{b}$ be two fuzzy numbers with membership functions $\mu_{\tilde{a}}$ and $\mu_{\tilde{b}}$, respectively. Then taking degree of uncertainty as the semantics of fuzzy number

$$\text{Pos}\ (\tilde{a} \star \tilde{b}) = \sup\{\min(\mu_{\tilde{a}}(x), \mu_{\tilde{b}}(y)), x, y \in \Re, x \star y\} \tag{1}$$

where $\star$ is any one of the relations $>, <, =, \leq, \geq$. Analogously if $\tilde{B}$ is a crisp number, say b, then

$$\text{Pos}\ (\tilde{a} \star b) = \sup\{\mu_{\tilde{a}}(x), x \in \Re, x \star b\} \tag{2}$$

On the other hand, necessity measure of an event $\tilde{a} \star \tilde{b}$ is a dual of possibility measure. The grade of necessity of an event is the grade of impossibility of the opposite event and is defined as

$$\text{Nes}\ (\tilde{a} \star \tilde{b}) = 1 - \text{Pos}\ (\overline{\tilde{a} \star \tilde{b}}) \tag{3}$$

where $\overline{\tilde{a} \star \tilde{b}}$ represents complement of the event $\tilde{a} \star \tilde{b}$.

Lemma 1 *If $\tilde{a} = (a_1, a_2, a_3)$ and $\tilde{b} = (b_1, b_2, b_3)$ be two TFNs where $0 < a_1 < a_2 < a_2$ and $0 < b_1 < b_2 < b_2$, then $Pos(\tilde{a} > \tilde{b})$ is given by*

$$Pos(\tilde{a} > \tilde{b}) = \begin{cases} 1 & a_2 \geq b_2 \\ \frac{a_3 - b_1}{(a_3 - a_2)(b_2 - b_1)} & a_2 < b_2 \text{ and } a_3 < b_1 \\ 0 & a_3 \leq b_1 \end{cases}$$

and $nes(\tilde{a} > \tilde{b})$ is given by

$$nes(\tilde{a} > \tilde{b}) = \begin{cases} 1 & b_3 \leq a_1 \\ \frac{a_2 - b_2}{a_3 - b_2 + a_2 - a_1} & b_2 \leq a_2 \text{ and } a_1 < b_3 \\ 0 & b_2 \leq a_2 \end{cases}$$

2.4 Credibility (Cr) Measure:

Using (2) and (3), credibility measure of an event $\tilde{a} \star \tilde{b}$ is denoted by Cr $(\tilde{a} \star \tilde{b})$ and is defined as

$$\text{Cr}\ (\tilde{a} \star \tilde{b}) = [\text{Pos}\ (\tilde{a} \star \tilde{b}) + \text{Nes}\ (\tilde{a} \star \tilde{b})]/2 \tag{4}$$

Lemma 2 *If $\tilde{a}$ and $\tilde{b}$ be two fuzzy numbers, then credibility measure of the event $\tilde{a} > \tilde{b}$ is denoted by $Cr(\tilde{a} > \tilde{b})$ and is given by*

$$Cr(\tilde{a} > \tilde{b}) = \begin{cases} 1 & b_3 < a_1 \\ \frac{1}{2}(1 + \frac{a_2 - b_2}{b_3 - b_2 + a_2 - a_1}) & b_2 \leq a_2 \text{ and } a_1 < b_3 \\ \frac{1}{2}(\frac{a_3 - b_1}{a_3 - a_2 + b_2 - b_1}) & a_2 < b_2 \text{ and } a_3 > b_1 \\ 0 & a_3 < b_1 \end{cases}$$

2.5 Graded Mean Integration Value (GMIV) of Fuzzy Number:

For the generalized fuzzy number $\tilde{a}$ with membership function $\mu_{\tilde{a}}(x)$, according to [30], the GMIV of $\tilde{a}$ is denoted by $P(\tilde{a})$ and defined as

$$P(\tilde{a}) = \int_0^1 x(1-w)L^{-1}(x) + wr^{-1}(x)dx / \int_0^1 xdx = 2\int_0^1 x(1-w)L^{-1}(x) + wr^{-1}(x)dx$$

where the pre-assigned parameter $w \in [0, 1]$ refers to the degree of optimism. $w = 1$ represents an optimistic point of view, $w = 0$ represents a pessimistic point of view and $w = 0.5$ indicates a moderately optimistic decision makers point of view, and membership function $\mu_{\tilde{a}}$ is given by

$$\mu_{\tilde{a}} = \begin{cases} Ł(x) & a \leq x \leq b \\ 1 & b \leq x \leq c \\ \mathrm{R(x)} & c \leq x \leq d \\ 0 & \text{otherwise} \end{cases}$$

Using this rule, GIVM of TFN is obtained and presented as

$$[(1 - w)a + 2b + wc]/3 \quad \text{where } (a, b, c) \text{ is a TFN .}$$

3 Model Formulation

3.1 Classical TSP for Minimum Total Travel Cost (Model1A):

In a classical two-dimensional TSP, TSP can be represented as graph $G = (V, E)$, where $V = 1, 2, \ldots, N$ is the set of nodes and E is the set of edges. A salesman has to travel N cities at minimum cost. In this tour, salesman starts from a city, visits all the cities exactly once and comes to the starting city using minimum cost. Let c_{ij} be the cost for travelling from ith city to jth city. Then the model is mathematically formulated as

$$\left.\begin{array}{ll} \text{Determine } x_{ij}, i = 1, 2, \ldots, N, j = 1, 2, \ldots, N. & \\ \text{to minimize } Z = \sum\limits_{i=1}^{N}\sum\limits_{j=1}^{N} x_{ij}c_{ij}, & \\ \text{subject to } \sum\limits_{i=1}^{N} x_{ij} = 1, & j = 1, 2, \ldots, N \\ \sum\limits_{j=1}^{N} x_{ij} = 1, & i = 1, 2, \ldots, N \\ \sum\limits_{i \in S}\sum\limits_{j \in S} x_{ij} \leq |S| - 1, \forall S \subset V, & \\ \qquad x_{ij} \in \{0, 1\}. & \end{array}\right\} \quad (5)$$

where $x_{ij} = 1$ if the salesman travels from city-i to city-j, otherwise $x_{ij} = 0$
Let $(x_1, x_2, \ldots, x_N, x_1)$ be a complete tour of a salesman, where $x_i \in \{1, 2, .., N\}$ for i = 1, 2,… ,N and all x_i's are distinct. Then the above model reduces to

$$\left.\begin{array}{l}\text{Determine a complete tour } (x_1, x_2, \dots, x_N, x_1)\\ \text{to minimize } Z = \sum_{i=1}^{N-1} c_{x_i,x_{i+1}} + c_{x_N,x_1}\end{array}\right\} \quad (6)$$

Classical TSP with time for minimum total travel time (Model1B): Let t_{ij} be the time for travelling from ith city to jth city. Following (6), this model is formulated as

$$\left.\begin{array}{l}\text{Determine a complete tour } (x_1, x_2, \dots, x_N, x_1)\\ \text{To minimize Z} = \sum_{i=1}^{N-1} t_{x_i,x_{i+1}} + t_{x_N,x_1}\end{array}\right\} \quad (7)$$

3.2 *Proposed Fuzzy Profit TSP with Fuzzy Return and Fuzzy Expenditure (Model2):*

Here, a model is considered where the salesman spends some time at each city to convince the customers and hence due to this, makes some uncertain returns and incurs some uncertain expenditure also. Amount of profit, which depends on uncertain returns from city and uncertain expenditures, is dependent on the duration of time he spends in the city. Let c_{ij} and t_{ij} be the cost and time, respectively, for travelling from ith city to jth city. t_i be the spent time in ith city. $\tilde{o}_i(t_i)$ and $\tilde{e}_i(t_i)$ are fuzzy output/return and fuzzy expenditure, respectively, in ith city where $\tilde{o}_i(t_i) = \tilde{a}_i + b_i t_i - c_i t_i^2$ and $\tilde{e}_i(t_i) = \tilde{e}_{0i} t_i$ for spending t time by the salesman in the ith city. Here, a constraint on total time used by the salesman is imposed. He can at most use H units of time for his total tour. So the model is mathematically formulated as

$$\left.\begin{array}{ll}\text{Determine } x_{ij}, \text{ and } t_i, i = 1,2,\dots,N, j = 1,2,\dots,N. & \\ \text{to maximize } Z = \sum_{i=1}^{N} (\tilde{o}_i(t_i) - \tilde{e}_i(t_i)) - \sum_{i=1}^{N}\sum_{j=1}^{N} x_{ij}c_{ij} & \\ \text{subject to } \sum_{i=1}^{N} x_{ij} = 1, & j = 1,2,\dots,N \\ \sum_{j=1}^{N} x_{ij} = 1, & i = 1,2,\dots,N \\ \sum_{i\in S}\sum_{j\in S} x_{ij} \le |S| - 1, \forall S \subset V & \\ & x_{ij} \in \{0,1\}. \\ \text{such that } \sum_{i=1}^{N}\sum_{j=1}^{N} t_{ij}x_{ij} + \sum_{i=1}^{N} t_i \le H & \end{array}\right\} \quad (8)$$

where $x_{ij} = 1$ if the salesman travels from city-i to city-j, otherwise $x_{ij} = 0$, H = total allowable time for the entire tour.

Let $(x_1, x_2, \ldots, x_N, x_1)$ be a complete tour for the salesman, where $x_i \in \{1, 2, \ldots, N\}$ for $i = 1, 2, \ldots,$ N and all x_i's are distinct. Then the above model reduces to

$$\left.\begin{array}{l} \text{Determine a complete tour } (x_1, x_2, \ldots, x_N, x_1) \text{ and } (t_1, t_2, \ldots, t_N) \\ \text{To maximize } Z = (\tilde{o}_1(t) + \tilde{o}_2(t) + \cdots + \tilde{o}_N(t)) - [\sum_{i=1}^{N-1} c_{x_i, x_{i+1}} \\ \qquad + c_{x_N, x_1} + (\tilde{e}_1(t) + \tilde{e}_2(t) + \cdots + \tilde{e}_N(t))] \\ \text{subject to } \sum_{i=1}^{N-1} t_{x_i, x_{i+1}} + t_{x_N, x_1} + (t_1 + t_2 + \cdots + t_N) \le H \end{array}\right\} \quad (9)$$

If $\tilde{a}_i$ and $\tilde{e}_{0i}$ are considered as TFNs (a_{i1}, a_{i2}, a_{i3}) and $(e_{0i1}, e_{0i2}, e_{0i3})$, respectively, then let $\tilde{o}_i(t) = (o_{i1}(t), o_{i2}(t), o_{i3}(t))$ where $o_{i1}(t) = a_{i1} + b_i t - c_i t^2$, $o_{i2}(t) = a_{i2} + b_i t - c_i t^2, o_{i3}(t) = a_{i3} + b_i t - c_i t^2$ and $\tilde{e}_i(t) = (e_{i1}(t), e_{i2}(t), e_{i3}(t)) = (e_{0i1} t, e_{0i2} t, e_{0i3} t)$, and hence the above model reduces to

$$\left.\begin{array}{l} \text{Determine a complete tour } (x_1, x_2, \ldots, x_N, x_1) \text{ and } (t_1 + t_2 + \cdots + t_N) \\ \text{To maximize } Z = (Z_1, Z_2, Z_3) \\ Z_1 = (o_{11}(t_1) + o_{21}(t_2) + \cdots + o_{N1}(t_N)) - [\sum_{i=1}^{N-1} c_{x_i, x_{i+1}} + c_{x_N, x_1} \\ \qquad + (e_{13}(t_1) + e_{23}(t_2) + \cdots + e_{N3}(t_N))] \\ Z_2 = (o_{12}(t_1) + o_{22}(t_2) + \cdots + o_{N2}(t_N)) - [\sum_{i=1}^{N-1} c_{x_i, x_{i+1}} + c_{x_N, x_1} \\ \qquad + (e_{12}(t_1) + e_{22}(t_2) + \cdots + e_{N2}(t_N))] \\ Z_3 = (o_{13}(t_1) + o_{23}(t_2) + \cdots + o_{N3}(t_N)) - [\sum_{i=1}^{N-1} c_{x_i, x_{i+1}} + c_{x_N, x_1} \\ \qquad + (e_{11}(t_1) + e_{21}(t_2) + \cdots + e_{N1}(t_N))] \\ \text{subject to } \sum_{i=1}^{N-1} t_{x_i, x_{i+1}} + t_{x_N, x_1} + (t_1 + t_2 + \cdots + t_N) \le H \end{array}\right\} \quad (10)$$

The defined single-objective fuzzy TSP is solved by the hybrid algorithm developed in Sect. 4.

4 Hybrid ACO-PSO System

Among the most popular nature inspired approaches, mimicking the behaviour of ant, a class of optimization algorithms named ACO has been developed by several authors [17–21]. Some PSO algorithms has been developed by several authors [22, 23] inspiration of behaviour laws of bird flocks, fish schools and human communities.

Both algorithms PSO and ACO have some merits and demerits in solving combinatorial optimization problems. Here, combining the features of these algorithms, a hybrid algorithm is proposed to solve TSP in some modified form. The algorithm is strong enough to solve single-objective fuzzy constraint TSP. In the algorithm initially, ant colony system is used to produce a set of paths (tours) for the salesman which is a set of potential solutions for the ACO part of the algorithm. For each path of the salesman, stay time at different cities to maximize the profit from the system is determined by the PSO algorithm. To do this for each path, a set of stay times at different nodes is randomly generated. Each set of stay time for the path is a particle for the PSO. A set of such particles is generated for a path which is initial swarm of the PSO for that particular path. PSO operations are made on this set of particle a finite number of times to determine the optimum stay time at different nodes for the path. In this way, PSO operation is done for every path of the salesman (ant) to determine optimal stay times at different nodes for the respective path. The hybrid ACO-PSO system is presented below. In the algorithm, τ_{ij} represents that the amount of pheromone lies on the path between node i and node j, $iter1$ and $iter2$ represents iteration counter, $maxiter$ and $maxgen$ represents maximum iteration number of the ACO algorithm and maximum iteration number in PSO part, n represents number of ants or population size and N represents number of nodes/cities in the problem. Here, a set of particles of size M is used.

4.1 Procedures for the Proposed Hybrid Algorithm

(a) **Representation**: Here, a complete tour on N cities represents a path of an ant, i.e. a potential solution for the ACO part of the algorithm. A 'n dimensional integer vector', $X_k = (x_{k1}, x_{k2}, \ldots, x_{kN})$, is used to represent a solution, where x_{k1}, x_{k2}, $\ldots$, x_{kN} represent N consecutive cities in a tour and $k = 1, 2, \ldots, n$ where n is the number of ants.

(b) **Pheromone Initialization**: As the aim of a TSP is to maximize the profit from a tour, i.e. minimize the cost of a tour, it is assumed that initial value of pheromone $\tau_{ij} = 1/c_{ij}^{1.5}$.

(c) **Path Construction**: To construct a path X_k for kth ant, the following steps are followed:

i. Let $NS = \{1, 2, \ldots, N\}$ and $l = 1$.

ii. x_{kl} = A random element from the set NS.

iii. Let $NS = NS - \{x_{kl}\}$

Algorithm 1 Group Creation(A, E, H)

1: **Begin**
2: Set $iter1 = 0$ and initialize $maxiter$, $maxgen$
3: Initialize pheromo τ_{ij} for $i = 1, 2, \ldots, N$ and $j = 1, 2, \ldots, N$.
4: Do
5: Construct path of n ants, i.e. create n tours $X_i = (x_{i1}, x_{i2}, \ldots, x_{iN}, x_{i1})$, $i = 1, 2, \ldots, n$ using τ_{ij}. /*For kth path $X_k = (x_{k1}, x_{k2}, \ldots, x_{kN}, x_{k1})$ determine proportion of optimum stay times at different cities $(OTP_{k1}, OTP_{k2}, \ldots, OTP_{kN})$ to maximize profit for the path using PSO. So $(OTP_{k1}, OTP_{k2}, \ldots, OTP_{kN})$ is a particle for the PSO*/
6: For each particle TP_{ki} do
7: initialize velocity V_{ki}.
8: End for
9: $iter2 = 0$
10: Do
11: For each particle TP_{ki} do
12: Find the personal best position $PBTP_{ki}$.
13: End for
14: Find the global best position $GBTP_k$ of the swarm.
15: For each particle TP_{ki} do
16: Update the velocity V_{ki}.
17: Update the position TP_{ki}.
18: End for
19: $iter2 = iter2 + 1$
20: while($iter2 < maxgen$).
21: Determine stay times of X_k from $GBTP_k$
22: Calculate profit for X_k.
23: End Do.
24: Made pheromone evaporation.
25: Updated pheromone for all the paths.
26: Find best solution.
27: $iter1 = iter1 + 1$
28: While ($iter1 < maxiter$).
29: Output: best solution.
30: **End**

iv. Let node i be the present position of the ant, i.e. $x_{kl} = i$. Then next node $j \in NS$ is selected by the ant with a probability p_{ij} given by the formula

$$p_{ij} = \frac{\tau_{ij}^{\alpha}}{\sum\limits_{j \in NS} \tau_{ij}^{\alpha}}$$

where α is a positive constant used to amplify the influence of pheromone concentrations. Roulette wheel selection process [31] is used for the purpose.

v. $l = l + 1$, $x_{kl} = j$.

vi. if $l < N$ goto step (iii). n-such paths are constructed for n ants.

(d) **For each path X_k do the following operations**:

i. **Swarm initialization**: For each tour, a set of proportion of stay times at different nodes is generated. Each set of proportion of stay times at different nodes is considered as a particle. A set of M such particles is randomly generated which is swarm for the path. PSO operations are made on this swarm to improve the profit from the path. For every path, total tour completion time H is given. Total stay time in different nodes due to kth path is $T = H - T_K$, where T_K is the travel time of path X_k. ith particle of the swarm is considered as N component vector $TP_{ki} = (TP_{ki1}, TP_{ki2}, \ldots, TP_{kiN})$, where each $TP_{ki} \in (0.0001, 1)$. TP_{ki} is ith particle for the swarm of path X_k. For each particle, velocity is initialized between V_{max} and V_{min}. For each TP_{ki}, stay times at different nodes are calculated as $t_{kij} = T \times (TP_{kij})/(TP_{ki1} + TP_{ki2} + \cdots + TP_{kiN})$ for j=1,2,...,N and i=1,2,...,M. so $t_{ki} = (t_{ki1}, t_{ki2}, \cdots, t_{kiN})$.

ii. **Find global best position**: Initially, profit of the path X_k due to different particles TP_{ki} is calculated. The particle which gives maximum profit is considered as initial value of $GBTP_k$. After iteration when particles get new positions, $GBTP_k$ is updated if a position gives better profit than $GBTP_k$.

iii. **Find personal best position**: Initial position of a particle is taken as initial personal best position $PBTP_{ki}$. After each iteration if new position gives better profit, then $PBTP_{ki}$ is replaced by new position.

iv. **Updating velocity**: For each particle (i.e. stay times at different nodes) of a path velocity is updated using the following equation:

$$V_{kij}(t+1) = w \times V_{kij}(t) + C_1 r_{1j(t)}[PBTP_{ki} - PPB_{ki}] + C_2 r_{2j(t)}[GBTP_k - PPB_{ki}]$$

where PPB_{ki} is the present position of ith particle of swarm k, $V_{kij}(t)$ is the velocity of particle i in jth dimension at time step t and $V_{kij}(t+1)$ is the velocity of particle i in jth dimension at time step $t+1$. C_1 and C_2 are the positive acceleration constants and here $C_1 = 1.49618$, $C_2 = 1.79618$ and $w = 0.7298$. $r_{1j}(t)$ and $r_{2j}(t) \sim U(0, 1)$ are the random values in [0, 1].

v. **Updating position**: Each particle of kth swarm is updated using the following equation:

$$PPB_{ki}(t+1) = PPB_{ki}(t) + V_{ki}(t+1)$$

(e) **Find fittest solution**: To find fittest solution X_k of n solutions, for maximization of fuzzy objective if $\tilde{Z}_i$ and $\tilde{Z}_j$ be the objectives of X_k and X_j, respectively, then X_k dominates X_j with respect this objective if $cr(\tilde{Z}_k < \tilde{Z}_j) > 0.5$ is described in Sect. 2.4. This is a valid comparison operator as $cr(\tilde{Z}_i < \tilde{Z}_j) + cr(\tilde{Z}_i \geq \tilde{Z}_j) = 1$.

(f) **Pheromone Evaporation**: For evaporation of pheromone, the following formula is used:

$$\tau_{ij} = (1 - \rho)\tau_{ij}$$

where ρ is in [0, 1]. The constant ρ specifies the rate at which pheromone evaporates, causing ants to forget previous decisions.

(g) **Pheromone Updating**: Once all ants have constructed their complete tour, pheromone is increased on the paths through which the ants move. Depending upon the nature of the problem after a complete tour, pheromone is increased using the following rules.

If $\tilde{P}_k$ be the cost of path X_k, then for this path $\tau_{x_{ki}x_{ki+1}}$ is increased by $1/\{GMIV(P_k)\}^{\beta}$, where GMIV is described in Sect. 2.5 and β is a parameter as mentioned above. If travelling costs are considered as TFN type, then $\tilde{P}_k$ is also a TFN and let $\tilde{P}_k = (P_{k1}, P_{k2}, P_{k3})$; then, according to the rule, $\tau_{x_{ki}x_{ki+1}}$ is increased by $1/\{(P_{k1} + 4P_{k2} + P_{k3})/6\}^{\beta}$.

5 Numerical Illustration

5.1 Input Test Data for Different Models (1A, 1B and 2):

The model is illustrated for ten cities (N = 10). The assumed values of travel costs and times between different cities are presented in Table 1 (times are presented under slash(/)). The fuzzy returns and expenditures are $\tilde{o}_i(t) = \tilde{a}_i + b_i t - c_i t^2$, $\tilde{e}_i(t) = \tilde{e_0}_i t$, respectively. Fuzzy constant values in $\tilde{o}_i(t)$ and $\tilde{e}_i(t)$ are presented in Table 2. For the model, the total tour time is limited to H hours.

Table 1 Travel costs/times between different cities

i/j	0	1	2	3	4	5	6	7	8	9
0	-/-	25/5	28/5	32/4	20/6	26/5	37/3	8/7	29/5	20/6
1	37/3	-/-	20/6	28/5	35/3	40/2	30/5	42/2	28/6	4/7
2	42/2	28/6	-/-	30/5	25/7	35/4	9/9	32/5	40/2	30/5
3	28/6	30/5	7/9	-/-	20/8	25/7	30/5	35/4	22/8	37/3
4	37/3	22/8	35/4	30/5	-/-	20/8	25/7	30/5	9/9	28/6
5	25/7	30/5	25/7	8/9	28/6	-/-	32/5	40/2	32/5	30/5
6	28/6	25/7	30/5	22/8	37/3	40/2	-/-	10/9	32/5	20/8
7	20/8	5/9	32/5	40/2	35/4	25/7	40/2	-/-	22/8	37/3
8	30/5	40/2	35/4	25/7	20/8	22/8	37/3	32/5	-/-	28/6
9	28/6	30/5	28/6	20/8	11/7	32/5	37/3	40/2	30/5	-/-

Table 2 Fuzzy returns($\tilde{a}_i$)'s constants $\tilde{a}_i, \tilde{b}_i, \tilde{c}_i$ and expenditure ($\tilde{e}_{0i}$) in different cities

i / j	0	1	2	3	4	5	6	7	8	9
$\tilde{a}_i$	45, 50, 55;	50, 60, 65;	35, 40, 45;	65, 70, 76;	75, 80, 85;	100, 110, 115;	85, 90, 95;	95, 100, 105;	55, 60, 65;	135, 140, 145
b_i	9.75	20	40	24	58.75	72	28.25	56	89	88
c_i	3	3	4	4	5	4	3	4	5	4
$\tilde{e}_{0i}$	2.5, 3, 3.5	3.5, 4, 4.5	2.5, 3, 3.4	6.5, 7, 7.5	4.5, 5, 5.6	5.5, 6, 6.7	4.5, 5, 5.6	5.5, 6, 6.7	3.5, 4, 4.5	3.5, 4, 4.5

5.2 Solutions of Different Defined TSPs:

The model 1 A and model 1B which are, respectively, the minimization of total travel cost and travel time (without stay times) are solved by ACO. These results are presented in Table 3. The imprecise Model2 is appropriately reduced to crisp representation using the expressions in Sects. 2 and 3. The reduced Model2 is solved for the fuzzy input data by the proposed HA outlined in Sect. 4. The optimum results are presented in Table 3. Here, H is total tour time in hours, PN model type, ST Stay time, TC Travel cost and TT Travel time.

5.3 TSP with Virtual Data Set

Here, we generate large TSP problems with virtual data set, travel costs (c_{ij}) and travel times (t_{ij}) for 20, 40, 60, 80 and 100 cities. These values are randomly generated within some specified values, (c_l, c_u) and (t_l, t_u) (using a relation with travel cost for the present problem). The above ranges differ for different problems. Random datasets are generated using rand() function in C++ language. For the present problem, t_{ij} s are obtained as $t_{ij} =$ **random no in** $[t_l, t_u] + (c_n/c_{ij})$, where $t_l = 0, t_u = 5, c_n = 200$. The returns and expenditures at different cities for large TSPs are assumed as the returns $o_i(t)s$ and expenditures $e_i(t)s$ at the ith city following the earlier expressions. Here, $\tilde{a}_i$s, b_is, c_is and $\tilde{e}_{i0}$s for first ten cities are given in Table 2 and for other cities, there values are repeated in the same order at the interval of 10 cities, i.e. values at (10th, 20th,...), (11th, 21st,...),... ,(29th, 39th,...) cities are same as the values at 0th, 1st, ... ,9th cities, respectively. With these input data, TSPs of sizes $(20 \times 20), (40 \times 40), \ldots, (100 \times 100)$ are solved by the proposed HA and the optimum results are presented in Table 4. Here, it is observed that the behaviours of ‘Return’, ‘Expenditure’ and ‘Profit’ for large-size problems are same as the (10×10) TSP (Table 3). For all problems, ‘Returns’ and ‘Expenditures’ increase with total tour time, but their rates of increase are such that the total profit increases initially and after sometime, it decreases. Thus the proposed hybrid algorithm works well even for larger TSPs.

5.4 Statistical Test

Performance of the proposed method is statistically tested running it 25 times and calculating the average value, standard deviation and percentage relative error according to optimal solution against virtual data problems given in Table 4. The percentage relative error is defined as

$$\text{Error}~(\%) = \frac{(\textit{average solution}-\textit{best known solution})}{\textit{best known solution}} \times 100.$$

Table 3 Problem no (PN), Total tour times (H hours), Optimum paths, Stay Time (ST hours), Travel Costs (TC in $), Travel Time (TT hours), Profit (in $) and GMIV values (GV) for different models

PN	H	Path	ST	TC	TT	Profit	GIVM of profit
1A	–	0, 7, 1, 9, 4, 8, 5, 3, 2, 6;	–	111	80	–	
1B	–	6, 4, 0, 2, 8, 1, 5, 7, 3, 9;	–	376	27	–	
2	40	5, 7, 3, 9, 6, 4, 0, 2, 8, 1;	2.79, 0.68, 0.35, 2.77, 0.29, 0.51, 2.10, 0.40, 2.71, 0.41	376	27	[965.77 1059.27 1152.04]	1059.14
2	50	8, 1, 5, 7, 3, 9, 6, 4, 0, 2;	2.73, 0.26, 2.46, 0.48, 0.19, 2.75, 0.64, 1.11, 2.12, 0.25	376	27	[1281.58 1375.08 1467.84]	1374.95
2	60	1, 0, 3, 9, 6, 5, 7, 4, 2, 8;	2.31, 4.86, 0.53, 8.68, 2.67, 5.64, 5.19, 3.67, 3.33, 6.11	373	27	[1505.89 1585.39 1562.42]	1568.31
2	70	6, 4, 2, 0, 8, 1, 5, 7, 3, 9;	1.80, 3.27, 3.55, 5.12, 6.46, 2.26, 6.23, 5.07, 1.51, 6.75	357	28	[1510.47 1778.47 1797.06]	1736.90
2	80	8, 6, 5, 1, 4, 2, 0, 7, 3, 9;	6.78, 3.44, 6.78, 1.45, 4.64, 2.56, 5.12, 5.21, 1.24, 6.78	334	36	[1511.23 1788.41 2038.28]	1783.85
2	90	8, 3, 6, 5, 7, 9, 4, 0, 1, 2;	6.53, 1.47, 3.25, 6.71, 5.39, 6.82, 5.67, 5.28, 2.52, 4.36	305	42	[1550.79 1837.14 2097.16]	1832.75
2	100	5, 1, 9, 4, 8, 2, 3, 0, 7, 6;	6.68, 1.92, 6.73, 4.75, 6.78, 2.62, 1.63, 5.16, 5.23, 4.51	235	54	[1605.16 1889.68 2146.80]	1885.11
2	110	7, 0, 1, 9, 4, 6, 5, 3, 2, 8;	5.21, 5.74, 2.59, 7.01, 5.24, 2.13, 7.38, 2.27, 4.43, 7.01	212	61	[1636.49 1926.49 2190.41]	1922.14
2	120	9, 4, 5, 3, 2, 6, 0, 7, 8, 1;	7.09, 4.57, 7.05, 1.81, 3.63, 3.45, 5.42, 5.57, 7.08, 2.32	157	72	[1702.57 1993.35 2256.92]	1988.81
2	130	7, 1, 9, 4, 8, 5, 3, 2, 6, 0;	5.64, 1.90, 7.49, 4.78, 7.48, 7.49, 2.66, 3.60, 3.29, 5.68	111	80	[1745.12 2045.64 2317.34]	2042.65
2	140	0, 7, 1, 9, 4, 8, 5, 3, 2, 6;	5.87, 6.63, 4.62, 7.78, 5.76, 7.52, 7.74, 3.34, 4.41, 6.33	111	80	[1686.04 2033.01 2303.77]	2020.69

Table 4 Total tour times (H hours), Stay Time (ST hours), Travel Costs (TC in $), Travel Time (TT hours) and Profit (in $) for different models

Sl no.	Problem size	H	TT	ST	TC	Profit	GIVM of profit
1	20	400	238	162	409	[6008.32 6237.19 6392.54]	6214.22
		500	336	164	327	[6153.75 6305.04 6462.21]	6307.04
		600	382	218	295	[5812.54 6102.82 6300.1]	6071.82
2	40	700	444	256	524	[9139.65 9360.5 9591.64]	9363.93
		800	534	266	451	[9221.39 9414.66 9701.41]	9445.82
		900	561	339	430	[8779.51 8974.91 9130.95]	8961.79
3	60	800	472	328	829	[12091.42 12288.71 12450.24]	12276.79
		900	558	342	698	[12231.21 12399.39 12621.66]	12417.42
		1000	644	356	· 681	[12136.44 12372.82 12654.8]	12388.02
5	80	1300	820	480	1173	[18183.76 18321.37 18587.29]	18364.14
		1400	904	496	999	[18312.35 18538.45 18675.78]	18508.86
		1500	954	546	1003	[18338.81 18549.28 18628.38]	18505.49
6	100	2700	1003	1797	1596	[26829.22 26989.49 27185.25]	27001.32
		2900	1778	1122	1553	[28209.71 28411.55 28643.66]	28421.64
		3000	1965	1035	1682	[28004.53 28236.42 28393.13]	28211.36

Table 5 Dispersion results of hybrid ACO-PSO

Instances	Best	Worst	Average	SD	Error (%)
10	2042	2020	2029.73	1.31	0.19
20	6307.04	6205.7	6287.43	1.31	0.52
40	9445.8	9378.54	9421.53	1.07	1.72
60	12417.42	12106.37	12203.65	2.13	1.47
80	18508.86	18034.51	18176.23	2.02	1.92
100	28421.64	27980.43	28265.15	1.08	1.17

Examining Table 5, it is concluded that the proposed Model2 which solved by hybrid ACO-PSO is effective for the most of instances. The above table shows that standard deviation (SD) low for 10, 20, 40 and 100 node problems excepts 80 node as well as percentage error propagate smoothly for every instances, i.e. not too much high.

6 Discussion

In Table 3, the results of model 1 A and model 1B are as per our expectations. When travel cost is minimized (Module-1A), the minimum travel cost is 111$ but required travel time is very high (80 h). On the contrary, for Module-1B, it is reversed, i.e. travel time is minimum (27 h) but the travel cost is highest (376$). For Model2 when total tour time (H) is fixed with variable stay times at different cities, TS tries to earn as much as possible allocating possible maximum stay time at stations out of the available total tour time as it is a problem of maximization of profit. In this process, TS adopts the minimum tour time accepting highest travel cost. This practice is followed upto H = 70 h. After this, as H increases, TS makes a trade-off between the savings due to stay and tour cost. To reduce the travel cost, TS increases the travel time, i.e. takes more time than the minimum travel time and the rest time (from the total tour time) as stay time at different stations, because more time is available now. By this arrangement, profit increases with H upto H = 130 h. After this, profit decreases when H is more because stay at different stations is not beneficial any more, and this is more clearly explained in Fig. 1.

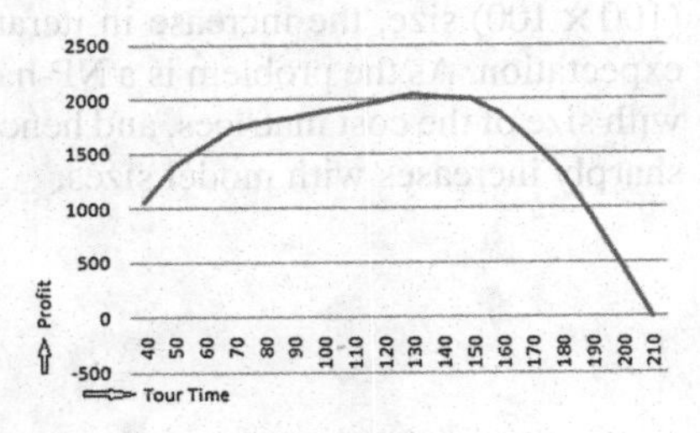

Fig. 1 Profit versus tour time for Model2

7 Model's Behavioural Studies

In this section, changing some model's parameters, the behaviour of the models is determined and presented graphically.

7.1 Profit Versus Tour Time:

For proposed fuzzy model, maximum profits are evaluated and plotted against different tour times in Fig. 1. Here, profits are maximum at $TT = 130$. It is interesting to note that for the model, as total tour time increases, profit also increases upto certain tour time and after that time profit decreases with the increase of tour time. In this process of evaluation, a TS first tries to make his total return ($\sum o_i$) maximum allotting the required total stay time for this from the tour time and then suitable tour path is selected against the travel time which is equal to available remaining time (i.e. *totaltourtime−totalstaytime*). Obviously, if the available travel time is more, TS adopts the path which gives comparatively minimum tour cost. However, when the total tour time is very high, TS adopts the lowest tour cost path and the rest time is spent as total stay time. As we know, if the stay time at each city excludes a certain limit, then expenditure is more than the return and due to these reasons, we observe that profits go down after certain tour time, and this study will be very helpful for the managers to fix/allot the tour times for TSs.

7.2 Model Size Versus Iteration Number:

Here, a study is made for the number of iterations required for optimum results against different sizes of cost matrices. For the Model2, optimum results were derived with $(5 \times 5), \ldots, (10 \times 10), (20 \times 20), \ldots, (60 \times 60), (100 \times 100)$ matrices and the corresponding iteration numbers were noted. These are plotted in Fig. 2 for Model2. It is to be noted that (10×10) matrix size onwards, the number of iterations required for optimum values increases sharply. However, when matrix size increases from (5×5) size to (10×10) size, the increase in iteration number is not much sharp. On the other hand, when matrix size increases from (80×80) size to (100×100) size, the increase in iteration number is very sharp. This is as per our expectation. As the problem is a NP-hard problem, the complexity and time increase with size of the cost matrices, and hence the number of iterations for optimum results sharply increases with model sizes.

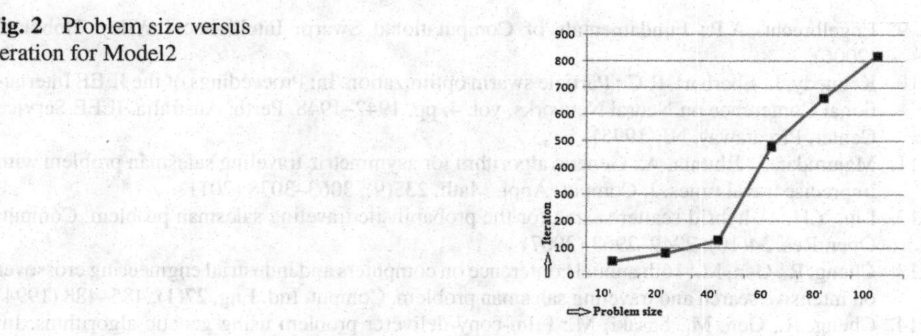

Fig. 2 Problem size versus iteration for Model2

8 Conclusion

In this investigation, a new type of TSP, i.e. a real-life TSP is proposed and solved. For a medical firm, normally a medical representative is asked to tour a number of town/sub-towns within a limited time and he/she fixes the path and stay at stations in such way that total outcome is maximum. Here, TSP model is formulated in fuzzy environment also with expenditures and returns as fuzzy. The models are illustrated with numerical examples and some interesting expected results are derived. For the first time, a hybrid algorithm—a combination of ACO and PSO—has been successfully applied for the above-mentioned TSP. Here, the model formulation and algorithm are quite general. These can be applied to formulate and solved solid TSPs and three-dimensional assignment problems. The present TSPs can also be extended to rough, fuzzy-rough, random, etc. environments.

References

1. Applegate, D.L., Bixby, R.E., Chvátal, V., Cook, W.: The Traveling Salesman Problem. Princeton University Press, Princeton (2007)
2. Fiechter, C.N.: A parallel tabu search algorithm for large traveling salesman problems. Discret. Appl. Math. **51**(3), 243–267 (1994)
3. Chang, T.S., WahWan, Y., Ooi, W.T.: A stochastic dynamic traveling salesman problem with hard time windows. Eur. J. Oper. Res. **198**(3), 748–759 (2009)
4. Chen, C.H., Liu, Y.C., Lin, C.J., Lin, C.T.: A hybrid of cooperative particle swarm optimization and cultural algorithm for neural fuzzy networks. In: FUZZ-IEEE, IEEE, 238–245 (2008)
5. Nagata, Y., Soler, D.: A new genetic algorithm for the asymmetric traveling salesman problem. Expert Syst. Appl. **39**(10), 8947–8953 (2012)
6. Lo, C.C., Hsu, C.C.: An annealing framework with learning memory. Trans. Syst. Man Cyber. Part A **28**(5), 648–661 (1998)
7. Michalewicz, Z.: Genetic Algorithms + Data Structures = Evolution Programs. Springer, New York (1992)
8. Bianchi, L., Gambardella, L.M., Dorigo, M.: An ant colony optimization approach to the probabilistic traveling salesman problem. In: Guervós, J.J.M., Adamidis, P., Beyer, H.-G., Fernández-Villacañas, J.-L., Schwefel, H.-P. (eds.) PPSN. Lecture Notes in Computer Science, vol. 2439, pp. 883–892. Springer, Berlin (2002)

9. Engelbrecht, A.P.: Fundamentals of Computational Swarm Intelligence. Wiley, Hoboken (2006)
10. Kennedy, J., Eberhart, R.C.: Particle swarm optimization. In: Proceedings of the IEEE International Conference on Neural Networks, vol. 4, pp. 1942–1948. Perth, Australia, IEEE Service Center, Piscataway, NJ (1995)
11. Majumdar, J., Bhunia, A.: Genetic algorithm for asymmetric traveling salesman problem with imprecise travel times. J. Comput. Appl. Math. **235**(9), 3063–3078 (2011)
12. Liu, Y.H.: A hybrid scatter search for the probabilistic traveling salesman problem. Comput. Oper. Res. **34**(10), 2949–2963 (2007)
13. Cheng, R., Gen, M.: 16th annual conference on computers and industrial engineering crossover on intensive search and traveling salesman problem. Comput. Ind. Eng. **27**(1), 485–488 (1994)
14. Cheng, R., Gen, M., Sasaki, M.: Film-copy deliverer problem using genetic algorithms. In: Proceedings of the 17th International Conference on Computers and Industrial Engineering Computers and Industrial Engineering **29**(14) 549–553 (1995)
15. Shi, X.H., Liang, Y.C., Lee, H.P., Lu, C., Wang, Q.X.: Particle swarm optimization-based algorithms for tsp and generalized tsp. Inf. Process. Lett. **103**(5), 169–176 (2007)
16. Moon, C., Kim, J., Choi, G., Seo, Y.: An encient genetic algorithm for the traveling salesman problem with precedence constraints. Eur. J. Oper. Res. **140**, 606–617 (2002)
17. Clerc, M., Kennedy, J.: The particle swarm—explosion, stability, and convergence in a multidimensional complex space. Trans. Evol. Comput. **6**(1), 58–73 (2002)
18. Cheng, C.B., Mao, C.P.: A modified ant colony system for solving the travelling salesman problem with time windows. Math. Comput. Model. **46**(910), 1225–1235 (2007)
19. Ghafurian, S., Javadian, N.: An ant colony algorithm for solving fixed destination multi-depot multiple traveling salesmen problems. Appl. Soft Comput. **11**(1), 1256–1262 (2011)
20. Focacci, F., Lodi, A., Milano, M.: A hybrid exact algorithm for the tsptw. INFORMS J. Comput. **14**(4), 403–417 (2002)
21. Bai, J., Yang, G.K., Chen, Y.W., Hu, L.S., Pan, C.C.: A model induced max-min ant colony optimization for asymmetric traveling salesman problem. Appl. Soft Comput. **13**(3) 1365–1375 (2013) Hybrid evolutionary systems for manufacturing processes
22. Leung, K.S., Jin, H., Xu, Z.: An expanding self-organizing neural network for the traveling salesman problem. Neurocomputing **62**, 267–292 (2004)
23. Chen, S.M., Chien, C.Y.: Parallelized genetic ant colony systems for solving the traveling salesman problem. Expert Syst. Appl. **38**(4), 3873–3883 (2011)
24. Dorigo, M., Gambardella, L.M.: Ant colonies for the travelling salesman problem. Biosystems **43**(2), 73–81 (1997)
25. López-Ibáñez, M., Blum, C.: Beam-aco for the travelling salesman problem with time windows. Comput. Oper. Res. **37**(9), 1570–1583 (2010)
26. Eberhart, R.C., Kennedy, J.: A new optimizer using particle swarm theory. In: Proceedings of the Sixth International Symposium on Micro Machine and Human Science, 39–43 (1995)
27. Petersen, H.L., Madsen, O.B.G.: The double travelling salesman problem with multiple stacks—formulation and heuristic solution approaches. Eur. J. Oper. Res. **198**(1), 139–147 (2009)
28. Ratnaweera, A., Halgamuge, S.K., Watson, H.C.: Self-organizing hierarchical particle swarm optimizer with time-varying acceleration coefficients. IEEE Trans. Evol. Comput. **8**(3), 240–255 (2004)
29. Zadeh, L.: Fuzzy sets. Inf. Control. **8**, 338–356 (1965)
30. Bianchi, L., Gambardella, L., Dorigo, M.: An ant colony optimization approach to the probabilistic traveling salesman problem. Parallel Problem Solving from Nature PPSN VII, vol. 2439, pp. 883–892. Springer, Berlin (2002)
31. Masutti, T.A.S., de Castro, L.N.: A self-organizing neural network using ideas from the immune system to solve the traveling salesman problem. Inf. Sci. **179**(10), 1454–1468 (2009)

Part VII
Wireless Sensor and Ad-hoc Networks

Mitigating Packet Dropping Problem and Malicious Node Detection Mechanism in Ad Hoc Wireless Networks

Shrikant V. Sonekar and Manali M. Kshirsagar

Abstract Lack of association among the nodes, mobility, device heterogeneity, bandwidth, multihop and battery power constraints in ad hoc wireless networks make the design of efficient routing protocols a major challenge. MANET is more susceptible to be attacked than wired network. Encryption, cryptography and authentication are the preventive measures that may be used at the initial level for sinking the attack possibilities, but these techniques are not capable to protect a network from newer attacks. Considering the parameters like bandwidth, energy, security, etc. many routing protocols categorized as reactive, proactive and hybrid have been proposed for communications; still an efficient communication and the dropping of packets are the major issues in MANET that need to be focused. Moreover, possibility of dropping a packet during transmission by malicious node cannot be ignored. In this paper, we have presented a practical approach for cluster head election and the possible probabilities. Issues with the packet dropping at localized level and maintaining the stability of the nodes have been explored. We also present a malicious node detection method. Preliminary simulation results have shown that proposed system reduces the impact of sequence number attack and packet dropping attack to very high extend.

Keywords MANET · Attacks · Packet dropping · Dirty packets · Cluster head

S.V. Sonekar (✉)
Research Scholar, Department of CSE, GHRCE, Nagpur, India
e-mail: srikantsonekar@gmail.com

M.M. Kshirsagar
Research Guide, Department of CSE, GHRCE, Nagpur, India
e-mail: manali_kshirsagar@yahoo.com

S. Das et al. (eds.), *Proceedings of the 4th International Conference on Frontiers in Intelligent Computing: Theory and Applications (FICTA) 2015*, Advances in Intelligent Systems and Computing 404, DOI 10.1007/978-81-322-2695-6_27

1 Introduction

MANET is a network formed by cooperation of mobile devices that communicate through wireless links when needed. The mobile devices are expected to work as routers and thus assist in route discovery and maintenance. The topology in MANET varies as the node moves in and out of the network [1], because of infrastructure less setup, inappropriate coordination, lack of association among the nodes and imperfect mechanism for authentication. Moreover, it is difficult to implement the cryptography-based mechanism because of scarce resources such as computational power and bandwidth. Local resources are owned and controlled by a node itself, whereas remote resources are owned and controlled by other nodes and those that can only be accessed through the network. Typically, accessing remote resources is more expensive than accessing local resources, since communication delays and the CPU overhead incurred to great extent in order to process communication protocols. Moreover, there is no guarantee that a routed communication path is free from malicious node. The radio channel which is used for communication is broadcast and is shared by all the nodes. Therefore, a malicious node can easily obtain the transmitted data. The directional antennas can be used to reduce the problem to some extent. There are two principal methods for dealing with malicious problems [2]. We can use malicious prevention technique to ensure that network will never enter it. Alternatively, we can allow the network to enter a malicious state, and then try to recover the same using malicious detection scheme. Prevention is commonly used if the probability that the network would enter an intrusion state is relatively high; otherwise, detection is more efficient. Note that a detection scheme requires overhead that includes not only the run-time cost of maintaining the required information and of executing the detection algorithm, but also the potential losses inherent in recovering from the intrusion. On the other hand, the routing protocols, such as AODV [3], DSR [4] and wireless MAC 802.11 [5] typically assume a trusted and supportive environment. Thus a malicious node can readily become router and dislocate network operations by disobeying the protocol specifications [6]. Moreover, dropping a packet during a transmission by the malicious nodes cannot be ignored. In order to extend the lifetime or to retain the battery backup, selfish node does not forward the packet and sometimes it drops the packet. In generic, at MAC or network layer packets can be dropped due to the following reasons:

- At MAC layer, the buffer size for packet transmission is limited, hence whenever new packet arrives from upper layer and if the buffer is full, then there is a possibility of dropping the packet.
- IEEE 802.11 protocol specification specified that when Request to Send frame reaches the maximum allowed number, the packet gets dropped [7].
- Owing to the transmission problems like interference, hidden and exposed terminal or high bit error rate, collision, etc. the packets may be dropped or get corrupted.

- Malicious node may alter the IEEE 802.11 MAC protocol's parameters to provoke packet dropping [8].
- If attempts of retransmission of packets reach the maximum allowed number, packets get dropped.

The rest of the paper is structured as: In the next section, we discuss route discovery and issues. Section 3 discusses cluster head formation concepts and state of art scenario. Proposed system for malicious node detection and probability scenarios are discusses in Sect. 4. Section 5 discusses the results obtained. Finally Sect. 6 draws conclusions and outlines for future work.

The aim of the paper is to study the issues involved in route discovery and the process to detect the malicious node in the network. The following areas are explored:

- Analysis of issues at localized level for packet dropping
- Cluster head formation, detection of malicious node and probability scenarios

2 Route Discovery and Issues

A route is considered to be active if it has an entry in the routing table that is marked as valid. When a node is trying to send a packet without knowing an active route, the source node broadcasts a route request (RREQ) message for searching a route to all its neighbours in the network [3]. When a RREQ packet arrives at an intermediate node, it could either forward the RREQ packet to another intermediate node or it prepared a route reply (RREP) packet if an available valid path to the destination exists in its cache. Ad hoc networks are dynamic in nature; nodes can join or leave the network at any moment. Hence it is difficult to control the individual nodes' movement by maintaining a stable neighbourhood topology [9]. Consider the nodes $n_0, n_1, n_2 \ldots n_N$ and R_m is the maximum range of the node, the comparative distance between nodes n_0 and n_1 is D(0, 1). If nodes communicates with each other without taking the help of any routing protocols then they are called neighbour nodes and the network topology for all the nodes is maintain if $D(0, 1) \leq R_m$. Let $Xn_0(t)$ and $Xn_1(t)$ be the positions of n_0 and n_1 nodes at time 't'. The distance between two nodes at time t is calculated as

$$D\{n_0, n_1\} = \|Xn_0(t) - Xn_1(t)\| \tag{1}$$

A communication link failure doesn't occur if $D\{n_0, n_1\} < R$ at time 't', R is radio range and $D\{n_0, n_1\}$ is the distance between two nodes, this condition exists and remains valid only for those nodes which are homogeneous in nature. Because of the multihoping and heterogeneity of the devices, it is difficult to maintain the homogenous nodes in ad hoc wireless networks. In AODV, each node must forward packets to the other node [10]. Sometimes accepting and forwarding RREQ

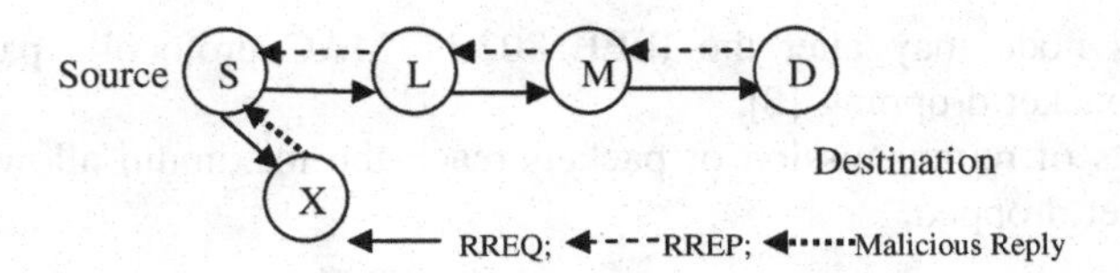

Fig. 1 Packet dropping at localized level

message depends on its remaining energy E_{Rem}. When it is lower than a threshold value θ ($E_{Rem} \leq \theta$), RREQ is dropped; otherwise, the message is forwarded to the next node. In Fig. 1, X is a malicious node, S is source node and D is destination node. The node S broadcasts RREQ packet to its one hop neighbour L. Node L forwards it to node M and finally reaches to node D. The node X disobeys this rule and shows that it has shortest path to reach to the destination and accordingly sends a RREP packet to source node S.

If node D sends a RREP to node S before the X's RREP then everything works well, otherwise node S assumes that there is a shortest path from node X to node D and it start sending all packets through node X, node X drops all the packets [11].

3 Cluster Head Formation and State of Art Scenario

Generally in ad hoc wireless networks the nodes are continuously moving and are not stable, this affects the efficiency of the protocol. Bandwidth reservations are of no use if the node mobility is very high. Moreover, MAC protocol does not influence the mobility of the nodes [12]. The cluster head is acknowledged by its own identification number and is responsible for running entire network. For cluster head election, different researchers have used different concepts. Proposed concept is based on the distance parameters. Consider two generic nodes L and M with x and y as the coordinators shown in Fig. 2, the distance between these two nodes with reference to x and y coordinators is calculated. Keeping node L as i and node M as j, compare xreference of i with xreference of j and yreference of i with yreference of j, set the value of threshold range R_{tr}, i.e. min-x-distance and min-y-distance. Carry out the same process for all the nodes, put coordinates x and y using 2D arrays.

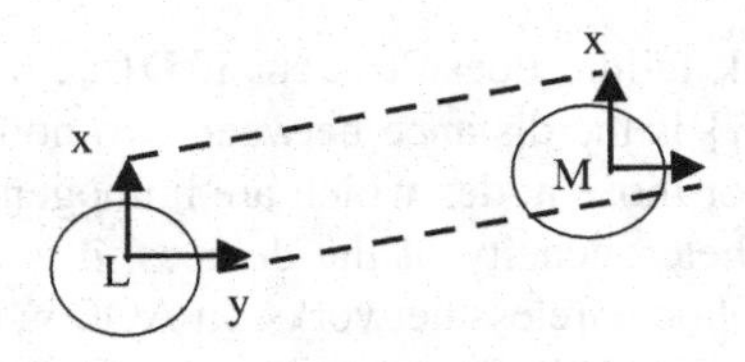

Fig. 2 Distance comparison between two nodes

$$\text{xreference} = \text{Amobile}[i][0] \quad (2)$$

$$\text{yreference} = \text{Amobile}[i][1] \quad (3)$$

$$\begin{aligned}\{(\text{xreference} - \text{Amobile}[j][0]) &= = \min - x - \text{distance})\ \&\& \\ (\text{yreference} - \text{Amobile}[j][1]) &= = \min - y - \text{distance})\}\end{aligned} \quad (4)$$

$$\begin{aligned}\{(\text{Amobile}[i][0] - \text{Amobile}[j][0]) &= = \min - x - \text{distance})\ \&\& \\ (\text{Amobile}[i][1] - \text{Mobile}[j][1]) &= = \min - y - \text{distance})\}\end{aligned} \quad (5)$$

where, xreference and yreference are the reference values of nodes L and M, respectively, min-x-distance and min-y-distance are the threshold values i.e. 300 specify in the pseudocode. The node which is eligible to become a cluster head has to calculate distance D_i of each node n_i from itself.

The node which is eligible to become a cluster head has to calculate distance D_i of each node n_i from itself.

$$D_i = \sqrt{(X_L - X_M)^2 + (Y_L - Y_M)^2} \quad (6)$$

Compare the distance D_i with R_{tr}. If $D_i \leq R_{tr}$ then find the node which has highest connectivity, sufficient energy level and signal strength for electing the cluster head. The pseudocode for cluster head formation is given as:-

```
Input: Nodes with initial energy
Output: Clusterhead with maximum residual energy
    # Calculation of clusterheads
    Create node= node_id;
    Set Channel=802.11;
    Set InitialEnergy= $inienergy;
    Set Residual Energy= $resenergy;
    Set radiorange= default;
    If(( node in radio range) && (next hop!=Null))
  {
    Capture data_load (node_all);
 Create node_Configure (rreq,rrep, tsend, tsend, trecv,
 tdrop, inienergy, resdualenergy);{   pkt_type;
    Time;
 tsend, trecv, tdrop, rrep, rreq;}}
    for (i=0; i<nn; i++)
    consumeenergy[i] = initialenergy[i]-finalenergy[i]
    totalenergy[i] = consumeenergy[i]
    if(maxenergy[i] < consumeenergy[i])
  { maxenergy = consumeenergy[i]
    node_id = i;} if (dist<300 && maxenergy[node_id]>
    energyneighbour[node_id])
  { maxenergy node is clusterhead ;}
```

3.1 State of Art Scenario

Each node maintains data_table for storing elected cluster head id and the key value generated by the cluster head. Elected cluster head sends *<CH id, Key Value>* to all members of the cluster. There is less probability of malicious node occurrence if this process is done before the transmission of packets. The cluster head creates the key value as, *Key=Utils.generateRSAKeyValue();* The generated key value is then converted into polynomial equations as *public static polynomial[] get polynomials (int cof[],Big Integer N, int n)*. The cluster head distributes this key value among the entire member through sendpoly () function.

Requesting all members of the cluster:-
Send REQUEST to all cluster members for maintaining *(<CH id, Key Value>)* in data_table. On receiving the REQUEST, cluster member [CM] sends REPLY message to the elected CH. CM[j] places *(<CH id, Key Value>)* in data_table [j].
Releasing the position of Cluster Head:-
The cluster head releases the head position when energy level or signal strength drops below the threshold or cluster head wants to release it voluntarily. Exiting cluster head send a time stamp RELEASE message to all cluster members. On receiving RELEASE message, CM[j] removes *(<CH id, Key Value>)* from its data_table and keeps only the exit time stamp of cluster head in data_table. The time stamp field gives the exact time when cluster head exits. This helps in finding the malicious node. The overhead here is it takes *2(N-1)* messages.

4 Proposed System and Probability Scenarios

Malicious node may participate in the route discovery phase but fails to forward the data packets to correct route. Each node is responsible for carrying out localized detection by overhearing ongoing transmissions and evaluating the behaviour of its neighbour node since the wireless medium is open in nature [13]. 'F_T' a failure tag is associated with all the nodes. 'F_T' consists of node id, counter value (CV) and MAC address. Moreover, malicious node forwards modified, fabricated or over-written packet to its neighbour node, such type of packet is known as dirty packet (DP).

In Fig. 3, node A can hear to its next hop X, node A knows that X's next hop is B. It then overhears the transmission between X and B using *carrier sense* mechanism. If node A does not hear the transmission after a time period t, a failure tag 'F_T' associated with X is increased and if it exceeds a predefined threshold value then node A sends a report packet to the source notifying about X's misbehaviour. The node has to forward the packet and supervise whether the neighbour forwards it or not. All the nodes maintain the suspicious node table. The table contains the suspicious node id and the number of suspicious entries [14]. The major issue with

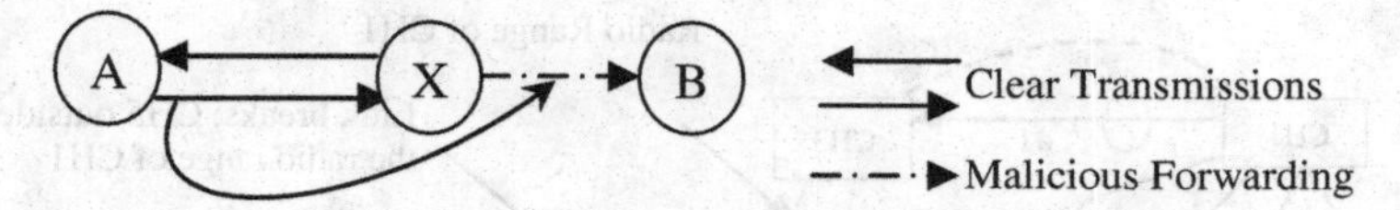

Acan hear the transmission in between X and B; A, X and B are intermediate nodes

Fig. 3 Overhearing of packet transmission by nodes

this scheme is the requirement of energy and the time for checking the transmission [15]. If E_F is the required energy for forwarding the packets and E_S is the required energy for supervision, remaining energy with the node is $E_{Rem} = E_F - E_S$. If T_{TXB} is the transmission time for packets to be sent from node X to B, then $T_{TXB} = T_{RB} - T_{SX}$, where T_{RB} is the exact time when node B received the packet and T_{SX} is the exact time when node X sends the packet. Es is the supervised energy for the T_{TXB}. If any node is behaving abnormally and the failure tag associated with node exceeds the threshold value then CH adds that node in the suspected node list and raises pre-alarm. CH also checks for the number of dirty packets sent by the node. CH asks information (*<CH id, Key Value>*) or (*<CH id, Key Value, Exit Time Stamp of Previous Cluster Head>*) from the suspected node. If the suspected node manages to respond appropriately, it is removed from the suspected node list and is reset normally. If suspected node does not respond appropriately, CH checks all the packets sent by the suspected node, based on the analysis cluster head declared node as malicious node. The proposed system declared a node X_i in the network as malicious,

$$\text{If } [(F_tCV(X_i) \geq \theta) \Lambda (DP(X_i) \geq \theta) \Lambda (\text{information doesn't match})] \quad (7)$$

where <CH id, Key Value, Exit Time Stamp of Previous Cluster Head> is information to be matched and 'θ' is the threshold value depends on network and type of Attacks.

In Fig. 4, the total distance of <CH_2, CH3, CH4> from CH_1 is calculated as, Distance (D) = $d_1 + d_3 + d_4$. We have not considered the cluster head CH_5, because CH_5 is not reachable from CH_1, i.e. d(CH1, CH5) = α. The distance of all the cluster heads from CH1 is calculated as $e(CH_1) = \max\{d\{(CH_1), (CH_2, CH_3, CH_4)\}, \forall i \in N\}$ [16]. Where 'e' is *eccentricity* and 'N' is network. If $E_{c(t)}$ is the energy consumed by CH_1 at time 't', E_D is the energy required to cover the distance 'D' and E_T is the energy required up to time 'T', then the remaining energy of CH_1 after time 't' is calculated as $E_{Rem} = \{[E_D + E_T] - E_{c(t)}\}$ (Table 1).

A global state, GS, is a collection of all the cluster heads in the network 'N', i.e. $GS = \{CH_1, CH_2 \ldots CH_n\}$ where n is the number of cluster heads in the network. The global state $GS = \{CH_1, CH_2, CH_3, CH_4, CH_5\}$ is consistent iff

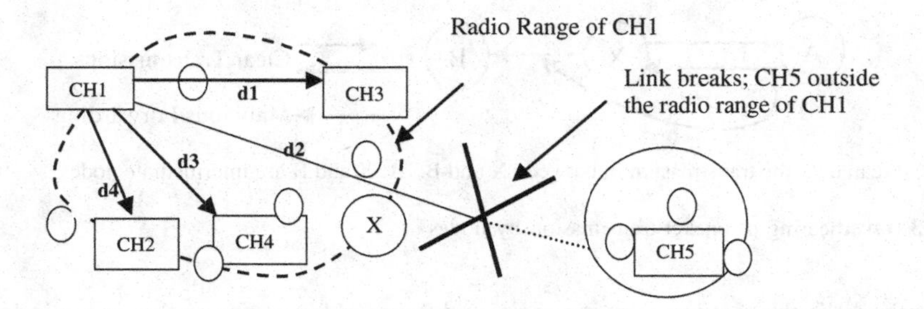

Fig. 4 Malicious node 'X' is trying to enter CH2, CH3, CH4 or CH5

Table 1 Probability scenarios of malicious node 'X' for Fig. 4

Scenarios	Conditions	Probability
Scenario I	If node X knows the radio range of CH1	$\text{Probability (P)} = \begin{cases} 1 & \text{for CH5} \\ 0 & \text{for CH2, CH3, CH4} \end{cases}$
Scenario II	If node X does not know the radio range of CH1, node X considers distance parameter	$\text{Probability (P)} = \begin{cases} 1 & \text{for CH5} \\ 0 & \text{for CH2, CH3, CH4} \end{cases}$
Scenario III	Otherwise	$\text{Probability (P)} = \begin{cases} 0.5 & \text{for CH5} \\ 0.5 & \text{for CH2, CH3, CH4} \end{cases}$

$$\forall i,\ \forall j\colon 1 \le i, j \le \text{n}::\textit{inconsistent}\ (CH_i, CH_j) = \Phi \tag{8}$$

Thus, in a consistent global state, for every received packet a corresponding send event is recorded in the global state. In an inconsistent global state, there is at least one packet whose receive event is recorded but its send event is not recorded in the global state.

5 Simulation and Results

We have used the random waypoint mobility model. It was first used in the evaluation of DSR. This model uses random speed uniformly distributed between [$\mathbf{V}_{\min}$, $\mathbf{V}_{\max}$]. As the node arrives at the new destination, it pauses for a selected pause time [T_{pause}] before starting to travel again. Krawetz [17] proposed the mobility metric to capture and quantify this nodal speed notion. The measure of relative speed between node A and B at time 't' is

$$\text{RS}(\text{A}, \text{B})^{\text{t}} = |VA(t) - VB(t)| \tag{9}$$

Table 2 Simulation parameters

Parameters	Value
Simulation duration	1000 s
Simulation area	1000 * 1000 m
Transmission range	250 m
Movement model	Random waypoint
Maximum speed	5–20 m/s
Traffic type	CBR (UDP)
Mac protocol	802.11
Routing protocol	AODV
Data payload	512 bytes
Host pause time	5 s

The choices of the simulator parameters that are presented in Table 2 consider accurateness and the throughput of the simulation. The experiments are carried out using network simulator (ns-2).

To evaluate the impact of sequence number attack, packet delivery ratio and routing packet dropped ratio metrics are chosen. The general observation shows that proposed algorithm keeps the delivery ratio higher around 60 %.

The throughput of the network is drastically reduced when AODV is under sequence number attack, as the node mobility increases the network topology changes making route requests more frequent, therefore malicious node has more chance to send more fake RREP packets. The black hole attack has a very big impact on decreasing the delivery ratio of the packets. In Fig. 5, we can observe that there is a greater decrease in the packet delivery when normal AODV is under attack. We observe that AODV performs much better in low mobility, as we raise the mobility rate the delivery of the packets drops slightly. The proposed algorithm manages to reduce the attack by around 50 %. As the node mobility increases, the participating node begins the route discovery processes more often and the malicious node can drop more routing packets (Figs. 6 and 7).

It has been observed that the additional routing overhead introduced by the attacking node reaches high when the size of the network is small and decreases as

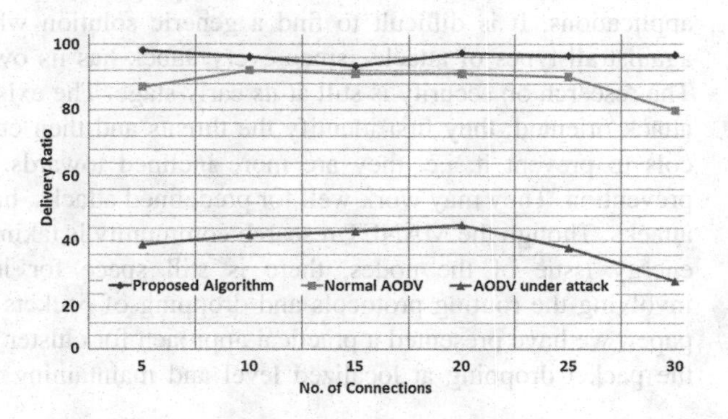

Fig. 5 Number of nodes versus delivery ratio (%) in sequence number attack

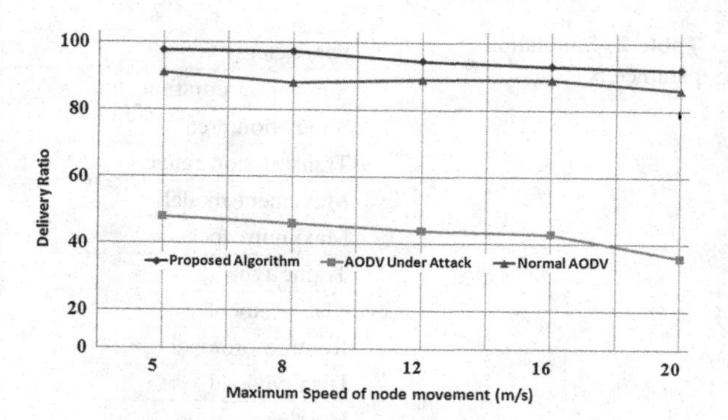

Fig. 6 Maximum speed of node movement versus delivery ratio (%) in sequence number attack

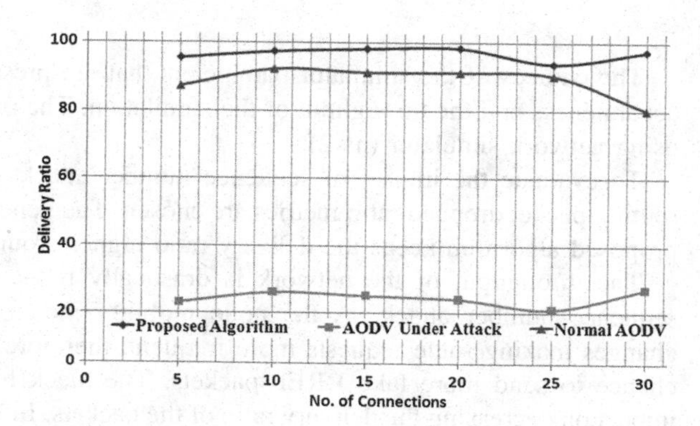

Fig. 7 Number of nodes versus delivery ratio (%) in dropping routing packet attack

the number of node connections increases. Simulations results show that proposed work manages to reduce the impact of dropping packet attacks to a great extend.

6 Conclusion and Outlines for Future Work

An ad hoc wireless network is an emerging research area with practical-based applications. It is difficult to find a generic solution which will work efficiently against all types of attacks, since every attack has its own distinct characteristics. The research on security is still at its early stage. The existing systems are typically attack oriented; they first identify the threats and then enhance the existing protocols to prevent it, i.e. they are more inclined towards the detection rather than prevention. They may work well for predefined attacks, but fail under unanticipated attacks. Though the MANET research community is taking great efforts to solve the energy issue of the nodes, there is still space for improvement, particularly involving the routing protocols and dropping of packets at localized level. In this paper, we have presented a practical approach for cluster head election. Issues with the packet dropping at localized level and maintaining the stability of the nodes have been explored. Based on the different parameters we declared the node as malicious. A comparative study of basic issues in MANET and probability scenarios is carried out. Simulation results have shown the realistic approach for finding the sequence number attack and dropping routing packet attack and they are reduced by around 50 %. In future, the research community may work on the loose clustering and incorporate this feature in ad hoc wireless networks for preserving the energy level of the cluster head. Moreover, there should be some mechanism which will restrict a malicious node to move in other part of the network.

References

1. Perkins, C.E.: Ad hoc Networking, pp. 198–264. Addison-Wesley, New York (2001)
2. Gupta, S.S., Ray, S.S., Mistry, O., Naskar, M.K.: A stochastic approach for topology management of mobile ad hoc networks. In: Asian International Mobile Computing Conference, pp. 90–99. Jadavpur University, Kolkata (2007)
3. Perkins, C.E., Royer, E.M., Das S.R.: Ad hoc on-demand distance vector routing. In: Proceedings of the 2nd IEEE Workshop on Mobile Computing and Applications, vol. 3, pp. 99–100. New Orleans, LA, USA (1999)
4. Johnson, D., Maltz, D.: In: Imielinski, T., Korth, H. (eds.) Dynamic Source Routing in Ad Hoc Wireless Networks, Mobile Computing. Kluwer (1996)
5. IEEE Std. 802.11: Wireless LAN Medium Access Control (MAC) and Physical Layer (PHY) Specifications (1997)
6. Yang, H., Luo, H., Ye, F., Lu, S., Zhang, L., Security in mobile ad hoc networks: challenges and solutions. IEEE Wirel. Commun. **11**(1), 38–47, UCLA Computer Science Department (2004)
7. IEEE 802.11: Wireless LAN Media Access Control (MAC) and Physical layer (PHY) Specifications. ANSI/EIEEE Std. 802.11 (1999)
8. Djahel, S., Nait-abdesselam, F., Zhang, Z.: Mitigating packet dropping problem in mobile ad hoc networks: proposals and challenges. IEEE Commun. Surv. Tutor. **13**, Fourth Quarter (2011)
9. Duarte-Melo, E.J., Liu, M.: Energy efficiency of many-to-one communication in wireless networks. In: Invited Paper in Proceeding of IEEE 45th Midwest Symposium on Circuits and Systems (MWSCAS'02), Volume 1,Tulsa (2002)
10. Murthy, S.R., Manoj, B.S.: Ad hoc Mobile Wireless Networks—Architectures and Protocols. Pearson Publication, New York (2014)
11. Marti, S., Giuli, T.J., Lai, K., Baker, M.: Mitigating routing misbehavior in mobile ad hoc networks. In: Proceedings of the 6th Annual International Conference on Mobile Computing and Networking, New York, NY, USA: ACM Press, pp. 255–265. Boston, MA (2000)
12. Wang, D.: Malicious node detection mechanism for wireless ad hoc network. Int. J. Secur. (IJS) **7**(1) (2013), School of Mathematical Sciences, Chuzhou University, China
13. Mishra, A., Nadkarni, K., Patcha, A., Virginia Tech: Intrusion detection in wireless ad hoc networks. IEEE Wirel. Commun. (2004)
14. Huang, Y., Lee, W.: A cooperative intrusion detection system for ad hoc networks. College of Computing, Georgia Institute of Technology

15. Sarkar, S.K., Basavaraju, T.G., Puttamadappa, C.: Ad hoc Mobile Wireless Networks-Principles, Protocols and Applications, pp. 267–270. Auerbach Publications, Taylor & Francis Group, Boca Raton (2012)
16. Dey, S.: Graph Theory with Applications, pp. 5–30. SPD (2012)
17. Krawetz, N.: Introduction to Network Security, pp. 5–13. Thomson Learning, Boston (2011)

Service Provisioning Middleware for Wireless Sensor Network

S. Sasirekha and S. Swamynathan

Abstract Earlier, wireless sensor network (WSN) applications tended to follow the traditional format of being specific to a device. Later, when applications evolved integrating heterogeneous devices, it rendered difficulty in enforcing a common standard among all the diverse devices. In order to handle this, a lot of WSN middleware platforms emerged to bind the application interaction with the heterogeneous devices through heterogeneous interfaces. This started increasing the service-based applications while decreasing the device-based applications. Apart from not only providing the classic task of reading the information from the sensor network, the middleware support were extended to address interoperability, management, security, and privacy. However, still there exists an important issue, which many of the existing middleware fail to address. For instance, the network design scenario varies depending on the application context. However, most of the existing middleware operate on the default network infrastructure and data dissemination protocol to collect the data and perform other tasks on the network. Therefore, there is a requirement to include support for customizing the network configuration for an application requirement with respect to its context. Hence, in this work, a service provisioning middleware based on service-oriented architecture (SOA) is proposed. To support network customization, in the middleware layer, a decision algorithm is proposed. It is used for generating the configuration file according to the application requirement. This service provisioning middleware would serve as a generic model for adapting to the required network environment.

Keywords Wireless sensor network · Service-oriented architecture · Middleware · Service provisioning

---

S. Sasirekha (✉)
SSN College of Engineering, Chennai 603110, India
e-mail: sasirekhas@ssn.edu.in

S. Swamynathan
Anna University, Chennai 600025, India
e-mail: swamyns@annauniv.edu

S. Das et al. (eds.), *Proceedings of the 4th International Conference on Frontiers in Intelligent Computing: Theory and Applications (FICTA) 2015*, Advances in Intelligent Systems and Computing 404, DOI 10.1007/978-81-322-2695-6_28

1 Introduction

Wireless sensor network (WSN) in the recent past has witnessed an explosive growth in connecting the digital realm to the physical world [1]. This is mainly due to the technology innovation in microprocessors, wireless communication, and its integration with the microelectromechanical system (MEMS). As WSN consents a fine-grained environment observation at an economical cost much lower than other known monitoring systems such as remote sensing, nowadays it plays a significant role in building large-sized WSNs for most currently evolving practical applications. These sensor networks provide robust service even in toxic and inaccessible regions to humans. WSN has great potential to efficiently plan and coordinate among the nodes to acquire information for real-time scenarios such as handling emergency, military, and disaster relief operations [2]. It brings a new and wide perspective for monitoring physical conditions like temperature, pressure, humidity, etc.

Nowadays, sensor nodes are available at low cost and are miniature in size. Due to their size, sensors have limited energy, storage, communication, and computation capabilities. Hence, they are usually deployed in large arrays to collaboratively extract the environmental data [2]. A WSN is designed to transmit the sensed data from an array of sensor nodes to a data repository on a server or directly to the user through wireless communication. When the sensor network is deployed in a hostile environment, it operates unattended for a long time with sensors equipped with limited battery power. Therefore in these situations, reducing the energy consumption to prolong the network lifetime stands as a critical issue. As most of the energy is spent on transmitting the data than other tasks in WSN, various communication protocols specific to WSN were proposed to handle efficient data transmission over the network. Some of them are LEACH (Low-Energy Adaptive Clustering Hierarchy), PEGASIS (Power-Efficient GAthering in Sensor Information Systems), TEEN (Threshold Efficient sensor Network protocol), MECN (Minimum Energy Communication Network), GAF (Geographic Adaptive Fidelity), and GEAR (Geographic and Energy-Aware Routing). Most of these solutions are dedicated to power efficient routing, focusing on short-range communication and adopt some aggregation mechanism to reduce the amount of data to be transmitted [3].

Some of the related works, as discussed in [4], highlight the importance of involving the application users in the WSN communication process. They also clearly state that it reduces the computation cost of the network with respect to application-specific optimization. Every application class satisfies a specific need and it requires a particular type of communication protocol. Thus, it is understood that the main issue is in providing the most suitable protocol to each application class. However, in general, it is difficult for application developers to choose the protocol that best meets their application needs. Therefore, in this work the main objective is to improve the performance of system by providing application users with a choice of the communication protocol, from which users can choose according to the application requirements [5, 6]. In spite of the advantages, the existing middleware [7–9] have been built with a high degree of dependency between applications and the underlying communication protocol. Accordingly, the framework also needs to afford a decision unit to provide a choice of suitable communication protocol.

Hence, in this paper a service provisioning middleware is proposed. It acts as a mediator between applications and the WSN. It helps in translating the application requirements to an efficient choice of network configuration and protocols. The middleware is generic enough to be used over a wide range of applications. The key focus of this middleware is interpreting the application needs and selecting the precise network protocol to configuration for performing efficient data aggregation. Besides, it also provides a high level graphical interface for subscribing, publishing, and collecting the sensor data. From the user's point of view, the system offers an abstraction layer to the applications developer from the sensor layer.

This paper is organized as follows: the related work is presented in Sect. 2; in Sect. 3 detailed descriptions of the different layers of the architecture are given; in Sect. 4, the operation of a middleware with a testing environment is illustrated; and finally in Sect. 5, some concluding remarks are given.

2 Related Works

In this section, the state-of-art WSN middleware are examined and discussed by classifying them as non-adaptable and self-adaptable middleware in the context of the proposed work. The classification was done based on the support for network reconfiguration at runtime. Here, some of the works related to self-adaptable middleware are discussed. Most of the self-adaptable middleware are based on the service-oriented concepts handling the heterogeneity and composition of the sensor nodes. They manage the adaptation issues based on the structural and behavioral reconfiguration actions. These existing service-oriented architecture (SOA)-based middleware can further be categorized based on the service deployment level. The services deployment basically can be on any of the three distinguished layers such as on the node, on the gateway, or on the base station layer. Middleware such as Tiny Web Services [10] and TinyWS [11] deploy services at the sensor nodes. They interact horizontally between the neighboring nodes and reform the structure according to the specification provided by the adaptive services. Tiny Web Services is an event-based middleware which provides mini Web Services operated by a small version of Transmission Control Protocol/Internet Protocol (TCP/IP) called µIP. TinyWS provides an opportunity to communicate directly with the sensor nodes without going through gateway by embedding Web Services (WS) on the sensor nodes using TCP/IP. Middleware Linking Applications and Networks (MiLAN) is another middleware for WSN, where a description of application requirement is received from the users and the best sensor protocol and network configuration is chosen. The work presented in [12] belongs to service deployment at the base station layer. It provides an Open Services Gateway initiative (OSGi)-based service-oriented middleware which uses a packet forwarder for communicating between the sensor nodes and base station. However, it requires a system administrator intervention to add a wrapper to communicate with the heterogeneous nodes. One other middleware as discussed in [13] provides a cross-level architecture. It provides a layered approach that deploys services in all the three layers and uses the DPWS for creating WS. However, no adaption is proposed to deal with service failures. At the base station level, the sensor network is also proposed as a database by deploying query proxies inside the network [14, 15]. In [16], a declarative language is proposed, which receives the query submitted by the user and submits it to the sensor networks. COUGAR provides a virtual database concept and a data centric routing approach [17].

All the above discussed middleware are advantageous in the defined context and aim at maximizing the network lifetime. However, the major shortcoming of the above discussed middleware is, they do not provide an abstract representation for the user to specify the application needs and how the sensor data has to be generated. Moreover, these middleware use eXtensible Markup Language (XML) and Simple Object Access Protocol (SOAP) standards for its communication, which stimulate an overhead and consume a large part of the device resources [18]. The main difference between the existing middleware and the proposed work is, here, a customary depiction of the application requirements and sensor data generated are provided. Also, the proposed system relies on XML and REpresentational State Transfer (REST) standards retaining the device resource for a much longer time. Hence, the proposed system aims at providing an adaptable WSN configuration and protocol support for every specific application requirement.

3 System Model for Data Provisioning Middleware

The proposed middleware serves as a uniform, standard, and abstract development model for developing WSN applications. This framework is developed by reflecting the concepts of service orientation specific to WSN [19]. The architecture includes three significant layers such as the sensor network layer, data processing layer, and the user interface layer as shown in Fig. 1. The sensor network layer is built by deploying several sensor nodes which have different sensing capabilities. Within the same network one powerful node with greater processing and storage capacity is deployed to act as sink node. The sensor network layer transmits the sensor data and the network management information to the outside entities via the sink node using the corresponding providers. These sensor data and other information received from the network layer are complied, classified, and stored for future use. Then the middleware-specific sink node in the network layer is used to pass request to the

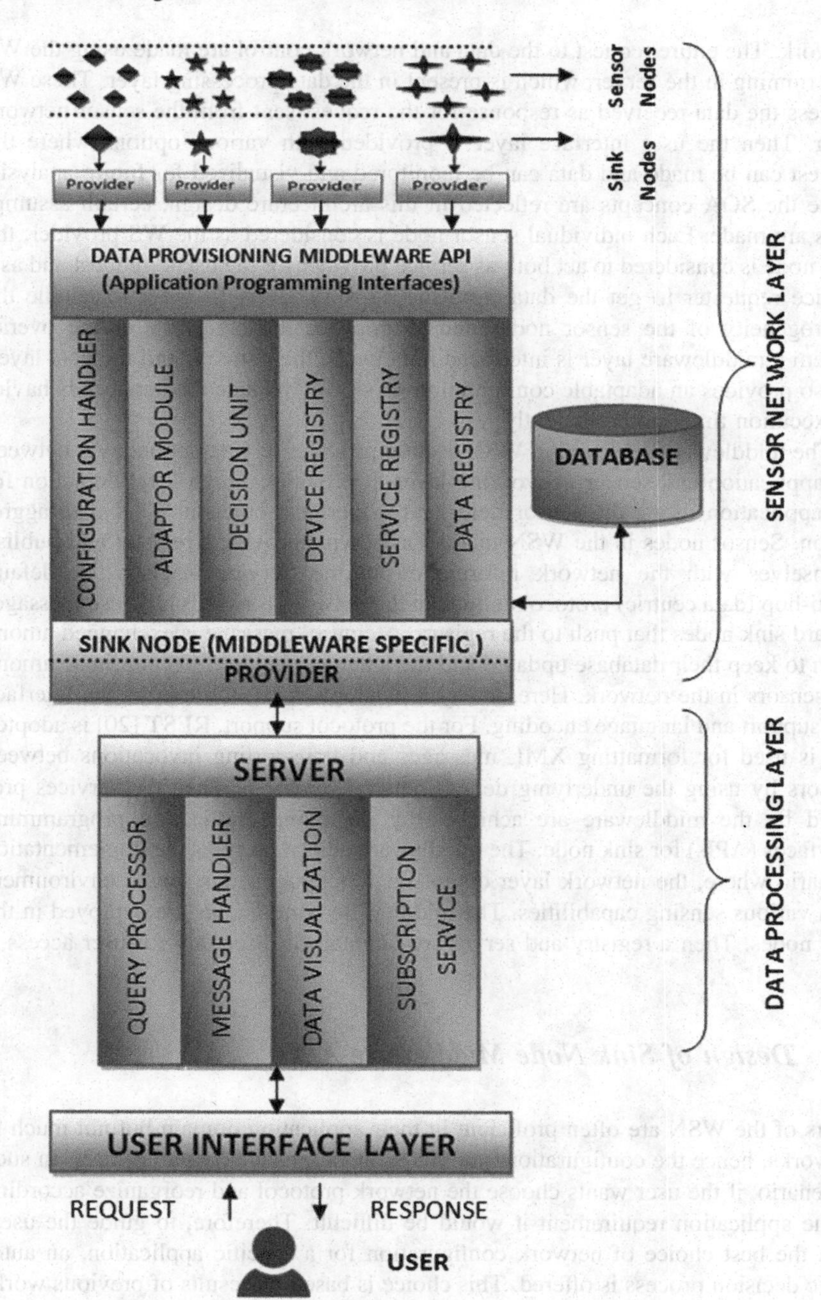

Fig. 1 Design of data provisioning middleware

network. The entire request to the data and network control are made using the WS [15] running in the server, which is present in the data processing layer. These WS process the data received as response for the request sent from the sensor network layer. Then the user interface layer is provided with various options where the request can be made and data can be monitored and visualized for future analysis. Since the SOA concepts are reflected in this architecture design, certain assumptions are made. Each individual sensor node is considered as the WS provider; the sink node is considered to act both as service provider for the users' request and as a service requester to get the data from the network layer. In order to handle the heterogeneity of the sensor nodes and to improve the efficiency of the overall system a middleware layer is interfaced in between the network and the data layer. It also provides an adaptable configuration service to manage the network behavior at execution time more efficiently.

The middleware deployed in WSN section provides an interaction layer between the application and sensor networking layer. It provides a high-level depiction for the application users, the sensor field, and to perform efficient WSN data aggregation. Sensor nodes in the WSN that act as service providers register and publish themselves with the network information in the Service Registry. A default multi-hop (data centric) protocol defined in the network forwards all these messages toward sink nodes that push to the registry. A control message is exchanged among them to keep their database updated and the latter can be used to coordinate among the sensors in the network. Here, the system adopts XML to describe the interface and support and language encoding. For the protocol support, REST [20] is adopted and is used for formatting XML messages and transporting invocations between sensors by using the underlying default network protocol. Then the services provided by the middleware are achieved by deploying application programming interfaces (APIs) for sink node. The middleware design portrays the implementation scenario where, the network layer comprises a heterogeneous sensor environment with various sensing capabilities. The middleware components are deployed in the sink nodes. Then a registry and server are maintained discretely for user access.

3.1 Design of Sink Node Middleware APIs

Users of the WSN are often proficient in their application domain but not much in networks, hence the configuration and WSN operations are preset [21, 22]. In such a scenario, if the user wants choose the network protocol and reorganize according to the application requirement it would be difficult. Therefore, to guide the users with the best choice of network configuration for a specific application, an automatic decision process is offered. This choice is based on results of previous works carried out in the field, reported by researchers. These works are studied and stored in the database. The adaptation module of the middleware holds this database with all the historical information about the different WSN configuration choices. In addition, the average values for the various performance metrics, such as delay and accuracy, for each executed application are included in the database for the corresponding network configuration. This information is used as the major input by the decision algorithm of decision unit to provide the best choice of network configuration. However, based on factors such as the number of sensors in the data source, number of sink nodes, and the choice of data delivery model, the decision algorithm decides the dissemination and network topologies as shown in Fig. 2.

For every request made by the application providers the decision unit API runs the decision algorithm. The decision algorithm generates the corresponding configuration file. The generated file is pushed to the configuration manager API. Receiving the configuration file, it processes them and using the adaptor module API, the configuration API controls the hardware and topology of each individual sensor network through its corresponding sink nodes. These API modules communicate among them with the XML messages to receive and process the application input messages. Also, the middleware has an XML-based API for sending

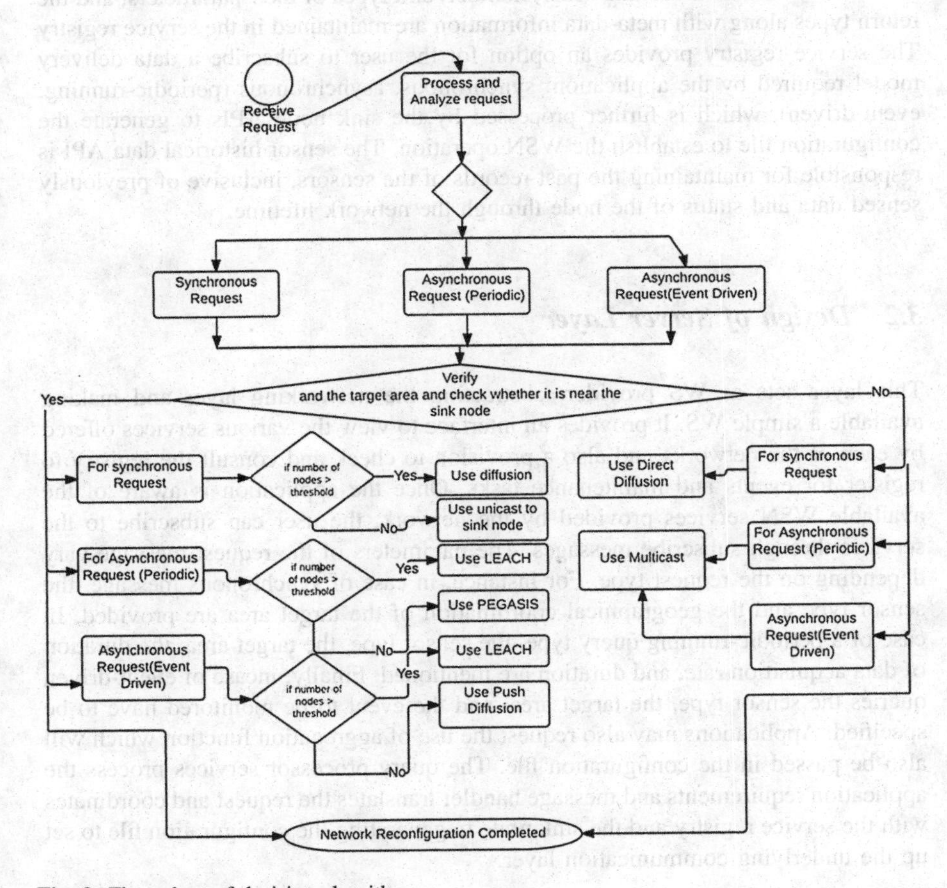

Fig. 2 Flow chart of decision algorithm

the configuration file to configuration handler of node level API and it activates the dissemination protocol according to the current WSN application. The decision algorithm decides the strategies of data dissemination and network topologies based on the request type submitted by the user, the size of the target area, and the expected number of data sources and sinks are considered. From the choice based on the previous works, the possible combination of network topology and dissemination strategy was identified. It was inferred that, in general, the network topology to support energy efficient data dissemination can be classified as flat and hierarchical. Further, according to the application requirement they are merged with dissemination algorithm such as direct diffusion, push diffusion, one-phase diffusion, unicast, multicast, and clustering with LEACH or PEGASIS.

Then the infrastructural information of the WSN and users are held in the registry. The various services related to the network are maintained in the form of various sub-registries. The aggregated network information from sink node service provisioning API is registered in the device registry. Service information such as the names of the methods, the order, number, and types of their parameters, and the return types along with meta-data information are maintained in the service registry The service registry provides an option for the user to subscribe a data delivery model required by the application: synchronous, asynchronous (periodic-running, event-driven), which is further processed by the sink node APIs to generate the configuration file to establish the WSN operation. The sensor historical data API is responsible for maintaining the past records of the sensors, inclusive of previously sensed data and status of the node through the network lifetime.

3.2 Design of Server Layer

This layer acts as WS provider, abstracting the networking layer and making available a simple WS. It provides an interface to view the various services offered by each of the networks and also a provision to check and consult the registry to register for events and maintenance tasks. Once the application is aware of the available WSN services provided by the network, the user can subscribe to the services through subscribe messages. The parameters of the request message vary depending on the request type. For instance, in case of synchronous message, the sensor type and the geographical coordination of the target area are provided. In case of a periodic-running query type, the sensor type, the target area, the duration of data acquisition rate, and duration are mentioned. Finally, incase of event-driven queries the sensor type, the target area, and the event to be monitored have to be specified. Applications may also request the use of aggregation function which will also be passed in the configuration file. The query processor services process the application requirements and message handler translates the request and coordinates with the service registry and the sink node in generating the configuration file to set up the underlying communication layer.

3.3 Configuration File Generation

As shown in Fig. 2, each application provider submits the request query to the decision algorithm along with various other parameters such as the size of the target area, the estimated number of data sources, and sinks. These parameters have a significant role in deciding the different strategies of data dissemination and network topologies. Initially the algorithm verifies for the request type submitted by the provider. The request query is analyzed and determined if the query type is a synchronous or asynchronous periodic or event-driven query. The network reorganization is mainly based on this request type. Next, for all the request types, the size of the target area and the number of nodes near the sink node is verified. It is analyzed for, if the number is greater than a defined threshold. In case it is large, for a synchronous request again, the total number of nodes in the target is verified against a defined threshold, if it is has a large number of nodes, diffusion dissemination strategy is chosen, otherwise unicast to sink node is chosen. Similarly, it is also verified for asynchronous periodic and event-driven query; in this case if the number of nodes is large, LEACH dissemination strategy is chosen, otherwise PEGASIS and push diffusion is chosen respectively. In case it is less, for a synchronous request the direct diffusion dissemination strategy is chosen. Similarly, for asynchronous periodic and event-driven query muticast dissemination strategy is chosen.

Further, the middleware passes the decision chosen to the network protocol, which in turn triggers the needed infrastructure for communication. Finally the message is propagated to sensor nodes, based on the chosen strategy for data dissemination. If the request sent from the application matches its data type it responds with the data delivery message.

Therefore the proposed middleware operates in a service-oriented way providing a flexible, adoptable, generic system for programming abstractions and positioning them right from the hardware platforms up to end-user applications in a heterogeneous WSN.

4 Testing Environment

A WSN environment is created with 4 Arduino Mega Board interfaced with Digi XBee using XBee shield to form a mesh network as shown in Fig. 3, to monitor the temperature and humidity of the testing environment. To build sensor network with a mixture of hardware platforms, Digi XBee is used to communicate because they are 802.15.4 compliant. Since tuning of many parameters of the XBee is very versatile, XBee is preferable to CC2420. Also, XBee module has an IEEE 64-bit address and has a provision to configure in the AT mode or API; as the aim is to

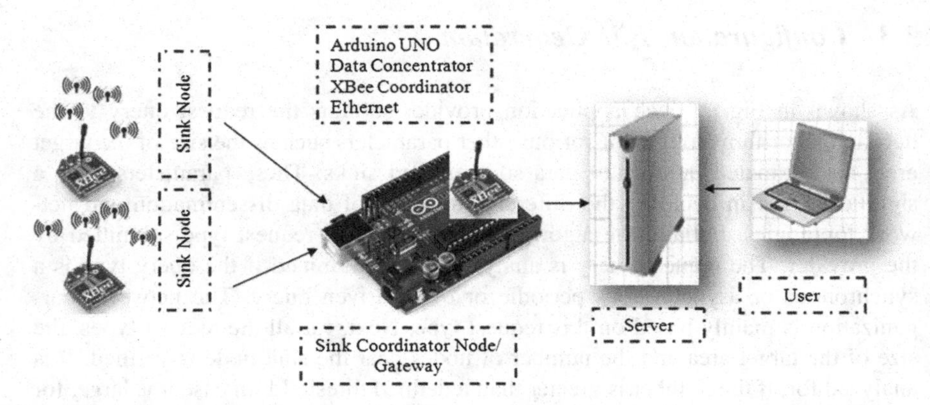

Fig. 3 Testing environment

configure the network from an API, we choose an API mode. The XBee modules come preloaded with the ZNet 2.5 firmware which implements the ZigBee protocol stack. XBee has its own microcontroller inside and for programming it uses the software called XCTU. This microcontroller is used to program the routing and other communication details of the XBee.

In order to make all the sensor platforms to communicate, some basic settings have to done in the network configuration level. First of all, for the radios to communicate, it is mandatory to set the same Personal Area Network (PAN) ID and sync them in the same channel. For the purpose of testing it is assumed that PAN ID 0x1234 and channel 0x0D are set for all the radios. Once the communication aspects are programmed among the various sensor platforms the sensor nodes are deployed. Then the middleware APIs are installed in the sensor and sink nodes to provide the abstraction. The necessary implementation codes for the sensors to implement them as services are defined. This implementation is platform-dependent, that is, the application programmer uses C++ to write Arduino programs. The users can access the WSN through the WS deployed in the server. The central server acts as a gateway to the WSN. The intelligence level of the WSN increases upstream in the network. Some of the data collected are shown in Fig. 4.

To address the improvements of the above stated middleware system, it is demonstrated that when using a network routing protocol chosen according to the application needs the network utilization is proven to be high. A comparison is run with middleware with fixed data centric protocol and an adaptive middleware where the network routing protocol changes according to the application need. It is initially fixed with multi-hopping flat network routing protocol, Sensor Protocols for Information via Negotiation (SPIN). The idea behind SPIN is to name the data using high-level descriptors or meta-data. Before transmission, meta-data are exchanged among sensors via a data advertisement mechanism. Each node upon receiving new data advertises it to its neighbors and interested neighbors, i.e., those that do not have the data, retrieve the data by sending a request message. The

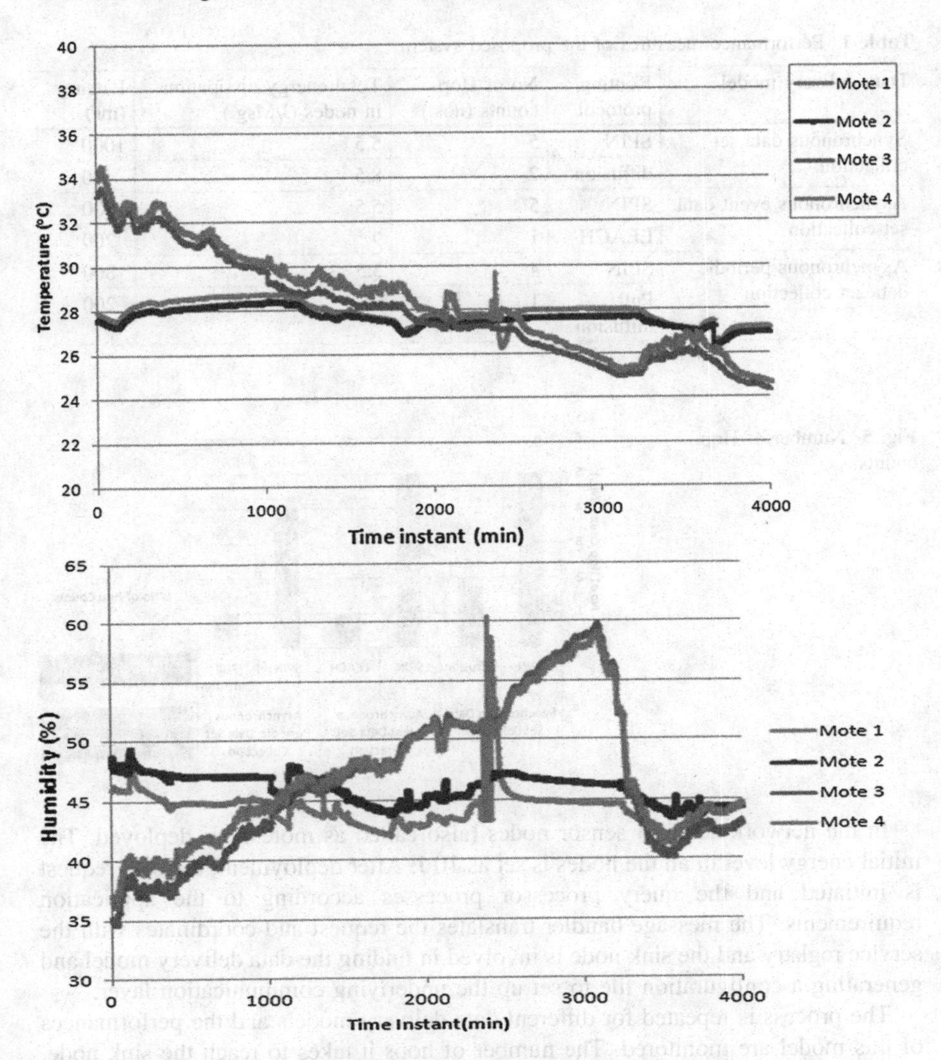

Fig. 4 Sample temperature and humidity data collected from 4 motes

adaptive network routing protocols for the data delivery model are chosen based on the decision algorithm explained in Configuration file generation section.

The results are generated for both the scenarios, the fixed routing protocol and adapted routing protocol for the three data delivery models—Synchronous, Asynchronous Event-Based, and Periodic Data Set models. The adaptive network routing protocols for the data delivery model are chosen based on the decision algorithm explained in Configuration file generation section.

Table 1 Performance measures of the proposed system

Data delivery model	Routing protocol	No of Hop counts (nos.)	Total energy dissipation in nodes (J/Msg.)	Latency (ms)
Synchronous data set collection	SPIN	5	5.5	1000
	diffusion	2	8.5	400
Asynchronous event data set collection	SPIN	5	5.5	1000
	LEACH	1	9	200
Asynchronous periodic data set collection	SPIN	4	5.5	800
	Pull diffusion	1	9	200

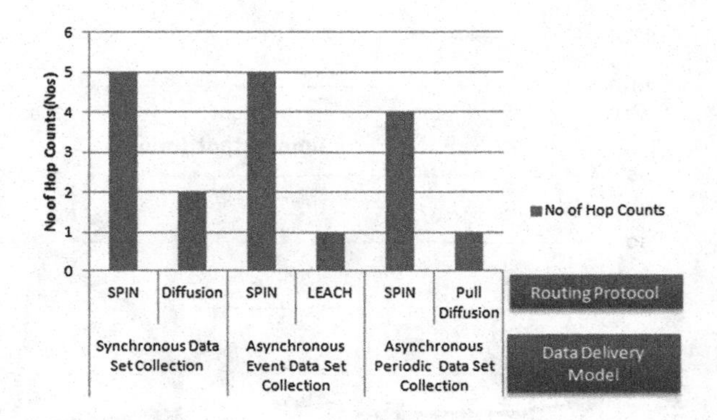

Fig. 5 Number of Hop counts

In the network setup, 4 sensor nodes (also called as motes) are deployed. The initial energy level in all the nodes is set as 10 J. After deployment, the user request is initiated and the query processor processes according to the application requirements. The message handler translates the request and coordinates with the service registry and the sink node is involved in finding the data delivery model and generating a configuration file to set up the underlying communication layer.

The process is repeated for different data delivery models and the performances of this model are monitored. The number of hops it takes to reach the sink node, total energy dissipated in the nodes while transmitting, and based on the hop count the latency is measured. The generated results are tabulated in Table 1 and graphically plotted as shown in Figs. 5, 6 and 7.

For simplicity of energy analysis, a first-order radio model [23, 24] is adopted. Energy consumption in circuitry for running the transmitter or receiver and in radio amplifier for wireless communication are $E_{circuitry} = 0.5$ J/msg and $E_{amplifier} = 0.5$ pJ/msg, respectively. The value of $E_{amplifier}$ is directly proportional to the square of transmission distance. Therefore, using the following formulas (1) and (2):

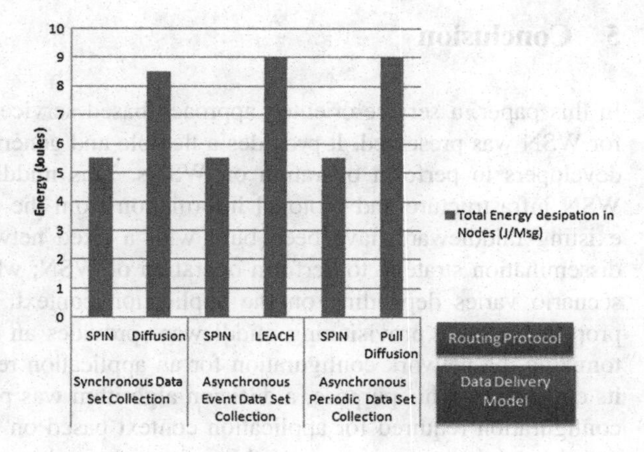

Fig. 6 Energy dissipation in the nodes

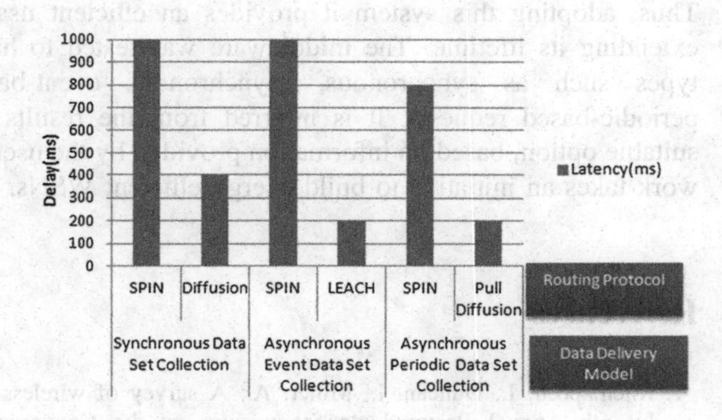

Fig. 7 Latency in transmission

The energy consumed for transmitting a single packet is

$$E_{\text{transmit}}(k,d) = E_{\text{circuitry}} \times k + E_{\text{amplifier}} \times k \times d^2 \quad (1)$$

The energy consumed for receiving single a packet is

$$E_{\text{recieve}}(d) = E_{\text{circuitry}} \times k \quad (2)$$

where k is the size of transmitted packets, and d is the distance between a transmitter and a receiver.

There is difference in the performance of the various data delivery models. The performance requirements of the model are application-specific. The user should be able to adapt or choose the required network routing protocol. The proposed middleware architecture supports the selection of required routing protocol for the data delivery model by the user. Hence the model is flexible to the user's specification. The analysis on the performance measures aids to infer that the overall system performance seems to be improved.

5 Conclusion

In this paper, a service-oriented approach-based service provisioning middleware for WSN was presented. It provides a flexible and generic platform for application developers to perform operation on WSNs. This middleware layer abstracts the WSN infrastructure and protocol information from the developer, as most of the existing middleware have been built with a fixed network topology and a data dissemination strategy to perform operation of WSN; whereas the network design scenario varies depending on the application context. Hence in this paper, the proposed service provisioning middleware provides an extended support for customizing the network configuration for an application requirement with respect to its context. For this purpose, a decision algorithm was proposed. It decides on the configuration required for application context based on the type of request query, the size of the target area, and the estimated number of data sources and sinks. Thus, adopting this system it provides an efficient usage of the WSN thereby extending its lifetime. The middleware was tested to handle three basic request types such as synchronous, asynchronous event-based, and asynchronous periodic-based requests. It is inferred from the results that it chooses the best suitable option, based on information provided by the users' interest. This proposed work takes an initiative to build energy efficient WSNs.

References

1. Khemapech, I., Duncan, I., Miller, A.: A survey of wireless sensor networks technology. In: 6th Annual Postgraduate Symposium on the Convergence of Telecommunications, Networking and Broadcasting, pp. 1–42. Liverpool, UK (2005)
2. Rawat, P., Singh, K.D., Chaouchi, H., Bonnin, J.M.: Wireless sensor networks: a survey on recent developments and potential synergies. J. Supercomputing **68/1**, 1–48 (2014)
3. Akkaya, K., Younis, M.: A survey on routing protocols for wireless sensor networks. Elsevier Ad Hoc Netw. J. **3**(3), 325–349 (2005)
4. Heidemann, J., Silva, F., Estrin, D.: Matching data dissemination algorithms to application requirements. In: Proceedings of the 1st International Conference on Embedded Networked Sensor Systems, pp. 218–229, USA (2003)
5. Heinzelman, W., Kulik, J., Balakrishnan, H.: Adaptive protocols for information dissemination in wireless sensor networks. In: Proceedings of the 5th Annual ACM/IEEE International Conference on Mobile Computing and Networking (MobiCom'99), pp. 1–15, WA (1999)
6. Wang, M., Cao, J., Li, J., Dasi, S.K.: Middleware for wireless sensor networks: a survey. J. Comput. Sci. Technol. **23**(3), 305–326 (2008)
7. Heinzelman, W., Murphy, A.L., Carvalho, H.S., Perillo, M.A.: Middleware to support sensor network applications. IEEE Netw. Mag. Special Issue **18**(1), 6–14 (2004)
8. Mohamed, N., Al-Jaroodi, J.: A survey on service-oriented middleware for wireless sensor networks. J. IEEE Int. Conf. Serv. Oriented Comput. Appl. **5/2**, 71–85 (2011)
9. Hadim, G. Mohamed, N.: Middleware challenges and approaches for wireless sensor networks. IEEE Distrib. Syst. Online **7/3**, 1–23 (2006)

10. Nissanka, B., Priyantha, A.K., Goraczko, M., Zhao F.: Tiny web services: design and implementation of interoperable and evolvable sensor networks. In: Proceedings of the 6th ACM Conference on Embedded Network Sensor Systems, pp. 253–266. ACM, USA (2008)
11. Othman, N.Y. Glitho, R.H., Khendek, F.: The design and implementation of a web service framework for individual nodes in sinkless wireless sensor networks. In: 12th IEEE Symposium on Computers and Communications, pp. 941–947 (2007)
12. Prinsloo J.M., Schulz C.L., Kourie, D.G., Theunissen, W.H.M.: Strauss, T., Van Den Heever, R. Grobbelaar, S.: A service-oriented architecture for wireless sensor and actor network applications. In: Proceedings of the 2006 Annual Research Conference of the South African Institute of Computer Scientists and Information Technologists on IT Research in Developing Countries (SAICSIT '06), pp. 145–154 (2006)
13. Leguay, J., Lopez-Ramos, M., Jean-Marie, K., Conan, V.: An efficient service oriented architecture for heterogeneous and dynamic wireless sensor networks. In: 33rd IEEE Conference on Local Computer Networks, pp. 740–747 (2008)
14. Devices Profile for Web Services (DPWS) specification. (2006). http://schemas.xmlsoap.org/ws/2006/02/devprof/
15. Bonnet, P., Gehrke, J.E., Seshadri, P.: Towards sensor database systems. In: Proceedings of the 2nd International Conference on Mobile Data Management, pp. 3–21, Hong Kong (2001)
16. Govindan, R., Hellerstein, J.M., Hong, W., Madden, S., Franklin, M., Shenker, S.: The sensor network as a database. USC Computer Science Department Technical Report, pp. 02–771, (2002)
17. Yao, Y., Gehrke, J.E.: The cougar approach to In-Network query processing in sensor network. Sigmod Record. **31**(3), 9–18 (2002)
18. Delicato, F., Pirmez, L., Pires, P., de Rezende, J.: Exploiting Web technologies to build automatic wireless sensor network. In: 8th IFIP IEEE International Conference on Mobile and Wireless Communication, vol. 211, pp. 99–114 (2006)
19. Delicato, F.C., Pires, P.F., Pirmez, L., Carmo, L.F.: A service approach for architecting application independent wireless sensor networks. Cluster Comput. **8/2–3**, 211–221 (2005)
20. Graham, S., et al.: Building Web services with Java: making sense of XML, SOAP, WSDL, and UDDI. Sams Publishing (2002)
21. Fok, C.-L., et al.: Adaptive service provisioning for enhanced energy efficiency and flexibility in wireless sensor networks. Sci. Comput. Program. **78/2**, 195–217 (2013)
22. Chandrakant, N., Tejas, J., Harsha, D., Deepa Shenoy, P., Venugopal, K.R., Patnaik, L.M., Chancellor, V: EMID: maximizing lifetime of wireless sensor network by using energy efficient middleware service. In: International Conference on Intelligent Information Networks, pp. 314–317 (2011)
23. Lindsay, S., Raghavendra, C.S., Sivalingam, K.M.: Data gathering in sensor networks using the energy delay metric. In: Proceedings of the 15th International Parallel & Distributed Processing Symposium, p. 188 (2001)
24. Tang, F., You, I., Gou, S., Gou, M., Ma, Y.: A chain-cluster based routing algorithm for wireless sensor networks. J. Intell. Manuf. **23**(4), 1305–1313 (2010)

A Novel MTC-RB Heuristic for Addressing Target Coverage Problem in Heterogeneous Wireless Sensor Network

Sonu Choudhary, R.S. Sharma and Sneha Shriya

Abstract Network life time is a crucial perspective in target specified applications of wireless sensor network. Expedient sensor scheduling and power management can efficaciously prolong network lifetime. Sensors are divided into cover sets in which all targets are covered by a set of sensors and targets are monitored by each cover set for a period of time. The coverage problem proved to be NP-Complete. In present study, a unified elucidation of prior algorithms is given an efficient and new heuristic is proposed. The proposed heuristic prioritizes sensors based on the value of heuristic function. Its shown analytically that proposed heuristic outperforms other heuristic in terms of solutions quality. Our experimental analysis over a randomly generated large set of problem instances demonstrate that proposed heuristic solution is close to the definite optimal solution.

Keywords Target coverage · Energy efficiency · Sensor cover · Sensor scheduling · Network lifetime · Wireless sensor network

1 Introduction

Recently, research area is greatly impacted by wireless sensor network. WSN constitute the base for a wide range of probable application which includes habitat monitoring, water monitoring, fire detection, chemical agent detection, health monitoring, etc. [7]. The wireless sensor network consists of a large number of small, light weight, energy constrained, and inexpensive devices called sensor nodes and a communication module. Sensor nodes are tiny electronic devices equipped with

S. Choudhary (✉) · R.S. Sharma · S. Shriya
Rajasthan Technical University, Kota, Rajasthan, India
e-mail: sonuuce15@gmail.com

R.S. Sharma
e-mail: rssharma@rtu.ac.in

S. Shriya
e-mail: sneha.shriya22@gmail.com

S. Das et al. (eds.), *Proceedings of the 4th International Conference on Frontiers in Intelligent Computing: Theory and Applications (FICTA) 2015*, Advances in Intelligent Systems and Computing 404, DOI 10.1007/978-81-322-2695-6_29

batteries, sensing hardware, processing, and storage resources, and these nodes are designed for specific application. In WSN, sensors are deployed randomly or predefined manner according to the requirement of environment and the application. Design of WSN is influenced by many factors, such as scalability, fault tolerance, network topology, hardware constraints, transmission modes, and power consumption. Among these factors, limited power supply is the most critical issue due to the sensors battery size and weight limitation. Thus the efficient utilization of available energy resources directly impacts the lifetime and the performance of wireless sensor network.

A sensor node can be in two state of monitoring: idle and active and in one of four state of communication: transmit, receive, idle, and sleep. In active mode, sensor continuously monitor environment or communicate with base station [3]. In sleep state, sensor cannot monitor or communicate. In large-scale networks, sensor nodes divided into cover sets. A cover set is characterized by active times during which the respective cover set continuously monitors its targets. Such deployment extend the life time of sensor networks. To maximize network lifetime, target coverage depends on scheduling active/sleep duration of sensors, while sensors of cover set continuously monitor all targets. Target coverage problem proved to be NP-Complete [4]. Approximate schemes and heuristics are proposed earlier. We propose an efficient and new heuristic. The proposed heuristic prioritizes sensors based on the value of heuristic function. It is shown analytically that proposed heuristic outperforms other heuristic in terms of solutions quality. Our experimental analysis over a randomly generated large set of problem instances demonstrate that proposed heuristic solution is close to the definite optimal solution.

2 Related Work

Coverage problem in WSN has been classified based on subject covered by sensors, as area or discrete points. In area coverage problem, the main objective is to cover given area under the condition that each portion of given area is covered by minimum one sensor at a time. In target coverage [2, 4, 10, 14], problem a set of targets is covered by sensors.

Target coverage problem aims to maximize lifetime of network by grouping sensor nodes into disjoint or non-disjoint cover sets and then schedule their active/sleep durations. In disjoint [10, 13] sensor cover, every sensor is a part of only one sensor cover and in non-disjoint [1, 4] sensor cover, each sensor can be a part of more than one sensor cover. Cardei et al. [4] proposes a greedy heuristic. In greedy heuristic priority is given to that sensor which covers maximum uncovered targets, maximum residual battery life, and critical target while generating cover sets. Zorbas et al. [14] propose algorithm in which deployed sensor nodes are grouped into four classes: Best, Good, Ok, and Poor. Recently Manju [1] proposes MCH. In MCH, sensors are prioritize according to maximum covered targets. In target coverage, two types of sensor networks homogeneous and heterogeneous are used.

Homogeneous WSN consist sensors with same energy and sensing range [1, 4]. Heterogeneous WSN consists sensors with different sensing range and computing power [9]. There are centralized and distributed algorithms for coverage problem in WSN. In centralized approach [1, 4], algorithms are executed on base station (BS). In distributed approach [5, 8, 10], there is no any central node and decision are taken by a number of sensor nodes in decentralized manner. Proposed heuristic use centralized approach.

3 Discussion

According to the literature survey, objective of all approaches is to maximize network lifetime. So, an efficient heuristic solution to maximize network lifetime of target coverage problem is proposed. This section introduce some definitions, preliminary knowledge and assumptions.

3.1 Coverage Matrix

Coverage matrix R is as

$$R_{ij} = \begin{cases} 1, & \text{if } \mathit{target}\ t_j\ \mathit{is\ covered\ by\ sensor}\ s_i \\ 0, & \text{otherwise.} \end{cases} \quad (1)$$

3.2 Sensor Cover

Given R, a sensor cover C is a n rows set of R such that for each column of j $R_{ij} = 1$. Sensor cover C is said to be a minimal sensor cover if sensor cover C', $C' \subseteq C$ if and only if $C' =$ C for any set cover.

3.3 Minimum Life Time of Cover Set C

The minimum lifetime of cover set C is the minimum available life time of sensor s_i.

$$max - lifetime(c) = Min_{s_i \in C^{b_i}} \quad (2)$$

3.4 Critical Target

Target t_j is a critical target, if number of sensors which cover target t_j and total battery life of covering sensors is minimum.

$$j = Min \sum_{i} R_{ij} \times b_i \tag{3}$$

3.5 Target Coverage

Let t_1,t_2..t_n targets and s_1,s_2..s_n sensors are randomly deployed. b_i is the battery life of each sensor s_i. Target t_j covered by a sensor s_i which is within the sensing range of s_i. Sensor cover C is a subset of sensors which collectively cover all targets.

$$c = s_i \quad | \; for\ each\ target\ t_j \tag{4}$$

Lifetime L(C) of sensor cover cannot exceed minimum battery life of $s_i Min_{s_i \in C^{b_i}}$. The objective of energy efficient target coverage problem is to maximize the sum of $L(C_k)$ for all cover sets in which the time allotted is not larger than battery life b_i for each sensor of cover sets.
S=s_1,s_2..s_n; set of n sensors.
$s_k = i| \;\; if\ target\ t_m\ covered\ by\ s_i$
Sensor cover matrix M is as follows.

$$M_{ij} = \begin{cases} 1, & \text{if } sensor\ s_i\ is\ in\ cover\ set\ s_i \\ 0, & \text{otherwise.} \end{cases} \tag{5}$$

Formulation of target coverage problem in linear programming is defined as follows.

$$Maximize\, L_1 + L_2 + ...L_x \tag{6}$$

$$Subject\ to \sum_{j=1}^{x} M_{ij} L_x \le b_i for\ all\ s_i \tag{7}$$

$$\sum_{i \in s_k} M_{ij} \ge 1 for\ all\ t_m \in T, j = 1, 2.....x \tag{8}$$

$$L_x \ge 0\, for\ all\ cover\ sets\ C_x \tag{9}$$

where x denotes number of set covers. First constraint ($\sum_{j=1}^{x} M_{ij} L_x \le b_i$) is stated that time allotted for all sensors $s_i \in$ C is not more than b_i. Where b_i denotes life time of sensor.

Second constraint ($\sum_{i \in s_k} M_{ij} \ge 1$) for j=1..x, all $t_m \in T$ stated that in a set cover C_j. Each target t_m is covered by minimum one sensor.

4 MTC-RB (Maximum Target Coverage with Residual Battery)

An efficient and new heuristic (MTC-RB) for target coverage is proposed. From experimental analysis, it is observed that parameter B plays an important role to getting approximate solution of definite optimal solution. Hence sensors are prioritized according to the maximum value of h (heuristic function). There are three important steps of heuristics in following manner.

4.1 Generate Sensor Cover

The heuristic generates sensor cover C by selecting sensor with highest value of h, where h is as follows.

$$h = c + l \tag{10}$$

where c is percentage of target covered by sensor s_i and l is life time of sensor s_i. We generate sensor cover by iteratively selecting sensors with higher value of h until all targets are covered by at least one sensor.

4.2 Assign Life Time to Cover

Some lifetime is assigned to a sensor cover generated in earlier step. We assign life time B'. Life time $B' =$ Min(B, max lifetime(C)) is added to the life time of sensor cover, and it is subtracted from battery life of each sensor $s_i \in C$, where B is a user-defined constant. This process ensure that sensor cover do not consume energy more than any sensor S_i of sensor cover. So sensors with remaining residual battery are available for other covers.

4.3 Modify Priorities of Sensors

To avoid repeated generation of sensor covers in successive iterations, the priority of sensor reduce after it used in a sensor cover. Priority of sensor is reduce in terms of reducing the value of h.

Sensor cover generation in MTC-RB is different from other heuristic. In proposed heuristic generated sensor, covers are nonminimal so we carried out an additional step to minimize the sensor cover. Process of minimizing cover set is to remove one sensor from acquired cover set and check whether it is a cover set or not, if it is a cover set then we remove the sensor from acquired cover set. By repeating the process with all sensors of sensor cover set, we get minimal cover.

Algorithm 1 MTC-RB algorithm

Assign life time to each sensor by random number generator within the range of 1 to 10.
$S = \emptyset$; S_c = sensors
while do$S_c \neq \emptyset$ do
  Step 1: Generate a sensor cover
  //generate a sensor cover C and return null if cover is not found//
  Initialize $C = \emptyset$, $T_{uncovered} = T$
  while do$(T_uncovered = \emptyset)$
    Select a sensor s with highest value of h
    $h = c + l$
    $C \leftarrow C \cup s$
    For all target t covered by sensor s
    $T_{uncovered} = T_{uncovered} - t$
  end while
  Step 2:Minimizes cover set C
  Step 3:Assign life time
  $L(C) \leftarrow B'$
  Where B'=Min(B,max lifetime(C))
  Step 4:Update
  For all $s_i \in C$ update $b_i \leftarrow b_i - L(C)$
  If $b_i = 0$ then $S_c \leftarrow S_c - s_i$
end while

5 Experimental Analysis

To evaluate performance of MTB-RB, Greedy, MCH, MTC-RB heuristic are implemented in c language. All experiments were carried out on a Pentium (4) 2.63 GHz host with 256 MB of RAM, running Window Xp/Linux operating systematic sensor network with target nodes and sensor nodes randomly deployed in sensing area of 800 × 800 m inside monitored area 1000 × 1000 m is simulated. In MTC-RB heuristic sensor, covers are non-disjoint in terms of residual battery life. It is assumed that sensors are heterogeneous in terms of energy and initially, they have different battery life and same sensing range. Coordinates of sensors and targets are generated by pseudo random number generator. It is also assumed that target is within the sensing range of sensor if Euclidean distance of target from sensor is less than equal 80 m. For experiment, number of targets in interval [20, 90] and number of sensors in [20, 150] are varied. Upper bound of network life time is calculated as $u = Min_j \sum_i R_{ij} b_i$.

5.1 Experiment 1

Experiments were performed on greedy [4] and MCH [1] heuristic for average value of battery ($b_i = 5$) and MTC-RB for different values of battery life varying within the range [1–10]. Greedy, MCH, MTC-RB were experimented for different value of B to find a solution which is closer to definite optimal solution. The experiment shows that network life time improved significantly with lesser values of B. We experimented for target varying between 20 and 90 with an increment of 10 and fix 150 sensor nodes. In this experiment, we take four different values of B. Average life time is calculated for 20 random problem instances. Result of experiment 1 is shown by Fig. 1.

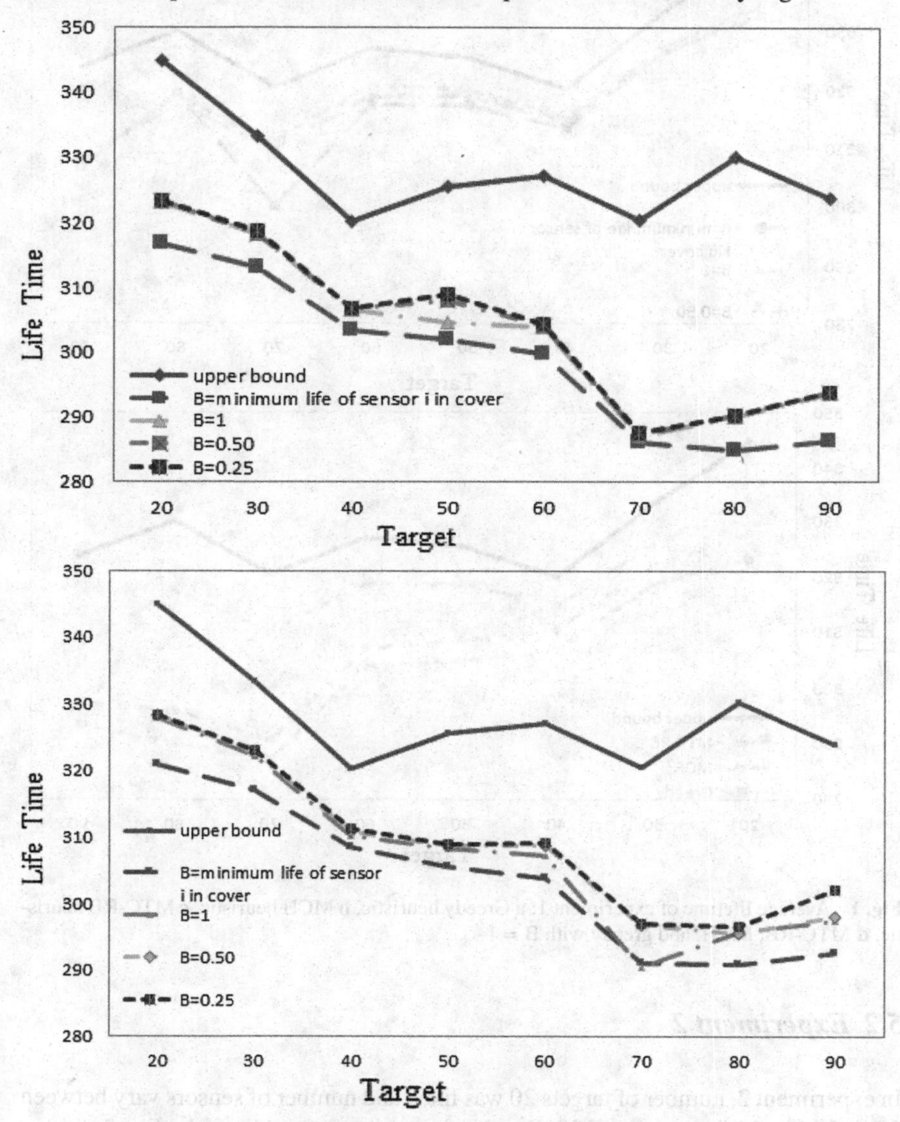

Optimal value of problem instance is not defined but upper bound is calculated as given above. Summary of experimental results is: for fixed number of sensors with sensing range, the lifetime of network increase as number of targets decreases. MTC-RB solution is close to upper bound and the performance of MTC-RB is better than MCH and greedy heuristic.

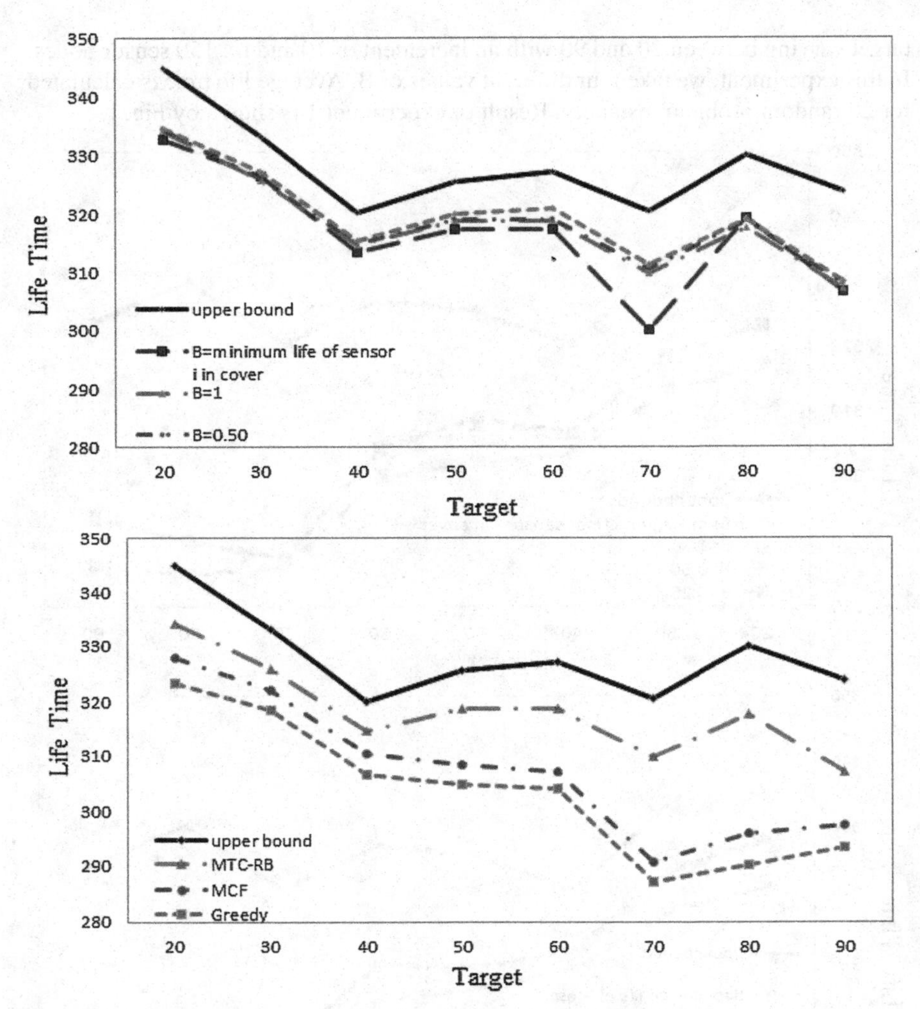

Fig. 1 Average lifetime of experiment 1; **a** Greedy heuristic, **b** MCH heuristic, **c** MTC-RB heuristic, **d** MTC-RB, MCH, and greedy with B = 1

5.2 Experiment 2

In experiment 2, number of targets 20 was fixed and number of sensors vary between 20 and 140 with increment of 20. Result of experiment 2 is shown by Fig. 2.

After analyzing experiment 1, 2 and it is observed that life time of sensor network increases with increase in density of sensor nodes because same target node is monitored by more number of sensors which form more set covers resulting in longer lifetime. Similarly, if number of target nodes increases, then target is monitored by less number of sensors which decrease number of set covers so network

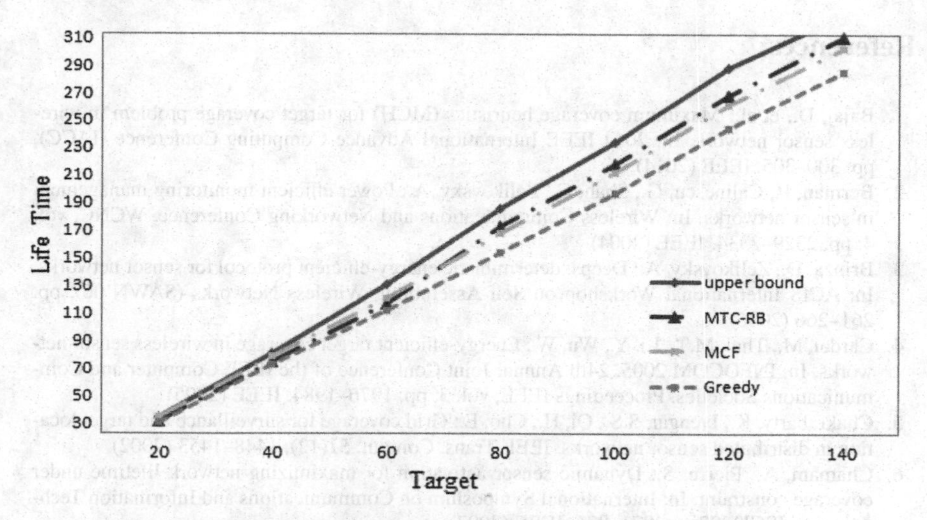

Fig. 2 Average life times obtain by MTC-RB, MCH, and greedy with B = 1

lifetime decreases. Choosing small B is better than higher B because for small value of B, a sensor can participate in more cover sets which results higher lifetime which we observed in Fig. 1 that MTC-RB provide solution close to the definite optimal solution.

6 Conclusion

This paper studies various centralized algorithms with homogeneous and heterogeneous network. A heuristic for target coverage problem in WSN with varying battery life is proposed. MTC-RB prioritizes sensors based on the value of heuristic function. It is shown analytically that proposed heuristic outperforms other heuristic in terms of solutions quality. To verify this approach, simulation results are presented. Proposed MTC-RB heuristic obtain 100 % coverage, because each target is covered by at least one sensor at a time. This paper addresses a basic target coverage problem in heterogeneous network. This study can be extend to other variants of target coverage problem as coverage with oriented sensing or adjustable sensing range [11, 12]. Effect of Q-coverage on network lifetime can also be analyzed. Scheduling active/sleep duration of individual sensors based on output of target coverage problem is also direction of future work [6, 13].

References

1. Bajaj, D., et al.: Maximum coverage heuristics (MCH) for target coverage problem in wireless sensor network. In: 2014 IEEE International Advance Computing Conference (IACC), pp. 300–305. IEEE (2014)
2. Berman, P., Calinescu, G., Shah, C., Zelikovsky, A.: Power efficient monitoring management in sensor networks. In: Wireless Communications and Networking Conference, WCNC, vol. 4, pp. 2329–2334. IEEE (2004)
3. Brinza, D., Zelikovsky, A.: Deeps: deterministic energy-efficient protocol for sensor networks. In: ACIS International Workshop on Self-Assembling Wireless Networks (SAWN'06), pp. 261–266 (2006)
4. Cardei, M., Thai, M.T., Li, Y., Wu, W.: Energy-efficient target coverage in wireless sensor networks. In: INFOCOM 2005. 24th Annual Joint Conference of the IEEE Computer and Communications Societies. Proceedings IEEE, vol. 3, pp. 1976–1984. IEEE (2005)
5. Chakrabarty, K., Iyengar, S.S., Qi, H., Cho, E.: Grid coverage for surveillance and target location in distributed sensor networks. IEEE Trans. Comput. **51**(12), 1448–1453 (2002)
6. Chamam, A., Pierre, S.: Dynamic sensor activation for maximizing network lifetime under coverage constraint. In: International Symposium on Communications and Information Technologies. ISCIT'07, pp. 971–976. IEEE (2007)
7. Chong, C.Y., Kumar, S.P.: Sensor networks: evolution, opportunities, and challenges. Proc. IEEE **91**, 1247–1256 (2003)
8. Cardei, I., Cardei, M.: Energy efficient connected coverage in wireless sensor networks. Int. J. Sens. Netw. **3**, 201–210 (2008)
9. Cardei, M., Wu, J., Lu, M., Pervaiz, M.O.: Maximum network lifetime with adjustable range. In: IEEE International Conference on Wireless and Mobile Computing, Networking and Communications (2005)
10. Cardei, M., Ding-Zhu, D.: Improving wireless sensor network lifetime through power aware organization. ACM Wirel. Netw. **11**, 333–340 (2005)
11. Dhawan, A., Vu, C.T., Zelikovsky, A., Li, Y., Prasad, S.K.: Maximum lifetime of sensor networks with adjustable sensing range. In: Seventh ACIS International Conference on Software Engineering, Artificial Intelligence, Networking, and Parallel/Distributed Computing, SNPD 2006, pp. 285–289. IEEE (2006)
12. Funke, S., Kesselman, A., Kuhn, F., Lotker, Z., Segal, M.: Improved approximation algorithms for connected sensor cover. Wirel. netw. **13**(2), 153–164 (2007)
13. Slijepcevic, S., Potkonjak, M.: Power efficient organization of wireless sensor networks. In: IEEE International Conference on Communications, ICC 2001, vol. 2, pp. 472–476. IEEE (2001)
14. Zorbas, D., Glynos, D., Kotzanikolaou, P., Douligeris, C.: B {GOP}: an adaptive algorithm for coverage problems in wireless sensor networks. In: 13th European Wireless Conference, EW (2007)

Part VIII
Quantum Dot Cellular Automata

2-Dimensional 2-Dot 1-Electron Quantum Cellular Automata-Based Dynamic Memory Design

Mili Ghosh, Debarka Mukhopadhyay and Paramartha Dutta

Abstract Quantum dot cellular automata (QCA) is supposed to be the most promising technology which is an efficient nanotechnology and overcomes the limitations of the CMOS technology. Thus this technology is considered as the most preferable replacement of CMOS. This paper presents a design methodology of dynamic memory using 2-dot 1-electron QCA. According to the best of our knowledge, this has not been reported yet. The proposed design is supposed to be an optimum design with respect to cell requirement.

Keywords Dynamic memory · 2-dot QCA · Memory loop · Majority voter · Planar crossing

1 Introduction

From the last few decades, the names of CMOS technology and the digital industry are taken synonymously. However, in recent times CMOS technology faces serious challenges due to rapid development in digital industry with increased cost [1]. QCA is one of the most promising emerging nanotechnologies which satisfies the said urge. Its high-density cost-effective capability with high-speed nanoscale architecture makes it an ideal replacement of CMOS technology in the near future. 2-dot 1-electron QCA is one of the recent structural forms of QCA which depicts binary

---

M. Ghosh (✉) · P. Dutta
Department of Computer & System Sciences, Visva Bharati University,
Santiniketan, India
e-mail: ghosh.mili90@gmail.com

P. Dutta
e-mail: paramartha.dutta@gmail.com

D. Mukhopadhyay
Department of Computer Science & Engineering, Bengal Institute
of Technology and Management, Santiniketan, India
e-mail: debarka.mukhopadhyay@gmail.com

S. Das et al. (eds.), *Proceedings of the 4th International Conference on Frontiers in Intelligent Computing: Theory and Applications (FICTA) 2015*, Advances in Intelligent Systems and Computing 404, DOI 10.1007/978-81-322-2695-6_30

information with the position of a single electron within a cell. As other forms of QCA, 2-dot 1-electron QCA transfers information using cell-to-cell interaction which follows the Coulomb's law. The advantages of 2-dot 1-electron QCA by the time are well known [2].

The domain of memory design using 4-dot 2-electron QCA has been widely explored. There exist numerous previous reportings in this domain such as [3–5]. In case of 4-dot 2-electron, the domain of parallel memory is also explored such as [6]. Here, we will focus only on the serial memory structures. The 2-dot 1-electron QCA is one of the latest domains of nanotechnology. So a few previous reportings exist in this domain such as [2, 7]. The basic components of any 2-dot 1-electron QCA are binary wire, majority voter gate, inverter, and planar crossing of wires as proposed in [2]. Clocking as stated in [8–10] is the most important aspect of any QCA design. In this article, we propose a design methodology of dynamic memory using 2-dot 1-electron QCA. The implementation of dynamic memory in 2-dot 1-electron QCA is completely planar. In [3], a square cell-based approach of memory design has been proposed. It basically uses a loop-based memory layout in 4-dot 2-electron QCA and enables data storage with the use of loopback construct. It consists of a majority voter gate and a feedback loop. Another reporting is done as in [4] known as H-memory architecture. The design resembles a binary tree structure with control circuit at each step. This design facilitates high density with equal access time. Memory designs using QCA using alternative clocking mechanism is reported in [5]. This memory structure provides a special arrangement for majority voter gate in which the signal wires behave differently with the clock phase. This approach uses the majority voter gate as a memory element and behaves accordingly as a shift register.

2 Basics of 2-Dot 1-Electron QCA

The base of QCA technology lies on the concept of cells. Cells are the building blocks of any QCA design. The most common form of QCA is 4-dot 2-electron QCA which contains four quantum dots and two electrons. The 2-dot 1-electron QCA consists of two quantum dots and one electron. The electron can tunnel between the quantum dots. 2-dot 1-electron QCA cells can align either horizontally or vertically as shown in Fig. 1. Figure 1 also shows how the electron alignment within a cell carries

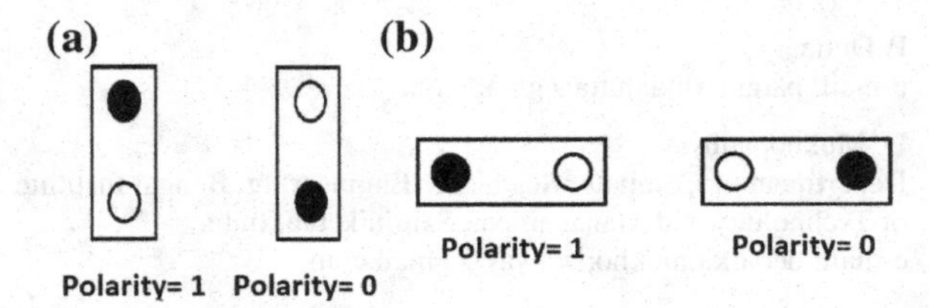

Fig. 1 Binary encoding in 2-dot 1-electron QCA cell **a** with vertical alignment and **b** with horizontal alignment

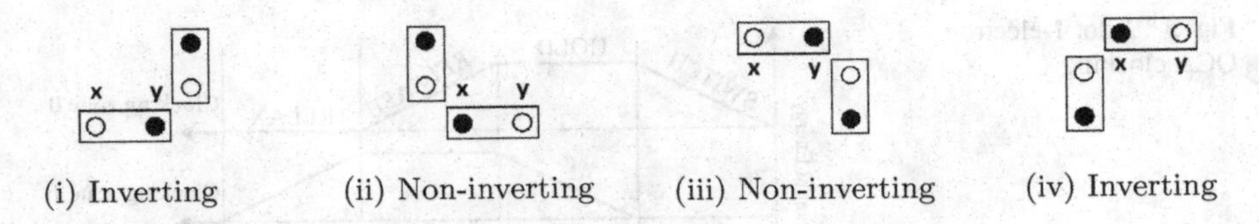

Fig. 2 Inversion by corner cell placement

binary information. This shows the most general convention. The basic components of any 2-dot 1-electron QCA architecture are wire, majority voter gate, and inverter as mentioned in [2]. The corner placement of 2-dot 1-electron QCA cells are shown in Fig. 2.

2.1 Clocking Mechanism

Clocking in case of QCA plays a significant role [11]. The purpose of clocking a QCA circuit is a bit different from that used in CMOS clocking. 2-dot 1-electron QCA uses the same clocking mechanism as general QCA clocking. QCA uses quasi-adiabatic clocking mechanism. In case of CMOS digital designs, clocking is used for the purpose of synchronization, whereas in case of QCA, clocking is used for controlling the data flow and supplying energy to weak input signals. QCA clocking has four phases: switch, hold, release, and relax. The electron potentials are initially low [12] and start to increase gradually in the switch phase. In the hold phase, the electron potentials are maximum and the cells do not have any definite polarity at this stage. During the release phase, the electron potentials start to lower and the cells gradually attain their polarity. During the relax phase, the electron potentials are minimum and the cells attain definite polarity. The actual computation of QCA is done in the release phase. Every QCA architecture is divided into four clock zones and each clocking zone consists of four phases. Each clock zone is $\frac{\pi}{2}$ out of phase with the next clock zone as shown in Fig. 3. The color scheme convention we use in this article to specify different clock zones is shown in Fig. 4.

3 Dynamic Memory

Dynamic memory is the most common type of memory used presently. Dynamic memory stores each bit of data in a separate unit. It is a kind of volatile memory, as it loses its data, once power is removed. Dynamic memory has to be refreshed periodically to maintain the data in the memory. Dynamic memory is preferred over static memory due to its structural simplicity. Thus, the basic characteristics of dynamic memory include

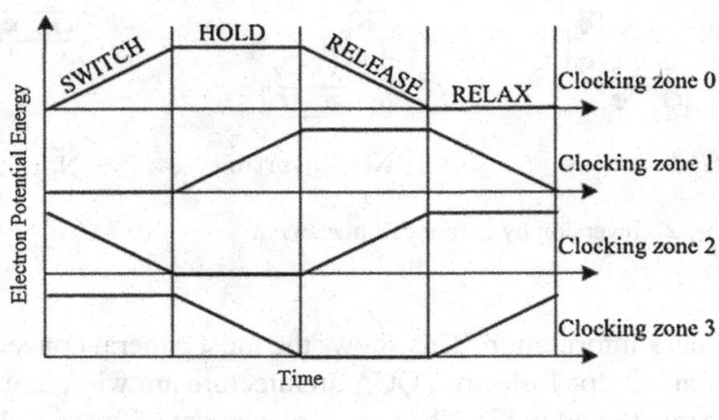

Fig. 3 2-dot 1-electron QCA clocking

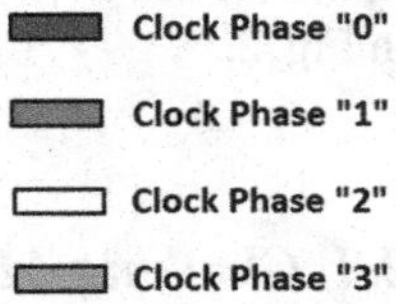

Clock Phase "0"
Clock Phase "1"
Clock Phase "2"
Clock Phase "3"

Fig. 4 Color code of different clock zones

1. Dynamic memory loses its data with removal of power.
2. Dynamic memory has to be refreshed periodically to retain the data in the memory.

3.1 Design of Dynamic Memory

A simple loop-based design of the dynamic memory is proposed in this article. As shown in Fig. 5, a feedback loop is incorporated to retain the data stored. We use a majority voter gate along with a feedback loop to serve the purpose of data. The 2-dot 1-electron QCA implementation is shown in Fig. 6. In QCA, to retain the state of a cell, one has to refresh it time to time. This ensures the dynamism in memory

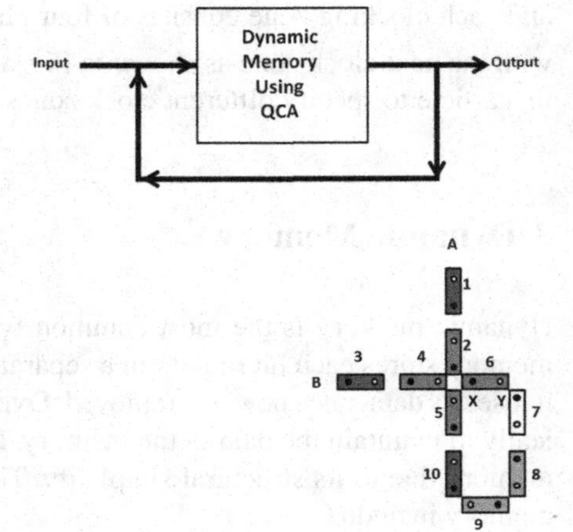

Fig. 5 Block diagram of dynamic memory

Fig. 6 Dynamic memory implementation using 2-dot 1-electron QCA with cell positions

realization. So we exploit this property of QCA to design dynamic memory unit. The majority voter gate in QCA has three input cells. In the proposed design, one of the inputs is used to feed the loopback data to the majority voter gate. The rest two inputs, denoted by A and B respectively, are assigned according to the user need. If we want to retain the previous input, we have to apply input 0 and 1 to the inputs of the proposed design A and B or the reverse. The third input of the majority voter gate in 2-dot 1-electron QCA is inverted naturally. So we apply the inverse of the previous data input to the third input of the majority voter gate in the design as shown in Fig. 6. In both cases, the majority voter gate will retain the previous data, which can be assured from Eq. 1. To write a new value say N to the memory, inputs A and B both are made equal to the value N. Then, the data value N persists in the memory as per Eq. 2.

$$\begin{aligned} M(1,0,Prev') = M(0,1,Prev) &= 1.0 + 0.(Prev')' + 1.(Prev')' \\ &= 1.0 + 0.Prev + 1.Prev \\ &= Prev \end{aligned} \tag{1}$$

$$\begin{aligned} M(N,N,Prev') &= N.N + N.(Prev')' + N.(Prev')' \\ &= N.N + N.Prev + N.Prev \\ &= N \end{aligned} \tag{2}$$

A multibit dynamic memory implementation can be done as shown in Fig. 7. A shown in the figure, separate dynamic memory 1 bit unit for each bit has to be incorporated. Thus to construct N bit dynamic memory unit we have to use N such 1 bit unit. The 2-dot 1-electron QCA implementation of N bit dynamic memory is shown in Fig. 8.

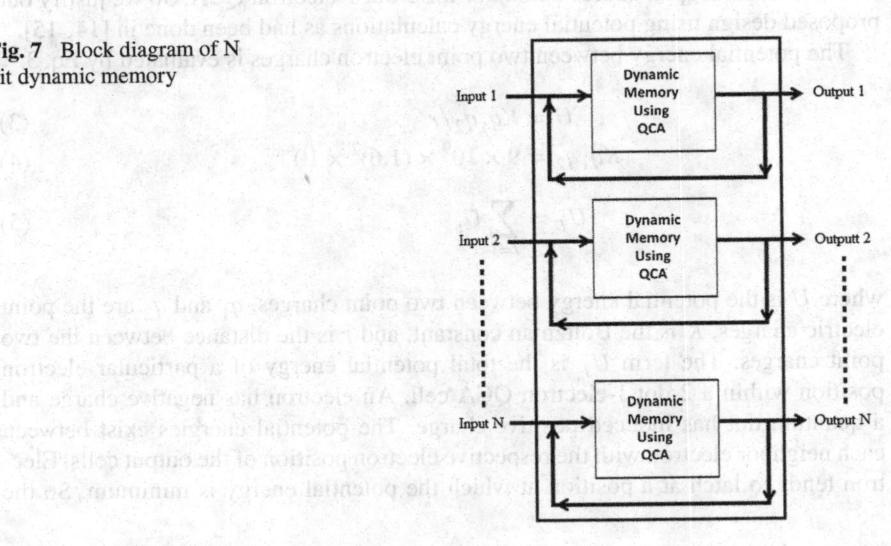

Fig. 7 Block diagram of N bit dynamic memory

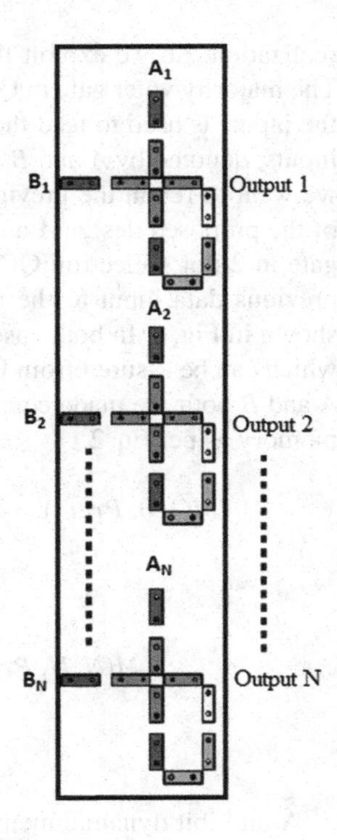

Fig. 8 N bit dynamic memory 2-Dot 1-electron QCA

3.2 Determination of Output State

As open-source simulator QCA Designer [13] of 4-dot 2-electron QCA, there does not exist such an open-source simulator for 2-dot 1-electron QCA. So we justify our proposed design using potential energy calculations as had been done in [14, 15].

The potential energy between two point electron charges is evaluated by Eq. 3.

$$U = Kq_1q_2/r \quad (3)$$

$$Kq_1q_2 = 9 \times 10^9 \times (1.6)^2 \times 10^{-38} \quad (4)$$

$$U_T = \sum_{t=1}^{n} U_t \quad (5)$$

where U is the potential energy between two point charges, q_1 and q_2 are the point electric charges, K is the Boltzman constant, and r is the distance between the two point charges. The term U_T is the total potential energy of a particular electron position within a 2-dot 1-electron QCA cell. An electron has negative charge and a quantum dot has induced positive charge. The potential energies exist between each neighbor electron with the respective electron position of the output cells. Electron tends to latch at a position at which the potential energy is minimum. So the

Table 1 Output state of dynamic memory

Cell	Electron position	Total potential energy	Comments
1	–	–	Input cell with polarity 0
2	–	–	Attains the polarity of cell 1
3	–	–	Input cell with polarity 1
4	–	–	Attains the polarity of cell 3
5 (*initial*)	–	–	Initially with polarity 0
6	x	-3.33×10^{-20} J	Electron will latch at position × due to less energy
6	y	-0.54×10^{-20} J	
7	–	–	Attains polarity of cell 6 due to cell placement as shown in Fig. 2(iii)
8	–	–	Attains the polarity of cell 7
9	–	–	Attains inverse polarity of cell 8 due to cell placement as shown in Fig. 2(i)
10	–	–	Attains polarity of cell 9 due to cell placement as shown in Fig. 2(ii)
5	–	–	Attains the polarity of cell 10

comparison of the total potential energies for the possible positions of an electron shows which electron position is more stable. The size of each 2-dot 1-electron QCA cell is 13 nm × 5 nm and the inter dot distance between two adjacent cells is 5 nm [2].

Figure 6 shows the dynamic memory design with 2-dot 1-electron QCA with cell numbers. The potential energy calculations are shown in Table 1.

4 Stability and Compactness of Proposed Design

Stability plays an important role to determine the acceptability and efficiency of a design. There are certain properties which ensure stability of a design. These are as follows:

1. Input signals of a Majority Voter gate must reach the inputs at the same time and with the same strength.
2. All the cells of a Majority Voter gate must be at the same clock zone.
3. The output of a Majority Voter gate must be taken off either at the same clock or at the next clock.

The proposed design met all the conditions and ensures stability as well. Another important measure of any 2-dot 1-electron QCA is compactness. 2-dot 1-electron QCA cells have rectangular structure. Let us assume that the cells have a size of $a \times b$. The design of dynamic memory as seen in Fig. 6 consists of 10 such cells. The area of coverage of the design is $12ab$ and the cell coverage area is $10ab$. Thus the ratio of coverage area is 10 : 12. So the index of compactness of the proposed design is 83 %. Thus the design facilitates with a high degree of compactness.

Robustness is another measure for any QCA architecture. To achieve robust signal transmission through a QCA architecture, it is desirable that more than one cell are incorporated per intermediate (non-input and non-output) clock zone. In [16], the maximum number of cells that can be accommodated within a clock zone has been derived. The reported design of dynamic memory consists of two clock zones which consists of only one cell as we can see in Fig. 6 cells 7 and 10. So the number of cells will be increased by two to achieve higher degree of robustness. However, the proposed design ensures stability as well as high degree of compactness.

5 Conclusion

We have proposed a design of dynamic memory using 2-dot 1-electron QCA. The design has been verified using potential energy calculations. We also give the stability measure and degree of compactness of the proposed design.

Acknowledgments The authors would like to thank the Department of Computer and System Sciences, Visva-Bharati, Santiniketan and the Department of Computer Science and Engineering, Bengal Institute of Technology and Management, Santiniketan for their infrastructure support. The authors gratefully acknowledge the support of INSPIRE Fellowship (Serial No. 1513–2012, INSPIRE Reg. No. IF31027) for pursuing Doctoral Research in any Indian University by Department of Science and Technology, Ministry of Science and Technology, Government of India.

References

1. International Technology Roadmap for Semiconductor (ITRS): http://www.itrs.net
2. Hook, L.R., IV, Lee, S.C.: Design and simulation of 2-d 2-dot quantum-dot cellular automata logic. IEEE Trans. Nanotechnol. **10**, 996–1003 (2011)
3. Berzon, D., Fountain, T.J.: A memory design in QCAs using the squares formalism. In: Proceedings of the Great Lakes Symposium on VLSI, pp. 166–169. IEEE, Mar 1999
4. Frost, S.E., Rodrigues, A.F., Janiszewski, A.W., Rausch, R.T., Kogge, P.M.: Memory in motion: a study of storage structures in QCA. In: Proceedings 1st Workshop on Non-Silicon Computing (2002)
5. Vankamamidi, V., Ottavi, M., Lombardi, F.: A Line-based parallel memory for QCA Implementation. IEEE Trans. Nanotechnol. **4**, 690–698 (2005)
6. Ottavi, M., Pontarelli, S., Vankamamidi, V., Salsano, A., Lombardi, F.: Design of a QCA memory with parallel read/serial write. In: Proceedings. IEEE Computer Society Annual Symposium on VLSI, pp. 292–294. IEEE, May 2005
7. Hook, L.R., IV, Lee, S.C.: Collinear 2-dot QCA nano-electronic wire structure to improve QCA computing device reliability. In: Proceedings SPIE Smart Structure Conference, vol. 1 (2011)
8. Vetteth, A., Walus, K., Dimitrov, V.S., Jullien, G.A.: Quantum-Dot Cellulur Automata of Flip Flop. ATIPS Laboratory, Alberta (2003)
9. Mukhopadhyay, D., Dinda, S., Dutta, P.: Designing and implementation of quantum cellular automata 2:1 multiplexer circuit. Int. J. Comput. Appl. **25**, 21–24 (2011)

10. Dutta, P., Mukhopadhyay, D.: New architecture for flip flops using quantum-dot cellular automata. In: Proceedings of the 48th Annual Convention of Computer Society of India. Vol II Advances in Intelligent Systems and Computing, vol. 249, pp. 707–714. Springer (2013)
11. Dutta, P., Mukhopadhyay, D.: A Study on energy optimized 4 dot 2 electron two dimensional quantum dot cellular automata logical reversible flip-flops. Microelectron. J. **46**, 519–530 (2015) (Elsevier)
12. Blum, K.: Density Matrix Theory and Applications. Springer Series on Atomic, Optical and Plasma Physics (2012)
13. Walus, K., Dysart, T., Jullien, G.A., Budiman, R.A.: QCA designer: a rapid design and simulation tool for quantum-dot cellular automata. IEEE Trans. Nanotechnol. **3**, 26–31 (2004)
14. Mukhopadhyay, D., Dutta, P.: Qca based novel unit reversible multiplexer. Adv. Sci. Lett. **16**, 163–168 (2012)
15. Ghosh, M., Mukhopadhyay, D., Dutta, P.: A 2 dot 1 electron quantum cellular automata based parallel memory. In: Information Systems Design and Intelligent Applications. Advances in Intelligent Systems and Computing, vol. 339, pp. 627–636. Springer (2015)
16. Huang, J., Momemzadeh, M., Lombardi, F.: Design of Sequential Circuits by Quantum-dot Cellular Automata. Microelectron. J. **38**, 525–537 (2007) (Elsevier)

Optimized Approach for Reversible Code Converters Using Quantum Dot Cellular Automata

Neeraj Kumar Misra, Subodh Wairya and V.K. Singh

Abstract Reversible logic has gained importance in the present development of low-power and high-speed digital systems in nanotechnology. In this manuscript, we have introduced and optimized the reversible Binary to Gray and Gray to Binary code converters circuit using two new types of reversible gates. Two new types of 3×3 reversible gates, namely BG-1 gate (Binary to Gray) and GB-1 gate (Gray to Binary), have been proposed to design converter circuits without any garbage outputs. In addition, useful theorems have been developed, associated with the number of gates, garbage outputs, constant input and quantum cost of the reversible converters. The QCA Designer v2.0.3 tool is used for simulation to test the workability of reversible code converters. The simulation results show that the design works correctly and extracted parameters are better than the previously reported designs. Area and lower bound parameter analysis also show that the design is based on the optimized approach.

Keywords Reversible logic · Code converter · BG-1 gate · GB-1 gate · Garbage output · Quantum cost · Quantum dot cellular automata

1 Introduction

Nowadays at the transistor level, the designer uses CMOS but it has physical limitations such as current leakage, short channel effect, power consumption, large layout area and delay [1–4]. It is predicted that these physical limitations of CMOS will result at the end of conventional CMOS. The vision is towards QCA (Quantum dot cellular automata). It principally focuses on high device density, low delay and low power circuits. QCA performs computation using electric or magnetic field

N.K. Misra (✉) · S. Wairya · V.K. Singh
Department of Electronics Engineering, Institute of Engineering and Technology, Lucknow, India
e-mail: neeraj.mishra3@gmail.com

S. Das et al. (eds.), *Proceedings of the 4th International Conference on Frontiers in Intelligent Computing: Theory and Applications (FICTA) 2015*, Advances in Intelligent Systems and Computing 404, DOI 10.1007/978-81-322-2695-6_31

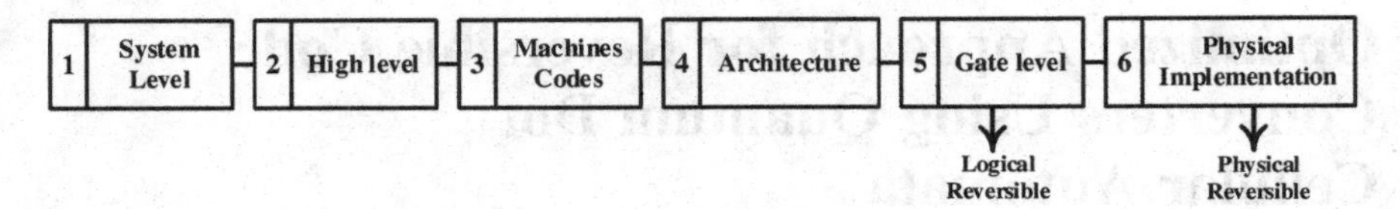

Fig. 1 Approach level of the digital hardware system

polarization [5]. The advancement in digital hardware systems and subsystems has emerged in higher computational density circuits and energy loss has been reduced.

In the early 1960s, Landauer [6] shows that each logic bit of information, KT ln2 joules of energy loss is due to missing bits of information. However, according to Bennett [7], there would be no energy loss and it can be avoided by using the digital hardware designed by reversible gates. Moreover, digital hardware systems are not only logically reversible but also physically reversible. In the design of logically reversible digital hardware system, gate level approach is used. The distinctive feature of the gate is one-to-one mapping between inputs and outputs and output logic can be recovered from input logic [8–11]. Physical reversibility of a digital system can be defined in a way such that if computation is in backward direction without energy loss, then that system design is fulfilling the criteria of physical reversibility. In the flow chart (shown in Fig. 1) of the designing approach, the various levels are shown; fifth level shows the logical reversibility and sixth level shows the physical reversibility.

The paper is organized as follows: Sect. 2 depicts the reversible logic and QCA terminology. Section 3 depicts the design of new code converters using two new types of 3×3 reversible gates. Sections 3.2 and 3.3 describe the Binary to Gray and Gray to Binary code converters with QCA block diagrams and layouts. Section 4 performs a simulation setup result and comparative performance analysis result with the existing approach for the proposed converters in terms of cell complexity, clock cycle and area occupancy. The last section concludes this paper.

2 Background Study

In this section, we present some preliminary knowledge about QCA, reversible logic and their behaviours along with lower bound parameters.

Basics of QCA: The fundamental unit of a QCA device is the QCA cell. These cells contain four quantum dots positioned near the corners of the cell, where another two free electrons can reside [5, 12–15]. Cells have two polarization states (Fig. 1a). These states permit the cells to represent binary value. By setting P = +1 polarization (binary 1), while a cell with P = −1 polarization (binary 0). The majority vote condition is most used frequently (shown in Fig. 2d). In this, the output is defined by Maj = (AB + BC + CA). If the set one input is at 1, it works as 2-input OR gate. If the set one input is at 0, it works as 2-input AND gate.

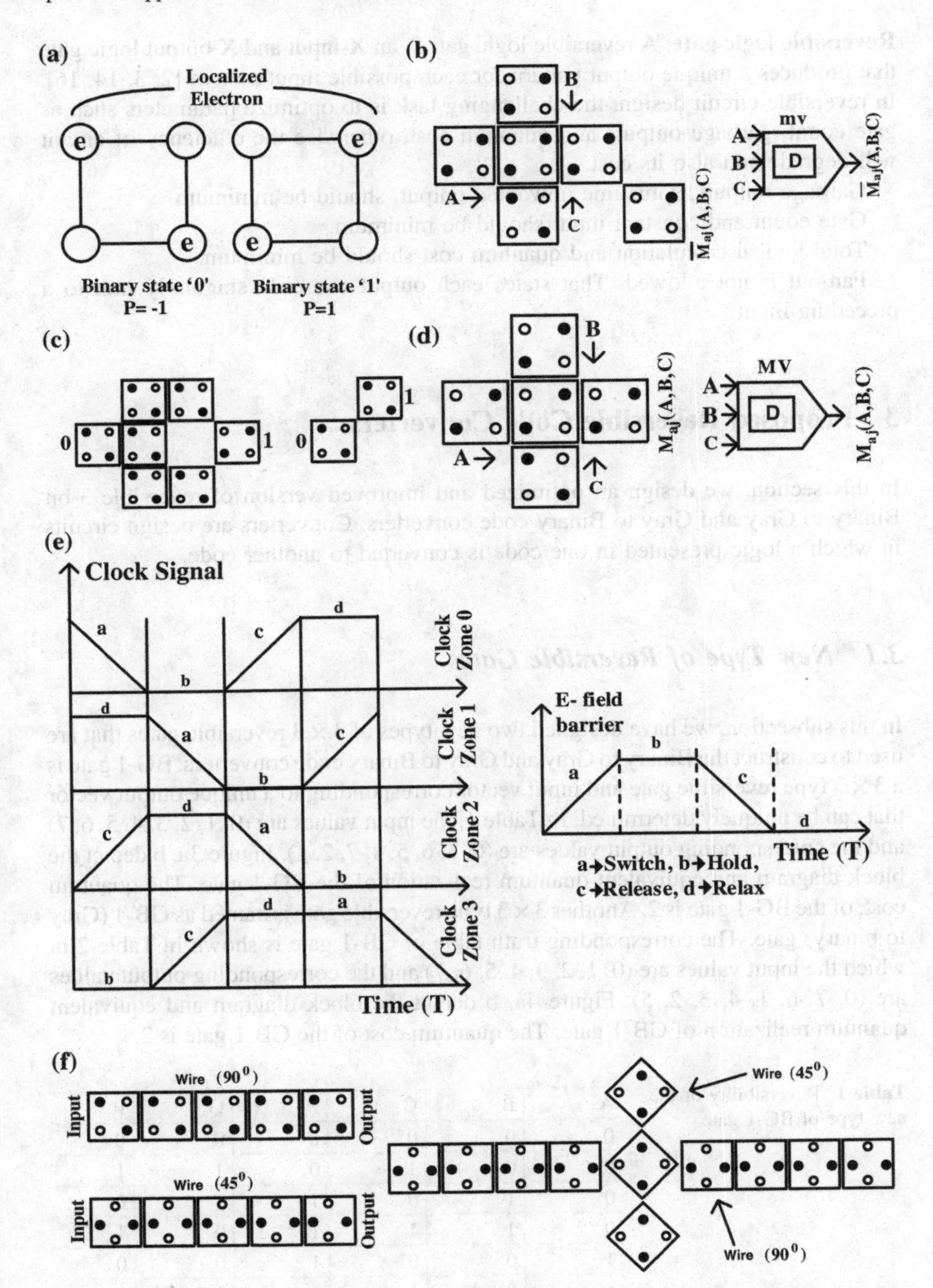

Fig. 2 **a** QCA cell with two binary interpretations of the states. **b** QCA minority voter. **c** QCA inverter. **d** QCA majority vote. **e** QCA phase and clock cycle. **f** QCA binary wire

Reversible logic gate: A reversible logic gate is an X-input and X-output logic gate that produces a unique output pattern for each possible input pattern [2, 3, 14, 16]. In reversible circuit design, the challenging task is to optimize parameters such as gate count, garbage outputs and quantum cost, otherwise the efficiency of circuit will degrade and also its cost.

Garbage outputs, sometime unwanted output, should be minimum.

Gate count and constant input should be minimum.

Total logical calculation and quantum cost should be minimum.

Fan-out is not allowed. That states each output can only singularly link to a preceding input.

3 Proposed Reversible Code Converters

In this section, we design an optimized and improved version of reversible 3-bit Binary to Gray and Gray to Binary code converters. Converters are design circuits in which a logic presented in one code is converted to another code.

3.1 New Type of Reversible Gates

In this subsection, we have designed two new types of 3 × 3 reversible gates that are used to construct the Binary to Gray and Gray to Binary code converters. BG-1 gate is a 3 × 3 type reversible gate and input vector corresponding to a unique output vector that can be uniquely determined. In Table 1, the input values are (0, 1, 2, 3, 4, 5, 6, 7) and the corresponding output values are (0, 3, 6, 5, 4, 7, 2, 1). Figure 3a, b depict the block diagram and equivalent quantum realization of the BG-1 gate. The quantum cost, of the BG-1 gate is 2. Another 3 × 3 type reversible gate is named as GB-1 (Gray to binary) gate. The corresponding truth table of GB-1 gate is shown in Table 2 in which the input values are (0, 1, 2, 3, 4, 5, 6, 7) and the corresponding output values are (0, 7, 6, 1, 4, 3, 2, 5). Figure 4a, b depict the block diagram and equivalent quantum realization of GB-1 gate. The quantum cost of the GB-1 gate is 2.

Table 1 Reversibility of the new type of BG-1 gate

A	B	C	P	Q	R
0	0	0	0	0	0
0	0	1	0	1	1
0	1	0	1	1	0
0	1	1	1	0	1
1	0	0	1	0	0
1	0	1	1	1	1
1	1	0	0	1	0
1	1	1	0	0	1

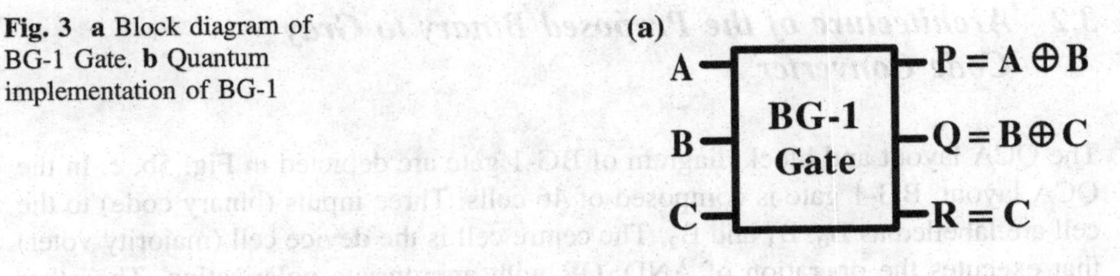

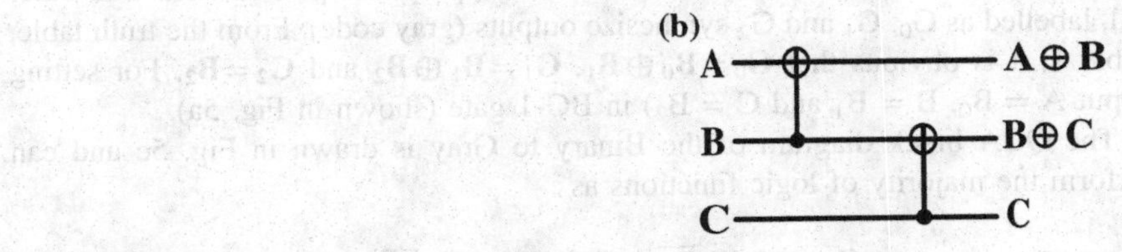

Fig. 3 **a** Block diagram of BG-1 Gate. **b** Quantum implementation of BG-1

Table 2 Reversibility of the new type of GB-1 gate

A	B	C	P	Q	R
0	0	0	0	0	0
0	0	1	1	1	1
0	1	0	1	1	0
0	1	1	0	0	1
1	0	0	1	0	0
1	0	1	0	1	1
1	1	0	0	1	0
1	1	1	1	0	1

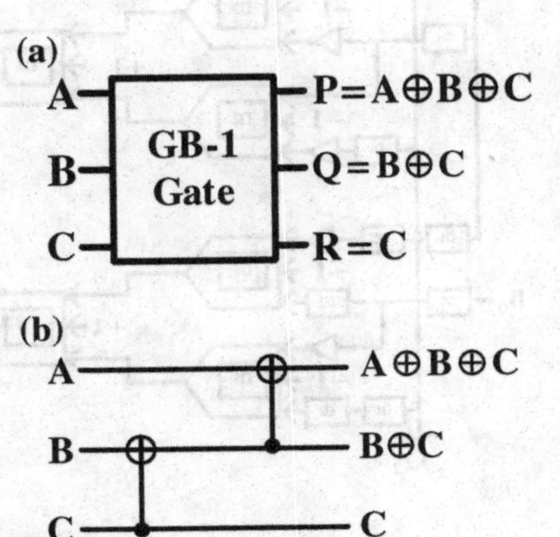

Fig. 4 **a** Block diagram of BG-1 Gate. **b** Quantum implementation of GB-1

3.2 *Architecture of the Proposed Binary to Gray Code Converter*

The QCA layout and block diagram of BG-1 gate are depicted in Fig. 5b, c. In the QCA layout, BG-1 gate is composed of 46 cells. Three inputs (binary code) to the cell are labelled as B_0, B_1 and B_2. The centre cell is the device cell (majority voter) that executes the operation of AND, OR with appropriate polarization. The other cell, labelled as G_0, G_1 and G_2 synthesize outputs (gray code). From the truth table, Table 1, it is obvious that $\mathbf{G_0 = B_0 \oplus B_1}$, $\mathbf{G_1 = B_1 \oplus B_2}$ and $\mathbf{G_2 = B_2}$. For setting (input A = B_0, B = B_1 and C = B_2) in BG-1 gate (shown in Fig. 5a).

The QCA block diagram of the Binary to Gray is drawn in Fig. 5c and can perform the majority of logic functions as

$$\mathbf{G_0 = M(M(\overline{B_1}, 0, B_0), M(B_1, 0, \overline{B_0}), 1),}$$
$$\mathbf{G_1 = M(M(\overline{B_1}, 0, B_2), M(B_1, 0, \overline{B_2}), 1),}$$
$$\mathbf{G_2 = B_2}$$

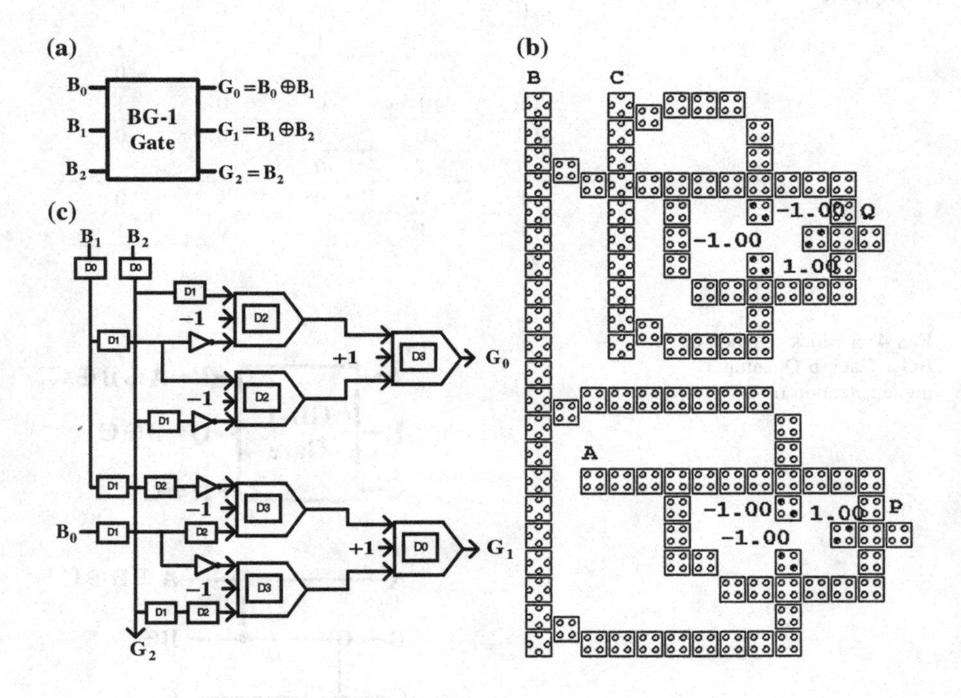

Fig. 5 **a** Block diagram of BG-1 to Gray converter. **b** QCA layout of Binary to Gray code converter. **c** QCA block diagram of Binary to Gray code converter

3.3 Architecture of the Proposed Gray to Binary Code Converters

The QCA layout and block diagram of GB-1 gate are depicted in Fig. 6b, c. The QCA layout of GB-1 gate is composed of 64 cells. Three gray code inputs G_0, G_1 and G_2 are applied instantly into the device cell (majority vote) that executes the operation of AND, OR with appropriate polarization. The other cell, labelled as B_0, B_1 and B_2 synthesize outputs (binary code). From the truth Table 2 we synthesis Boolean expressions as $\mathbf{B_0 = G_0 \oplus G_1 \oplus G_2}$, $\mathbf{B_1 = G_1 \oplus G_2}$ and $\mathbf{B_2 = G_2}$. For setting (input A = G_1, B = G_2 and C = G_0) in GB-1 gate (shown in Fig. 6a). The QCA block diagram of the Binary to Gray converter is drawn in Fig. 6c and can perform the majority and minority logic functions as

$$\mathbf{B_0 = M_{aj}(M_{aj}(M_{aj}(G_1, G_2, 1), \overline{M}_{aj}(G_1, G_2, 0), G_0, 0), \overline{M}_{aj}((M_{aj}(G_1, G_2, 1), \overline{M}_{aj}(G_1, G_2, 0), 0), G_0, 0), 0)}$$

$$\mathbf{B_1 = M_{aj}(M_{aj}(G_1, G_2, 1), \overline{M}_{aj}(G_1, G_2, 0), 0), \ B_2 = G_2}$$

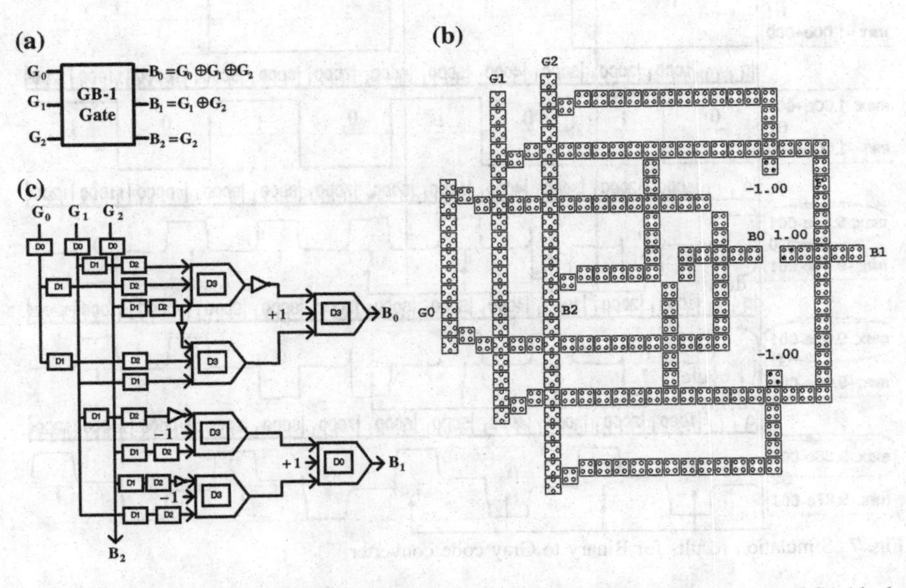

Fig. 6 **a** Block diagram of GB-1. **b** QCA layout of Gray to Binary code converter. **c** QCA block diagram of Gray to Binary code converter

4 Simulation Setup Results

The two different code converters have been laid out and simulated by using QCADesigner v2.0.3 tool. In this simulation, we used the bistable approximation and the following set parameters: (number of samples 12800, radius of effect 65 nm, clock amplitude factor 2, convergence tolerance 0.001000, relative permittivity 12.9, clock high and low are $9.8e^{22}$ and $3.8e^{23}$, respectively, maximum iterations per sample 100, cell size 18 nm × 18 nm and dot diameter 5 nm) (Figs. 7 and 8).

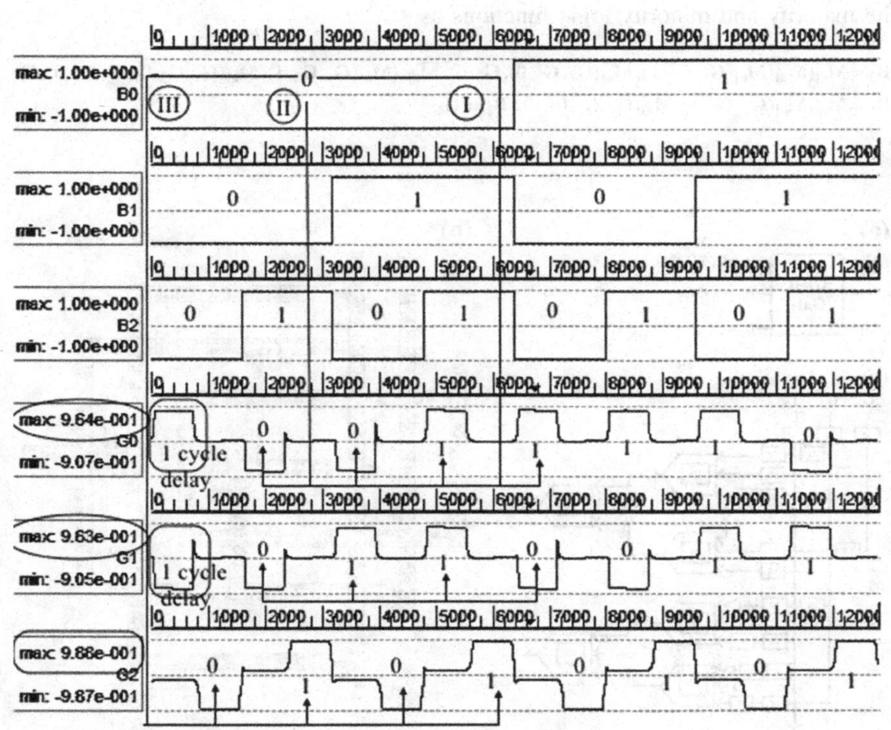

Fig. 7 Simulation results for Binary to Gray code converter

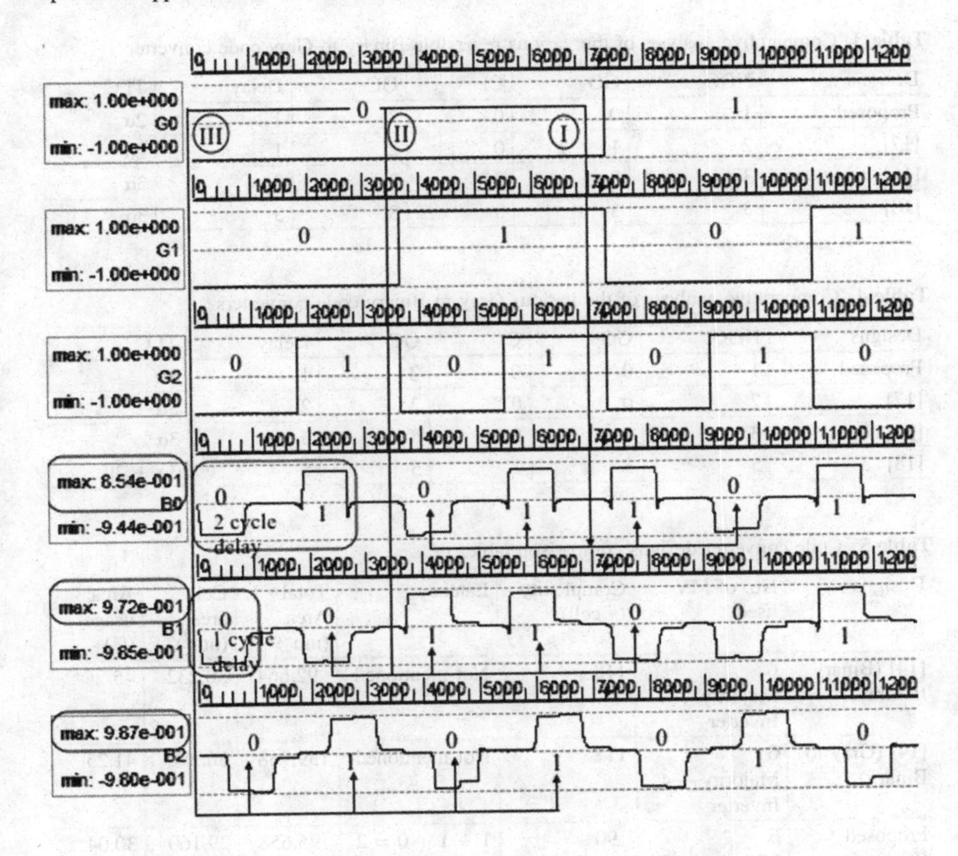

Fig. 8 Simulation results for Gray to Binary code converter

5 The Comparative Performance Tables 3, 4 and 5 of New Designs

The comparative performance Tables 3, 4 and 5 sketches the performance of our proposed designs with existing designs in terms of number of majority gate (MV), cell count and latency. The QCA layouts used the designs of L-shaped wire, majority vote (MV) and inverters. The stability and robustness of these designs can be tested based on cell count, area and latency.

Table 3 Comparative analysis of the various reversible Binary to Gray code converter

Designs	NOG	GO	CI	QC	Delay	TLC
Proposed	1	0	0	2	1	2α
[17]	2	1	0	2	1	2α
[16]	3	0	0	3	3	3α
[18]	3	3	0	3	3	3α

Table 4 Comparative analysis of the various Gray to Binary code converters

Designs	NOG	GO	CI	QC	Delay	TLC
Proposed	1	0	0	2	1	3α
[17]	2	0	0	2	2	3α
[16]	3	0	0	3	3	3α
[18]	5	3	2	5	5	3α + 2β

Table 5 Code converter designs comparison table

Designs	No. of MV used	Complexity (# cell)	Latency	Total Area (nm^2)	Cell area (nm^2)	Area usage (%)
[14] (Binary to Gray)	6 Majority + 4 Inverter	118	Not mentioned	92,664	38,232	45
[14] (Gray to Binary)	6 Majority + 4 Inverter	112	Not mentioned	139,968	36,288	41.25
Proposed (Binary to Gray)	6 Majority + 7 Inverter	90	1 + 1 + 0 = 2	95,658	29,160	30.04
Proposed (Gray to Binary)	6 Majority + 9 Inverter	194	2 + 2 + 0 = 4	280,171	62,856	22.43

6 Conclusion

This paper has presented the QCA implementations of a Binary to Gray and Gray to Binary code converters using two new types of reversible gates. For the first time ever proposed in the literature with an optimal approach and using emerging nanotechnology archetype QCA Designer tool. The new types of gates have a capability to optimize reversible parameters for the design of converters. The nanoscale Binary to Gray converter design requires, area (proposed design has 95,658 nm^2 compared with 38.232 nm^2 [17]), number of MVs used (proposed design has 6Majority + 3inverters compared with 6Majority + 4inverter [17]) and complexity (proposed design requires 46 cells compared with 118 [17]), whereas the nanoscale Gray to Binary converter requires, area (proposed design has 20,736 nm^2 compared with 36.288 nm^2 [17], number of MVs used (proposed design has 6Majority + 1inverter compared to 6Majority + 4inverter [17]) and complexity (proposed 64 cells compared with 112 [17]). Moreover, we have verified the functionality of the proposed converters in QCA-Designer tool. Finally, converter circuit reveals the major QCA advantages such as less area utilized, high device density and latency. The proposed design outperforms the existing ones in terms of scalability and efficiency. The proposed converter circuits will be useful in many applications such as inside computer operations, communication systems many digital circuits, etc.

References

1. Wairya, S., Nagaria, R.K., Tiwari, S.: Performance analysis of high speed hybrid CMOS full adder circuits for low voltage VLSI design. VLSI Des. **7** (2012)
2. Misra, N.K., Wairya, S., Singh, V.K.: Preternatural low-power reversible decoder design in 90 nm technology node. Int. J. Sci. Eng. Res. **5**(6), 969–978 (2014)
3. Misra, N.K., Mukesh, M.K., Wairya, S., Kumar, A.: Cost efficient design of reversible adder circuits for low power applications. Int. J. Comput. Appl. **117**(19), 37–45 (2015)
4. Misra, N.K., Wairya, S., Singh, V.K.: An inventive design of 4*4 bit reversible NS gate. In: Recent Advances and Innovations in Engineering (ICRAIE), pp. 1–6, IEEE (2014)
5. Walus, K., Dysart, T.J., Jullien, G., Budiman, R.A.: QCADesigner: A rapid design and simulation tool for quantum-dot cellular automata. IEEE Trans. Nanotechnol. **3**(1), 26–31 (1973)
6. Landauer, R.: Irreversibility and heat generation in the computing process. IBM J. Res. Dev. **5**(3), 183–191 (1961)
7. Bennett, C.H.: Logical reversibility of computation. IBM J. Res. Dev. **17**(6), 525–532 (1973)
8. Saravanan, P., Kalpana, P.: A novel and systematic approach to implement reversible gates in quantum dot cellular automata. Quantum **2**(5), 15 (2013)
9. Misra, N.K., Wairya, S., Singh, V.K.: Approaches to design feasible error control scheme based on reversible series gates. Eur. J. Sci. Res. **129**(3), 224–240 (2015)
10. Saravanan, M., Manic, K.S.: Energy efficient code converters using reversible logic gates. In: IEEE International Conference on Green High Performance Computing (ICGHPC), pp. 1–6 (2013)
11. Misra, N.K., Wairya, S., Singh, V.K.: Evolution of structure of some binary group-based Nbit comparator. NTO-2 N Decoder Byreversible Technique (2014)
12. Angizi, S., Alkaldy, E., Bagherzadeh, N., Navi, K.: Novel robust single layer wire crossing approach for exclusive or sum of products logic design with quantum-dot cellular automata. J. Low Power Electron. **10**(2), 259–271 (2014)
13. Sarker, A., Bahar, A.N., Biswas, P.K., & Morshed, M.: A novel presentation of Peres Gate (Pg) in Quantum-Dot Cellular Automata (QCA). Eur. Sci. J. **10**(21) (2014)
14. Thapliyal, H., Ranganathan, N.: Conservative QCA gate (CQCA) for designing concurrently testable molecular QCA circuits. In: 22nd International Conference on IEEE VLSI Design, pp. 511–516 (2009)
15. Sen, B., Dutta, M., Goswami, M., Sikdar, B.K.: Modular Design of testable reversible ALU by QCA multiplexer with increase in programmability. Microelectron. J. **45**(11), 1522–1532 (2014)
16. Haghparast, M., Hajizadeh, M., Hajizadeh, R., Bashiri, R.: On the synthesis of different nanometric reversible converters. Middle-East J. Sci. Res. **7**(5), 715–720 (2011)

17. Das, J.C., De, D.: Reversible binary to grey and grey to binary code converter using QCA. IETE J. Res. (ahead-of-print) 1–7 (2015)
18. Saligram, R., & Rakshith,T. R.: Novel code converter employing reversible logic. Int. J. Comput. Appl. **781** (2012)

Design of a Logically Reversible Half Adder Using 2D 2-Dot 1-Electron QCA

Kakali Datta, Debarka Mukhopadhyay and Paramartha Dutta

Abstract Among the emerging technologies in the nanotechnology domain, quantum-dot cellular automata (QCA) is an important name. It overcomes the serious technical limitations of CMOS. In this article, we have used two-dimensional 2-dot 1-electron QCA cells, a variant of two-dimensional 4-dot 2-electron QCA cells, to design a logically reversible half adder. Potential energies have been computed to substantiate and analyse the proposed design. The issues related to energy and power dissipation have also been discussed. Finally, our proposed architecture using 2-dot 1-electron QCA cells has been compared with the architecture using 4-dot 2-electron QCA cells.

Keywords QCA · Majority voter · Reversible half adder · Coulomb's repulsion

1 Introduction

The prevailing CMOS technology has been under challenge by the emerging QCA technology as regards to space and power requirements are concerned. CMOS technology has reached its scaling limit due to quantum-mechanical effects. Moreover, off-state leakage current demands more power consumption. Lent and Tougaw [1] proposed the technology of quantum-dot cellular automata (QCA). The research community is attracted by the extremely small feature sizes, ultra-low power

K. Datta (✉) · P. Dutta
Department of Computer & System Sciences, Visva-Bharati University, Santiniketan, India
e-mail: kakali.datta@visva-bharati.ac.in

P. Dutta
e-mail: paramartha.dutta@gmail.com

D. Mukhopadhyay
Department of Computer Science & Engineering, Bengal Institute of Technology and Management, Santiniketan, India
e-mail: debarka.mukhopadhyay@gmail.com

S. Das et al. (eds.), *Proceedings of the 4th International Conference on Frontiers in Intelligent Computing: Theory and Applications (FICTA) 2015*, Advances in Intelligent Systems and Computing 404, DOI 10.1007/978-81-322-2695-6_32

consumption, concept simplicity and potential for real-world applications in the near future. Among the top six emerging technologies in future computers, QCA provides a solution at nano-scale and also offers a new method of computation, information transfer and transformation.

The advantage of QCA is that when an input is brought to an edge of a QCA clock, it is evaluated and output is obtained at another clocking edge. Thus, no power lines need to be routed internally. Moreover, QCA system needs the minimum energy to add to lift the electrons from their ground state. The QCA system does away with off-state leakage current, dimensional restrictions, and degraded switching performance, which are the inherent drawbacks of the CMOS systems. The advantage of 2-dot 1-electron QCA is that the number of dots and the number of electrons per cell is half of that in case of 4-dot 2-electron QCA. Moreover, out of the six possible configurations only two are admissible [2]. Also, as binary information passes from cell to cell using the inter-cellular interaction obeying the Coulomb's principle, the complexity of wiring is reduced.

In the following Sects. 2 through 6, we have gone through a quick review of two-dimensional 2-dot 1-electron QCA, proposed a design of the logically reversible half adder, and then we have verified the outputs using potential energy calculations and finally we have analysed our proposed design. In Sect. 7, we have made a comparative study of the proposed design with the existing design using 4-dot 2-electron QCA. Finally, we have discussed the amount of energy and power required to operate the proposed architecture in Sect. 8.

2 2-D 2-Dot 1-Electron QCA

A variant of two-dimensional 4-dot 2-electron QCA is the two-dimensional 2-dot 1-electron QCA. Here, the cells are rectangular (horizontally or vertically oriented) with two holes (or dots) at the two ends. One free electron may tunnel through between these two quantum dots. The structure of the 2-dot 1-electron QCA cells and their polarities are shown in Fig. 1. The position of the electron within a cell represents binary information which passes from cell to cell using the inter-cellular interaction obeying the Coulomb's principle.

2-dot 1-electron QCA clocking is somewhat different from the CMOS clocking in the sense that CMOS clocking synchronizes the operations, whereas 2-dot 1-electron QCA clocking not only determines the direction of the signal flow but also supplies energy to weak input signals so that they can propagate through the entire archi-

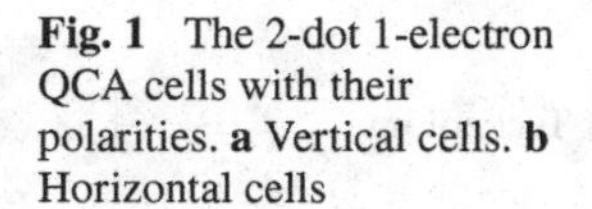

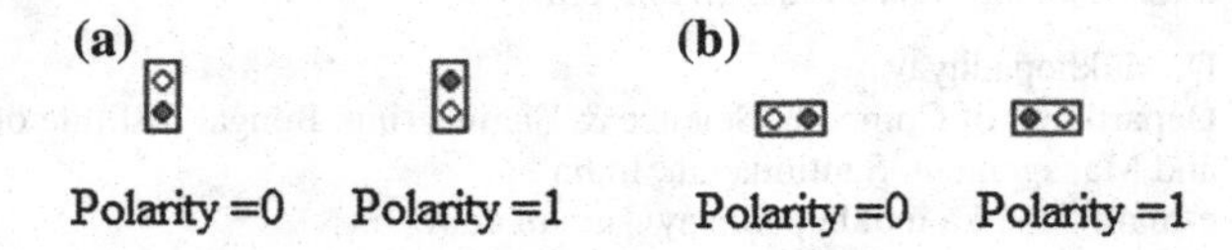

Fig. 1 The 2-dot 1-electron QCA cells with their polarities. **a** Vertical cells. **b** Horizontal cells

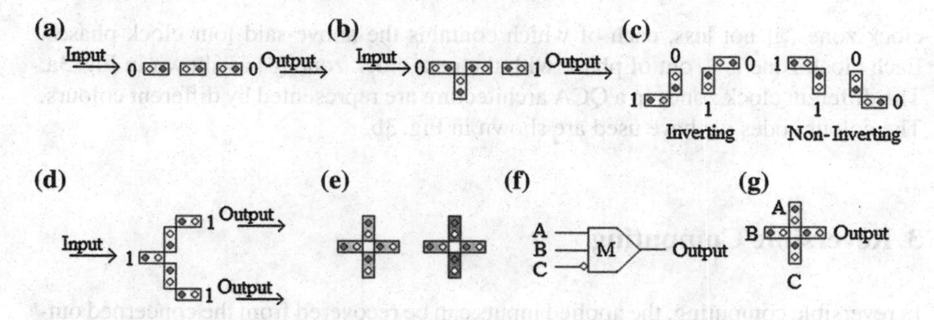

Fig. 2 The 2-dot 1-electron **a** binary wire **b** inverter with differently oriented cell **c** inverter at turnings **d** fan-out **e** planar wire crossing **f** majority voter **g** implementation of majority voter

tecture [3, 4]. In [5], we get a lucid description of the clocking mechanism and the different building blocks namely the binary wire (Fig. 2a), the inverter using a cell of different orientations in between two cells of a binary wire (Fig. 2b), the inverters by turning at corners (Fig. 2c), a fan-out gate (Fig. 2d), crossing wires (Fig. 2e), the majority voter gate (Fig. 2f) and its implementation (Fig. 2g).

The clocking of QCA follows the quasi-adiabatic clocking mechanism, which consists of four phases namely switch, hold, release, and relax. Initially, when the potential energy of electron is low [6] and the electron is not capable of tunnelling between quantum dots, it has a definite polarity. With the beginning of the switch phase, the potential energy of electrons starts to rise and at the end of this phase the electron attains its maximum potential energy. During the hold phase, the electron maintains its maximum potential energy and becomes completely delocalized loosing its polarity. In the release phase, the potential energy of the electron starts to lower and the cell gradually moves towards a definite polarity. During the last phase, i.e. the relax phase, the electron has minimum energy and is too weak to tunnel between the dots. Thus cell attains a definite polarity. Each QCA architecture comprises four

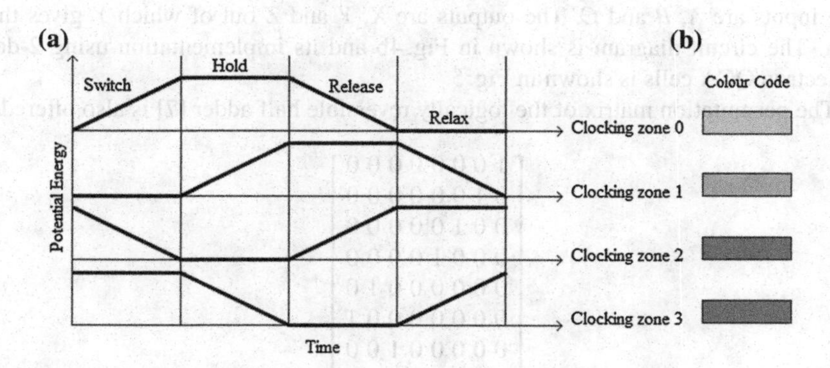

Fig. 3 **a** The 2-dot 1-electron clocking **b** Colour codes of the clock phases

clock zones, if not less, each of which contains the above-said four clock phases. Each clock zone is $\frac{\pi}{2}$ out of phase with the next clock zone [3] as shown in Fig. 3a. The different clock zones in a QCA architecture are represented by different colours. The colour codes we have used are shown in Fig. 3b.

3 Reversible Computing

In reversible computing, the applied inputs can be recovered from the concerned outputs. Hence ideally, the reversible circuits do not lose information. Reversible logic has the ability to reduce the dissipation of power to a great extent and can be particularly used in ultra-low power logic circuit design. Energy dissipation results in irreversible computation due to information loss. In irreversible logic, at the room temperature, heat generated due to loss of one bit of information is very small, but when the number of bits involved is significant, the amount of heat dissipated here assumes such a dimension that it affects the performance and reduces the lifetime of the system. On the contrary, no information is ever lost in the case of reversible computing paradigm. So we can recover any earlier stage by computing backward. In the present scope, our discussion is confined to logical reversibility. This logic is to use a one-to-one mapping between the input and output vectors. The NOT operation is a classical example of reversible logic where the input can be recovered by the output. Reversible computing may be categorized as logical reversibility or thermodynamic reversibility. In this paper, we are considering logical reversibility only. Thermodynamic reversibility is beyond the scope of this paper.

4 Logically Reversible Half Adder

The block diagram of the proposed logically reversible half adder is shown in Fig. 4a. The inputs are A, B and C. The outputs are X, Y and Z out of which Y gives the sum. The circuit diagram is shown in Fig. 4b and its implementation using 2-dot 1-electron QCA cells is shown in Fig. 5.

The permutation matrix of the logically reversible half adder [7] is also offered

$$\begin{bmatrix} 1&0&0&0&0&0&0&0 \\ 0&1&0&0&0&0&0&0 \\ 0&0&1&0&0&0&0&0 \\ 0&0&0&1&0&0&0&0 \\ 0&0&0&0&0&0&1&0 \\ 0&0&0&0&0&0&0&1 \\ 0&0&0&0&0&1&0&0 \\ 0&0&0&0&1&0&0&0 \end{bmatrix}$$

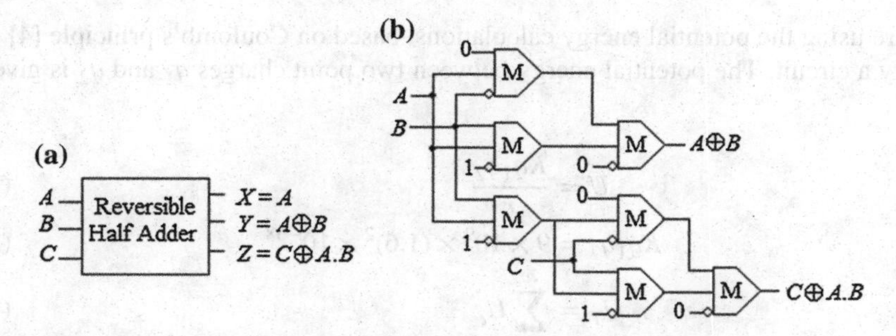

Fig. 4 Logically reversible half adder **a** block diagram **b** circuit diagram

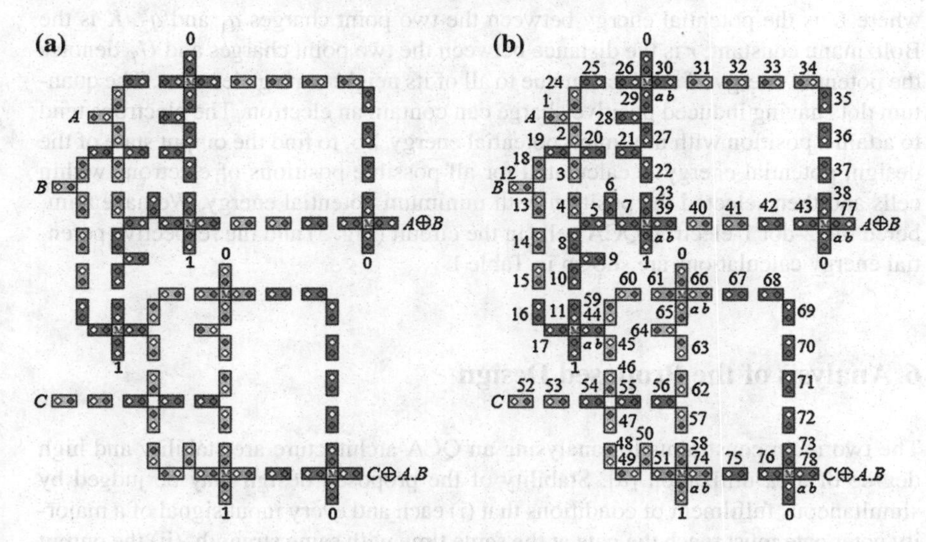

Fig. 5 **a** The implementation of 2-dot 1-electron QCA logically reversible half adder **b** Numbering of the cells of the same circuit

This matrix has only one in each row and each column. Moreover, its determinant evaluates nonzero value. Thus, it ensures the reversible properties.

5 Determination of Output Energy State

QCA designer, an open-source simulator software is available for 4-dot 2-electron-based QCA circuits. Unfortunately, no such open-source simulator that simulates 2-dot 1-electron QCA is available till date. We have to use some standard mathematical functions in order to justify our proposed 2-dot 1-electron QCA architecture. Here, we are using the potential energy calculations, based on Coulomb's principle [4] to verify a circuit. The potential energy between two point charges q_1 and q_2 is given by

$$U = \frac{Kq_1q_2}{r} \tag{1}$$

$$Kq_1q_2 = 9 \times 10^9 \times (1.6)^2 \times 10^{-38} \tag{2}$$

$$U_T = \sum_{t=1}^{n} U_t \tag{3}$$

where U is the potential energy between the two point charges q_1 and q_2, K is the Boltzmann constant, r is the distance between the two point charges and U_T denotes the potential energy of an electron due to all of its neighbouring electrons. The quantum dots having induced positive charge can contain an electron. The electrons tend to attain a position with minimum potential energy. So, to find the output state of the design, potential energy is calculated for all possible positions of electrons within cells and then selected the position with minimum potential energy. We have numbered the 2-dot 1-electron QCA cells in the circuit (Fig. 5) and the respective potential energy calculations are shown in Table 1.

6 Analysis of the Proposed Design

The two main constraints for analysing an QCA architecture are stability and high degree of area utilization [8]. Stability of the proposed design may be judged by simultaneous fulfilment of conditions that (i) each and every input signal of a majority voter gate must reach the gate at the same time with same strength, (ii) the output of a majority voter gate must be taken off at the same clock phase or at the next clock phase and (iii) every cell of a majority voter gate must be at the same clock. It is needless to say that it is clocking only that is responsible for governing the stability of the circuit. Figure 5 satisfies the above conditions thereby ensuring that the proposed design is stable.

Let a 2-dot 1-electron QCA cell be of size $p \times q\,\text{nm}^2$ with p and q representing length and breadth of a 2-dot 1-electron QCA cell. Figure 5 shows that we required 85 cells. The effective area covered is $85pq\,\text{nm}^2$. The area covered by the design is $(10p + 9q) \times (13p + 10q)\,\text{nm}^2$. Thus the area utilization ratio is $85pq : (10p + 9q) \times (13p + 10q)$. Thus high degree of area utilization is met.

Table 1 Output state of 2-dot 1-electron QCA logically reversible half adder architecture

Cell number	Electron position	Total potential energy	Comments
1	–	–	Input cell *A* which is equal to the *X* output
12	–	–	Input cell *B*
52	–	–	Input cell *C*
2–5	–	–	Attains the inverse polarity of cell 5 (Fig. 2c)
6–7	–	–	Attains the inverse polarity of cell 4 (Fig. 2b)
8–9	–	–	Attains the inverse polarity of cell 4 (Fig. 2b)
10-11	–	–	Attains the inverse polarity of cell 8 (Fig. 2b)
13–17	–	–	Attains the polarity of cell 12 (Fig. 2c)
18	–	–	Attains the inverse polarity of cell 12 (Fig. 2c)
19–23	–	–	Attains the inverse polarity of cell 18 (Fig. 2c)
24	–	–	Attains the inverse polarity of cell 1 (Fig. 2c)
25–26	–	–	Attains the inverse polarity of cell 24 (Fig. 2c)
27–28	–	–	Attains the inverse polarity of cell 22 Fig. 2c)
29	–	–	Attains the inverse polarity of cell 22 (Fig. 2b)
30	*a*	-3.328×10^{-20} J	Electron will latch at position *a* due to less
	b	-0.537×10^{-20} J	energy
31–38	–	–	Attains the same polarity of cell 30 (Fig. 2c)
39	*a*	14.102×10^{-20} J	Electron will latch at position *b* due to less
	b	1.368×10^{-20} J	energy

(continued)

Table 1 (continued)

Cell number	Electron position	Total potential energy	Comments
40–43	–	–	Attains the same polarity of cell 39 (Fig. 2a)
44	*a* *b*	14.102×10^{-20} J 1.368×10^{-20} J	Electron will latch at position *b* due to less energy
45–49	–	–	Attains the same polarity of cell 44 (Fig. 2c)
50–51	–	–	Attains the inverse polarity of cell 49 (Fig. 2c)
53-58	–	–	Attains the polarity of cell 52 (Fig. 2c)
59	–	–	Attains the inverse polarity of cell 44 (Fig. 2b)
60–61	–	–	Attains the inverse polarity of cell 59 (Fig. 2c)
62–64	–	–	Attains the inverse polarity of cell 56 (Fig. 2c)
65	–	–	Attains the inverse polarity of cell 63 (Fig. 2c)
66	*a* *b*	3.329×10^{-20} J 0.537×10^{-20} J	Electron will latch at position *b* due to less energy
67–73	–	–	Attains the polarity of cell 66 (Fig. 2c)
74	*a* *b*	14.102×10^{-20} J 1.368×10^{-20} J	Electron will latch at position *b* due to less energy
75–76	–	–	Attains the same polarity of cell 74 (Fig. 2a)
77	*a* *b*	-6.905×10^{-20} J -0.294×10^{-20} J	Electron will latch at position *a* due to less energy. Thus we get output *Y*.
78	*a* *b*	3.329×10^{-20} J 0.537×10^{-20} J	Electron will latch at position *b* due to less energy. Thus we get output *Z*.

Table 2 Comparison of existing 4-dot 2-electron QCA logically reversible half adder design with the proposed design in 2-dot 1-electron QCA

	Proposed in [7]	Proposed in this paper
Number of cells	142	85
Area covered (in nm^2)	193752	40075
Number of majority voter gates	7	7
Number of clock phases	9	8
Energy required to run	Relatively higher (due to 284 electrons)	Less (due to 85 electrons)

7 Comparative Study

In Table 2, we have compared the proposed design of the logically reversible half adder with the existing design using logically reversible half adder [7] with respect to the number of cells required, the area covered, the number of gates, the number of clock phases and energy required. Here, we have considered $p = 13$ nm and $q = 5$ nm [2].

8 Energy and Power Dissipation in the Proposed Design

From [9], we get the expressions for the following parameters: E_m, the minimum energy to be supplied to the architecture with N cells; E_{clk}, the clock signal energy; E_{diss}, energy dissipation from the architecture with N cells; ν_2, frequency of dissipation energy; τ_2, time to dissipate into the environment to come to the relaxed state; ν_1, incident energy frequency; τ_1, time required to reach the quantum level n from quantum level n_2; τ, time required by the cells in one clock zone to switch from one to the next polarization; t_p, propagation time through the architecture; and $\nu_2 - \nu_1$, the differential frequency. We have calculated the same in Table 3 for our proposed design of the logically reversible half adder. $\hbar$ is the reduced Plank's constant, m is the mass of an electron, a^2 the area of a cell, N is the number of cells in the architecture and k is the number of clock phases used in the architecture. Here we have considered $n = 10$ and $n_2 = 2$.

Table 3 Different parameter values of our findings for 2-dot 1-electron QCA logically reversible half adder

Parameters	Value
$E_m = E_{clock} = \dfrac{n^2\pi^2\hbar^2 N}{ma^2}$	6.059×10^{-18} J
$E_{diss} = \dfrac{\pi^2\hbar^2}{ma^2}(n^2 - 1)N$	6.005×10^{-18} J
$\nu_1 = \dfrac{\pi\hbar}{ma^2}(n^2 - n_2^2)N$	8.780×10^{15} Hz
$\nu_2 = \dfrac{\pi\hbar}{ma^2}(n^2 - 1)N$	9.054×10^{15} Hz
$(\nu_1 - \nu_2) = \dfrac{\pi\hbar}{ma^2}(n^2 - 1)N$	0.274×10^{15} Hz
$\tau_1 = \frac{1}{\nu_1} = \dfrac{ma^2}{\pi\hbar(n_1^2 - 1)}N$	0.252×10^{-16} s
$\tau_2 = \frac{1}{\nu_2} = \dfrac{ma^2}{\pi\hbar(n_1^2 - 1)}N$	2.455×10^{-16} s
$\tau = \tau_1 + \tau_2$	4.987×10^{-16} s
$t_p = \tau + (k - 1)\tau_2 N$	14.81×10^{-16} s

9 Conclusion

In this article, we have proposed a design of logically reversible half adder using 2-dot 1-electron QCA cells. As no open-source simulator is available for 2-dot 1-electron QCA, justification is done here using potential energy calculations. It is justified that the logically reversible half adder using 2-dot 1-electron QCA cells is much more cost effective with respect to size and power consumption than the same using 4-dot 2-electron QCA cells.

Acknowledgments The authors would like to thank the Department of Computer and System Sciences, Visva-Bharati, Santiniketan and the Department of Computer Science and Engineering, Bengal Institute of Technology and Management, Santiniketan for their infrastructure support. The authors greatly acknowledge the support of Ms. Mili Ghosh, INSPIRE Fellow IF31027 and Ms. Sunanda Mondal, Research Scholar, for their active support and help.

References

1. Lent, C., Tougaw, P.: A device architecture for computing with quantum dots. Proc. IEEE **85**, 541–557 (1997)
2. Iv, L.R.H., Lee, S.C.: Design and simulation of 2-D 2-Dot quantum-dot cellular automata logic. IEEE Trans. Nanotechnol. **10**, 996–1003 (2011)

3. Mukhopadhyay, D., Dinda, S., Dutta, P.: Designing and implementation of quantum cellular automata 2:1 multiplexer circuit. Int. J. Comput. Appl. **25**, 21–24 (2011)
4. Dutta, P., Mukhopadhyay, D.: Quantum cellular automata based novel unit reversible multiplexer. Adv. Sci. Lett. **5**, 163–168 (2012)
5. Datta, K., Mukhopadhyay, D., Dutt, P.: Design of n-to-2^n Decoder using 2-Dimensional 2-Dot 1-Electron Quantum Cellular Automata. In: National Conference on Computing, Communication and Information Processing, pp. 77–91 (2015)
6. Blum, K.: Density Matrix Theory and Applications. Springer Series on Atomic, Optical and Plasma Physics. Springer, Berlin (2012)
7. Mondal, S., Mukhopadhyay, D., Dutt, P.: A Novel Design of a Logically Reversible Half Adder using 4-Dot 2-Electron QCA. In: National Conference on Computing, Communication and Information Processing, pp. 123–130 (2015)
8. Ghosh, M., Mukhopadhyay, D., Dutta, P.: A 2 dot 1 electron quantum cellular automata based parallel memory. In: Information Systems Design and Intelligent Applications. Advances in Intelligent Systems and Computing, vol. 339, pp. 627–636. Springer India (2015)
9. Mukhopadhyay, D., Dutta, P.: A study on energy optimized 4 dot 2 electron two dimensional quantum dot cellular automata logical reversible flipflops. Microelectron. J. **46**(6), 519–530 (2015)

Part IX
Fuzzy Sets and Systems

Intuitionistic Fuzzy Multivalued Dependency and Intuitionistic Fuzzy Fourth Normal Form

Asma R. Shora, Afshar Alam and Ranjit Biswas

Abstract Intuitionistic fuzzy databases are used to handle imprecise and uncertain data as they represent the membership, nonmembership, and hesitancy associated with a certain element in a set. This paper presents the Intuitionistic Fuzzy Fourth Normal Form to decompose the multivalued dependent data. A technique to determine Intuitionistic Fuzzy multivalued dependencies by working on the closure of dependencies has been proposed. We derive the closure by obtaining all the logically implied dependencies by a set of Intuitionistic Fuzzy multivalued dependencies, i.e., Inference Rules. A complete set of inference rules for the Intuitionistic Fuzzy multivalued dependencies has been given along with the derivation of each rule. These rules help us to compute the dependency closure and we further use the same for defining the Intuitionistic Fuzzy Fourth Normal Form.

Keywords Intuitionistic fuzzy normal forms · Intuitionistic fuzzy multivalued dependency · 4NF-IF

1 Introduction

Intuitionistic fuzzy sets are similar to fuzzy sets, except for the fact that the membership value of every element in the set is not equal to its nonmembership value and that there is a factor of hesitancy associated with it. Intuitionistic fuzzy

A.R. Shora (✉)
Fairfield Institute of Management Technology, New Delhi, India
e-mail: asmashora.sch@jamiahamdard.ac.in

A. Alam · R. Biswas
Department of Computer Science, Jamia Hamdard University, Hamdard Nagar, New Delhi, India
e-mail: aalam@jamiahamdard.ac.in

R. Biswas
e-mail: rbiswas@jamiahamdard.ac.in

S. Das et al. (eds.), *Proceedings of the 4th International Conference on Frontiers in Intelligent Computing: Theory and Applications (FICTA) 2015*, Advances in Intelligent Systems and Computing 404, DOI 10.1007/978-81-322-2695-6_33

databases are special relational databases that can contain crisp, fuzzy, and intuitionistic fuzzy values. Like any other database, the complexity of the table increases as the size and dimensions increase, giving rise to anomalies. In order to avoid this, the normalization /decomposition of an Intuitionistic Fuzzy relation into sub–relations must be performed. The decomposition has to be lossless and thus the data dependencies need to be identified and preserved during this process.

The intuitionistic fuzzy database models [13] are based on the relational theory and thus they inherit several concepts and operations from the relational database models [1–3]. Intuitionistic fuzzy logic, along with the relational databases, provides a concrete platform to handle complexity and uncertainty of data. Most of the information we humans deal with everyday, is ubiquitous. Real-world systems work on information which is mostly incomplete, inconsistent, uncertain and imprecise [4]. Incomplete information has missing links or data that makes it difficult to interpolate the information in between. Atanasov [5, 6] introduced IF sets as a mathematical infrastructure to handle imprecise and vague data. IFS are often considered to be the extensions of the Fuzzy sets.

Suppose A is an intuitionistic fuzzy set and X € A, such that $A=\{(x, \mu_A(x), \nu_A(x))\}$, where, $\mu_A(x)$ is the membership degree and $\nu A(x)$ is the nonmembership degree for x belonging to A. The degree of hesitation is calculated as

$$\Pi_A(x) = 1 - \{\mu_A(x) + \nu_A(x)\}$$

such that

$$\mu_A(x) + \nu_A(x) + \Pi_A(x) = 1$$

Fuzzy set is a special intuitionistic fuzzy set with $\Pi_A(x)=0$ and thus $\nu_A(x) = 1 - \mu_A(x)$. There are a number of modifications [7, 8] proposed over intuitionistic fuzzy sets. Intuitionistic fuzzy sets find a great scope in fields involving decision theory, e.g., expert systems, medical diagnosis, machine learning, match predictions, etc. [9–12]. Similar database systems have been developed in order to represent and handle incomplete [14–19] information in relational databases. Intuitionistic fuzzy databases were introduced by Supriya, Biswas, and Roy [13]. An intuitionistic fuzzy database is a set of relations where each relation R is a subset of the cross product $2^{D1} \times 2^{D2} \times \ldots \times 2^{Dm}$, where $2^{Di} = P(Di)$ and P(Di) is the power set of Di (domain). In the subsequent sections of this paper, we introduce the multivalued data dependencies in IF databases. This is in continuation with the previous work done on Intuitionistic Fuzzy Normal Forms [28]. We derive the dependency closure by obtaining all the logically implied dependencies by a set of intuitionistic fuzzy multivalued dependencies, i.e., Inference Rules. The inference rules for Intuitionistic Fuzzy multivalued dependencies will be explored and the derivation of each rule will be given. The same rules are used to define the Intuitionistic Fuzzy Fourth Normal Form.

2 Data Dependencies in Intuitionistic Fuzzy Databases

Both Fuzzy and Intuitionistic Fuzzy relations and dependencies have been studied in detail [20–24, 29]. On the basis of dependencies among data, intuitionistic fuzzy normal forms were defined [25–28]. Intuitionistic fuzzy multivalued dependency, a general case of an IF functional dependency is determined using the following measure.

2.1 The IF Equality Measure

In order to be able to compare two or more attribute dependencies, we need to have a measure. The comparison operator used here is the IF equality operator that we defined previously in [28]. The crisp set elements are compared using syntactic measure while as non-crisp sets are compared semantically as the exact measure of equality is not known. We can only determine the extent to which the sets are 'equal' (similar). The equality measure plays an important role in defining dependencies among the attributes and tuples. The IF measure EQUAL (IEQ), which is a resemblance relation over U to compare two entities, *such that*

$$\text{IEQ UxU}, \mu_{\text{IEQ}} - - > [0,1], \nu_{\text{IEQ}} - > [0,1]$$

Being a resemblance relation, IEQ holds the following properties:

$$\text{IEQ}(a, a) = 1 \quad \text{(reflexivity)}$$
$$\text{IEQ}(a, b) = \text{IEQ}(b, a) \quad \text{(symmetry)}$$

The generalized IEQ measure over a given domain can be defined as $\text{IEQ}(t1, t2) = \{(t1, t2), \mu\text{IEQ}(t1, t2), \nu\text{IEQ}(t1, t2)\}$, where $\text{IEQ}(t_1, t_2)$ yields an IF tuple *such that*

$$\mathbf{\mu_{IEQ}(t_1, t_2) = \min\{\mu_{IEQ}(t_1[A_1]), t_2[A_1]), \ldots, \mu_{IEQ}(t_1[A_n]), t_2[A_n])\}}$$
$$\mathbf{\nu_{IEQ}(t_1, t_2) = \max\{\nu_{IEQ}(t_1[A_1]), t_2[A_1]), \ldots, \nu_{IEQ}(t_1[A_n]), t_2[A_n])\}}.$$

The same definition can be extended for a domain with more than two tuples $(t_1, \ldots, t_n)$. The IEQ measure gives us a semantic equality measure of two or more IF sets, which is used for defining relationships and intuitionistic fuzzy dependencies among data.

3 Intuitionistic Fuzzy Multivalued Dependency (IFMVD)

Multivalued dependency (MVD) is a condition that arises in a database when the tuples of a relation represent more than one many-to-many relationships [30]. Thus, certain attributes become independent of one another, and their values must appear in all combinations. MVDs in the fuzzy data model have been studied defined [22].

We introduce the IFMVD, based on the IF equality measure as follows:

Definition 3.1 Let there be a universe, where X, Y, and Z $\in$ R.

The IFMVD, $\mathbf{X} - > - >_{(\mu,\nu)} \mathbf{Y}$ holds in an intuitionistic fuzzy relation R, if for tuples t_1 and t_2 of R, there exists another tuple t_3 in R, *such that* $IEQ(t_1[X], t_2[X], t_3[X])$, which means

$$\begin{aligned}
&\mu_{IEQ}(t_1[X], t_2[X], t_3[X]) <= \max\{\min(\mu_{IEQ}(t_1[Y], t_3[Y]), \mu_{IEQ}(t_2[Z], t_3[Z]))\\
&\min(\mu_{IEQ}(t_2[Y], t_3[Y]), \mu_{IEQ}(t_1[Z], t_3[Z]), \mu_{IEQ}(t_1[Y], t_2[Y], t_3[Y]), \mu_{IEQ}(t_1[Z], t_2[Z], t_3[Z])\},\\
&\nu_{IEQ}(t_1[X], t_2[X], t_3[X]) > \min\{\max(\nu_{IEQ}(t_1[Y], t_3[Y]), \nu_{IEQ}(t_2[Z], t_3[Z])),\\
&\max(\nu_{IEQ}(t_2[Y], t_3[Y]), \nu_{IEQ}(t_1[Z], t_3[Z]), \nu_{IEQ}(t_1[Y], t_2[Y], t_3[Y]), \nu_{IEQ}(t_1[Z], t_2[Z], t_3[Z])\},\\
&\text{where, } Z = R - XY \text{ and } 0 > \mu > 1 \& 0 > \nu > 1
\end{aligned} \tag{1}$$

4 Inference Rules

Let there be a relation R and let there be tuples $t_1,\ldots,t_4$ belonging to R. The following inference rules hold for the IFMVD:

1. IF − REPLICATION: $(\text{If } X - >_{\mu\nu} Y, \text{ then } X - > - >_{\mu\nu} Y)$

Proof (by contradiction) Let $X - > - >_{\mu\nu} Y$ does not hold, then

$$\begin{aligned}
&\mu_{IEQ}(t_1[X], t_2[X], t_3[X]) > \max\{\min(\mu_{IEQ}(t_1[Y], t_3[Y]), \mu_{IEQ}(t_2[Z], t_3[Z])),\\
&\min(\mu_{IEQ}(t_2[Y], t_3[Y]), \mu_{IEQ}(t_1[Z], t_3[Z]), \mu_{IEQ}(t_1[Y], t_2[Y], t_3[Y]), \mu_{IEQ}(t_1[Z], t_2[Z], t_3[Z])\},\\
&\nu_{IEQ}(t_1[X], t_2[X], t_3[X]) < = \min\{\max(\nu_{IEQ}(t_1[Y], t_3[Y]), \nu_{IEQ}(t_2[Z], t_3[Z])),\\
&\max(\nu_{IEQ}(t_2[Y], t_3[Y]), \nu_{IEQ}(t_1[Z], t_3[Z]), \nu_{IEQ}(t_1[Y], t_2[Y], t_3[Y]), \nu_{IEQ}(t_1[Z], t_2[Z], t_3[Z])\}
\end{aligned}$$

□

Since, $X - >_{\mu\nu} Y$ holds in R, by the definition of IFFD [28], we have

$$\begin{aligned}
&\mu_{IEQ}(t_1[X], t_2[X], t_3[X]) <= \max\{\min(\mu_{IEQ}(t_1[Y], t_3[Y]), \mu_{IEQ}(t_2[Z], t_3[Z])),\\
&\min(\mu_{IEQ}(t_2[Y], t_3[Y]), \mu_{IEQ}(t_1[Z], t_3[Z]), \mu_{IEQ}(t_1[Y], t_2[Y], t_3[Y]), \mu_{IEQ}(t_1[Z], t_2[Z], t_3[Z])\}
\end{aligned}$$

This contradicts our assumption. Hence, $\mathbf{X->->_{\mu\nu}Y}$ **holds in R**

2. IF – COMPLEMENTATION: (If $X->->_{\mu\nu}Y$, then $X->->_{\mu\nu}Z$, where $Z = R - XY$)

Proof Since, $X->->_{\mu\nu}Y$ holds in R,

$$\mu_{IEQ}(t_1[X], t_2[X], t_3[X]) <= \max\{\min(\mu_{IEQ}(t_1[Y], t_3[Y]), \mu_{IEQ}(t_2[Z], t_3[Z])),$$
$$\min(\mu_{IEQ}(t_2[Y], t_3[Y]), \mu_{IEQ}(t_1[Z], t_3[Z]), \mu_{IEQ}(t_1[Y], t_2[Y], t_3[Y]), \mu_{IEQ}(t_1[Z], t_2[Z], t_3[Z])\},$$
$$\nu_{IEQ}(t_1[X], t_2[X], t_3[X]) > \min\{\max(\nu_{IEQ}(t_1[Y], t_3[Y]), \nu_{IEQ}(t_2[Z], t_3[Z])),$$
$$\max(\nu_{IEQ}(t_2[Y], t_3[Y]), \nu_{IEQ}(t_1[Z], t_3[Z]), \nu_{IEQ}(t_1[Y], t_2[Y], t_3[Y]), \nu_{IEQ}(t_1[Z], t_2[Z], t_3[Z])\}$$

$$\text{Hence, } \mu_{IEQ}(t_1[X], t_2[X], t_3[X]) <= \max\{\min(\mu_{IEQ}(t_1[Z], t_3[Z]), \mu_{IEQ}(t_2[Y], t_3[Y])),$$
$$\min(\mu_{IEQ}(t_2[Z], t_3[Z]), \mu_{IEQ}(t_1[Y], t_3[Y]), \mu_{IEQ}(t_1[Z], t_2[Z], t_3[Z]), \mu_{IEQ}(t_1[Y], t_2[Y], t_3[Y])\},$$
$$\nu_{IEQ}(t_1[X], t_2[X], t_3[X]) > \min\{\max(\nu_{IEQ}(t_1[Z], t_3[Z]), \nu_{IEQ}(t_2[Y], t_3[Y])),$$
$$\max(\nu_{IEQ}(t_2[Z], t_3[Z]), \nu_{IEQ}(t_1[Y], t_3[Y]), \nu_{IEQ}(t_1[Z], t_2[Z], t_3[Z]), \nu_{IEQ}(t_1[Y], t_2[Y], t_3[Y])\}$$

i.e., $\mathbf{X->->_{\mu\nu}Z}$ **holds in R** □

3. IF – REFLEXITIVITY (IF $Y \underline{c} X$, then $Y->->_{\mu\nu}X$)

Proof (by contradiction) Let $Y->->_{\mu\nu}X$ does not hold in R, i.e.,

$$\mu_{IEQ}(t_1[X], t_2[X], t_3[X]) > \max\{\min(\mu_{IEQ}(t_1[Y], t_3[Y]), \mu_{IEQ}(t_2[Z], t_3[Z])),$$
$$\min(\mu_{IEQ}(t_2[Y], t_3[Y]), \mu_{IEQ}(t_1[Z], t_3[Z]), \mu_{IEQ}(t_1[Y], t_2[Y], t_3[Y]), \mu_{IEQ}(t_1[Z], t_2[Z], t_3[Z])\},$$
$$\nu_{IEQ}(t_1[X], t_2[X], t_3[X]) < \min\{\max(\nu_{IEQ}(t_1[Y], t_3[Y]), \nu_{IEQ}(t_2[Z], t_3[Z])),$$
$$\max(\nu_{IEQ}(t_2[Y], t_3[Y]), \nu_{IEQ}(t_1[Z], t_3[Z]), \nu_{IEQ}(t_1[Y], t_2[Y], t_3[Y]), \nu_{IEQ}(t_1[Z], t_2[Z], t_3[Z])\}$$

Since, $Y \underline{C} X$, i.e., a trivial dependency, therefore Y-> ->X exists, according to Eq. (1)

This contradicts our assumption. **Hence,** $\mathbf{Y->->_{\mu\nu}X}$ □

4. IF – AUGMENTATION: (If Z C W and $X->->_{\mu\nu}Y$, then $XW->->_{\mu\nu}YZ$)

Proof Since Z is a subset of W, by applying Rule c, we get: $W->->_{\mu\nu}Z$

Also, $X->->_{\mu\nu}Y$ holds in R, we have

$$\begin{aligned}&\mu_{IEQ}(t_1[X], t_2[X], t_3[X]) <= \max\{\min(\mu_{IEQ}(t_1[Y], t_3[Y]), \mu_{IEQ}(t_2[Z], t_3[Z])),\\&\min(\mu_{IEQ}(t_2[Y], t_3[Y]), \mu_{IEQ}(t_1[Z], t_3[Z]), \mu_{IEQ}(t_1[Y], t_2[Y], t_3[Y]), \mu_{IEQ}(t_1[Z], t_2[Z], t_3[Z])\},\\&\nu_{IEQ}(t_1[X], t_2[X], t_3[X]) > \min\{\max(\nu_{IEQ}(t_1[Y], t_3[Y]), \nu_{IEQ}(t_2[Z], t_3[Z])),\\&\max(\nu_{IEQ}(t_2[Y], t_3[Y]), \nu_{IEQ}(t_1[Z], t_3[Z]), \nu_{IEQ}(t_1[Y], t_2[Y], t_3[Y]), \nu_{IEQ}(t_1[Z], t_2[Z], t_3[Z])\}\end{aligned} \quad (2)$$

Also,

$\mu_{IEQ}(t_1[W], t_2[W], t_3[W]) <= \max\{\min(\mu_{IEQ}(t_1[Z], t_3[Z]), \mu_{IEQ}(t_2[R-WZ], t_3[R-WZ])), \mu_{IEQ}(t_2[Z], t_3[Z], t_3[Z]), \mu_{IEQ}(t_1[R-WZ], t_2[R-WZ], t_3[R-WZ])\}$

$\nu_{IEQ}(t_1[W], t_2[W], t_3[W]) > \min\{\max(\nu_{IEQ}(t_1[Z], t_3[Z]), \nu_{IEQ}(t_2[R-WZ], t_3[R-WZ])), \nu_{IEQ}(t_2[Z], t_3[Z], t_3[Z]), \nu_{IEQ}(t_1[R-WZ], t_2[R-WZ], t_3[R-WZ])\}$

From (1) and (2), we have

$\min\{\mu_{IEQ}(t_1[X], t_2[X], t_3[X]), \mu_{IEQ}(t_1[W], t_2[W], t_3[W])\} <= \min[\max\{\min(\mu_{IEQ}(t_1[Y], t_3[Y]), \mu_{IEQ}(t_2[R-XY], t_3[R-XY])), \min(\mu_{IEQ}(t_2[Y], t_3[Y]), \mu_{IEQ}(t_1[R-XY], t_3[R-XY])), \mu_{IEQ}(t_1[Y], t_2[Y], t_3[Y]), \mu_{IEQ}(t_1[R-XY], t_2[R-XY], t_3[R-XY]), \max\{\min(\mu_{IEQ}(t_1[Z], t_3[Z]), \mu_{IEQ}(t_2[R-WZ], t_3[R-WZ])), \min(\mu_{IEQ}(t_2[Z], t_3[Z]), \mu_{IEQ}(t_1[R-WZ], t_3[R-WZ])), \mu_{IEQ}(t_1[Z], t_2[Z], t_3[Z]), \mu_{IEQ}(t_1[R-WZ], t_2[R-WZ], t_3[R-WZ]), \max\{[\nu_{IEQ}(t1[X], t2[X], t3[X]), \nu_{IEQ}(t1[W], t2[W], t3[W])\} > \max[\min\{\max(\nu_{IEQ}(t_1[Y], t_3[Y]), \nu_{EQ}(t_2[R-XY], t_3[R-XY])), \max(\nu_{IEQ}(t_2[Y], t_3[Y]), \nu_{IEQ}(t_1[R-XY], t_3[R-XY])), \nu_{IEQ}(t_1[Y], t_2[Y], t_3[Y]), \nu_{IEQ}(t_1[R-XY], t_2[R-XY], t_3[R-XY]), \min\{\max(\nu_{IEQ}(t_1[Z], t_3[Z]), \nu_{IEQ}(t_2[R-WZ], t_3[R-WZ])), \max(\nu_{IEQ}(t_2[Z], t_3[Z]), \nu_{IEQ}(t_1[R-WZ], t_3[R-WZ])), \nu_{IEQ}(t_1[Z], t_2[Z], t_3[Z]), \nu_{IEQ}(t_1[R-WZ], t_2[R-WZ], t_3[R-WZ])$

i.e.,

$\mu_{IEQ}(t_1[XW], t_2[XW], t_3[XW]) <= \max\{\min(\mu_{IEQ}(t_1[YZ], t_3[YZ]), \mu_{IEQ}(t_2[R-XYZW], t_3[R-XYZW])), \min(\mu_{IEQ}(t_2[YZ], t_3[YZ]), \mu_{IEQ}(t_1[R-XYZW], t_3[R-XYZW])), \mu_{IEQ}(t_1[YZ], t_2[YZ], t_3[YZ], \mu_{IEQ}(t_1[R-XYZW], t_2[R-XYZW], t_3[R-XYZW])$
$\nu_{IEQ}(t_1[XW], t2[XW], t_3[XW]) > \min\{\max(\nu_{IEQ}(t_1[YZ], t_3[YZ]), \nu_{IEQ}(t_2[R-XYZW], t_3[R-XYZW])), \max(\nu_{IEQ}(t_2[YZ], t_3[YZ]), \nu_{IEQ}(t_1[R-XYZW], t_3[R-XYZW])), \nu_{IEQ}(t_1[YZ], t_2[YZ], t_3[YZ], \nu_{IEQ}(t_1[R-XYZW], t_2[R-XYZW], t_3[R-XYZW])$

i.e., $\mathbf{XW->->_{\mu\nu}YZ}$

5. IF − TRANSITIVITY(If $X->->_{\mu\nu}Y$ and $Y->->_{\mu\nu}Z$ then $X->->_{\mu\nu}Z$, where X,Y,Z are disjoint sets and R = XYZW)

Proof Since, $X->->_{\mu\nu}Y$ for t_1, t_2, and t_3 in R,

$\mu_{IEQ}(t_1[X], t_2[X], t_3[X]) <= \max\{\min(\mu_{IEQ}(t_1[Y], t_3[Y]), \mu_{IEQ}(t_2[Z], t_3[Z])), \min(\mu_{IEQ}(t_2[Y], t_3[Y]), \mu_{IEQ}(t_1[Z], t_3[Z]), \mu_{IEQ}(t_1[Y], t_2[Y], t_3[Y]), \mu_{IEQ}(t_1[Z], t_2[Z], t_3[Z])\}$,
$\nu_{IEQ}(t_1[X], t_2[X], t_3[X]) > \min\{\max(\nu_{IEQ}(t_1[Y], t_3[Y]), \nu_{IEQ}(t_2[Z], t_3[Z])), \max(\nu_{IEQ}(t_2[Y], t_3[Y]), \nu_{IEQ}(t_1[Z], t_3[Z]), \nu_{IEQ}(t_1[Y], t_2[Y], t_3[Y]), \nu_{IEQ}(t_1[Z], t_2[Z], t_3[Z])\}$

Suppose there exist a tuple t_4 in R.
Since $X - > - >_{\mu\nu} Y$ and $Y - > - >_{\mu\nu} Z$ holds in R,
We have, $\mu_{IEQ}(t_1[X], t_3[X], t_4[X]) < = \mu_{IEQ}(t_1[Y], t_3[Y], t_4[Y])$

$$\nu_{IEQ}(t_1[X], t_3[X], t_4[X]) > \nu_{IEQ}(t_1[Y], t_3[Y], t_4[Y]) \quad (3)$$

$$\mu_{IEQ}(t_1[Y], t_3[Y], t_4[Y]) < = \max\{\min(\mu_{IEQ}(t_1[Z], t_3[Z]), \mu_{IEQ}(t_2[XW], t_3[XW]))$$

□

Let,
$$\mu_{IEQ}(t_1[Z], t_2[Z], t_3[Z]), \mu_{IEQ}(t_1[XW], t_2[XW], t_3[XW]) = \max\{A1, A2, A3, A4\}$$

$$\nu_{IEQ}(t_1[Y], t_3[Y], t_4[Y]) > \min\{\max(\nu_{IEQ}(t_1[Z], t_3[Z]), \nu_{IEQ}(t_2[XW], t_3[XW]))$$

Let,

$$\nu_{IEQ}(t_1[Z], t_2[Z], t_3[Z]), \nu_{IEQ}(t_1[XW], t_2[XW], t_3[XW]) = \min\{A1', A2', A3', A4'\} \quad (4)$$

Case: Let A1 be the highest and A1' be the lowest

$$\mu_{IEQ}(t_1[X], t_3[X], t_4[X]) < = \min(\mu_{IEQ}(t_1[Z], t_4[Z]), \mu_{IEQ}(t_3[XW], t_4[XW])$$
$$\nu_{IEQ}(t_1[X], t_3[X], t_4[X]) > \max(\nu_{IEQ}(t_1[Z], t_4[Z]), \nu_{IEQ}(t_3[XW], t_4[XW])$$

i.e., $\mu_{IEQ}(t_1[X], t_3[X], t_4[X]) < = \mu_{IEQ}(t_1[Z], t_4[Z])$

$\nu_{IEQ}(t_1[X], t_3[X], t_4[X]) > \nu_{IEQ}(t_1[Z], t_4[Z])$ and $\mu_{IEQ}(t_3[X], t_4[X])$ and $\mu_{IEQ}(t_3[W], t_4[W])$
$\nu_{IEQ}(t_3[X], t_4[X])$ and $\nu_{IEQ}(t_3[W], t_4[W])$

(5)

$$\mu_{IEQ}(t_1[X], t_3[X], t_4[X]) < = \mu_{IEQ}(t_3[X], t_4[X])$$
$$\nu_{IEQ}(t_1[X], t_3[X], t_4[X]) > \nu_{IEQ}(t_3[X], t_4[X]) \quad (6)$$

is trivially true, using (6), we get

$$\mu_{IEQ}(t_1[X], t_3[X], t_4[X]) < = \mu_{IEQ}(t_3[Y], t_4[Y])$$
$$\nu_{IEQ}(t_1[X], t_3[X], t_4[X]) > \nu_{IEQ}(t_3[Y], t_4[Y]) \quad (7)$$

From (5) through (7), we get

$$\mu_{IEQ}(t_1[X], t_3[X], t_4[X]) < = \min(\mu_{IEQ}(t_1[Z], t_4[Z]), \mu_{IEQ}(t_3[YW], t_4[YW])$$
$$\nu_{IEQ}(t_1[X], t_3[X], t_4[X]) > \max(\nu_{IEQ}(t_1[Z], t_4[Z]), \nu_{IEQ}(t_3[YW], t_4[YW])$$

i.e., $\mathbf{X -> -->_{\mu\nu} Z}$ holds in R

Similar proof exists for rest of the cases.

6. IF – COALSESCENCE(If $X - > - >_{\mu\nu} Y$ and $Z - - >_{\mu\nu} W$, where $W \underline{C} Y$, and Y and Z are disjoint, then $X \text{--}>_{\mu\nu} W$)

Proof (by contradiction) Let, $X - >_{\mu\nu} W$ does not hold in R i.e.,

$$\mu_{IEQ}(t_1[X], t_2[X]) > \mu_{IEQ}(t_1[W], t_2[W])$$
$$\nu_{IEQ}(t_1[X], t_2[X]) < = \nu_{IEQ}(t_1[W], t_2[W])$$

Since, $X - > - >_{\mu\nu} Y$ holds, we have

$$\mu_{IEQ}(t_1[X], t_2[X], t_3[X]) < = \max\{\min(\mu_{IEQ}(t_1[Y], t_3[Y]), \mu_{IEQ}(t_2[Z], t_3[Z])),$$
$$\min(\mu_{IEQ}(t_2[Y], t_3[Y]), \mu_{IEQ}(t_1[Z], t_3[Z]), \mu_{IEQ}(t_1[Y], t_2[Y], t_3[Y]), \mu_{IEQ}(t_1[Z], t_2[Z], t_3[Z])\},$$
$$\nu_{IEQ}(t_1[X], t_2[X], t_3[X]) > \min\{\max(\nu_{IEQ}(t_1[Y], t_3[Y]), \nu_{IEQ}(t_2[Z], t_3[Z])),$$
$$\max(\nu_{IEQ}(t_2[Y], t_3[Y]), \nu_{IEQ}(t_1[Z], t_3[Z]), \nu_{IEQ}(t_1[Y], t_2[Y], t_3[Y]), \nu_{IEQ}(t_1[Z], t_2[Z], t_3[Z])\}$$

□

Case: Let $\min(\mu_{IEQ}(t_1[Y], t_3[Y]), \mu_{IEQ}(t_2[Z], t_3[Z]))$ be the maximum and $\max(\nu_{IEQ}(t_1[Y], t_3[Y]), \nu_{IEQ}(t_2[Z], t_3[Z]))$ be the lowest in the above equations.

$$\mu_{IEQ}(t_1[X], t_2[X], t_3[X]) < = \min\{\mu_{IEQ}(t_1[Y], t_3[Y]), \mu_{IEQ}(t_2[R - XY], t_3[R - XY])\}$$
$$\nu_{IEQ}(t_1[X], t_2[X], t_3[X]) > \max\{\nu_{IEQ}(t_1[Y], t_3[Y]), \nu_{IEQ}(t_2[R - XY], t_3[R - XY])\} \quad (8)$$

Now since, $W \underline{C} Y$, we have $Y - > - >_{\mu\nu} W$ (by reflexivity)
Hence, $\mu_{IEQ}(t_1[Y], t_3[Y]) < = \mu_{IEQ}(t_1[W], t_3[W])$

$$\nu_{IEQ}(t_1[Y], t_3[Y]) > \nu_{IEQ}(t_1[W], t_3[W]) \quad (9)$$

From (7) and (8), we have: t_2 does not agree with t_3 on W.

This contradicts the fuzzy multivalued dependency $Z - > - >_{\mu\nu} W$ because $Y \Lambda Z$ is an empty set, which means: $Z\underline{C}(R\text{–}XY)$ and t_2 and t_3 agree on Z. So, our assumption is wrong.

Hence,
$\mu_{IEQ}(t_1[X], t_2[X]) > \mu_{IEQ}(t_1[W], t_2[W]) \;\&\; \nu_{IEQ}(t_1[X], t_2[X]) <= \nu_{IEQ}(t_1[W], t_2[W])$

Therefore, **X** $\rightarrow_{\mu\nu}$ **W** holds in R. All other rules follow from the above proved inference rules.

IF-UNION

$$\left(\text{If } X->->_{\mu\nu} Y \text{ and } X->->_{\mu\nu} Z, \text{ then } X->->_{\mu\nu} YZ\right)$$

IF-DECOMPOSITION

$$\left(\text{if } X->->_{\mu\nu} Y \text{ and } X->->_{\mu\nu} Z, \text{ Then } X->->_{\mu\nu} Y \prod Z, X->->_{\mu\nu} Y-Z \text{ and } X->->_{\mu\nu} Z-Y\right)$$

IF-PSEUDOTRANSITIVITY

$$(\text{If } X->->_{\mu\nu} Y \;\&\; YW->->_{\mu\nu} Z, \text{ then } XW->->_{\mu\nu} Z-YW$$

5 Intuitionistic Fuzzy Fourth Normal Form (4NF-iF)

The previous normal forms, i.e., the 2NF (IF), 3NF (IF), and BCNF (IF) are defined on the basis of IF functional dependencies. The IF fourth normal form is based on the IFMVDs present in an IF relation. The formal definition of an IF multivalued dependency IFMVD is

Definition 5.1 $X->->_{(\mu,\nu)} Y$ is an IF multivalued dependency if and only if

$$\begin{aligned}
&\mu_{IEQ}(t_1[X], t_2[X], t_3[X]) <= \max\{\min(\mu_{IEQ}(t_1[Y], t_3[Y]), \mu_{IEQ}(t_2[Z], t_3[Z])),\\
&\min(\mu_{IEQ}(t_2[Y], t_3[Y]), \mu_{IEQ}(t_1[Z], t_3[Z]), \mu_{IEQ}(t_1[Y], t_2[Y], t_3[Y]),\\
&\mu_{IEQ}(t_1[Z], t_2[Z], t_3[Z])\}, \nu_{IEQ}(t_1[X], t_2[X], t_3[X]) > \min\{\max(\nu_{IEQ}(t_1[Y], t_3[Y]),\\
&\nu_{IEQ}(t_2[Z], t_3[Z])), \max(\nu_{IEQ}(t_2[Y], t_3[Y]), \nu_{IEQ}(t_1[Z], t_3[Z]),\\
&\nu_{IEQ}(t_1[Y], t_2[Y], t_3[Y]), \nu_{IEQ}(t_1[Z], t_2[Z], t_3[Z])\}
\end{aligned} \quad (10)$$

where X,Y, and Z belong to R and $t_1, t_2, \ldots, t_n$ are the tuples of R, Z = R –XY, and $0 > \mu > 1$ & $0 > \nu > 1$

Definition 5.2 A relation R is in **fourth normal Form (4NF)** if and only if

(a) R is in BCNF(IF)
(b) R has no Intuitionistic Fuzzy Multivalued dependencies.

Suppose we have a relation R with a multivalued dependency, **X** $->->_{(\mu,\nu)}$ **Y**. The IFMVD can be removed by decomposing R into: **R1(R−Y)** and **R2(X U Y)**

Example 5.1 The following Table 1 is an example of an un-normalized IF table

Table 1 PATIENT_MOD

Pat_No	Name	Age	Cons_Date	Symptom
1	Sara	32	2/1/12	Obesity, severe skin streaks
2	Maya	40	1/11/11	Obesity
3	Sahil	29	22/4/11	Obesity, moderate insulin resistance
4	Kapila	42	21/02/12	Obesity, severe insulin resistance

In Table 1, we have a given set of patients in an IF relation PATIENT_MOD {Pat_No, Age, Name, Cons_date, Address, Symptom}, where Pat_No, Name, and Pat_No, Cons_date are the (0.9, 0.1) candidate keys. The following set of IFFDs exist.

Pat_No → (0.8, 0.1) Age, Pat_No, Name → (0.9, 0.1), Cons_date, Cons_-date → (0.8, 0.2) Name

The symptoms are quantified using tolerance relations, for example, one symptom skin streaks can be represented as in Table 2. Similar tables can be made for other symptoms.

For the 2NF (IF) check, we have to remove the partial key dependencies (if any). Since there is a partial dependence of a nonkey attribute, Age on the part of the key Pat_No, PATIENT_MOD is not in 2- NF (IF), i.e., Pat_No → (0.8, 0.1) Age. Thus, we need to decompose the relation PATIENT_MOD into: PATIENT_MOD (Pat_No, Name, Cons_date, Address, Symptom) and PATIENT_MOD1 (Pat_No, Age).

Since no transitive dependencies exist, both relations are in 3NF (IF). The dependency Pat_No → (0.8, 0.1) Age does not violate any BCNF (IF) rules as Pat_No is a (0.8, 0.1)–key for the relation PATIENT_MOD1.

Similarly, (Pat_No, Name $\rightarrow_{(0.9,0.1)}$ Cons_date, Address) is acceptable. But the dependency Cons_date $\rightarrow_{(0.8,0.2)}$ Name creates a violation as Cons_date is not a key attribute so there is a need to decompose the relation **PATIENT_MOD**.

The final decomposition is as follows:

PATIENT_MOD (Pat_No, Cons_date, Address,Symptom)
PATIENT_MOD1 (Pat_No, Age)
PATIENT_MOD2 (Cons_date, Name)

Table 2 PATIENT_SYMPTOM

Skin streaks	Mild	Moderate	Severe	Absent
Mild	(1.0, 0.0, 0.0)	(0.8, 0.2, 0.0)	(0.9, 0.1, 0.0)	(0.2, 0.7, 0.1)
Moderate	(0.8, 0.2, 0.0)	(1.0, 0.0, 0.0)	(0.7, 0.2, 0.1)	(0.4, 0.5, 0.1)
Severe	(0.9, 0.1, 0.0)	(0.7, 0.2, 0.1)	(1.0, 0.0, 0.0)	(0.77, 0.03, 0.2)
Absent	(0.2, 0.7, 0.1)	(0.4, 0.5, 0.1)	(0.77, 0.03, 0.2)	(1.0, 0.0, 0.0)

Even though **PATIENT_MOD** is in BCNF, an IFMVD exists, i.e., Pat_No $\rightarrow\rightarrow_{(0.9,0.1)}$ symptom. The IFMVD can be removed by decomposing **PATIENT_MOD** into

PATIENT_MOD3 (Pat_No, Cons_date, Address)
PATIENT_MOD4 (Pat_No, Symptom)

Thus the final decomposition is:

PATIENT_MOD1 (Pat_No, Age), **PATIENT_MOD2** (Cons_date, Name), **PATIENT_MOD3** (Pat_No, Cons_date, Address), and **PATIENT_MOD4** (Pat_No, Symptom).

6 Conclusion

The determination of intuitionistic fuzzy multivalued dependency needs a measure. The operator we used is the IF equality measure. This measure helps us compare and represent the dependencies, according to the IF schema design constraints. The operator is used to compute inference rules and dependency closure based on which, the intuitionistic fuzzy fourth normal form is defined. The multivalued dependencies are to be preserved while decomposing the relation into the IF fourth normal form. We have proposed a simple way of normalizing an IF database into intuitionistic fuzzy fourth normal form by exploring intuitionistic fuzzy data dependencies.

References

1. Codd, E.F.: A relational model of data for large shared data banks. Commun. ACM **13**(6), 377–387 (1970)
2. Zadeh, L.A.: Fuzzy sets. Inf. Control **8**, 338–353 (1965)
3. Xu, Z.S., Yager, R.R.: Some geometric aggregation operators based on intuitionistic fuzzy sets. Int. J. Gen Syst. **35**, 417–433 (2006)
4. Pons, O., et al.: Dealing with disjunctive and missing information in logic fuzzy databases. Int. J. Uncertainty Fuzziness Knowl. Based Syst. **4**(2), 177–201 (1996)
5. Atanassov, K.: Intuitionistic fuzzy sets. Fuzzy Sets Syst. **20**, 87–96 (1986)
6. Atanassov, K.: More on intuitionistic fuzzy-sets. Fuzzy Sets Syst. **33**, 37–45 (1989)
7. Atanassov, K.T., Gargov, G.: Interval-valued intuitionistic fuzzy sets. Fuzzy Sets Syst. **31**, 343–349 (1989)
8. Bustince, H., Burillo, P.: Vague sets are intuitionistic fuzzy sets. Fuzzy Sets Syst. **79**, 403–405 (1996)
9. Szmidt, E., Kacprzy,J.: Intuitionistic fuzzy sets in some medical applications. In: Proceeding of International Conference on Computational Science, Part–II, pp. 263–271(2001)
10. Sanchez, E.: Solutions in Composite Fuzzy Relation Equations: Application to Medical Diagnosis In Brouwerian Logic. Fuzzy Automata and Decision Processes, pp. 221–234, (Elsevier, New York 1977)

11. Chountas, P., et al.: On intuitionistic fuzzy expert systems with temporal parameters. Comput. Intell. Theory Appl. **38**, 241–249 (2006)
12. Jun, Y.: Intuitionistic fuzzy finite state machines. J. Appl. Math. Comput. **17**(1–2), 109–120 (2005)
13. De, S.K.,Biswas, R., Roy, A. R.: Intuitionistic fuzzy database. In: Second International Conference on IFS, NIFS, vol. 4(2), pp. 43–31, Sofia (1998)
14. Imielinski, T., Lipski, W.: Incomplete information in relational databases. J. ACM **31**(4), 761–791 (1984)
15. Laurent, D., Spyratos, N.: Partition semantics for incomplete information in relational databases. In: SIGMOD, ACM record, pp. 66–73, New York (1988)
16. Lipski, W.: On semantic issues connected with incomplete information databases. ACM Trans. Database Syst. **4**(3), 262–296 (1979)
17. Liu, K.C., Sunderraman, R.: A generalized relational model for indefinite and maybe information. IEEE Trans. Knowl. Data Eng. **3**(1), 65–76 (1991)
18. Ola, A., Ozsoyoglu, G.: Incomplete relational database models based on intervals. IEEE Trans. Knowl. Data Eng. **5**(2), 293–308(1993)
19. Vassiliou, Y.: Functional dependencies and incomplete information. In: Proceeding of Sixth International Conference on VLDB, pp. 260–269, Canada (1980)
20. Hamouz, S.A., Biswas, R.: Fuzzy functional dependencies in relational databases. Int. J. Comput. cogn. **4**(1) (2006)
21. Cubero, J.C., et al.: Computing fuzzy dependencies with linguistic label. Stud. Fuzziness Soft Comput. **34**, 368–382, Springer (1999)
22. Raju, K.V.S.V.N., Majumdar, A.K.: Fuzzy functional dependencies and lossless join decomposition of fuzzy relational database systems. ACM Trans. Database syst. **13**, 129–166 (1988)
23. Deschrijver, G., Kerre, E.E.: On the composition of intuitionistic fuzzy relations. Fuzzy Sets Syst. **136**, 333–361(2003)
24. Kumar, D.A., et al.: A method of intuitionistic fuzzy functional dependencies in relational databases. Eur. J. Sci. Res. **29**(3), 415–425 (2009)
25. Alam, M.A., Ahmad, S., Biswas, R.: Normalization of intuitionistic fuzzy relational database. NIFS **l0**(1), 1–6(2004)
26. Hussain, S., Alam, M.A., Biswas, R.: Normalization of intuitionistic fuzzy relational database into second normal form—2NF (IF). Int. J. Math. Sci. Eng. Appl. (IJMSEA) **3**(3), 87–96 (2009)
27. Hussain, S., Alam, M.A.: Normalization of intuitionistic fuzzy relational database into third normal form—3NF (IF). Int. J. Math. Sci. Eng. Appl. (IJMSEA), **4**(1), 151–157 (2010)
28. Shora, A.R., Alam, M.A.: Data dependencies and normalization of intuitionistic fuzzy databases. In: Advanced Computing, Networking and Informatics, Vol1, Smart Innovation, Systems and Technologies, vol. 27, pp. 309–318, Springer (2014)
29. Jyothi, S., Babu, M.S.: Multivalued dependencies in fuzzy relational databases and lossless join decomposition. Fuzzy Sets Syst. **88**, 315–332 (1997)
30. Silberschatz, A., Korth, S.: Database System Concepts, 5th edn, pp. 295. (McGraw-Hill, New York 2006)

Parameter Reduction of Intuitionistic Fuzzy Soft Sets and Its Related Algorithms

Sumonta Ghosh and Sujit Das

Abstract Contribution of intuitionistic fuzzy soft set (IFSS) in uncertain real-life applications is inevitable. Computation with IFSS may be complicated by the use of less important parameters. However, there has been a little focus on parameter reduction of IFSSs. In this paper, we introduce two different parameter reduction algorithms in IFSSs to satisfy the different needs of decision makers. The first algorithm is based on selection of a set of parameters whose combined contribution is less important in the decision-making process. The second approach selects parameter(s) which has less deviation in comparison to the other parameters. Finally, the proposed algorithms have been demonstrated using illustrative numerical examples. This study also preserves the decision abilities while reducing the redundant parameters.

Keywords Intuitionistic fuzzy soft set • Parameter reduction • Redundant parameter

1 Introduction

Molodtsov (1999) introduced soft set in [1] as a generic mathematical tool to deal with uncertain problems which are difficult to solve using traditional mathematical tools. Finding the necessity of incorporating fuzzy set with soft set, Maji et al. [2] introduced fuzzy soft set (FSS) in 2001. Since its introduction, FSS theory has been successfully applied in many different fields such as decision making, data analysis,

---

S. Ghosh (✉)
Department of MCA, Bengal College of Engineering and Technology, Durgapur 12, India
e-mail: mesumonta@gmail.com

S. Das
Department of C.S.E., Dr. B. C. Roy Engineering College, Durgapur 6, India
e-mail: sujit_cse@yahoo.com

S. Das et al. (eds.), *Proceedings of the 4th International Conference on Frontiers in Intelligent Computing: Theory and Applications (FICTA) 2015*, Advances in Intelligent Systems and Computing 404, DOI 10.1007/978-81-322-2695-6_34

forecasting, simulation, optimisation, texture classification, etc. FSS and its various generalisations like intuitionistic fuzzy soft set (IFSS) [13], hesitant fuzzy soft set (HFSS) [14–16], etc. are mainly based on the set of objects and parameters, where parameters are the descriptive properties of the objects. Parameters have an important role for representing the objects. At the same time, unnecessary use of parameters may result in unexpected situations. Thus there is a definite need to avoid the unnecessary parameters with the assurance that the reduction of redundant parameters will not influence the application. Some significant efforts have already been made towards parameter reduction in soft sets. Chen et al. [3] pointed out that the outcome of soft set reduction presented in [4] was incorrect and then proposed a new idea of parameter reduction in the context of soft sets. Kong et al. [5] introduced the concept of normal parameter reductions in soft sets and FSSs, which solved the problem of suboptimal choice and added parameter set of soft sets. Ma et al. [6] proposed a novel parameter reduction algorithm for soft sets. A set of four heuristic algorithms for parameter reduction in the framework of interval-valued fuzzy soft sets was proposed by Ma et al. [7]. However, a very few works have been done on parameter reduction of IFSS, which is very much significant to apply IFSS in decision-making paradigm [17–19].

In this study, we focus on parameter reduction of IFSSs and propose two algorithmic approaches by ensuring certain decision abilities. Two numerical examples have been used to demonstrate the proposed approaches. To accomplish this, the remaining of this paper is organised as follows. Section 2 recalls some basic ideas relevant to this study. The proposed approaches and corresponding numerical examples are demonstrated in Sect. 3 followed by conclusions in Sect. 4.

2 Preliminaries

In this section, we review some basic notions of fuzzy soft set, intuitionistic fuzzy set, and intuitionistic fuzzy soft set.

Definition 1 [2] Let E be a set of parameters, which are fuzzy words or sentences involving fuzzy words and U be the universe of discourse. Also, assume that the set of all fuzzy subsets of U is denoted by $FS(U)$ and $A \subseteq E$. The pair $\left(\widehat{F}_{\{A\}}, E\right)$ is described as a fuzzy soft set (FSS) over U, where $\widehat{F}_{\{A\}}$ is a mapping, given by $\widehat{F}_{\{A\}} : E \to FS\left(U\right)$. If $e \notin A$, then $\widehat{F}_{\{A\}}(e) = \widehat{\varnothing}$, where $\widehat{\varnothing}$ denotes null fuzzy set.

Example 1 Let $U = \{z_1, z_2, z_3, z_4, z_5\}$ be the set of five diseases and $E = \{p_1, p_2, p_3, p_4, p_5\}$ be the set of five symptoms. Let $A = \{p_1, p_2, p_3, p_5\} \subseteq E$. Now suppose that

$\widehat{F}_{\{A\}}(p_1)=\{z_1/0.3, z_2/0.5, z_3/0.9, z_4/0.8\}$, $\widehat{F}_{\{A\}}(p_2)=\{z_2/0.8, z_3/0.2, z_4/0.7\}$, $\widehat{F}_{\{A\}}(p_3)=\{z_1/0.6, z_2/0.2, z_3/0.8\}$, $\widehat{F}_{\{A\}}(p_5)=\{z_1/0.6, z_2/0.8, z_3/0.4, z_4/0.8\}$.

Then the FSS is given by

$$(\widehat{F}_{\{A\}},E)=\begin{Bmatrix}(p_1,\{z_1/0.3, z_2/0.5, z_3/0.9, z_4/0.8\}),\\(p_2,\{z_2/0.8, z_3/0.2, z_4/0.7\}),\\(p_3,\{z_1/0.6, z_2/0.2, z_3/0.8\}),\\(p_4,\{\widehat{\varnothing}\}),\\(p_5,\{z_1/0.6, z_2/0.8, z_3/0.4, z_4/0.8\})\end{Bmatrix}$$

Definition 2 [8] Let E be a set of parameters, U be an initial universe, and the set of all intuitionistic fuzzy sets of U is denoted by $IFS(U)$. Let $A \subseteq E$. The pair $(\widehat{F}_{\{A\}}, E)$ is described as an intuitionistic fuzzy soft set (IFSS) over U, where $\widehat{F}_{\{A\}}$ is a mapping, given by $\widehat{F}_{\{A\}}: E \rightarrow IFS(U)$. If $e \notin A$, then $\widehat{F}_{\{A\}}(e)=\widehat{\varnothing}$, where $\widehat{\varnothing}$ is null intuitionistic fuzzy set.

Example 2 Let U, A and E are same like as Example 1. Let us take

$$\widehat{F}_{\{A\}}(p_1)=\{z_1/(0.3,0.6), z_2/(0.5,0.4), z_3/(0.9,0.1), z_4/(0.8,0.1)\},$$
$$\widehat{F}_{\{A\}}(p_2)=\{z_2/(0.8,0.1), z_3/(0.2,0.7), z_4/(0.7,0.3)\},$$
$$\widehat{F}_{\{A\}}(p_3)=\{z_1/(0.6,0.2), z_2/(0.2,0.7), z_3/(0.8,0.2)\},$$
$$\widehat{F}_{\{A\}}(p_5)=\{z_1/(0.6,0.3), z_2/(0.8,0.1), z_3/(0.4,0.4), z_4/(0.8,0.1)\}.$$

Then the IFSS is given below.

$$(\widehat{F}_{\{A\}},E)=\begin{Bmatrix}(p_1,\{z_1/(0.3,0.6), z_2/(0.5,0.4), z_3/(0.9,0.1), z_4/(0.8,0.1)\}),\\(p_2,\{z_2/(0.8,0.1), z_3/(0.2,0.7), z_4/(0.7,0.3)\}),\\(p_3,\{z_1/(0.6,0.2), z_2/(0.2,0.7), z_3/(0.8,0.2)\}),\\(p_4,\widehat{\varnothing}),\\(p_5,\{z_1/(0.6,0.3), z_2/(0.8,0.1), z_3/(0.4,0.4), z_4/(0.8,0.1)\})\end{Bmatrix}$$

Definition 3 [9] An intuitionistic fuzzy set (IFS) A on universe X is defined as $A=\{x, \mu_A(x), \nu_A(x): x \in X\}$. Here, the functions $\mu_A: X \rightarrow [0,1]$ and $\nu_A: X \rightarrow [0,1]$ reflect, respectively, the degree of membership and non-membership of the element x to the set A such that for every $x \in X$, $0 \le \mu_A(x)+\nu_A(x) \le 1$. The function $\pi_A: X \rightarrow [0,1]$, given by $\pi_A(x)=1-(\mu_A(x)+\nu_A(x)), x \in X$ is termed as intuitionistic fuzzy index. It represents the degree of hesitation or indeterminacy of the element x to the set A.

For a fixed $x_1 \in X$, an object $\{\mu_A(x_1), \nu_A(x_1)\}$ is normally called intuitionistic fuzzy value (IFV) or intuitionistic fuzzy number (IFN). An important problem is the comparison of IFNs. In order to compare two IFNs, Chen and Tan [10] introduced the score function (or net membership) as $S(x_1) = \mu(x_1) - \nu(x_1)$ and Hong and Choi [11] defined the accuracy function as $H(x_1) = \mu(x_1) + \nu(x_1)$, where x_1 is an IFV. Xu [12] used both the score and accuracy functions to form the order relations between any pair of IFVs (x_1 and x_2) as given below:

$$\begin{aligned} &\text{if } (S(x_1) > S(x_2)), \text{ then } x_1 > x_2;\\ &\text{if } (S(x_1) == S(x_2)), \text{ then}\\ &(1) \text{ if } (H(x_1) == H(x_2)), \text{ then } x_1 = x_2;\\ &(2) \text{ if } (H(x_1) < H(x_2)), \text{ then } x_1 < x_2; \end{aligned}$$

3 Parameter Reduction Algorithm

In this section, we propose two algorithmic approaches to reduce the unnecessary parameter(s) of an IFSS. This section has also provided two illustrative examples to show the effectiveness of the proposed approaches. Let $U = \{h_1, h_2, \ldots, h_n\}$ be the set of objects and $E = \{e_1, e_2, \ldots, e_m\}$ be a set of parameters. An IFSS $(\widehat{F}_{\{A\}}, E)$ is defined over U.

Algorithm 1

Step 1 Score values for each of the entries of the IFSS $(\widehat{F}_{\{A\}}, E)$ are calculated and denoted as $S(h_i, e_j), \forall i, j$.

Step 2 Aggregated score $\widehat{S}(h_i)$ for each object $h_i, i = 1, 2, \ldots, n$ is computed as $\widehat{S}(h_i) = \sum_{j=1}^{m} S(h_i, e_j)$.

Step 3 Any two subsets of the parameter set $E = \{e_1, e_2, \ldots, e_m\}$, $A = \{e_1', e_2', \ldots, e_p'\} \subset E$ and $B = \{e_1'', e_2'', \ldots, e_q''\} \subset E$ are selected, where $A, B \neq \Phi$, $A \cap B = \Phi$ and $A \cup B \neq E$.

Step 4 Compute the scores of each object $h_i, i = 1, 2, \ldots, n$ corresponding to the parameter set A and B, i.e. $\widehat{S}_{A \subset E}(h_i)$ and $\widehat{S}_{B \subset E}(h_i) \forall h_i \in U$ are computed.

Step 5 If $\widehat{S}_{A \subset E}(h_i) + \widehat{S}_{B \subset E}(h_i) = 0$, then $E - (A \cup B)$ is chosen as the reduction parameter set of the IFSS $(\widehat{F}_{\{A\}}, E)$.

Example 3 Let $(\widehat{F}_{\{A\}}, E)$ be an IFSS shown in Table 1. Let $E = \{e_1, e_2, \ldots, e_8\}$ and $U = \{h_1, h_2, \ldots, h_6\}$.

Table 1 Tabular representation of IFSS $(\widehat{F}_{\{A\}}, E)$

	e_1	e_2	e_3	e_4	e_5	e_6	e_7	e_8
h_1	(0.6, 0.3)	(0.7, 0.2)	(0.4, 0.2)	(0.6, 0.3)	(0.5, 0.4)	(0.3, 0.5)	(0.7, 0.1)	(0.1, 0.7)
h_2	(0.5, 0.4)	(0.3, 0.5)	(0.5, 0.3)	(0.2, 0.5)	(0.2, 0.6)	(0.35, 0.2)	(0.6, 0.1)	(0.4, 0.3)
h_3	(0.5, 0.35)	(0.4, 0.3)	(0.5, 0.45)	(0.6, 0.1)	(0.55, 0.3)	(0.45, 0.25)	(0.5, 0.2)	(0.15, 0.55)
h_4	(0.6, 0.4)	(0.2, 0.5)	(0.45, 0.3)	(0.5, 0.3)	(0.15, 0.6)	(0.05, 0.15)	(0.15, 0.35)	(0.5, 0.4)
h_5	(0.45, 0.3)	(0.6, 0.2)	(0.6, 0.35)	(0.45, 0.15)	(0.2, 0.5)	(0.3, 0.4)	(0.05, 0.25)	(0.4, 0.5)
h_6	(0.5, 0.2)	(0.4, 0.1)	(0.5, 0.25)	(0.2, 0.15)	(0.25, 0.6)	(0.15, 0.35)	(0.5, 0.45)	(0.4, 0.6)

Table 2 Score of each entry and object for the IFSS $(\widehat{F}_{\{A\}}, E)$

	e_1	e_2	e_3	e_4	e_5	e_6	e_7	e_8	$\widehat{S}(h_i)$
h_1	0.3	0.5	0.2	0.3	0.1	−0.2	0.6	−0.6	1.2
h_2	0.1	−0.2	0.2	−0.3	−0.4	0.15	0.5	0.1	0.15
h_3	0.15	0.1	0.05	0.5	0.25	0.2	0.3	−0.45	1.1
h_4	0.2	−0.3	0.15	0.2	−0.45	−0.1	−0.2	0.1	−0.4
h_5	0.15	0.4	0.25	0.3	−0.3	−0.1	−0.2	−0.1	0.4
h_6	0.3	0.3	0.25	0.05	−0.35	−0.2	0.05	−0.2	0.2

From Table 1 of IFSS $(\widehat{F}_{\{A\}}, E)$, we first calculate score values of each object $h_i, i=1,2,\ldots,6$ corresponding to the parameters $E=\{e_1, e_2, \ldots, e_8\}$ which is defined as $\widehat{S}(h_i)=\sum_{j=1}^{8} S(h_i, e_j)$. Then the objects are ranked using score and accuracy values (Table 2). The rank is found as $\{h_1, h_3, h_5, h_6, h_2, h_4\}$. If we consider $A=\{e_1, e_3\}$ and $B=\{e_5, e_8\}$, where $A, B \subset E$, $A \cup B \neq E$, then we have $\widehat{S}_A(h_1)=0.5$ and $\widehat{S}_B(h_2)=-0.5$. Addition of these two score values yield $\widehat{S}_A(h_1)+\widehat{S}_B(h_1)=0$. So the parameter set $\{e_1, e_3, e_5, e_8\}$ can be reduced and the reduced set of parameters will be $E'=\{e_2, e_4, e_6, e_7\}$. Next score values $\widehat{S}(h_i)$ of each object $h_i, i=1,2,\ldots,6$ are computed using the reduced parameter set $E'=\{e_2, e_4, e_6, e_7\}$. The result is shown in Table 3. Then the objects are ranked again based on the score values and we find the rank similar to the previous one.

Algorithm 2

Step 1 Score values for each of the entries of the IFSS $(\widehat{F}_{\{A\}}, E)$ are calculated and denoted as $S(h_i, e_j), \forall i, j$.

Step 2 Aggregated score $\widehat{S}(h_i)$ for each object $h_i, i=1,2,\ldots,n$ is computed as
$$\widehat{S}(h_i)=\sum_{j=1}^{m} S(h_i, e_j).$$

Step 3 Calculate maximum score deviation ε with the parameter set $E=\{e_1, e_2, \ldots, e_m\}$, where $\varepsilon=\min\{\left|\widehat{S}(h_i)-\widehat{S}(h_j)\right|\}, i, j=1,2,\ldots, n, i \neq j$

Table 3 Reduced parameter set and score of each object

	e_2	e_4	e_6	e_7	$\widehat{S}(h_i)$
h_1	0.5	0.3	−0.2	0.6	1.2
h_2	−0.2	−0.3	0.15	0.5	0.15
h_3	0.1	0.5	0.2	0.3	1.1
h_4	−0.3	0.2	−0.1	−0.2	−0.4
h_5	0.4	0.3	−0.1	−0.2	0.4
h_6	0.3	0.05	−0.2	0.05	0.2

Step 4 Select the parameter $e_k \in E$ which satisfies the following conditions

(a) $S(h_i, e_k) \leq S(h_i, e_j), j \neq k, j = 1, 2, \ldots, m, \forall i$

(b) $\max |S(h_i, e_k) - S(h_i, e_j)| < \varepsilon, \ j \neq k, \ i = 1, 2, \ldots, m, \ \forall i.$

Step 5 Select maximum number of parameter set A which satisfy Step 4.

Step 6. Compute $E - A$ as the reduced set of parameter of IFSS.

Example 4 Let $(\widehat{F}_{\{A\}}, E)$ be an IFSS, where the set of objects $U = \{h_1, h_2, \ldots, h_4\}$ and the set of parameters $E = \{e_1, e_2, \ldots, e_5\}$. From Table 4 of IFSS $(\widehat{F}_{\{A\}}, E)$, we first calculate score values of each object $h_i, i = 1, 2, \ldots, 4$ corresponding to the parameters $E = \{e_1, e_2, \ldots, e_5\}$ which is defined as $\widehat{S}(h_i) = \sum_{j=1}^{5} S(h_i, e_j)$. Then the objects are ranked using score and accuracy values. The rank is found as $\{h_2, h_4, h_3, h_1\}$.

The maximum deviation is computed as $\varepsilon = \min\{1.5, 0.3, 1.25, 1.20, 0.25, 0.95\} = 0.25$. We consider the parameter $e_1 \in E$, where $S(h_i, e_1) \leq S(h_i, e_j), \ \forall i, j$ and $\max |S(h_i, e_1) - S(h_i, e_j)| < \varepsilon, \ \forall i, j$ (Table 5). Thus the reduction of the parameter e_1 will have no effect in the ranking of objects, which is shown in Table 6. So we get the reduced set of parameters as $E - \{e_1\}$.

Table 4 Tabular representation of IFSS $(\widehat{F}_{\{A\}}, E)$

	e_1	e_2	e_3	e_4	e_5
h_1	(0.5, 0.4)	(0.6, 0.4)	(0.45, 0.3)	(0.5, 0.4)	(0.6, 0.4)
h_2	(0.5, 0.1)	(0.5, 0.05)	(0.6, 0.1)	(0.6, 0.2)	(0.7, 0.2)
h_3	(0.5, 0.35)	(0.4, 0.2)	(0.5, 0.25)	(0.5, 0.3)	(0.55, 0.3)
h_4	(0.65, 0.3)	(0.5, 0.1)	(0.55, 0.1)	(0.5, 0.15)	(0.65, 0.2)

Table 5 Score of each entry and object for the IFSS $(\widehat{F}_{\{A\}}, E)$

	e_1	e_2	e_3	e_4	e_5	$\widehat{S}(h_i)$
h_1	0.1	0.2	0.15	0.1	0.2	0.75
h_2	0.4	0.45	0.5	0.4	0.5	2.25
h_3	0.15	0.2	0.25	0.2	0.25	1.05
h_4	0.35	0.4	0.45	0.35	0.45	2.0

Table 6 Reduced parameter set and score of each object

	e_2	e_3	e_4	e_5	$\widehat{S}(h_i)$
h_1	0.2	0.15	0.1	0.2	0.65
h_2	0.45	0.5	0.4	0.5	1.85
h_3	0.2	0.25	0.2	0.25	0.9
h_4	0.4	0.45	0.35	0.45	1.65

4 Conclusion

Parameter reduction mainly avoids the redundant parameter(s). The decision with the minimum subset of parameters remains same as the entire set of parameters. This is the main significance of parameter reduction procedure. In this paper, we have proposed two parameter reduction algorithms applicable for IFSSs. Two numerical examples are given to clarify the applicability of those proposed algorithms. In future, researchers may investigate the parameter reduction of IFSSs for other uncertain environment to satisfy the various concerns of decision makers.

References

1. Molodtsov, D.: Soft set theory-first results. Comput. Math Appl. **37**(4/5), 19–31 (1999)
2. Maji, P.K., Biswas, R., Roy, A.R.: Fuzzy soft sets. J. Fuzzy Math. **9**(3), 589–602 (2001)
3. Chen, D., Tsang, E.C.C., Yeung, D.S., Wang, X.: The parameterization reduction of soft sets and its applications. Comput. Math Appl. **49**(5–6), 757–763 (2005)
4. Maji, P.K., Roy, A.R.: An application of soft sets in a decision making problem. Comput. Math Appl. **44**, 1077–1083 (2002)
5. Kong, Z., Gao, L., Wang, L., Li, S.: The normal parameter reduction of soft sets and its algorithm. Comput. Math Appl. **56**(12), 3029–3037 (2008)
6. Ma, X.Q., Sulaiman, N., Qin, H.W., Herawan, T., Zaino, J.M.: A new efficient normal parameter reduction algorithm of soft sets. Comput. Math Appl. **62**(2), 588–598 (2011)
7. Ma, X., Qin, H., Sulaiman, N., Herawan, T., Abawajy, J.H.: The parameter reduction of the interval-valued fuzzy soft sets and its related algorithms. IEEE Trans. Fuzzy Syst. **22**(1), 57–71 (2014)
8. Maji, P.K., Biswas, R., Roy, A.R.: Intuitionistic fuzzy soft sets. J. Fuzzy Math. **9**(3), 677–692 (2001)
9. Atanassov, K.: Intuitionistic fuzzy sets. Fuzzy Sets Syst. **20**, 87–96 (1986)
10. Chen, S.M., Tan, J.M.: Handling multi criteria fuzzy decision-making problems based on vague set theory. Fuzzy Sets Syst. **67**, 163–172 (1994)

11. Hong, D.H., Choi, C.H.: Multi criteria fuzzy decision-making problems based on vague set theory. Fuzzy Sets Syst. **114**, 103–113 (2000)
12. Xu, Z.: Intuitionistic preference relations and their application in group decision making. Inf. Sci. **177**, 2363–2379 (2007)
13. Das, S., Kar, S.: Group decision making in medical system: an intuitionistic fuzzy soft set approach. Appl. Soft Comput. **24**, 196–211 (2014)
14. Das, S., Kar, S.: The hesitant fuzzy soft set and its application in decision making. In: International Conference on Facets of Uncertainties and Applications, Springer Proceedings in Mathematic and Statics, vol. 125, pp. 235–247 (2013)
15. Broumi, S., Smarandache, F.: New operations over interval valued intuitionistic hesitant fuzzy set. Math. Stat. **2**(2), 62–71 (2014)
16. Wang, J.Q., Li, X.E., Chen, X.H.: Hesitant fuzzy soft sets with application in multicriteria group decision making problems. Sci. World J. (2015). Article ID 806983
17. Das, S., Kar, S., Pal, T.: Group decision making using interval-valued intuitionistic fuzzy soft matrix and confident weight of experts. J. Artif. Intell. Soft Comput. Res. **4**(1), 57–77 (2014)
18. Das, S., Kar, M.B., Pal, T., Kar, S.: Multiple attribute group decision making using interval-valued intuitionistic fuzzy soft matrix. In: Proceedings of IEEE International Conference on Fuzzy Systems (FUZZ-IEEE), pp. 2222–2229. Beijing (2014). doi:10.1109/FUZZ-IEEE.2014.6891687
19. Das, S., Kar, S.: Intuitionistic multi fuzzy soft set and its application in decision making. In: Maji, P. et al (eds.) Fifth International Conference on Pattern Recognition and Machine Intelligence (PReMI). Lecture Notes in Computer Science, 8251, pp. 587–592 (2013)

Correlation Measure of Hesitant Fuzzy Linguistic Term Soft Set and Its Application in Decision Making

Debashish Malakar, Sulekha Gope and Sujit Das

Abstract Hesitant fuzzy linguistic term set (HFLTS) provides a new and more powerful technique to represent the qualitative judgments of experts. Molodtsov [13] proposed soft set theory which is used as a general mathematical tool for dealing with uncertainty. In this paper, we introduce hesitant fuzzy linguistic term soft set (HFLTSS) by combining HFLTS with soft sets. HFLTSS can be used to provide approximate description of an object using linguistic terms. This study also proposes hesitant fuzzy linguistic term soft matrix (HFLTSM) to represent HFLTSS. As an extension of correlation measures in HFLTS, we introduce correlation coefficient in the context of HFLTSS. Finally, the applicability of correlation coefficient in the framework of HFLTSS has been demonstrated using a real-life decision-making problem.

Keywords Hesitant fuzzy linguistic term set • Hesitant fuzzy linguistic term soft set • Hesitant fuzzy linguistic term soft matrix • Correlation coefficient

1 Introduction

In real-life decision-making problems, experts often prefer to express their opinions using qualitative approach rather than quantitative approach, since qualitative approach resembles human's inherent reasoning capability. Linguistic terms are

---

D. Malakar (✉) · S. Gope
Department of Computer Application, Asansol Engineering College, Asansol 5, India
e-mail: debashish.malakar@gmail.com

S. Gope
e-mail: sulekhagope81@gmail.com

S. Das
Department of C.S.E., Dr. B. C. Roy Engineering College, Durgapur 6, India
e-mail: sujit_cse@yahoo.com

S. Das et al. (eds.), *Proceedings of the 4th International Conference on Frontiers in Intelligent Computing: Theory and Applications (FICTA) 2015*, Advances in Intelligent Systems and Computing 404, DOI 10.1007/978-81-322-2695-6_35

used to model the qualitative approach. Zadeh [2] introduced the fuzzy linguistic approach in 1975. The author used linguistic variables to represent qualitative information, whose values are not numbers but words or sentences. The fuzzy linguistic approach considers only one linguistic term to represent the value of a linguistic variable. In this age of uncertainty, use of single linguistic term may not be suitable to properly represent an expert's opinion, where multiple number linguistic terms may be more comfortable. To cope up with this kind of situation, Rodríguez et al. [3] introduced the concept of hesitant fuzzy linguistic term set (HFLTS), which provides a new and more powerful technique to represent experts' qualitative judgments. Since its introduction in 2012, HFLTS has attracted the attention of many researchers. Liao et al. [4] proposed distance and similarity measures for HFLTSs and applied them in multi-criteria decision-making (MCDM) problem. Wei et al. [5] studied aggregation theory in the context of HFLTS. Using optimistic and pessimistic attitude of decision makers, Chen and Hong [6] proposed a decision-making approach based on HFLTS. In [7], Liu and Rodríguez used fuzzy envelope to propose a new representation of HFLTS. Motivated by the works found in [8–10], Liao et al. [11] emphasized on determining correlation measure and correlation coefficient of HFLTS. However, a few works have been done on the parameterization of HFLTS.

Correlation measures between two fuzzy sets find the degree of dependency between them. Since its introduction, many researchers have contributed on correlation measures of different types of fuzzy sets, such as fuzzy correlation and correlation coefficients [8, 12], intuitionistic fuzzy correlation and correlation coefficients [9, 13], hesitant fuzzy correlation and correlation coefficients [10, 14], and correlation measures and correlation coefficients of HFLTSs [11]. However, no efforts have been found for measuring the correlation coefficients of HFLTSSs.

In this paper, we extend the concept of HFLTS [3] with soft set [1] to introduce hesitant fuzzy linguistic term soft set (HFLTSS). HFLTSSs are specifically useful in complex and uncertain situations, where one hesitates to forward his/her opinion due to lack of experience or knowledge. HFLTSS is more flexible to represent the judgment of a group of decision makers in multiple attribute decision-making problems maintaining anonymity using linguistic terms. HFLTSS allows decision makers to select a subset of attributes as per their own intuition in an unbiased manner. We have developed hesitant fuzzy linguistic term soft matrix (HFLTSM) to represent a HFLTSS. Motivated by the correlation measurement approaches for HFLTSs found in [11], we have investigated HFLTSS-based correlation measurements. A real-life example has been presented to demonstrate the applicability of correlation measurements in the framework of HFLTSMs.

The rest of this paper is structured as follows. Section 2 presents some basic concepts related to hesitant fuzzy set (HFS), hesitant fuzzy soft set (HFSS), HFLTS, and correlation measures of HFLTSs. In Sect. 3, we introduce the concepts of

HFLTSS, HFLTSM, information measure, correlation measure, and correlation coefficient of HFLTSSs. Section 4 illustrates the applicability of the proposes ideas in decision-making paradigm. Finally, Sect. 5 gives a brief conclusion.

2 Preliminaries

This section briefly describes the ideas relevant with this paper. First, we discuss briefly HFS, HFSS, and HFLTS. Then correlation measures of HFLTSs are discussed.

2.1 *HFS, HFSS, HFLTS*

HFS is an extension of fuzzy sets. When membership degrees are assigned to the elements of a fuzzy set, decision makers often hesitate among a set of possible values. HFS can be used to model this kind of uncertainties originated by the hesitation.

Definition 1 [15]. Let X be a fixed set, a HFS M on X is defined in terms of a function $h_M(x)$ that returns a subset of [0, 1] when applied to X. This can be represented using the mathematical expression: $M=\{\langle x, h_M(x)\rangle \mid x\in X\}$, where $h_M(x)$ is a set of values in [0, 1], representing the possible membership degrees of the element $x\in X$ to the set M. For convenience, Xia and Xu [16] defined hesitant fuzzy element (HFE) as $h=h_M(x)$.

It is noted that the number of values in different HFEs may be different. A HFE $h_M(x)$ with k number of values can be defined as $h_M^k(x)$, where k is a positive integer. Here, the membership degree of an element for a set is represented by several possible values between 0 and 1. HFSs are mainly used in situations, where one has hesitation in giving his/her preferences over objects in a decision-making [17–19] process.

Example 1 Let $X=\{x_1, x_2, x_3\}$ be a fixed set. $h_M(x_1)=\{0.2, 0.3\}$, $h_M(x_2)=\{0.6, 0.4, 0.6\}$, and $h_M(x_3)=\{0.4, 0.5, 0.7\}$ be the HFEs of $x_i (i=1, 2, 3)$ to the set M. Then the HFS M is considered as

$$M=\{(x_1, \{0.2, 0.3\}), (x_2, \{0.6, 0.4, 0.6\}), (x_3, \{0.4, 0.5, 0.7\})\}.$$

Definition 2 [20]. Let $\widehat{H}(X)$ be the set of all HFSs of X, where X is a fixed set. Let E be a set of parameters and $A\subseteq E$. A HFSS over X is defined by the set of ordered pairs $(\widehat{F}_{\{A\}}, E)$, where $\widehat{F}_{\{A\}}$ is a mapping given by $\widehat{F}_{\{A\}}: E\to \widehat{H}(X)$. For any parameter,

$e \in A, \widehat{F}_{\{A\}}(e)$ is a hesitant fuzzy subset of X and may be considered as e-elements or e-approximate elements in the HFSS. Clearly, $\widehat{F}_{\{A\}}(e)$ can be written as a HFS such that $\widehat{F}_{\{A\}}(e) = \{\langle x, h_E(x)\rangle | x \in X\}$, where $h_E(x)$ is a set of values in [0, 1].

Example 2 Let X be set of three computer, i.e., $X = \{x_1, x_2, x_3\}$. Let $E = \{e_1, e_2, e_3\} =$ {cheap, latest version, high processing}. Suppose that
$\widehat{F}_{\{A\}}(e_1) = (\{x_1, (0.2, 0.6)\}, \{x_2, (0.5, 0.6, 0.7)\}, \{x_3, (0.4, 0.5, 0.2)\})$,
$\widehat{F}_{\{A\}}(e_2) = (\{x_1, (0.2, 0.7, 0.6)\}, \{x_2, (0.2, 0.4)\}, \{x_3, (0.3, 0.1)\})$, *and*
$\widehat{F}_{\{A\}}(e_3) = (\{x_1, (0.2, 0.5, 0.6)\}, \{x_2, (0.8, 0.1)\}, \{x_3, (0.2, 0.2)\})$

Thus HFSS is represented as

$$(\widehat{F}_{\{A\}}, E) = \left\{ \begin{array}{l} \langle e_1, (\{x_1, (0.2, 0.6)\}, \{x_2, (0.5, 0.6, 0.7)\}, \{x_3, (0.4, 0.5, 0.2)\})\rangle \\ \langle e_2, (\{x_1, (0.2, 0.7, 0.6)\}, \{x_2, (0.2, 0.4)\}, \{x_3, (0.3, 0.1)\})\rangle \\ \langle e_3, (\{x_1, (0.2, 0.5, 0.6)\}, \{x_2, (0.8, 0.1)\}, \{x_3, (0.2, 0.2)\})\rangle \end{array} \right\}.$$

The use of linguistic approach is important in real-life problems, because many problems can be well described using qualitative approach rather that quantitative one. Fuzzy linguistic approach [2] can be used to model the qualitative aspects in linguistic terms using linguistic values and variables.

Definition 3 [3]. Let $x_i \in X$, $i = 1, 2, \ldots, n$ be fixed and $S = \{s_t | t = -\tau, \ldots, -1, 0, 1, \ldots, \tau\}$ be a linguistic term set. A hesitant fuzzy linguistic term set (HFLTS) on X, denoted by H_S, is an ordered finite subset of the consecutive linguistic terms of S. In mathematical terms, $H_s = \{\langle x_i, h_s(x_i)\rangle | x_i \in X\}$, where $h_s(x_i)$ is a set of some values in the linguistic term set S and can be expressed as $h_s(x_i) = \{s_{\phi_l}(x_i) | s_{\phi_l}(x_i) \in S, l = 1, \ldots, L\}$ with L being the number of linguistic terms in $h_s(x_i)$. $h_s(x_i)$ denotes the possible degrees of linguistic variable x_i to the linguistic term set S. For convenience, $h_s(x_i)$ is called the hesitant fuzzy linguistic element (HFLE) and H_S is the set of all HFLEs.

Example 3 Let S be a linguistic term set as shown below.
$S = \{s_0$: nothing, s_1: very low, s_2: low, s_3: medium, s_4: high, s_5: very high, s_6: perfect$\}$.
Then $H_s = \{s_3\}$, $H_S^1 = \{s_2, s_3\}$, and $H_S^2 = \{s_3, s_4, s_5\}$ are three HFLTSs on S.

Rodrıguez et al. [3] defined the following operations on HFLTSs. Results of these operations are shown based on the values of H_s, H_S^1, and H_S^2 as given in Example 3.

(1) $H_S^C = S - H_S = \{s_i | s_i \in S \text{ and } s_i \notin H_S\} = \{s_0, s_1, s_2, s_4, s_5, s_6\}$
(2) $H_S^1 \cap H_S^1 = \{s_i | s_i \in H_S^1 \text{ and } s_i \in H_S^2\} = \{s_3\}$
(3) $H_S^1 \cup H_S^1 = \{s_i | s_i \in H_S^1 \text{ or } s_i \in H_S^2\} = \{s_2, s_3, s_4, s_5\}$

2.2 Correlation Measure of HFLTS

This section briefly describes informational energy, correlation, and correlation coefficients concerned with HFLTSs. In the following definitions, we consider $S = \{s_t | t = -\tau, \ldots, -1, 0, 1, \ldots, \tau\}$ as a linguistic term set.

Definition 4 [11]. For a HFLTS, $H_s = \{\langle x_i, h_s(x_i)\rangle | x_i \in X\}$ with $h_s(x_i) = \{s_{\varphi_l}(x_i) | s_{\varphi_l}(x_i) \in S, l = 1, \ldots, L\}$, the information energy of H_S is defined as

$$E(H_s) = \sum_{i=1}^{n}\left(\frac{1}{L_i}\sum_{l=1}^{L_i}\left(\frac{\varphi_l(x_i)}{2\tau+1}\right)^2\right).$$

where $\varphi_l(x_i)$ is the index of the lth smallest linguistic term in HFLE $h_s(x_i)$, L_i is the number of linguistic terms in $h_s(x_i)$, and n is the cardinality of X.

Definition 5 [11]. The correlation between two HFLTSs H_s^1 and H_s^2 is introduced using the information energy of HFLTS. For two HFLTSs $H_s^1 = \{\langle x_i, h_s^1(x_i)\rangle | x_i \in X\}$ and $H_s^2 = \{\langle x_i, h_s^2(x_i)\rangle | x_i \in X\}$ with $h_s^k(x_i) = \{s_{\varphi_k^l}(x_i) | s_{\varphi_k^l}(x_i) \in S, l = 1, \ldots, L_i\}$, $k = 1, 2$, the correlation between H_s^1 and H_s^2 is defined as

$$C(H_s^1, H_s^2) = \sum_{i=1}^{n}\left(\frac{1}{L_i}\sum_{l=1}^{L_i}\left(\frac{|\varphi_l^1(x_i)|}{2\tau+1}\cdot\frac{|\varphi_l^2(x_i)|}{2\tau+1}\right)\right).$$

Definition 6 [11]. For two HFLTSs $H_s^1 = \{\langle x_i, h_s^1(x_i)\rangle | x_i \in X\}$ and $H_s^2 = \{\langle x_i, h_s^2(x_i)\rangle | x_i \in X\}$ with $h_s^k(x_i) = \{s_{\varphi_k^l}(x_i) | s_{\varphi_k^l}(x_i) \in S, l = 1, \ldots, L_i\}, k = 1, 2,$ the correlation coefficient between H_s^1 and H_s^2 is defined as below:

$$\rho_1(H_s^1, H_s^2) = \frac{C(H_s^1, H_s^2)}{\left(E(H_s^1).E(H_s^2)\right)^{1/2}} = \frac{\sum_{i=1}^{n}\left(\frac{1}{L_i}\sum_{l=1}^{L_i}\left(\frac{|\varphi_l^1(x_i)|}{2\tau+1}\cdot\frac{|\varphi_l^2(x_i)|}{2\tau+1}\right)\right)}{\left(\sum_{i=1}^{n}\left(\frac{1}{L_i}\sum_{l=1}^{L_i}\left(\frac{|\varphi_l^1(x_i)|}{2\tau+1}\right)^2\right)\cdot\sum_{i=1}^{n}\left(\frac{1}{L_i}\sum_{l=1}^{L_i}\left(\frac{|\varphi_l^2(x_i)|}{2\tau+1}\right)^2\right)\right)^{1/2}}$$

Here, L_i is the maximum number of linguistic terms in $H_s^1(x_i)$ and $H_s^2(x_i)$ (the shorter is extended till equal length by repeating the final linguistic term), and n is the cardinality of X.

3 HFLTSS, HFLTSM, and Correlation Coefficient

In this section, we introduce HFLTSS and HFLTSM. We also propose information measure, correlation, and correlation coefficient of HFLTSSs, which can be considered as an extended study of the work done by Liao et al. [11] in the context of HFLTS.

Definition 7 Let $\widehat{H}(x)$ be the set of all hesitant fuzzy linguistic term set of X, where X is a fixed set. Let E be a set of parameters and $A \subseteq E$. A HFLTSS over X is defined by the set ordered pairs $(\widehat{F}, A)$ where $\widehat{F}$ is a mapping given by $\widehat{F}: A \rightarrow \widehat{H}(x)$. For any parameter, $e \in A$, $\widehat{F}(e)$ is a hesitant fuzzy linguistic term subset of X and may be considered as e-elements or e-approximate elements in HFLTSS. Clearly, $\widehat{F}(e)$ can be written as a hesitant fuzzy linguistic term set such that $\widehat{F}(e) = \left\{ \langle x, \widehat{h}_E(x) \rangle | x \in X \right\}$, where $\widehat{h}_E(x)$ is a set of values in hesitant fuzzy linguistic term set H_s and can be expressed as $\widehat{h}_E(x) = \{ s_{\delta_l}(x) | s_{\delta_l}(x) \in S, l = 1, \ldots, L \}$ with L is the number of linguistic term in $\widehat{h}_E(x)$. $\widehat{h}_E(x)$ denotes the possible degrees of linguistic variable x to the linguistic term set S.

Example 4 Consider $U = \{h_1, h_2, \ldots, h_6\}$ be a set of six houses and $A \subseteq E$ be a set of parameters, where $E = \{e_1, e_2, e_3, e_4\} =$ {'expensive', 'beautiful', 'wooden', 'in the green surrounding'}. A HFLTSS $(\widehat{F}, A)$ describes the attractiveness of house. Linguistic term sets for the five parameters $E = \{e_1, e_2, e_3, e_4\}$ are given below:

$$S_1 = \left\{ \begin{array}{l} s_{-3} = \text{very expensive}, \ s_{-2} = \text{expensive}, \ s_{-1} = \text{a little expensive}, \\ s_0 = \text{medium}, \ s_1 = \text{a little cheap}, \ s_2 = \text{cheap}, \ s_3 = \text{very cheap} \end{array} \right\}$$

$$S_2 = \left\{ \begin{array}{l} s_{-3} = \text{least beautiful}, \ s_{-2} = \text{less beautiful}, \ s_{-1} = \text{a little beautiful}, \ s_0 = \text{beautiful}, \\ s_1 = \text{more beautiful}, \ s_2 = \text{most beautiful}, \ s_3 = \text{outstanding beautiful} \end{array} \right\}$$

$$S_3 = \left\{ \begin{array}{l} s_{-3} = \text{very expensive wooden}, \ s_{-2} = \text{expensive wooden}, \ s_{-1} = \text{a little expensive wooden}, \\ s_0 = \text{medium expensive wooden}, \ s_1 = \text{a little cheap wooden}, \ s_2 = \text{cheap wooden}, \\ s_3 = \text{very cheap wooden} \end{array} \right\}$$

$$S_4 = \left\{ \begin{array}{l} s_{-3} = \text{full green surrounding}, \ s_{-2} = \text{front side green surrounding}, \\ s_{-1} = \text{a little green surrounding}, \ s_0 = \text{medium green surrounding}, \\ s_1 = \text{a big green surrounding}, \ s_2 = \text{back side green surrounding}, \\ s_3 = \text{less green surrounding} \end{array} \right\}$$

If we consider

$F(e_1) = \{\langle h_1, (s_{-1}, s_0, s_2)\rangle, \langle h_2, (s_{-2}, s_0)\rangle, \langle h_3, (s_1, s_2, s_3)\rangle, \langle h_4, (s_2)\rangle, \langle h_5, (s_{-1}, s_1, s_2)\rangle, \langle h_6, (s_3, s_{-1}, s_2)\rangle\}$,
$F(e_2) = \{\langle h_1, (s_{-1}, s_2)\rangle, \langle h_2, (s_{-3}, s_1, s_0)\rangle, \langle h_3, (s_1)\rangle, \langle h_4, (s_2, s_2, s_1)\rangle, \langle h_5, (s_{-1})\rangle, \langle h_6, (s_{-3}, s_{-1}, s_2)\rangle\}$,
$F(e_3) = \{\langle h_1, (s_{-1}, s_2, s_{-2})\rangle, \langle h_2, (s_{-2}, s_2, s_3)\rangle, \langle h_3, (s_1, s_{-3})\rangle, \langle h_4, (s_2, s_3, s_{-1})\rangle, \langle h_5, (s_{-1})\rangle, \langle h_6, (s_{-3}, s_{-2})\rangle\}$,
$F(e_4) = \{\langle h_1, (s_{-2}, s_3, s_2)\rangle, \langle h_2, (s_{-1}, s_1, s_1)\rangle, \langle h_3, (s_{-3})\rangle, \langle h_4, (s_3, s_1)\rangle, \langle h_5, (s_{-1}, s_1, s_{-2})\rangle, \langle h_6, (s_2)\rangle\}$,

Then the HFLTSS is represented as

$$(\widehat{F}, A) = \left\{\begin{array}{l} (e_1, \{\langle h_1, (s_{-1}, s_0, s_2)\rangle, \langle h_2, (s_{-2}, s_0)\rangle, \langle h_3, (s_1, s_2, s_3)\rangle, \langle h_4, (s_2)\rangle, \langle h_5, (s_{-1}, s_1, s_2)\rangle, \langle h_6, (s_3, s_{-1}, s_2)\rangle\}) \\ (e_2, \{\langle h_1, (s_{-1}, s_2)\rangle, \langle h_2, (s_{-3}, s_1, s_0)\rangle, \langle h_3, (s_1)\rangle, \langle h_4, (s_2, s_2, s_1)\rangle, \langle h_5, (s_{-1})\rangle, \langle h_6, (s_{-3}, s_{-1}, s_2)\rangle\}) \\ (e_3, \{\langle h_1, (s_{-1}, s_2, s_{-2})\rangle, \langle h_2, (s_{-2}, s_2, s_3)\rangle, \langle h_3, (s_1, s_{-3})\rangle, \langle h_4, (s_2, s_3, s_{-1})\rangle, \langle h_5, (s_{-1})\rangle, \langle h_6, (s_{-3}, s_{-2})\rangle\}) \\ (e_4, \{\langle h_1, (s_{-2}, s_3, s_2)\rangle, \langle h_2, (s_{-1}, s_1, s_1)\rangle, \langle h_3, (s_{-3})\rangle, \langle h_4, (s_3, s_1)\rangle, \langle h_5, (s_{-1}, s_1, s_{-2})\rangle, \langle h_6, (s_2)\rangle\}) \end{array}\right\}.$$

Definition 8 HFLTSS can be represented well by HFLTSM. If $X = \{x_1, x_2, \ldots, x_n\}$ and $E = \{e_1, e_2, \ldots, e_m\}$,, then HFLTSM is defined as $F = (f_{ij})_{m \times n}$, where $f_{ij} = \widehat{h}_E(x_i, e_j)$, $i = 1, 2, \ldots, n, j = 1, 2, \ldots, m$. Here, $\widehat{h}_E(x_i, e_j)$ is a set of values in hesitant fuzzy linguistic term set H_S and can be expressed as $\widehat{h}_E(x_i, e_j) = \{s_{\delta_l}(x_i, e_j) | s_{\delta_l}(x_i, e_j) \in S, l = 1, \ldots, L\}$ with L is the number of linguistic term in $\widehat{h}_E(x_i, e_j)$. Table 1 presents a HFLTSM.

Definition 9 Informational energy $I_{\text{HFLTSS}}(A)$ for HFLTSS $(\widehat{F}, A)$ is defined as

$$I_{\text{HFLTSS}}(\widehat{F}, A) = \sum_{j=1}^{m} \left(\sum_{i=1}^{n} \frac{1}{L_i} \left(\sum_{l=1}^{L_i} \left(\frac{(S_{A_{\delta(l)}}(x_i, e_j))}{2\tau + 1} \right)^2 \right) \right).$$

Definition 10 For two HFLTSSs $(\widehat{F}, A)$ and $(\widehat{G}, B)$,, the correlation between $(\widehat{F}, A)$ and $(\widehat{G}, B)$ is defined by

$$C_{\text{HFLTSS}}\{(\widehat{F}, A), (\widehat{G}, B)\} = \sum_{j=1}^{m} \left(\sum_{i=1}^{n} \frac{1}{L_i} \left(\sum_{l=1}^{L_i} \frac{(S_{A_{\delta(l)}}(x_i, e_j))}{2\tau + 1} \cdot \frac{(S_{B_{\delta(l)}}(x_i, e_j))}{2\tau + 1} \right) \right).$$

Table 1 Tabular representation of HFLTSM

	e_1	e_2	...	e_m
x_1	$\widehat{h}_E(x_1, e_1)$	$\widehat{h}_E(x_1, e_2)$	...	$\widehat{h}_E(x_1, e_m)$
x_2	$\widehat{h}_E(x_2, e_1)$	$\widehat{h}_E(x_2, e_2)$	...	$\widehat{h}_E(x_2, e_m)$
⋮	⋮	⋮	⋮	⋮
x_n	$\widehat{h}_E(x_n, e_1)$	$\widehat{h}_E(x_n, e_2)$	...	$\widehat{h}_E(x_n, e_m)$

Definition 11 Correlation coefficient between two HFLTSS $(\widehat{F},A)$ and $(\widehat{G},B)$ is given as

$$\rho_{\text{HFLTSS}}\{(\widehat{F},A),(\widehat{G},B)\} = \frac{C_{\text{HFLTSS}}\{(\widehat{F},A),(\widehat{G},B)\}}{\left(C_{\text{HFLTSS}}\{(\widehat{F},A),(\widehat{F},A)\}.C_{\text{HFLTSS}}\{\ (\widehat{G},B),(\widehat{G},B)\}\right)^{\frac{1}{2}}}$$

$$= \frac{\sum_{j=1}^{m}\left(\sum_{i=1}^{n}\frac{1}{L_i}\left(\sum_{l=1}^{L_i}\frac{(S_{A_{\delta(l)}}(x_i,e_j))}{2\tau+1}.\frac{(S_{B_{\delta(l)}}(x_i,e_j))}{2\tau+1}\right)\right)}{\left(\sum_{j=1}^{m}\left(\sum_{i=1}^{n}\frac{1}{L_i}\left(\sum_{l=1}^{L_i}\left(\frac{(S_{A_{\delta(l)}}(x_i,e_j))}{2\tau+1}\right)^2\right)\right)\right)^{\frac{1}{2}}\cdot\left(\sum_{j=1}^{m}\left(\sum_{i=1}^{n}\frac{1}{L_i}\left(\sum_{l=1}^{L_i}\left(\frac{(S_{B_{\delta(l)}}(x_i,e_j))}{2\tau+1}\right)^2\right)\right)\right)^{\frac{1}{2}}}.$$

Example 5 Consider two HFLTSSs $(\widehat{F},A)$ and $(\widehat{G},B)$ as given below:

$$(\widehat{F},A) = \left\{\begin{array}{l} \{\langle h_1,(s_{-1},s_0,s_2)\rangle,\langle h_2,(s_{-2},s_1,s_0)\rangle,\langle h_3,(s_1,s_2,s_3)\rangle,\langle h_4,(s_2,s_0,s_1)\rangle\} \\ \{\langle h_1,(s_{-1},s_1,s_2)\rangle,\langle h_2,(s_{-3},s_1,s_0)\rangle,\langle h_3,(s_1,s_1,s_3)\rangle,\langle h_4,(s_2,s_2,s_1)\rangle\} \\ \{\langle h_1,(s_{-1},s_2,s_{-2})\rangle,\langle h_2,(s_{-2},s_2,s_3)\rangle,\langle h_3,(s_1,s_{-2},s_{-3})\rangle,\langle h_4,(s_2,s_3,s_{-1})\rangle\} \\ \{\langle h_1,(s_{-2},s_3,s_2)\rangle,\langle h_2,(s_{-1},s_1,s_1)\rangle,\langle h_3,(s_{-1},s_2,s_{-3})\rangle,\langle h_4,(s_2,s_3,s_1)\rangle\} \end{array}\right\}$$

$$(\widehat{G},B) = \left\{\begin{array}{l} \{\langle h_1,(s_{-1},s_0,s_2)\rangle,\langle h_2,(s_{-2},s_1,s_3)\rangle,\langle h_3,(s_1,s_2,s_3)\rangle,\langle h_4,(s_1,s_{-2},s_3)\rangle\} \\ \{\langle h_1,(s_{-1},s_1,s_2)\rangle,\langle h_2,(s_{-3},s_1,s_0)\rangle,\langle h_3,(s_2,s_3,s_{-3})\rangle,\langle h_4,(s_2,s_2,s_1)\rangle\} \\ \{\langle h_1,(s_{-1},s_3,s_2)\rangle,\langle h_2,(s_{-2},s_2,s_{-3})\rangle,\langle h_3,(s_1,s_{-2},s_{-3})\rangle,\langle h_4,(s_{-2},s_3,s_1)\rangle\} \\ \{\langle h_1,(s_{-2},s_{-3},s_2)\rangle,\langle h_2,(s_{-1},s_1,s_1)\rangle,\langle h_3,(s_1,s_2,s_3)\rangle,\langle h_4,(s_2,s_3,s_1)\rangle\} \end{array}\right\}$$

Informational energy, correlation, and correlation coefficient of two HFLTSSs $(\widehat{F},A)$ and $(\widehat{G},B)$ are computed as $I_{\text{HFLTSS}}(\widehat{F},A) = 1.14$, $I_{\text{HFLTSS}}(\widehat{G},B) = 1.37$, and $\rho_{\text{HFLTSS}}\{(\widehat{F},A),(\widehat{G},B)\} = 0.73$.

4 Decision Making Using Correlation Coefficient of HFLTSSs

Suppose an investment company wants to select a car manufacturer among a set of four manufacturers given by $X = \{X_1, X_2, X_3, X_4\}$ = {Maruti Suzuki, Hyundai, Toyota, Honda} for the purpose of investment. Initially, the company randomly selects a set of four cars from each of the manufactures for its internal policy purpose and investigates a set of four attributes, given by $E=\{E_1,E_2,E_3,E_4\}=$ {Fuel economy, Price, Comfort, Design} for each of the randomly selected cars. The company has its own requirement regarding the various cars and their attributes, which is represented by a HFLTSM as shown in Table 2. Information about the selected cars of the different car manufacturers are given using HFLTSMs shown in Tables 3, 4, 5 and 6. Information about the various cars has been taken randomly for the experimentation purpose only, which might not reflect the reality. We present the attribute values in terms of linguistic expression. Each attribute can be considered as a linguistic variable and the corresponding linguistic term sets are given below:

$$E_1=\{s_{-3}=\text{verylow},\ s_{-2}=\text{low},\ s_{-1}=\text{a little low},\ s_0=\text{normal},\ s_1=\text{a little high},\ s_2=\text{high},\ s_3=\text{very high}\},$$
$$E_2=\{s_{-3}=\text{verylow},\ s_{-2}=\text{low},\ s_{-1}=\text{a little low},\ s_0=\text{normal},\ s_1=\text{a little high},\ s_2=\text{high},\ s_3=\text{very high}\},$$
$$E_3=\{s_{-3}=\text{verylow},\ s_{-2}=\text{low},\ s_{-1}=\text{a little low},\ s_0=\text{normal},\ s_1=\text{a little high},\ s_2=\text{high},\ s_3=\text{very high}\},$$
$$E_4=\{s_{-3}=\text{worst},\ s_{-2}=\text{below averge},\ s_{-1}=\text{average},\ s_0=\text{above agerage},\ s_1=\text{good},\ s_2=\text{better},\ s_3=\text{best}\}$$

Table 2 Requirement of the investment company using HFLTSM

	E_1	E_2	E_3	E_4
Car 1	$\{s_2,s_0,s_{-1}\}$	$\{s_1,s_{-2}\}$	$\{s_3,s_2,s_1\}$	$\{s_2,s_1,s_0\}$
Car 2	$\{s_2,s_{-1}\}$	$\{s_1,s_0,s_{-3}\}$	$\{s_3,s_1\}$	$\{s_2\}$
Car 3	$\{s_3,s_2,s_{-2}\}$	$\{s_1,s_1,s_{-1}\}$	$\{s_{-1},s_{-3}\}$	$\{s_3,s_1\}$
Car 4	$\{s_2,s_{-2},s_{-1}\}$	$\{s_2,s_{-2}\}$	$\{s_1,s_{-2},s_{-3}\}$	$\{s_3,s_2,s_{-1}\}$

Table 3 Information of Maruti Suzuki $\{X_1\}$ using HFLTSM

	E_1	E_2	E_3	E_4
Celerio	$\{s_3,s_2,s_{-1}\}$	$\{s_0,s_{-1},s_{-2}\}$	$\{s_3,s_2,s_1\}$	$\{s_{-1},s_{-2}\}$
Swift	$\{s_2,s_{-1},s_{-2}\}$	$\{s_1,s_{-3}\}$	$\{s_1,s_{-1},s_{-2}\}$	$\{s_2\}$
DZire	$\{s_3,s_{-2}\}$	$\{s_2,s_{-1}\}$	$\{s_1,s_{-3}\}$	$\{s_3,s_{-1}\}$
Ciaz	$\{s_1,s_{-2}\}$	$\{s_2,s_{-3}\}$	$\{s_{-3},s_{-2}\}$	$\{s_3,s_2,s_{-1}\}$

Table 4 Information of Hyundai $\{X_2\}$ using HFLTSM

	E_1	E_2	E_3	E_4
Xcent	$\{s_3,s_0,s_{-1}\}$	$\{s_0,s_{-1},s_{-2}\}$	$\{s_2,s_{-1}\}$	$\{s_{-1},s_{-2}\}$
EON	$\{s_{-1},s_{-2}\}$	$\{s_1,s_{-2},s_{-3}\}$	$\{s_0,s_{-1},s_{-2}\}$	$\{s_2\}$
i10	$\{s_3,s_1,s_{-2}\}$	$\{s_2,s_{-1}\}$	$\{s_{-1},s_{-3}\}$	$\{s_{-1},s_{-3}\}$
Santa Fe	$\{s_2,s_1,s_{-2}\}$	$\{s_{-2},s_{-3}\}$	$\{s_2,s_{-3}\}$	$\{s_2,s_{-1}\}$

Table 5 Information of Toyota $\{X_3\}$ using HFLTSM

	E_1	E_2	E_3	E_4
Yaris	$\{s_3, s_2\}$	$\{s_0, s_{-1}, s_{-2}\}$	$\{s_3, s_2, s_1\}$	$\{s_{-1}, s_{-2}\}$
Corolla	$\{s_2, s_0, s_{-1}\}$	$\{s_0, s_{-1}, s_{-3}\}$	$\{s_1, s_{-1}, s_{-2}\}$	$\{s_2\}$
Camry	$\{s_3, s_{-2}\}$	$\{s_2, s_1, s_{-1}\}$	$\{s_{-1}, s_{-3}\}$	$\{s_3, s_2, s_{-1}\}$
Prius	$\{s_{-1}, s_{-2}\}$	$\{s_2, s_{-3}\}$	$\{s_{-2}, s_{-3}\}$	$\{s_2, s_{-1}, s_{-2}\}$

To exhibit the selection of the investment company, this study computes the

Table 6 Information of Honda $\{X_4\}$ using HFLTSM

	E_1	E_2	E_3	E_4
Brio	$\{s_2, s_{-1}\}$	$\{s_{-1}, s_{-2}\}$	$\{s_3, s_2, s_1\}$	$\{s_{-2}, s_{-1}\}$
Amaze	$\{s_0, s_{-1}, s_{-2}\}$	$\{s_2, s_1, s_{-3}\}$	$\{s_{-2}, s_{-1}\}$	$\{s_3, s_2\}$
City	$\{s_3, s_1, s_{-2}\}$	$\{s_3, s_2, s_{-1}\}$	$\{s_2, s_1, s_{-3}\}$	$\{s_3, s_0, s_{-3}\}$
Mobilio	$\{s_2, s_1, s_{-2}\}$	$\{s_2, s_{-3}\}$	$\{s_{-2}, s_{-3}\}$	$\{s_3, s_2, s_{-1}\}$

Table 7 Correlation coefficient of HFLTSMs

	Maruti Suzuki	Hyundai	Toyota	Honda
Investment company	0.633	0.52	**0.677**	0.638

correlation coefficient of each HFLTSM given in Tables 3, 4, 5 and 6 with the HFLTSM given in Table 2. The result is shown in Table 7. Since the correlation coefficient of HFLTSM (Table 5) for the car manufacturer Toyota $\{X_3\}$ with HFLTSM (Table 2) for the investment company is more, Toyota $\{X_3\}$ will be the optimal choice for the investment purpose.

5 Conclusion

Molodtsov [1] introduced soft set theory, which was later combined with many other mathematical models for dealing with uncertainty. Rodríguez et al. [3] introduced the concept of HFLTS which provides a new and more powerful technique to represent experts' qualitative judgments. In this paper, we have introduced the concept of HFLTSS. The HFLTSS is a combination of HFLTS and soft set. HFLTSM has been developed to represent HFLTSS. This study has investigated correlation coefficient measurements in the context of HFLTSS. Finally, real-life examples are given to validate the proposed ideas. In future, researchers might investigate the various properties of HFLTSS and apply them in uncertain decision-making problems, where group decision making is crucial due to lack of information, expertise domain of the experts, risk amendment, etc.

References

1. Molodtsov, D.: Soft set theory-first results. Comput. Math Appl. **37**(4–5), 19–31 (1999)
2. Zadeh, L.A.: The concept of a linguistic variable and its applications to approximate reasoning-Part I. Inf. Sci. **8**, 199–249 (1975)
3. Rodrıguez, R.M., Martínez, L., Herrera, F.: Hesitant fuzzy linguistic terms sets for decision making. IEEE Trans. Fuzzy Syst. **20**, 109–119 (2012)
4. Liao, H.C., Xu, Z.S., Zeng, X.J.: Distance and similarity measures for hesitant fuzzy linguistic term sets and their application in multi-criteria decision making. Inf. Sci. **271**, 125–142 (2014)
5. Wei, C.P., Zhao, N., Tang, X.J.: Operators and comparisons of hesitant fuzzy linguistic term sets. IEEE Trans. Fuzzy Syst. **22**(3), 575–585 (2014)
6. Chen, S.M., Hong, J.A.: Multicriteria linguistic decision making based on hesitant fuzzy linguistic term sets and the aggregation of fuzzy sets. Inf. Sci. **286**, 63–74 (2014)
7. Liu, H.B., Rodríguez, R.M.: A fuzzy envelope of hesitant fuzzy linguistic term set and its application to multicriteria decision making. Inf. Sci. **258**, 220–238 (2014)
8. Yu, C.H.: Correlation of fuzzy numbers. Fuzzy Sets Syst. **55**, 303–307 (1993)
9. Gerstenkorn, T., Manko, J.: Correlation of intuitionistic fuzzy sets. Fuzzy Sets Syst. **44**, 39–43 (1991)
10. Chen, N., Xu, Z.S., Xia, M.M.: Correlation coefficients of hesitant fuzzy sets and their application to clustering analysis. Appl. Math. Model. **37**, 2197–2211 (2013)
11. Liao, H., Xu, Z.S., Zeng, X.J., Merigó, J.M.: Qualitative decision making with correlation coefficients of hesitant fuzzy linguistic term sets. Knowl.-Based Syst. **76**, 127–138 (2015)
12. Liu, S.T., Gao, C.: Fuzzy measures for correlation coefficient of fuzzy numbers. Fuzzy Sets Syst. **128**, 267–275 (2002)
13. Mitchell, H.B.: A correlation coefficient for intuitionistic fuzzy sets. Int. J. Intell. Syst. **19**, 483–490 (2004)
14. Xu, Z.S., Xia, M.M.: On distance and correlation measures of hesitant fuzzy information. Int. J. Intell. Syst. **26**, 410–425 (2011)
15. Torra, V.: Hesitant fuzzy sets. Int. J. Intell. Syst. **25**, 529–539 (2010)
16. Xu, Z.S., Xia, M.M.: Hesitant fuzzy information aggregation in decision making. Int. J. Approximate Reasoning **52**(3), 395–407 (2011)
17. Das, S., Kar, S.: Group decision making in medical system: an intuitionistic fuzzy soft set approach. Appl. Soft Comput. **24**, 196–211 (2014)
18. Das, S., Kar, S., Pal, T.: Group Decision Making using Interval-valued Intuitionistic Fuzzy Soft Matrix and Confident Weight of Experts. J. Artif. Intell. Soft Comput. Res. **4**(1), 57–77 (2014)
19. Das, S., Kar, M.B., Pal, T., Kar, S.: Multiple attribute group decision making using interval-valued intuitionistic fuzzy soft matrix. In: Proceedings of IEEE International Conference on Fuzzy Systems (FUZZ-IEEE), pp. 2222–2229, Beijing (2014) doi:10.1109/FUZZ-IEEE.2014.6891687
20. Das, S., Kar, S.: The hesitant fuzzy soft set and its application in decision making, In: International Conference on Facets of Uncertainties and Applications, 2013, Springer Proceedings in Mathematics and Statistics, vol. 125, pp. 235–247 (2013)

Constrained Solid Travelling Salesman Problem Solving by Rough GA Under Bi-Fuzzy Coefficients

Samir Maity, Arindam Roy and Manoranjan Maiti

Abstract In this paper, a Rough Genetic Algorithm (RGA) is proposed to solve constrained solid travelling salesman problems (CSTSPs) in crisp and bi-fuzzy coefficients. In the proposed RGA, we developed a 'rough set based selection' (7-point scale) technique and 'comparison crossover' with new generation dependent mutation. A solid travelling salesman problem (STSP) is a tavelling salesman problem (TSP) in which, at each station, there are a number of conveyances available to travel to another station. The costs and risk/discomforts factors are in the form of crisp, bi-fuzzy in nature. In this paper, CSTSPs are illustrated numerically by some standard test data from TSPLIB using RGA. In each environment, some statistical significance studies due to different risk/discomfort factors and other system parameters are presented.

Keywords Rough selection · RGA · Generation dependent mutation · STSP · CSTSP

S. Maity (✉)
Department of Computer Science, Vidyasagar University, Medinipur 721102, West Bengal, India
e-mail: maitysamir13@gmail.com

A. Roy
Department of Computer Science, P.K. College, Contai, Purba Medinipur 721401, West Bengal, India
e-mail: royarindamroy@yahoo.com

M. Maiti
Department of Applied Mathematics, Vidyasagar University, Medinipur 721102, West Bengal, India
e-mail: mmaiti2005@yahoo.co.in

S. Das et al. (eds.), *Proceedings of the 4th International Conference on Frontiers in Intelligent Computing: Theory and Applications (FICTA) 2015*, Advances in Intelligent Systems and Computing 404, DOI 10.1007/978-81-322-2695-6_36

1 Introduction

In optimization, the TSP is one of the most intensively studied problems. A TSP is to find a possible tour along which a travelling salesman (TS) visits each city exactly once for a given list of cities and back to the starting city, so that total cost spent/distance covered is minimal. TSP is a well-known NP-hard combinatorial optimization problem [1]. Different types of TSPs have been solved by the researchers during last two decades. These are TSPs with time windows [2], stochastic TSP [3], double TSP [4], asymmetric TSP [5], TSP with precedence constraints [6].

In our day to day life, for travelling a set of conveyances may be available at each city. In that case, a salesperson has to design his/her tour for minimum cost maintaining the TSP conditions and using the suitable conveyances at different cities. This problem is called Solid Travelling Salesman Problem (STSP). Travelling cost from one city to another city depends on the types of conveyances, condition of roads, geographical areas, weather condition at the time of the travel, etc., so there always prevail some uncertainties/vagueness.

As optimization of bi-fuzzy objective is not well defined. We use fuzzy possibility and necessity based approaches [7, 8] to represent and to solve the bi-fuzzy CSTSP. There are many types of GA developed by the researchers such as Pareto GA [9], Localized GA (LGA) [10], Adaptive GA (AGA) [11], Enhance GA [12], Efficient GA [13] etc., which are used to get the optimal solutions in different research areas. Last and Eyal [14] and Roy et al. [15] improved the performance of GAs by providing a new, fuzzy based extension of the Life Time feature. They used a Fuzzy Logic Controller, to adapt the crossover probability as a function of the chromosomes 'age'. In GA, a set of potential solutions to the problem is formed by the random generations.

For CSTSPs solution, a RGA is proposed here with rough set based age dependent selection. We extended the imprecise selection technique with age as rough variable and it divided as 7-point scale as the linguistic variables Very Very Low (VVL), Very Low (VL), Low (L), Medium (M), High (H), Very High (VH) and Very Very High (VVH). For the first time from the trust measure of rough variable in 7-point scale also. A crossover technique, comparison crossover [7] and a virgin generation dependent random mutation are also implemented in the present algorithm. The proposed algorithm is tested with standard data set from TSPLIB [16] against the classical GA which is the combination of Roulette Wheel Selection (RWS), cyclic crossover and random mutation and hence the efficiency of the new algorithm is established.

This paper is organized as follows: In Sect. 2, we describe mathematical preliminaries. In Sect. 3, RGA is presented. Now in Sect. 4, we compute the complexity of the RGA. Again in Sect. 5 different kinds of CSTSP are given . We illustrate the above problems using some empirical data with discussion in Sect. 6. Finally, in Sect. 7, we conclude the paper with the scope of future development.

2 Mathematical Preliminaries

2.1 Bi-Fuzzy Variable

Definition 1 A fuzzy fuzzy (Fu-Fu) variable ξ is a fuzzy variable with fuzzy parameters.

Theorem 2.1 ([8]) *Assume that $\tilde{\tilde{c}}_{ij}$ is LR bi-fuzzy variable, for any $\theta \in \Theta$. Then the membership function of $\tilde{\tilde{c}}_{ij}(\theta)$ is*

$$\mu_{\tilde{\tilde{c}}_{ij}(\theta)}(t) = \begin{cases} L(\frac{c_{ij}(\theta)-t}{\alpha^c_{ij1}}) & \text{if } c_{ij}(\theta) \geq t, \alpha^c_{ij1} \geq 0, \\ R(\frac{t-c_{ij}(\theta)}{\beta^c_{ij1}}) & \text{if } c_{ij}(\theta) \leq t, \beta^c_{ij1} \geq 0 \end{cases} \quad (1)$$

where the fuzzy vector $(c_{ij}(\theta))_{n\times 1} = (c_{i1}(\theta), c_{i2}(\theta), c_{i3}(\theta), \ldots c_{in}(\theta))^T$ is also LR fuzzy variable with membership function as follows,

$$\mu_{c_{ij}}(t) = \begin{cases} L(\frac{c_{ij}-t}{\alpha^c_{ij2}}) & \text{if } c_{ij} \geq t, \alpha^c_{ij2} \geq 0, \\ R(\frac{t-c_{ij}}{\beta^c_{ij2}}) & \text{if } c_{ij} \leq t, \beta^c_{ij2} \geq 0 \end{cases} \quad (2)$$

and $\alpha_{ij1}, \alpha_{ij2}, \beta_{ij1}, \alpha_{ij2}$, are the left and right spread of $\tilde{\tilde{c}}_{ij}$, i = 1, 2 …, m, j = 1, 2 …, n, the reference function L, R:[0,1]→ [0,1] satisfies that L(1) = R(1) = 0, L(0) = R(0) = 1, and it is monotone function.Thus $Pos\{\omega|Pos\{\{\tilde{\tilde{c}}_{ij}(\theta)^T \geq f_i\} \geq \delta_i\} \geq \gamma_i$ is equivalent to $c^T_i x + R^{-1}(\delta_i)\beta^{cT}_{i1} x + R^{-1}(\gamma_{i1})\beta^{cT}_{i2} x \geq f_i, i = 1, 2, \ldots, m$ where $\delta_i, \gamma_i \in [0, 1]$ are predetermined confidence levels.

2.2 Rough Set

Rough Variable [18]: Let a rough variable ξ is a measurable function from the rough space $(\Lambda, \Delta, \kappa, \Pi)$ to the set of real numbers. i.e. for every Borel set of $\mathfrak{R}$, $\{\lambda \in \Lambda | \xi(\lambda) \in B\} \in \kappa$.

Trust Measure [18]: Let $(\Lambda, \Delta, \kappa, \Pi)$ be a rough space. The trust measure of event A is denoted by Tr$\{A\}$ and defined by Tr$\{A\} = \frac{1}{2}(\underline{Tr}\{A\} + \overline{Tr}\{A\})$. Upper and lower trust measures of event A are respectively defined by $\overline{Tr}\{A\}) = \frac{\Pi\{A\}}{\Pi\{\kappa\}}$, and $\underline{Tr}\{A\} = \frac{\Pi\{A \cap \Delta\}}{\Pi\{\Delta\}}$. When enough information about the measure Π is not available, it may be treated as the Lebesgue measure.

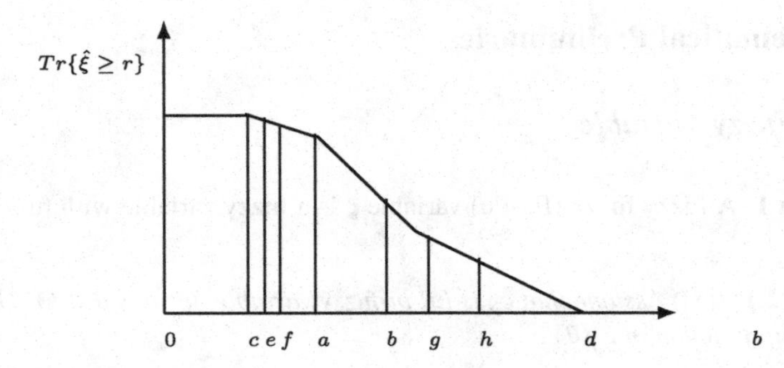

Fig. 1 $Tr\{\hat{\xi} \geq r\}$ function curve

Here first time the trust measure for 7-point scale of the rough event $\hat{\xi} \geq r$, $Tr\{\hat{\xi} \geq r\}$ and its function curve (Fig. 1) is presented, where r is a crisp number, $\hat{\xi}$ is a rough variable given by $\hat{\xi} = ([a, b], [c, d]), 0 \leq c \leq e \leq f \leq a \leq b \leq g \leq h \leq d$.

$$Tr\{\hat{\xi} \geq r\} = \begin{cases} 0 & for \quad d \leq r \\ \frac{(d-r)}{4(d-c)} & \text{for } h \leq r \leq d \\ \frac{(d-r)}{4(d-c)} + \frac{(h-r)}{4(h-e)} & \text{for } g \leq r \leq h \\ \frac{1}{4}(\frac{(d-r)}{(d-c)} + \frac{(h-r)}{(h-e)} + \frac{(g-r)}{(g-f)}) & \text{for } b \leq r \leq g \\ \frac{1}{4}(\frac{(d-r)}{(d-c)} + \frac{(h-r)}{(h-e)} + \frac{(g-r)}{(g-e)} + \frac{(b-r)}{(b-e)}) & \text{for } a \leq r \leq b \\ \frac{1}{4}(\frac{(d-r)}{(d-c)} + \frac{(h-r)}{(h-e)} + \frac{(g-r)}{(g-e)} + 1) & \text{for } f \leq r \leq a \\ \frac{1}{4}(\frac{(d-r)}{(d-c)} + \frac{(h-r)}{(h-e)} + 2) & \text{for } e \leq r \leq f \\ \frac{1}{4}(\frac{(d-r)}{(d-c)} + 3) & \text{for } c \leq r \leq e \\ 1 & \text{for } r \leq c. \end{cases} \quad (3)$$

3 Proposed RGA

We propose a new algorithm RGA using the rough set based selection (7-point), comparison crossover and generation dependent random mutation. The proposed RGA and its procedures are presented below:

3.1 Representation

Here a complete tour of N cities represents a solution. So an N dimensional integer vector $X_i = (x_{i1}, x_{i2}, \ldots, x_{iN})$ and $Y_i = (v_{i1}, v_{i2}, \ldots, v_{iP})$ are used to represent a solution, where $x_{i1}, x_{i2}, \ldots, x_{iN}$ represent N consecutive cities in a tour. Population size number of such solutions $X_i = (x_{i1}, x_{i2}, \ldots, x_{iN})$, and $Y_i = (v_{i1}, v_{i2}, \ldots, v_{iP})$, $i = 1, 2, \ldots, M$, are randomly generated by random number generator.

3.2 Rough Set Based Selection

The above M such solutions have fitnesses represented by $f(x_i)$ of the *ith* chromosomes. At the time of initialization, each chromosome age is defined as null. Now in every generation the age is counted as using the following mechanism :

If $avgfit \geq f(x_i)$, $age(x_i) = avg(age) + \frac{k*(avgfit - f(x_i))}{(avgfit - minfit)}$

If $avgfit \leq f(x_i)$, $age(x_i) = \frac{avg(age)}{2} + \frac{k*(f(x_i) - avgfit)}{(maxfit - avgfit)}$

where avgfit is the average fitness values, maxfit and minfit are maximum and minimum fitness values of the last generation and $k = \frac{maxfit + minfit}{2}$. Also avg(age) means the average age of the set of chromosomes. Here the maximum and minimum ages depend on the requirement of the problems.

Now since age calculated as crisp values, we construct the common rough values form it,

Rough Age = $([r_1 * \text{avg age}, r_2 * \text{avg age}], [r_3 * \text{avg age}, r_4 * \text{avg age}])$,

where $r_1 = \frac{Max\ Age - Avg\ Age}{Avg\ Age}$, $r_2 = \frac{Max\ Age + Min\ Age}{2}$, $r_3 = \frac{Max\ Age - Min\ Age}{2}$, $r_4 = \frac{Avg\ Age - Min\ Age}{Avg\ Age}$.

According to the extended age of the chromosome, it belongs to the any one of the common rough age corresponding p_c are created of each chromosome as VVL, VL, L, M, H, VH, VVH. The common rough age ([a,b],[c,d]) is extended to $0 \leq c \leq e \leq f \leq a \leq b \leq g \leq h \leq d$ and is described as below,

Table 1 Rough extended trust based linguistic

Gene	VVY	VY	Y	M	O	VO	VVO
VVY	VVL	VL	VL	L	VL	VL	VVL
VY	VL	VL	L	M	L	VL	VL
Y	VL	L	L	H	L	L	VL
M	L	M	H	VH	H	M	L
O	VL	L	L	H	L	L	VL
VO	VL	VL	L	M	L	VL	VL
VVO	VVL	VL	VL	VL	VL	VL	VVL

$$Age = \begin{cases} Very\ Very\ Young(VVY)\ for\ c \le age < e \\ Very\ Young(VY) \quad for\ e \le age < f \\ Young(Y) \quad for\ f \le age < a \\ Middle(M) \quad for\ a \le age \le b \\ Old(O) \quad for\ b < age \le g \\ Very\ Old(VO) \quad for\ g < age \le h \\ Very\ Very\ Old(VVO)\ for\ h < age \le d \end{cases} \tag{4}$$

3.3 Comparison Crossover

(a) Determination Probability of Crossover (p_c): Probability of crossover (p_c), for a pair of chromosomes (X_i,X_j) is determined as below:

A. p_cs for rough set based age selection
(i) At first age levels, (VVY, VY, M, O, VO, VVO) of X_i and X_j are determined by making trust measure of rough values w.r.t their ages in common rough age region given in Eqs. (3) and (4).
(ii) After determination of age intervals of the chromosomes their crossover probabilities are determined as linguistic variables (VVL, Vl, L, H, VH, VVH) as in Fig. 1 using rough trust measure which is presented in Table 1 and trust levels are given as Eq. (3).
(b) Crossover Mechanism: Here we used crossover method, we choose two individuals(parents) to produce new individuals(child's). To get optimal result of a TSP, we take a tour from one node(city)to next node(city) with minimum cost/value. we construct the crossover mechanism according Maity et al. [7].

3.4 Generation Dependent Random Mutation

(a) Selection for mutation: For each solution of P(t), generate a random number r from the range [0,1]. If $r < p_m$ then the solution is taken for mutation where p_m be the probability of mutation.
(b) Generation Oriented Mutation (Variable Method): Here we model a new form of mutation mechanism where probability of mutation (p_m) are determined as follows

$$p_m = \frac{k}{\sqrt{Current\ generation\ number}},\ k \in [0,1].$$

So, here proposed mutation mechanism follow the real world demand and p_m decreases smoothly as generation may increases. After calculating the p_m, then mutation operation done as conventional random mutation.

3.5 Algorithm of RGA

Input: max_ gen, pop_ size, Max_age, Min_ age, Problem Data (cost matrix, risk matrix).

Output: The optimum and near optimum solutions.

1. **Start**
2. g ← 0 // g: generation number
3. **Initialize P(g)**
4. **Evaluate f(P(g))**;
5. **while** (g ≤ *max_gen*)
6. Evaluate the average fitness
7. **if** average fitness > current fitness
8. age(x_i)=avg(age)+$\frac{k*(avgfit-f(X_i))}{(avgfit-minfit)}$
9. **else**
10. age(x_i)=$\frac{avg(age)}{2}$ + $\frac{k*(f(X_i)-avgfit)}{(maxfit-avgfit)}$
11. **if** (age(x_i)> maximum age)
12. age(x_i)= maximum age
13. **else if** (age(x_i)< minimum age)
14. age(x_i)= minimum age
15. Determine average age
16. Determine common rough age
17. Developed VVY, VY, M, O, VO, VVO
18. **for** each pair of parents **do**
19. Trust based p_c created
20. **end for**
21. **for** i=1 to Pop Size//**Comparison crossover**
22. Choose pair of chromosomes according to p_c
23. Randomly generate node between 1 to N (say a_r)
24. Replace a_r at first place of each parents
25. Determine value of each corresponding node
26. **for** j=1 to N
27. Compare minimum value
28. Check the node existence in child
29. Concatenated node to the child (offspring)
30. **end for**
31. Replace a_r at end place of each parents
32. Compare minimum value from end of the parents
33. Repeat step 26 to step 30
34. Replace the child's in offspring's set
35. **end for**
36. $p_m=\frac{k}{g}$, k∈[0,1]
37. **for** i=0 to *pop_size*
38. Select chromosome depending p_m

39. Randomly select two different nodes in [1,N]
40. Swap the places of the selected two nodes
41. **end for**
42. Store the new off springs into offspring set
43. **Reproduce a new P(g)**
44. **Evaluate f(P(g))**
45. Store the local optimum and near optimum solutions
46. g ← g+1
47. **endwhile**
48. Store the global optimum and near optimum results
49. **End Algorithm**.

3.6 Termination Criteria

RGA (Rough set based) algorithm is terminated any one of the following conditions is satisfied (which over is earlier):
(a) the best solution does not improve within 20 consecutive generations
(b) number of generations reaches user defined iterations (generations).

4 Complexity of the RGA

Time Complexity: The time complexities of selection operator, crossover operator and mutation operator in genetic algorithm are O(MN), $O(Mp_cN^2)$, $O(Mp_mN^2)$ respectively. While M is the initial size of the population. Normally $p_c > p_m$, so $O(Mp_cN^2) > O(Mp_mN^2) > O(MN)$. Here s_0 is the number of generations (outer iterations), so the time complexity of the outer loop is $O(s_0MN^2)$. The time complexity of initial population generation and fitness function calculation are $O(MN) < O(s_0MN^2)$. Now $O(MN) < O(s_0MN^2)$. So the time complexity of the GA is $O(s_0MN^2)$.
Space Complexity: Genetic algorithm needs to save the population, so it needs MN of the space. Normally M>N, so GA space complexity is O(MN).

5 Proposed CSTSP

5.1 Classical TSP with Risk/Discomfort Constraints (CTSP)

Let $c(i,j)$ be the cost for travelling from ith city to jth city and $r(i,j)$ be the risk/discomfort factor level in travelling from ith city to jth city. Then the problem can be mathematically formulated as:

$$\left.\begin{aligned} &\text{Minimize} \quad Z = \sum_{i \neq j} c(i,j) x_{ij} \\ &\text{subject to} \quad \sum_{i=1}^{N} x_{ij} = 1 \text{ for } j = 1, 2, \ldots, N \\ &\sum_{j=1}^{N} x_{ij} = 1 \text{ for } i = 1, 2, \ldots, N \\ &\sum_{i=1}^{N} \sum_{j=1}^{N} r(i,j) x_{ij} \leq r_{max} \\ &where\ x_i \neq x_j, i, j = 1, 2 \ldots, N. \end{aligned}\right\} \quad (5)$$

where x_{ij} is the decision variable and $x_{ij} = 1$ if the salesman travels from city-i to city-j, otherwise $x_{ij} = 0$ and r_{max} is the maximum risk/discomfort factor that should be maintained for the entire tour to avoid unwanted situation. Then the above CTSP reduces to

$$\left.\begin{aligned} &\text{determine a complete tour} \quad (x_1, x_2, \ldots, x_N, x_1) \\ &\text{to minimize} \quad Z = \sum_{i=1}^{N-1} c(x_i, x_{i+1}) + c(x_N, x_1) \\ &\text{subject to} \quad \sum_{i=1}^{N-1} r(x_i, x_{i+1}) + r(x_N, x_1) \leq r_{max} \\ &where\ x_i \neq x_j, i, j = 1, 2 \ldots, N. \end{aligned}\right\} \quad (6)$$

5.2 STSP with Risk/Discomfort Constraints (CSTSP)

Let $c(i,j,k)$ be the cost for travelling from ith city to jth city using kth type conveyance and $r(i,j,k)$ be the risk/discomfort factor in travelling from ith city to jth using kth type conveyance. Then the salesman has to determine a complete tour $(x_1, x_2, \ldots, x_N, x_1)$ and corresponding conveyance types $(v_1, v_2, \ldots, v_P)$ to be used for the tour, where $x_i \in \{1, 2, \ldots N\}$ for $i = 1, 2, \ldots, N$, $v_i \in \{1, 2, \ldots P\}$ for $i = 1, 2, \ldots, N$ and all x_i are distinct. Then the problem can be mathematically formulated as:

$$\left.\begin{aligned} &\text{to minimize} \quad Z = \sum_{i=1}^{N-1} c(x_i, x_{i+1}, v_i) + c(x_N, x_1, v_l), \\ &\text{subject to} \quad \sum_{i=1}^{N-1} r(x_i, x_{i+1}, v_i) + r(x_N, x_1, v_l) \leq r_{max}, \\ &where\ x_i \neq x_j, i, j = 1, 2 \ldots N, \quad v_i, v_l \in \{1, 2 \ldots, orP\} \end{aligned}\right\} \quad (7)$$

5.3 CSTSP in Bi-Fuzzy Environment (BFCSTSP)

In the above problem Eq. (7), if costs and risk/discomfort factors are bi-fuzzy variables, i.e., $\tilde{\tilde{c}}(i,j,k)$ and $\tilde{\tilde{r}}(i,j,k)$ respectively, risk/discomfort limit r_{max} is also bi-fuzzy number $\tilde{\tilde{r}}_{max}$, then the above problem reduces to according Theorem 2.1 and [8]. The objective function for Pos-Pos are equivalently written as below:

$$\left.\begin{array}{c} \text{minimize } f \\ Pos\{\theta|Pos\{|\tilde{\tilde{C}}(\theta)^T x \leq f\} \geq \delta\} \geq \gamma \\ Pos\{\theta|Pos\{|\tilde{\tilde{R}}(\theta)^T x \leq \tilde{\tilde{r}}_{max}(\theta)^T\} \geq \theta\} \geq \eta \end{array}\right\} \quad (8)$$

The objective function for Nes-Nes are equivalently written as below:

$$\left.\begin{array}{c} \text{minimize } f \\ Nes\{\theta|Nes\{|\tilde{\tilde{C}}(\theta)^T x \leq f\} \geq \delta\} \geq \gamma \\ Nes\{\theta|Nes\{|\tilde{\tilde{R}}(\theta)^T x \leq \tilde{\tilde{r}}_{max}(\theta)^T\} \geq \theta\} \geq \eta \end{array}\right\} \quad (9)$$

The Eq. (8) transformed as

$$\left.\begin{array}{c} \text{minimize } c^T - L^{-1}(\delta)\alpha_1^{cT} - L^{-1}(\gamma)\alpha_2^{cT} \\ \text{s.t. } r_{max} - R^T + R^{-1}(\theta)\beta_1^{r_{max}} + L^{-1}(\theta)\alpha_1^{r_{max}T} + L^{-1}(\eta)(\alpha_2^{RT} + \beta_2^{r_{max}} \geq 0 \end{array}\right\} \quad (10)$$

The objective function for Nes-Nes Eq. (9) are equivalently written as below:

$$\left.\begin{array}{c} \text{minimize } c^T + R^{-1}(1-\delta)\beta_1^{cT} + R^{-1}(1-\gamma)\beta_2^{cT} \\ \text{s.t. } r_{max} - R^T - L^{-1}(1-\eta)(\alpha_2^{r_{max}} + \beta_2^{RT}) \\ -L^{-1}(1-\theta)\alpha_1^{r_{max}} - R^{-1}(\theta)\beta_1^{RT} \geq 0 \end{array}\right\} \quad (11)$$

where $\alpha_1^c, \alpha_2^c, \alpha_1^R, \alpha_2^R, \alpha_1^{r_{max}}, \alpha_2^{r_{max}}, \beta_1^c, \beta_2^c, \beta_1^R, \beta_2^R, \beta_1^{r_{max}}, \beta_2^{r_{max}}$ are corresponding left and right spreads of the reference function of LR fuzzy numbers and $\theta, \eta, \delta, \gamma$ are predetermined confidence levels.

6 Numerical Experiments

6.1 Testing for RGA

Test the standard TSP problems from TSPLIB [16]. Table 2 gives the results of along RGA with the standard GA comparison in terms of total cost and iterations. Here standard GA (SGA) is the combinations of RW selection, cyclic crossover with well

Table 2 Results for standard TSP problem (TSPLIB)

Instances	Result	RGA Cost avg	RGA Iteration	RGA SD error (%)	SGA Cost avg	SGA Iteration	SGA SD error (%)
fri26	937	937	**43**	0.73	937	269	2.65
		938.74		0.87	940.71		3.46
bays29	2020	2020	**53**	1.48	2020	451	2.81
		2023.4		1.65	2027.79		3.21
bayg29	1610	1610	**62**	0.45	1610	378	3.57
		1611.52		0.98	1615.71		2.63
dantzig42	699	699	**140**	1.72	699	612	3.27
		700.35		0.07	704.75		2.87
eil51	426	426	**79**	0.68	426	341	2.01
		427.38		1.17	429.38		2.78
berlin52	7542	7542	**120**	1.62	7542	526	4.31
		7548.75		0.63	7562.29		2.57
st70	675	675	**154**	1.38	675	813	2.4
		676.25		1.01	679.45		4.25
eil76	538	538	**113**	0.97	538	457	2.47
		540.73		0.69	543.27		1.64
pr76	108159	108159	**151**	1.05	108159	410	2.13
		108180.34		0.74	108243.39		4.06
rat99	1211	1211	**135**	1.34	1211	328	3.63
		1213.76		0.57	1217.43		3.36
kroa100	21282	21282	**262**	1.78	21282	285	4.73
		21284.75		1.05	21289.9		3.65

known random mutation. Also the parameters such as p_c and p_m are taken as RGA. We have taken the result under 25 independent runs the best optimal solution with standard deviation (SD) and error are presents here.

6.2 CSTSP with Risks/Discomforts Constraint in Crisp values

Now for a CSTSP, where we consider three types of conveyances as Eq. (7). The cost matrix for the CSTSP and corresponding risk/discomfort matrix are represented below:

Table 3 Input data: crisp CSTSP

	Crisp cost matrix(5 *5) with three conveyances				
i/j	1	2	3	4	5
1	∞	(35,36,27)	(18,39,30)	(20,33,34)	(30,21,62)
2	(35,26,17)	∞	(40,21,32)	(18,29,10)	(35,26,37)
3	(38,30,29)	(17,58,34)	∞	(12,25,14)	(42,25,46)
4	(28,20,11)	(10,22,14)	(17,8,29)	∞	(30,19,24)
5	(17,15,9)	(42,23,34)	(35,36,37)	(20,31,43)	∞
	Crisp risks/discomforts matrix(5*5) with three conveyances				
i/j	1	2	3	4	5
1	∞	(.69,.68,.75)	(.84,.63,.7)	(.82,.7,.71)	(.72,.8,.42)
2	(.67,.76,.84)	∞	(.61,.8,.7)	(.83,.73,.92)	(.67,.76,.65)
3	(.63,.71,.73)	(.83,.44,.67)	∞	(.89,.76,.86)	(.59,.76,.75)
4	(.73,.81,.9)	(.9,.78,.86)	(.84,.93,.72)	∞	(.71,.82,.77)
5	(.84,.86,.92)	(.59,.78,.67)	(.66,.65,.64)	(.82,.71,.59)	∞

6.3 CSTSP in Bi-Fuzzy Environments (BFCSTSP)

Here we took the cost and risk/discomfort factors as bi-fuzzy for the CSTSP as Eqs. (10) and (11). Also we consider three types of conveyances. Here we use triangular LR Fu-Fu variables where (ξ, α, β) is LR fuzzy variable with known left and right spreads. Also ξ is a triangular fuzzy variable connect with the corresponding components in Table 3.

Here predetermined confidence levels $\delta = \eta = 0.9$, $\theta = \eta = 0.95$ and reference function L(x) $=$ R(x) $= 1 -$ x are taken. Left and right spreads of the LR fuzzy numbers are given in the Table 5. The result are shown in Table 6 as decision maker (DM) choose for both case optimistic DM (ODM) and pessimistic DM (PDM) are consider (Table 4).

Table 4 Results of CSTSP in crisp environment

Algorithm	Path(vehicle)	Cost	Risk	R_{max}
RGA	3(1)-4(1)-2(2)-5(3)-1(1)	70	4.50	4.75
	2(3)-1(1)-3(1)-4(2)-5(2)	89	4.52	
	1(1)-3(1)-2(2)-5(1)-4(3)	92	4.3	
	1(2)-5(1)-4(1)-2(1)-3(3)	93	4.54	
	4(1)-5(2)-1(2)-2(2)-3(2)	127	4.16	
SGA	2(3)-3(3)-1(3)-5(1)-4(2)	142	4.7	4.75
RGA	4(2)-5(3)-2(1)-1(1)-3(2)	138	4.25	4.25
	1(2)-3(2)-2(2)-5(1)-4(3)	154	3.68	4.00
	4(3)-5(1)-2(3)-3(3)-1(2)	160	3.71	

Table 5 Input data: BFCSTSP

i/j	1	2	3	4	5
1		$(\xi, 2, 2)$	$(\xi, 3, 3)$	$(\xi, 4, 4)$	$(\xi, 5, 5)$
	∞	$(\xi, 4, 4)$	$(\xi, 5, 5)$	$(\xi, 6, 6)$	$(\xi, 3, 3)$
		$(\xi, 1, 1)$	$(\xi, 2, 2)$	$(\xi, 7, 7)$	$(\xi, 4, 4)$
2	$(\xi, 3, 3)$		$(\xi, 5, 5)$	$(\xi, 1, 1)$	$(\xi, 2, 2)$
	$(\xi, 4, 4)$	∞	$(\xi, 7, 7)$	$(\xi, 3, 3)$	$(\xi, 4, 4)$
	$(\xi, , 2)$		$(\xi, 6, 6)$	$(\xi, 8, 8)$	$(\xi, 3, 3)$
3	$(\xi, 6, 6)$	$(\xi, 1, 1)$		$(\xi, 5, 5)$	$(\xi, 1, 1)$
	$(\xi, 8, 8)$	$(\xi, 4, 4)$	∞	$(\xi, 4, 4)$	$(\xi, 2, 2)$
	$(\xi, 7, 7)$	$(\xi, 3, 3)$		$(\xi, 6, 6)$	$(\xi, 9, 9)$
4	$(\xi, 6, 6)$	$(\xi, 3, 3)$	$(\xi, 5, 5)$		$(\xi, 6, 6)$
	$(\xi, 4, 4)$	$(\xi, 7, 7)$	$(\xi, 3, 3)$	∞	$(\xi, 4, 4)$
	$(\xi, 5, 5)$	$(\xi, 3, 3)$	$(\xi, 6, 6)$		$(\xi, 7, 7)$
	$(\xi, 4, 4)$	$(\xi, 2, 2)$	$(\xi, 8, 8)$	$(\xi, 9, 9)$	
5	$(\xi, 3, 3)$	$(\xi, 7, 7)$	$(\xi, 7, 7)$	$(\xi, 5, 5)$	∞
	$(\xi, 6, 6)$	$(\xi, 6, 6)$	$(\xi, 6, 6)$	$(\xi, 6, 6)$	
	Bi-fuzzy risks/discomforts matrix(5 *5) with three conveyances				
i/j	1	2	3	4	5
1		$(\xi, .12, .12)$	$(\xi, .13, .13)$	$(\xi, .14, .14)$	$(\xi, .15, .15)$
	∞	$(\xi, .02, .02)$	$(\xi, .03, .03)$	$(\xi, .04, .04)$	$(\xi, .05, .05)$
		$(\xi, .07, .07)$	$(\xi, .04, .04)$	$(\xi, .06, .06)$	$(\xi, .08, .08)$
2	$(\xi, .16, .16)$		$(\xi, .17, .17)$	$(\xi, .01, .01)$	$(\xi, .11, .11)$
	$(\xi, .24, .24)$	∞	$(\xi, .16, .16)$	$(\xi, .17, .17)$	$(\xi, .21, .21)$
	$(\xi, .14, .14)$		$(\xi, .06, .06)$	$(\xi, .1, .1)$	$(\xi, .2, .2)$
3	$(\xi, .06, .06)$	$(\xi, .18, .18)$		$(\xi, .03, .03)$	$(\xi, .04, .04)$
	$(\xi, .13, .13)$	$(\xi, .11, .11)$	∞	$(\xi, .16 ,.16)$	$(\xi, .05, .05)$
	$(\xi, .16, .16)$	$(\xi, .22, .22)$		$(\xi, .25, .25)$	$(\xi, .01, .01)$
4	$(\xi, .07, .07)$	$(\xi, .13, .13)$	$(\xi, .15, .15)$		$(\xi, .26, .26)$
	$(\xi, .04, .04)$	$(\xi, .07, .07)$	$(\xi, .13, .13)$	∞	$(\xi, .14, .14)$
	$(\xi, .05, .05)$	$(\xi, .06, .06)$	$(\xi, .14, .14)$		$(\xi, .2, .)$
	$(\xi, .11, .11)$	$(\xi, .2, .2)$	$(\xi, .19, .19)$	$(\xi, .18, .18)$	
5	$(\xi, .03, .03)$	$(\xi, .1, .1)$	$(\xi, .13, .13)$	$(\xi, .12, .12)$	∞
	$(\xi, .05, .05)$	$(\xi, .06, .06)$	$(\xi, .17, .17)$	$(\xi, .16, .16)$	

6.4 CSTSP for Virtual Data

Here CSTSP are solved by RGA with large scale data which are randomly generated for different cities and the results are presented below in Table 7.

Table 6 Optimum results of BFCSTSP

DM	Path(vehicle)	Obj value	Risk	R_{max}
ODM	3(2)-1(1)-4(2)-5(3)-2(3)	63.5	4.37	4.5
PDM	5(1)-1(2)-4(2)-3(3)-2(3)	68.9	4.48	4.5
ODM	3(1)-4(3)-2(1)-5(3)-1(1)	72.5	4.2	4.5
PDM	4(3)-1(2)-5(1)-3(2)-2(1)	79.4	4.32	4.5
ODM	1(1)-4(1)-2(2)-5(3)-3(3)	81.5	4.03	4.5
PDM	1(3)-5(3)-3(3)-2(2)-4(3)	96.5	4.23	4.5
ODM	2(3)-1(1)-5(3)-3(3)-4(2)	93.5	3.78	4.25
PDM	5(3)-2(1)-4(2)-3(3)-1(2)	118.2	3.81	4.25
ODM	4(3)-1(2)-5(3)-3(2)-2(2)	102.2	3.6	4
PDM	2(1)-4(3)-1(2)-5(3)-3(3)	129.5	3.91	4

Table 7 Results of CSTSP for different cities

Instances (cities)	Costs	R_{max}
15×15	142	5.5
20×20	196	6.5
25×25	244	7.5
30×30	273	9.5
35×35	398	11.25
40×40	446	13.0
45×45	518	15.5
50×50	692	18.0
80×50	1145	23.7
100×50	1468	32.5
150×50	2354	41.4
200×50	3623	73.9

6.5 Statistical Test for RGA

To compare the efficiency of the developed algorithm, another two established heuristic technique Fuzzy age based GA (FGA developed by Last et al. [14] and used by Roy et al. [15]) and SGA are used. Here 100 independent run for individual algorithm are considered.

For calculation of different steps of ANOVA easily, we subtract 50 (with out lose of generality) from each numbers and the Table 8 is reduced to Table 9 where X1, X2 and X3 represents RGA, FGA and SGA respectively.

Here, total sample size of each algorithm is equal and say, I = 11 (problems) and number of algorithm is, J = 3. Mean of the sample means, $\bar{\bar{X}} = 16.82$.

Table 8 Number of win for different algorithms

Problem	1	2	3	4	5	6	7	8	9	10	11
RGA	84	91	78	90	71	79	87	75	97	77	81
FGA	67	76	63	82	57	68	63	59	71	69	64
SGA	59	43	56	41	51	57	49	37	56	52	55

Table 9 Subtracted table from Table 8

Problem	1	2	3	4	5	6	7	8	9	10	11	Mean
X_1	34	41	28	40	21	29	37	25	47	27	31	$\overline{X}_1 = 32.74$
X_2	17	26	13	32	7	18	13	9	21	19	14	$\overline{X}_2 = 17.18$
X_3	9	−7	6	−9	1	7	−1	−13	6	2	5	$\overline{X}_3 = 0.55$

Table 10 ANOVA summary table (data taken from Table 9)

Source of variation	Sum of square	df	Mean of square	F
Between groups	$SS_B = 5701.19$	$J - 1 = 2$	$MS_B = \frac{SS_B}{J-1} = 2850.6$	$\frac{MS_B}{MS_W} = 31.28$
Within groups	$SS_W = 1640$	$J(I - 1) = 20$	$MS_W = \frac{SS_W}{J(I-1)} = 91.11$	
Total	$SS_T = 7341.19$	$IJ - 1 = 32$		

Here, critical F values, $F_{0.05(2,20)} \approx 3.57$. As the computed F (cf. Table 10) is higher than the standard critical F values (= 3.57) for 0.05 level of significance, it may be inferred that there is a significant differences between the groups. Scheffe's multiple comparison F- test is done for this purpose to find out whether RGA & SGA and/or RGA & FGA are significant. For the first pair i.e., for RGA & SGA, calculated F value is given by $F = \frac{(\overline{X}_1 - \overline{X}_3)^2}{MS_W(\frac{1}{I} + \frac{1}{J})} = 26.80$. Similarly, for the second pair i.e., for RGA and FGA, calculated F = 6.26. As both calculated F values are greater than the standard value (3.57), there is significant difference between RGA & SGA and also RGA & FGA. It can be concluded that RGA is better compared with the other two algorithms

7 Conclusion

In this paper, a new algorithm for GA, called RGA is proposed and illustrated in CSTSP formulated in different environments. In RGA, a new rough 7-point age based selection and comparison crossover are used along with virgin generation dependent random mutation. CSTSPs are recently introduced in the area of TSPs and regarded as highly NP-hard combinatorial optimization problems. Such CSTSPs are here formulated in crisp and bi-fuzzy costs and risk/discomfort levels and solved by the proposed RGA. Here, development of RGA is in general form and it can be applied in other discrete problems such as network optimization, graph theory, solid transportation problems, vehicle routing, VLSI chip design, etc. In spite of the better results by RGA, there is a lot of scope for development in RGA, specially for the CSTSPs. In the three dimensional TSPs with conveyances, we have assigned a conveyance arbitrarily during each crossover and mutation for the optimum selection of the routes. This is a limitation of the present CSTSPs.

References

1. Lawler, E.L., Lenstra, J.K., Rinnooy Kan, A.H.G.: The TSP: G.E. Re Guided Tour of Combinatorial Optimization. Wiley, New York (1985)
2. Focacci, F., Lodi, A., Milano, M.: A hybrid exact algorithm for the TSPTW. Inf. J. Comput. **14**(4), 403–417 (2002)
3. Chang, T., Wan, Y., Tooi, W.: A stochastic dynamic travelling salesman problem with hard time windows. Eur. J. Oper. Res. **198**(3), 748–759 (2009)
4. Petersen, H.L., Madsen, O.B.G.: The double travelling salesman problem with multiple stack—formulation heuristic solution approaches. Eur. J. Oper. Res. **198**, 339–347 (2009)
5. Majumder, S.K., Bhunia, A.K.: Genetic algorithm for asymmetric traveling salesman problem with imprecise travel times. J. Comput. Appl. Math. **235**(9), 3063–3078 (2011)
6. Moon, C., Ki, J., Choi, G., Seo, Y.: An efficient genetic algorithm for the TSP with precedence constraints. EJOR **140**, 606–617 (2002)
7. Maity, S., Roy, A., Maiti, M.: A modified genetic algorithm for solving uncertain CSTSP. Comput. Ind. Eng. **83**, 273–296 (2015)
8. Xu, J., Zhou, X.: Fuzzy-Like Multiple Objective Decision Making, Studies in Fuzziness and Soft Computing. Springer, New York (2011)
9. Ni, H., Wang, Y.: Stack index tracking by pareto efficient GA. Appl. Soft Comput. **13**(12), 4519–4535 (2013)
10. Ursani, Z., Essam, D., Cornforth, D., Stocker, R.: Localized genetic algorithm for vehicle routing problem with time windows. Appl. Soft Comput. **11**(8), 5375–5390 (2011)
11. Neungmatcha, W., Sethanan, K., Gen, M., Theerakulpisut, S.: Adaptive genetic algorithm for solving sugarcane loading stations with multi-facility services problem. Comput. Electron. Agric. **98**(10), 85–99 (2013)
12. Huang, K.Y.: An enhance classification method comparing a GA, rough set theory and modified PBMF-index function. Appl. Soft Comput. **12**(10), 46–63 (2012)
13. Enigin, O., Ceran, G., Yilmaz, M.K.: An efficient GA for hybrid flow shop scheduling with multiprocessor task problem. Appl. Soft Comput. **11**(3), 3056–3065 (2011)
14. Last, M., Eyal, S.: A fuzzy-based lifetime extension of genetic algorithms. Fuzzy Sets Syst. **149**, 131–147 (2005)
15. Roy, A., Kar, S., Maiti, M.: A production inventory model with stock dependent..: a fuzzy genetic algorithm with varying population size approach. Comput. Ind. Eng. **57**, 1324–1335 (2009)
16. TSPLIB. http://comopt.ifi.uniidelberg.de/software/TSPLIB95/

Part X
Routing and Traffic Grooming

A Novel Multi-Criteria Multi-Agent-Based Routing Strategy Based on Tarantula Mating Behavior

Susmita Bandyopadhyay and Arindam Kumar Chanda

Abstract This paper proposes a routing strategy based on the mating behavior of a species of spider, Tarantula, in which the female Tarantula sometimes eats the male Tarantula just after the mating for food or genetic need. This behavior has been used in a multi-criteria multi-agent-based routing strategy. A hierarchical structure of agents has been considered where the worker agents at the leaf level calculate shortest paths, congestion in a path, number of intermediate nodes, and identify deadlock condition in the network. A master agent at the top of the hierarchy controls them. Fuzzy orientation has been given to calculate fuzzy edge lengths of network instance while finding shortest path and in fuzzy weight calculation in PROMETHEE multi-criteria outranking method. A network instance has been used in order to implement the strategy as proposed in this research study.

Keywords Tarantula mating behavior • Routing strategy • Multi-criteria decision analysis • Fuzzy • Hierarchical agent system

1 Introduction

In this section, a brief overview of the research studies in the area of multi-criteria optimization approaches and multi-agent systems are presented. An agent is a computational system which is long-lived, has goals, self-contained, autonomous, and capable of independent decision making. Among the benchmark multi-agent technologies, the significant ones are GAIA [1], ROADMAP [2], PROMETHEUS [3], PROSA [4], and ADACOR [5]. The strategy as proposed in this paper has also used multi-criteria decision analysis technique [6]. Some of the frequently applied

S. Bandyopadhyay (✉)
University of Burdwan, Burdwan, West Bengal
e-mail: bandyopadhyaysusmita2010@gmail.com

A.K. Chanda
G. B. Pant Govt. Engineering College, New Delhi, India

S. Das et al. (eds.), *Proceedings of the 4th International Conference on Frontiers in Intelligent Computing: Theory and Applications (FICTA) 2015*, Advances in Intelligent Systems and Computing 404, DOI 10.1007/978-81-322-2695-6_37

MCDA techniques include AHP (Saaty [7]), ELECTRE I, II, III, IV [8–11], PROMETHEE [12].

Among the agent-based technologies, as applied in manufacturing, the holonic approaches have been well-practiced. The other architectures include PROSA [13], ADACOR [14], and HCBA [15]. Authors combining agent-based approach with various other approaches include the work of Sinha et al. [16] (Particle Swarm Optimization), Shi and Qian [17] (Artificial Immune Algorithm), Hsieh [18] (Petri Net), Leung et al. [19] (Ant Colony Algorithm), Lee et al. [20] (Simulated Annealing), Shao et al. [21] (Genetic Algorithm), Lo´pez-Ortega and Villar-Medina [22] (Artificial Neural Network).

The agent-based applications are observed in various areas such as, flow shop scheduling [23], shop floor scheduling for ship building yards [24], process planning and scheduling [25], freight transport [26], travelling salesman problem, [27] and so on. Among the research studies on routing in manufacturing applying agent-based and multi-criteria approaches, the significant ones include the work of Maione and Naso [28] (dealt with task controlling mechanism), Mikler et al. [29] (applied in traffic management), Eltarras and Eltoweissy [30] (dealt with dynamic addressing problem).

2 The Proposed Strategy

This paper has used a hierarchical structure of agents (Fig. 1) and PROMETHEE multi-criteria outranking method with a fuzzy orientation. The leaf level of the hierarchy contains worker agents. Each of the worker agents performs a particular task. The worker agents considered in this research study are (1) shortest path agent, (2) congestion agent, (3) deadlock agent, and (4) hops agent. The master agent takes the final decision from top of the hierarchy. After performing the task, each of the worker agents is killed by the master agent after taking the result of the performed task from the worker agent. Thus, the hierarchical structure does not exist after all the tasks are performed by all the worker agents. The final decision is taken by the master agent based on PROMETHEE multi-criteria decision analysis technique

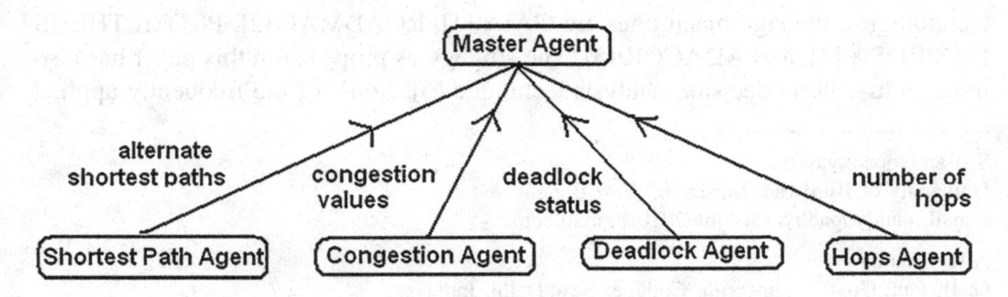

Fig. 1 Hierarchy of agents

based on the information provided by the worker agents. The master agent gets notification after killing each of the worker agents.

In dynamic environment, by the time when a particular job reaches a particular node, there is a chance that the previously determined optimum path may not be optimum anymore because of congestion, deadlock, and so on. Thus, instead of finding the entire point-to-point optimum path, it may be better to choose the next best node to route a job toward the destination. The analogy of the interesting mating behavior of Tarantula with the idea in this research study can be described in Fig. 2.

The routing strategy considered in this research study finds the next optimum neighboring node through agent-based technique, instead of finding the entire source-to-destination path. Thus, the worker agents and then the master agent are active and will start functioning whenever there will be a need to route a job to the next optimum neighboring node and whenever a new job enters the system. The master agent invokes and initiates the actions of the worker agents, just like the female spider chooses a male spider for mating. The worker agents, after performing their tasks, return the results to the master agent, just like the male spider transfers the genetic material to the female spider during mating. The master agent kills the worker agents after receiving the results from the worker agents, just like the female spider kills and eats the male spider after mating. The master agent gets the notification of the killing of the worker agents, just like the female spider takes the male spider as food. The various functions as performed by various worker agents and the master are described in the following subsections.

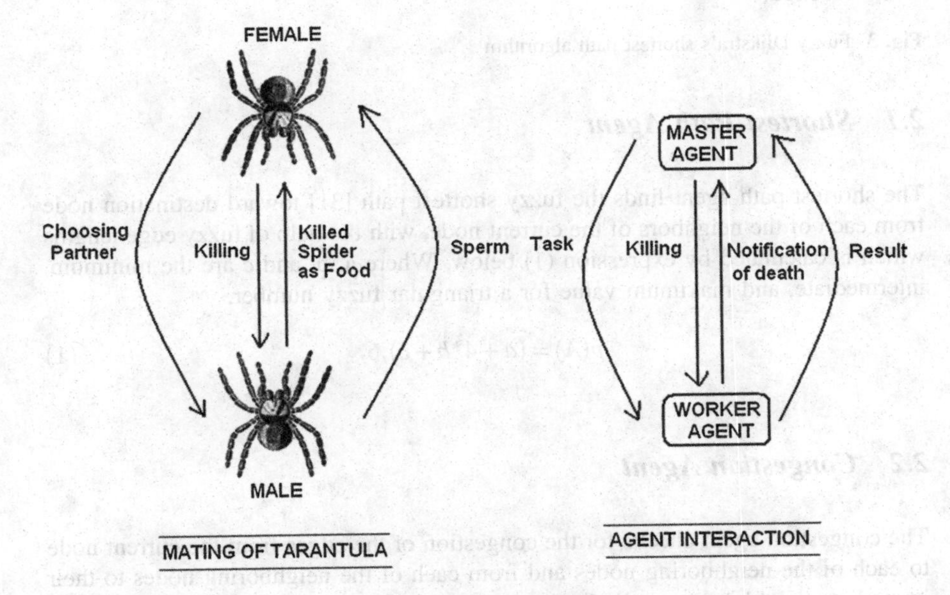

Fig. 2 Analogy of mating behavior of tarantula with idea in this research study

```
Input: Set of edges (E), Set of vertex (V), d[], source node (s), destination node (d)
Output: Shortest Path from s to d of length contained in v[d]

1. Initialize:
   For each edge e ∈ E
      Set Fuzzy Number W_e = (a_e + 4*b_e + c_e) / 6
   End For
   Set d[s] ← 0 and d[i] ← ∞, ∀ i, i ≠ s
   Set permanent node p ← s
2. For each neighbor j ∈ neighbor(p)
      Set v[j] ← minimum {d[j], d[p]+c_p,j}
      Set pred[j] ← p
   End For
3. Find minimum among non-permanent node:
   Set min ← ∞
   For each vertex v ∈ V
      If v is not permanent node Then
         If d[v] < min Then
            Set min ← d[v]
            Set j' ← v
         End If
      End If
   End For
4. Set next permanent node p ← j'
5. Repeat steps 2 – 4 until all the nodes become permanent or the destination is
   reached.
```

Fig. 3 Fuzzy Dijkstra's shortest path algorithm

2.1 *Shortest Path Agent*

The shortest path agent finds the fuzzy shortest path [31] toward destination node from each of the neighbors of the current node, with the help of fuzzy edge lengths which is calculated by expression (1) below, Where a, b, and c are the minimum, intermediate, and maximum value for a triangular fuzzy number.

$$p(A) = (a + 4*b + c)/6 \tag{1}$$

2.2 *Congestion Agent*

The congestion agent checks for the congestion of the edges from the current node to each of the neighboring nodes and from each of the neighboring nodes to their immediate neighbors on their shortest paths toward destination (Fig. 4). Although

Fig. 4 Algorithm for finding congestion

```
For each neighbor i
        Record number of jobs on arc (src, i)
End For
For each neighbor i
        Find the nextnode n of i on SP(i)
        Record number of jobs on arc (i, n)
End For
```

congestion can be represented by more than one factor, but in this research study, congestion is represented by the number of jobs traveling on a particular edge.

2.3 *Deadlock Agent*

The deadlock agent (Fig. 5) checks whether the neighbors of the current node faces any immediate cyclic path. Let the current node be c and the neighbors of c be x, y, and z. If the immediate neighbors of two or more neighbors of x, y, z are same, then the algorithm marks those neighbors (x and/or y and/or z) as unsafe, otherwise they are safe (Fig. 3).

2.4 *Hops Agent*

The hops agent finds the intermediate nodes on a path toward destination (Fig. 6). Thus, for each of the shortest path from the neighboring nodes toward destination, there is a particular number representing the number of intermediate nodes on the way. The target is to choose that particular neighbor as the better node which will have least number of hops since the greater the number of hops, greater is the

Fig. 5 Algorithm for deadlock

```
For each neighbor n1
    For each neighbor n2
        If n1 != n2 Then
            If neighbor[n1] != neighbor[n2] Then
                Safe[n1] = 1
                Safe[n2] = 1
            End If
        End If
End For
```

```
For each neighbor i
    Find shortest path SP(i) to destination d from i
    Set count[i] ← 0
    For each node from n on SP(i)
        Set count[i] ← count[i] + 1
    End
End For
```

Fig. 6 Algorithm for number of hops

chance of facing more congestion, more deadlock, and more blockage at the nodes due to loaded buffers.

2.5 *Master Agent*

The master agent takes the final decision to choose the optimum neighboring node to route a job next by PROMETHEE [32] multi-criteria decision analysis technique based on the information provided by the worker agents (Fig. 7)—shortest path data from shortest path agent, congestion data from congestion agent, deadlock data from deadlock agent, number of hops from hops agent, and between the current node and the destination node on each of the alternate paths through the neighbors.

```
1. Input a network with fuzzy edge values
2. Input source and destination nodes s and d respectively
3. Find neighbor N of s
4. For each neighbor n ∈ N
       Find shortest path p from n to d
       Store p in PATH
   End For
5. For each neighbor n ∈ N
       Find path p ∈ PATH FROM N
       Find number of jobs on edge (s, n)
       Find number of jobs on edge (n, i), i: neighbor of n on p
       Check whether there is any deadlock cycle with n
       Calculate number of hops from n on p
   End For
6. Apply PROMETHEE to find the best neighbor n
```

Fig. 7 Algorithm for master agent

3 Results and Discussion

The multi-agent multi-criteria approach, as proposed in this research study, has been applied on a network example (Fig. 8) which has been taken based on the book by Herrmann [33]. The relevant details are shown in Fig. 9. Next, a total of four decision makers are assumed and they all assign their own preference to the five criteria. The five criteria are: (1) Path length, (2) Number of jobs traveling on the edge from current node (node 3) to immediate neighbors (nodes 2, 5). The respective edges in this example are: 3-2 and 3-5, (3) Number of jobs traveling on the edge between the immediate neighbors (nodes 1, 3) and their neighbors on their respective shortest paths, (4) Deadlock status of each of the neighbors, and (5) Number of intermediate nodes or hops.

Figure 10 shows the preferences as provided by the four decision makers (DM), the fuzzy weights as calculated for the above five criteria, and the values of the

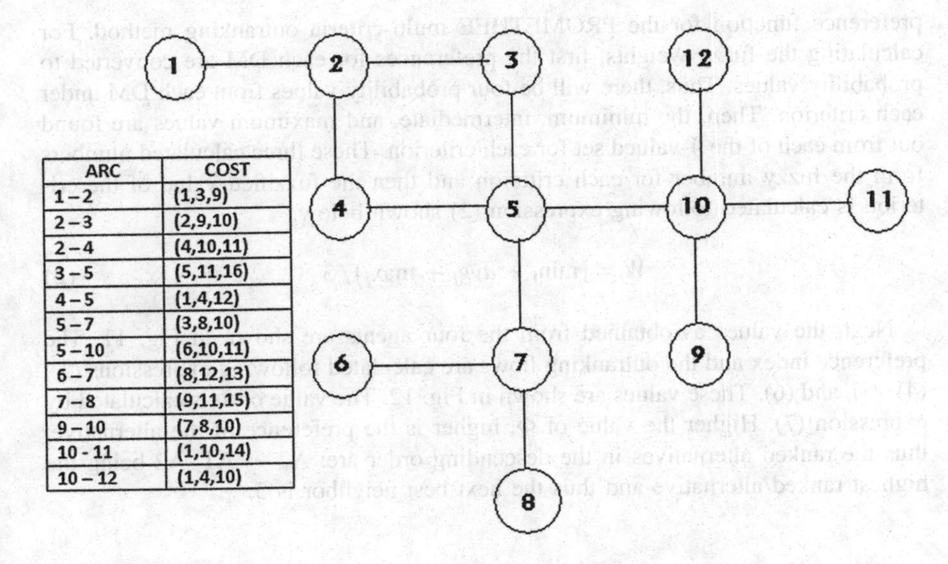

ARC	COST
1 – 2	(1,3,9)
2 – 3	(2,9,10)
2 – 4	(4,10,11)
3 – 5	(5,11,16)
4 – 5	(1,4,12)
5 – 7	(3,8,10)
5 – 10	(6,10,11)
6 – 7	(8,12,13)
7 – 8	(9,11,15)
9 – 10	(7,8,10)
10 - 11	(1,10,14)
10 – 12	(1,4,10)

Fig. 8 Network diagram

Source node 3; Destination node: 12
Number of neighbors of source node 3: 2 => 2, 5

Alternate paths through neighbors

Alternates	Paths
A1	3 – 2 – 4 – 5 – 10 – 12
A2	3 – 5 – 10 – 12

Congestion Weights

6	1	5	4	3
7	2	2		

Fig. 9 Relevant details

Preference for 5 Criteria from Four Decision Makers (DM)

DM	C1	C2	C3	C4	C5
1	2	5	1	4	3
2	1	3	2	4	5
3	4	2	3	1	3
4	5	2	1	4	3

Weights for 5 Criteria

Sl. No.	Fuzzified Weights
1	0.222222
2	0.222222
3	0.133333
4	0.111111
5	0.177778

Values of Preference Function

ALTERNATIVES					
A1, A2	-11	-1	-2	0	-2
A2, A1	11	1	2	0	2

Fig. 10 Weights, preferences

preference function for the PROMETHEE multi-criteria outranking method. For calculating the fuzzy weights, first the preferences for each DM are converted to probability values. Thus, there will be four probability values from each DM under each criterion. Then, the minimum, intermediate, and maximum values are found out from each of the 4-valued set for each criterion. These three calculated numbers form the fuzzy number for each criterion and then the fuzzified value of the criterion is calculated following expression (2) shown below.

$$W_j = \left(\min_j + avg_j + \max_j\right)/3 \tag{2}$$

Next, the values as obtained from the four agents are shown in Fig. 11. The preference index and the outranking flows are calculated following expressions (3), (4), (5), and (6). These values are shown in Fig. 12. The value of Φ is calculated by expression (7). Higher the value of Φ, higher is the preference of the alternative, thus the ranked alternatives in the descending order are: A2 → A1, A2 being the highest ranked alternative and thus the next best neighbor is 5.

Path length	Number of jobs from source node to neighbors	Number of jobs from neighbors to neighbors of neighbors	Deadlock status	Number of hops
36	3	4	0	5
25	2	2	0	3

Fig. 11 Data values from agents

Fig. 12 Preference index and outranking flows

Preference Index

ALTERNATIVES	
A1, A2	-3.28889
A2, A1	3.28889

Outranking Flows

Alternatives	Φ^+	Φ^-	Φ
A1	-3.28889	3.28889	-6.577778
A2	3.28889	-3.28889	6.577778

$$\pi(a,b) = \sum_{j=1}^{C} W_j P_j(a,b) \tag{3}$$

$$\pi(b,a) = \sum_{j=1}^{C} W_j P_j(b,a) \tag{4}$$

$$\phi^+(a) = \frac{1}{(n-1)} \sum_{x \in A} \pi(a,x) \tag{5}$$

$$\phi^-(a) = \frac{1}{(n-1)} \sum_{x \in A} \pi(x,a) \tag{6}$$

$$\phi(a) = \phi^+(a) - \phi^-(a) \tag{7}$$

4 Conclusion

This paper focuses on routing problem in manufacturing networks. Instead of establishing a point-to-point connection between source and destination nodes, this paper proposes to route a job or message to the next optimal neighboring node. For this purpose, a hierarchical multi-agent-based system with multi-criteria decision technique PROMETHEE is considered with a master agent and several worker agents. The master agent takes all these inputs from the worker agents and selects the best immediate neighbor using a multi-criteria outranking method known as PROMETHEE. The entire idea is based on the mating behavior of a species of spider known as Tarantula. Specific example has been considered to implement the proposed strategy. However, bigger network and more number of criteria can be added in future in order to make the strategy more applicable and realistic.

References

1. Wooldridge, M., Jennings, N.R., Kinny, D.: The Gaia methodology for agent-oriented analysis and design. Auton. Agent. Multi-Agent Syst. **3**(3), 285–312 (2000)
2. Juan, T., Pearce, A., Sterling, L.: ROADMAP: extending the Gaia methodology for complex open systems. In: Gini, M., Ishida, T., Castelfranchi, C., Johnson, W.L. (eds.) Proceedings of the First International Joint Conference on Autonomous Agents and Multiagent Systems (AAMAS'02), pp. 3–10. ACM Press (2002)
3. Padgham, L., Winikoff, M.: Developing Intelligent Agent Systems—A Practical Guide. (Wiley, Chichester 2004)
4. Van Brussel, H., Wyns, J., Valckenaers, P., Bongaerts, L.: Reference architecture for holonic manufacturing systems: PROSA. Comput. Ind. **37**(3), 255–274 (1998)
5. Leitaõ, P., Colombo, A., Restivo, F.: ADACOR: A collaborative production automation and control architecture. IEEE Intell. Syst. **20**(1), 58–66 (2005)
6. Lootsma, F.A.: Multi-criteria Decision Analysis via Ratio and Difference Judgement. (Kluwer Academic Publishers, Netherlands 1999)
7. Saaty, T.L.: The Analytic Hierarchy Process: Planning, Priority Setting, Resource Allocation. McGraw-Hill, New York (1980)
8. Roy, B.: The outranking approach and the foundations of electre methods. In: Bana e Costa, C. A. (ed.) Readings in Multiple Criteria Decision Aid, pp. 155–183. Springer, Berlin (1990)
9. Roy, B.: The outranking approach and the foundation of ELECTRE methods. Theor. Decis. **3**, 149–173 (1991)
10. Roy, B.: Decision science or decision-aid science? Eur. J. Oper. Res. **2**, 184–203 (1993)
11. Roy, B.: Multicriteria Methodology for Decision Aiding. Kluwer Academic Publishers, Dordrecht, Holland (1996)
12. Brans, J.P., Vincke, Ph: PROMETHEE: A new family of outranking methods in MCDM. Manage. Sci. **6**, 647–656 (1985)
13. Van, Brussel H., Wyns, J., Valckenaers, P., Bongaerts, L.: Reference architecture for holonic manufacturing systems: PROSA. Comput. Ind. **37**(3), 255–274 (1998)
14. Leitão, P., Colombo, A., Restivo, F.: ADACOR: a collaborative production automation and control architecture. IEEE Intell. Syst. **20**(1), 58–66 (2005)
15. Chirn, J., McFarlane, D.: A component-based approach to the holonic control of a robot assembly cell. In: Proceedings of the IEEE 17th International Conference on Robotics and Automation, ICRA (2000)
16. Sinha Ashesh K., Aditya, H.K., Tiwari, M.K, Chan, F.T.S.: Agent oriented petroleum supply chain coordination: co-evolutionary particle swarm optimization based approach. Expert Syst. Appl. **38**(5), 6132–6145 (2011)
17. Xuhua, Shi, Feng, Qian: A Multi-Agent immune network algorithm and its application to murphree efficiency determination for the distillation column. J. Bionic Eng. **8**(2), 181–190 (2011)
18. Fu-Shiung, Hsieh: Design of reconfiguration mechanism for holonic manufacturing systems based on formal models. Eng. Appl. Artif. Intell. **23**(7), 1187–1199 (2010)
19. Leung, C.W., Wong, T.N., Mak, K.L., Fung, R.Y.K.: Integrated process planning and scheduling by an agent-based ant colony optimization. Comput. Ind. Eng. **59**(1), 166–180 (2010)
20. Lee W.C., Chen, S.K., Wu, C.C.: Branch-and-bound and simulated annealing algorithms for a two-agent scheduling problem. Expert Syst. Appl. **37**(9), 6594–6601 (2010)
21. Xinyu, Shao, Xinyu, Li, Liang, Gao, Chaoyong, Zhang: Integration of process planning and scheduling-a modified genetic algorithm-based approach. Comput. Oper. Res. **36**(6), 2082–2096 (2009)
22. Omar, López-Ortega, Israel, Villar-Medina: A multi-agent system to construct production orders by employing an expert system and a neural network. Expert Syst. Appl. **36**(2), 2937–2946 (2009)

23. Pedro, Gómez-Gasquet, Carlos, Andrés, Francisco-Cruz, Lario: An agent-based genetic algorithm for hybrid flowshops with sequence dependent setup times to minimise makespan. Expert Syst. Appl. **39**(9), 8095–8107 (2012)
24. Choi, H.S., Park, K.H.: Shop-floor scheduling at shipbuilding yards using the multiple intelligent agent system. J. Intell. Manuf. **8**(6), 505–515 (1997)
25. Xinyu, Li, Liang, Gao, Xinyu, Shao: Anactive learning genetic algorithm for integrated process planning and scheduling. Expert Syst. Appl. **39**(8), 6683–6691 (2012)
26. Nilesh, Anand, Mengchang, Yang: Duin J.H.R. van, Tavasszy Lori: GenCLOn: An ontology for city logistics. Expert Syst. Appl. **39**(15), 11944–11960 (2012)
27. Sorin, Llie, Costin, Bădică: Multi-agent approach to distributed ant colony optimization. Sci. Comput. Program. **78**(6), 762–774 (2013)
28. Guido, Maione, David, Naso: A soft computing approach for task contracting in multi-agent manufacturing control. Comput. Ind. **52**(3), 199–219 (2003)
29. Mikler, A.R., Honavar, V, Wong, J.S.K.: Autonomous agents for coordinated distributed parameterized heuristic routing in large dynamic communication networks. J. Syst. Softw. **56**, 231–246 (2001)
30. Ramy, Eltarras, Mohamed, Eltoweissy: Associative routing for wireless sensor networks. Comput. Commun. **34**(18), 2162–2173 (2011)
31. Deng, Y., Chen, Y., Zhang, Y., Mahadevan, S.: Fuzzy Dijkstra algorithm for shortest path problem under uncertain environment. Appl. Softw. Comput. **12**(3), 1231–1237 (2012)
32. Alessio, Ishizaka, Philippe, Nemery: Multi-Criteria Decision Analysis: Methods and Software. Wiley, UK (2013)
33. Herrmann, J.W. (ed.): Handbook of Production Scheduling. (Springer, New York 2006)

A Tree Based Multicast Routing Protocol Using Reliable Neighbor Node for Wireless Mobile Ad-Hoc Networks

Yadav Ajay Kumar and Tripathi Sachin

Abstract In this paper, we have focused on multicast routing which is one of the critical issues in mobile ad-hoc network. A wireless ad-hoc network is a self-configuring and dynamic network consisting of mobile devices. Mobile devices can move independently without any help of the base station and network infrastructure. There are various constraints in mobile ad-hoc networks, such as the dynamic nature of the topology, scarcity of bandwidth, and energy constraint in the network which is also responsible for the short lifetime of the nodes. These constraints lead to design a routing protocol for multicasting, which is energy efficient, scalable, durable, and simple. The multicast routing protocols have the potential to reduce communication costs and saving the network resources by broadcasting the same data packet to a group of receivers at a time. Since multicast ad-hoc on demand distance vector (MAODV) routing protocol has several limitations, e.g., low-packet delivery ratio, high-packet delivery delay, and high energy consumption due to link failure, node failure, mobility, and dynamic topology of wireless nodes. For group communication, multicast routing in MANETs needs to establish the reliable link between the neighbor nodes. So, in this paper, a mesh-based multicast routing protocol using reliable neighbor node for mobile ad-hoc networks has been proposed to overcome the limitations as mentioned above. The proposed scheme consider only the reliable nodes for routing process and discard the node having reliably pair factor less than the threshold value. The reliability pair factors are calculated based on the energy level of the nodes. The simulation results illustrate that the proposed protocol outperforms MAODV and EMAODV.

Keywords Wireless ad-hoc networks (MANETs) · Mobility · Multicast routing · Routing table · Neighbor node

---

Y.A. Kumar (✉) · T. Sachin
Department of Computer Science and Engineering,
Indian School of Mines, Dhanbad 826004, Jharkhand, India
e-mail: ajay.csjmku@cse.ism.ac.in
URL: http://www.ismdhanbad.ac.in

T. Sachin
e-mail: var_1285@yahoo.com

S. Das et al. (eds.), *Proceedings of the 4th International Conference on Frontiers in Intelligent Computing: Theory and Applications (FICTA) 2015*, Advances in Intelligent Systems and Computing 404, DOI 10.1007/978-81-322-2695-6_38

1 Introduction

A wireless ad-hoc network (WANET) is a decentralized type of wireless network. The network is called ad-hoc because it does not rely data packet on a pre-existing infrastructure [15]. Wireless mobile ad-hoc networks are self-configuring, dynamic networks in which nodes are free to move. Wireless mobile ad-hoc networks is useful in various fields as military battlefields where sharing of information is very important, emergency searches, rescues operation, and classroom where rapid deployment and quick reconfigurable infrastructure requires. The primary objective of a MANETs routing protocol is to set up an efficient routes between the communicating nodes to attain the following: reduce delivery time of the data packet, reduce packet losses, provide more stable connectivity, and diminish routing control overheads [5, 13, 17]. These constraints put in to force to construct a simple, scalable, robust, and energy efficient routing protocol in multicast environment. Multicast routing in MANETs for group communication requires to create a reliable communication links among the neighboring hops. Multicast is a technique to transmit a packets to a group of zero or more receivers identified by a single destination address [18]. The multicast routing protocols has been classified in to three categories: (1) Tree-based multicast routing protocols, in tree-based routing protocols only one route exists between a pair of nodes and hence these protocols are efficient in terms of the number of link transmissions. The examples of tree-based multicast routing protocols are ad-hoc multicast routing (AMRoute), ad-hoc multicast routing protocol utilizing increasing id-numbers (AMRIS), multicast ad-hoc on demand distance vector routing protocol (MAODV), and entropy-based MAODV.

Multicast ad-hoc on demand distance vector (MAODV) [10] is well known tree-based multicast routing protocol. The MAODV multicast routing protocol discover the multicast routes on demand using route discovery mechanism. When a mobile node wish to flow a data packet to a group of receivers then it generate the route request (*RQ*) message but it does not have the route to that group of receivers. Only a member of desired multicast receivers may respond to route request (*RQ*) message.

Entropy-based MAODV (EMAODV) [4] is also tree-based multicast routing protocol. In this entropy, concepts are used to develop an analytical modeling, and choose the long-life multicast routing as per entropy metric. This development reduces the number of route renewal as well as assurances the route stability in dynamic mobile networks, but it increases difficulty of route developments. (2) Mesh-based multicast routing protocols, in mesh-based approaches, there are more than one path between a pair of nodes are available. Mesh-based multicasting is better suited for highly dynamic topologies, simply due to the redundancy associated with this approach [8]. Typical mesh-based protocols includes centered protocol for on demand multicast routing protocol (ODMRP) [12] and its variations (Patch ODMRP) [11], Pool ODMRP [2], PDAODMRP [3], and EODMRP [14], core-assisted mesh protocol (CAMP) [6], and (3) hybrid-based multicast routing protocols. For a comprehensive review, the reader is referred to [9].

This paper proposes a mesh-based multicast routing protocol using reliable neighbor node selection approach for mobile ad-hoc networks [1]. It is observed that the existing multicast routing protocols have several limitations such as link failures, node failures, node mobility, dynamic topology, low reliability, and low scalability. The proposed protocol is designed on considering these limitations and trying to overcomes these. The proposed protocol operates in several phases: (1) Computation of reliability pair factor using energy level of the node. (2) Remove the neighbor nodes that have reliability pair factor (F_{RP}) lesser than the threshold reliability pair factor (F_{RP}^{th}) value. (3) Discovery of multicast routes using request packets (RQ) and reply packets (RP). (4) Multicast route has been allocated based on the maximum value of reliability pair factor of a path and data packet has been transmitted from source node to a group of receivers. In this way, the proposed protocol avoiding the unreliable node to participates in routing process, and thus we can save the various network resources and hence the efficiency of the protocol will increased.

The rest of the paper is organized as follows: Sect. 2 present description of our proposed protocol, Sect. 3 includes performance evaluation and analysis through relevant simulation results and compare to multicast ad-hoc on demand distance vector routing protocol (MAODV) and entropy-based multicast ad-hoc on demand distance vector routing protocol (EMAODV), and finally Sect. 4 provides conclusions.

2 Proposed Mesh-Based Multicast Routing Protocol Using Reliable Neighbor Node Selection Scheme

In this section, a reliable neighbor nodes selection mesh-based multicast routing has been proposed. In this scheme, we have randomly deployed few mobile nodes in a specific area associated with some initial energy and also define the reliability pair factor (F_{RP}^{th}) = 2.8 as a threshold value. Since multicast routing in MANETs for a group communication needed the establishment of reliable communication link between the neighbor nodes. So in this paper, we have designed a reliable neighbor node selection mesh-based multicast routing protocol to send a multicast data packet from source node to a group of destination node by observing the inborn limitations of existing multicast routing protocols such as the packet delivery ratio is low, packet delivery delay is high, and energy consumption is also high due to link failure, node failure, mobility and dynamic topology of a wireless nodes.

2.1 Reliable Neighbor Node Selection Scheme

A reliable neighbor node selection scheme is used for create multicast routes in MANETs with the help of a set of reliability pairs factor [1]. Reliability pair factor (F_{RP}) defines the link connectivity status. When we calculate the reliability pair factor (F_{RP}) of a node, we consider the several parameters, i.e., it is a function of energy level of two nodes involves in the communication, distance between the nodes and mobility states of the nodes. The procedure to compute the reliability pair factor (F_{RP}) involves the several steps: (1) Use of energy model to identify reliability pair among neighbor nodes. (2) Use of mobility model to find the change in nodes location with the help of its average velocity and the direction of movement. (3) Calculation of (F_{RP}) using the models given in steps 1–2.

2.2 Energy Model

In order to compute the reliability pair factor (F_{RP}), we need to know the remaining battery energy of a node. The energy model is given as follows, if F is the full battery energy of a node, then rest battery energy of node p at a time t, ($E_p^{rem}(t)$) is given by

$$E_p^{rem}(t) = E_p^{rem}(t-1) - E_p^{Tr} \times k(t-1,t) - E_p^{Rc} \times k(t-1,t) \quad (1)$$

where E_p^{Tr} is amount of energy required to transmit a bit data, k is the number of bit transmitted from time $t-1$ to t and E_p^{Rc} is amount of energy required receive k bit data from time $t-1$ to t. At time $t=0$, $E_p^{rem}(t) = \mathrm{F}$, i.e., at initial time the rest energy of a node is equal to full battery energy of a node.

2.3 Reliability Pair Factor

The successful transmission of the data packets from source node to a set of destination node is directly proportional to the minimum energy level of either nodes and inversely proportional to the distance between them [1]. We define F_{RP} to account for successful transmission between the nodes using above discussed models. F_{RP} at time T = 0 and t is given by

$$F_{RP} = k\frac{min(E_P^{rem}, E_Q^{rem})}{d_{(PQ,t)}} \quad (2)$$

where k is proportionality constant. Equation (2) are used to calculate the reliability pair factor (F_{RP}) between the neighbor nodes that is established multicast route between source node to a group of destination nodes.

2.4 Route Discovery Phase

In Fig. 1, when source node want S to send a data packet to a group of destination nodes it broadcast a request packet (RQ) across the network, and the nodes which are coming in its transmission range have received the request packet. While the request packet (RQ) broadcasted across the network, it carried out the several components in its header: $\{S_{addr}, MC_{addr}, Seq_{no}, R_{flag}, P_{info}, F_{RP}^{th}\}$, where S_{addr} refers the address of the source node, MC_{addr} is a group of destination address, Seq_{no} is originated at source to establish a path and help in identification of duplicate RQ packets, R_{flag} is one bit flag identifier if $R_{flag} = 1$ implies RQ packets, P_{info} is used to store the node address, while traveling from source node to a group of destination nodes, and F_{RP}^{th} is a threshold value of the reliability, this is used to compare the reliability pair factor of a connection.

Let us consider a network, in which few mobile nodes are randomly deployed in a specific area as shown in Fig. 3. Each node in the network maintains two database tables as neighboring information table (NI) and routing table (RT). The neighboring information (NI) table stores the neighbor node address that are coming in a transmission range of particular node and reliability pair factor (F_{RP}) of respective nodes, and the second table is used to store the multicast group address MC_{addr} and next hop address ($Nexthop$) for forwarding the data packet to next hop.

Suppose the source node S want to transmit a multicast data packet to a group of destination node say X, Y and Z. When source node S broadcast a request packet (RQ) across the network suppose the node A, B and C are coming in its transmission range and receive the request packet (RQ). When node A, B, and C receives the request packet (RQ) it compares its reliability pair factor (F_{RP}) with threshold reliability

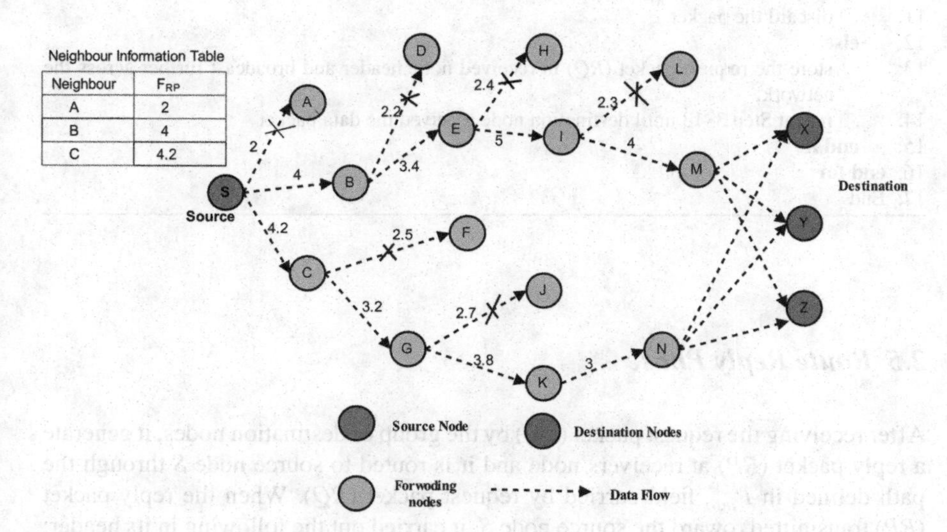

Fig. 1 Route discovery process

pair factor (F^{th}_{RP}). The node A discards the request packet (RQ) because its reliability pair factor is less than threshold reliability pair factor(F^{th}_{RP}) as recorded in neighbor information table (NI) shown in Fig. 3. While in case of node B and C receive and store the request packet (RQ) in its header because the reliability pair factor (F_{RP}) of these node are more than threshold reliability pair factor (F^{th}_{RP}). Further node B and C broadcast the request packet (RQ) to its neighboring nodes D, E, and F, G, respectively. Node D discards the request packet and node E receives and stores the request packet (RQ) because its reliability pair factor F_{RP} is more than threshold reliability pair factor (F^{th}_{RP}). In the same way, the node F discards the data packet and the node G receives and stores in it's header. This process is carried out until multicast group member receive the request packet (RQ). Finally, multicast tree has been generated to send the data packet from source node to a group of receivers as shown in Fig. 4. The route discovery process shown in Algorithm 1.

Algorithm 1 Route discovery process

1: Begin
2: Initialize $F^{th}_{RP} = 2.8$, $RQ = \{S_{addr}, MC_{addr}, \_, \_, \_, \_\}$ and other fields of the request packet at source node S.
3: Source node S broadcast the request packet (RQ) to its neighboring nodes those are coming in its transmission range.
4: **for** every request packet (RQ) arrived at its neighboring node **do**
5:     **if** there is no neighbor of a particular node which broadcast the request packet (RQ) **then**
6:         go to Step 17.
7:     **else**
8:         compare the reliability pair factor (F_{RP}) with threshold reliability pair factor(F^{th}_{RP}) of a received node which are coming in its transmission range.
9:     **end if**
10:     **if** $F_{RP}(Rec\ node) < F^{th}_{RP}\ value$ **then**
11:         discard the packet.
12:     **else**
13:         store the request packet (RQ) in received node header and broadcast further across the network.
14:         repeat Step 3–13 until destination node received the data packet
15:     **end if**
16: **end for**
17: End

2.5 Route Reply Phase

After receiving the request packet (RQ) by the group of destination nodes, it generate a reply packet (RP) at receivers node and it is routed to source node S through the path defined in P_{info} field carried by request packet (RQ). When the reply packet (RP) transmitted toward the source node S, it carried out the following in its header: $\{S_{addr}, RPath_{info},\ and\ MC_{addr}\}$, where S_{addr} refers source address, $RPath_{info}$ field

Fig. 2 Route reply process

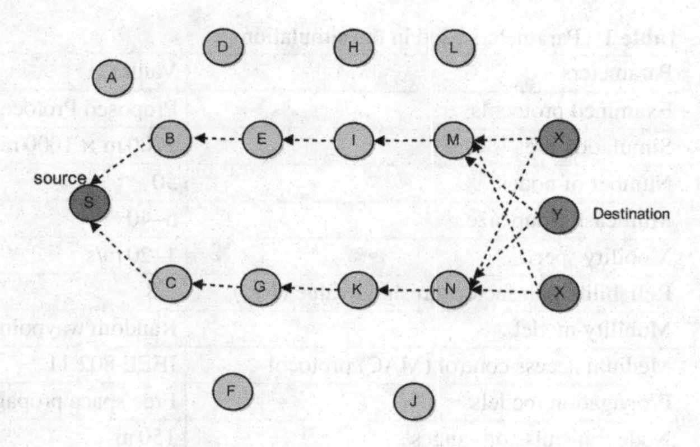

refers the reverse path to carry the reply packet (*RP*) from group of destination nodes to source node *S*, and MC_{addr} refers the group of destination address.

Let us consider the group of destination node *X*, *Y*, and *Z* generates the reply packet (*RP*) and send to its upstream nodes *M* and *N*, respectively, as shown in Fig. 2. After receiving the request packet (*RQ*) by the node *M* it again send the reply packet to its neighboring node, say *I* receives and stores in it's header. This forwarding operation is carried out until the source node *S* has received the reply packet. In the same way, the upstream node *N* also send a reply packet *RP* toward the source node *S* until received the reply packet (*RP*). Finally, the source node *S* receive two reply packet (*RP*) via two different routes.

3 Performance Evaluation

In this segment, we will introduce the parameters used in our simulation. In addition, we will compare the performance of the proposed RNSMMRP with some existing routing protocols.

3.1 Simulation Setup

The simulation is shown using NS-2 (Network Simulation 2, version 3.5) [16]. An ad-hoc network is generated in the area of 1000 m × 1000 m square meter. It is consist of 50 mobile nodes that are randomly deployed in the network. We have used random waypoint model as mobility model. In random waypoint model, each node randomly selects the moving direction, and when hitting the simulation area boundary, it back and continues to move. We have used free space propagation model [7] in our simulations. The free space model basically represents the communication range as a circle around the transmitter. Every node is assumed to have its remaining

Table 1 Parameters used in the simulations

Parameters	Values
Examined protocols	Proposed Protocol, EMAODV, and MAODV
Simulation area	1000 m × 1000 m
Number of nodes	50
Multicast group size	5–40
Mobility speed	1–20 m/s
Reliability pair factor threshold value (F_{RP}^{th})	2.8
Mobility model	Random waypoint model
Medium access control (MAC) protocol	IEEE 802.11
Propagation models	Free space propagation model
Node transmission ranges	150 m
Data packet size	512 bytes

battery energy in two ranges, low range, and high range. The higher value for the low range is L_{th}. A node has full battery energy, F. E_P^{Tr} is the amount energy required to transmit a bit of information and E_P^{Rc} is the amount of energy required to receive a bit of information. We used constant bit rate (CBR) as the traffic type. In CBR model, the source transmits a certain number of fixed-size packets. The parameters used in the simulations are listed in Table 1. The performance evaluation metrics used in the simulations are:

1. **Packet delivery ratio (PDR)**: It defines the number of packet received at receivers to the number of packets sent from source.

$$Packet\ delivery\ ratio = \frac{\sum Number\ of\ packet\ receive}{\sum Number\ of\ packet\ send} \tag{3}$$

2. **Packet delivery delay**: It defines the average time taken to transmit the predefined number of packets from source node to multicast receivers for various group sizes.

3.2 Simulation Outcome and Explication

We have evaluated the performance of the proposed scheme in different network scenarios in terms of node moving speed, packet delivery ratio, and packet delivery delay. The effect of mobility speed on proposed protocol, EMAODV and MAODV has been analyzed. These protocols have been simulated for packet delivery ratio (*PDR*) and packet delivery delay. We have try to show the performance of the proposed protocols under various mobility speeds. The explication of performance parameters are given in this segment.

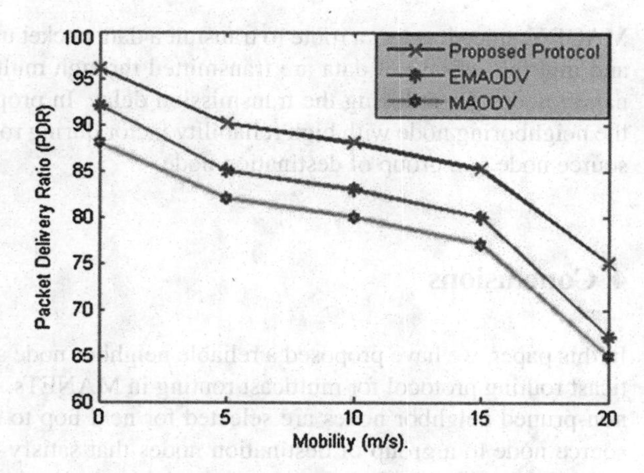

Fig. 3 Packet delivery ratio versus mobility

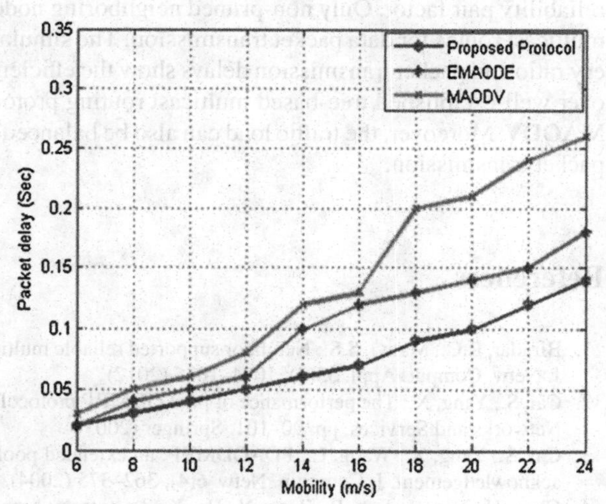

Fig. 4 Packet delivery delay versus mobility

Figure 3 shows that the effect of node mobility on packet delivery ratio for proposed protocol, EMAODV and MAODV. As mobility speed of the nodes increases, the packet delivery decreases. The proposed protocol perform better in compare to EMAODV and MAODV due to multiple paths and considering the routes to send the data packet based on reliability pair factor scheme, i.e., those mobile nodes having low reliability pair factor will not share in routing process. The MAODV and EMAODV do not consider the reliability pair factor scheme to find the non-pruned node to transmit the data packet.

From Fig. 4 it is clear that if mobility speed of the node increases the packet delivery delay also increases, due to very high-mobility speed communication link may be failures this will cause the delay in packet delivery. The time require to transmit the multicast data in the protocol proposed is lesser than that of EMAODV and

MAODV, since it select a route to transmit a data packet using the more reliable link and multiple stream of data are transmitted through multipath to a group of destination nodes for reducing the transmission delay. In proposed scheme selection of the neighboring node with high reliability factor during routing the data packet from source node to a group of destination node.

4 Conclusions

In this paper, we have proposed a reliable neighbor node selection mesh-based multicast routing protocol for multicast routing in MANETs. In this scheme, only those non-pruned neighbor nodes are selected for next hop to send the data packet from source node to a group of destination nodes that satisfy certain threshold value of reliability pair factor. Only non-pruned neighboring nodes are used to create reliable multicast routes for data packet transmission. The simulation results for packet delivery ratio and packet transmission delays show the efficiency of the developed scheme over well-established tree-based multicast routing protocols such as EMAODV and MAODV. Moreover, the traffic load can also be balanced using multipath during data packet transmission.

References

1. Biradar, R.C., Manvi, S.S.: Neighbor supported reliable multipath multicast routing in manets. J. Netw. Comput. Appl. **35**(3), 1074–1085 (2012)
2. Cai, S., Yang, X.: The performance of pool ODMRP protocol. In: Management of Multimedia Networks and Services, pp. 90–101. Springer (2003)
3. Cai, S., Yang, X., Wang, L.: PDAODMRP: an extended pool ODMRP based on passive data acknowledgement. J. Commun. Netw. **6**(4), 362–375 (2004)
4. Chen, H., Yan, Z., Sun, B., Zeng, Y., He, X.: An entropy-based long-life multicast routing protocol in maodv. In: ISECS International Colloquium on Computing, Communication, Control, and Management, CCCM 2009, vol. 1, pp. 314–317. IEEE (2009)
5. Das, S.K., Tripathi, S., Burnwal, A.: Intelligent energy competency multipath routing in wanet. In: Information Systems Design and Intelligent Applications, pp. 535–543. Springer (2015)
6. Garcia-Luna-Aceves, J., Madruga, E.L.: The core-assisted mesh protocol. IEEE J. Sel. Areas Commun. **17**(8), 1380–1394 (1999)
7. Iskander, M.F., Yun, Z.: Propagation prediction models for wireless communication systems. IEEE Trans. Microw. Theory Tech. **50**(3), 662–673 (2002)
8. Jetcheva, J.G., Johnson, D.B.: Adaptive demand-driven multicast routing in multi-hop wireless ad hoc networks. In: Proceedings of the 2nd ACM International Symposium on Mobile Ad Hoc Networking and Computing, pp. 33–44. ACM (2001)
9. Junhai, L., Danxia, Y., Liu, X., Mingyu, F.: A survey of multicast routing protocols for mobile ad-hoc networks. IEEE Commun. Surv. Tutorials **11**(1), 78–91 (2009)
10. Kharraz, M.A., Sarbazi-Azad, H., Zomaya, A.Y.: On-demand multicast routing protocol with efficient route discovery. J. Netw. Comput. Appl. **35**(3), 942–950 (2012)
11. Lee, M., Kim, Y.K.: Patch odmrp: an ad-hoc multicast routing protocol. In: 15th International Conference on Information Networking, Proceedings, pp. 537–543. IEEE (2001)

12. Lee, S.J., Su, W., Gerla, M.: On-demand multicast routing protocol in multihop wireless mobile networks. Mob. Netw. Appl. **7**(6), 441–453 (2002)
13. Manvi, S., Kakkasageri, M.: Multicast routing in mobile ad hoc networks by using a multiagent system. Inf. Sci. **178**(6), 1611–1628 (2008)
14. Oh, S.Y., Park, J.S., Gerla, M.: E-ODMRP: enhanced ODMRP with motion adaptive refresh. J. Parallel Distrib. Comput. **68**(8), 1044–1053 (2008)
15. Remondo, D., Niemegeers, I.G.: Ad hoc networking in future wireless communications. Comput. Commun. **26**(1), 36–40 (2003)
16. The Network Simulator 2. http://www.isi.edu/nsnam/ns/index.html/
17. Yadav, A.K., Tripathi, S.: Load balanced multicast routing protocol for wireless mobile ad-hoc network. In: 2015 Third International Conference on Computer, Communication, Control and Information Technology (C3IT), pp. 1–6. IEEE (2015)
18. Zeng, Q.A., Agrawal, D.P.: Introduction to wireless and mobile systems. CENGAGE Learning (2002)

Traffic Grooming in Hybrid Optical-WiMAX Mesh Networks

Deepa Naik, Soumyadeb Maity and Tanmay De

Abstract In this work, a static lightpath approach is used for traffic grooming in order to minimize the bandwidth blocking probability in Wavelength Division Multiplexed Passive Optical Networks (WDM-PON). In WDM-PON, each subscriber is assigned a distinct wavelength as the transmission channel. It offers greatest flexibility, because of its point to point (P2P) technology. Each wavelength channel supports the traffic in Gbps range. In practice, individual traffic request from subscribers needs bandwidth in Mbps range. To utilize the network resources efficiently in WDM-PON, traffic grooming technique is implemented. The proposed algorithm, here is for the traffic grooming and routing in hybrid networks. In this network optical and World Interoperable Microwave Access (WiMAX) architectures are integrated. The proposed algorithm improves the bandwidth blocking probability rate for both upstream and downstream traffic requests. The performance of our algorithm is evaluated using simulation results. The simulation and result shows that the proposed algorithm provides better result than existing shortest path algorithm.

Keywords Optical line terminal (OLT) · Optical network units (ONUs) · Wavelength division multiplexed passive optical networks (WDM-PON) · Traffic grooming

D. Naik (✉) · S. Maity · T. De
Department of Computer Science and Engineering, National Institute of Technology, Durgapur, India
e-mail: naiksavantdeepa@gmail.com

S. Maity
e-mail: soumyamaity121@gmail.com

T. De
e-mail: tanmayd12@gmail.com

S. Das et al. (eds.), *Proceedings of the 4th International Conference on Frontiers in Intelligent Computing: Theory and Applications (FICTA) 2015*, Advances in Intelligent Systems and Computing 404, DOI 10.1007/978-81-322-2695-6_39

1 Introduction

Passive Optical Network (PON) consists of optical line terminal (OLT) and a number of optical network units (ONUs) at customer premises. In WDM-PON, each ONU uses the multiple wavelengths. Here traffic request is carried out in all optical channel from source node to destination node is known as lightpath [1]. Traffic grooming in hybrid networks comprises of three stages. They are grooming, routing, and wavelength assignment. Grooming multiplexes the multiple low rate traffics into one high capacity wavelength channel. Routing allocates the route from source node to destination node through a lightpath. Finally, a dedicated wavelength is assigned to established lightpath.

In this work, the unicast traffic is static in nature, i.e., source, destination, and data rate are known beforehand. The proposed Hybrid Unicast Traffic Grooming (HUTG) algorithm reduces the number of wavelengths by sharing of the links among various unicast sessions and minimize the bandwidth blocking probability of traffic requests.

The rest of the paper is organized as follows: Sect. 2 presents the previous works. In Sect. 3, problem is defined in formal notation. The proposed approach is presented in Sect. 4. The complexity analysis and results are described in Sect. 5 and Sect. 6 respectively. Finally, conclusion and future work is drawn in Sect. 7.

2 Previous Works

Traffic grooming in WDM Hybrid Optical-WiMAX Networks in a mesh topology is relatively new and only limited works are done. The traffic grooming problem can be broadly classified as static traffic grooming and dynamic traffic grooming. In static traffic grooming, the set of traffic requests from a source node to destination node is known a priori. In dynamic traffic grooming, traffic connection requests arrive and depart dynamically and both routing and wavelength assignment are dynamically done as per incoming connection requests. B. Mukherjee and Zhu in [2] studied the node architecture for WDM optical network and proposed traffic grooming using the integer linear programming (ILP). They proposed two algorithms: Maximizing Resource Utilization (MRU) and Maximizing the Single-Hop Traffic (MST) algorithms with the objective of maximizing the network throughput. The authors in [3] propose a multi-objective evolutionary algorithm with three objective maximising the traffic throughput, minimizing the number of transceivers, and minimising the average propagation delay or average hop counts. In [4], De et al. has considered traffic grooming problem for multi-hop traffic requests with the objective of maximizing the network throughput. They also considered the traffic grooming as similar to clique partitioning problem and proposed a heuristic algorithm called Traffic Grooming based on Clique Partitioning (TGCP). In [5], Zyane et al. the performance of Dijkstra's shortest path algorithm is investigated in terms of blocking probability, which leads to more unbalanced routes over the network links. Ahmed et al. in [6] proposed an architecture for Resilient Ethernet Passive Optical-WiMAX hybrid network and suggested the routing algorithm for ring topology.

3 Problem Description

Traffic grooming is a technique used to multiplex the low speed connection requests on the available wavelengths, maintaining the lightpath capacity and other network resource constraints. We have considered an optical and WiMAX integrated hybrid network for traffic grooming. Hybrid network employs, optical network as an back haul and WiMAX as an front end. The ONUs with integrated WiMAX Base station are connected to OLT node. Each link from OLT to ONUs are connected by an individual wavelength channel in the upward and downward directions. Each node has tunable transmitter and receiver. In our approach connection request can be served through single-hop or multi-hop. In single-hop, the request served in a single lightpath between the source and destination. On other hand multi-hop, the request can be groomed with multiple lightpaths. Initially the requests are served with single-hop and the leftover requests are served with multi-hop. The least congested shortest route (i.e., the maximum number of wavelength available shortest path) is selected in order to route the traffic requests. The main objective is to minimize the bandwidth blocking probability rate of the traffic requests for a given network.

Assumptions:
Following assumptions are made in this work.

1. Traffic demand is static in nature, i.e., traffic demand is known in advance.
2. Each network node can act as both source node as well as destination nodes.
3. Granularity of low speed traffic request is $x \in \{1, 3, 12\}$, which means that bandwidth capacity can be one of the OC-1, OC-3 and OC-12 between various source and destination nodes.
4. Wavelength converters are not used here.
5. Each unicast sessions may be supported by multi-hop light paths.
6. WDM-PON creates a point to point link between OLT and ONUs, So each ONU can operate at a rate up to the full bit rate of wavelength channel.
7. WDM-PON employs separate wavelength channel from OLT to ONUs (called downstream direction) and ONUs to OLT (called up stream direction). Former is called downlink wavelength and later is called uplink wavelength.
8. The number of transceivers at each node and number of wavelengths on each link is fixed.
9. Each wavelength channel has capacity c.

We introduce the following notations to define the problem:
T: set of n number of unicast session requests, $T = \{t_1, t_2, t_3, \ldots, t_n\}$.
ω_i^j: presents the current status of the free bandwidth, the ith wavelength channel

used by jth path. Initially $\omega_i^j = c$.
w: number of wavelength channel in each link.
x: sorted traffic request in descending order of their the bandwidth requirement.
tr_i: total number of transmitter at node i.
rec_i: total number of receiver at node i.
tr_s: at present, number of transmitter available at source node s.
rec_d: at present, number of receiver available at destination node d.
c: channel capacity.
s: source node.
d: destination node.
b_i: bandwidth requested by unicast session request i.
p_i: it is 1 if bandwidth b_i of the request i is satisfied, otherwise it is 0.

Objective: The objective here is to minimize the bandwidth blocking probability of the traffic requests in a given network. i.e.,

$$Minimize \left[\frac{\sum_{i=1}^{n} b_i - \sum_{i=1}^{n} (b_i \times p_i)}{\sum_{i=1}^{n} b_i} \right] \quad (1)$$

Constraints:
We assume the following constraints in this work.

1. Cumulative bandwidth of all sub-wavelengths request established must be less than or equal to the total wavelength capacity of all the lightpath established, i.e., capacity of the virtual topology.
2. The number of lightpath originating from a node cannot be more than the number of transmitter at the node.
3. The number of lightpath terminating at a node cannot be more than the number of receivers at the node.

4 Proposed Approach

In this section, an algorithm called Hybrid Unicast Traffic Grooming (HUTG) is proposed in order to groom the traffic requests between node pairs. The bandwidth demands keep growing day by day, as the higher number of number of users and higher number of bandwidth requirement per user is increased. The transmission rate on a wavelength channel is expected to reach OC-768 (40 Gbps) in near future. The traffic request from individual user possibly varies from OC-1 (51.84 Mbps) to OC-192 (10 Gbps). Hence this mismatch between the traffic demand per user and wavelength channel transmission capacity leads to inefficiency in utilizing the network resources. This disparity motivated us to develop the algorithm which leads to better utilization of network resources and minimize the bandwidth blocking probability in optical-wireless networks. The basic idea of this approach is to set up lightpath and groom the common source node and common destination node pair, which has the largest amount of traffic requirements. The traffic requests are in the form of three tuple $[s, d, b_i]$, where s is the source node, d is the destination node, b_i is the bandwidth request between the source-destination pair. The each node act as a source or destination node depending on the uplink/downlink direction of the traffic flow. The proposed algorithm is divided into two phases similar to existing algorithm in [7]. In first phase of the proposed approach, traffic requests are generated and groomed. The groomed requests are sorted in descending order of their traffic requirements. In second phase, the ligthpath between source-destination node pairs is established and assignment of wavelength is done.

In WDM static traffic grooming, the main problem is to find the optimal route for the arrived requests. The light path should take the same wavelength along its path. This constraint is known as wavelength continuity constraint. In shortest-path algorithm the route is selected depending on the hop count. When the request arrives, a free wavelength is searched along the corresponding route to establish the connection. The request is blocked, if the network resources are not available for corresponding requests. We have considered the eight node network, shown in Fig. 1, to illustrate the working principal of proposed algorithm HUTG. If there is a request, arrived from routing node 0 and access node OLT, then according to (Shortest Path Traffic Grooming) SPT algorithm the shortest path is 0-1-2-OLT between the node pair 0 and OLT. The wavelength is not available along this path and hence the request gets blocked. In case of HUTG the request is served along the path of available wavelength depending on the network status. There are total eight paths available to serve the same request. But only two paths 0-4-3-OLT and 0-1-4-3-OLT have the wavelengths available along their traversal. The algorithm selects the shortest path 0-4-3-OLT to serve the request.

Example: We have considered an eight node network as shown in Fig. 1 to illustrate the working principle of HUTG algorithm. We have assumed the physical link is bidirectional. In this simulation, each wavelength capacity is assumed OC-48 and allowed traffic bandwidth requests were assumed to be OC-1, OC-3 and OC-12. The traffic matrices are randomly generated. These matrices are shown in Tables 1, 2, and 3 respectively. The groomed traffic bandwidth requests are shown in Table 4.

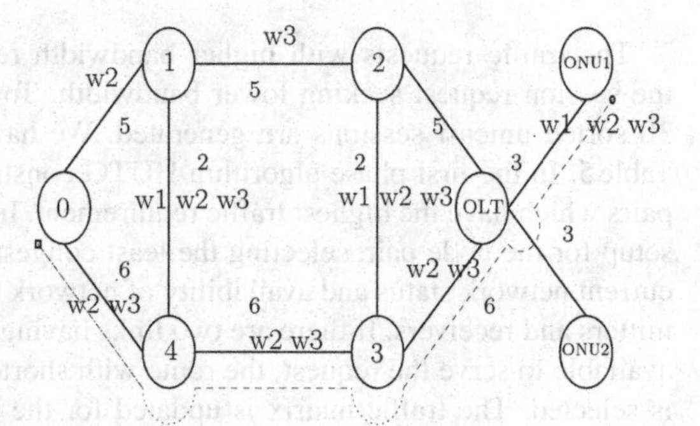

Fig. 1 Least conjested shortest Path from node 0 to ONU1

Algorithm 1: Hybrid Unicast Traffic Grooming (HUTG)

Input : $T[s, d, b_i]$, A Traffic matrix of Unicast requests $T = [t_1, t_2, \ldots, t_n]$, where $t_i = [s, d, b_i]$ with a wavelength capacity c
Output: bandwidth blocking probability of unicast sessions

```
1  Sort the traffic matrix T according to the Bandwidth requirement from decending order of
   their traffic requests and place in x, where x contains sorted request list.
2  for i = 1 to w do
3      for j = 1 to E do
4          ω_i^j = c
5      /* Each edge is assigned Channel capacity c */
6  while x ≠ NULL do
7      s ← t_i;   /* Add t_i = (s, d, b_i) into s */
8      path = LeastCongestedShortestPath(s, d, b_i)   /* Generate the least
       congested shortest path (Free wavelength available shortest
       path) */
9      for j = 1 to |path| do
10         if (tr_s > 0) and (rec_d > 0) then
11             /* check the transmitter and receiver availability
               status */
12             if ω_k^j ≥ b_i then
13                 Continue   /* check the wavelength availability along
                   the path */
14     if j > |path| then
15         Assign wavelength to t_i
16         for j = 1 to |path| do
17             ω_k^j = ω_k^j − b_i
18             tr_s = tr_s − 1   /* updates the transmitter at source node
               */
19             rec_d = rec_d − 1   /* updates the receiver at destination
               node */
20             T[s, d] = 0   /* update the traffic matrix */
21         Break
22     x = x − s   delete the request which is satisfied from list x.
23 end
```

The traffic requests with higher bandwidth requirement are groomed prior to the session request seeking lower bandwidth. To sort the sessions in this fashion, 30 sorted unicast sessions are generated. We have shown only for ten, shown in Table 5. In the first phase algorithm HUTG construct the sorted list x for the node pairs which have the highest traffic requirement. In the second phase the ligthpath is setup for the node pair, selecting the least congested shortest path according to the current network status and availability of network resources like wavelengths, transmitters and receivers. If there are two links having the same number of wavelengths available to serve the request, the route with shortest distance from the source node is selected. The traffic matrix is updated for the served requests and status of the

Function Least Congested Shortest Path(s,d,b_i)

```
Input  : G(V,E) and request t_i[s, d, b_i]
Output: Least Congested Shortest Path
1  for k = 1 to |w| do
2      a=v_0,v_1,v_2, ...v_n
3      z=v_0,v_1,v_2, ...v_n
4      weight(v_i,v_j)   /* weight between two vertices */
5      for i = 1 to |n| do
6          L(v_i)=∞
7      L(s_i)=0;   /* Initialize all distance from source will be
       infinity except source */
8      S=s    /* S is initialized with source node */
9      while z ≠ S do
10         for j = 1 to |n| do
11             if (v_j  ∉  in  S) then
12                 if ((ω_k^j(S_j, v_j) > 0) then
13                     if (L(v_j) + weight(S_j, v_j) < L(v_j) then
14                         L(v_j)=L(z)+weight(S_j, v_j)   /* This adds a vertex to
                           S with minimal label and updates the label of
                           vertices not in S*/
15                         p = v_j;
16         S = S ⋃ p    /* Add path to S *///
17     if L(d) ≠ ∞ then
18         break
19 return (S)   /* returns least congested shortest path and
   wavelength indexed */
```

Table 1 OC-1 traffic for 8-node network

Index	0	1	2	3	4	OLT	ONU1	ONU2
0	0	0	3	3	0	3	0	0
1	0	0	0	0	0	3	3	0
2	0	3	0	3	3	0	0	0
3	0	0	0	0	0	0	0	0
4	0	0	0	0	0	0	0	0
OLT	0	3	0	0	0	0	3	0
ONU1	0	0	3	0	0	0	0	0
ONU2	0	0	3	1	0	0	0	0

network resources are also updated. Finally, the process is repeated until the list x is empty or the availability of network resources is exhausted.

This ensures network load is distributed. The result shows that shortest path algorithm is not clearly optimal when performance is measured in terms of bandwidth

Table 2 OC-3 traffic for 8-node network

Index	0	1	2	3	4	OLT	ONU1	ONU2
0	0	0	0	0	4	0	0	0
1	4	0	4	0	0	0	0	0
2	0	0	0	0	0	0	0	0
3	0	0	0	0	0	0	0	0
4	0	4	0	0	0	0	0	0
OLT	0	0	0	0	0	0	0	0
ONU1	4	0	0	0	0	0	0	0
ONU2	0	0	0	0	0	0	4	0

Table 3 OC-12 traffic for 8-node Network

Index	0	1	2	3	4	OLT	ONU1	ONU2
0	0	4	0	0	0	0	4	4
1	0	0	0	4	4	0	0	4
2	4	0	0	0	0	4	4	4
3	4	4	4	0	0	4	4	4
4	4	0	4	4	0	4	4	4
OLT	4	0	4	4	4	0	0	4
ONU1	4	4	0	4	4	4	0	4
ONU2	4	4	0	0	4	4	0	0

Table 4 Groomed requests for 8 node network

Index	0	1	2	3	4	OLT	ONU1	ONU2
0	0	48	3	3	12	3	48	48
1	12	0	12	48	48	3	3	48
2	48	3	0	3	3	48	48	48
3	48	48	48	0	0	48	48	48
4	48	12	48	48	0	48	48	48
OLT	48	3	48	48	48	0	3	48
ONU1	48	48	3	48	48	48	0	48
ONU2	48	48	3	1	48	48	12	0

blocking probability and wavelength utilization. Our approach leads to better performance in the network with high network connectivity.

Table 5 Ten sorted request for the 8 node network

Index	Source (S)	Destination (D)	Bandwidth (OC)
0	0	1	48
1	0	ONU1	48
2	1	3	48
4	1	4	48
5	1	ONU2	48
6	2	0	48
7	2	OLT	48
8	2	ONU1	48
9	2	ONU2	48
10	3	0	48

5 Complexity Analysis

Let us assume N is the total number of nodes in the optical network and M is the total number of node pair having nonzero traffic requests. Let W is the total number of wavelength supported by a single fiber link. In first phase, to sort the nonzero traffic requests worst case time complexity is $O(Mlog_2M)$. In second phase, i.e., after sorting the requests to find the path using least congested shortest path algorithm, the worst case time complexity is $O(WN^2)$ and to update the wavelength matrix as well as transmitter receiver status, the worst case time complexity is $O(N^2)$ and $O(N)$. The worst case time complexity of second phase is $O(WN^2)$. Therefore the worst case time complexity of the HUTG algorithm for M requests is $O(Mlog_2M) + O(MWN^2)$.

6 Results

In this section, results obtained using Hybrid Unicast Traffic Grooming (HUTG) algorithm are presented and discussed. The main objective is to minimize the bandwidth blocking probability of the traffic requests using traffic grooming technique and simultaneously reducing the number of wavelength channel requirements.

Figure 2 shows the results of both the algorithms HUTG with single-hop and (Shortest Path Traffic grooming) SPT. They demonstrate the relationship between the number of wavelengths required and the bandwidth blocking probability of the network. Here, total bandwidth requirement for all the request is OC-2000 with fixed transceivers set to 7. When the number of wavelengths increased to 7 the bandwidth blocking probability is reduced to zero. In Fig. 3 the simulation results show that the bandwidth blocking probability of the proposed approach is significantly reduced with increase in the number of wavelength as compared to SPT. When the number

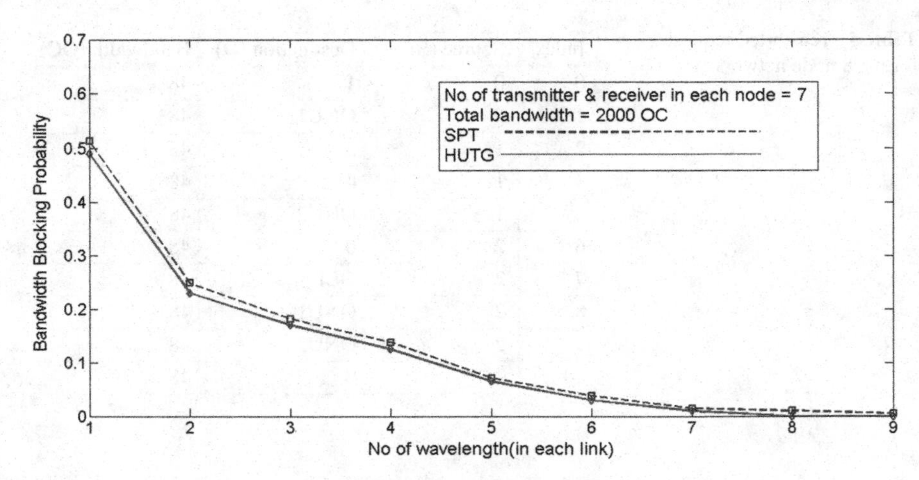

Fig. 2 Relationship of total number of wavelength consumed versus bandwidth blocking probability rate for 8-node network (HUTG with single-hop)

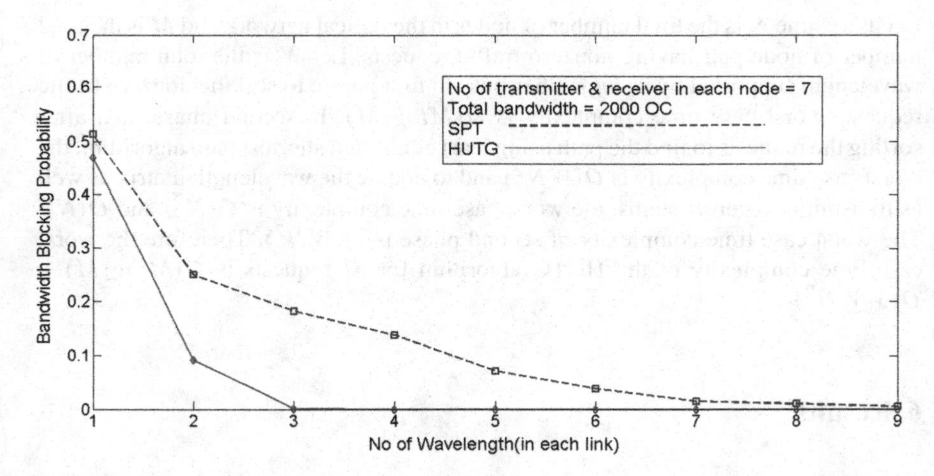

Fig. 3 Relationship of total number of wavelength consumed versus bandwidth blocking probability rate for 8-node network (HUTG with multi-hop)

of wavelengths are increased to 3, the bandwidth blocking probability is reduced to zero in HUTG algorithm with multi-hop. Here, both number of wavelengths required as well as bandwidth blocking probability are reduced by HUTG algorithm than the existing SPT algorithm.

7 Conclusion and Future Work

In this work, the problem of grooming for unicast traffic in Hybrid networks is studied. The Shortest Path algorithm (Dijkstra) makes one single path, which may be more congested. It leads to congestion in the network and requests get blocked. To solve this problem the proposed algorithm HUTG is used keeping the congestion in mind during routing. The simulation results show that HUTG leads to better performance in order to reduce the bandwidth blocking probability and balancing network load. HUTG algorithm is based on static lightpath traffic grooming approach which can be extended using light-trail instead of lightpath concept. The proposed algorithm can be further extended for dynamic traffic grooming in WiMAX to Wi-Fi networks in future.

References

1. Chiu, A.L., Modiano, E.H.: Traffic grooming algorithms for reducing electronic multiplexing costs in WDM ring networks. J. Lightwave Technol. **18**(1), 2–12 (2000)
2. Zhu, K., Mukherjee, B.: Traffic grooming in an optical WDM mesh network. IEEE J. Sel. Areas Commun. **20**(1), 122–133 (2002)
3. Prathombutr, P., Stach, J., Park, E.: An algorithm for traffic grooming in WDM optical mesh networks with multiple objectives. In: (ICCCN) Proceedings of the 12th International Conference on Computer Communications and Networks, pp. 405–411 (2003)
4. De, T., Pal, A., Sengupta, I.: Traffic grooming, routing, and wavelength assignment in an optical WDM mesh networks based on clique partitioning. Photonic Netw. Commun. **20**(2), 101–112 (2010)
5. Zyane, A., Guennoun, Z., Taous, O.: Performance evaluation of shortest path routing algorithms in wide all-optical WDM networks. In: (ICMCS) International Conference on Multimedia Computing and Systems, pp. 831–836 (2014)
6. Ahmed, A., Shami, A.: RPR-EPON-WiMAX hybrid network: a solution for access and metro networks. J. Opt. Commun. Netw. **4**(3), 173–188 (2012)
7. Yoon, Y.R., Lee, T.J., Chung, M.Y., Choo, H.: Traffic grooming based on shortest path in optical WDM mesh networks. In: CCS Computational Science, pp. 1120–1124. Springer, Heidelberg (2005)

Part XI
Authentication and Network Security

Host-Based Intrusion Detection Using Statistical Approaches

Sunil Kumar Gautam and Hari Om

Abstract An intrusion detection system (IDS) detects the malicious activities, running in the system that may be a single system or a networked system. Furthermore, the intrusion-based systems monitor the data in a system against the suspicious activities and also secure the entire network. Detection of malicious attacks with keeping acceptability of low false alarm rate is a challenging task in intrusion detection. In this paper, we analyze the three statistical approaches namely principal component analysis (PCA), linear discriminant analysis (LDA), and naive Bayes classifier (NBC), employed in host-based intrusion detection systems (HIDS) and we detect the accuracy rate using these approaches.

Keywords Principal component analysis (PCA) · Linear discriminant analysis (LDA) · Naive bayes classifier (NBC) · Host-based intrusion detection systems (HIDSs)

1 Introduction

The Internet technologies provide fantastic facilities for information gathering, retrieving, and exchanging data among users [1]. The Internet provides several user-friendly facilities to human beings, such as e-business, financial security, e-education, health sector, etc. Internet indeed has become an integral part of human life in most of the activities [2]. The society's dependency has increased on Internet due to the expansion of local area network (LAN), metropolitan area network (MAN), and wide area network (WAN). Due to extensive usage of Internet, the

S.K. Gautam (✉) · H. Om
Department of Computer Science & Engineering, Indian School of Mines, Dhanbad 826004, India
e-mail: gautamsunil.cmri@gmail.com

H. Om
e-mail: hariom4india@gmail.com

S. Das et al. (eds.), *Proceedings of the 4th International Conference on Frontiers in Intelligent Computing: Theory and Applications (FICTA) 2015*, Advances in Intelligent Systems and Computing 404, DOI 10.1007/978-81-322-2695-6_40

network traffic as well as various kinds of attacks such as Internet fraud, hacking, intrusion, etc. are taking place. Several researchers have developed techniques to protect system resources against these suspicious activities, such as firewall, user's authentication, data encryption techniques, etc. These techniques provide primary level security to the network, but they are not highly reliable [3]. Researchers have also developed intrusion detection systems (IDSs), which usually protect the entire network from hackers as well as viruses. The IDSs are mainly categorized into two types namely host-based intrusion detection system (HIDS) and network-based intrusion detection system (NIDS). The former is applicable only in a local system, i.e., single system, and it examines the local data stored in that system. The latter analyzes the entire data packet of the network and detects the malicious code [4–6]. There are various statistical approaches such as support vector machine (SVM), principal component analysis (PCA), independent component analysis (ICA), naive Bayes classifier (NBC), singular value decomposition (SVD), linear discriminant analysis (LDA), etc., for detecting the false alarm rate for intrusion, where false alarm rate refers to the probability of falsely rejecting the null hypothesis for a particular test. In this paper, we use three feature reduction approaches for intrusion detection, i.e., PCA, LDA, and NBC, in the HIDS and analyze their performance.

The rest of the paper is organized as follows. In Sect. 2, we discuss LDA and in Sect. 3 we discuss PCA. The NBC is discussed in Sects. 4 and 5 provides the experimental methodology. Section 6 describes the datasets and in Sect. 7 we discuss the anomaly detection rate. Finally, in Sect. 8 we draw conclusion.

2 Linear Discriminant Analysis

The LDA is a traditional statistical approach that is generally used for data dimensionality reduction and data classification. It is applied to convert the high-dimensional data into low-dimensional data without loss of generality [7]. It provides a precise accuracy of intrusion detection. It has been found useful in many fields such as face recognition, speech recognition, intrusion detection, etc. [8]. The LDA can be used to identify the anomalies in large datasets. It is basically different from the PCA approach because the PCA is useful in extracting features from a huge dataset, not in class discrimination. In other words, the LDA finds the project axes on which the data points of different classes are far from each other, while the data points of the same class are closer to each other. The PCA approach does not provide such flexibility [2, 9]. Besides, the PCA approach is not suitable for large and nonlinear datasets [10]. The LDA approach can be used in both types of intrusion, i.e., NIDS and HIDS. In this paper, we employ LDA in HIDS.

Let D be an n-dimensional dataset, i.e., $D = \{x_1, \ldots, x_n\}$ that is assigned to k different classes, represented by $C_i, i = 1, 2, \ldots, k$. Here, we take two classes, i.e., normal and test classes. The process of finding intrusion in HIDS using LDA [14, 15] is described as follows.

Step 1: Find mean value of each class C_i and complete dataset as follows:

$$\mu_i = \frac{1}{n_i} \sum_{x \in C_i} x \tag{1}$$

$$\mu = \frac{1}{n} \sum_{j=1}^{n} x_j \tag{2}$$

Step 2: Find project matrix, represented by S_b, and between-class scatter matrix as follows:

$$S_b = \sum_{i=1}^{k} n_i (\mu - \mu_i)(\mu - \mu_i)^t \tag{3}$$

Step 3: Find within-class scatter matrix, represented by S_w, as follows:

$$S_w = \sum_{i=1}^{k} \sum_{x \in C_i} (x - \mu_i)(x - \mu_i)^t \tag{4}$$

Step 4: Find eigen value as follows:

$$W^* = S_w^{-1} (\mu_1 - \mu_2) \tag{5}$$

Here μ_1 and μ_2 are mean values of the aforesaid two classes.
Step 5: Sort eigen values and then find project matrix:

$$y_i = W^t \times x \tag{6}$$

Step 6: Calculate major linear discriminant value to detect maximum deviations, denoted by l_m, as follows:

$$l_m = \sum_{i=1}^{q} \frac{y_i^2}{w_i} \tag{7}$$

Step 7: Calculate minor linear discriminant value to detect minimum deviations with lowest value, denoted by l_n, as follows:

$$l_n = \sum_{i=m-r+1}^{m} \frac{y_i^2}{w_i} \tag{8}$$

The minor linear discriminant detects some attacks, which do not exist in correlation model. Thus, we select two threshold values for identification of the attacks namely major threshold value, represented by T_{lq}, and minor threshold value, represented by T_{lr}. The intrusion occurs only if the following two conditions are satisfied:

$$\sum_{i=1}^{q} \frac{y_i^2}{w_i} > T_{lq} \tag{9}$$

$$\sum\nolimits_{i=m-r+1}^{m} \frac{y_i^2}{w_i} > T_{lr} \tag{10}$$

3 Principal Component Analysis (PCA)

Due to advancements in Internet technologies, the amount of data generated by a system is very large. In order to efficiently handle such high-dimensional data, we should reduce its dimensionality without losing its important information. The PCA approach helps reducing the data dimensionality and it is widely used in several applications such as intrusion detection, face recognition, voice recognition, pattern recognition, etc. The PCA is a statistical approach applied on large datasets to detect anomalies. It is also called as Karhunen–Loeve transform, which is the basis for data compression especially JPEG standard. It reduces the complexity and converts a large dataset into small datasets without losing any important property of the original dataset [1, 4, 7]. It is basically a data reduction technique, not an intrusion detection technique. Mathematically, it de-correlates the correlated random variables as the linear combination of the original variables, used to express the data in a reduced form. The first principal component is the projection on the direction in which the variance of the projection is maximized. The second principal component is the linear combination of the original variables with the second largest variance and orthogonal to the first principal component, and so on [11, 12]. The process of finding intrusion in HIDS using LDA is described as follows:

Step 1: We take a set of observation X_1, X_2 … X_n, where each observation is represented by a vector of length m. We can represent matrix $X_{m \times n}$, as follows:

$$\begin{bmatrix} x_{11} & x_{12} & \dots & x_{1n} \\ x_{21} & x_{22} & \dots & x_{2n} \\ \dots & \dots & \dots & \dots \\ x_{m1} & x_{m2} & \dots & x_{m} \end{bmatrix} \tag{11}$$

$$= [X_1 \quad X_2 \quad \dots \quad X_{mn}]$$

Step 2: Find average μ of matrix $X_{m \times n}$ as follows:

$$\mu = \begin{bmatrix} x_1 \\ \dots \\ x_j \\ \dots \\ x_m \end{bmatrix} = \begin{bmatrix} \frac{1}{n}\sum_{i=1}^{n} x_{1i} \\ \dots \\ \frac{1}{n}\sum_{i=1}^{n} x_{ji} \\ \dots \\ \frac{1}{n}\sum_{i=1}^{n} x_{mi} \end{bmatrix} \tag{12}$$

Step 3: Calculate deviation $\varnothing_i$ from the average as follows:

$$\varnothing_i = [x_i - \mu] \tag{13}$$

Step 4: Calculate covariance matrix $V_{m\times m}$ of the dataset using deviation as follows:

$$V_{m\times m} = \frac{1}{n}\sum_{i=1}^{n}(X_i - \mu)(X_i - \mu)^t \tag{14}$$

Step 5: Covariance matrix $V_{m\times m}$ reduces the dataset dimensionality. So we can easily compute the eigen values and eigen vectors of the covariance matrix and also arrange them in ascending order as given bellow:

$$(\lambda_1, \omega_1), (\lambda_2, \omega_2), \ldots, (\lambda_m, \omega_m)$$

Step 6: Let sample principal component of observation X be $y_1, y_2, \ldots, y_m$. Then we have

$$y_i = w^m(X - \bar{\mu}), i = 1, 2, \ldots, m \tag{15}$$

Step 7: Sum of square of each partial principal component score is equal to the principal component score, i.e.,

$$\sum_{i=1}^{m}\frac{y_i^2}{\lambda_i} = \frac{y_1^2}{\lambda_1} + \frac{y_2^2}{\lambda_2} + \cdots + \frac{y_m^2}{\lambda_m} \tag{16}$$

This equation is the Mahalanobis distance of the dataset X.

Step 8: Calculate major principal component, denoted by c_1, to detect maximum deviations with highest value as follows:

$$c_1 = \sum_{i=1}^{q}\frac{y_i^2}{\lambda_i} \tag{17}$$

Step 9: Find minor principal component to detect minimum deviations with lowest value, denoted by c_2, as follows:

$$c_2 = \sum_{i=m-r+1}^{m}\frac{y_i^2}{\lambda_i} \tag{18}$$

The minor principal component detects some attacks, which do not exist in correlation model. Thus, we select two threshold values for identification of the attacks namely major threshold value, represented by T_q, and minor threshold value, represented by T_r. If any attack occurs, then it must satisfy the following two conditions:

$$\sum_{i=1}^{q} \frac{y_i^2}{\lambda_i} > T_q \tag{19}$$

$$\sum_{i=m-r+1}^{q} \frac{y_i^2}{\lambda_i} > T_r \tag{20}$$

4 Naïve Bayes Classifier

The NBC is based on probability model and it is efficient in classifying the attributes. This technique, however, works on assumptions that if the assumptions are correct, then it works very well; otherwise it does not [13]. It is not so much feasible to miscellaneous datasets [16]. In this paper, we implement naïve Bayes classification on Weka tools, and thus we get corrected classified instances.

5 Experimental Methodology

In this paper, we have taken the data generated by a personal computer for evaluating the performance of the LDA and PCA. These data are log files containing normal datasets and test datasets. These datasets are generated from a standalone personal computer without connecting it to Internet, with the following configuration:

- Processor—Intel (R) Core (TM) i3;
- Memory—4.00 GB RAM;
- Operating System—Microsoft Window 8; and
- Architecture—64 bit Operating System.

We have taken the following attributes from the system generated data.

a. *Datagrams Received/Delivered per second*
   It is the rate at which the input IP datagrams have successfully been delivered to IP user protocols, including ICMP.
b. *Datagram per second*
   It is the overall transmission rate for IP datagrams being sent and received over the network interfaces per second (sum of IP datagrams sent/s and IP datagrams received/s).

c. *Datagram received per second*
   It refers to the rate at which the IP datagrams are received from the network interfaces. This counter does not include the datagrams forwarded to another server.
d. *File Write Operation per second*
   It refers to the combined rate (in incidents per second) of file system write requests to all devices on the computer, including requests to write data into the file system cache.
e. *Committed Bytes in use*
   The committed memory is the physical memory in use for which the space has been reserved in paging file so that it can be written to disk.
f. *User Time*
   The percentage of time for which a process executes in user mode is the user time.
g. *Segment per second*
   It refers to the rate per second of TCP segments that are sent or received using the TCP protocol. It is in addition of the segments received per second and the segments send per second.
h. *Processor Privileged time*
   It is same as the processor time, but counts only the kernel mode processor cycles.

6 Datasets

In this section, we describe the process of generating the dataset. The following steps have been followed to generate the data. Figure 1 shows the process of the system's generated datasets.

Step1: Computer's Start-Menu → Settings → Control Panel → Administrative Tools and then double click on Computer Management
Step2: Explore the performance and then select Data Collector Sets → right click on User Defined and finally choose new → Create new Collector Set
Step3: Provide a name for Data Collector Set and then click on OK
Step4: Finally, user can add or remove the counters by going through the properties of the created New Data Collector Set
Step5: The generated log file can be found by going through the following menu: Performance → Reports → User Defined and then select created log-file
Step6: The log files which are comma separated value (.csv) format can be easily exported to Microsoft Excel

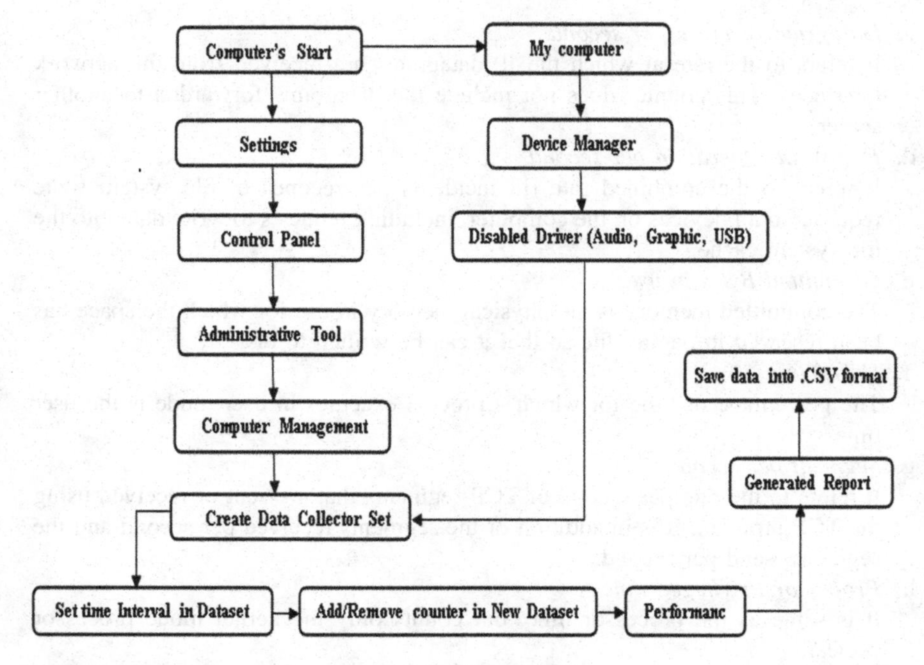

Fig. 1 Process to generate datasets for LDA, PCA, and NBC

Step7: Using this process, we collect the performance logs files for every day. It is the training dataset, which is shown in Table 1

Step8: Now, disable the following drivers: Audio driver, Graphic driver, USB driver, and again collect the performance log files using steps 1 to 7 above. It is the tested data, which is shown in Table 2

7 Detection False Alarm Rate

In this section, we have analyzed the false alarm rate and detection rate used by confusion matrix. The simulation has been done on MATLAB and after performing experiments, we have analyzed the results based on training (normal) dataset and tested (intruded) dataset classified into normal and abnormal classes. The confusion matrix is a set of actual class and predicted class shown in Table 3. Both these classes contain normal class and anomaly class.

Table 1 Training dataset

D.g. Rec. Del./s	D.g./s	D.g. Rec./s	File wt. Op./s	Committed bytes in use	User time	Segment/s	Pro. pri. time
10.49531	8.36454	13.24786	40.23469	35.94079	5.32469	0.00000	2.80230
11.47996	11.47996	11.47996	17.21994	36.56161	21.52234	0.00000	0.77939
2.01888	6.056647	6.056647	33.31156	37.02617	22.86933	42.43424	1.97324
19.93784	28.90986	28.90986	141.5586	37.27585	3.50498	17.95173	0.77919
9.92481	17.86466	17.86466	113.1428	37.29353	3.10138	37.86034	6.22710
10.07906	15.11859	15.11859	573.4986	37.40068	13.7801	41.83300	6.22508
7.93749	9.92187	9.92187	340.32014	37.42835	4.65076	2.02232	1.97493
5.04168	27.22512	27.22512	218.80935	37.44587	6.30219	0.00000	3.49883
6.94568	28.77499	28.77499	113.11553	37.44048	3.48830	16.94285	1.94651
3.02613	17.14812	17.14812	105.91493	37.47542	3.54629	31.53968	6.62731
4.96336	9.92673	9.92673	88.34794	37.49529	1.16330	10.70318	6.23417
3.02639	3.02639	3.02639	133.16123	37.50328	4.33458	97.38710	3.95405
2.97850	2.97850	2.97850	34.74918	37.50493	0.77566	44.90650	1.55934
5.65423	9.32657	4.72698	69.60123	37.50721	1.57611	3.98393	2.72338
10.91216	10.91216	10.91216	130.94651	37.48927	3.09999	4.97871	2.33369
0.36469	7.13465	2.36463	67.72533	37.48564	1.94534	37.45454	2.76802
10.09207	10.09207	10.09207	114.04045	37.47736	2.75956	26.86964	0.77749
0.99220	0.99220	0.99220	93.26764	37.47723	1.55032	17.90992	1.16601
0.00000	6.96461	0.00000	37.85281	48.57050	1.16721	5.97615	1.55628
3.03355	25.89929	3.98368	12.11785	48.16886	3.94986	18.20134	1.18496
5.98640	3.02946	2.98826	25.86400	49.50400	6.62576	60.86182	2.33850
0.00000	5.96862	5.98215	6.97014	49.50537	14.39920	42.83985	4.28084
0.00000	29.87204	4.04191	67.72898	49.50620	8.55731	2.98518	0.38871
13.93927	15.92869	26.89150	6.97438	49.45230	12.86475	0.00000	0.77735
0.00000	1.01116	5.05694	13.14495	49.45312	4.27936	4.97902	3.11175
1.01226	3.98434	8.98149	6.97073	49.46271	14.77012	11.34951	3.16338
13.93927	0.00000	5.97520	4.97972	49.48716	19.09495	0.00000	2.72889
3.98367	0.00000	3.02975	12.94671	49.47983	1.97466	10.95512	1.16709
3.98017	2.02426	0.00000	7.07728	49.48497	0.38900	35.82162	3.10950
6.98324	0.99595	1.99572	11.95191	49.48901	9.73096	82.80132	6.23509
2.02205	0.00000	16.93189	4.98035	49.49366	1.58029	3.12463	1.57972
1.99174	0.99566	0.00000	5.97578	49.50839	2.72894	1.99174	1.16703
0.99644	3.03358	15.92878	21.23393	49.50003	17.50069	19.92890	1.94619
0.00000	2.98828	1.01030	6.96964	49.50051	10.11934	0.00000	1.58029
29.87203	0.99697	11.95185	13.95305	49.54790	4.27955	0.99800	0.38985
0.99563	1.99173	6.96461	11.95855	49.59748	15.01229	38.82972	1.94452
7.97071	6.06760	25.89929	4.04663	49.52373	5.06786	0.00000	2.80230
0.00000	0.00000	3.02946	6.96461	49.53297	7.00325	0.00000	0.77939
1.01027	0.99611	5.96862	37.85281	49.14264	12.45698	42.43424	12.45698

Table 2 Tested dataset

D.g. Rec. Del./s	D.g./s	D.g. Rec./s	File wt. Op./s	Committed bytes in use	User time	Segment/s	Pro. Pri. time
0.00000	0.99996	0.00000	6.12993	48.72842	16.26893	0.00000	2.36677
0.00000	0.00000	0.99996	2.99984	49.13995	1.17181	0.00000	1.17000
1.99981	6.99935	0.00000	2.98784	49.13734	0.00000	0.00000	0.38907
1.00857	5.04285	6.99935	2.99989	49.13566	0.00000	0.00000	1.58145
5.99966	0.99673	5.04285	1.99991	49.13734	5.85907	0.00000	1.55641
1.99983	1.00015	0.99673	24.99772	49.13801	0.78797	0.99604	0.77849
0.00000	1.00000	1.00015	7.05999	49.29697	20.31188	0.00000	0.00000
0.00000	0.99994	1.00000	5.98037	49.28994	20.70197	0.00000	2.76853
0.99992	0.00000	0.99994	3.00043	49.28974	42.18534	0.00000	0.38879
5.99966	0.00000	0.00000	3.00000	49.29007	28.51424	0.00000	1.16728
0.99986	0.00000	1.99996	2.99982	49.28953	25.38932	4.98038	0.38910
0.99999	0.99994	15.99080	16.99971	49.28712	23.82691	0.00000	0.39499
1.99997	7.99955	12.99892	1.99996	49.2665	24.60811	0.00000	1.94539
0.99985	8.99930	7.99955	2.99986	49.24668	23.82691	0.00000	0.00000
0.99986	3.99993	8.99930	1.99987	49.24568	26.95174	0.00000	0.38909
0.99999	6.99903	3.99993	5.99989	49.24246	13.28035	0.00000	1.18607
2.46575	5.99995	6.99903	2.99975	49.24367	0.39060	0.00000	1.16723
3.26475	3.99994	5.99995	1.99988	49.24394	0.00000	1.99331	0.38908
3.99344	25.89704	25.89704	18.92476	43.37004	0.39499	1.99080	0.00000
20.9163	1.99196	1.99196	3.98393	43.37141	0.00000	0.00000	1.18610
3.03641	0.00000	0.00000	15.93712	43.36997	0.00000	1.01213	0.00000
13.9452	16.19375	16.19375	7.08477	43.37018	0.00000	0.00000	0.38910
7.97177	5.97617	5.97617	42.82923	43.36956	0.39535	0.00000	0.00000
0.00000	0.00000	0.00000	13.94499	43.37210	0.38908	0.00000	0.39541
8.09993	1.99120	1.99120	4.97800	43.37073	0.00000	0.00000	0.38904
1.99065	4.04854	4.04854	14.16990	43.37025	0.00000	0.00000	0.00000
1.99196	0.00000	0.00000	5.97657	43.37449	0.00000	0.00000	0.79066
0.00000	3.98439	3.98439	11.95318	43.37292	0.38910	12.95004	0.00000
16.19375	2.02446	2.02446	7.08561	43.37360	1.55639	0.00000	0.77815
5.97617	1.99190	1.99190	12.94734	43.37490	0.39541	0.00000	0.38910
0.00000	1.99194	1.99194	6.97180	43.37333	0.00000	0.00000	0.39542
1.99120	0.00000	0.00000	31.87445	43.36614	0.00000	3.98468	0.77889
4.04854	3.03608	3.03608	12.14431	43.36860	0.00000	0.00000	0.38932
1.99215	9.96075	9.96075	5.97645	43.36991	0.00000	0.00000	1.16649
0.00000	1.99213	1.99213	14.94099	43.37182	0.38909	0.00000	0.39576
3.98439	32.87120	22.91023	11.95316	43.37600	0.00000	0.00000	0.38907
2.02446	4.04910	4.04910	4.04910	43.37853	0.38911	0.00000	0.77889
1.99190	1.99395	1.99395	14.95463	43.38018	0.79084	3.98468	0.00000
1.99194	25.89704	25.89704	18.92476	43.37004	0.39499	43.37004	0.76985

Table 3 Confusion matrix

Actual class	Predicted class	
	C	NC
C	TN	FP
NC	FN	TP

Table 4 Performance analysis

Analysis technique	Detection rate (%)	False alarm rate (%)
LDA	94.86	5.14
PCA	89.72	10.28
NBC	92.58	7.42

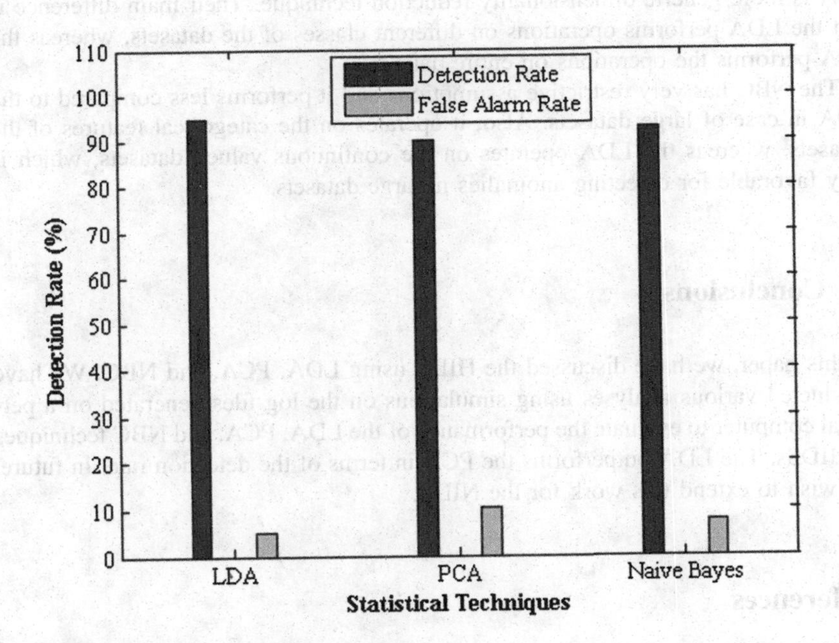

Fig. 2 Performance of detection rate

Here,

C—Anomaly Class	NC—Normal Class
TN—True Negative	FP—False Positive
FN—False Negative	TP—True Positive

True Negative: Percentage of valid records that are correctly classified.
False Positive: Percentage of records that are incorrectly classified as positive.
False Negative: Percentage of records that are incorrectly classified as negative.
True Positive: Percentage of records that are correctly classified as positive.

$$\textbf{Accuracy} = \frac{TN + TP}{TN + TP + FN + FP}$$

$$\textbf{Recall}(R) = \frac{TP}{TP + FN}$$

$$\textbf{Precision}(P) = \frac{TP}{TP + FP}$$

In this paper, we have observed that the LDA outperforms the PCA and naïve Bayes classifier. From our experiment results shown in Fig. 2 and in Table 4, we have observed that the LDA is more specialized generative method, whereas the PCA is more generic dimensionality reduction technique. Their main difference is that the LDA performs operations on different classes of the datasets, whereas the PCA performs the operations on entire datasets.

The NBC has very restrictive assumptions and it performs less compared to the LDA in case of large datasets. Also, it operates on the categorical features of the datasets, whereas the LDA operates on the continuous valued datasets, which is very favorable for detecting anomalies in large datasets.

8 Conclusions

In this paper, we have discussed the HIDS using LDA, PCA, and NBC. We have conducted various analyses using simulations on the log files generated on a personal computer to evaluate the performance of the LDA, PCA, and NBC techniques in HIDSs. The LDA outperforms the PCA in terms of the detection rate. In future, we wish to extend this work for the NIDS.

References

1. Singh, S., Silakari, S.: Generalized Discriminant Analysis Algorithm for Feature Reduction in Cyber-Attack Detection System. arXiv preprint arXiv:0911.0787 (2009)
2. Tan, Z., Jamdagni, A., He, X., Nanda, P.: Network intrusion detection based on LDA for payload feature selection. In: IEEE GLOBECOM Workshops, pp. 1545–1549 (2010)
3. He, X., Yan, S., Hu, Y., Niyogi, P., Zhang, H.J.: Face recognition using Laplacian faces. IEEE Trans. Pattern Anal. Mach. Intell. **27**(3), 328–340 (2005)
4. Aydın, M.A., Zaim, A.H., Ceylan, K.G.: A hybrid intrusion detection system design for computer network security. Comput. Electr. Eng. **35**(3), 517–526 (2009)
5. Gascon, H., Orfila, A., Blasco, J.: Analysis of update delays in signature-based network intrusion detection systems. Comput. Secur. **30**(8), 613–624 (2011)
6. Moskovitch, R., Pluderman, S., Gus, I., Stopel, D., Feher, C., Parmet, Y., Elovici, Y.: Host based intrusion detection using machine learning. In: IEEE Intelligence and Security Informatics, pp. 107–114 (2007)

7. Delac, K., Grgic, M., Grgic, S.: Independent comparative study of PCA, ICA, and LDA on the FERET data set. Int. J. Imaging Syst. Technol. **15**(5), 252–260 (2005)
8. Kasliwal, B., Bhatia, S., Saini, S., Kumar, C.A.: A hybrid anomaly detection model using G-LDA. In: IEEE International Advance Computing Conference (IACC), 2014, pp. 288–293 (2014)
9. Patcha, A., Park, J.M.: An overview of anomaly detection techniques: existing solutions and latest technological trends. Comput. Netw. **51**(12), 3448–3470 (2007)
10. Mechtri, L., Tolba, F.D., Ghoualmi, N.: Intrusion detection using principal component analysis. In: Second International Conference on Engineering Systems Management and Its Applications (ICESMA), 2010, pp. 1–6 (2010)
11. Labib, K., Vemuri, V.R.: An application of principal component analysis to the detection and visualization of computer network attacks. Ann. Télécommun. **61**(1–2), 218–234 (2006)
12. Om, H., Hazra, T.: Statistical techniques in anomaly intrusion detection system. Int. J. Adv. Eng. Technol. **5**(1), 387–398 (2012)
13. Hsu, C.C., Huang, Y.P., Chang, K.W.: Extended Naive Bayes classifier for mixed data. Expert Syst. Appl. **35**(3), 1080–1083 (2008)
14. Imran, H.M., Abdullah, A.B., Hussain, M., Palaniappan, S., Ahmad, I.: Intrusions detection based on optimum features subset and efficient dataset selection. Int. J. Eng. Innovative Technol. **2**, 265–270 (2012)
15. Martínez, A.M., Kak, A.C.: PCA versus LDA. IEEE Trans. Pattern Anal. Mach. Intell. **23**(2), 228–233 (2001)
16. Mukherjee, S., Sharma, N.: Intrusion detection using naive Bayes classifier with feature reduction. Procedia Technol. **4**, 119–128 (2012)

Secure Remote Login Scheme with Password and Smart Card Update Facilities

Rajeev Kumar, Ruhul Amin, Arijit Karati and G.P. Biswas

Abstract Smart card and password-based user authentication scheme is popular for accessing remote services from the remote server over insecure communication. In this regard, numerous user authentication protocols have been proposed in the literature. However, we have observed that still none of the protocols provide complete facilities such as password change process, password recover process, and smart card revocation process to the registered user. The main aim of this paper is to design a secure user authentication protocol which provides complete facilities to the registered user. The security analysis of the protocol is presented which confirms that the same protocol is secure against various common attacks. Our protocol is not only provide complete facilities to the registered user, but also provides session key agreement as well as mutual authentication between the U_i and RS. The performance of the proposed scheme is relatively better than existing related schemes.

Keywords Authentication · Security attacks · Security attributes · Smart card

1 Introduction

In the client–server environment, many password-based user authentication schemes using smart card are widely used to exchange confidential data by encrypting using the generated session key. The concept of the remote user authentication scheme was

R. Kumar (✉) · R. Amin · A. Karati · G.P. Biswas
Department of Computer Science and Engineering, Indian School of Mines, Dhanbad, Jharkhand, India
e-mail: rajeev.kumar@nic.in

R. Amin
e-mail: amin_ruhul@live.com

A. Karati
e-mail: arijit.karati@gmail.com

G.P. Biswas
e-mail: gpbiswas@gmail.com

S. Das et al. (eds.), *Proceedings of the 4th International Conference on Frontiers in Intelligent Computing: Theory and Applications (FICTA) 2015*, Advances in Intelligent Systems and Computing 404, DOI 10.1007/978-81-322-2695-6_41

first proposed by Lamport [24] based on the one-way hash function and thereafter many password-based authentication protocols [1, 2, 4, 6, 7, 10, 14–16, 18–21] have been proposed in the literature. It has been observed that many schemes [17, 23, 27] suffer from smart card stolen attack resulting in offline password guessing. In 2007, Wang et al. [29] illustrated that the schemes [27, 32] cannot withstand forgery attack, offline password guessing attack, and denial-of-service attack and presented better solutions to fix the problems. Later, Awasthi et al. [11] pointed out that Shen et al.'s scheme [26] cannot withstand user impersonation attack, and also proposed an improved protocol. In 2012, Wang et al. [28] pointed out that the Yang et al. [31] and Hsieh-Leu [17] schemes are insecure against smart card loss attack and subsequently proposed an efficient scheme to thwart the problems of smart card security breach. In 2013, Ruhul et al. [10] pointed out that the scheme [28] suffers from offline identity-password guessing attack, user-server impersonation attack, and also proposed an improved protocol to fix the Wang et al. [28] scheme. In this paper, we have designed a secure user authentication protocol using smart card, which is not only efficient in terms of complexities, but also provides essential facilities to the registered user.

1.1 Road Map of the Paper

In Sect. 2, we address our proposed protocol and the security analysis of the proposed scheme appears in Sect. 3. Section 4 measures the performance of this protocol and the conclusion is given in Sect. 5. We complete the paper with several related references.

2 Proposed Protocol

This section presents the smart card-based user authentication and key agreement protocol using the cryptographic one-way hash function. We have listed all the notations used throughout this paper in Table 1. Our proposed protocol has several processes such as registration, login and authentication, password change, password recovery, and smart card revocation, each of which is discussed below.

2.1 Registration Process

In this phase, a new user chooses an identity ID_i and sends registration message $\langle ID_i, MOB \rangle$ along with personal credential information UC_i to the remote server through the insecure channel in order to get remote services, where MOB is the valid mobile number of the user. After getting the registration message, the RS computes $Reg_i = h(ID_i \parallel X_{rs})$ and sends it to the mobile number securely. Then the remote

Table 1 List of notations used

Symbol	Description
SC	Smart card
U_i	User/client
RS	Remote server
$\mathcal{A}$	Attacker/adversary
ID_i	Identity of user U_i
PW_i	User's password
X_s	Secret key of the RS
R_i	Random number generated by the user
R_{rs}	Random number generated by the RS
SK	Session key of the protocol
$\oplus$	The bitwise exclusive or operation
$\parallel$	The concatenation operation
$h(\cdot)$	One-way hash function, $h : (0,1)^* \rightarrow (0,1)^n$

server maintains a table (say, User Table) containing attributes $\langle ID_i, MOB\rangle$ and personal credentials.

After receiving Reg_i, the U_i computes $A_i = h(ID_i \parallel PW_i \parallel Reg_i)$, $B_i = Reg_i \oplus PW_i$ and issues a smart card after storing $\langle ID_i, A_i, B_i\rangle$ into the memory of smart card. After that, the U_i sends $\langle ID_i, SCN_i, B_i\rangle$ to the remote server through open networks.

After receiving the message, the RS first checks whether the ID_i exists in the database. If it exists, the RS further stores $\langle SCN_i, B_i\rangle$ into the user table (See Table 2) and completes the registration process, where SCN_i is the unique number of the smart card.

2.2 Login and Authentication Process

In this process, U_i and RS perform several operations in order to negotiate a session key over insecure networks. The steps of this process are discussed below.

Table 2 User table

Identity	Parameter	Mobile number	SCN	User credential
ID_1	B_1	9804557	SCN_1	UC_1
ID_2	B_2	9868754	SCN_2	UC_2
ID_3	B_3	9878712	SCN_3	UC_3
ID_4	B_4	9878712	SCN_4	UC_4
..	..	....	....	....
ID_n	B_n	6668712	SCN_n	UC_n

Step 1. U_i initially inserts the smart card into the card reader and provides high entropy password PW_i. The term high entropy means that the attacker cannot guess user's password in polynomial time. After getting PW_i, the card reader computes $Reg_i^* = B_i \oplus PW_i$, $A_i^* = h(ID_i \parallel PW_i \parallel Reg_i^*)$ and verifies whether $A_i^* = A_i$ or not. If the verification holds, U_i has provided correct PW_i; otherwise, it aborts the connection. After that, the card reader generates a 128-bit random number R_i and computes $C_i = h(ID_i \parallel R_i \parallel PW_i)$, $D_i = R_i \oplus PW_i$. Finally, the card reader sends $\langle ID_i, C_i, D_i \rangle$ to the remote server over insecure networks.

Step 2. After receiving the login message, *RS* first checks whether ID_i exists in the database. If it does not exist, *RS* immediately rejects the login message; otherwise, the *RS* securely sends one-time password *(OTP)* to the registered mobile number. On receiving *OTP*, the U_i submits it to the *RS* and after that *RS* checks the *OTP* verification. If the verification does not hold, it immediately rejects the session; otherwise, it computes $Reg_i^* = h(ID_i \parallel X_s)$, $PW_i^* = B_i \oplus Reg_i^*$, $R_i^* = D_i \oplus PW_i$, $C_i^* = h(ID_i \parallel R_i^* \parallel PW_i^*)$ and checks the condition $C_i^* \stackrel{?}{=} C_i$. If the condition does not hold, *RS* rejects the session; otherwise, it believes that the sender of the login message is registered.

Step 3. The *RS* now generates a 128-bit random number R_{rs} and computes $G_i = R_i \oplus R_{rs}$, $F_i = h(ID_i \parallel R_i^* \parallel R_{rs} \parallel Reg_i)$, and then sends $\langle F_i, G_i \rangle$ to the user through open networks.

Step 4. After receiving the message $\langle F_i, G_i \rangle$, the user computes $R_{rs} = G_i \oplus R_i$, $F_i^* = h(ID_i \parallel R_i \parallel R_{rs} \parallel Reg_i)$ and checks the correctness as $F_i^* \stackrel{?}{=} F_i$. If it does not hold, it terminates the connection; otherwise, the user believes that the remote server is authentic. Now, both parties compute the session key $SK = h(ID_i \parallel Reg_i \parallel R_i \parallel R_{rs})$ and start secure communication in future.

2.3 Password Change Process

It is possible that the password of the user may be leaked by someone and hence the user should change the password. In any password-based authentication system, the protocol should provide password change facility. The description of password change procedure is as follows.

Step 1. The U_i initially inserts the smart card into the card reader and provides high entropy password PW_i. After getting PW_i, the card reader computes $Reg_i^* = B_i \oplus PW_i$, $A_i^* = h(ID_i \parallel PW_i \parallel Reg_i^*)$ and verifies $A_i^* \stackrel{?}{=} A_i$. If the verification holds, the card reader asks to enter new password PW_i^{new} to the user.

User U_i/Smartcard
Inserts SC and provides PW_i
SC computes $Reg_i = B_i \oplus PW_i$, $A_i^* = h(ID_i \parallel PW_i \parallel Reg_i)$
If($A_i^* \neq A_i$)
Rejects Else
Computes $B_i^{new} = Reg_i \oplus PW_i^{new}$, $A_i^{new} = h(ID_i \parallel PW_i^{new} \parallel Reg_i)$.
Replace $\langle A_i, B_i \rangle$ with new information $\langle A_i^{new}, B_i^{new} \rangle$ respectively.

Fig. 1 Password change process

Step 2. After getting the PW_i^{new}, the card reader computes $B_i^{new} = Reg_i \oplus PW_i^{new}$, $A_i^{new} = h(ID_i \parallel PW_i^{new} \parallel Reg_i)$ and replaces the old information $\langle A_i, B_i \rangle$ with the new information $\langle A_i^{new}, B_i^{new} \rangle$ into the memory of smart card. Thus, the U_i can easily change his/her password without taking any assistance from the remote server.

We further describe password change phase in Fig. 1.

2.4 Forgot Password Recover Process

It is a common problem with many users that they forget their password due to either accessing several web servers or rarely use them. Therefore, it is an important desirable property of the authentication protocol to provide password recover facility. The proposed protocol provides password recover facility to the user. We further describe this process in Fig. 2.

Step 1. User initially submits $\langle ID_i, MOB \rangle$ to the remote server.

Step 2. On getting it, *RS* checks whether the ID_i and the corresponding *MOB* exists in the user table. If it exists, the *RS* computes $PW_i = B_i \oplus Reg_i$; otherwise, it terminates the request.

Step 3. Finally, the *RS* sends PW_i to the registered *MOB* securely. Thus, our protocol provides password recover facility to the registered user.

User U_i/Smartcard	Remote Server RS
U_i submits $\langle ID_i, MOB \rangle$	Checks ID_i and it's corresponding MOB
	If correct, Compute $PW_i = B_i \oplus Reg_i$
	←Sends securely PW_i to the registered MOB

Fig. 2 Forgot password recovery process

User U_i/Smartcard	Remote Server RS
U_i submits $\langle ID_i, MOB\rangle$	
Along with personal credentials	Checks U_i's legitimacy using personal credential If correct, Compute $Reg_i = h(ID_i \parallel X_s)$ $\leftarrow$Sends securely Reg_i to the registered MOB
U_i computes $B_i = Reg_i \oplus PW_i^{new}$ $A_i = h(ID_i \parallel PW_i^{new} \parallel Reg_i)$ Embeds $\langle ID_i, A_i, B_i\rangle$ in smart card Sends ID_i, SCN_i, B_i to the RS	
	Checks existence of ID_i. then updates user table with new SCN_i, B_i

Fig. 3 Smart card revocation process

2.5 Smart Card Revocation Process

It is possible that the user's smart card may have got lost. As a result, the user is unable to access the remote server. In order to get new smart without re-registration, our protocol provides smart card revocation facility to the registered user. We further describe this process in Fig. 3.

Step 1. The user submits $\langle ID_i, MOB\rangle$ along with personal credential information to the remote server.

Step 2. After receiving the request, *RS* first verifies the legitimacy of the user based on the personal credential along with $\langle ID_i, MOB\rangle$. If the verification holds, *RS* computes $Reg_i = h(ID_i \parallel X_s)$ and sends it to the registered mobile number securely.

Step 3. After getting Reg_i information, U_i computes $A_i = h(ID_i \parallel PW_i^{new} \parallel Reg_i)$, $B_i = Reg_i \oplus PW_i^{new}$.

Step 4. Then the U_i produces a new smart card after storing $\langle ID_i, A_i, B_i\rangle$ into the memory of smart card. Finally, U_i sends $\langle ID_i, SCN_i, B_i\rangle$ to the remote server through open networks.

Step 5. After receiving the message, *RS* first checks whether the ID_i exists in the database. If it exists, *RS* further stores $\langle SCN_i, B_i\rangle$ into the user table.

3 Security Analysis and Discussion

In this section, we cryptanalyze the proposed protocol against various attacks. It is well known that an attacker has extreme capabilities over insecure networks, this means he/she can trap, delete, regenerate, reroute the login-reply message and try to authenticate him/herself to the server or the user for retrieving the confidential information(s). In this paper, we have assumed that the user always uses high entropy password which cannot be guessed in polynomial time. The high entropy password includes lower–upper character, special symbol, and numeric character and it should not be trouble-free for guessing the high entropy password [3].

3.1 Smart Card Stolen Attack

We assumed that the attacker $\mathcal{A}$ has obtained the legal user smart card and extracted smart card information $\langle ID_i, A_i, B_i \rangle$ using power analysis attack [22, 25], where $A_i = h(ID_i \parallel PW_i \parallel Reg_i)$, $B_i = Reg_i \oplus PW_i$. Then $\mathcal{A}$ tries to extract confidential information $\langle PW_i, X_s \rangle$. $\mathcal{A}$ is unable to extract $P\langle W_i, X_s \rangle$ using A_i, as it is protected by the cryptographic one-way hash function. Furthermore, he/she cannot extract PW_i without the knowledge of Reg_i from B_i. Similarly, $\mathcal{A}$ is unable to extract PW_i after intercepting login message during protocol execution.

The above discussion not only resists the smart card stolen attack, but also provides strong security protection on the user's password.

3.2 User Impersonation Attack

In this attack, $\mathcal{A}$ tries to impersonate as a valid user after intercepting login request message. We suppose that $\mathcal{A}$ has trapped $\langle ID_i, C_i, D_i \rangle$ during protocol execution and tries to compute another valid message, where $C_i = h(ID_i \parallel R_i \parallel PW_i)$, $D_i = R_i \oplus PW_i$. However, $\mathcal{A}$ is unable to compute valid $\langle C_i, D_i \rangle$ without knowledge of $\langle Reg_i, PW_i \rangle$. Therefore, we claim that the proposed protocol can withstand user impersonation attack.

3.3 Server Impersonation Attack

Resembling user impersonation attack [8], the attacker $\mathcal{A}$ tries to impersonate as valid server after providing valid reply message $\langle F_i, G_i \rangle$ to the user, where $G_i = R_i \oplus R_{rs}$, $F_i = h(ID_i \parallel R_i^* \parallel R_{rs} \parallel Reg_i)$. It is noticeable that the $\mathcal{A}$ cannot compute valid reply message without knowledge of Reg_i. Therefore, $\mathcal{A}$ cannot impersonate as valid remote server.

3.4 Session Key Computation Attack

In this attack model, $\mathcal{A}$ attempts to compute valid session key during protocol execution, so that $\mathcal{A}$ can extract confidential information after decrypting the ciphertext.

Therefore, $\mathcal{A}$ needs the session key $SK = h(ID_i \parallel Reg_i \parallel R_i \parallel R_{rs})$ parameters. It is noted that the computation of the session key of the proposed protocol depends on the non-invertible cryptographic one-way hash function. We have demonstrated earlier that $\mathcal{A}$ cannot compute Reg_i from the protocol description. Additionally, the random numbers $\langle R_i, R_{rs} \rangle$ used in our protocol are secure. Therefore, $\mathcal{A}$ is not able to compute valid session key of the proposed protocol.

3.5 Known Session Specific Temporary Information Attack

We assume that the short-term information $\langle R_i, R_{rs} \rangle$ used in our protocol has leaked in some way and been obtained by the attacker. Now $\mathcal{A}$ tries to compute valid session key of the proposed protocol. Since the session key depends on another secret information Reg_i, the attacker is not able to launch such type of attack [5, 9].

3.6 Mutual Authentication

Our protocol provides mutual authentication property between U_i and RS. After getting the login message, RS verifies the legitimacy of U_i, and U_i also verifies the authenticity of RS after receiving the reply message. Therefore, the proposed protocol provides mutual authentication property.

3.7 Fast Error Detection

During the login phase, the smart card first verifies the authenticity of U_i based on the user's PW_i. It is noted that if U_i provides wrong password by mistake or nonregistered user tries to access stolen smart card, the system quickly detects U_i. Therefore, the protocol provides fast error detection mechanism which reduces network congestion as well as communication and computation cost of the protocol.

4 Performance Comparison

This section evaluates the performance of the proposed scheme with several related competitive schemes. In order to measure performance of the protocol, computation and communication cost are the two main important attributes and should be as minimum as possible. In Table 3, we have provided computation and communication cost of the proposed scheme with several related schemes, where T_h and $T_{\oplus}$ indicate hash function and X-or operations respectively. It is noticeable that the

Table 3 Computation cost and communication cost comparison of the proposed scheme with related schemes

Scheme	Communication cost	Computation cost
Ruhul et al. [10]	640 bits	$14T_h + 5T_e$
Saru et al. [23]	768 bits	$15T_h + 18T_{\oplus}$
Wang et al. [29]	768 bits	$11T_h + 13T_{\oplus}$
Chou et al. [13]	1080 bits	$29T_h + 24T_{\oplus}$
Chang et al. [12]	768 bits	$12T_h + 6T_{\oplus}$
Wen and Li [30]	1152 bits	$25T_h + 22T_{\oplus}$
Our	640 bits	$10T_h + 7T_{\oplus}$

communication cost is reasonable compared to the other schemes, but the computation cost is better. The proposed scheme is not only relatively better in terms of complexity, but also provides strong security protection against common attacks.

5 Conclusion

This paper contributes an efficient and practical smart card-based user authentication along with session key agreement protocol usable in client–server environment. The protocol not only negotiates session key agreement, but also provides most important desirable facilities such as password change process, password recovery process, and smart card recovery process to the registered user. The security analysis ensures that the proposed protocol provides good security protection against several common attacks and achieves mutual authentication property. The performance of the proposed scheme is also relatively better than existing schemes. The overall efficiency of the proposed protocol claims that it should be applicable in the client–server environment.

References

1. Amin, R.: Cryptanalysis and an efficient secure id-based remote user authentication scheme using smart card. In: IJCA. vol. 75, pp. 1149–1157. Citeseer (2013)
2. Amin, R., Biswas, G.P.: Anonymity preserving secure hash function based authentication scheme for consumer usb mass storage device. In: IEEE 2015 Third International Conference on Computer, Communication, Control and Information Technology (C3IT), pp. 1–6. (2015)
3. Amin, R., Biswas, G.P.: Cryptanalysis and design of a three-party authenticated key exchange protocol using smart card. Arab. J. Sci. Eng. 1–15 (2015). http://dx.doi.org/10.1007/s13369-015-1743-5
4. Amin, R., Biswas, G.P.: Design and analysis of bilinear pairing based mutual authentication and key agreement protocol usable in multi-server environment. Wireless Pers. Commun. 1–24 (2015)

5. Amin, R., Biswas, G.P.: An improved rsa based user authentication and session key agreement protocol usable in tmis. J. Med. Syst. **39**(8), 79 (2015). http://dx.doi.org/10.1007/s10916-015-0262-y
6. Amin, R., Biswas, G.P.: A novel user authentication and key agreement protocol for accessing multi-medical server usable in tmis. J. Med. Syst. **39**(3), 1–17 (2015)
7. Amin, R., Biswas, G.P.: Remote access control mechanism using rabin public key cryptosystem. In: Information Systems Design and Intelligent Applications, pp. 525–533. Springer (2015)
8. Amin, R., Biswas, G.P.: A secure light weight scheme for user authentication and key agreement in multi-gateway based wireless sensor networks. Ad Hoc Netw. (2015)
9. Amin, R., Biswas, G.P.: A secure three-factor user authentication and key agreement protocol for tmis with user anonymity. J. Med. Syst. **39**(8), 78 (2015). http://dx.doi.org/10.1007/s10916-015-0258-7
10. Amin, R., Maitra, T., Rana, S.P.: An improvement of wang. et. al.s remote user authentication scheme against smart card security breach. Int. J. Comput. Appl. **75**(13), 37–42 (2013)
11. Awasthi, A.K., Srivastava, K., Mittal, R.: An improved timestamp-based remote user authentication scheme. Comput. Electr. Eng. **37**(6), 869–874 (2011)
12. Chang, Y.F., Tai, W.L., Chang, H.C.: Untraceable dynamic-identity-based remote user authentication scheme with verifiable password update. Int. J. Commun. Syst. **27**(11), 3430–3440 (2014)
13. Chou, J.S., Huang, C.H., Huang, Y.S., Chen, Y.: Efficient two-pass anonymous identity authentication using smart card. IACR Cryptology ePrint Archive **2013**, 402 (2013)
14. Giri, D., Maitra, T., Amin, R., Srivastava, P.: An efficient and robust rsa-based remote user authentication for telecare medical information systems. J. Med. Syst. **39**(1), 1–9 (2015)
15. He, D., Kumar, N., Chilamkurti, N.: A secure temporal-credential-based mutual authentication and key agreement scheme with pseudo identity for wireless sensor networks. Information Sciences 321, 263–277 (2015), security and privacy information technologies and applications for wireless pervasive computing environments
16. He, D., Kumar, N., Chilamkurti, N., Lee, J.H.: Lightweight ecc based rfid authentication integrated with an id verifier transfer protocol. J. Med. Syst. **38**(10), 116 (2014)
17. Hsieh, W.B., Leu, J.S.: Exploiting hash functions to intensify the remote user authentication scheme. Comput. Secur. **31**(6), 791–798 (2012)
18. Islam, S.H.: A provably secure id-based mutual authentication and key agreement scheme for mobile multi-server environment without esl attack. Wireless Pers. Commun. **79**(3), 1975–1991 (2014)
19. Islam, S.H.: Design and analysis of a three party password-based authenticated key exchange protocol using extended chaotic maps. Inf. Sci. **312**, 104–130 (2015)
20. Islam, S., Gosta Pada Biswas, K.K.C.: Cryptanalysis of an improved smartcard-based remote password authentication scheme. Inf. Sci. Lett. **3**(1), 35–40 (2014)
21. Islam, S., Khan, M.K., Obaidat, M., Muhaya, F.: Provably secure and anonymous password authentication protocol for roaming service in global mobility networks using extended chaotic maps. Wireless Pers. Commun. 1–22 (2015)
22. Kocher, P., Jaffe, J., Jun, B.: Differential power analysis. In: Advances in Cryptology CRYPTO99. pp. 388–397. Springer (1999)
23. Kumari, S., Khan, M.K., Li, X.: An improved remote user authentication scheme with key agreement. Comput. Electr. Eng. **40**(6), 1997–2012 (2014)
24. Lamport, L.: Password authentication with insecure communication. Commun. ACM **24**(11), 770–772 (1981)
25. Messerges, T.S., Dabbish, E., Sloan, R.H., et al.: Examining smart-card security under the threat of power analysis attacks. Comput. IEEE Trans. **51**(5), 541–552 (2002)
26. Shen, J.J., Lin, C.W., Hwang, M.S.: Security enhancement for the timestamp-based password authentication scheme using smart cards. Comput. Secur. **22**(7), 591–595 (2003)
27. Ku, W.C.: S.M.C. Weakness and improvement of an efficient password based remote user authentication scheme using smart cards. IEEE Trans. Consum. Electron. **50**(1), 204–207 (2004)

28. Wang, D., Ma, C.G., Zhang, Q.M., Zhao, S.: Secure password-based remote user authentication scheme against smart card security breach. J. Netw. **8**(1), 148–155 (2013)
29. Wang, X.M., Zhang, W.F., Zhang, J.S., Khan, M.K.: Cryptanalysis and improvement on two efficient remote user authentication scheme using smart cards. Comput. Stan. Interfaces **29**(5), 507–512 (2007)
30. Wen, F., Li, X.: An improved dynamic id-based remote user authentication with key agreement scheme. Comput. Electr. Eng. **38**(2), 381–387 (2012)
31. Yang, G., Wong, D.S., Wang, H., Deng, X.: Two-factor mutual authentication based on smart cards and passwords. J. Comput. Syst. Sci. **74**(7), 1160–1172 (2008)
32. Yoon, E.J., Ryu, E.K., Yoo, K.Y.: Further improvement of an efficient password based remote user authentication scheme using smart cards. Consum. Electron. IEEE Trans. **50**(2), 612–614 (2004)

Generation and Risk Analysis of Network Attack Graph

Keshav Prasad, Santosh Kumar, Anuradha Negi and Aniket Mahanti

Abstract Attack graph describes how an attacker can compromise with network security. To generate the attack graph, we required system as well as vulnerability information. The system information contains scanned data of a network, which is to be analyzed. The vulnerability data contain information about, how exploits can be generated due to multiple vulnerabilities and what effects can be of such exploitation. Multihost multistage vulnerability analysis (MulVAL) tool is used for generating attack graph in this work. MulVAL generated graphs are logical attack graphs based on logical programming and based on dependencies among attack goal and configuration information. The risk of network attack graph is measured through graph topology theoretic properties (connectivity, cycles, and depth), and analysis of possible attacks paths is carried out in this paper.

Keywords Attack graph · Logical attack graph · Network security metrics · Risk analysis · Attack paths

1 Introduction

It is an emerging issue to control network security with growth of network size and numbers of vulnerabilities. The dynamic nature of the attacks and network size are the current challenges for the attack graph generation and analysis of attack graph.

K. Prasad (✉) · S. Kumar · A. Negi
Department of Computer Science & Engineering, Graphic Era University, Dehradun, India
e-mail: kainulyk@gmail.com

S. Kumar
e-mail: amu.santosh@gmail.com

A. Negi
e-mail: anuradhanegi40@gmail.com

A. Mahanti
Department of Computer Science, University of Auckland, Auckland, New Zealand
e-mail: a.mahanti@auckland.ac.nz

S. Das et al. (eds.), *Proceedings of the 4th International Conference on Frontiers in Intelligent Computing: Theory and Applications (FICTA) 2015*, Advances in Intelligent Systems and Computing 404, DOI 10.1007/978-81-322-2695-6_42

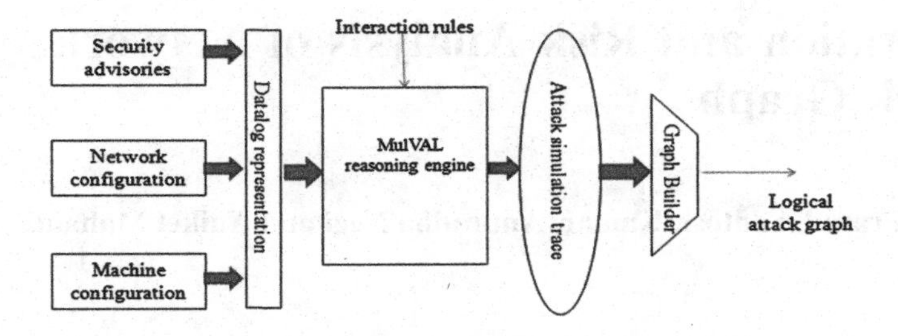

Fig. 1 Logical attack graph generator [1]

Therefore, it is essential to integrate very fast and accurate security approaches to evaluate the existing security state of the network. For this purpose, attack graph [1, 2] is mostly used technique. An attack graph describes how an attacker can gain privileges on a system by linking multiple vulnerabilities [3]. How, a logical attack graph can be generated, the required approach is shown in Fig. 1. There are four phases for logical attack graph generation, phase 1 Data log representation, phase 2 interaction rule with MulVAL and XSB reasoning engine, phase 3 attack simulation traces, and finally phase 4 graph builder as show in Fig. 1.

Phase 1 Datalog Representation It is obtained with integration of three components which are major inputs for an attack graph generator:

First component security advisories: which consists firewall rules, i.e., allow the connection, allow the connection with security check, and block the connection.
Second component network Configuration: which provides description about routers, switches, and other network devices which are involved to control the entire network.
Third component machine configuration: which describes about platform of host machine.

Phase 2 Interaction Rule with MulVAL and XSB Reasoning Engine MulVAL [4, 5] is an end-to-end framework and reasoning system that conducts multihost, multistage vulnerability analysis of a network. The MulVAL reasoning engine takes encoded information as an input from datalog representation and produces complete logical attack graph along with the basic network topology information for analyzing the whole network. Figure 1 shows the complete framework of the MulVAL system. MulVAL system is able to import data from the vulnerability scanners such as Nessus [6], Oval and vulnerability database NVD [7]. The reasoning engine consists of collection of Datalog rules that capture behavior of the operating system and interaction between various components in the network. MulVAL is an integrated system of MulVal software, XSB, and graphviz. XSB is a reasoning system and logical programming of attacks is done in XSB.

Phase 3 Attack Simulation Traces The result of this phase details the vulnerable IP addresses in the network and list of vulnerabilities present in those IP addresses.

Phase 4 Graph builder The attack simulation traces are fed into Graphviz to obtain logical attack graph in various formats.

Logical Attack Graph

Logical attack graph describes logical dependencies among network attack goals and configuration information. A logical attack graph always has size polynomial to the network, which is to be analyzed. The edges of the network attack graphs specify causal relations between network configurations and attackers potential privileges.

Figure 2 shows an example of a simple logical attack graph, nodes p1 and p2 are called as privilege nodes, while nodes e1, e2, and e3 are called exploit nodes. The nodes c1, c2, c3, c4, c5, c6, and c7 are referred to as fact/configuration nodes. In essence, the exploit nodes represent the host machines in the network that can be attacked, while the configuration nodes are sets of configurations (such as firewall rules and IDS configurations) that those hosts have been configured with. The privilege nodes represent the privileges that those host machines are capable of running. In the representation of logical attack graphs, the privilege nodes are treated as OR nodes, which can be satisfied, if any of the child node is true. Exploit nodes are treated as AND nodes which only be satisfied when all of its children are satisfied.

In this paper, we have used Ubuntu 12.04 platform for generation and MATLAB for risk analysis of attack graph. Rest of the paper is organized as follows. Section 2 describes graph theoretic properties-based security metrics [8, 9, 10]. The generated graph analysis and possible attack paths in attack graph [11] are analyzed in Sect. 3. Finally, Sect. 4 concludes the work.

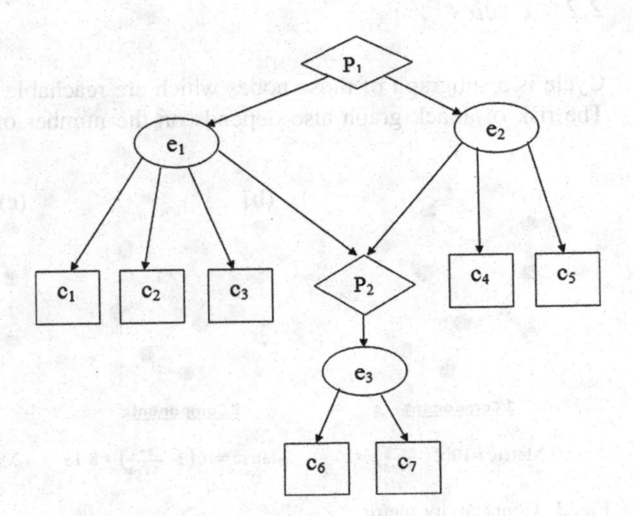

Fig. 2 Logical attack graph

2 Graph Theoretic Properties

Risk analysis over attack graph can be done using graph theoretic properties [8]. Graph theoretic properties only can be applied on directed graph. These properties are useful to determine risk associated with the attack graph. We have calculated the risks on the base of common vulnerability scoring system (CVSS) [12] scale of (0, 10) where 0 signify most secure and 10 signify least secure.

2.1 *Connectivity*

Connectivity defines the number of connected components in a graph. The worst case (least secure) is when a graph has only single components and best case (most secure) is when the graph is totally disconnected. If w is the number of components in the graph and d is the most possible number of components (total number of nodes) in the graph then connectivity metric is given by equation:

$$\text{Connectivity metric} = 10\left(1 - \frac{w-1}{n-1}\right) \tag{1}$$

An example of the connectivity metric with various components is shown in Fig. 3. In this example, we have found that the connectivity score decreases as the number of connected components increases. Therefore, it is proved by statistical results obtained that single-connected component has the highest connectivity score which signifies high risk.

2.2 *Cycles*

Cycle is a subgraph of those nodes which are reachable to each others in a graph. The risk of attack graph also depends on the number of cycles in the graph. If a

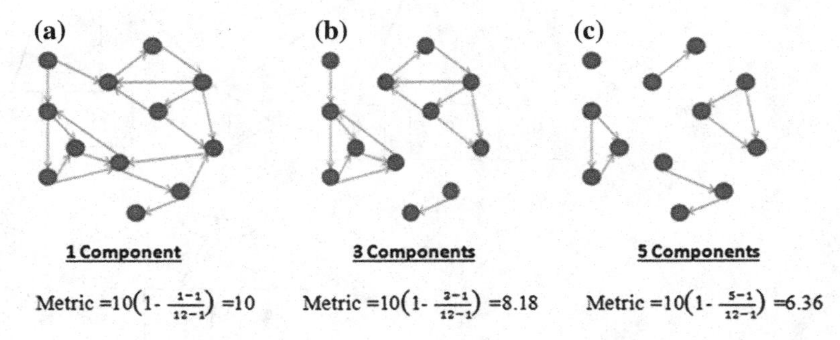

Fig. 3 Connectivity metric

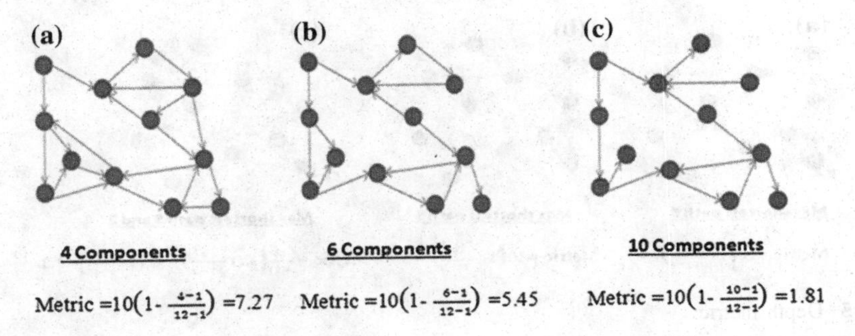

Fig. 4 Cycles metric

number of cycles are less than an attack graph is more risky and if cycles are more then attack graph is more secure. If c is the number of cycles in the graph and d is the most possible numbers of cycles in the graph then cycles metric is given by equation:

$$\text{Cycles metric} = 10\left(1 - \frac{c-1}{d-1}\right) \tag{2}$$

An example of the cycles metric values with various components are shown in Fig. 4. In this example, we have found that the cycle metric score decreases as the number of connected components increases. Therefore, it is proved by statistical results obtained that if the number of cycles in the graph are less then cycles score is high (less secure) and when the number of cycles are more then cycles score is less (more secure).

2.3 Depth

Depth of the graph is the length of maximum shortest path in the graph. It is also known as graph diameter. If n is the total number of components, d is the total number of nodes, c is number of nodes in different components, and s is the depth of the components. Depth metric is calculated by equation:

$$\text{Depth metric} = \frac{10}{nd}\sum_{i}^{n} c_i\left(1 - \frac{s_i}{c_i - 1}\right) \tag{3}$$

Figure 5 shows an example of depth metric with the help of 3 graphs. In this example, the every graph having equal number of nodes, but different number of diameters and components. Figure 5a, b have single connection with different diameter. Figure 5c has two component with different diameters. Result of metric concludes that if attack graph diameter is higher then network is more secure.

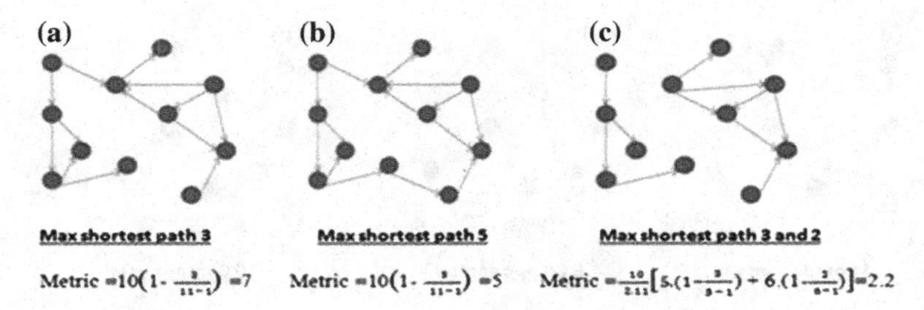

Fig. 5 Depth metric

2.4 Combined Score

It describes the combined score of all considered metrics, i.e., connectivity, cycles, and depth metrics of considered graph. The combined score can be calculated by the following formula

$$\text{Combined score} = 10\sqrt{\frac{\sum_{i=1}^{n} (s_i)^2}{\sum 10^2}} \tag{4}$$

where n is number of considered metrics and s is the individual score of metrics.

Figure 6a has 1 connectivity component, 4 cycles, and 9 diameter; Fig. 6b has 1 connectivity component, 6 cycles, and 6 diameter; and Fig. 6c has 1 connectivity component, 10 cycles, and 6 diameter. Combined score of Fig. 6c is the lowest among the three graph, therefore, this is the more secure attack graph as compare to Fig. 6a, b.

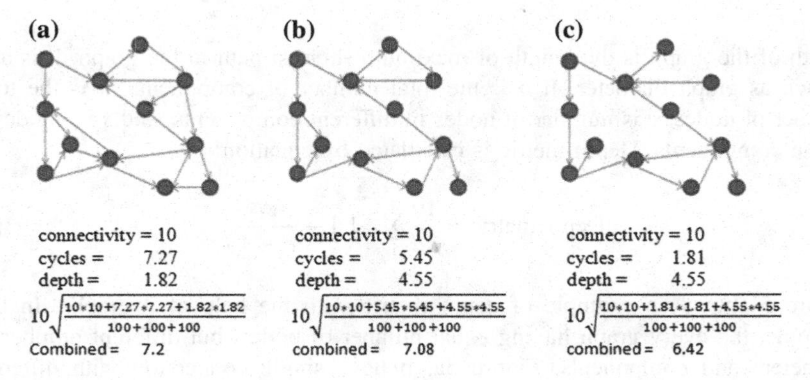

Fig. 6 Combined score metric

3 Results Analysis

The Ubuntu version 12.04 platform is used for logical attack graph generator. We have used Nessus vulnerabilities scanner for scanning the networks. Section 1 of this paper describes the used methodology to generate the logical attack graph. It consists four phases which are shown in Fig. 1. In the present work, the network configuration is done with the creation of a set of host access control list (hostname1, hostname2, protocol, port) and Datalog tuples as input to the MulVAl reasoning engine. The vulnerabilities are obtained with the creation of set of vulExists and vulProperty.

Figure 7a, b represent logical attack graphs generated by MulVAL. This graph has three types of nodes, diamonds nodes are privileged nodes, oval nodes are exploiting nodes, and rectangle nodes are fact or configuration nodes. The logical attack graph shown in Fig. 7a has 12 nodes, 1 connectivity component, 9 cycles and

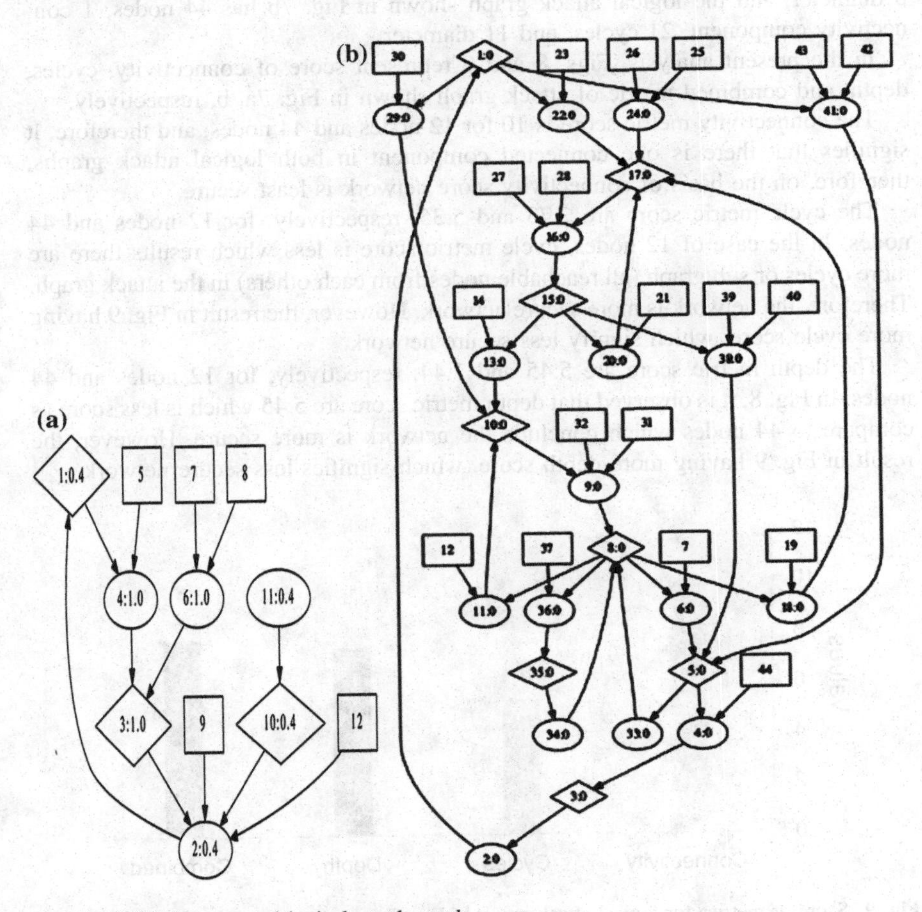

Fig. 7 MulVAL generated logical attack graph

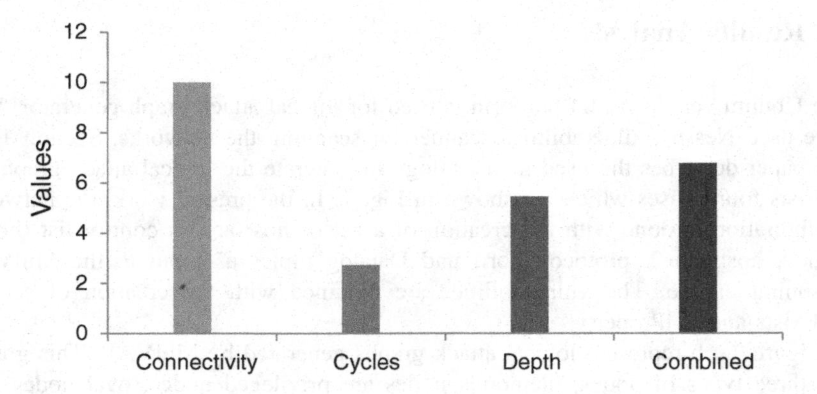

Fig. 8 Score of metrics for logical attack graph of Fig. 7a

5 diameter, and the logical attack graph shown in Fig. 7b has 44 nodes, 1 connectivity component, 21 cycles, and 11 diameter.

In the present analysis, Figs. 8 and 9 represent score of connectivity, cycles, depth, and combined metric of attack graph shown in Fig. 7a, b, respectively.

The connectivity metric score is 10 for 12 nodes and 44 nodes, and therefore, it signifies that there is one connected component in both logical attack graphs, therefore, on the basis of connectivity score network is least secure.

The cycle metric score are 2.73 and 5.35, respectively, for 12 nodes and 44 nodes. In the case of 12 nodes, cycle metric score is less which results there are more cycles or sub graph (all reachable nodes from each others) in the attack graph. Therefore, the network is more secure network. However, the result in Fig. 9 having more cycle score, which signify less secure network.

The depth metric score are 5.45 and 7.44, respectively, for 12 nodes and 44 nodes. In Fig. 8, it is observed that depth metric score are 5.45 which is less score as compare to 44 nodes which conclude the network is more secure. However, the result in Fig. 9 having more depth score, which signifies less secure network.

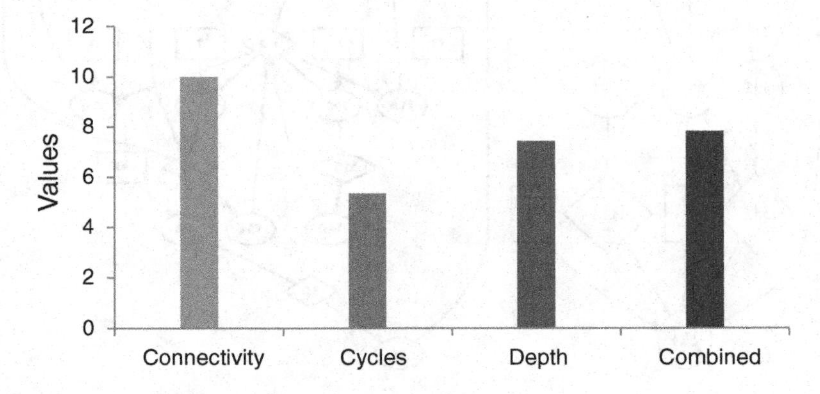

Fig. 9 Score of metrics for logical attack graph of Fig. 7b

The combined scores are 6.76 and 7.83, respectively, for 12 nodes and 44 nodes. In Fig. 8, it is observed that combined scores is 6.76 which is less score as compare to 44 nodes which shows the network is more secure. However, the result in Fig. 9 having more combined score, which signify less secure network.

3.1 Attack Paths Analysis

If we calculate all attack paths in the logical attack graph then we can prevent the attack, which could be happen on the network. In logical attack, graph fact nodes are source nodes for the attacker that means fact nodes (which describe firewall and machine configuration) has to be satisfied for any attack. If fact nodes not satisfied, then no attack is possible. Attacker goal is to gain privileges in the network host machine, therefore, privilege nodes in the logical attack graphs are the destination nodes. In Fig. 2 e1, e2, and e3 are exploiting nodes. Since exploit nodes are AND nodes, so all children nodes must be true for the satisfaction of exploiting node. The exploit node e1 will be true if children c1, c2, c3, and p2 are true. The exploit node e2 will be true if p2, c4, and c5 are true. The exploit node e3 will be true if c6 and c7 are true. Privileges are OR nodes so if any child is true, then condition is true. All attack paths for graph in Fig. 2 are:

(i) e3 –> p2
(ii) e3 –> p2 –> e1 –> p1
(iii) e3 –> p2 –> e2 –> p1

There are three attack paths for a graph given in Fig. 2, but these attacks are only possible when e1, e2, and e3 are logically true.

4 Conclusion

The modeling of logical attack graph gives opportunities to analyze the network with more efficient ways. This work describes how MulVAL runs efficiently for networks with numbers of hosts, and it has discovered an interesting security problem in a real network. We examined risk associated with the network attack graph using graph theoretic properties (connection, cycle, and depth) on the scale of (0–10), where score 0 is fully secure and 10 is worst secure. The analysis has been done on a network with different number of hosts (different numbers of nodes) and the result shows how risk factor heavily depends on depth and cycles present in the graph. All attack paths in logical attack graph also calculated in this work; once we know attack paths, we can prevent them by changing system or firewall setting.

References

1. Ou, X., Boyer, W., McQueen, M.: A Scalable Approach to Attack Graph Generation. ACM (2006)
2. Ingols, K., Lippmann, R., Piwowarsi, K.: Practical attack graph generation for network defense. In: 22nd Annual Conference on Computer Security Application, pp. 121–130 (2006)
3. Zhang, S., Caragea, D., Ou, X.: An Empirical Study on Using the National Vulnerability Database to Predict Software Vulnerabilities. Database and Expert Systems Applications, pp. 217–231. Springer, Berlin (2011)
4. Ou, X., Appel, A.W.: A logic-programming approach to network security analysis. USENIX Security (2005)
5. Ou, X., Govindavajhala, S., Appel, A.W.: MulVAL: a logic-based network security analyzer. In: 14th Usenix Security Symposium (2005)
6. Nessus security scanner. http://www.nessus.org
7. NIST, NVD. http://nvd.nist.gov/cvss.cfm
8. Noel, S., Jajodia, S.: Metrics suite for network attack graph analytics. In: Proceedings of the 9th Cyber and Information Security Research Conference. Oak Ridge National Laboratory, Tennessee (2014)
9. Wang, L., Islam, T., Long, T., Singhal, A.: An Attack Graph-Based Probabilistic Security Metric. Data and Applications Security, pp. 283–296. Springer, Berlin (2008)
10. Wang, L., Singhal, A., Jajodia, S.: Measuring the overall security of network configurations using attack graphs. Lecture Notes in Computer Science, vol. 4602, pp. 98–112. Springer, New York (2007)
11. Williams, L., Lippmann, R., Ingols, K.: GARNET—a graphical attack graph and reachability network evaluation tool. In: Proceedings of the 5th International Workshop. Springer, Cambridge (2008)
12. Common Vulnerability Scoring System (CVSS). http://www.first.org/cvss

Unsupervised Spam Detection in Hyves Using SALSA

Mohit Agrawal and R. Leela Velusamy

Abstract With the escalation in popularity of social networking sites such as Twitter, Facebook, LinkedIn, MySpace, Google+, Weibo, and Hyves, the rate of spammers and unsolicited messages has increased significantly. Spamming agents can be automated spam bots or users. The main objective of this paper is to propose an unsupervised approach to detect spam content messages. In this paper, stochastic approach for link-structure analysis (SALSA) algorithm is used to classify a message being spam or not-spam. The dataset from the popular Dutch social networking site named Hyves has been obtained and tested with different performance measures namely true positive rate, false positive rate, accuracy, and time of execution, and it is found that this mechanism outperforms the previously existing unsupervised author-reporter model for spam detection based on HITS.

Keywords HITS • Markov chain • Naïve Bayes • SALSA • Spam bot • Web 2.0

1 Introduction

Today, social networking sites have become an important platform service regarding information sharing among people [1–3]. With the introduction of web 2.0 along with its user friendly features, lots of global social networking sites namely Twitter, Facebook, LinkedIn, MySpace etc., along with its local invariants such as Weibo, and Hyves, have grown exponentially. The features and understanding of web 2.0 has been highlighted by Murugesan [4]. This growth and popularity attracts a huge count of spammers performing unsolicited activities.

M. Agrawal (✉) · R. Leela Velusamy
National Institute of Technology, Tiruchirappalli, Tamil Nadu, India
e-mail: mohit.agrawal619@gmail.com

R. Leela Velusamy
e-mail: leela@nitt.edu

S. Das et al. (eds.), *Proceedings of the 4th International Conference on Frontiers in Intelligent Computing: Theory and Applications (FICTA) 2015*, Advances in Intelligent Systems and Computing 404, DOI 10.1007/978-81-322-2695-6_43

Besides spamming, there are many other problems which surround social networking sites in an adverse manner such as Sybil attack, link farming, and phishing.

Spammers are e-criminals whose motives can be either personal or business oriented. Initially, e-mails were the main target of spamming but with the advancement in its spam filtering techniques it can filter almost 95 % of the unsolicited messages. Lack of security measures in the social networking sites has made it an active area of interest for researchers. To tackle the problem of spamming, many supervised and unsupervised approaches have been proposed in the past. But regular improvement of spammers to these spamming activities made the researchers to find new ways to improve the existing mechanisms through further modification. In this paper, improvement of the existing unsupervised author-reporter model for spam detection [16] has been proposed.

Stochastic approach for link-structure analysis (SALSA) is an efficient and widely used algorithm for ranking, which was introduced by Lempel and Morgan [5]. SALSA is a combination of Hyperlink-Induced Topic Search (HITS) and PageRank algorithm. Apart from inheriting features from Page rank and HITS algorithm, SALSA uses random walk through chains of authorities and hubs based on the theory of Markov chain. Initially, Random walk was used by Google for the purpose of ranking pages and later SALSA algorithm has been used for recommending the reliable account that can be followed in Twitter [6]. The spam messages are generated by different spammers across the world to perform some unsolicited task. A latest research by Cisco Systems found the volume of spam originating from different countries and the survey found India at the top with 13.7 % [7]. Because of the dynamic nature of spam, most of the sites rely on their users to fight against them. For this reason, users are provided with some options for reporting spam. This paper highlights the benefits of unsupervised approach over supervised approach for detecting spam content. Also, the improved performance of the proposed algorithm over the existing unsupervised algorithm namely author-reporter model with respect to different parameters is highlighted.

Rest of the paper includes the following divisions. The existing mechanism for spam detection has been discussed in Sect. 2. Section 3 describes proposed mechanism for spam detection using SALSA Algorithm. In Sect. 4, the proposed algorithm has been compared with existing mechanism based on unsupervised methodology followed by conclusion in Sect. 5.

2 Related Work

In a survey on the existing solutions of spam detection in social media, Heymann et al. [8] presented the application differences in social media to that of e-mail and web spams. The antispam methods highlighted in this paper are identification based, rank based, and limit based.

A topology-based spam detection in web has been proposed by Castillo et al. [9]. This scheme combines link-based and content-based features. Base classifier based on decision tree along with bagging has been used by the author. For smoothing purpose, author used clustering technique, propagation, and stacked graphical learning. Further 45 unique features has been used and found that the classifier can detect up to 88.4 % of spams.

Zeng et al. [10] proposed a spammer filtering method based on supervised machine learning technique to detect spammer in Weibo. Message content and user's behavior has been used to train SVM classifier. Information gain and chi-squared method are the two famous feature selection methods that have been used to highlight the important features in detection of spammers. Benevenuto et al. [11] proposed a method for detecting spammer on Twitter using tweet content and user behavior as the decision parameters. SVM is used along with fivefold cross validation for the classification purpose. In this method, 70 % of spammers are correctly identified.

Spam detection based on random boost has been proposed by DeBarr et al. [12]. This model has been generated using TF/IDF feature vectors and classifiers based on medoids for training purpose. Random boost has been found to be taking one fourth of the time with the same accuracy of random forest.

Chu et al. [13] proposed a method for differentiating Human, Bot, and Cyborg. Twitting Behavior, content, and properties of the account are the parameters based on which the method differentiated human and bot. Weka tool has been used with tenfold cross validation, and it is found that the proposed method has a true positive rate of 96 % on an average.

A generic statistical approach has been put forward by Ahmed and Abulaish [14]. Naive Bayes, Jrip, and J48 are the classification algorithms used for detection purpose, and Markov clustering algorithm has been used to identify the features and spam campaigns. The author used 14 statistical features and obtained a detection rate of 0.957 and 0.964 in Twitter and Facebook respectively. Wang et al. [15] proposed a SVM-based approach to classify a spam with a benign micro blog. Social network features and textual features are combined and composed to a feature vector. The influential words are extracted using chi-square algorithm.

The first unsupervised method for detecting spam in social network has been introduced by Bosma et al. [16]. The framework is based on HITS web link analysis, and user reports are used for classifying spam messages. The spam score and hub score are calculated based on link between reporters and messages. Spam score describes the likeliness of a message to be spam and hub score shows the trust worthiness of a reporter.

Najork et al. [17, 18] compared the effectiveness of HITS and SALSA with different parameters such as MRR, MAP, NDCG, and found SALSA to outperform HITS algorithm. They also found sampling of nodes and edges made SALSA more effective.

3 Proposed Work

This section elaborates the proposed detection model and its functioning. The working of the proposed model in different scenario has been described. Further this section is divided into two different sections. Section 3.1 focuses on the operational details of proposed model followed by implementation and observation of the proposed algorithm in Sect. 3.2.

3.1 Detection Model

Spams are dynamic in nature and most of the research on its detection is based on supervised approaches. Static nature of these approaches needs retraining to detect a new kind of spams. The main underlying base of this paper is to detect spam using unsupervised mechanism in order to detect spam content without retraining the model every time. The framework is basically inspired by SALSA. Initially, the algorithm divides the set of contents (*C*) and reporters (*R*) into a bipartite graph connected by edges (*E*) as shown in Fig. 1.

G(*R*, *C*, *E*) denotes a bipartite graph having partition *R* and *C*, with the set of directed edges from *R* to *C* denoted by *E*. The directed edge from *R* to *C* indicates a spam report issued by reporter for specific content/message. Using the basic framework of SALSA, the algorithm performs a random walk through the chain of content. Each step of this mechanism is composed of forward and backward link traversal. The content set *C* of bipartite graph represents the messages that have been reported by users from reporter set *R*. The weight of the edge from reporter_i to content_j is represented by $w_{i,j}$. The edge from reporter set to content set represent the issued report that has been initialized with some uniform weight $w_{i,j}$. Further, the content set has been divided into subcomponents using random walk through the chain of contents for the graph shown in Fig. 1a. The result of random walk through chain of content is shown in Fig. 1b. A random walk represents mathematical

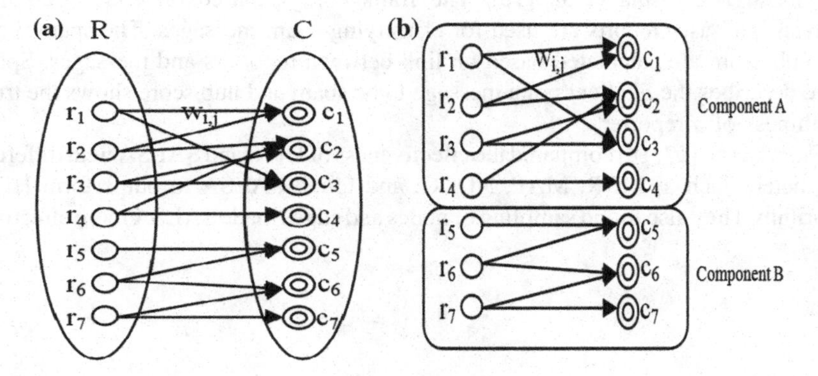

Fig. 1 Bipartite graph to represent **a** example of proposed model, **b** working of model

formalization of a path that consists of a chain of random steps which finds different connected components of content set using the common content between the reporters as shown in Fig. 1b.

After dividing the dataset into different components, the in-degree for each content has been calculated using number of inward directed edges to a content c_i and represented as indeg (c_i). The total in-degrees in a component is calculated as the sum of in-degrees of all c_i in that component and it is represented as comp_indeg. Then the normalized in-degree can be calculated by dividing the in-degree of each content c_i to the total in-degrees in the component containing c_i. Size of a component defines the number of contents in a component and represented as comp_size. Similarly, total_size represents the size of total content set C and it can be calculated as $|C|$. The ratio between comp_size to total_size defines the relative size of a component.

$$\text{Spam\_score}(c_i) = (\text{indeg}(c_i)/\text{comp\_indeg}) \times (\text{comp\_size}/\text{total\_size}) \quad (1)$$

The spam score for each content c_i has been calculated using Eq. 1. The working of the algorithm for calculating spam score for content c_2 in Fig. 2 is as follows:

Indeg (c_2) = Number of incoming edges to $c_2 = 3$.
comp_indeg = Number of incoming edges to component $A = 8$
comp_size = Number of contents in component $A = 4$
total_size = $|C| = 4 + 3 = 7$
Using Eq. 1, the spam score for c_2 can be calculated as:
spam_score (c_2) = (3/8) × (4/7)

Similarly, spam scores for each content can be calculated using Algorithm 1.

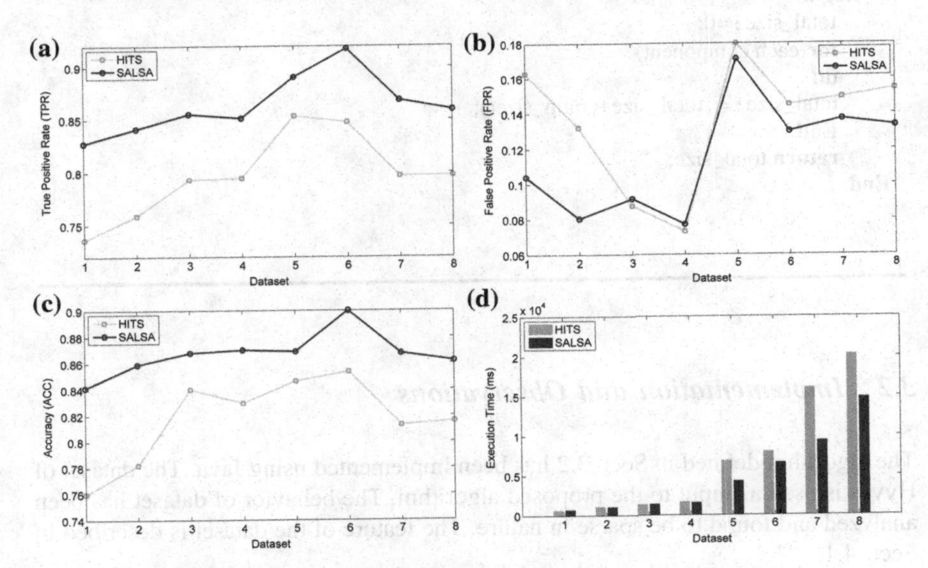

Fig. 2 Obtained results with different size dataset **a** TPR, **b** FPR, **c** accuracy, **d** time

Algorithm 1. Pseudocode to calculate spam score.

```
Input:  Bipartite graph of user and messages with spam reports.
Output: Spam score of each message.

Begin
   call calculate_size(); // To find total size of content set.
   for (each component)
   do
        comp_indeg : =0;
        for (each Vertex n in component c)
        do
           for (each edge e in incoming_edge(ni))
           do
              comp_indeg : =comp_indeg+edge_weight(ei);
           End
        End
        for(each content n in component c)
        do
           normalized_indeg(ni) : =  indeg(ni)/comp_indeg;
             //where n1, n2, n3 represents content 1, 2, 3 respectively
           relative_size : = comp_size/total_size;
           spam_score(ni) : =  normalized_indeg(ni)*relative_size;
        End
        return spam_score(ni);
   End
End.


Calculate_size()
Begin
   total_size : =0;
   for(each component)
   do
   total_size : =  total_size+comp_size();
   End
   return total_size;
End
```

3.2 *Implementation and Observations*

The algorithm defined in Sect. 3.2 has been implemented using Java. The dataset of Hyves is used as input to the proposed algorithm. The behavior of dataset has been analyzed and found to be sparse in nature. The feature of the dataset is described in Sect. 4.1.

In the collected dataset, it has been analyzed that more than 90 % of the data has been reported by only one user. About 5 % of data is reported by two users. Rest 5 % of the data contains 3 or 4 user spam reports. After input, the dataset has been converted into directed bipartite graph. The content set of bipartite graph has been broken into different connected components, and spam score has been calculated using Eq. 1. The result obtained from the proposed algorithm is described in Sect. 4.3.

4 Experiment and Result Analysis

4.1 Data Collection

The dataset used in this research consists of 28,998 reports, 13,188 messages, and 9,491 users from the social networking site Hyves. The collected dataset has been crawled for the period of January 2010 to January 2011. The collected dataset contains two subdivisions as test set and training set. From the previous works, it has been found that the training set consists of the oldest messages used for training of semisupervised and supervised spam detection. Similarly test set consists of latest messages of the dataset on which the machine learning approach has been tested. After the collection of dataset, it has been fragmented into different size as shown in Table 1, to test the working of proposed model with varying size of dataset. The result is discussed in the Sect. 4.3.

4.2 Performance Metrics

For evaluation purpose, a matching matrix illustrated in Table 2 is considered where true positive (TP) represents the correctly identified spam messages and false Positive (FP) represents the incorrectly identified spam messages. True negative (TN) represents correctly identified not-spam messages and false negative (FN) represents the

Table 1 Fragmentation description

Dataset	Content percentage (%)	Description
1	25	Test set
2	50	
3	75	
4	100	
5	25	Training set
6	50	
7	75	
8	100	

Table 2 Example of matching matrix

	Actual	
	Spam	Not spam
Outcome		
Spam	TP	FP
Not spam	FN	TN

incorrectly identified not-spam messages. According to the matching matrix, a set of other metrics is considered for comparison of performance and efficiency, between proposed method and existing approach.

$$\text{True Positive Rate, TPR} = \frac{TP}{TP + FN} \quad (2)$$

The ratio between correctly identified spam to the total spam reports is termed as true positive rate, and it is calculated using Eq. 2. False positive rate can be defined as the ratio between incorrectly identified spam reports to that of total identified not-spam reports, and it is calculated using Eq. 3. Accuracy defines the ratio between correctly identified messages to total identified messages, and it is calculated using Eq. 4. The time taken by the algorithm for detection of spam messages is defined as time of execution.

$$\text{False Positive Rate, FPR} = \frac{FP}{FP + TN} \quad (3)$$

$$\text{Accuracy, ACC} = \frac{TP + TN}{TP + FN + FP + TN} \quad (4)$$

4.3 Results and Comparison

The observed true positive (TP), false positive (FP), false negative (FN), and true negative (TN) values for both algorithms has been tabulated below in Table 3 and matching matrix has been prepared. Using metrics defined in Sect. 4.2, true positive

Table 3 Observed TP, FP, FN, and TN

Dataset	TP		FP		FN		TN	
	HITS	SALSA	HITS	SALSA	HITS	SALSA	HITS	SALSA
1	894	1004	50	32	320	210	257	275
2	1355	1501	66	42	430	284	456	480
3	1847	1991	70	67	481	337	661	664
4	2179	2334	73	76	563	408	913	910

Table 4 Statistical performance comparison of competing algorithms

Algorithm used	Average TPR (%)	Average FPR (%)	Average ACC (%)
HITS	79.71	13.54	81.75
SALSA	86.38	11.55	87.08

rate (TPR), false positive rate (FPR), and accuracy (ACC) has been calculated and compared with the existing unsupervised spam detection approach and the results are made known. The statistical comparison of competing algorithms has been tabulated below in Table 4.

True positive rate of both algorithms has been compared with the varying size of datasets and plotted as shown in Fig. 2a. The graph clearly explains the improvement of detection rate achieved using proposed model because of the effectiveness of SALSA in scoring the content highlighted by Najork et al. [17, 18]. Similarly Fig. 2b illustrates the false positive rate of both algorithms with varying size of datasets and proposed approach is found to perform better for most of the test cases.

Figure 2c points up the comparison of accuracy achieved using existing algorithm and proposed model. From the plotted graph, it can be observed that SALSA-based spam detection outperforms the previously proposed mechanism because computationally light nature of SALSA makes it similar to weighted in/out degree ranking. Unlike HITS, the proposed approach calculates spam score based on relative scoring which makes is more efficient.

Author-reporter model calculates both hub score and spam score for each reporter and content, respectively. On the other hand, the proposed work calculates only spam score (authority score) for detection of spam content messages. The convergence of both hub and spam score in HITS-based model is one of the reason of its slow execution compared to SALSA-based detection. It can be seen that SALSA takes 34.92 % of execution time less than that of HITS. The comparison of execution time with varying size of datasets is shown in Fig. 2d.

5 Conclusion and Future Work

Due to the ever existing popularity of social networking sites, a significant growth in the rate of spammers is bound to happen hence the researchers have to keep finding ways to improve the existing techniques to detect these unsolicited messages. To overcome the drawbacks of supervised approaches, an unsupervised technique to filter spam content messages has been proposed. SALSA algorithm is a link analysis algorithm which has been used for the first time in context of spam detection in this paper.

A comparative analysis has been done between the proposed approach and existing author-reporter model proposed by Bosma et al. [16] using parameters such as true positive rate, false positive rate, accuracy, and time of execution. Results from the above experiment illustrate SALSA-based detection performs better than

HITS-based author-reporter model. Further testing of the proposed model with other social network dataset such as Facebook and Twitter, to check its behavior with different datasets will be done later. Some more features will be added to the proposed model to improve its efficiency and detection rate. In future, the behavior of the reporters can be analyzed to find the affect of their false reporting.

References

1. Subrahmanyam, K., Reich, S.M., Waechter, N., Espinoza, G.: Online and offline social networks: use of social networking sites by emerging adults. J. Appl. Dev. Psychol. **29**(6), 420–433 (2008)
2. Lin, K.-Y., Lu, H.-P.: Why people use social networking sites: an empirical study integrating network externalities and motivation theory. Comput. Hum. Behav. **27**(3), 1152–1161 (2011)
3. Brandtzaeg, P.B., Heim, J.: Why people use social networking sites. In: Online Communities and Social Computing, pp. 143–152. Springer, Berlin (2009)
4. Murugesan, S.: Understanding Web 2.0. IT Prof. **9**(4), 34–41 (2007)
5. Lempel, R., Moran, S.: SALSA: the stochastic approach for link-structure analysis. ACM Trans. Inf. Syst. TOIS **19**(2), 131–160 (2001)
6. Gupta, P., Goel, A., Lin, J., Sharma, A., Wang, D., Zadeh, R.: Wtf: the who to follow service at twitter. In: Proceedings of the 22nd International Conference on World Wide Web, pp. 505–514 (2013)
7. Cisco 2011 Annual Security Report
8. Heymann, P., Koutrika, G., Garcia-Molina, H.: Fighting spam on social web sites: a survey of approaches and future challenges. Internet Comput. IEEE **11**(6), 36–45 (2007)
9. Castillo, C., Donato, D., Gionis, A., Murdock, V., Silvestri, F.: Know your neighbours: web spam detection using the web topology. In: ACM SIGIR, pp. 423–430 (2007)
10. Zeng, Z., Zheng, X., Chen, G., Yu, Y.: Spammer detection on Weibo social network, pp. 881–886 (2014)
11. Benevenuto, F., Magno, G., Rodrigues, T., Almeida, V.: Detecting spammers on twitter. In: Collaboration, Electronic Messaging, Anti-Abuse and Spam Conference (CEAS), vol. 6, p. 12 (2010)
12. DeBarr, D., Wechsler, H.: Spam detection using random boost. Pattern Recognit. Lett. **33**(10), 1237–1244 (2012)
13. Chu, Z., Gianvecchio, S., Wang, H., Jajodia, S.: Detecting automation of twitter accounts: are you a human, bot, or cyborg? IEEE Trans. Dependable Secure Comput. **9**(6), 811–824 (2012)
14. Ahmed, F., Abulaish, M.: A generic statistical approach for spam detection in online social networks. Comput. Commun. **36**(10–11), 1120–1129 (2013)
15. Wang, K., Wang, Y., Li, H., Xiong, Y., Zhang, X.: A new approach for detecting spam microblogs based on text and user's social network features. In: 4th International Conference on Wireless Communications, Vehicular Technology, Information Theory and Aerospace and Electronic Systems (VITAE), pp. 1–5 (2014)
16. Bosma, M., Meij, E., Weerkamp, W.: A framework for unsupervised spam detection in social networking sites. In: Advances in Information Retrieval, pp. 364–375. Springer, Berlin (2012)
17. Najork, M.A.: Comparing the effectiveness of HITS and SALSA. In: Proceedings of the Sixteenth ACM Conference on Information and Knowledge Management, pp. 157–164 (2007)
18. Najork, M., Gollapudi, S., Panigrahy, R.: Less is more: sampling the neighborhood graph makes salsa better and faster. In: Proceedings of the Second ACM International Conference on Web Search and Data Mining, pp. 242–251 (2009)

Cryptanalysis of a Chaotic Map-Based Authentication and Key Agreement Scheme for Telecare Medicine Information Systems

Sandip Roy and Santanu Chatterjee

Abstract User authentication and privacy is quite essential in telecare medicine information systems (TMIS) for a secure and efficient access of the healthcare services. Very recently, in 2014, Li et al. proposed an efficient *chaotic maps and smart cards* based password authentication and key agreement scheme TMIS (Journal of Medical Systems). In this paper, we analyze that though the Li et al. scheme is computationally efficient, it has several security weaknesses. As for example, it has design flaws in both login authentication phase and in password change phase. Moreover, it cannot resist denial-of-service attack and adopts incorrect strategy in design of server status table. As a result, the Li et al. scheme is not suitable for practical applications. Finally, we hint at some possible improvements that can be adopted by their scheme to make it more secured against various possible known attacks.

Keywords Chebyshev chaotic maps · Remote user authentication · Smart card · Denial-of-servive attack · Telecare medicine information system (TMIS)

1 Introduction

A secure communication among connected patients, doctors, and the telecare server is an essential requirement in TMIS. Further, medical electronic records of a patient should be well protected by TMIS because it directly involves with patient privacy. Password authentication based on smart card along with user anonymity is very popular mechanism to check the authenticity of genuine login users and remote servers [1–7].

S. Roy (✉)
Department of Computer Science and Engineering, Asansol Engineering College, Asansol 713305, West Bengal, India
e-mail: sandiproy9500@gmail.com

S. Chatterjee
Defence Research and Development Organization, Hyderabad 500 069, India
e-mail: santanu.chatterjee@rcilab.in

S. Das et al. (eds.), *Proceedings of the 4th International Conference on Frontiers in Intelligent Computing: Theory and Applications (FICTA) 2015*, Advances in Intelligent Systems and Computing 404, DOI 10.1007/978-81-322-2695-6_44

Chaotic map-based cryptography has attracted wide attention of the researchers as its shows better performance than that of traditional cryptography. A number of chaotic map and smart cards based user authentication schemes with user anonymity for TMIS have been proposed in recent times [8–11]. Cryptography using chaotic maps provides the semigroup property and is more efficient than cryptography using modular exponential computations and scalar multiplications on elliptic curve [12, 13].

Guo et al. [8] is the first to propose a low cost and efficient chaotic map-based user authentication scheme that provides key agreement between the user and medical server. In 2013, Hao et al. [9] identified that the Guo et al. scheme suffers from two weaknesses such as lack of proper user anonymity and inefficiency of the double secret keys. The enhanced scheme proposed by Hao et al. addressed and removed these pitfalls. Later, Lee et al. analyzed that both Guo et al. and Hao et al. authentication schemes violate the contributory property of key agreements [11]. In 2014, Jiang et al. [10] investigated that the Hao et al. scheme is vulnerable to the stolen smart card attack. Applying this attack, the identity of a legal user might be disclosed to an adversary, thus it cannot provide necessary privacy to a genuine user. Then, they proposed a comparatively robust chaotic map-based user authentication and key agreement scheme with strong anonymity.

Very recently, in 2014, Li et al. proposed a secure chaotic maps and smart cards based password authentication and key agreement scheme with user anonymity for TMIS [14]. Li et al. found a serious security problem in the Jiang et al. and Lee et al. chaotic map-based authentication schemes proposed earlier. Li et al. investigated that, in those two schemes, a registered user's secret parameters may be intentionally exposed to many nonregistered users and this problem causes the service misuse attack. He proposed some enhancements over these two schemes to prevent the shortcomings. In this paper, we analyze that the Li et al. scheme has several security weaknesses: (1) it has design flaws in login and authentication phases, as the scheme is unable to detect the correct password input (2) it has design flaws in password change phase and (3) inefficiency in password change phase can incur denial-of-service attack. Using a detailed security analysis, we show that our improved scheme can successfully resist the security weaknesses of the Li et al. scheme. Finally, through proper security analysis and comparison, we show that our scheme has much better security than the existing schemes in the literature.

The remainder of this paper is organized as follows. In Sect. 2 we discuss the fundamentals Chebyshev polynomial and chaotic map operation. In Sect. 3, we review all the phases of the Li et al. scheme. In Sect. 4, we perform a detailed cryptanalysis of the Li et al. scheme and identify their security pitfalls. In Sect. 5, as a future research direction, we suggest some remedial improvements on this scheme in order to make it more secured against those security weaknesses.

2 Mathematical Preliminaries

In this section, we discuss some fundamental concepts on Chebyshev polynomial and chaotic map. For more details, please refer to [13, 15].

2.1 Chebyshev Chaotic Maps

Definition 1 Let n be an integer ≥ 0, and let x be a variable taking value over the interval $[1, 1]$. Chebyshev polynomial of degree n $T_n(x) : [-1, 1] \rightarrow [-1, 1]$ is defined as

$$T_n(x) = cos(narccosx), \qquad x \in [-1, 1],$$
$$or$$
$$T_n(x) = cosn\theta, \quad x = cos\theta, \quad \theta \in [0, \pi]$$

From the above definition, the recursive relation of Chebyshev polynomial is defined as

$$T_n(x) = 2.x.T_{n-1}(x) - T_{n-2}(x), \quad \text{for any n} \geq 2$$
$$T_0(x) = 1 \quad \text{and}$$
$$T_1(x) = x$$

Definition 2 The Chebyshev polynomial satisfies the important semigroup property:

$$\begin{aligned} T_r(T_s(x)) &= cos(rcos^{-1}(scos^{-1})(x))) \\ &= cos(rscos^{-1}(x)) \\ &= T_{rs}(x) \\ &= T_s(T_r(x)) \end{aligned}$$

The semigroup property of Chebyshev polynomial also holds on the interval $(-\infty, +\infty)$, as mentioned by [15], given below:
$T_n(x) = (2.x.T_{n-1}(x) - T_{n-2}(x))$ mod p, where $n \geq 2$, $x \in (-\infty, +\infty)$, and p is a large prime number.
Obviously,
$T_r(T_s(x)) \equiv T_{rs}(x) \equiv T_s(T_r(x))$ mod p

3 Review of Li's Authentication Scheme for TMIS

In this section, first we briefly review the Li et al. scheme [14] and then analyze its security weaknesses. The proposed scheme also comprises four phases, namely, parameter generation phase, registration phase, authentication phase, and password change phase.

3.1 Parameter Generation Phase

In this phase, the remote server S generates a random number mk as the master key and one-way hash functions $h(\cdot)$ and $H(\cdot)$.

3.2 Registration Phase

In this phase, the user U performs the following steps with S through a secure channel:

1. U chooses identity ID, password PW, and a random number b. Next, U computes and sends $h(ID||h(PW||b))$ and ID to S.
2. Next, S generates a random number r and computes $IM_1 = IM_3 = h(mk) \oplus r$, $IM_2 = IM_4 = h(mk||r) \oplus ID, D1 = h(ID||mk) \oplus h(ID||h(PW||b))$. Then S stores $(IM_1, IM_2, IM_3, IM_4, D_1, h(\cdot),$
$H(\cdot))$ into user's smart card. Moreover, S maintains a status table for a registration service, where every row contains U's ID, latest random number r and status bit or login bit ($LB = 0$ indicates that U is not logged into S, 1 otherwise).
3. Finally, U computes $D_2 = h(ID||PW) \oplus b$ and stores D_2 into smart card.

3.3 Authentication Phase

In this phase, User U and server S mutually authenticate each other, and establish a session key.

To prevent denial-of-service (DoS) attacks, the scheme is composed of two cases. Case 1: the latest random numbers kept by U and S are identical. Case 2: the latest random numbers kept by U and S are different from each other.

In case 1, the following steps will be executed between user U and server S.

1. U inserts the smart card and inputs ID and password PW. Further, smart card generates a random number u and computes $b = h(ID||PW) \oplus D_2$, $K = D_1 \oplus h(ID||h(PW||b)) = h(ID||mk)$, $T_u(K)$, and $X_1 = h(K||IM_1||IM_2||T_u(K)||T_1)$, where T_1 is the current time stamp. U sends the login request $M_1 = IM_1, IM_2$, $Tu(K), X_1, T_1$ to S.

2. Server S receives message M_1 from U and checks if it came within allowable time delay limit or not. If it does not hold, S rejects this service request. Otherwise, S computes $r' = IM_1 \oplus h(mk)$ and $ID' = IM_2 \oplus h(mk||r')$ and checks if computed (ID, r) equals maintained (ID, r). If (ID, r) is found, S computes $K' = h(ID'||mk)$ and checks if computed $h(K||IM_1||IM_2||T_u(K)||T_1)$ equals received hash value X_1 or not. If there is a mismatch, S rejects this service request. Otherwise, set login bit of status table $LB = 1$. Moreover, S generates two random numbers r_{new} and v and computes $IM_1^* = h(mk) \oplus r_{new}$, $IM_2^* = h(mk||r_{new}) \oplus ID'$, $T_v(K')$, $sk = H(T_u(K)), T_v(K'), T_v(T_u(K)))$, $Y_1 = IM_1^* \oplus h(sk||T_1)$, $Y_2 = IM_2^* \oplus h(sk||T_2)$ and $Y3 = h(sk||IM_1^*||IM_2^*||T_v(K')$ $||T_2)$, here, T_2 is the current time stamp of S and sk is the session key established between U and S. Finally, S replies $M_2 = Y_1, Y_2, Y_3, T_v(K), T_2$ to U.
3. User U receives message M_2 from server S and checks whether message transmission delay is within valid time limit or not. Then, U computes $sk' = H(T_u(K)$, $T_v(K'), T_u(T_v(K')))$, $IM_{1new}^* = Y_1 \oplus h(sk'||T_1)$ and $IM_{2new}^* = Y_2 \oplus h(sk'||T_2)$. Next, U checks if computed $h(sk'||IM_{1new}^*||IM_{2new}^*||T_v(K')||T_2)$ equals received Y_3. U rejects this service request for a mismatch. Otherwise, U successfully authenticates S and the smart card replaces original IM_1, IM_2 with IM_{1new}^* and IM_{2new}^*, respectively. Also IM_3 and IM_4 are replaced with IM_1 and IM_2 for next login.
4. In addition, U computes a hash parameter $X_2 = h(IM_{1new}^*||IM_{2new}^*||T_u(T_v(K'))$ $||sk'||T_3)$ and sends a mutual authentication message $M_3 = X_2, T_3$ to server S, where T_3 is the current time stamp of the reply.
5. Upon receiving M_3 from U, server S checks whether message transmission delay is within valid time limit or not. If it holds, S checks if computed $h(IM_{1new}^*||IM_{2new}^*$ $||T_v(T_u(K))||sk||T_3)$ equals X_2 or not. S rejects user login request for a mismatch of these values. Otherwise, S allows the user U to login into the server S and random number r is updated with r_{new}.

Operations in Case 2 is almost the same as explained in Case 1 except the fact that IM_3 and IM_4 are used instead of IM_1 and IM_2 in message M_1 and hash parameter X_1 in authenticaiton phase. For details, see [14].

3.4 Password Change Phase

In this phase, a valid user U updates his old password PW into new password PW' as follows:

1. U inserts the smart card and inputs identity ID, original password PW and a new password PW'.
2. The smart card computes $b = D_2 \oplus h(ID||PW)$, $D_1' = D_1 \oplus h(ID||h(PW||b)) \oplus h(ID||h(PW'||b))$, and $D_2' = h(ID||PW') \oplus b$.
3. Finally, replacing D_1 with D_1' and D_2 with D_2' on the smart card.

4 Cryptanalysis of the Li et al. Scheme

In this section, we investigated the security of the Li et al. scheme and found that this scheme is insecure to the following attacks:

4.1 *Inefficient Password Change Phase*

In the Li et al. scheme, during the password change phase, user smart card does not verify the correctness of existing password. It simply accepts the old password PW and new password PW_{new} from the user. However, it is quite possible that a user may temporarily forget the original old password or may unknowingly enter wrong old password during password change phase. Also, by mistake, user may use one account password into another account. This may cause the denial-of-service scenario, where a user will no longer login to the server using the same device. Suppose, by mistake, at step 1 of "Password Change Phase" of this scheme, a user U inputs the wrong password PW^* instead of original password PW. Here, $PW^* \neq PW$. Now the smart card updates password as follows: Smart card SC does not verify whether the input password PW^* is right or wrong and computes D_1' and D_2' as given in step 2 and step 3.

$$\begin{aligned} b^* &= D_2 \oplus h(ID||PW^*), \quad \text{where } PW^* \neq PW \\ &= h(ID||PW) \oplus b \oplus h(ID||PW^*) \end{aligned}$$

Clearly, $b^* \neq b$, as $PW^* \neq PW$.
Again,

$$\begin{aligned} D_1' &= D_1 \oplus h(ID||h(PW^*||b^*)) \oplus h(ID||h(PW_{new}||b^*)) \\ &= h(ID||mk) \oplus h(ID||h(PW||b)) \oplus h(ID||h(PW^*|| \\ &\quad b^*)) \oplus h(ID||h(PW_{new}||b^*)) \\ &\neq h(ID||mk) \oplus h(ID||h(PW_{new}||b^*)), \\ &\quad \text{as } PW^* \neq PW \text{ and } b^* \neq b. \end{aligned}$$

Moreover,

$$\begin{aligned} D_2' &= h(ID||h(PW_{new}) \oplus b^* \\ &\neq h(ID||h(PW_{new}) \oplus b, \text{ as } b^* \neq b \end{aligned}$$

Clearly, D_1', D_2', and b^* are updated and written into the smart card SC with wrong values. Note that the user U is completely unaware of this fact and later on he/she will not be allowed to login into the system, even if correct new password PW_{new} is entered. This will initiate a permanent denial-of-service, where a legitimate user can never establish an authorized session with the server using the same smart card. Hence, in presence of such a fatal error in the password change phase, user will have to change the smart card and execute the registration process in a fresh way.

4.2 Fails to Protect Denial-of-Service Attack Due to Inefficiency in Password Change Phase

In Li's protocol, an attacker can launch a kind of denial-of-service attack, if he/she manages to get temporary access to a legal user's smart card SC. The remaining process of the attack is described below:

1. The adversary collects a legal user U's smart card, inserts it into a card reader and initiates password change.
2. The adversary inputs two distinct 128 bit random numbers n_1 and n_2.
3. The smart card computes $D_1^* = D_1 \oplus h(n_1) \oplus h(n_2)$. Finally, SC replaces D_1 with D_1^*.

Now, the legal user U fails to make a proper login into the server S. The details are given below:

1. U inserts the smart card and inputs ID and password PW. Further, smart card generates a random number u and computes $b = h(ID||PW) \oplus D_2$.
2.

$$\begin{aligned} K^* &= D_1^* \oplus h(ID||h(PW||b)) \\ &= D_1 \oplus h(n_1) \oplus h(n_2) \oplus h(ID||h(PW||b)) \\ &= h(ID||mk) \oplus h(ID||h(PW||b)) \oplus h(n_1) \oplus h(n_2) \oplus \\ &\quad h(ID||h(PW||b)) \\ &= h(ID||mk) \oplus h(n_1) \oplus h(n_2) \\ &\neq h(ID||mk), \text{ as } n_1 \neq n_2 \end{aligned}$$

Note that server S generates $K = h(ID||mk) \neq k^*$ as generated by user U.

3. User U sends the login message $M_1 = \{IM_1, IM_2, T_u(K), X_1^*, T_1\}$ to server S, where $X_1^* = h(K^*||IM_1 ||IM_2||T_u(K^*)||T_1)$.
4. Server S generates $K = h(ID||mk)$, where $K \neq k^*$
5. S computes hash value $h(K||IM_1||IM_2||T_u(K)||T_1) \neq X_1^*$, as $K \neq K^*$. Obviously, for this mismatch, S rejects the login request.

Hence, even if user U enters a valid identity ID and password PW, he/she cannot make a successful login into the server. Adversary can successfully execute this attack, because there is no verification of old password before update of D_1 and D_2.

4.3 Design Flaws in Authentication Phase

Suppose, due to mistake or unknowingly, a legal user enters a wrong password. In the Li et al. scheme, smart card SC does not verify the correctness of input password in authentication phase. As a result, even if a legal user U enters his/her password incorrectly by mistake, no password verification is performed at the smart card reader and login request is forwarded to the server. Though, server S rejects this login request but it consumes unnecessarily extra communication and computational overheads during the authentication phase. So, we consider that the design of the authentication is inefficient. The details are as follows:

1. A legal user U inserts the smart card and inputs ID, wrong password PW^*, random number u.
   (a) U computes $b^* = h(ID||PW^*) \oplus D_2$, $K^* = D_1 \oplus h(ID||h(PW^*||b)) \neq h(ID||mk)$, $T_u(K^*)$ and $X_1^* = h(K^*||IM_1||IM_2||T_u(K^*)||T_1)$, where T_1 is the current time stamp.
   (b) U sends the login request $M_1 = IM_1, IM_2$, $Tu(K^*), X_1^*, T_1$ to S. Note that no password verification is executed at this time.
2. Next, S checks whether $T_2 - T_1 \leq \triangle T$ holds or not, where T_2 is the current time stamp. If it does not hold, S rejects this service request.
3. Server S then computes $r' = IM_1 \oplus h(mk)$ and $ID' = IM_2 \oplus h(mk||r')$ and checks if computed (ID', r') equals maintained (ID, r) or not. Though U sent wrong password PW^*, this computed (ID', r') will be found in the server status table, as he/she sent corrent identity ID.
4. S computes $K = h(ID||mk)$ and using K, generates $h(K||IM_1||IM_2||T_u(K)||T_1)$.
5. Finally, S checks if generated $h(K||IM_1||IM_2 ||T_u(K)||T_1) \stackrel{?}{=} X_1^*$. Obviously, this login request will be rejected, as because these hash values are not equal.

Such wrong design causes unnecessary extra communication and computational overheads on the server. It can be avoided if both the user's identity and password verification take place at the smart card reader. Moreover, the user U is totally unaware of the fact that he/she entered his/her password incorrectly in this phase. So, he/she might think the server S as a malicious server, but S is actually an honest server.

4.4 Failure Againt Denial-of-Service Attack Due to Design Flaws in Server Status Table

We analyze that the Li et al. protocol can cause denial-of-service attack due to design flaws in server status table. The details of the attack are as follows:

1. A legal user U inserts his/her smart card SC and enter ID, PW, and random number b. Then U sends a valid login message to the server, as mentioned in step 1 of the authentication phase.
2. Server S checks validity of message transmision delay and generates (ID', r') pair using received IM_1 and IM_2.
3. If (ID', r') pair is found in the server status table, S checks the validity of the user login message as given in step 2 of the authentication phase. If verification holds, then proceed to step 3, otherwise reject the login request.
4. S sets status table login bit $LB = 1$ and generates new random number r_{new}. Further, as given in step 3, step 4, and step 5, U and S mutually autheticate each other and server updates r with r_{new} in its status table.
5. Suppose that, in a different session, the same legal user U with identity ID tries to login in the server again. Then server S generates (ID', r'_{new}) pair using received IM^*_{1new} and IM^*_{2new} and checks if stored $(ID', r'_{new}) \stackrel{?}{=} (ID, r_{new})$.
6. If pair (ID, r_{new}) is found, then server checks the value of login bit LB in the status table. Note that LB is already set to 1 because of the earlier login request made by the same user U. Now, as $LB = 1$, the server will think that this second login request made by user U is a replayed message sent by an adversary. As a consequesnce, server S will reject this valid login request made by a legal user U, so, user U with identity ID will never be able to login again into the system.

So, these design flaws in the server status table cause denial-of-service attack.

5 Discussion and Future Work

From the cryptanalysis of the Li et al. scheme discussed in Sect. 4, it is obvious that their scheme becomes impractical, inefficient, and insecure due to the fact that design of login and authentication phase does not verify the authenticity of the user ID and password before sending the request to the medical server. It can be avoided if both the users identity and password verification take place at the smart card reader, as mentioned in the Das et al. scheme [16].

Second, password change phase suffers from a serious design problem in that a user no longer be able to login to the server using the same smart card device. So, before initiating the password chage phase, the card reader must check the authenticity of the user by checking the correctness of the password, as mentioned in the Mishra et al. scheme [17].

Third, as already discussed, an adversary can launch DoS attack by preventing a genuine user from making a successful login into the system. To prevent an adversary from doing so, the scheme should verify the correctness of the existing password before a user can initiate the password change phase in the Mishra et al. scheme [17].

Fourth, once a user makes a successful login into the system, the login bit LB is set to one permanently in server status table. As a result, a valid user cannot make a successsful login into the system in future. So, to address this problem, instead of login bit, if a user can ingeniously use an increment counter variable (verified from server side) for each successful login into the medical server.

In future, we aim to propose an enhanced scheme that will provide unconditional security against possible known security attacks and threats.

References

1. Chang, C.C., Lee, C.Y.: A smart card-based authentication scheme uing user identify cryptography. Int. J. Netw. Secur. **15**(2), 139–147 (2013)
2. Das, A.K.: Improving identity-based random key establishment scheme for large-scale hierarchical wireless sensor networks. Int. J. Netw. Secur. **14**(1), 121 (2012)
3. He, D., Zhao, W., Wu, S.: Security analysis of a dynamic ID-based authentication scheme for multi-server environment using smart cards. Int. J. Netw. Secur. **15**(5), 350–356 (2013)
4. Kar, J.: ID-based deniable authentication protocol based on diffie-hellman problem on elliptic curve. Int. J. Netw. Secur. **15**(5), 357–364 (2013)
5. Lee, C.C., Lou, D.C., Li, C.T., Hsu, C.W.: An extended chaotic-maps-based protocol with key agreement for multiserver environments. Nonlinear Dyn. **76**(1), 853–866 (2014)
6. Li, C.T., Hwang, M.S.: An efficient biometrics-based remote user authentication scheme using smart cards. J. Netw. Comput. Appl. **33**(1), 15 (2010)
7. Lee, C.C., Li, C.T., Hsu, C.W.: A three-party password-based authenticated key exchange protocol with user anonymity using extended chaotic maps. Nonlinear Dyn. **73**(1–2), 125–132 (2013)
8. Guo, C., Chang, C.C.: Chaotic Maps-Based PasswordAuthenticated Key Agreement Using Smart Cards. Commun. Nonlinear Sci. Numer. Simul. **18**(6), 1433–1440 (2013)
9. Hao, X., Wang, J., Yang, Q., Yan, X., Li, P.: A chaotic map based authentication scheme for telecare medicine information systems. J. Med. Syst. **37**(2), 9919 (2013)
10. Jiang, Q., Ma, J., Lu, X., Tian, Y.: Robust Chaotic mapbased authentication and key agreement scheme with strong anonymity for telecare medicine information systems. J. Med. Syst. **38**(2), 12 (2014)
11. Lee, T.F.: An efficient chaotic map-based authentication and key agreement scheme using smartcards for telecare medicine information systems. J. Med. Syst. **37**(6), 9985 (2013)
12. Kocarev, L., Tasev, Z.: Public-key encryption based on Chebyshev maps. Proc. Int. Symp. Circ. Syst. **3**:III-28–III-31 (2003)
13. Bergamo, P., DArco, P., Santis, A., Kocarev, L.: Security of public-key cryptosystems based on Chebyshev polynomials. IEEE Trans. Circ. Syst.-I **52**:1382–1393 (2005)
14. Li, C.T., Lee, C.C., Weng, C.Y.: A secure chaotic maps and smart cards based password authentication and key agreement scheme with user anonymity for telecare medicine information systems. J. Med. Syst. **38**, 77 (2014). doi:10.1007/s10916-014-0077-2
15. Zhang, L.: Cryptanalysis of the public key encryption based on multiple chaotic systems. Chaos. Soliton. Fract. **37**(3), 669–674 (2008)

16. Das, A.K., Goswami, A.: An enhanced biometric authentication scheme for telecare medicine information systems with nonce using chaotic hash function. J. Med. Syst. **38**, 27 (2014). doi:10.1007/s10916-014-0027-z
17. Mishra, D., Srinivas, J., Mukhopadhyay, S.: A secure and efficient chaotic map-based authenticated key agreement scheme for telecare medicine information systemss. J. Med. Syst. **38**, 120 (2014). doi:10.1007/s10916-014-0120-3

Improving the Accuracy of Intrusion Detection Using GAR-Forest with Feature Selection

Navaneeth Kumar Kanakarajan and Kandasamy Muniasamy

Abstract Intrusion detection systems (IDS) are designed to detect malicious activities in a large-scale infrastructure. Many classification methods have been proposed to improve the classification accuracy of IDS. In this paper, we have applied greedy randomized adaptive search procedure with annealed randomness—Forest (GAR-Forest), a novel tree ensemble technique, with feature selection to improve classification accuracy of IDS. GAR-forest uses metaheuristic GRASP with annealed randomness to increase the diversity of ensemble. We used NSL-KDD datasets to study the classification accuracy of GAR-forest for both binary and multi-class classification problems. The results show that GAR-forest performs better when compared with random forest, C4.5, naive Bayes and multilayer perceptron for binary and multi-class classification problem achieving 82.3989 and 77.2622 % accuracy, respectively, while classifying test data. We have also applied feature selection procedures, such as information gain, symmetrical uncertainty and correlation-based feature subset, to select relevant features for improving the accuracy of GAR-forest. GAR-forest with symmetrical uncertainty yields 85.0559 % accuracy using 32 features for binary classification problem and information gain yields accuracy of 78.9035 % using 10 features for multi-class classification problem. GAR-forest is found to be relatively much faster than multilayer perceptron though it is slower than naive Bayes, random forest and C4.5 algorithm. The metaheuristic GRASP procedure enables GAR-forest to reach the global optimal solution which greedy deterministic approaches fail to reach.

Keywords Feature selection · GAR-forest · GRASP metaheuristic · Intrusion detection system · Random forest

N.K. Kanakarajan (✉) · K. Muniasamy
TIFAC-CORE in Cyber Security, Amrita Vishwa Vidyapeetham, Coimbatore
641 112, India
e-mail: navaneeth031@gmail.com

K. Muniasamy
e-mail: kandamuniasamy@yahoo.com

S. Das et al. (eds.), *Proceedings of the 4th International Conference on Frontiers in Intelligent Computing: Theory and Applications (FICTA) 2015*, Advances in Intelligent Systems and Computing 404, DOI 10.1007/978-81-322-2695-6_45

1 Introduction

Advancement in the field of communication resulted in the proliferation of networked systems. This leads to the creation of wide variety of attacks, which can use the vulnerabilities of networks and hosts to perform malicious activities. Intrusion detection is a security measure that helps to protect systems from potential abuses. Much has been done to increase the classification accuracy of the intrusion detection system.

Hota et.al. [7] have studied decision tree techniques: C4.5, ID3 (Iterative Dichotomizer 3), CART (Classification and Regression Tree), REP (Reduced Error Pruning) tree and decision table on NSL-KDD datasets. This work uses 20 % of NSL-KDD training dataset containing multiple classes (Normal, DoS, Probe, U2R, R2L) for building the models. The study reports that an accuracy of 99.56 % was achieved for C4.5 learning model when all available features are considered. Furthermore, feature selection procedures information gain, relief, correlation-based feature subset (CFS) and symmetrical uncertainty were studied on the C4.5 model. An accuracy of 99.68 % was achieved with information gain feature selection procedure using 17 features.

Hota et al. [6] studied multilayer perceptron, decision table, C4.5, random forest, REP tree classification techniques for both binary (Normal, Anomaly) and multi-class classification using 20 % NSL-KDD datasets. Random forest achieves high accuracy of 99.73 and 99.65 % for binary and multi-class classifiers, respectively. Information gain feature selection procedure was applied with random forest for binary classification problem and an accuracy of 99.76 % was achieved using 15 features.

Tesfahun et.al. [10] addressed the issue of class imbalance in NSL-KDD training dataset by applying synthetic minority oversampling technique (SMOTE) on training dataset to improve the prediction of minority classes (R2L, U2R). With random forest as classifier, information gain as feature selection procedure and SMOTE generated NSL-KDD as training data, the experiments achieved 96.3 % detection rate for R2L class (compared to 96.1 % for original NSL-KDD + Random Forest) and 96.2 % detection rate for U2R class (compared to 59.6 % for original NSL-KDD + Random Forest).

The major contribution of this work is the application of a novel tree ensemble technique GAR-forest [2] with feature selection to improve the classification accuracy of intrusion detection system. While random forest [1] selects best attribute based on information gain score among a randomly selected subset of attributes for splitting the node in a decision tree, GAR-forest employs the metaheuristic greedy randomized adaptive search procedure (GRASP) to arrive at a set of randomized adaptive solutions from which an attribute is selected randomly for splitting the node. The level of randomness and the number of attributes available for random selection decreases as the tree is constructed from top to bottom.

This work [2] shows that the best solution selection method employed by random forest in each step during ensemble construction need not achieve global optimum result. The metaheuristic GRASP with annealed randomness procedure overcomes this problem by employing a randomized solution selection method which selects solutions from a larger search space. The advantage of GAR-forest is that it is relatively faster than multilayer perceptron (MLP) [9]. Furthermore, the construction of individual trees in GAR-forest ensemble can be parallelized resulting in fast learning. However, it is slower than random forest, C4.5 [5] and naive Bayes [5] algorithms. Since GAR-forest is an ensemble method, it solves the problem of overfitting found in C4.5 method.

This work compares the accuracy of GAR-forest with random forest, C4.5, naive Bayes and MLP in classifying attacks for binary and multi-class classification problem. We used complete NSL-KDD dataset containing 125973 training instances and 22544 test instances for study. Feature selection procedures such as information gain [5], symmetrical uncertainty [8] and CFS [4] are applied with GAR-forest to enhance the classification accuracy.

The rest of the paper is organized as follows. Section 2 describes proposed system, Sect. 3 describes experiments and results, and Sect. 4 concludes the work and gives the future line of research.

2 Proposed System

The proposed system shown in Fig. 1 describes our machine learning methodology. The training data is fed into feature selection module where the relevant features are extracted according to the chosen procedure. The training data is reconstructed with the extracted relevant features and then used to build GAR-forest machine learning model. Similarly, the test data is reconstructed with relevant features and then passed into the classification module where GAR-forest is used for classification.

2.1 Dataset

NSL-KDD [11] datasets are used for studying the classification accuracy of the above-proposed method. NSL-KDD datasets are preferred over the KDDCup99 dataset as they do not contain unwanted duplicates. The datasets comprised 41 features and a binary or multi-class label. For this study, we have used complete NSL-KDD two-class dataset (Normal, Anomaly) and NSL-KDD multi-class dataset (Normal, DoS, Probe, R2L, U2R). The statistics of both datasets are shown in Tables 1 and 2, respectively.

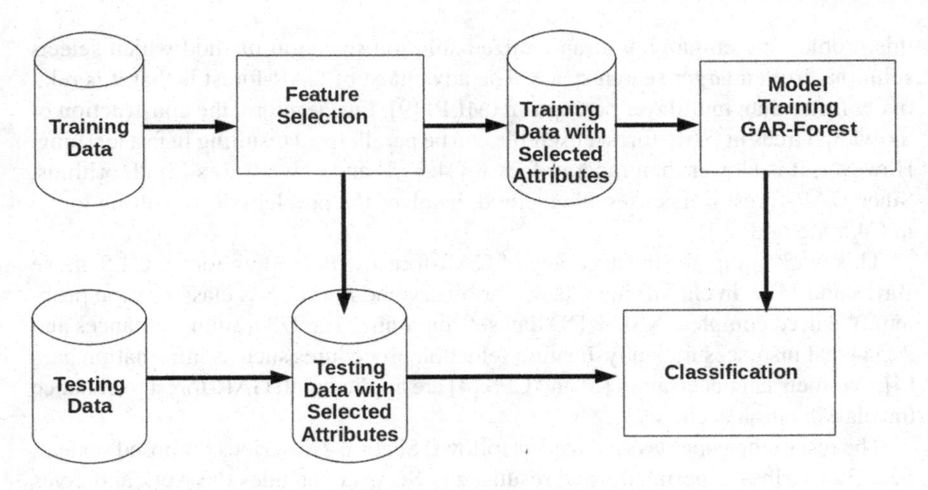

Fig. 1 Machine learning methodology

Table 1 NSL-KDD two-class dataset statistics

Dataset	Normal	Anomaly	Total
KDDTrain+	67343	58630	125973
KDDTest+	9711	12833	22544

Table 2 NSL-KDD multi-class dataset statistics

Dataset	Normal	DoS	Probe	U2R	R2L	Total
KDDTrain+	67343	45927	11656	52	995	125973
KDDTest+	9711	6775	2421	67	3570	22544

2.2 Feature Selection

Feature selection is a well-defined approach through which the statistical irregularities can be removed from dataset by selecting salient features. Feature selection methods identify relevant features which are highly correlated to the class distribution. Many studies had been done regarding the impact of feature selection on a classifier's performance and it is proven that well-selected features can significantly improve classification accuracy and reduce time for training the model. As [7] states, "An optimal feature set will depend on data, processing goal, and the selection criteria being used". In this work, we have studied information gain, symmetrical uncertainty and CFS with GAR-forest to improve the classification accuracy.

Information gain (InfoGain) selection method ranks the attribute according to the information gain measure which reflects the purity of the subsets obtained after splitting the set based on that attribute. CFS selection method ranks the subset of attributes which are highly correlated with the class, yet uncorrelated with other attributes in the subset. Symmetrical uncertainty selection method ranks the features in the same way as information gain but addresses the information gain measure's bias towards features with more values.

2.3 GAR-Forest Classifier

The proposed methodology employs a tree ensemble technique GAR-forest which incorporates a metaheuristic GRASP to generate randomized adaptive solutions. During the construction of each tree in the ensemble, the attribute based on which the node has to be split is chosen from a set of randomized adaptive solutions called candidate solution set *CSS*. Given a set of *M* attributes $F_1, F_2, \ldots, F_M$, the set *CSS* is defined by

$$CSS = \{F_i | InfoGain(F_i) \geq \alpha * MaxGain + (1 - \alpha) * MinGain\} \quad (1)$$

where
$InfoGain(F_i)$ = Information Gain of feature F_i,
α = ranges from 0 to 1,
$MaxGain$ = Maximum of $InfoGain(F_i)$,
$MinGain$ = Minimum of $InfoGain(F_i)$ and
i = varies from $1, 2, \ldots, M$.
The α value determines the cardinality of *CSS*. It is calculated by

$$\alpha = 1 - \left(\frac{N_{node}}{N}\right)^{\tau} \quad (2)$$

where
N = total number of instances in training dataset,
N_{node} = number of instances available at the node and
τ = controls the rate at which α varies from 0 to 1.

A key feature of GAR-forest tree construction method is the randomness of solutions generated decreases during the construction of the tree being maximum at the root and minimum at the leaves. From (2), it can be seen that at the root of a tree $N_{node} = N$ resulting in $\alpha = 0$ and hence the set *CSS* includes all available attributes for the selection of a solution at random. As the tree gets constructed from top down α reaches 1 and *CSS* will include the best attribute which will be used to split the node.

The algorithm for choosing the attribute for splitting the node is given below. Let *M* be the total number of attributes, F_i denotes *i*th attribute where $i = 1, 2, 3 \ldots, M$ then at each node of the tree, then

Step 1: Calculate α given by (2).
Step 2: Calculate $InfoGain(F_i)$ where $i = 1, 2, 3 \ldots, M$.

Step 3: Find *MaxGain* and *MinGain*.
Step 4: Construct *CSS* given by (1).
Step 5: Randomly choose a feature from *CSS* for splitting the node.

The determination of split point for numerical attributes has to be done before the attribute is considered for node splitting. While random forest selects best split point for splitting the values into two subsets, GAR-forest employs the above procedure to select the split point. The dataset is sorted out based on the numerical attribute in consideration. The algorithm for choosing split point is as follows. Let SP_i be the ith split point, where $i = 1, 2, 3 \ldots, N_{node}$, then

Step 1: Calculate α given by (2).
Step 2: Calculate $InfoGain(SP_i)$, where $i = 1, 2, 3 \ldots, N_{node}$.
Step 3: Find *MaxGain* and *MinGain*.
Step 4: Construct $CSS = \{SP_i | InfoGain(SP_i) \geq \alpha * MaxGain + (1 - \alpha) * MinGain\}$.
Step 5: Randomly choose a split point from *CSS*.

3 Experiments and Results

The experimental work is carried out using WEKA [3] data mining tool. The implementation of GAR-forest tree ensemble is done in JAVA by modifying the existing random forest source code from WEKA. The modification is done on the modules which calculate the splitting criterion of a node and numerical attribute. The evaluation parameters [6] such as accuracy, true positive rate (TPR), false positive rate (FPR), precision and F-score are considered for illustrating the classification accuracy of GAR-forest.

This work [2] cites that the classification accuracy of GAR-forest is better when the τ value is closer to 1. Hence, experiments have been conducted with τ value varied from 0.8 to 1.2 for both binary and multi-class classification problem. Tenfold cross validation is used for evaluating the classification accuracy. The results obtained confirmed the same. Table 3 displays the accuracy of GAR-forest for the corresponding τ value.

Four kinds of experiments have been conducted to illustrate the classification accuracy of GAR-forest tree ensemble. For all the experiments, the τ value is set

Table 3 Impact of τ value on classification accuracy of GAR-forest

Classification	Evaluation criteria	Accuracy (%)				
		τ value				
		0.8	0.9	1.0	1.1	1.2
Binary	10-fold cross validation	99.8452	99.8873	99.8936	99.8698	99.8865
Multi-class	10-fold cross validation	97.7694	99.796	99.87	95.78	94.8013

Table 4 Comparison of performance measures of binary classifiers

Classifier	Evaluation criteria	Accuracy (%)	TPR	FPR	Precision	F-score
Naive bayes	Cross validation	97.1121	0.971	0.032	0.972	0.971
MLP	Cross validation	97.9527	0.98	0.022	0.98	0.98
C4.5	Cross validation	99.7817	0.998	0.002	0.998	0.998
Random forest	Cross validation	99.8905	0.999	0.001	0.999	0.999
GAR-forest	Cross validation	99.8936	0.999	0.001	0.999	0.999
Naive bayes	KDDTest+	74.4544	0.745	0.2	0.822	0.739
MLP	KDDTest+	73.8733	0.739	0.214	0.796	0.735
C4.5	KDDTest+	81.5339	0.815	0.146	0.858	0.815
Random forest	KDDTest+	80.18	0.802	0.157	0.85	0.801
GAR-forest	KDDTest+	82.3989	0.824	0.143	0.858	0.824

Table 5 Comparison of performance measures of multi-class classifiers

Classifier	Evaluation criteria	Accuracy (%)	TPR	FPR	Precision	F-score
Naive bayes	Cross validation	95.4117	0.954	0.017	0.967	0.959
MLP	Cross validation	98.075	0.981	0.019	0.98	0.98
C4.5	Cross validation	99.7619	0.998	0.002	0.998	0.998
Random forest	Cross validation	99.796	0.999	0.001	0.999	0.999
GAR-forest	Cross validation	99.87	0.999	0.001	0.999	0.999
Naive bayes	KDDTest+	73.7624	0.738	0.164	0.801	0.714
MLP	KDDTest+	71.7619	0.718	0.197	0.737	0.666
C4.5	KDDTest+	74.929	0.749	0.167	0.802	0.697
Random forest	KDDTest+	75.692	0.757	0.128	0.8	0.723
GAR-forest	KDDTest+	77.2622	0.773	0.161	0.827	0.72

to 1 for calculating α as in (2). In Experiment-I and II, GAR-forest is compared with random forest, C4.5, naive Bayes and MLP. Both random forest and GAR-forest are not pruned. The ensemble size is set to 100.

In Experiment-I, binary classification problem is studied using NSL-KDD two-class dataset. KDDTrain+ dataset is used for training the classification models. All features were considered for training. Both tenfold cross validation (CV) and KDDTest+ dataset is used to evaluate the classification accuracy of the trained models. In Table 4, GAR-forest displays higher accuracy for both evaluation criteria. KDDTest+ GAR-forest has higher accuracy and TPR and lower FPR than other methods implying better classification accuracy.

In Experiment-II, multi-class classification problem is studied using NSL-KDD multi-class dataset. All features were considered for training. The results from Table 5 display GAR-forest has higher accuracy and TPR compared to random forest.

Table 6 Comparison of performance measures of feature selection procedures for binary classification

Method	Features	Accuracy (%)	TPR	FPR	Precision	F-score
GAR-forest	All	82.3989	0.824	0.143	0.858	0.824
GAR-forest + InfoGain	8	83.6409	0.836	0.133	0.866	0.837
GAR-forest + CFS	19	82.9755	0.83	0.149	0.847	0.83
GAR-forest + Symmetrical Uncertainty	32	85.0559	0.851	0.122	0.875	0.851

Table 7 Comparison of performance measures of feature selection procedures for multi-class classification

Method	Features	Accuracy (%)	TPR	FPR	Precision	F-score
GAR-forest	All	77.2622	0.773	0.161	0.827	0.72
GAR-forest + InfoGain	10	78.9035	0.798	0.174	0.824	0.799
GAR-forest + CFS	18	77.9454	0.779	0.186	0.816	0.779
GAR-forest + Symmetrical Uncertainty	23	77.6038	0.789	0.181	0.819	0.789

In Experiment-III, feature selection procedures such as InfoGain, symmetrical uncertainty and CFS are carried out on NSL-KDD two-class datasets. The selected features are used to process KDDTrain+ and KDDTest+ datasets which are used for training and testing, respectively. For each feature selection procedure, the number of features for which the classifier gives maximum classification accuracy and the corresponding results is displayed in Table 6. It can be seen that GAR-forest + symmetrical uncertainty with 32 features yields higher accuracy, TPR and lower FPR among other methods.

Similar procedure is carried out using NSL-KDD multi-class datasets. In Experiment-IV, results are displayed in Table 7. The results show that GAR-Forest+ InfoGain with 10 features yields higher accuracy and TPR among all the procedures.

4 Conclusion

In this work, we have used GAR-forest tree ensemble to improve the classification accuracy of intrusion detection system. The method was applied for both binary and multi-class classification problem. NSL-KDD datasets such as KDDTrain+ and KDDTest+ were used for training and testing the classifier. The experimental results show that for test data classification GAR-forest performs better than random forest, C4.5, naive Bayes and multilayer perceptron achieving 82.3989 and 77.2622 %

accuracy for binary and multi-class classification problem, respectively. Furthermore, feature selection procedures have been employed to improve the performance of GAR-forest. Results show that for binary classification problem, symmetrical uncertainty with 32 features yields 85.0559 % accuracy and for multi-class classification problem, information gain with 10 features yields 78.9035 % accuracy. We conclude that GAR-forest tree ensemble performs with high accuracy in classifying threats. The major drawback of GAR-forest is the construction of candidate solution set from the scratch at each step of tree construction. This results in a computational overload during training phase. Furthermore, the memory required to store the model is large compared to C4.5 and naive Bayes classifiers.

In future, the relation between the parameters α and τ which ensures annealed randomness of the GRASP metaheuristic will be studied to fine tune the performance of GAR-forest. Furthermore, the impact of ensemble size on the classification accuracy of GAR-forest will be studied.

References

1. Breiman, L.: Random forests. Mach. Learn. **45**(1), 5–32 (2001)
2. Díez-Pastor, J.F., García-Osorio, C., Rodríguez, J.J.: Tree ensemble construction using a GRASP-based heuristic and annealed randomness. Inf. Fusion **20**, 189–202 (2014)
3. Hall, M., Frank, E., Holmes, G., Pfahringer, B., Reutemann, P., Witten, I.H.: The WEKA data mining software: an update. ACM SIGKDD Explor. Newsl. **11**(1), 10–18 (2009)
4. Hall, M.A., Smith, L.A.: Feature subset selection: a correlation based filter approach (1997)
5. Han, J., Kamber, M., Pei, J.: Data Mining: Concepts and Techniques, 3rd edn. Morgan Kaufmann Publishers Inc., San Francisco (2011)
6. Hota, H., Shrivas, A.K.: Data mining approach for developing various models based on types of attack and feature selection as intrusion detection systems (IDS). In: Intelligent Computing, Networking, and Informatics, pp. 845–851. Springer (2014)
7. Hota, H., Shrivas, A.K.: Decision tree techniques applied on NSL-KDD data and its comparison with various feature selection techniques. In: Advanced Computing, Networking and Informatics-Volume 1, pp. 205–211. Springer (2014)
8. Press, W.H., Teukolsky, S., Vetterling, W., Flannery, B.: Numerical Recipes in C, vol. 1, p. 3. Cambridge University Press, Cambridge (1988)
9. Pujari, A.K.: Data Mining Techniques. Universities Press (2001)
10. Tesfahun, A., Bhaskari, D.L.: Intrusion detection using random forests classifier with SMOTE and feature reduction. In: 2013 International Conference on Cloud and Ubiquitous Computing and Emerging Technologies (CUBE), pp. 127–132. IEEE (2013)
11. The NSL-KDD Data Set. http://nsl.cs.unb.ca/NSL-KDD/

Part XII
Text Processing

Sentiment Analysis with Modality Processing

Surabhi Jain, Louella Mesquita Colaco and Okstynn Rodrigues

Abstract Online retailers selling goods have desired to know what the customers think about their products. Sentiment analysis has enabled the people to do so by allowing them to rate their products as positive or negative. In this paper, we present a novel method for sentiment analysis of sentences with modalities which helps to determine the polarity of a phrase that describes opposite opinions. The model determines the sentiment orientation of the sentence and the type of modality observed in the sentence. This model was implemented on customer reviews of four different products used by them and the result shows that 72 % of documents were correctly classified.

Keywords Opinion mining · Sentiment analysis · Modality · Natural language processing

1 Introduction

In today's world online shopping is a boon to the people worldwide, as it allows consumers to buy goods or services from a seller over the Internet on the click of a button. These electronic retailers get the feedback from the users, who have purchased the products. This feedback is of benefit to both the retailers and consumers. In case of customers, it helps in decision making whereas for product manufacturers

S. Jain (✉) · L.M. Colaco · O. Rodrigues
Department of Information Technology, Padre Conceicao College of Engineering, Goa University, Goa, India
e-mail: jsurbhi23@gmail.com

L.M. Colaco
e-mail: lmesquita@rediffmail.com

O. Rodrigues
e-mail: mecta2k7@gmail.com

S. Das et al. (eds.), *Proceedings of the 4th International Conference on Frontiers in Intelligent Computing: Theory and Applications (FICTA) 2015*, Advances in Intelligent Systems and Computing 404, DOI 10.1007/978-81-322-2695-6_46

it helps in knowing the efficiency, reliability of their products, and areas of improvement to make their business profitable.

Sentiment analysis has been focused upon in recent years. It aims to determine the attitude of a writer with respect to some topic or the overall contextual polarity of a document. The attitude may be his or her judgment or evaluation [1]. A basic task in sentiment analysis is to classify the polarity of a given text at the document level, sentence, or feature level as positive, negative, or neutral. This paper focuses on sentiment analysis at the sentence level. Here the sentences dealt with, are sentences with modality. These sentences always contain one or more modal verbs. Modal verbs are frequently used in English language, some of them are can, could, may, might, ought to, will, would, want, and should.

In this research, we study the problem of determining polarity of customer reviews of products sold online as positive or negative. Given a set of customer reviews for a particular product, the tasks involves: (1) Identifying and selecting feedback with modalities; (2) For each sentence, identifying the features of product that customers have expressed their opinions on; (3) Determining the polarity of the sentence at run time.

This paper is organized as follows. Related work on sentiment analysis with modality is described Sect. 2. Proposed method and proposed algorithm for identifying polarity of a sentence is described in Sect. 3. Section 4 provides experimental results followed by Sect. 5 which concludes the paper followed by references.

2 Related Works

This work is closely related to "Semantic analysis of sentences with modality" [2] by Liu et al. In [2], a supervised learning approach is studied to classify the sentiment orientation of sentences. To build a feature vector for each sentence, two groups of features, i.e., modality-specific features and general linguistic features are considered. Modality features like modality sequence and category of modality were used as part of the feature set. They designed and implemented their own classification strategy which made use of all the modality and linguistic features. The experiments were carried on customer reviews on four categories of products such as vacuum, cell phone, hair care, and mattress. The results have shown good performance.

Our work focuses on the following.

- To determine the polarity of any type of sentences with modalities, specifically customer reviews.
- The classification strategy used is different from the work done in [2].
- The type of modality is also concerned along with its polarity of sentence.

A more closely related work is [3], in which the authors specify the modality features and linguistic features in detail. General linguistic features such as opinion words (adjectives, adverbs, or nouns), negation words (such as not, never, hardly, no), relative location with adversative conjunctions (used in compound sentences), and punctuations (such as '?', '!') are considered. Along with linguistic features, modality features were also used to construct a classifier. They built a classifier based on SVMs to determine whether an opinion expressed in a sentence with modality is positive, negative, or neutral. From the results of both two-class classification and three-class classification, the method demonstrates significant improvement over the other three traditional methods(lexicon-based strategy, standard SVM classifier, and naive Bayes strategy).

3 The Proposed System

The system architecture is shown in Fig. 1. The inputs to the system are product reviews of four different categories. The system performs tokenization on the input data to convert it into a string of tokens. It then pre-processes the data. Preprocessing comprises of stop word removal and stemming.

3.1 *Preprocessing*

Stop Word Removal
Stops words are words of lesser importance, and are filtered to reduce the size of dataset and to keep only relevant data.

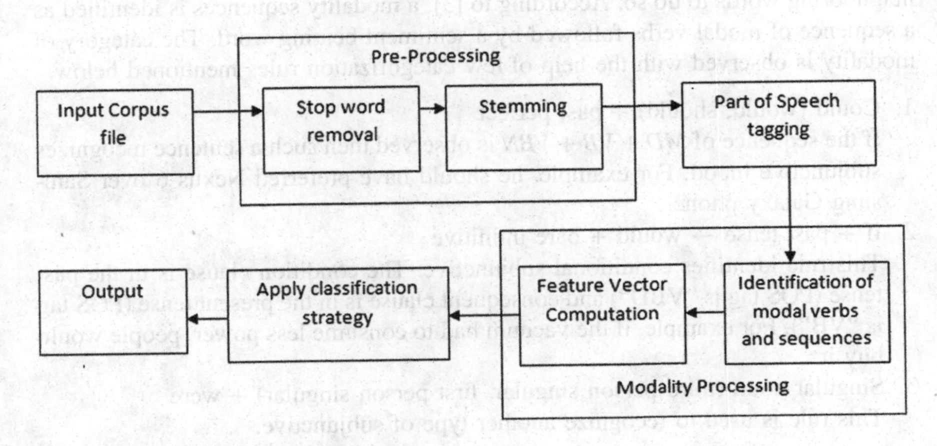

Fig. 1 Proposed system architecture

Stemming
Stemming is a technique which stems the suffixes of the words and stores it in its root form. For example, the words connection, connects, connected, and connecting all can be stemmed to the word 'connect'. We use Porter Stemmer algorithm [4], the most widely used algorithm for English language. It is a process for reducing derived words to their word stem. Porter stemming algorithm is based on a series of cascaded rewrite rules.

3.2 Part-of-Speech Tagging (POS)

Product features are usually nouns or noun phrases in review sentences. Thus the part-of-speech tagging is crucial. Stanford parts-of-speech tagger [5] is used to parse each review to produce the part-of-speech tag for each word (whether the word is a noun, modal verb, verb, and adjective). The tagger identifies simple nouns and verb groups.

3.3 Modality Processing

This is the heart of the architecture. It has three sub steps (as shown in the Fig. 1):

Identification of modal verbs and sequences
This step is used to identify the modal verbs in the input corpus. The modal verbs concerned are can, could, may, might, ought to, will, would, want, and should(as mentioned before).

A modal verb alone cannot determine the polarity of a sentence; it needs the neighboring words to do so. According to [3], a modality sequences is identified as a sequence of modal verbs followed by a sentiment bearing word. The category of modality is observed with the help of few categorization rules mentioned below.

1. Could (would, should) + past perfect
   If the sequence of *MD* + *VB* + *VBN* is observed then such a sentence recognizes subjunctive mood. For example, he should have preferred Nexus 6 over Samsung Galaxy phone.
2. If + past tense → would + bare infinitive
   This rule identifies conditional subjunctive. The condition clause is in the past tense (POS tag is "VBD") and consequent clause is in the present tense (POS tag is "VB"). For example, if the vacuum had to consume less power, people would buy it.
3. Singular noun (third person singular, first person singular) + were
   This rule is used to recognize another type of subjunctive.

4. You should (May) + VB
   This rule is used to recognize deontic modality. For example, you should buy an SLR canon camera.
5. I will (would) + VB
   This rule is used to recognize dynamic modality, with which the speaker expresses his/her willingness to do something. The subject of this type of sentence is "I", and the modality auxiliary word is "will" or "would". For example, i will compare the features of Samsung Edge with Sony xperia Z.

Feature vector computation
We use Delta TF-IDF [6], a widely used technique to calculate weight word scores before classification. It is computed by executing the following steps.

- Find the inverse document frequency (idf) with respect to each word of the category.
- Find the term frequency (tf) with respect to each word of the category. Term frequency is the number of times a term occurs in that document.
- Calculate the weight of the word by multiplying tf * idf values of that word with respect to the category and store it in the database.

3.4 Classification Strategy

In this section, we introduce the classification strategy used to determine the polarity of the sentence.

Decision list classifier
A decision list classifier is a machine learning technique used widely to determine the sentiment orientation of sentences. It is a collection of individual classification rules that collectively forms a classifier [7]. The process works as follows, first we identify the type of modality observed in the query sentence. Second, the weight of the test query is calculated with respect to each product category (cell phones, jewelry, software, and watches) for both positive and negative polarity. The maximum weight among all the possibilities is considered and the polarity associated with it is identified as the polarity of the test query.

3.5 Polarity Detection

This step predicts the sentiment orientation of an opinion sentence. It identifies the type of modality observed in the query sentence. This research work focuses on subjunctive mood and one particular type of modality, i.e., event modality along with its subtypes. The event modality is classified as deontic and dynamic modalities.

3.6 Proposed Algorithm

(1) Input the data corpus for preprocessing (for example Customer Reviews).
(2) The Preprocessing step includes:
Stop word removal—The stop words are eliminated from the corpus file.
Stemming—The Porter Stemmer Algorithm is used to remove the suffixes from the words.
(3) Then apply Stanford parts-of-Speech Tagger to the preprocessed data from the above step.
(4) Perform Modality Processing on the tagged data, which has two substeps:
Identify modal verbs and sequences with the help of categorization rules as mentioned in Sect. 3.3.
Evaluate the Feature Vector Computation using Delta TF-IDF.
(5) Apply Decision List as a Classification Strategy to determine the polarity of the sentence.
(6) Examine the results obtained on test query sentences.

4 Experimental Evaluation

4.1 Dataset Resources

The dataset used is a SNAP web dataset. This dataset consists of amazon reviews which include product, user information, ratings, and plain text review. The customer reviews of four product categories: cell phones and its accessories, software, jewelry, and watches were chosen.

4.2 Results and Observations

The total numbers of sentences considered were 5000 sentences from which 1000 sentences were detected as sentences with modalities. These sentences were identified with the help of Stanford POS tagger. A supervised machine learning approach was followed which included training and testing phase. The data from all the four categories was trained. The testing phase included untrained 100 random sentences of any of the four categories. The result shows that out of 100 sentences,

Table 1 Training data for customer review corpus

	Categories	Cell phones	Watches	Software	Jewelry
Training data of customer review corpus	Number of documents	240	260	260	240

Table 2 Testing dataset for customer review corpus

Test dataset (100 random documents)	1. Number of miscalculations with respect to customer review corpus	28
	2. Accuracy rate (%)	72

the number of documents correctly classified were 72. Table 1 shows the number of trained documents in each category. Table 2 shows the accuracy rate obtained is 72 %.

The correct sentiment orientation of the sentences is identified based on human interpretation.

5 Conclusion

In this paper, we implement a model to determine the sentiment orientation of sentences with modalities. A detailed study on modality and its types was done with the help of [8]. The linguistic and general features were taken from [2] and [3] and then the categorization rules were used accordingly. A decision list classifier was used to train the dataset. Our experimental results indicate that the proposed technique is quite effective in determining whether the polarity of a sentence is positive or negative.

References

1. Sentiment analysis. In: Wikipedia, The Free Encyclopedia. http://en.wikipedia.org/w/index.php?title=Sentiment_analysis&oldid=637966670. Accessed 13 Dec 2014
2. Liu, Y., Yu, X., Chen, Z., Liu, B.: Sentiment analysis of sentences with modalities. In: Proceedings of the International Workshop on Mining Unstructured Big Data using Natural Language Processing (Unstructured NLP 2013), pp. 39–44 (2013)
3. Liu, Y., Yu, X., Chen, Z., Liu, B.: Sentence level sentiment analysis in the presence of modalities. In: Proceedings of 15th international Conference, CICLing 2014, Part II Kathmandu, Nepal, 6–12 April 2014
4. Karaa, W.B.A., Gribaa, N.: Information retrieval with porter stemmer: a new version for English. In: Nagamalai, D. et al. (eds) Advances in Computational Science, Engineering and Information Technology, pp. 243–254. Springer International Publishing, Switzerland, AISC 225 (2013)

5. Toutanova, K., Klein, D., Manning, C., Singer, Y.: Feature-rich part-of-speech tagging with a cyclic dependency network. In: Proceedings of the Conference of the North American Chapter of the Association for Computational Linguistics on Human Language Technology, vol. 1, p. 173 (2003)
6. Martineau, J., Finin, T.: Delta tfidf: an improved feature space for sentiment analysis. In: Proceedings of the 3rd AAAI International Conference on Weblogs and Social Media, pp. 258–261 (2009)
7. Decision List. In: Encylopedia of Machine Learning, Springer US, pp. 261 (2010)
8. Palmer, F.: Mood and Modality. Cambridge University Press (2001)

Query-Based Extractive Text Summarization for Sanskrit

Siddhi Barve, Shaba Desai and Razia Sardinha

Abstract Sanskrit consists of lots of literature available in the form of epics, stories, puranas, Vedas, and many more. Most of the Sanskrit documents have been digitized and made available online. Searching for the required information from the plentiful documents available is a tedious task. Automatic summarization serves the purpose in such situations. Many tools for summarization have been developed for English and foreign languages. The research for such kind of tools in Sanskrit is under exploration. In this paper, we propose three query-based summary generation methods to obtain extractive summary for single document written in Sanskrit. The methods are based on average term frequency-inverse sentence frequency, the VSM (Vector Space Model) and a graph-based technique using PageRank. All the techniques are compared and evaluated on the basis of performance.

Keywords Summarization · Query-Based · Sanskrit · Extractive · Single document · Sentence similarity · tf.sf · Vector space model · Pagerank

1 Introduction

In everyday life, we see different kinds of summarizations such as newspaper headlines, trailer of a movie, menu card, gadget specifications, abstracts of research papers, and many more. Due to time pressure, an individual is interested in condensed

---

The erratum of this chapter can be found under 10.1007/978-81-322-2695-6_62

S. Barve (✉) · S. Desai · R. Sardinha
Department of Information Technology, Padre Conceicao College of Engineering,
Goa University, Goa, India
e-mail: barvesiddhi@gmail.com

S. Desai
e-mail: shaba.desai@gmail.com

R. Sardinha
e-mail: razia.sardinha@gmail.com

S. Das et al. (eds.), *Proceedings of the 4th International Conference on Frontiers in Intelligent Computing: Theory and Applications (FICTA) 2015*, Advances in Intelligent Systems and Computing 404, DOI 10.1007/978-81-322-2695-6_47

representation of any information which he wants to read, watch, or listen to. Summaries help the reader to obtain the relevant information. Document summaries assist user to decide on whether the document he has been reading is useful for him or not by eliminating the need to read the entire document. Automatic text summarization is the process of reducing a text document with a computer program in order to create summaries that retains the most important points in the original document.

Various kinds of views for characterizing summarizers depending on the various parameters are described in [1]. It includes summaries on the basis of the form of the summary we recognize, level of processing, the purpose the summary serves, the audience we recognize, span of processed text, language, and genre.

A variety of document summarization methods have been developed. Authors of [1] review research on automatic summarization over the past few years. It mentions salient notations and developments and seeks to assess the state-of-art for this challenging Natural Language Processing (NLP) task.

Authors of [2] explain various features for extractive text summarization which includes content word (keyword) feature, title word feature, sentence location feature, sentence length feature, proper noun feature, upper-case word feature, cue-phrase feature, biased word feature, pronoun, sentence-to-sentence cohesion, sentence-to-centroid cohesion, occurrence of nonessential information, and discourse analysis.

The goal of this paper is to adapt an approach, which has been applied to summarize English documents using query-based single document summarization methodologies in an attempt to develop and experiment with Sanskrit documents and compare the summaries to learn their efficiency. The system takes as an input a single document and the query specified in Sanskrit and bids to provide plausible summary for the given query. The reason behind choosing Sanskrit as a language for summarization is its vast and rich literature. Generating summaries for such kind of texts should definitely help the system users to conserve resources such as time.

The dataset used is a collection of Sanskrit articles collected from Sanskrit Wikipedia, a free encyclopedia also known as "sawiki" [3] and the set of queries collected from Sanskrit experts. The summary length is between 20–30 % of the total text and should not exceed 30 %.

2 Related Work

Lots of work has taken place in the area of text summarization with respect to Indian languages. Authors of [4] present comparison of various text summarization methods seen in Indian languages. Here, summarization techniques for Tamil, Kannada, Odia, Bengali, Punjabi, and Gujarati are taken into consideration. Authors of [5] have text summarizer for Tamil which uses graph-based approach for the entire document where each vertex is represented as a sentence and edges show connectivity between the sentences. Sentences are ranked and summary is formed. Authors of [6] have another method for Tamil that uses sentence scoring techniques used to select highly ranked sentences. Authors of [7] present a summarization approach for Kannada where sentences are scored based on line score and sentence score. Authors of [8] state method for summarization based on keyword extraction using tf-idf and GSS coefficients.

Authors of [9] describe about Odia text summarizer using stemmer where each sentence is ranked by assigning weight value to each word of sentence. Authors of [10] present a text summarizer for Bengali where thematic terms and positional scores are considered. Authors of [11] concentrate on feature selection and weight learning for Punjabi text summarization system. The important sentences are selected on the basis of statistical and linguistic features of a sentence.

3 System Description

Figure 1 shows a general framework for the Sanskrit text summarization. Preprocessing module, sentence extraction, and ranking are core of the system.

3.1 Sanskrit Document

To demonstrate the summary generated, we have downloaded 50 articles from Sanskrit Wikipedia which form the dataset. The articles cover random topics in Sanskrit. Since the system is a single document summarization, only one document will be taken into consideration at a time.

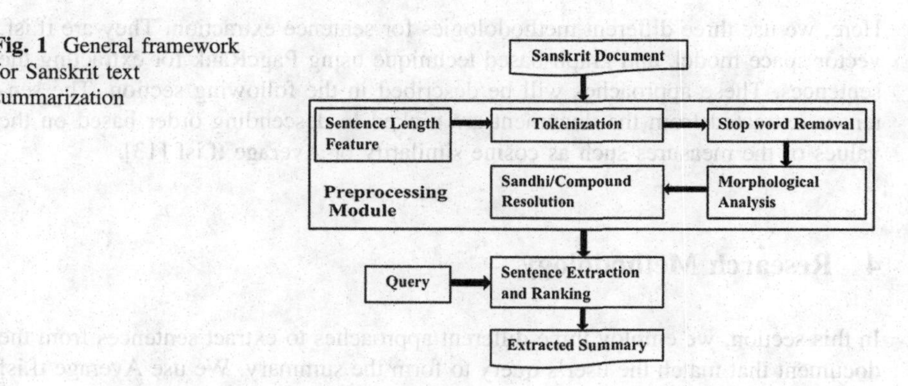

Fig. 1 General framework for Sanskrit text summarization

3.2 Preprocessing Module

- Sentence Length Feature
  The very short and very long sentences are eliminated from the document.
- Tokenization
  Tokenization includes splitting the sentence into words. Here, we tokenize the sentence using space as a delimiter.
- Stop-word Removal
  A list of commonly occurring words is supplied for the removal.
- Morphological analysis [17]
  This step aims at analyzing Sanskrit words giving its nominal stem or verbal stem along with its various linguistic features such as gender, number, case, person, etc.
- Resolving Compound and Sandhi [12, 17]
  This aspect focuses on identifying whether an encountered word is a compound or sandhi and then resolving the word accordingly.

3.3 Query

A query for each document has been formed separately depending on the content. The set of queries used in this process are formed by considering into account the frequently occurring nonstop words from each of the document.

3.4 Sentence Extraction and Ranking

Here, we use three different methodologies for sentence extraction. They are tf.isf, vector space model, and graph-based technique using PageRank for extracting the sentences. These approaches will be described in the following section. The sentences extracted from the document are ranked in descending order based on the values of the measures such as cosine similarity or Average tf.isf [13].

4 Research Methodology

In this section, we employ three different approaches to extract sentences from the document that match the user's query to form the summary. We use Average tf.isf [13], vector space model [14] and graph-based approach [15, 16].

4.1 Term Frequency. Inverse Sentence Frequency (tf.isf)

This technique is the convention of tf-idf [14]. Since the summarization is for a single document, here we work on sentential level rather than document and retrieve the sentences similar to the query based on the predefined threshold [16]. The computation is also similar to tf-idf and computed as follows:

$$tf.isf_{(w,s)} = tf_{(w,s)} * isf_{(w,s)} \quad (1)$$

Here, $tf.isf_{(w,\ s)}$ is the number of times the word w occurs in sentence s.

$$isf_w = \log\left(\frac{|S|}{sf_w}\right) \quad (2)$$

Here, isf_w is the inverse sentence frequency, sentence frequency sf_w is the number of sentences in which word w occurs, and S is the total number of sentences in the document.

For each sentence, Average tf.isf is computed by

$$Avg - tf.isf = \sum_{i=1}^{w(s)} \frac{tf.isf(s)}{W(s)} \quad (3)$$

Here, $W(s)$ is the number of words in the sentence s.

Later, the sentences are arranged in the descending order of Average tf.isf and the top-ranked sentences are selected as a part of summary.

4.2 Vector Space Model

The vector space model is also applied at a sentential level. Each sentence in the document and the respective query is represented as a vector. Each term is assigned a weight to reflect its importance. The weighing scheme used here makes use of Term Frequency (*tf*) and Inverse Sentence Frequency (*isf*) [12]. The weight $W(w,\ s)$ of word w in sentence s can be computed as

$$W(w,s) = tf(w,s) * isf(w,s) \quad (4)$$

Here, $tf(w,\ s)$ is given by Eq. (1) and $isf(w,\ s)$ by Eq. (2).

After computing weight of each term using tf.isf, we proceed toward finding the sentences similar to the query. This procedure is done by using cosine similarity between the sentence and the query using tf.isf as weights.

$$\cos\theta = \frac{s \cdot q}{\|s\|\|q\|} = \frac{\sum_{i=1}^{n} s_i \times q_i}{\sqrt{\sum_{i=1}^{n} (s_i)^2} \times \sqrt{\sum_{i=1}^{n} (q_i)^2}} \tag{5}$$

Here, s is the sentence and q is the query. Again, the sentences are ranked in descending order of cosine similarity and top-ranked sentences form the part of summary.

4.3 *Graph-Based Approach*

In this approach, we construct a graph of the document to be summarized [15, 16]. We represent each sentence as a vertex and similarity between them as an edge. The similar sentences are considered to be connected with each other. Each sentence is compared to every other sentence except itself.

All the sentences in the document graph are initialized to the PageRank of 1 and every sentence is evaluated by applying PageRank. After the computation of PageRank values is complete, the query is supplied and the sentences matching the query along with PageRank values are obtained by cosine similarity measure. The sentences are then ranked by descending PageRank values and top-ranked sentences are selected to be a part of summary. The PageRank is given by the formula

$$PR(V_i) = (1 - d) + d^* \sum_{V_j \in In(V_i)} \frac{PR(V_j)}{\mathrm{Out}(V_j)} \tag{6}$$

Here, $PR(V_i)$ is PageRank of vertex V_i, In (V_i) is all the predecessor vertices to node V_i, Out(V_i) is set of vertices that V_i points to, and d is the dampening factor which is between 0 and 1.

5 Experiments and Results

Before we discuss about the experiments and the results, let us consider an example to demonstrate how the query-based extractive text summarization for Sanskrit works.

1. Assume that the user selects a single document titled “कदली” in Table 1 [3] from the set of documents available.
2. The preprocessing steps are performed on the selected document. The steps include sentence length feature, tokenization, stop word removal, morphological analysis, and sandhi and compound resolution.

Table 1 Raw Sanskrit document

आम्रं पनसं, कदलीफलं च प्रमुखफलत्रयत्वेन परिगण्यन्ते । एवम् एव कदली-आम्र-उदुम्बर तिन्त्रिणीवृक्षाः प्रमुखाः वृक्षाः इति उच्यन्ते ।एतान् अधिकृत्य काचित विचित्रा कथा केषुचित प्रदेशेषुश्रूयते ।एते पञ्च वृक्षाः अपि कस्मिंश्चित् काले मनुष्यरुपेण सहोदर्यः आसन् ।बहुकालं यावत् तासां विवाहः न जातः ।कदाचित् देवः प्रत्यक्षीभूय 'विवाहम् इच्छन्ति वा?' इति ताः पृष्टवान् ।तदा चतस्रः सहोदर्यः विवाहम् अङ्घीकृतवत्यः । किन्तु अन्तिमा सहोदरी न अङ्गीकृतवती । सा सन्तानमात्रं प्रार्थितवती । अनन्तरं देवः ताः सर्वाः वृक्षरुपेण परिवर्त्य - "यः एतान् वृक्षान् आरोहति सः एव पतिः' इति उक्तवान् । अतः एव विवाहम् अङ्गीकृतवतः वृक्षान सर्वे आरोढुं शक्नुअन्ति चेदपि कदलीवृक्षं कोऽपि आरोढुं न शक्नोति । कदली इति संस्कृतभाषया, हिन्दिभाषया केला इति, आंगलभाषया बनाना प्लाण्टन्, आडम्स् आपलि इत्यपि उच्यते । काभिश्चित् भारतीयभाषाभिः कदलीइत्येव उच्यते ।तमिळु मलयाळभाषया च वाळ्:ऐ इति, तेलुगुभाषया अरटि इति च उच्यते । बौद्धाः कदलीवृक्षं पवित्रं मन्यन्ते । बौद्धसाहित्ये उक्तं मोचापानं कदलीफलेन एव सज्जीक्रियिते । अस्माकं पुराणे श्रूयते यत् -हनुमान् हिमालयप्रान्ते कदलीवने आसीत् इति । कदलीवृक्षस्य पुष्पं, शलाटुः, फलं च आहाररुपेण उपयुज्यते । कदलीवृक्षस्य शलाटुः हरितवर्णीयः भवतिी तस्य फलं हरितं,पीतं, पीतहरितमिश्रितं, रक्तवर्णीयं वा भाति । तदा आकारेण लघु बृहत् चापि भवति । कदल्यां ५,००० प्रभेदाः सन्ति इति श्रूयते । पच्चेकदली, बूदकिदली, वृक्षकदली, रसकदली, एलाकदली, अरण्यकदली इति अष्ट, दश वा विधाः प्रसिद्धाः सन्ति । कदलीपर्णं बृहदाकारकं पञ्चषपादमतिदीर्धम्, अधिकविशालं च भवति । दक्षिणाभारते केषुचित स्थलेषु भोजनार्यम् एतस्य उपयोगः क्रियते । जनाः विवाहादिशुभकार्येषु वितानं मण्डपं द्वारं च पुष्पगुच्छसहितेन कदलीवृक्षेण अलङ्कुर्वन्ति । एतं शुभसङ्केतं मन्येन्ते जनाः । पक्वानि कदलीफलानि कदलीपुष्पम् यदा शलाटुः पक्वं भवति तदा वृक्षं कर्तयन्ति । कर्ततिवृक्षस्य प्रकाण्डं परितः स्थितिभ्यः कन्देभ्यः नूतनसस्यानि उत्पद्यन्ते । एवं कदिलीसन्तानः वर्धते ।

Below is an illustration of morphological analysis and sandhi/compound resolution for the document "कदली."

- **Morphological Analysis**
  Analysis for the word "आम्रं" follows:
  आम्रम्=आम्र{पुं}{2;एक}
  आम्र{नपुं}{1;एक}
  आम्र{नपुं}{2;एक}
- **Sandhi/Compound Resolution**
  Suppose if the encountered word, "रक्तवर्णीयम्" is compound, then the split can be obtained as रक्तवर्णीयम्= रक्त-वर्णि-इयम्.
  The word obtained could also be a Sandhi; if it is Sandhi, then the splits are obtained as चेदपि= चेत्+अपि.
  After preprocessing is carried out, the user is prompted to enter the query corresponding to the document. The query is "वृक्षाः कदली शलाटुः फलं."

3. Next, we use sentence extraction techniques along with the query to find most relevant sentences matching the query.
4. Now, the system is ready to select 20–30 % of top-ranked sentences for summary generation.
5. Finally, the top-ranked sentences are fused together to generate summary. The summary generated by using the three techniques is illustrated in Tables 2, 3, and 4.

The results generated using all the three techniques, i.e., Average tf.isf, VSM, and graph-based technique were presented to a set of human experts in Sanskrit for review and rating. We explained the experts about the purpose of our system and

Table 2 Summary generated by average tf.isf approach

आम्रं पनसं, कदलीफलं च प्रमुख फल तरयत्वेन परगिणयन्ते । एवम् एव कदली-आम्र-उदुम्बर तन्त्रिणीवृक्षाः प्रमुखाः वृक्षाः इति उच्यन्ते । अतः एव विवाहम् अङ्गीकृतवतः वृक्षान सर्वे आरोढुं श क्नुअन्ति चेत् अपि कदलीवृक्षं कः अपि आरोढुं न शक्नोति। कदली इति संस्कृतभाषया, हिन्दिभाषया केला इति, आंगलभाषया बनाना प्लाण्टन्, आडम्स् आपलि इति अपि उच्यते । अस्माकं पुराणे श्रूयते यत् हनुमान् हिमालयप्रान्ते कदलीवने आसीत् इति। प-च्चेकदली, बृदकिदली, वृक्ष-कदली, रस-कदली, एला-कदली, अरण्य-कदली इति अष्ट, दश वा विधाः प्रसिद्धाः सन्ति।

Table 3 Summary generated by graph-based approach

कदलीवृक्षस्य शलाटुः हरितिवर्णीयः भवति। कदलीपर्णं बृहदाकारकं पञ्चषप अदमतिदीर्धम्, अधिक विशालम् च भवति। बौद्धाः कदली-वृक्षं पवित्रं मन्यन्ते। कदलीवृक्षस्य पुष्पं, शलाटुः, फलं च आहाररुपेण उपयुज्यते । काभिः चित् भारतीयभाषाभिः कदली इति एव उच्यते । पक्वानि कदलीफलानि कदली-पुष्पम् यदा शलाटुः पक्वं भवति तदा वृक्षं कर्तयन्ति।

Table 4 Summary generated by VSM approach

कदली इति संस्कृतभाषया, हिन्दिभाषया केला इति, आंगलभाषया बनाना प्लाण्टन्, आडम्स् आपलि इति अपि उच्यते। काभिःचित् भारतीयभाषाभिः कदली इति एव उच्यते । कदलीवृक्षस्य पुष्पं, शलाटुः, फलं च आहाररुपेण उपयुज्यते । पक्वानि कदलीफलानि कदली-पुष्पम् यदा शलाटुः पक्वं भवति तदा वृक्षं कर्तयन्ति । एवम् एव कदली-आम्र-उदुम्बर तन्त्रिणीवृक्षाः प्रमुखाः वृक्षाः इति उच्यन्ते । कदलीवृक्षस्य शलाटुः हरितिवर्णीयः भवति।

Table 5 Evaluation scale for summarization

Scale measure	Score	Clarification
Poor	0	Summary is poor having no relation with the document
Below average	1	Summary is below average and away from main idea of the document
Average	2	Summary is average, i.e., it just meets the expectations of the user
Fairly good	3	Summary is convincing and contains main idea of the document
Very good	4	The summary is of good quality and the results obtained are very good
Excellent	5	The quality of the summary is excellent and the results are best and convey idea of the document

supplied them with the text documents, respective queries, and the summaries generated using the above three techniques. We asked them to read the documents and the corresponding summaries and then provide a score from 0–5 as shown in Table 5.

Thirty percentage of the reviewers rated Average tf.isf to be fairly good, 26 % of them rated it as very good, 14 % rated as excellent, and 28 % rated it for average evaluation. In case of VSM, 42 % were for a score = very good, 24 % assigned score = excellent, 18 % gave score = fairly good, and 14 % average. Coming to

Fig. 2 Results of evaluation

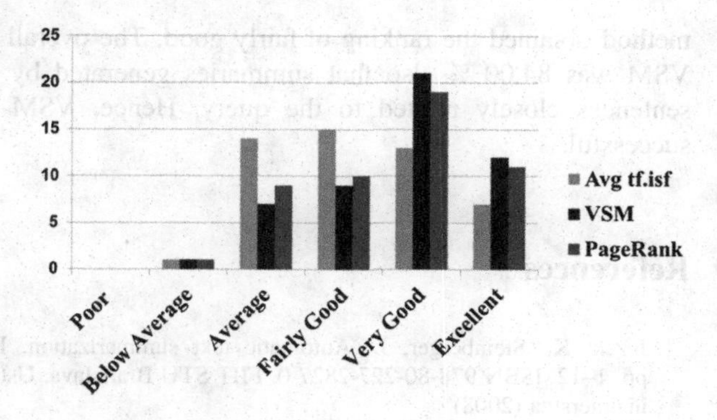

Table 6 Performance of query-based summarization for Sanskrit

Methods	Scale measure and score						
	Poor (%)	Below average (%)	Average (%)	Fairly good (%)	Very good (%)	Excellent (%)	Performance (%)
Avg tf.isf	0.0	2.0	28.0	30.00	26.00	14.00	70.00
VSM	0.0	2.0	14.0	18.00	42.00	24.00	84.00
Graph-based	0.0	2.0	18.0	20.00	38.00	22.00	80.00

PageRank, 38 % of reviewers gave score = very good, 22 % assigned score = excellent, 20 % rated as score = fairly good, and 18 % of them as average. Not a single reviewer gave score = poor to the summaries generated using the above mentioned approaches.

This indicates that the quality of the summary generated is quite acceptable. From the above evaluation scale and graph, we can predict that VSM and PageRank perform well as compared to Average tf.isf.

From the results obtained after the evaluation, we assumed that the overall performance is based on, "Excellent," "Very Good," and "Fairly Good" results. The overall performance is sum of these three scores and is close to 80 %. The results are shown in the form of graph in Fig. 2 and performance is displayed in Table 6.

6 Conclusion

In this work, we compared three query-based extractive text summarization methods; one based on Average tf.isf another on VSM and third one a graph-based technique using PageRank. The results obtained for all the three methods were presented to Sanskrit experts for the purpose of evaluation. Out of the three methods, majority of the summaries generated by VSM and graph-based PageRank methods got the ranking of very good, while those generated by the Average tf.isf method obtained the ranking of fairly good. The overall performance obtained for VSM was 84.00 % also that summaries generated by it contained most of the sentences closely related to the query. Hence, VSM-based method was most successful.

References

1. Jezek, K., Steinberger, J.: Automatic text summarization. In: Snasel, V. (ed.) Znalosti, pp. 1–12. ISBN 978-80-227-2827-0, FIIT STU Brarislava, Ustav Informatiky a softveroveho inzinierstva (2008)
2. Gupta, V., Lehal, G.S.: A survey of text summarization extractive techniques. J. Emerg. Technol. Web Intell. **2**(*3*), 258–268 (2010) doi:10.4304/jetwi.2.3.258-268
3. Sanskrit Wikipedia. https://sa.wikipedia.org/. Accessed Aug 2014
4. Dhanya, P.M., Jathavedan, M.: Article: comparative study of text summarization in indian languages. Int. J. Comput. Appl. **75**(6), 17–21 (2013)
5. Sankar K., Vijay Sundar Ram R., Devi, S.L.: Text extraction for an agglutinative language. In: Vijayanand, K., Ramamoorthy, L. (eds.) Language in India, vol. 11, (2011) ISSN 1930-2940
6. Perumal, K., Chaudhuri, B.B.: Language independent sentence extraction based text summarization. In: Proceedings of ICON-2011: 9th International Conference on Natural Language Processing (2011)
7. kallimani, J.S., Srinivasa, K.G.: Information retrieval by text summarization for an Indian regional language. In: IEEE (2010)
8. Jayashree, R., Srikanta Murthy, K., Sunny, K.: Document summarization in Kannada using keyword extraction. CS & IT-CSCP (2011)
9. Balabantaray, R.C., Sahoo, B., Sahoo, D.K., Swain, M.: Odia text summarization using stemmer. Int. J. Appl. Inf. Syst. **1**(3) (2012). ISSN: 2249-0868
10. Sarkar, K.: Bengali text summarization by sentence extraction. In: Proceedings of International Information Management (ICBIM-2012), pp. 233–245, NIT Durgapur
11. Gupta, V., Lehal, G.S.: Features selection and weight learning for Punjabi text summarization. Int. J. Eng. Trends Technol. **2**(2) (2010)
12. Goyal, P., Huet, G., Kulkarni, A., Scharf, P., Bunker, P.: A distributed platform for Sanskrit processing. In: Proceedings of COLING 2012: technical papers, pp. 1011–1028 (2012)
13. Azmi-Murad, M.A., Martin, T.P.: Similarity-based estimation for document summarization using fuzzy sets. Int. J. Comput. Sci. Secur. **1**, 1–12 (2007)
14. Christopher, D.M., Prabhakar, R., Hinrich, S.: Introduction to Information Retrieval. Cambridge University Press (2008)
15. Mihalcea, R.: Graph-based ranking algorithms for sentence extraction, applied to text summarization. In: Proceedings of the ACL 2004 on Interactive poster and demonstration sessions, ACLdemo'04 (2004)
16. Thakkar, K.S., Dharaskar, R.V., Chandak, M.B.: Graph-based algorithms for text summarization. In: IEEE (2010)
17. University of Hyderabad. http://sanskrit.uohyd.ernet.in/scl/#. Accessed Dec 2014

Code Obfuscation by Using Floating Points and Conditional Statements

Chandan Kumar Behera and D. Lalitha Bhaskari

Abstract Code obfuscation is one of the trickiest methods for software code transformation, and proves to be one of the major sources for malwares as well as legitimate software. There are many techniques for code obfuscation, which can be used to manipulate the source code, so that it becomes hard to analyze for the adversary. In this paper, the conditional code obfuscation using floating point arithmetic and control statements is discussed, where the 'if condition' has been used several times in place of 'if-else if' conditions. The objective of the proposed conditional code obfuscation method is to keep the code simple, without major modification, so that the adversary will be unable to identify the difference between the original code and the obfuscated code, but the code will be highly obfuscated.

Keywords Code obfuscation · Conditional statements · Floating point numbers and precisions · Radix conversion

1 Introduction

Protection of software is gradually a more important constraint for software development in industries. This protection problem is basically harder than other security problems. When one has the adversary for full access to the chosen software or hardware and can examine, or modify it, then no piece of software can be protected for a long period of time [2].

For analysing any software or understanding the executable files, reverse engineering has to be done and one of the examples of simple form of protection against

C.K. Behera (✉) · D.L. Bhaskari
Department of Computer Science and Systems Engineering, Andhra University, Visakhapatnam, India
e-mail: ckb.iitkgp@gmail.com

D.L. Bhaskari
e-mail: lalithabhaskari@yahoo.co.in

S. Das et al. (eds.), *Proceedings of the 4th International Conference on Frontiers in Intelligent Computing: Theory and Applications (FICTA) 2015*, Advances in Intelligent Systems and Computing 404, DOI 10.1007/978-81-322-2695-6_48

reverse engineering attacks is obfuscation, which modifies the program and as well as make it harder for the adversary to understand or analyse [9]. In the beginning, this technique is developed for automatically creating multiple transformations of the same program, and by that each version will be difficult to analyse and modify for some more time. This means, the concept of code obfuscation makes the code more difficult for understanding, so that it will not be advisable to go for code tampering without proper understanding of the code [5].

The technique 'obfuscation' helps for manipulating source code to make it harder to analyze and more difficult to understand for the attackers [13]. It is a general technique used for protecting software against malicious reverse engineering also. This approach can focus on changing a specific feature of the code (e.g., complexity). But, the aim of the code obfuscation techniques is to prevent malicious users by disclosing the properties of the original source program [16, 20].

Typical code obfuscation techniques include splitting of codes into smaller pieces, merging pieces of unrelated codes, randomizing the code placement, mangling of data structures, obfuscates the literal strings of a program, merging local integers, use of random dead codes [3], inserting dead variables, reordering of instructions, parameter reordering, transparent branch insertion, variable renaming, variable reassigning, field assignments, aggressive methods renaming, renaming of registers, duplication of registers, promoting primitive registers, reorders the constants in the bytecode and assigns random keys to them, arbitrarily marks all the basic bytecode blocks in the program with either 0 or 1, mapping of bytecode instructions to source code line numbers, removal of local variable tables in the bytecode that store the local variable names in the source code, also various techniques using opaque predicates (like branch insertion) array folding, array splitting, constant unfolding, introducing bogus control flow, control flow obfuscation, control flow flattening, breaking abstraction boundaries, false refactoring, etc [1, 6, 11, 22].

Some more techniques that can be used in higher level languages, mostly in object-oriented programming, are splitting or merging of classes, finding of inner classes (if available or not and then use obfuscation there even), new obfuscated names for methods and classes in a random fashion, encrypts class files and causes them to be decrypted at runtime, converting functions into inline methods at runtime, interleave methods (i.e., will have the same signature), use of more methods having the same names (overload names), takes a class and replaces all the fields with fields of the objects belonging to the same class, converting the fields of a class to public, splits all of the non-static methods into a static method, open all the classes for modification, group the classes for modifying the original structure, selects a random method from the class or a random basic block from a method (i.e., a copy of the basic block will be created and some additional malicious code will be added in the new basic block, by which the values of the local variables might be changed and the basic blocks will be bypassed at the time of execution), etc. [7, 19, 21].

In this paper, the discussion is basically about the conditional statement. There are many ways to use obfuscation techniques by using conditional statements like, reversing the 'if' and 'else' conditions, using negation of the condition with swapping the blocks, introducing 'if condition' which will be never or always true, breaking of the condition into nested conditions. Conversion from 'if–else if' conditions to switch cases, copying the block of code by inserting the conditions with different syntax, etc.

2 Proposed Methodology

Generally, use of conditional statements and loops in a program is common. Mostly, for optimizing a program, the programmers give the stress toward the loops. According to this thought, usually the obfuscation techniques can also be implied on conditions, wherever the conditions are used. Normally, after obfuscating a program, the code will be lengthier as well as difficult to understand. But, in the proposed logic, neither the code will be lengthier nor the code will look difficult to understand. By this reason, the code can be ignored easily by the adversary. In this paper, our idea is to use the conditional statement and fractional numbers to obfuscate the code. The difficulty with floating point variables is the rules of algebra, which are not properly applicable like integers. Mostly, programmers use standard algebraic rules for floating point arithmetic operations. This may create some undesired results in the programs, in some cases. In our paper, in place of 'if else' or 'if-else if–else' conditions, we use 'if condition' several times. In the proposed method, the use of 'if-else or if-else if-else' conditions will be replaced by several 'if conditions', as the obfuscated code and the original code look similar, but the obfuscated code will work differently, because of floating point operations, like addition, comparison, etc.

In the contemporary programs, it will be beneficiary, if the reader understands the use of floating point arithmetic suitably. To represent the real numbers, most floating point formats use scientific notations and few bits to represent the mantissa and a smaller number of bits to represent exponent. And so, the floating point numbers can represent with some significant digits and becomes a big impact on floating point arithmetic operations' execution. To easily observe the impact of limited precision arithmetic in programs, we apply decimal floating point format with mantissa and exponent. There is no dilemma for both mantissa and exponents, if signed also. As the floating point calculation is limited to some significant digits during the computation, so it truncates last digit of the smaller number, which is the result of not as much of accurate as it would be. Extra digits available during computation are known as guard digits (or guard bits in the case of binary format) [15, 17, 18] and it can enhance accuracy at the time of computation.

The accuracy loss during a single computation usually is not enough to worry, unless it concerns greatly, on the accuracy of the results of the computations. However, the computed value which is the result of a sequence of floating point operations can accumulate the error and greatly affect the computation itself [3, 10].

In case of addition of two numbers with different signs, the accuracy of the result may be less than the precision, what is available in the floating point format. Also, it is known that comparing floating point numbers is very dangerous. Because of inaccuracies present in floating point computation, it should be taken care, while comparing two floating point variables [12]. It is identified that, binary floating point computations produce same result, but may differ in their least significant bits. By the test of equality, if and only if all the bits of two operands are exactly same, then also it is not necessarily true for two different floating point computations. There are some reasons followed for this discussion [4].

2.1 Different Pattern Generation

In this paper, some proper fractions have been used for generating different patterns. Those patterns with bitwise representation and by rounding-off the digits, sometimes the result will seem to be undesirable. But, actually this is not at all undesirable, but properly calculated and unique. So, before execution of the code, it can be identified that, which condition which is going to be true, by that, which function is going to be executed. As the result is not undesirable and looks very simple, the readers may not pay that much attention to the conditional statements.

2.2 Pattern Repetition

The second concerned point is that, the repetition of patterns. As the remainder will be not zero, during the conversion of the proper fractional number into binary format, which is an infinite string of zeroes and ones. Obviously, the string will be getting a repetition of digits, which can be considered as a kind of pattern. Because of limited memory space, it is not possible to store the whole string. So, the exact visualization of the value in a variable is not possible, because of the reason that limited bytes are allowed to the variables.

2.3 "if" Conditions

The third concerned point is that it is not possible to use any other code, in between the 'if and else' conditions. But, in place of that the 'if condition' can be used several times, without any problem at all. Because of this reason, we can include some junk code in between the code, to make the program more obfuscated or to divert the adversary from our proposed conditional code obfuscation technique.

2.4 Floating Points

Although, the floating point numbers are perfectly suitable for many applications, but the dynamic range is somewhat small for many scientific applications. In computations, the limited precision format may introduce some serious exception, which will help for different obfuscation techniques.

According to Fig. 1, if we consider res = c, and if res is c ± e, (where e is a bit value), then by the simple bitwise comparison of the variables res and c argue that res < c, if c is greater than res, but less than res ± e. However, in most of the cases, the value of res will be equal to the value of c. The snap shot of execution of this program shown in Fig. 2.

Therefore, it is always mandatory that the comparison of two floating point numbers should be considered according to their ranges, regardless of actual comparison. So, by trying to compare between two floating point numbers can directly lead to an exception.

So we can state a rule, as while comparing two floating point numbers, it's always better to compare the range of one variable to the range of another variable with the values of both the variables plus or minus a small value, considering as some exception. The reason behind this is, while storing a single or double precision value to memory, the value come down to the appropriate size, before storing it and always works in the extended precision format [14].

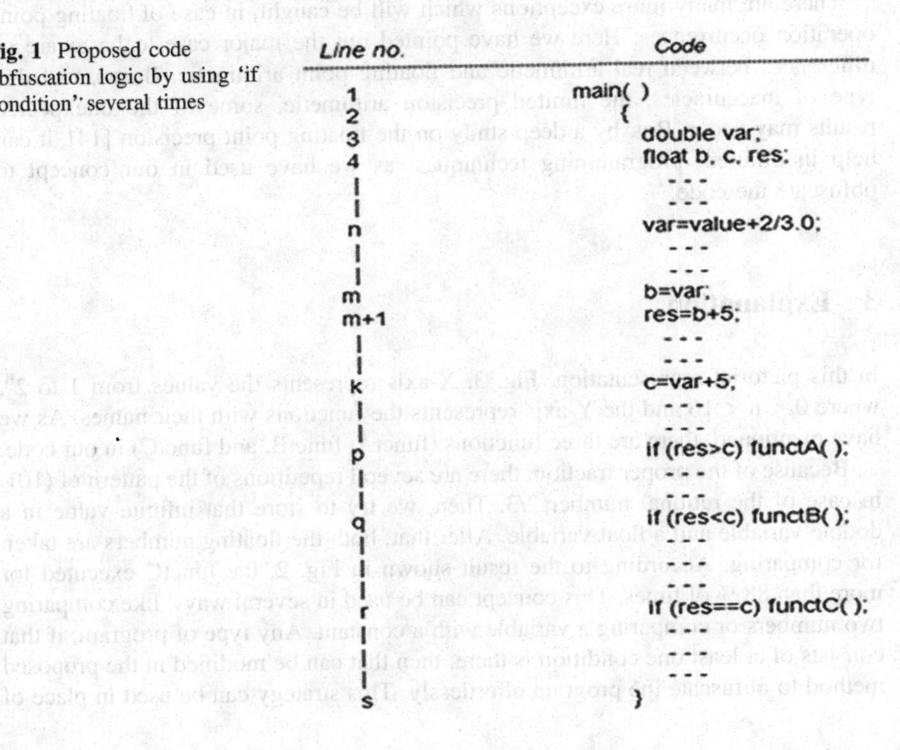

Line no.
Code
1 main( )
2 {
3 double var;
4 float b, c, res;
n var=value+2/3.0;
m b=var;
m+1 res=b+5;
k c=var+5;
p if (res>c) functA( );
q if (res<c) functB( );
r if (res==c) functC( );
s }

Fig. 1 Proposed code obfuscation logic by using 'if condition' several times

```
0: functC     1: functC     2: functA     3: functC     4: functB
5: functB     6: functB     7: functB     8: functC     9: functC
10: functC    11: functA    12: functA    13: functA    14: functA
15: functA    16: functC    17: functC    18: functC    19: functC
20: functC    21: functC    22: functC    23: functC    24: functC
25: functC    26: functC    27: functB    28: functB    29: functB
30: functB    31: functB    32: functC    33: functC    34: functC
35: functC    36: functC    37: functC    38: functC    39: functC
40: functC    41: functC    42: functC    43: functC    44: functC
45: functC    46: functC    47: functC    48: functC    49: functC
50: functC    51: functC    52: functC    53: functC    54: functC
55: functC    56: functC    57: functC    58: functC    59: functA
60: functA    61: functA    62: functA    63: functA    64: functC
65: functC    66: functC    67: functC    68: functC    69: functC
70: functC    71: functC    72: functC    73: functC    74: functC
75: functC    76: functC    77: functC    78: functC    79: functC
80: functC    81: functC    82: functC    83: functC    84: functC
85: functC    86: functC    87: functC    88: functC    89: functC
90: functC    91: functC    92: functC    93: functC    94: functC
95: functC    96: functC    97: functC    98: functC    99: functC
100: functC   101: functC   102: functC   103: functC   104: functC
105: functC   106: functC   107: functC   108: functC   109: functC
110: functC   111: functC   112: functC   113: functC   114: functC
115: functC   116: functC   117: functC   118: functC   119: functC
120: functC   121: functC   122: functC   123: functB   124: functB
125: functB   126: functB   127: functB   128: functC
```

Fig. 2 Snapshot to demonstrate the calling of functions

There are many more exceptions which will be caught, in case of floating point operation occurrences. Here we have pointed out the major case to be aware of differences between real arithmetic and floating point arithmetic. Because of this type of inaccuracies, the limited precision arithmetic, some of the unexpected results may occur. But, by a deep study on the floating point precision [14], it can help in different programming techniques, as we have used in our concept to obfuscate the code.

3 Explanation

In this pictorial representation, Fig. 3, X-axis represents the values from 1 to 2^n, where $0 < n < 16$ and the Y-axis represents the functions with their names. As we have mentioned, there are three functions (functA, functB, and functC) in our code.

Because of the proper fraction, there are several repetitions of the pattern of $(10)_2$ in case of the rational number 2/3. Then, we try to store that infinite value in a double variable and a float variable. After that, both the floating numbers are taken for comparing. According to the result shown in Fig. 2, the functC executed for more than 88 % of times. This concept can be used in several ways, like comparing two numbers or comparing a variable with a constant. Any type of program, if that consists of at least one condition is there, then that can be modified in the proposed method to obfuscate the program effortlessly. This strategy can be used in place of

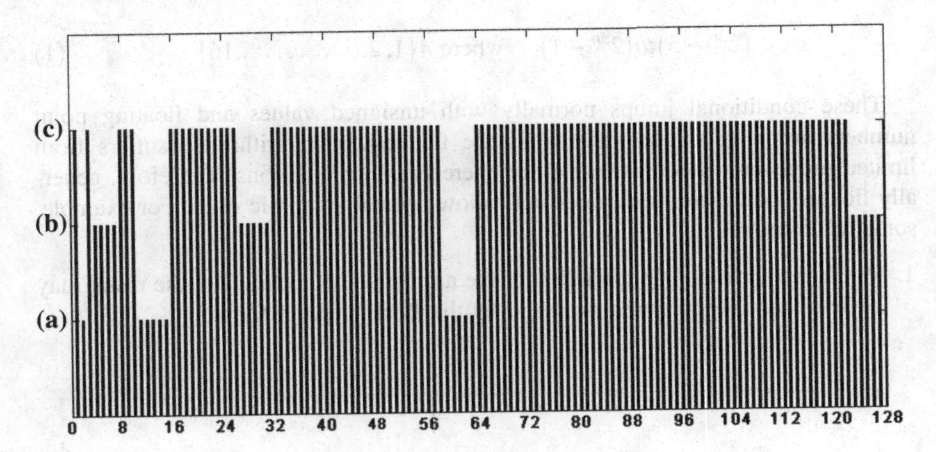

Fig. 3 Result of the execution of program in Fig. 1

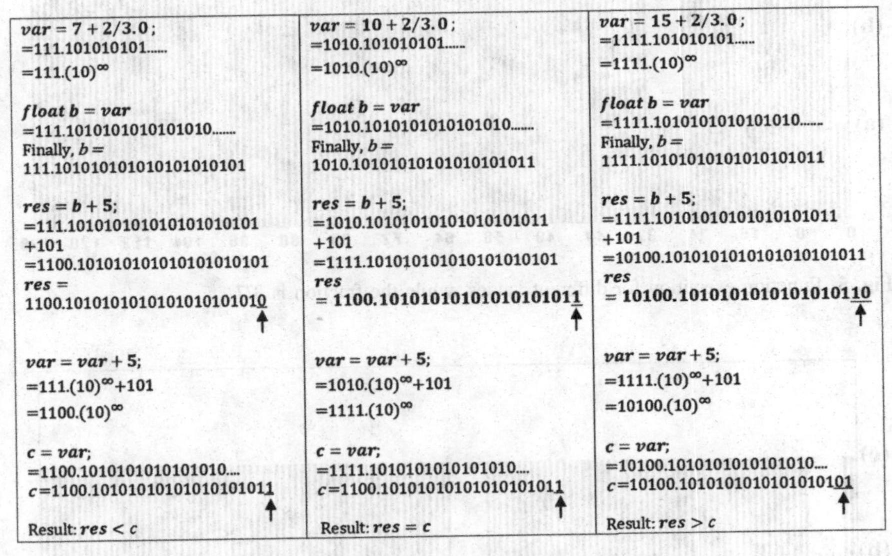

var = 7 + 2/3.0;
=111.101010101......
=111.(10)∞
float b = var
=111.1010101010101010.......
Finally, b =
111.10101010101010101010101
res = b + 5;
=111.10101010101010101010101
+101
=1100.10101010101010101010101
res =
1100.101010101010101010101010
var = var + 5;
=111.(10)∞+101
=1100.(10)∞
c = var;
=1100.1010101010101010....
c=1100.101010101010101010111
Result: res < c
var = 10 + 2/3.0;
=1010.101010101......
=1010.(10)∞
float b = var
=1010.1010101010101010.......
Finally, b =
1010.1010101010101010101011
res = b + 5;
=1010.1010101010101010101011
+101
=1111.10101010101010101010101
res
= 1100.10101010101010101010111
var = var + 5;
=1010.(10)∞+101
=1111.(10)∞
c = var;
=1111.1010101010101010....
c=1100.10101010101010101010111
Result: res = c
var = 15 + 2/3.0;
=1111.101010101......
=1111.(10)∞
float b = var
=1111.1010101010101010.......
Finally, b =
1111.1010101010101010101011
res = b + 5;
=1111.1010101010101010101011
+101
=10100.1010101010101010101011
res
= 10100.101010101010101010110
var = var + 5;
=1111.(10)∞+101
=10100.(10)∞
c = var;
=10100.1010101010101010....
c=10100.1010101010101010101
Result: res > c

Fig. 4 Function execution for different values, while the fraction is 2/3

or with inserting some dead code or XOR operation or as well as doing some bit wise operations as shown in Fig. 4.

In the code shown in Fig. 1, if the rational number 2/3 is replaced by 3/7, then the functB will be not executed at all (Fig. 5). The functA will be called, when the value will be between 1 to 2 or 27 to 31 or 251 to 255, and so on. Here the pattern is $(001)_2$, i.e., the functA executes, when the value will be with the range as follows:

$$\left(2^{3n}-5\right) to \left(2^{3n}-1\right) \quad \text{where } n\{1,2,\ \ldots\ldots\ldots\ldots 16\} \tag{1}$$

These conditional jumps normally with unsigned values and floating point numbers are with signed values. As the floating point arithmetic suffers from limited precision, inaccuracies may be there in the calculations. Therefore, generally floating point arithmetic does not follow normal algebraic rules. For example, some of them are

1. While the addition or subtraction of the numbers, the accuracy of the result may be less than the precision provided by the floating point format.

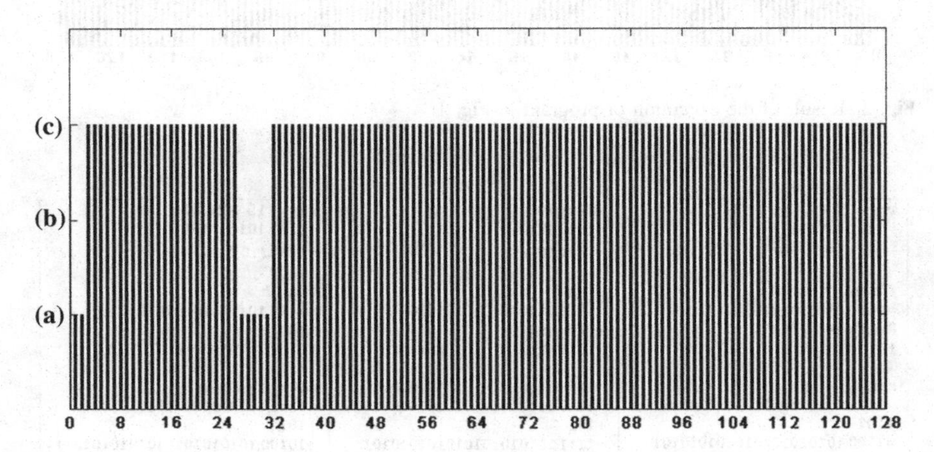

Fig. 5 Function execution for different values, while the fraction is 3/7

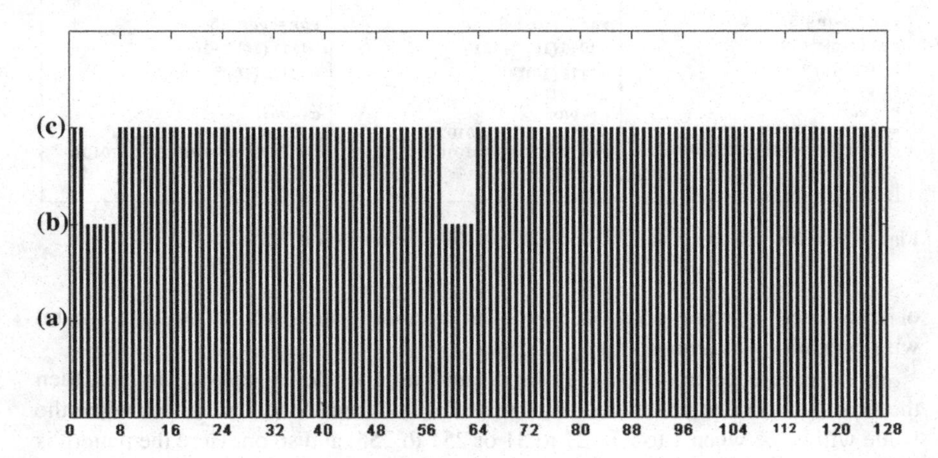

Fig. 6 Function execution for different values, while the fraction is 1/7

2. When comparing two floating point numbers, the result may confuse, because of the exception on computations. Therefore, the range of the variable should be taken care, while comparing the floating point variables.

Similarly, if the proper fraction number 2/3 is replaced by 1/7, then the functA will not execute (Fig.6). But, the functB will be called, when the value will be from 3 to 7 or 59 to 63 or 507 to 511, and so on. Here, the pattern is $(011)_2$ (Fig. 5).

4 Conclusion

The software program's code may be seemed to be simple and understandable, but the reality is different. Some programs may be tricky and thereby confuse the adversary. Code obfuscation is one type of method that is used intentionally to make the programming code harder to understand. In this paper, the binary floating point arithmetic and control statement obfuscation technique are used, which can be implemented in any high level programming language program, as a wide adopted predictable result and also independent of both architecture and format of binary floating point numbers. It is observed that, by implementing this technique, neither the size of the obfuscated program increases nor it looks much different than the original code. Commonly, several times use of 'if' condition looks similar to the use of 'if-else if-else' conditional statements, but, while going through the code, the depth of obfuscation can be understood. Another important point is that the adversary should not able to identify the output as well as the executed function, based on the input values. The presented explanation will help the reader to understand the conditional code obfuscation with better clarity. There are several software engineering code techniques to measure the effect of code obfuscation, in terms of the complexity, modularity or the size of obfuscated code. The proposed methodology will not affect much regarding these issues. It is suggested to use this strategy with other obfuscation techniques, to make it more effective and efficient. As future work, this concept can be combined with other obfuscation techniques, followed by finding the deobfuscating method of this conditional code obfuscation technique.

References

1. Anckaert, B., Madou, M., Sutter, B.D., Bus, B.D., Bosschere, K.D., Preneel, B.: Program obfuscation: a quantitative approach. Proceedings of the 2007 ACM workshop on Quality of protection, USA, pp. 15–20 (2007)
2. Balakrishnan, A., Schulze, C.: Code obfuscation: literature survey, Technical report, Computer Science Department, University of Wisconsin, Madison, USA (2005)
3. Beaucamps P., Filiol E.: On the possibility of practically obfuscating programs—towards a unified perspective of code protection. J. Comput. Virol. 2(4), WTCV'06 Special Issue (2006)
4. Brisebarre, N., Louvet, N., Martin E.D., Ercegovac, M.D.: Implementing decimal floating-point arithmetic through binary: some suggestions, 21st IEEE International

Conference on Application-specific Systems Architectures and Processors, pp. 317–320 (2010)
5. Ceccato, M., Penta, M.Di., Falcarin, P., Ricca, F., Torchiano, M., Tonella, P.: A family of experiments to assess the effectiveness and efficiency of source code obfuscation techniques. Empirical Softw. Eng. **19**(4), 1040–1074 (2014)
6. Chang, H., Atallah M.: Protecting software code by guards. ACM Workshop on Security and privacy in digital rights management, DRM'01, pp. 160–175 (2002)
7. Collberg C., Thomborson, C., Low D.: A taxonomy of obfuscating transformations, Technical report, Department of Computer Science, The University of Auckland, NewZealand (1997)
8. Doornik, J.A.: Conversion of high-period random numbers to floating point, ACM transactions on modeling and computer simulation, **17**(1) (2007)
9. Drape, S.: Intellectual property protection using obfuscation, Proceedings of SAS, pp. 133–144 (2009)
10. Eisenbarth, T., Koschke, R., Simon, D.: Locating features in source code. IEEE Trans. Softw. Eng. **29**(3), 210–224 (2003)
11. Ernst, M.: Quickly detecting relevant program invariants. In: 22nd International Conference on Software Engineering, pp. 449–458 (2000)
12. Fan, W., Lei, X., An, J.: Obfuscated malicious code detection with path condition analysis. J. Netw. **9**(5), 1208–1214 (2014)
13. Kulkarni, A., Metta, R.: A new code obfuscation scheme for software protection, service oriented system engineering (SOSE). In: 2014 IEEE 8th International Symposium, pp. 409–414 (2014)
14. Lauha, J.: The neglected art of fixed point arithmetic, Seminar presentation, Assembly (2006)
15. Lumetta, S.S.: Example: Bit sliced comparison, Introduction to Computer Engineering lecture, pp. 21-26 (2012)
16. Luo, H., Jianqin, J., Qingkai, Z.: Code obfuscation techniques based on software protection. Comput. Eng. **32**(11) (2006)
17. Mann R., How to program an 8-bit microcontroller using C Language, Journal of Atmel Applications, **3**(4),13–16 (2015)
18. Overton, M.L.: Floating point representation (1996)
19. Popov, I.V., Debray, S.K., Andrews, G.R.: Binary obfuscation using signals, Proceedings of 16th USENIX Security Symposium, Article-19 (2007)
20. Sharif, M., et al.: Impeding malware analysis using conditional code obfuscation. In: Network and Distributed System Security Symposium (2008)
21. Wroblewski, G.: General method of program code obfuscation. PhD thesis, Wroclaw University of Technology (2002)
22. You, I., Yim, K.: Malware obfuscation techniques: a brief survey, International Conference on Broadband, Wireless Computing, Communication and Applications, IEEE Computer Society. pp. 297–300 (2010)

Privacy Preserving Spam Email Filtering Based on Somewhat Homomorphic Using Functional Encryption

Sumit Jaiswal, Subhash Chandra Patel and Ravi Shankar Singh

Abstract With the advent of cloud computing, there has been a recent trend of delegating the computation (of a specific function) from the client to the third party (cloud). The privacy condition in this scenario requires that the third party should be able to perform the computation of the required specific function over the private (encrypted) data, but learn nothing else about the data (apart from the specific computation performed on that specific encrypted data). We take into account this concern in scenario of *proxy email server* (semi honest) which is required to perform specific functions (filtering encrypted spam mails) with minimal or no knowledge about the input data. We also highlight the limitations and challenges over the set of functions that can be performed by proxy email server in context of functional encryption.

Keywords Proxy email · Functional encryption · Master secret function key · Spam filtering · Input privacy · Cloud · Minimum disclosure

1 Introduction

With the advent of cloud computing, computing is witnessing the change how we perceive the services. Clients have the advantage to delegate their data and applications to third party without any need to setup the required hardware on their own or

S. Jaiswal (✉) · S.C. Patel · R.S. Singh
IIT (BHU), Varanasi, India
e-mail: sumit.rs.cse13@iitbhu.ac.in
URL: http://www.iitbhu.ac.in

S.C. Patel
e-mail: scpatel.rs.cse@iitbhu.ac.in

R.S. Singh
e-mail: ravi.cse@iitbhu.ac.in

S. Das et al. (eds.), *Proceedings of the 4th International Conference on Frontiers in Intelligent Computing: Theory and Applications (FICTA) 2015*, Advances in Intelligent Systems and Computing 404, DOI 10.1007/978-81-322-2695-6_49

software licensing fees [2]. This saves the significant amount of resource and time from client point of view. But with this paradigm transformation comes the security issues also [1]. On one hand where Cloud computing enables the client to remotely execute the desired functionality, there are several issues and trade offs the client has to pay in order to get the task executed by the third party. The privacy of inputs and efficiency of the delegated computation is one of the main concern from client point of view. The client wishes the third party to execute computation of some specific function over the data inputs but wishes not to disclose the information (even minimal) about the data inputs [10]. At the end, the third party should not be able to learn any additional information about the data other then the computation itself. In simple terms, client does not want to sacrifice the secrecy of its data still while transferring the computation over data to some remote third party.

2 Previous Related Work in Spam Mail Detection

Previous related work in identification of spam characteristics of mail was mainly dependent on signature pattern analysis of emails based on identification of having atleast one signature indicative of spam [19]. While some other techniques [21] of spam detection use the text clustering methods based on vector space model. They compute disjoint clusters automatically by extracting the clusters. While some other methods use negative selection methods by discovering unknown temporal patterns [20]. These methods do not take into account the encrypted nature of mails in scenario of cloud computing as well as the confidentiality of emails. These methods include constant updating the database of signature of spam mails on the server which in turn increases the complexity and load over the filtering server. These methods involve probability factor as the limiting factor of success where the server's ability to detect and flag any mail as spam depends upon the attributes the mail posses which relates its closeness to database of spam mail signatures.

3 Preliminaries

In this section we briefly review the preliminary concepts and definitions related:

3.1 Public Key Encryption

Given the security parameter λ, message space M, ciphertext space C, conventional public key cryptosystem [4] $\mathcal{E}$, consists of three algorithms: $KeyGen_{\mathcal{E}}$, $Encrypt_{\mathcal{E}}$, $Decrypt_{\mathcal{E}}$ such that:

1. $\textbf{KeyGen}_{setup}(1^{\lambda}) \rightarrow (PK, SK)$: Taking into account the security parameter λ, the *randomized key generation algorithm* generates PK (which is publicly announced) and computes a matching secret key SK (which is held secret by owner.)
2. $\textbf{Encrypt}_{PK}(m) \rightarrow c$: The *encryption algorithm* takes input string m and encodes it to random string c using the public key PK. This encryption algorithm maps a value m from message space M to a value c from a ciphertext space C.
3. $\textbf{Decrypt}_{SK}(c) \rightarrow m$: Using the *decryption algorithm* and the corresponding secret key SK for matching public key PK, the owner of SK (holding secret key) can deterministically compute the input m (corresponding pre-image of c).
   All of the 3 algorithms: $\textbf{KeyGen}_{setup}$, $\textbf{Encrypt}_{PK}$, $\textbf{Decrypt}_{SK}$ must be efficient i.e. run in polynomial time corresponding to λ i.e. polynomial in a security parameter (where λ signifies the bit length of keys PK, SK.)

3.2 Homomorphic Encryption

A homomorphic encryption scheme [6] $\mathcal{E}$ contains the algorithms $\textbf{KeyGen}_{setup}$, $\textbf{Encrypt}_{PK}$, $\textbf{Decrypt}_{SK}$ as mentioned above in public key cryptosystem scheme. It has an additional fourth algorithm $\textbf{Evaluate}_{\mathcal{E}}$ as:

$$\textbf{Evaluate}_{\mathcal{E}} f(c_1, c_2) \rightarrow c_3$$

where c_1, c_2 are encryptions of m_1, m_2 under public key encryption:

$$\textbf{Encrypt}_{PK}(m_1) \rightarrow c_1$$

$$\textbf{Encrypt}_{PK}(m_2) \rightarrow c_2$$

The fourth $\textbf{Evaluate}_{\mathcal{E}}$ algorithm described above takes into input set of ciphertexts c_1, c_2 in function f and outputs the resultant ciphertext c_3 such that when c_3 is decrypted, outputs m_3

$$m_3 \leftarrow \textbf{Decrypt}_{\mathcal{E}}(SK, c_3)$$

which is similar to the result of applying the function f' (homomorphic equivalent of function f) over the plaintexts m_1, m_2 i.e.:

$$f'(m_1, m_2) \rightarrow m_3$$

The function f' can be as simple as multiplication and addition. Considering the recent advances in homomorphic encryption, other complex functions can be achieved too (but it is still computationally expensive).

3.3 Functional Encryption

Given the message space M, ciphertext space C, security parameter λ and a specific functionality F where F describes the function of a plaintext that can be learned from ciphertext. Functional Encryption [8] consist of four algorithms: *Setup*, *Keygen*, *Encrypt*, *Decrypt*:

1. **Setup**$(1^\lambda) \rightarrow (pk, msk)$: The setup algorithm takes into input the security parameter λ and randomly generates pair of keys pk and msk i.e. public key and master secret key respectively.
2. **Keygen**$(msk, f) \rightarrow sk[f]$: The keygen algorithm when given input msk (master secret key generated in setup step) and a description of *specific function* (which needs to be computed over the ciphertext) generates a key *specific* to the functionality f and denoted by $sk[f]$.
3. **Encrypt**$(pk, x) \rightarrow c$: When given a message x and the public key pk (generated in setup step), Encrypt algorithm generates a ciphertext c corresponding to the message x.
4. **Decrypt**$(sk[f], c) \rightarrow y$: The decrypt algorithm when given (key specific to the function f) and corresponding ciphertext c (generated in Encrypt step above) outputs y which is *equivalent* to $f(x)$.

4 Privacy Preserving "Partial" Delegation of Work

As the client delegates [10] the computational work to the third party, there is a strong requirement to maintain the privacy of data input without sacrificing the complexity of the computation. In order to maintain the privacy of data, the input data must be encrypted in order to do so, but this affects the ability of third party to process the data in encrypted form. Recent advances in homomorphic encryption [12] address this issue. Using Fully Homomorphic Encryption, data can be processed in encrypted form. However, despite of its limitation over its practical application in real life applications, it seems to be prospective solution in the coming time.

But Fully Homomorphic Encryption computes the operation over the encrypted data as a whole. Given the pair of ciphertexts, it is possible to compute the operation over the ciphertexts which outputs the resultant ciphertext which is *equivalent* to result when applied the *same* operation over the corresponding plaintexts. Thus the output ciphertext (when derived from parent ciphertexts) contains the *whole* information about the corresponding resultant plaintext (when decrypted).

What if we want to delegate the processing work (a specific function) [7] to third party and wish to process the input ciphertexts *partially* (i.e. a specific operation over it) while still maintaining the privacy of inputs? This represents a slightly different problem (as compared to homomorphic encryption) i.e. when outsourcing the computation to the third party, we wish the third party to perform the specific function over the given ciphertext rather than outputting the whole ciphertext (corresponding to resultant plaintext). The situation demands that the ciphertext should be sent in clear to third party which output the specific computation (partial homomorphic in nature) over the ciphertext which is *equivalent* to the specific operation getting performed over the plaintext.

Any calculation of specific function over the encrypted data is desirable in scenarios like Proxy spam email filter. The requirement is that delegation of spam detection work should be *somewhat* homomorphic in nature (rather than fully homomorphic). We analyse the scenario of (untrusted) proxy email server, requiring to output the specific function of the plaintext rather than outputting fully plaintext (ciphertext output of homomorphic encryption) in clear.

In Ideal case, the requirement is that we would like to allow the third party to learn about the *specific function* of the ciphertext, but learns nothing else about knowledge [3] about the plaintext (apart from what implies from the function of plaintext). This notion of security can be achieved using functional encryption. We discuss this notion of security in scenario of proxy spam email filter which represents the "selective disclosure of knowledge" [17] although it maintains privacy preserving.

5 Privacy Preserving Secure Computation via Proxy Spam Email Filter

We assume the scenario: proxy spam email filtering server where the mail administrator Bob uses the notion of functional encryption by setting up a Functional Encryption scheme $\mathcal{E}$ using the following steps:

- Bob randomly generates pair of keys pk and msk using security parameter λ, chooses the encryption algorithm and broadcasts the public key pk (keeping the msk i.e. master secret key secret).
- Bob decides upon a function f (to be delegated to proxy mail server) and generates a corresponding secret key $sk[f]$ for function f using msk where f is user defined function such that:

$$f(x) = \begin{cases} 1 \text{ if } x \text{ belongs to spam} \\ 0 \text{ otherwise} \end{cases} \tag{1}$$

- Bob communicates the $sk[f]$ to the proxy email server to perform the operation.
- Alice transmits the encrypted mail to Bob using the public key pk which passes through spam email server
- The proxy email server performs the operation to compute $sk[f]$ to test if the encrypted message is spam (without learning anything more about the plaintext apart from what is obvious from the $sk[f]$)

The important property is that *distribution* of $sk[f]$ to interested parties and choice of f decides exactly *who* is authorized to decrypt the information implied by $sk[f]$

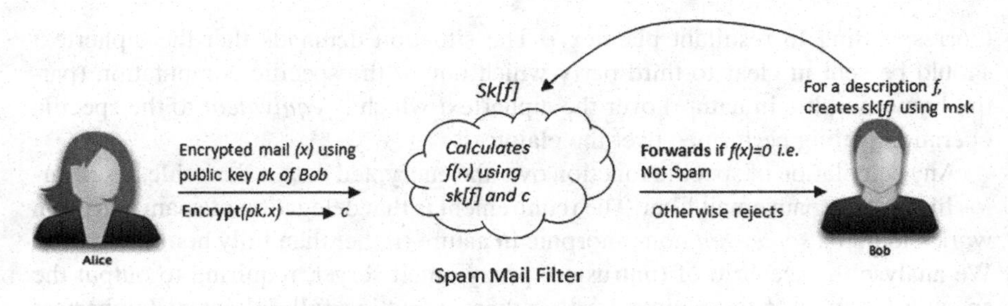

Fig. 1 Filtering of email based on *spam* function

and determines *how much* amount of plaintext information can be accessed. Thus the scenario discussed represents the ideal case of simple functional encryption applicable in scenario of proxy spam mail filter. However, there are other security issues which come into play while deciding the key exchange and trust issues over the functioning of spam mail detection. In this scenario we assume that the proxy mail server is semi-honest (i.e. *honest but curious*) adversary.

6 Security Analysis and Conclusions

There is indeed an important necessity to make the whole process (i.e. output preprocessing by email proxy server) *non-interactive* in nature with only one single key exchange step and with no other extra input. However, there exist other *interactive* methods of different computational complexity trade off's. Apart from this, there is a need of broad set of functions over which the functional encryption is applicable while considering the scenario of proxy mail server. It is yet to be seen whether other set of functions similar to Attribute based access control methods [15, 16], partial information disclosure [17], identity based encryption [13, 14] scheme can be developed over the functional encryption.

Another important security issue is that: suppose there are n different proxy email servers having n different secret keys $sk[f_1], sk[f_2], sk[f_3] \ldots sk[f_n]$, we assume that even if the attacker gets the possession of all those n different keys, the attacker can learn nothing about the decryption of cipheretxt c apart from what is implied from functions of plaintext x i.e. $f_1(x), f_2(x), f_3(x) \ldots f_n(x)$.

We describe the scheme as *partial homomorphic* because if the scheme was fully homomorphic, the proxy email server can run the spam function $sk[f]$ on encrypted mail, but will only learn the encrypted output of function f i.e. it can not learn whether if the encrypted email output was spam. In short, fully homomorphic encryption scheme can tag encrypted email message either 'spam' or 'not spam' but can not decide it. So functional encryption is more powerful notion in this scenario than fully homomorphic encryption scheme. However, we observe that constructing fully functional encryption scheme is still an open problem as compared to fully homomorphic encryption scheme.

References

1. Zissis, D., Lekkas, D.: Addressing cloud computing security issues. Futur. Gener. Comput. Syst. **28**(3), 583–592 (2012)
2. Armbrust, M., Fox, A., Griffith, R., Joseph, A.D., Katz, R., Konwinski, A., Zaharia, M.: A view of cloud computing. Commun. ACM **53**(4), 50–58 (2010)
3. Brassard, G., Chaum, D., Crpeau, C.: Minimum disclosure proofs of knowledge. J. Comput. Syst. Sci. **37**(2), 156–189 (1988)
4. Rivest, R.L., Shamir, A., Adleman, L.: A method for obtaining digital signatures and public-key cryptosystems. Commun. ACM **21**(2), 120–126 (1978)
5. Diffie, W., Hellman, M.E.: New directions in cryptography. Inf. Theory, IEEE Trans. **22**(6), 644–654 (1976)
6. Gentry, C.: A fully homomorphic encryption scheme. Doctoral dissertation, Stanford University (2009)
7. Gennaro, R., Gentry, C., Parno, B.: Non-interactive verifiable computing: outsourcing computation to untrusted workers. In: Advances in Cryptology CRYPTO, pp. 465-482. Springer, Berlin (2010)
8. Boneh, D., Sahai, A., Waters, B.: Functional encryption: definitions and challenges. In: Theory of Cryptography, pp. 253-273. Springer, Berlin (2011)
9. Sahai, A., Waters, B.: Fuzzy identity-based encryption. In: Advances in Cryptology EUROCRYPT 2005, pp. 457–473. Springer, Berlin (2005)
10. Chung, K.M., Kalai, Y., Vadhan, S.: Improved delegation of computation using fully homomorphic encryption. In: Advances in Cryptology CRYPTO 2010, pp. 483–501. Springer, Berlin (2010)
11. Waters, B.: Efficient identity-based encryption without random oracles. In: Advances in Cryptology EUROCRYPT 2005, pp. 114–127. Springer, Berlin (2005)
12. Gentry, C., Halevi, S.: Implementing Gentrys fully-homomorphic encryption scheme. In: Advances in Cryptology EUROCRYPT 2011, pp. 129-148. Springer, Berlin (2011)
13. Boneh, D., Franklin, M.: Identity-based encryption from the Weil pairing. In: Advances in Cryptology CRYPTO 2001, pp. 213–229. Springer, Berlin (2001)
14. Garg, S., Gentry, C., Halevi, S., Raykova, M., Sahai, A., Waters, B. Candidate indistinguishability obfuscation and functional encryption for all circuits. In: 2013 IEEE 54th Annual Symposium on Foundations of Computer Science (FOCS), pp. 40–49. IEEE (2013)
15. Goyal, V., Pandey, O., Sahai, A., Waters, B. Attribute-based encryption for fine-grained access control of encrypted data. In: Proceedings of the 13th ACM Conference on Computer and Communications Security, pp. 89–98. ACM (2006)
16. Bethencourt, J., Sahai, A., Waters, B. Ciphertext-policy attribute-based encryption. In: IEEE Symposium on Security and Privacy, 2007, SP'07, pp. 321–334. IEEE (2007)
17. Goldreich, O., Micali, S., Wigderson, A.: Proofs that yield nothing but their validity or all languages in NP have zero-knowledge proof systems. J. ACM (JACM) **38**(3), 690–728 (1991)
18. Boneh, D., Raghunathan, A., Segev, G.: Function-private identity-based encryption: hiding the function in functional encryption. In: Advances in Cryptology CRYPTO 2013, pp. 461–478. Springer, Berlin (2013)
19. Sobel,W.: System utilizing updated spam signatures for performing secondary signature-based analysis of a held e-mail to improve spam email detection, uS Patent 7,293,063 (6 Nov 2007)
20. Ma, W., Tran, D., Sharma, D.: A novel spam email detection system based on negative selection. In: Fourth International Conference on Computer Sciences and Convergence Information Technology, 2009. ICCIT'09, pp. 987–992 (2009) doi:10.1109/ICCIT.2009.58
21. Sasaki, M., Shinnou, H., Spam detection using text clustering. In: International Conference on Cyberworlds, 2005, pp. 4, pp.–319 (2005). doi:10.1109/CW.2005.83

Part XIII
Image Processing

Digital Image Watermarking Scheme Based on Visual Cryptography and SVD

Ajay Kumar Mallick, Priyanka and Sushila Maheshkar

Abstract Imperceptibility, robustness and security are the most important parameters of a watermarking scheme and tradeoff between these parameters is a challenging task. In this paper, we propose a watermarking scheme to balance the tradeoff between security and imperceptibility by using visual cryptography and singular value decomposition (SVD). Watermark image is encoded into *n* shares using visual cryptography of shares scheme. Among the *n* shares any two shares belonging to qualified set of shares are chosen and embedded into U and V component of SVD. The unused shares are with the authorized owner which can be used to extract the whole watermark. This can be done by superimposing shares or bitwise logical AND operation with the binary shares extracted from U and V component of watermarked image. Moreover the U and V component of SVD are exploited to enhance the imperceptibility. Experimental results show that the proposed technique is imperceptible, secure and robust to the common image processing and other attacks.

Keywords Visual cryptography · SVD · Secret sharing schemes · PSNR · MSE

1 Introduction

With the rapid development of multimedia and internet technologies, protecting the copyright of digital media has become an important issue [1]. Digital image watermarking has emerged as one of the solutions for this. It has received lot of

A.K. Mallick (✉) · Priyanka · S. Maheshkar
Dept. of Computer Science and Engineering, Indian School of Mines, Dhanbad, India
e-mail: mallickajay6@gmail.com

Priyanka
e-mail: priyankasingh401@gmail.com

S. Maheshkar
e-mail: sushila_maheshkar@yahoo.com

S. Das et al. (eds.), *Proceedings of the 4th International Conference on Frontiers in Intelligent Computing: Theory and Applications (FICTA) 2015*, Advances in Intelligent Systems and Computing 404, DOI 10.1007/978-81-322-2695-6_50

attention for image copyright protection in the last few decades [2]. Watermarking is defined as the practice of imperceptibly altering a Work to embed a message about that Work [3]. Watermark can be extracted from the watermarked images to prove the copyright. Important characteristics of watermarking schemes are invisibility, robustness, security, and capacity [4]. Digital watermarking algorithms are classified into two main categories namely spatial domain and frequency domain watermarking [5].

A versatile technique known as visual cryptography is one of the solutions to achieve security. It can represent the watermark or secret message by several different shares of binary images. It is difficult to detect any clues about a watermark from individual shares. The secret message is revealed when parts or all of these shares are aligned and stacked together. Visual cryptography scheme eliminates complex computation problem in decryption process, and the secret message can be restored by stacking operation [6].

Naor and Shamir's [7] were the first to propose an encoding scheme to share a binary image into two shares. Hwang [8] proposed the typical idea for a digital image watermarking based on the visual cryptography. The method uses visual threshold scheme defined by Naor–Shamir. Chang et al. [9] suggested spatial-domain image hiding schemes to hide a binary image into two meaningful shares. These two secret shares are embedded into two gray-level cover images. To decode the hidden messages, embedding images can be superimposed. For balancing the performance between pixel expansion and contrast, Liguo Fang [10] recommend a (2, n) scheme based on combination. Threshold visual secret sharing schemes having mixed XOR and OR operation with reversing and based on binary linear error correcting code was suggested by Xiao-qing and Tan.

This paper proposes a secure watermarking scheme based on SVD and visual cryptography for gray-scale images. In the proposed technique public share image will be open to the public, but only the owner, who uses the private sharing image of the watermark, can retrieve the watermark for ownership. If there is any dispute then the private sharing image can be provided to the arbitrator to resolve the ownership issue more convincingly. Organization of the paper is as follows: Sect. 2 explains the proposed watermarking technique, Sect. 3 discusses experimental results, and Sect. 4 concludes the paper.

2 Proposed Work

This section describes the proposed digital image watermarking scheme. Watermarking process is done in two phases: embedding and extraction.

2.1 Embedding

Proposed algorithm embeds the watermark using visual cryptography and SVD. 4×4 block-wise SVD is applied on the cover image. U and V components are used to embed the shares generated from the original watermark. The watermark image is encoded into n shares in such a manner that each original pixel appears in n modified versions, one for each share. Each share is a collection of m black and white sub pixels, which are printed in close proximity to each. The pixels in each share are encoded from the two permutation matrices C0 and C1. To share white pixel C0 matrices is chosen whereas to share black pixel C1 is chosen. Smaller pixel expansion results in smaller size of the share. Among the n shares any two share belonging to qualified set having $k \geq 3$ are chosen and embedded into U and V component of SVD transformation of blocks of cover image. The unused shares are with the authorized owner and among the two shares, only one of the shares with the owner can reveal the watermark when superimposed or bitwise logical AND operation with the binary shares extracted from U and V component of SVD transformation. Thus, k out of n shares is maintained. To accommodate a bonded change in elements of SVD Transformed coefficient, a threshold is defined while the embedding of watermark. The overall procedure of embedding and extraction is shown in Figs. 1 and 2, respectively.

To insert bit value 0 in U component of block we use the following Eq. (1) and interchange $u_{2,1}$ and $u_{3,1}$ after values are evaluated.

$$f(u_{2,1}, u_{3,1}) = \begin{cases} -\left(||u_{2,1}| + \frac{\mathrm{thr} - ||u_{2,1}| - |u_{3,1}||}{2}\right), -\left(||u_{3,1}| - \frac{(\mathrm{thr} - ||u_{2,1}| - ||u_{3,1}||)}{2}\right); \\ \quad \text{if } 0 < |u_{2,1}| - |u_{3,1}| < \mathrm{thr} \\ -\left(||u_{2,1}| + \frac{\mathrm{thr} + ||u_{2,1}| - |u_{3,1}||}{2}\right), -\left(||u_{3,1}| - \frac{(\mathrm{thr} + ||u_{2,1}| - |u_{3,1}|)}{2}\right); \\ \quad \text{if } -\mathrm{thr} < |u_{2,1}| - |u_{3,1}| < 0 \\ \text{exchange } u_{2,1} \text{ and } u_{3,1}; \text{if } |u_{2,1}| - |u_{3,1}\} < 0 \text{ and } |u_{2,1}| - |u_{3,1}| < -\mathrm{thr} \end{cases} \tag{1}$$

To insert bit value 0 in V component of block use the above Eq. (2) and interchange $v_{1,2}$ and $v_{1,3}$ after values are evaluated.

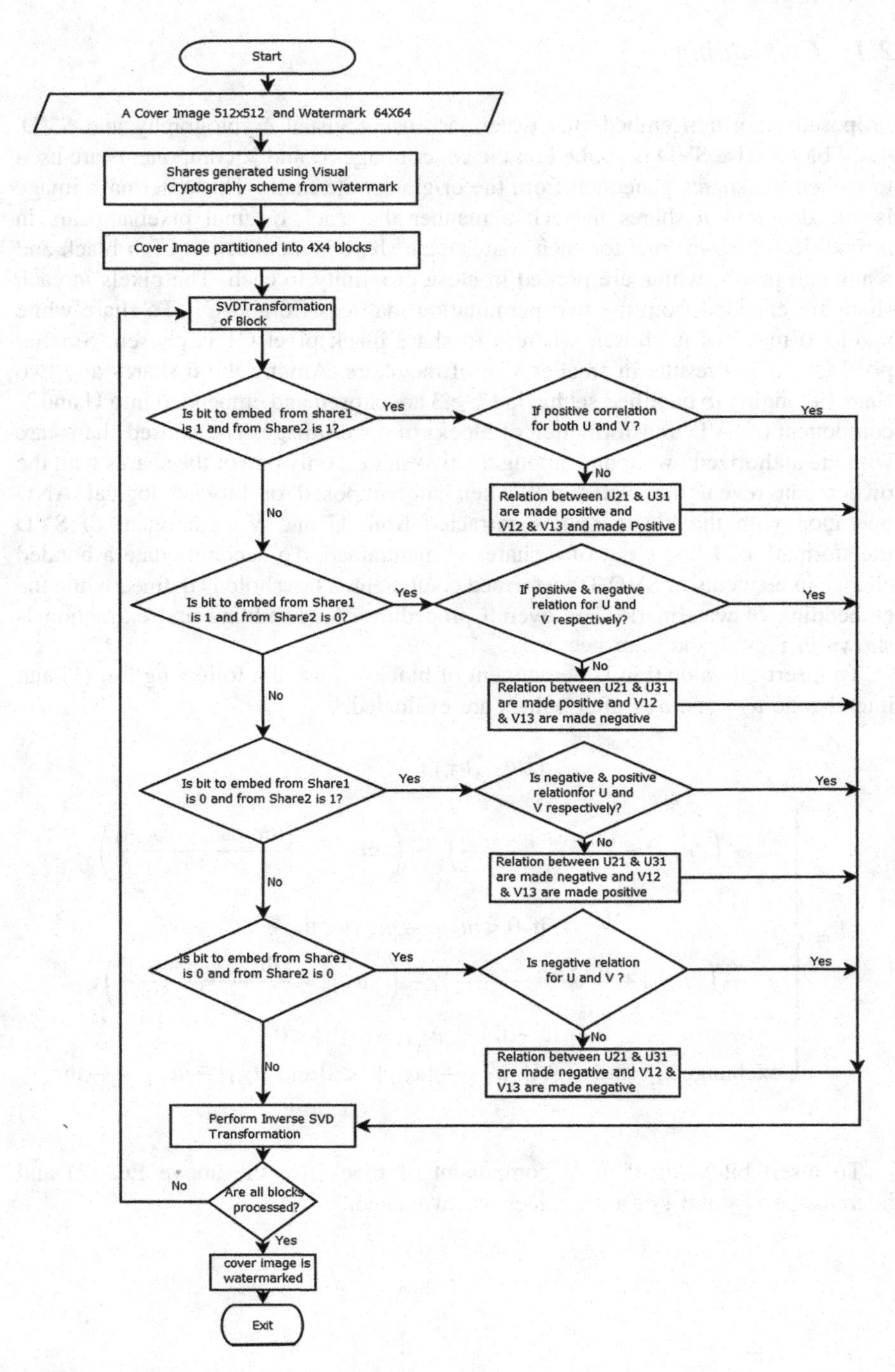

Fig. 1 Embedding process

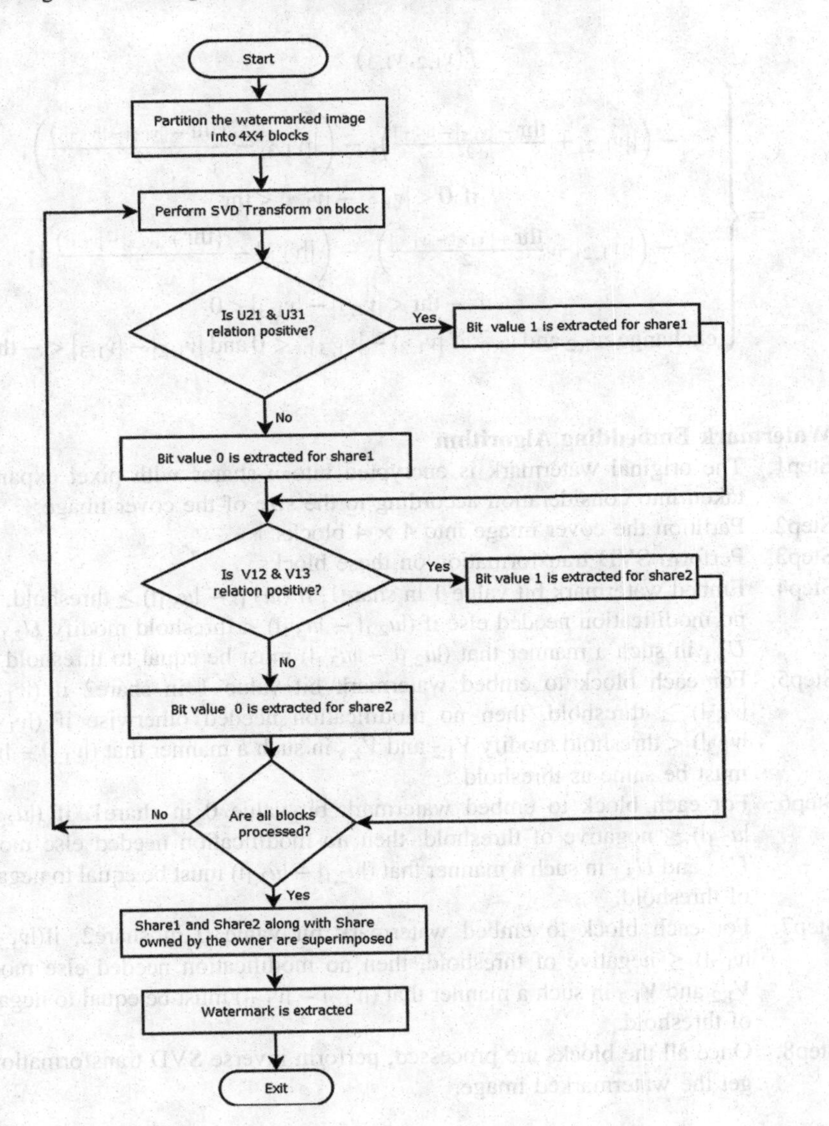

Fig. 2 Extraction process

$$f(v_{1,2}, v_{1,3}) = \begin{cases} -\left(||v_{1,2}| + \frac{\text{thr} - ||v_{1,2}| - |v_{1,3}||}{2}\right), -\left(||v_{1,3}| - \frac{(\text{thr} - ||v_{1,2}| - ||v_{1,3}||)}{2}\right); \\ \quad \text{if } 0 < |v_{1,2}| - |v_{1,3}| < \text{thr} \\ -\left(||v_{1,2}| + \frac{\text{thr} + ||v_{1,2}| - |v_{1,3}||}{2}\right), -\left(||v_{1,3}| - \frac{(\text{thr} + ||v_{1,2}| - |v_{1,3}|)}{2}\right); \\ \quad \text{if } -\text{thr} < |v_{1,2}| - |v_{1,3}| < 0 \\ \text{exchange } v_{1,2} \text{ and } v_{1,3}; \text{if } |v_{1,2}| - |v_{1,3}\} < 0 \text{ and } |v_{1,2}| - |v_{1,3}| < -\text{thr} \end{cases} \quad (2)$$

Watermark Embedding Algorithm

Step1. The original watermark is encrypted into n shares with pixel expansion taken into consideration according to the size of the cover image.

Step2. Partition the cover image into 4 × 4 blocks.

Step3. Perform SVD transformation on these blocks.

Step4. Embed watermark bit value 1 in share1, if $(|u_{2,1}| - |u_{3,1}|) \geq$ threshold, then no modification needed else if $(|u_{2,1}| - |u_{3,1}|) <$ threshold modify $U_{2,1}$ and $U_{3,1}$ in such a manner that $(|u_{2,1}| - |u_{3,1}|)$ must be equal to threshold.

Step5. For each block to embed watermark bit value 1 in share2 if $(|v_{1,2}| - |v_{1,3}|) \geq$ threshold, then no modification needed otherwise if $(|v_{1,2}| - |v_{1,3}|) <$ threshold modify $V_{1,2}$ and $V_{1,3}$ in such a manner that $(|v_{1,2}| - |v_{1,3}|)$ must be same as threshold.

Step6. For each block to embed watermark bit value 0 in share1, if $(|u_{2,1}| - |u_{3,1}|) \leq$ negative of threshold, then no modification needed else modify $U_{2,1}$ and $U_{3,1}$ in such a manner that $(|u_{2,1}| - |u_{3,1}|)$ must be equal to negative of threshold.

Step7. For each block to embed watermark bit value 0 in share2, if$(|v_{1,2}| - |v_{1,3}|) \leq$ negative of threshold, then no modification needed else modify $V_{1,2}$ and $V_{1,3}$ in such a manner that $(|v_{1,2}| - |v_{1,3}|)$ must be equal to negative of threshold.

Step8. Once all the blocks are processed, perform inverse SVD transformation to get the watermarked image.

2.2 Extraction

To extract the watermark first the watermarked image is partitioned into blocks and SVD is applied. If $|u_{2,1}| - |u_{3,1}|$ has a positive relation then bit value 1 is extracted otherwise bit value 0 extracted to construct share1. Similarly, if $(|v_{1,2}| - |v_{1,3}|)$ has a positive relation then bit value 1 is extracted otherwise value 0 is extracted to construct share2 bit. Shares belonging to qualified set owned by the owner are superimposed with extracted share to reveals the watermark.

Watermark Extraction Algorithm

Step1. Partition the watermarked image into blocks and apply SVD transform.

Step2. For each block if $(|u_{2,1}| - |u_{3,1}|)$ if a positive relation then bit value 1 is extracted otherwise bit value 0 extracted to construct share1. Similarly, if $(|v_{1,2}| - |v_{1,3}|)$ if a positive relation then bit value 1 is extracted otherwise value 0 is extracted to construct share2 bit.

Step3. Superimposition or bitwise logical AND operation of Extracted share and shares belonging to qualified set owned by the owner. The resultant image reveals the watermark.

3 Result Analysis and Discussion

This section presents experimental results of the proposed technique. Several Matlab simulations were performed on set of gray-scale images (Lena, Pepper, Boat and Texture) of size 512 × 512 and a binary watermark image of size 64 × 64 as shown in Fig. 3.

Watermark after pixel expansion during share generation enlarges to 128 × 128 bits as shown in Fig. 4.

The Shares generated and there specific combination during superimposition gives encrypted watermark hidden in the shares is demonstrated in Table 1.

Fig. 3 Test images. **a** Lena, **b** pepper, **c** boat, **d** texture, **e** watermark

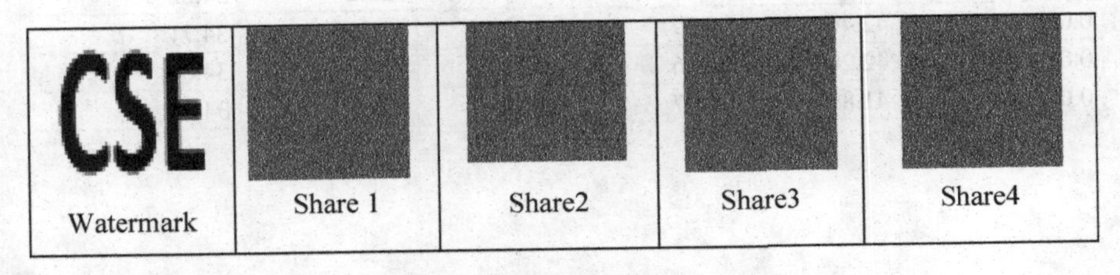

Fig. 4 Watermark image and shares generated using visual cryptography

Table 1 Share superimposed to obtain watermark

Shares	Watermark obtained
Share1, Share2 and Share3	Yes
Share1, Share2 and Share4	Yes
Share1, Share2, Share3 and Share4	Yes
Share1 and Share2	No

The watermark of dimension 64 × 64 is encrypted into four shares and by superimposing the specific combination in shares the watermark is obtained is shown in Fig. 5.

While embedding the watermark, various thresholds are considered showed variation in the PSNR values of different standard images as shown in Table 2. On increasing the threshold the invisibility of watermark degraded.

In Fig. 6, We can observe that the cover image and watermarked image have hardly any visual difference. Extracted watermark for various cover images is also shown.

We have compared the results of our proposed technique with the existing scheme proposed by Chang et al. [10]. Figure 7 shows comparison between proposed and existing scheme. From Fig. 7, it is evident that proposed scheme has better PSNR values and is more secure.

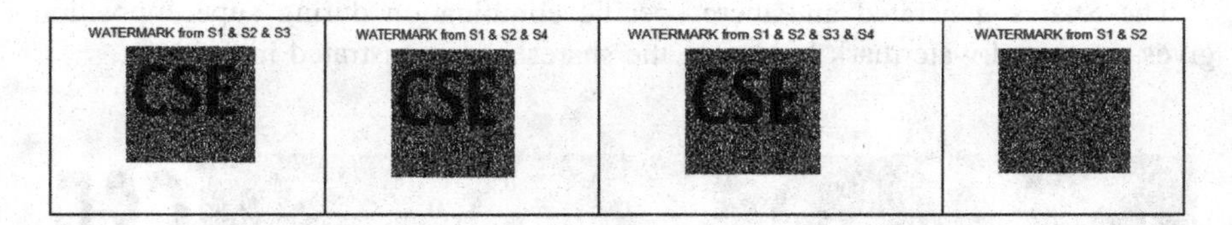

Fig. 5 Recovered watermark by AND operation bitwise of qualified set of shares

Table 2 PSNR value of proposed technique for different test images

Threshold	Lena	Pepper	Boat	Texture	Livingroom
0.002	43.12	37.22	36.94	32.43	34.94
0.004	43.08	36.72	36.81	32.42	34.82
0.009	42.46	36.59	36.59	32.29	34.71
0.012	42.20	36.26	36.31	32.18	34.60
0.015	41.89	36.07	36.08	32.13	34.57

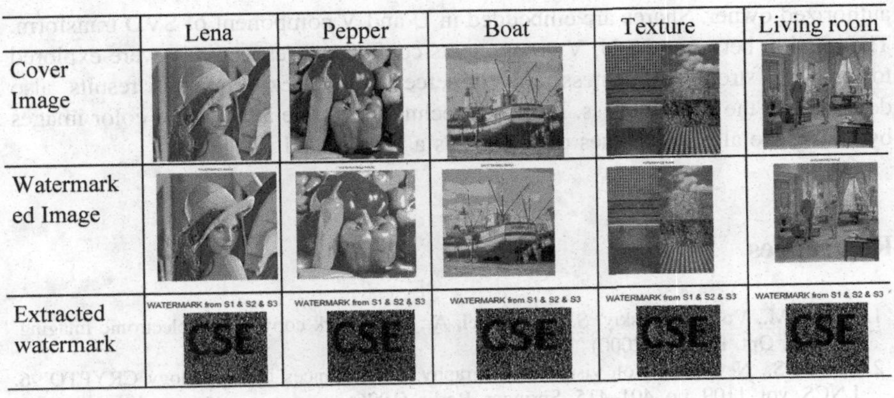

	Lena	Pepper	Boat	Texture	Living room
Cover Image					
Watermarked Image					
Extracted watermark	WATERMARK from S1 & S2 & S3 CSE	WATERMARK from S1 & S2 & S3 CSE	WATERMARK from S1 & S2 & S3 CSE	WATERMARK from S1 & S2 & S3 CSE	WATERMARK from S1 & S2 & S3 CSE

Fig. 6 PSNR value of proposed technique for different 512 × 512 images

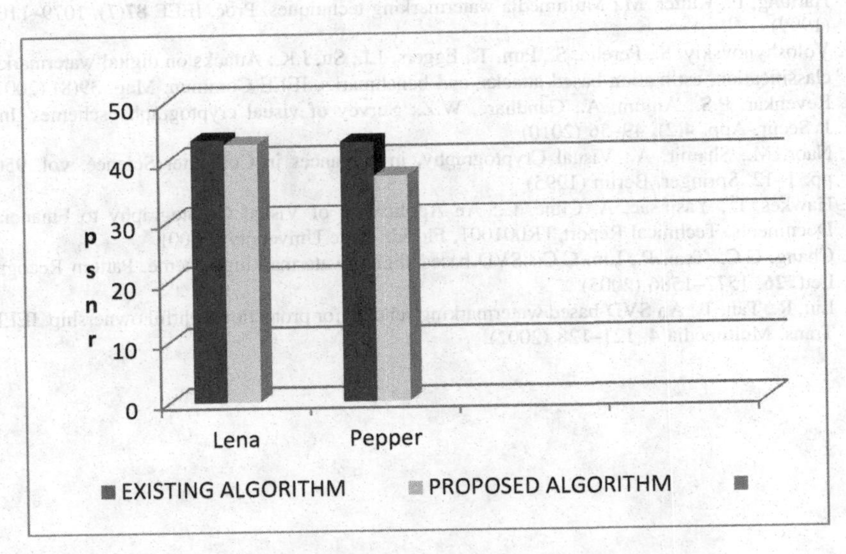

Fig. 7 Comparison between proposed and existing scheme for threshold $T = 0.0$

4 Conclusions

The paper presents an efficient scheme to enhance the security but not at the cost of imperceptibility. It encrypts the watermark into number of shares using visual cryptography. One of the share which act as a key is with the authorized owner. Key share play a vital role in watermark regeneration. In the proposed scheme even if the watermark is extracted, it is almost impossible even for the most sophisticated cryptanalysts to decrypt the watermark without the key share owned by the authorized owner. Shares are embedded in U and V component of SVD transform. The relation between the U, V components coefficients respectively were explored to provide strong robustness and imperceptibility. Experimental results also demonstrate the effectiveness. Proposed technique can be adapted for color images by taking the all three planes of images as a gray scaled image.

References

1. Kutter, M., Voloshynovskiy, S.V., Herrigel, A.: Watermark copy attack, electronic imaging. Int. Soc. Opt. Photon. (2000)
2. Droste, S.: New results on visual cryptography. In: Advances in Cryptology CRYPTO'96. LNCS, vol. 1109, pp. 401–415. Springer, Berlin (1996)
3. Cox, I.J., Miller, M.L., Bloom, J.A.: Digital Watermarking. Morgan Kaufmann Publishers Inc., San Francisco (2002)
4. Hartung, F., Kutter, M.: Multimedia watermarking techniques. Proc. IEEE **87**(7), 1079–1107 (1999)
5. Voloshynovskiy, S., Pereira, S., Pun, T., Eggers, J.J., Su, J.K.: Attacks on digital watermarks: classification, estimation based attacks, and benchmarks. IEEE Commun. Mag. **39**(8) (2001)
6. Revenkar, P.S., Anjum, A., Gandhare, W.Z.: Survey of visual cryptography schemes. Int. J. Secur. App. **4**(2), 49–56 (2010)
7. Naor, M., Shamir, A.: Visual Cryptography, in Advances in Computer Science, vol. 950, pp. 1–12. Springer, Berlin (1995)
8. Hawkes, L., Yasinsac, A. Cline, C.: An Application of Visual Cryptography to Financial Documents. Technical Report TR001001, Florida State University (2000)
9. Chang, C.C., Tsai, P., Lin, C.C.: SVD-based digital watermarking scheme. Pattern Recogn. Lett. **26**, 1577–1586 (2005)
10. Liu, R., Tan, T.: An SVD-based watermarking scheme for protection rightful ownership. IEEE Trans. Multimedia **4**, 121–128 (2002)

Compression of Hyper-Spectral Images and Its Performance Evaluation

K. Subhash Babu, K.K. Thyagharajan and V. Ramachandran

Abstract Effective lossy algorithms for compressing hyperspectral images using singular value decomposition (SVD) and discrete cosine transform (DCT) have been presented in this paper. A Hyperspectral image consists of a number of bands and each band contains some specific information. This paper suggests two compression algorithms that are used to compress the hyperspectral images by considering hyperspectral image data, band-by-band and applying compression to each band employing SVD and DCT. The compression performance of the resultant images are evaluated using various objective image quality metrics and are found to be attractive for hyperspectral image compression.

Keywords Hyperspectral images · SVD · DCT · Image compression · Image quality measures

1 Introduction

Hyperspectral images are images obtained from airborne/space-borne sensors which are then sent to the base station for further processing. Systems onboard usually will have limitations by way of storage, the power to process the data, and consumption of power. Such systems employ compression algorithms that have

K. Subhash Babu (✉)
Department of Computer Applications, SSN College of Engineering, Chennai, India
e-mail: subashbabuk@gmail.com

K.K. Thyagharajan
Department of Computer Science & Engineering, RMD Engineering College, Chennai, India
e-mail: kkthyagharajan@yahoo.com

V. Ramachandran
Department of Computer Science & Engineering, NIT Nagaland, Dimapur, Nagaland, India
e-mail: rama5864@gmail.com

S. Das et al. (eds.), *Proceedings of the 4th International Conference on Frontiers in Intelligent Computing: Theory and Applications (FICTA) 2015*, Advances in Intelligent Systems and Computing 404, DOI 10.1007/978-81-322-2695-6_51

good compression performance with low complexity [1]. Since the available bandwidth and onboard storage is limited in aircraft/satellite, it is necessary to compress the image data. The compression techniques are classified as lossless and lossy. The lossy compression achieves higher compression ratios as compared to lossless compression algorithms [2]. Various image compression algorithms are found in the literature for both lossy and lossless compression techniques. Most of the image compression techniques are transform-based. An improved version of EZW (embedded image coding using zerotrees of wavelet coefficients) algorithm has been suggested by Cheng for compressing the AVIRIS (airborne visible/infrared imaging spectrometer) images and performance is evaluated using compression ratio for both lossless and lossy compression [1].

A Joint KLT (Karhunen–Loeve Transform) Lasso algorithm for compressing hyperspectral images was proposed by Simplice A. Alissou and Ye Zhang [2]. A compression method that employs a hybrid transform comprising of Karhunen–Loeve transform (KLT) and discrete wavelet transform (DWT) for integers, was suggested by Cheng. [3]. A compression algorithm for hyperspectral images, with low complexity, based on distributed source coding multilinear was suggested by Nian and Wan [4]. In [5], another version of lossy image compression technique emphasizes the usage of singular value decomposition (SVD) and the difference reduction of wavelets. SVD is a numerical technique which is used to diagonalize matrices [6]. Medical images were compressed by discrete cosine transform (DCT) spectral similarity strategy in [7]. A hybrid approach for image compression using neural networks, vector quantization, and DCT is suggested in [8] to achieve good energy compaction. In [9], for colour image compression, DCT technique is used and in order to avoid recurring compressions of analogous blocks, fractal image compression is used. A modified coding framework based on H.264/SVC is employed in [10]. The same can be applied in the compression of hyperspectral images. Various other methods also have been suggested in the literature. Most of them are complex and are not usually recommended for hyperspectral image compression.

In this paper, two lossy image compression algorithms, namely SVD and DCT were studied and suggested to compress the hyperspectral images by extracting important information band-by-band. SVD is used in the image compression techniques to achieve high quality in the images with less computational complexity [11]. When the signals are highly correlated, DCT is the most sought after technique for obtaining a compact compressed image. In such cases, the performance of DCT is in par with that of the optimum transform KLT [7]. The organization of the paper is as follows: Sect. 2 discusses SVD for image compression followed by DCT in Sect. 3, then experimental results are discussed in Sect. 4, and conclusion is given in Sect. 5.

2 Hyperspectral Image Compression Using SVD

The block diagram in Fig. 1 shows our lossy compression algorithm based on SVD, employed on each band of the hyperspectral image. Singular value decomposition (SVD) technique is popular in various image-processing applications [12]. SVD can be performed on any general matrix of $M \times N$ size. SVD is a numerical technique that performs diagonalization on matrices. The pixel values of the image at the image coordinates are stored in a matrix and operations are performed on them. Although SVD compression offers good image quality, the compression ratio it gives is quite low. When the image is transformed using SVD, it is not compressed, but the data is made up of product expansion using singular values. The first singular value has the maximum information about the image and the information gradually comes down as we proceed further with subsequent singular values. This implies that if we consider only the first few singular values to represent the image and ignore the rest, we are not losing much of information in the image and there will not be much difference between the reconstructed image and the original image [6]. The more singular values we take into account, the better is its quality and worse is its compression and vice versa. The purpose of SVD is to factor matrix A into product of three matrices U, S, and V such that

$$A = USV^T \tag{1}$$

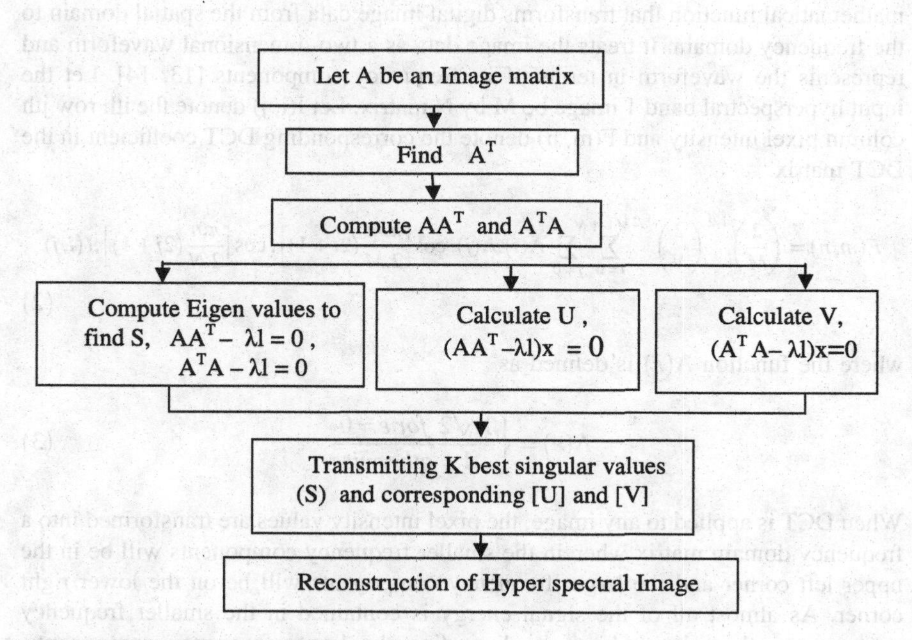

Fig. 1 Block diagram for SVD image compression

In the above equation, U and V are orthonormal matrices and S is a diagonal matrix which contains singular values. The singular values $\sigma 1 > \sigma 2 > \cdots > \sigma n$ appear in descending order along the main diagonal of S.

The SVD algorithm has been implemented in Matlab by including K larger values (and dropping the smaller values), to produce the final compressed version. The following algorithm used in MATLAB to read the hypercube band-by-band. The computational complexity is $O(N^3)$.

Algorithm for SVD

```
Step-1 Read each band of the Hyperspectral image to be
       compressed using multibandread function in Matlab
Step-2 Find the Eigenvalues of the matrix, formulate the
       Eigenvectors, normalize the vector, generate U,D,V
       vectors using SVD function in Matlab
Step-3 Retain the first K larger singular values to get the
       reconstructed image
```

3 Hyperspectral Image Compression Using DCT

DCT (discrete cosine transform) is one of the modern data compression techniques for compressing the multimedia data. The discrete cosine transform (DCT) is a mathematical function that transforms digital image data from the spatial domain to the frequency domain. It treats the image data as a two-dimensional waveform and represents the waveform in terms of its frequency components [13, 14]. Let the input hyperspectral band 1 image be M by N matrix. Let I(i, j) denote the ith row jth column pixel intensity and F(m, n) denote the corresponding DCT coefficient in the DCT matrix.

$$F(m,n) = \left(\frac{2}{M}\right)^{1/2}\left(\frac{2}{N}\right)^{1/2}\sum_{i=0}^{M-1}\sum_{j=0}^{N-1}\Lambda(i).\Lambda(j).\cos\left[\frac{\pi.m}{2.M}(2i+1)\right].\cos\left[\frac{\pi.n}{2.N}(2j+1)\right].I(i,j) \tag{2}$$

where the function $\Lambda(\varepsilon)$ is defined as

$$\Lambda(\varepsilon) = \begin{cases} 1/\sqrt{2} & for\ \varepsilon = 0 \\ 1 & otherwise \end{cases} \tag{3}$$

When DCT is applied to any image, the pixel intensity values are transformed into a frequency domain matrix wherein the smaller frequency components will be in the upper left corner and the larger frequency components will be on the lower right corner. As almost all of the signal energy is contained in the smaller frequency components, there is no harm in discarding the larger frequency components.

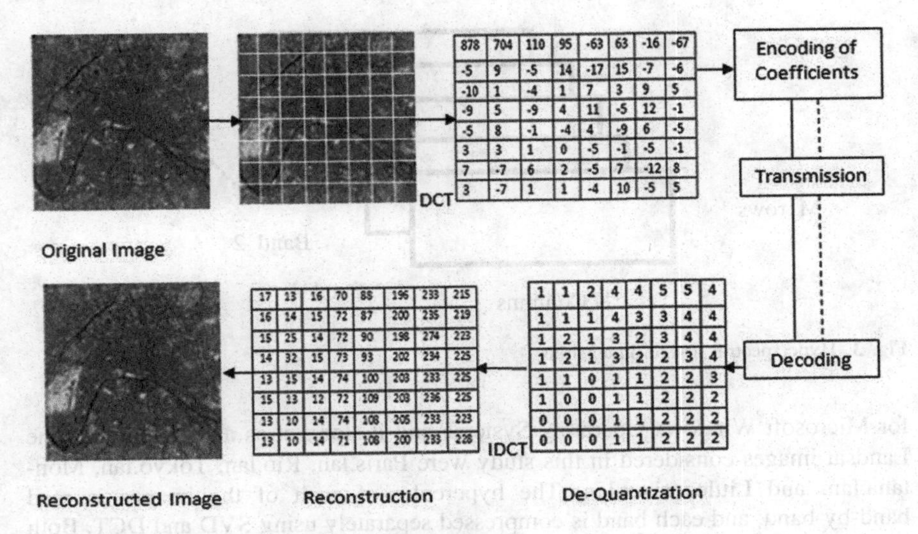

Fig. 2 Block diagram for DCT image compression

This may result in a very slight distorted image but an advantageous compressed image. The input to DCT is 8 × 8 matrix of integers, which are gray scale values of each pixel. The 8 bit pixels have values from 0 to 255 [15]. The block diagram for DCT based hyperspectral image compression is given Fig. 2, with details of typical pixel values. The image is divided into 8 × 8 blocks resulting in 64 pixels.

The following algorithm is used to compress each band of the hyperspectral image using DCT. The computational complexity of this algorithm is $O(N^2)$.

Algorithm for DCT

```
Step-1 Read each band of the Hyperspectral image to be
       compressed using multibandread function in matlab
Step-2 Apply the discrete cosine transform(DCT) on first 8 × 8
       block of the image
Step-3 Apply Step-2 to the remaining 8 × 8 blocks of the
       entire image using blockproc function in matlab
Step-4 Apply chosen 8 × 8 mask to all the blocks
Step-5 Obtain the DCT compressed image
```

4 Experimental Results

The hyperspectral images are compressed using SVD and DCT and the image quality is estimated using various image quality metrics. The hyperspectral images used in this study are Landsat images which are in .lan format. The lan file extension is associated with the ERDAS image geospatial data authoring software

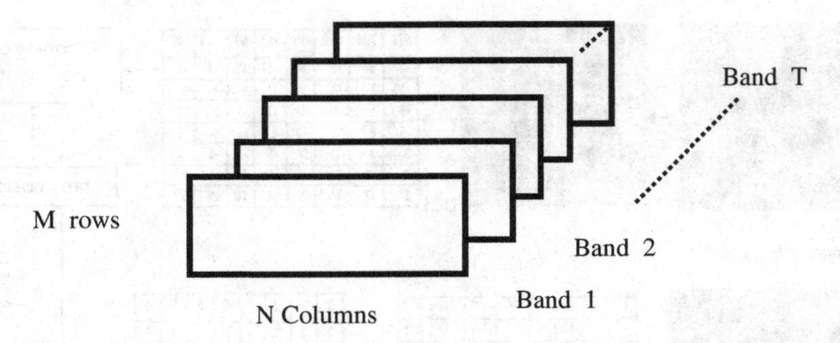

Fig. 3 Hyperspectral image dimensions

for Microsoft Windows Operating Systems and it stores it as a raster image. The Landsat images considered in this study were Paris.lan, Rio.lan, Tokyo.lan, Montana.lan, and Littlecoriver.lan. The hypercube of each of the images is read band-by-band, and each band is compressed separately using SVD and DCT. Both the methods are lossy methods and they are applied on 2D images. The hyperspectral image consists of M rows, N columns, and T bands as shown in Fig. 3. The original hyperspectral images considered in this study were given in .jpg format from Fig. 4a–e which are of size 512 × 512 with 7 bands.

4.1 Image Quality Measures Used in This Study

The compressed images are then evaluated for performance using the metrics like mean square error (MSE), peak signal to noise ratio (PSNR), normalized cross-correlation (NCC), structural content (SC), maximum difference (MD), and normalized absolute error (NAE). Table 1 lists the size of the images before and after the SVD and DCT compression on storing the images in .jpg format. Table 2 lists the values of various quality measures that are estimated after the SVD and DCT compression.

Mean Square Error (MSE)
Mean square error measures the error with respect to the center of the image values. It is calculated as average of cumulative squared value of the error difference between the original and the compressed image.

$$MSE = \frac{1}{MN} \sum_{i=1}^{M} \sum_{j=1}^{N} (I(i,j) - C(i,j))^2 \tag{4}$$

where I(i, j) and C(i, j) represent two images of size M × N, I is the original image and C is the compressed image. A lower MSE value indicates less error in the compressed image [16, 17].

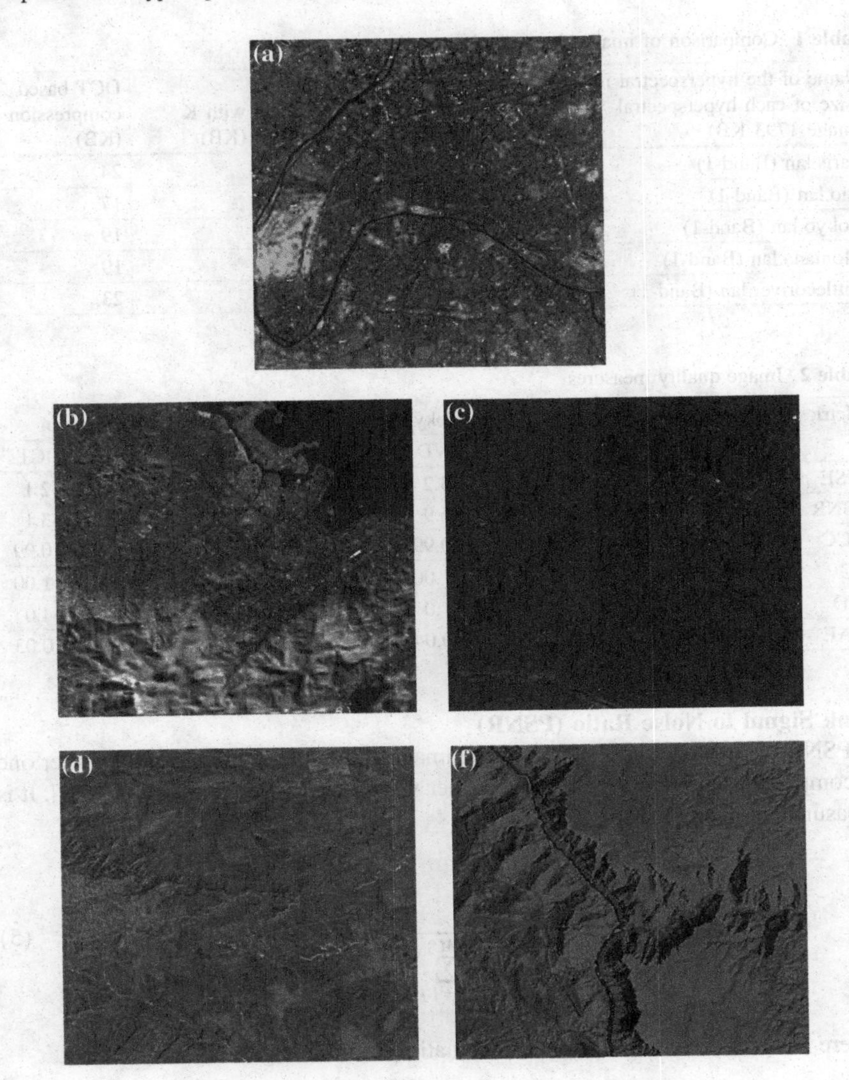

Fig. 4 **a** Paris.lan. **b** Rio.lan. **c** Tokyo.lan. **d** Montana.lan. **e** Littlecoriver.lan

Table 1 Comparison of image size (.jpg format)

Name of the hyperspectral image (size of each hyperspectral image 1793 KB)	Original image size (KB)	SVD based compression with K value = 100 (KB)	DCT based compression (KB)
Paris.lan (Band-1)	46	40	24
Rio.lan (Band-1)	26	24	17
Tokyo.lan (Band-1)	32	28	19
Montana.lan (Band-1)	28	25	19
Littlecoriver.lan (Band-1)	37	35	23

Table 2 Image quality measures

Metrics	Paris.lan		Rio.lan		Tokyo.lan		Montana.lan		Coriver.lan	
	SVD	DCT	SVD	DCT	SVD	DCT	SVD	DCT	SVD	DCT
MSE	41.3	54.7	9.0	10.6	13.2	15.5	9.4	12.7	22.2	32.1
PSNR	32.0	30.8	38.6	37.9	36.9	36.23	38.4	37.1	34.6	33.1
NCC	0.99	0.99	0.99	0.99	0.99	0.99	0.99	0.99	0.99	0.99
SC	1.00	1.00	1.00	1.00	1.00	1.00	1.00	1.00	1.00	1.00
MD	63.0	95.0	30.0	61.0	31.0	97.0	38.0	94.0	44.0	64.0
NAE	0.04	0.05	0.03	0.03	0.04	0.04	0.02	0.02	0.03	0.03

Peak Signal to Noise Ratio (PSNR)

A PSNR of zero can be obtained if one image is completely white and another one is completely black (or vice versa). Higher value implies better image [16, 17]. It is measured in decibels (db).

$$PSNR = 20\log_{10}\left(\frac{R^2}{\frac{1}{MN}\sum_{i=1}^{M}\sum_{j=1}^{N}(I(i,j)-C(i,j))^2}\right) \tag{5}$$

where R is the maximum possible fluctuation in the image data or R is 255.

Normalized Cross Correlation (NCC)

NCC gives the structural content of the image [17, 18]

$$NCC = \sum_{i=1}^{M}\sum_{j=1}^{N} I(i,j)C(i,j) / \sum_{i=1}^{M}\sum_{j=1}^{N} I(i,j)^2 \tag{6}$$

Structural Content (SC)

It is used to measure the image similarity. Consider the image is represented as M × N matrix, the formula for finding structural content variation factor is

$$SC = \sum_{i=1}^{M}\sum_{j=1}^{N} I(i,j) / \sum_{i=1}^{M}\sum_{j=1}^{N} C(i,j) \quad (7)$$

where SC is the structural content factor which is the ratio between the sum of pixel values of original image I(i, j) before compression and the sum of the pixel values of the degraded image C(i, j) after compression [18].

Maximum Difference (MD)
It is a measure to find the difference in content between the original and compressed image [18]. The higher the difference, the lesser will be the image quality.

$$MD = (|I(i,j) - C(i,j)|) \quad (8)$$

Normalized Absolute Error (NAE)
It is also used to find the difference between the original and reconstructed image [18]. Lesser the error, higher will be the quality.

$$NAE = \sum_{i=1}^{M}\sum_{j=1}^{N} |I(i,j) - C(i,j)| / \sum_{i=1}^{M}\sum_{j=1}^{N} |C(i,j)| \quad (9)$$

The images used in this study are as follows:
The size of the original hyperspectral images was 1793 KB. But when it is decomposed band-by-band, it is found that there is a good amount of decrease in the file sizes. According to the results obtained, it is evident from Table 2 that the MSE and MD vary considerably, but the values of PSNR, NCC, SC, and NAE were not affected significantly.

4.2 *Compression Ratio (CR)*

Using DCT Method

Compression Ratio using DCT method depends on the choice of the mask for each 8 × 8 block. Table 3 shows three different masks and the corresponding compression ratios. It is easy to observe that the compression depends upon the number of 1's in the mask—the more 1's we have in the mask the less will be the compression ratio. Correspondingly, image quality measures also will be affected. It is assumed that the mask is chosen ahead of compression and is known aprior to the decompression of the algorithm. The values that are to be made available for decompression are the nonzero values of the DCT coefficients and the size of the original image.

Table 3 Three different masks and the corresponding compression ratio using DCT

Mask 1								Mask 2								Mask 3							
1	1	1	1	1	0	0	0	1	1	1	1	0	0	0	0	1	1	1	0	0	0	0	0
1	1	1	1	0	0	0	0	1	1	1	0	0	0	0	0	1	1	0	0	0	0	0	0
1	1	1	0	0	0	0	0	1	1	0	0	0	0	0	0	1	0	0	0	0	0	0	0
1	1	0	0	0	0	0	0	1	0	0	0	0	0	0	0	0	0	0	0	0	0	0	0
1	0	0	0	0	0	0	0	0	0	0	0	0	0	0	0	0	0	0	0	0	0	0	0
0	0	0	0	0	0	0	0	0	0	0	0	0	0	0	0	0	0	0	0	0	0	0	0
0	0	0	0	0	0	0	0	0	0	0	0	0	0	0	0	0	0	0	0	0	0	0	0
0	0	0	0	0	0	0	0	0	0	0	0	0	0	0	0	0	0	0	0	0	0	0	0
Compression ratio																							
4.27								6.4								10.67							

Table 4 Three different values of K and the corresponding CR using SVD

Value of K	Compression ratio
50	5.115
100	2.5575
150	1.7050

Using SVD Method

As explained in Sect. 2, the compression ratio depends on the value of K which represents the number of singular values that are being considered for the calculation and the size of the image. The higher the value of K the lower will be the compression ratio. Of course the value of K correspondingly affects the image quality parameters. Table 4 shows CR for three values (50, 100, and 150) of K.

5 Conclusion

This paper has proposed a technique of decomposing the hyperspectral image into individual bands and compression algorithms have been applied to these individual images. The paper presented two methods for band-by-band compression of hyperspectral images using SVD and DCT techniques. Although these techniques achieve compression for images in each band with considerable reduction in size, the study found that not all the suggested objective quality metrics perform well in terms of estimating the quality of the compressed hyperspectral images. Also the quality of the image to a great extent depends on the type of the application in which it is used. The hyperspectral image compression comprising of many bands of images using DCT and SVD algorithms is T times complexity of single image where T is number of bands. Further methods could be explored to combine the band-wise compression of hyperspectral images and to achieve better quality along with the desired compression.

References

1. Cheng, K.J., Dill, J.: Hyperspectral images lossless compression using the 3D binary EZW algorithm. In: Proceedings of SPIE 8655 on Image Processing: Algorithms and Systems, vol. XI, pp. 865515–22, 19 Feb 2013
2. Alissou, S.A., Zhang, Y.: Hyperspectral data compression using lasso algorithm for spectral decorrelation. In: Proceedings of SPIE 9124 on Satellite Data Compression, Communications, and Processing, vol. X, pp. 91240A-9. 22 May 2014
3. Cheng, K., Dill, J.: An improved EZW hyperspectral image compression. J. Comput. Commun. **2**, 31–36 (2014)
4. Nian, Y., He, M., Wan, J.: Low-complexity compression algorithm for hyperspectral images based on distributed source coding. Math. Probl. Eng. **2013**(Article ID 825673), 7 p. (2013)
5. Anbarjafari, G., et al.: Lossy image compression using singular value decomposition and wavelet difference reduction. Digital Sig. Process. (Impact Factor: 1.92), 117–123 (2013)
6. Jayaraman, S., Esakirajan, S., Veerakumar, T.: Digital Image Processing, pp. 209–221. Tata McGrawHill Education Private Ltd., New Delhi (2009)
7. Wu, Y.-G., Tai, S.-C.: Medical image compression by discrete cosine transform spectral similarity strategy. IEEE Trans. Inf. Technol. Biomed. **5**(3), 236–243 (2001)
8. El Zorkany, M.: A hybrid compression technique using neural network and vector quantization with DCT. Adv. Intell. Syst. Comput. **233**, 233–244 (2014)
9. Rawat, C.S., Meher, S.: A hybrid image compression scheme using DCT and fractal image compression. Int. Arab J. Inf. Technol. **10**(6), 553–562 (2013)
10. Balaji, L., Thyagharajan, K.K.: H.264/SVC mode decision based on mode correlation and desired mode list. Int. J. Autom. Comput. **11**(5), 510–516 (2014)
11. Kahu, S., Rahate, R.: Image compression using singular value decomposition. Int. J. Adv. Res. Technol. **2**(8), 244–248 (2013)
12. Sadek, R.A.: SVD based image processing applications: state of the art, contributions and research challenges. Int. J. Adv. Comput. Sci. Appl. **3**(7), 26–34 (2012)
13. Zhou, X.: Research on DCT-based image compression quality. In: Cross Strait Quad-Regional Radio Science and Wireless Technology Conference (CSQRWC), vol. 2, pp. 1490–1494 (2011)
14. Watson, A.B.: Image Compression Using the Discrete Cosine Transform. In: Mathematica Journal, 4(1), pp. 81-88.(1994)
15. Cabeen, K., Gent, P.: Image compression and the discrete cosine transform. In: Math 45, pp. 1–11. College of the Redwoods
16. Maruthi, R., Dr. Sankarasubramanian, K.: Assessing the blurred image quality using some uni-variate and bi-variate measures. In: IJCECA-SERC-DST'ISSN 0974- 4983, vol. 02, Issue No 03, Spring Edition 2010, pp. 32–38 (2010)
17. Naidu, V.P.S., Raol, J.R.: Pixel level image fusion using wavelets and PCA. Def. Sci. J. **58**(3), 338–352 (2008)
18. Desai, D., Kulkarni, L.: A quantitative comparitive study of analytical and iterative reconstruction technique. Int. J. Image Process. (IJIP) **4**, 307–319 (2010)

Hand Gesture Recognition for Sign Language: A Skeleton Approach

Y.H. Sharath Kumar and V. Vinutha

Abstract In this work, we propose two novel approaches to classify sign images based on their skeletons. In the first approach, the distance features are extracted from the endpoints and junction points remarked on the skeleton. The extracted features are aggregated by the use of interval-valued type data and are saved in the knowledge base. A symbolic classifier has been used for the purpose of classification. In second approach, the concept of spatial topology is combined with the symbolic approach for classifying the sign gesture skeletons. From the end points and junction points of skeletons, triangles are generated using Delaunay Triangulation; and for each triangle, features like lengths of each side and angles are extracted. These extracted features of each skeleton of signs are clumped and represented in the form of interval-value type data. A suitable symbolic classifier is designed for the purpose of classification. Experiments are conducted on our own real dataset to evaluate the performance of two approaches. The experimental results disclose the success of the proposed classification approach.

Keywords Threshold-based segmentation • Skeleton extraction • Symbolic classifiers • Delaunay triangle

1 Introduction

A sign language, a mode of communication for deaf and dumb people to express and emphasize their views through a set of hand gestures and facial expressions to the community. The main purpose of sign language recognition (SLR) system is to provide systematic and an accurate mechanism to convert sign gestures into text or

Y.H. Sharath Kumar (✉) · V. Vinutha
Department of Computer Science and Engineering, Maharaja Institute of Technology, Belawadi, Srirangapatna Tq, Mandya, Karnataka, India
e-mail: sharathyhk@gmail.com

V. Vinutha
e-mail: aasta130591@gmail.com

S. Das et al. (eds.), *Proceedings of the 4th International Conference on Frontiers in Intelligent Computing: Theory and Applications (FICTA) 2015*, Advances in Intelligent Systems and Computing 404, DOI 10.1007/978-81-322-2695-6_52

voice format so that communication between impaired (deaf and dumb) people and hearing society can be more convenient. In terms of grammar and syntax, a sign language may be completely different from that of spoken language. It can be categorized into two subgroups: hand performed (movements and hand orientations, shapes) signs and nonmanual signs (facial expressions and body postures). The sign languages are not standard and universal; their grammar differs from country to country. American Sign Language (ASL), Chinese Sign Language (CSL), British Sign Language (BSL), and Indian Sign Language (ISL) are such languages to mention a few. Some countries share a bunch of similarities among them, while some differ from each other. About 3 % of the population in India are lacking in verbal capability. On the other hand, from the human aspect the most basic and realistic way to interact with the computer is through speech and gesture unites. Thus, the creative work undertaken for sign language and gesture recognition is likely focuses on shifting the paradigm from winning point of click Therefore SLR is esteemed to be noted areas of research in human–computer interaction (HCI) that has attracted more and more interest to researchers to work in HCI society.

Until now, SLR has been classified into two classes: one is based on the device used to capture gestures, i.e., data gloves-based SLR and another one is vision-based SLR. In the first case, the users are given a pair of gloves rigged with sensors and cables to transfer the exact hand movement to computer but these devices are too expensive with limited hand movements. In the second case, people can interact mutually with the computer in a friendlier manner which is just a goal of HCI, i.e., the process is carried out in two steps: First is to capture hand sign image with the help of camera and then analyze sign image in order to classify hand gesture and translate the recognized sign into voice/text form. However, due to the current computer vision techniques, i.e., large amount of technical hitch on image understanding are still open or need to be upgraded, the use of colored gloves or other visual abet is necessary for vision-based form or wide-reaching vocabulary SLR task.

2 Related Works

Vision-based SLR is more realistic and user friendly than datagloves-based one. Meantime, it faces more challenges and seduces more researchers working on this topic. Different researchers have used different varieties of paradigms for recognition of various hand gestures. Here we have reviewed some of these approaches.

Shreyashi et al. [1] presented a real-time system in order to identify gestures from ISL. The sign images are segmented using HSV color model, for this segmented image the principle component analysis (PCA) features are extracted and fed into distance-based classifier. Finally, recognized gesture is converted into text or voice format. Rohit et al. [2] designed a gesture recognition system capable of recognizing the most ambiguous static single-handed gestures. The sign image is segmented using YCbCr color model. Contour tracing descriptor is used to relate gesture contours and the classifiers like K-nearest neighbor (KNN) and support vector machine (SVM) is used for classification. Sakshi et al. [3] introduced a method for an application which would help in recognizing 9 different ISL alphabet signs. Here the sign of right- and left-hand side are used for experimentation. The scale invariance Fourier transform (SIFT) is used as features for matching between the different sign gesture. Joyeeta et al. [4] has designed a system in which the first step is to distinguish between the nonskin-colored pixels and skin-colored pixel in the given image; next step is to obtain the desired portion of hand till wrist then the extracted features calculate Eigenvalues and Eigenvectors. Finally, classification is done using Eigenvalue weighted Euclidean distance classifier. Joyeeta et al. [5] presented a novel approach for recognizing ISL by considering a continuous video sequences. In preprocessing stage, skin filtering is performed to extract skin-colored portions from nonskin-colored region after the histogram matching is done between the sign gestures. Karishma et al. [6] have proposed a system which automatically recognizes ISL. For segmented sign image, the Hu invariant moment and structural shape descriptors and their combinations are extracted and fed into multiclass SVM (MSVM). In Bhuyan [7] system determine VOPs (video object plane) for different hand pose from input video sequence using Hausdorff tracker detect the changing hand posture from one frame to other frame then extract the key VOPs by measuring the shape similarities using Hausdorff distance measure that eliminates the dispensable frames. These frames are fed as input to recognition module where recognition of particular sign gesture is achieved by comparing the input state sequence with the states of individual FSM (finite state machine). However, instead of comparing all representative FSMs, differentiate only the first state of input to all the FSMs using ART (Angular Radial transform) shape descriptor. This speeds up recognition task and thus efficiently the gesture is recognized. In Subha et al. [8], 32 combinations of binary number sign are developed using only right-hand palm images. These images are processed using Canny Edge detection to extract the outline of palm. Jaya et al. [9] presented a method for hand gesture recognition using Microsoft Kinect sensors.

From the above survey, we have overviewed that some researchers have considered only few alphabets and some have taken all 26 sign Alphabets and have conducted experimentation using different classification technique (like k-NN classifier, Euclidian weighted distance classification method, Bayer's classifier, etc.) and recognized American Sign Language (ASL) or Indian Sign Language (ISL) alphabets. Only Static gestures are taken and recognized in (Joyeeta et al. [4]). Intensely, it is been noticed from all the previous work that authors have not worked on special characters such as hand gestures indicating different activities like daily routine activities, stationary objects, action, action resembling family members, etc., recognizing the special characters and signs is one of the challenging task.

According to the recent progress in the area of symbolic data analysis, real-time objects can be described in a better way using symbolic data, which is nothing but the supplement of traditional crisp data (Billard and Diday [10]). Symbolic data are visible in forms of discrete, continuous, ratio absolute, interval-valued data, and multivalued. The conception of symbolic data analysis has been widely studied in the field of classification as it is been proved theoretically and practically that the classification paradigms based on symbolic data outstrip conventional classification techniques. In skeleton-based sign language representation, the skeletal distance between end points and junction points of sign gesture of each class possesses significant discrepancy. Therefore, we interpreted that it would be more relevant to capture these dissimilarities and are instantiated as interval-valued features in order to furnish an effective representation for sign gesture classification on our own real set of data. Based on the information collected from intensive literature review, it is learnt that no work has been reported so far to classify the gestures using symbolic data classifier. This work is carried out by fixing inter valued feature by considering minimum and maximum of feature value sets.

This chapter is organized as follows. Section 3 presents proposed method with the process of skeleton pruning as a preprocessing step. Section 3.3 presents the proposed two alternative approaches for classification of skeletonized image. Section 4 presents the database used for experimentation. Section 5 presents the experimentation carried out using different proposed models. Finally conclusions are drawn in Sect. 6.

3 Proposed Model

The proposed method encompasses of five sections: Segmentation, extraction of skeleton, Feature reduction, Representation and Classification as shows in Fig. 1. The sign gestures are segmented based on thresholding methods and then we need to generate skeleton for each of the segmented sign gestures and prune the skeleton applying discrete curve evolution (DCE) method. Symbolic classifier has been adopted in order to represent the distance features between junction points and endpoints of skeletonized sign. Finally, the gestures are classified to particular class and are identified in text or speech form.

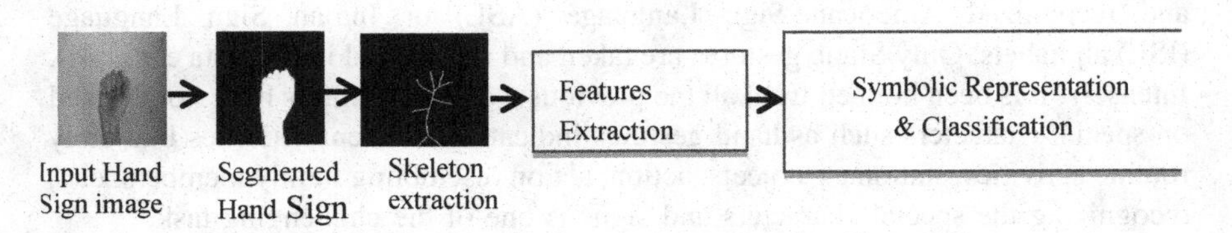

Fig. 1 Block diagram of the proposed method

3.1 Segmentation

Initially, segmentation is carried out applying threshold-based segmentation algorithm [11]. Thresholding is one of the simple technique by which the object of interest can be extracted from its background. A given image transfigured to HSV planes and an intensity histogram corresponding to each channel is extracted. The histogram intensity values correlate with two dominant regions belonging to background wherein sign gestures are identified. On the basis of intensity values, the sign gesture is segmented. Figure 2 shows the results of segmentation using the threshold-based method on a few sets of images.

3.2 Skeleton Pruning

For skeleton generation, we adopt the work proposed by Bai et al. [12] generating skeleton for each of the segmented hand gesture. Initially, discrete curve evolution (DCE) is used to make the polygon structure as simpler as possible for each of the skeletonized gesture. Skeletons are pruned in such a way that only branches ending at convex DCE vertices will remain. The skeleton pruning guarantees to preserve the topology of the gesture and it is robust to outliers (noise) and borderline distortion. The advantage of using DCE is the fact that it outlooks whether shape components is relevant or not based on context features. It repeatedly removes least appropriate polygon vertices, where the relevance measure is estimated with respect to the actual partially simplified prior versions of the polygon. Therefore, the left out skeleton branches are decided in the context of the entire shape. We can obtain equivalent skeletons of only relevant branches, provided none of the branches are missing. Same threshold or appropriate stopping norm of the DCE is needed for simplification purpose. Figure 3 displays detected endpoints and junction points for given sign gesture skeleton.

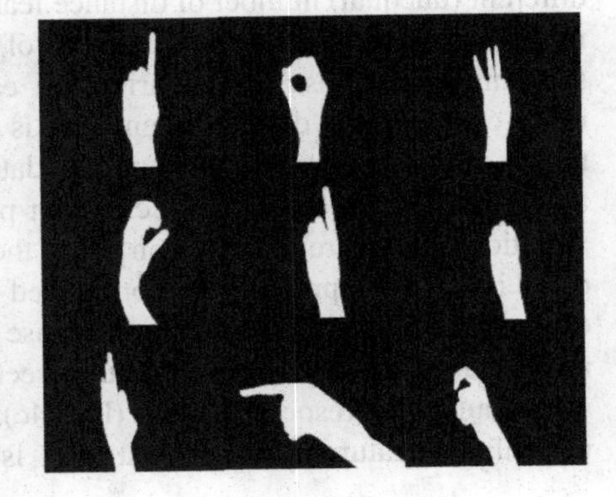

Fig. 2 Segmentation results on sign gesture images

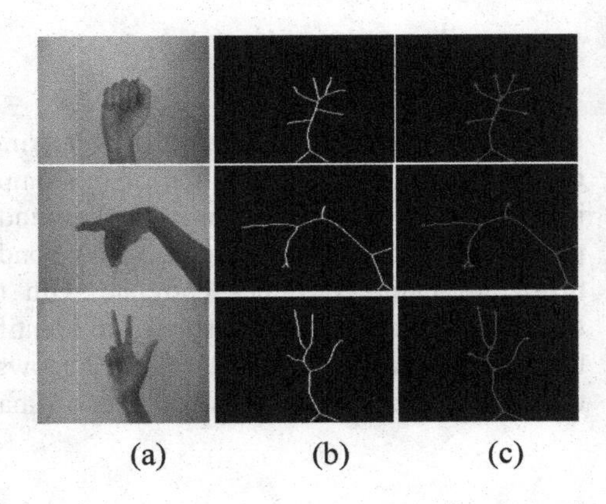

Fig. 3 Skeleton generation **a** Sign images **b** Skeletons generated **c** Skeletons with detected endpoints and junction points

3.3 Proposed Classification Models

As each signs is represented by its respective skeleton, the problem of sign classification has thus become a representation and classification of skeletons. In this section, we propose three different approaches for representation and classification of skeletons.

Symbolic Classification Model.
In this section, we extend a method to classify signs using a symbolic approach. From the obtained skeleton endpoints and junction points, distance features are extracted. The features are further normalized using discrete cosine transform (DCT). The minimized features are aggregated and instantiated through interval-valued type data and are saved on to the knowledge base. A symbolic classifier has been mostly used for the purpose of classification.

Representation and Classification.
In this section, we calculate the Euclidean distance between the detected end points and junction points on the skeleton. It is to be noticed that the skeletons of samples of signs of same class have large intraclass variations and also the number of endpoints and junction points detected vary from skeleton to skeleton, thereby resulting with different (unequal) number of distance features. Hence we endorse applying DCT to normalize and extract a fixed number of features. Further, as sign skeletons have substantial intraclass dissimilarities in each subgroup, preserving the variations using conventional data representation is quite difficult. Hence, the proposed solution is intended to use unconventional data called symbolic data [10]. In this work, the representation of symbolic classifier proposed by [13] is adopted to impact the variations of feature assimilation using the interval-valued feature vectors as elucidated in Fig. 4a represents the normalized DCT feature data of N classes. Each class of normalized feature data depicts the use of interval-valued features (Fig. 4b). It is noted that unlike conventional feature vectors are stored in the knowledge base as a substitute of the respective class (Fig. 4c). During sign classification process, each normalized feature value of test sign is compared with the respective interval

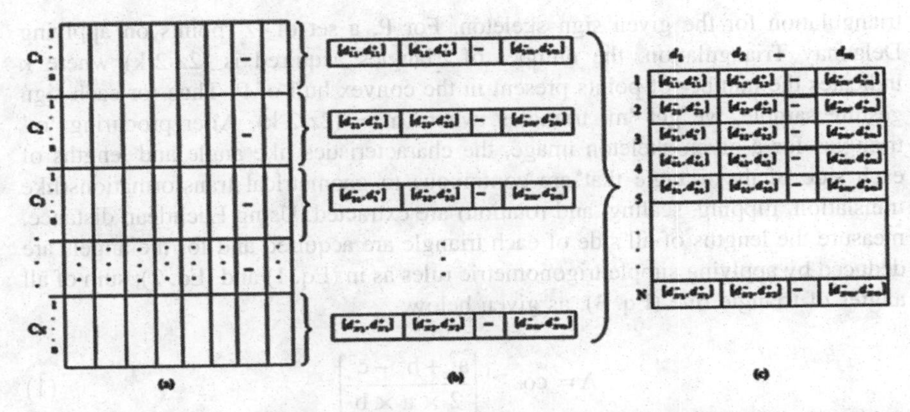

Fig. 4 Symbolic representation of feature data **a** Feature matrix **b** Interval-valued representation of class **c** Interval feature matrix for all the classes

representatives of all the classes (Fig. 4c) to review whether the feature value of the test sign lies within the respective interval of the class representatives or not. Such features are taken as acceptance count. A test sign sample with maximum acceptance count is said to belong to a particular class category.

Delaunay Triangulation (DT) Model.
In this section, we extend a method to classify a hand gesture using Delaunay Triangles. For the skeletonized sign, identify the endpoints and junction points. Based on the detected points generate the triangles. Later on for each of triangles extricate certain features (length and orientation) that are represented by interval-valued data. For the purpose of classification, we outline a voting-based classifier.

Representation and Classification.
The proposed sign classification system represents signs symbol in terms of their skeletons. A set of detected junctions and end points for each of the skeletonized gestures are represented by coordinate's values (x, y) indeed the Delaunay triangles are computed for available coordinates (x, y) values. Figure 5 elucidate Delaunay

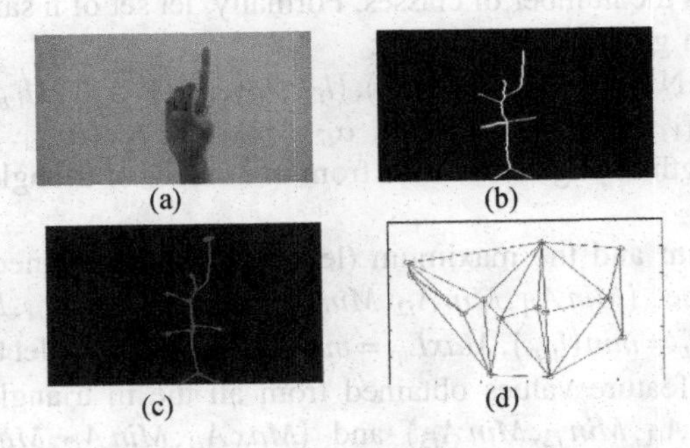

Fig. 5 Delaunay triangulation for the given sign skeleton **a** Sign image, **b** Sign Skeleton **c** Detected endpoints and junction points **d** Triangles obtained after applying Delaunay triangulation

triangulation for the given sign skeleton. For P, a set of 'z' points on applying Delaunay Triangulation, the number of triangles acquired is (2z-2-k) where k indicates the number of points present in the convex hull of P. Thus for each sign gesture sample, we get 'm' triangles, where m = (2z-2-k). After procuring 'm' triangles for a given skeleton image, the characteristics like angle and lengths of each side of the triangle that are continuous to geometrical transformations(like translation, flipping, scaling, and rotation) are extracted. Using Euclidean distance, measure the lengths of all side of each triangle are acquired and for the angels are deduced by applying simple trigonometric rules as in (Eq. 1) and (Eq. 2), sum of all angles of triangle rule (Eq. 3), as given below:

$$A = \cos^{-1}\left[\frac{a^2 + b^2 - c^2}{2 \times a \times b}\right] \tag{1}$$

$$C = \sin^{-1}\left[\frac{a \times \sin(A)}{c}\right] \tag{2}$$

$$B = (180 - A - C) \tag{3}$$

Here the lengths of each sides of triangle are represented by a, b, and c and it is procured by calculating the distance between the Delaunay triangles coordinates values and A, B, and C denotes the opposite internal angles of the sides of triangle, respectively. Once the lengths and the angles of all sides of the triangle are obtained, utterly we can see (m × 6 matrix) 6 m numbers of features representing a single sign gesture in a sample. Therefore, there are 6 mn (m × n × 6) number of features representing a class Cj with n samples as shown in Fig. 6.

In this model also, as done in previous model, we assimilate the length and angles of all triangles of an interval-valued type feature that belong to a sample in the structure where m × 6 stimulating features are transmuted into 1 × 6 interval-valued type features. Comprehension is done by examining the minimum and maximum lengths and angles of all m triangles of a sample. Now the feature matrix is of size (N × n) × 6, subsisting of 6 intervalued type features for N × n sample, where N is the number of classes. Formally, let set of n samples be [S1, S2, S3,…., Sn] of sign gesture class say Cj;

j = 1, 2, 3,……N $F_i = [(l_{i\,11}, l_{i\,12}, l_{i13}), (l_{i21}, l_{i22}, l_{ti23}) \ldots\ldots (l_{i\,m1}, l_{im2}, l_{i\,m3})]$ and let and $Th_t = [(\alpha_{t\,11},\ \alpha_{t\,12},\ \alpha_{t\,13}), (\alpha_{t\,21},\ \alpha_{t22},\ \alpha_{t23}) \ldots\ldots (\alpha_{t\,m1},\ \alpha_{t\,m2},\ \alpha_{t\,m3})]$ be set of features (length & angle) obtained from m number of triangles in sign sample Si of class Cj.

Let the minimum and the maximum (length) feature obtained from all the m triangles of Si be $(Min\,A_{i1}, Min\,A_{i2}, Min\,A_{i3})$ and $(Max\,A_{i1}, Min\,A_{i2}, Min\,A_{i3})$, respectively. $MinL_{ip} = min(l_{ikp}), MaxL_{ip} = max(l_{ikp})$. Similarly, let the minimum and maximum (angle) feature values obtained from all the m triangles of Si be and, respectively, $(Min\,A_{i1}, Min_{\ i2}, Min\,A_{i3})$ and $(Max\,A_{i1}, Min\,A_{i2}, Min\,A_{i3})$, $MinA_{ip} = min(\alpha_{ikp}), MaxA_{ip} = max(\alpha_{ikp})$ where k = 1, 2, 3…., m; i = 1, 2, 3….n, and p = 1, 2, 3.

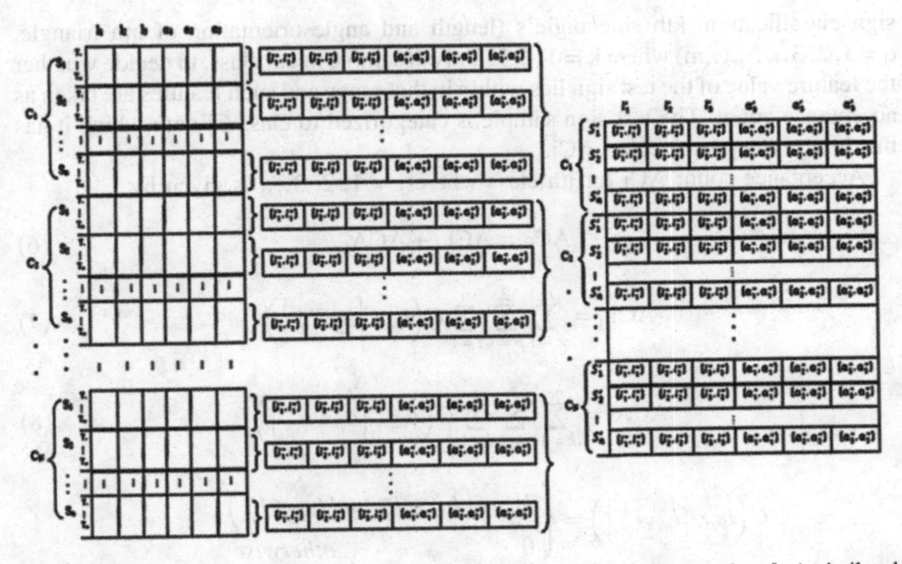

Fig. 6 Feature matrix representation **a** Coventional feature matrix representation, **b** Assimilated length and angles features, and **c** Symbolic feature matrix

Now, its recommended for each sample Si, the varying features are represented in the interval-valued features form

$$\left(\left[l_{ip}^{-}, l_{ip}^{+}\right]\right), \left(\left[\alpha_{ip}^{-}, \alpha_{ip}^{+}\right]\right) \tag{4}$$

Where $l_{ip}^{-} = MinL_{ip}$ and $l_{ip}^{+} = MaxL_{ip}\alpha_{ip}^{-} = MinA_{ip}$ and $\alpha_{ip}^{+} = MaxA_{ip}$. The interval $\left[l_{ip}^{-}, l_{ip}^{+}\right]$ and $\left[\alpha_{ip}^{-}, \alpha_{ip}^{+}\right]$ represent the upper and lower limits of the length and angle feature values of the sample Si . Now, the formation of class Cj of reference sign symbol is represented by the use of interval type and is given by Eq. 5 where i = 1, 2,…,n; j = 1, 2,…, N, and p = 1,2,3.

$$\mathrm{RF_{ji}}\left\{\left(\left[l_{ip}^{-}, l_{ip}^{+}\right]\right), \left(\left[\alpha_{ip}^{-}, \alpha_{ip}^{+}\right]\right)\right\} \tag{5}$$

In this section for classifying the signs, we have adopted the symbolic classifier [13] with a slight moderation. A set of m features of an unknown sign of test sample are matched between the interval-valued type features of the respective symbolic reference sample RF ji during classification process, which are then saved on to the knowledge base to determine efficiently.

Let $FA_t = [(l_{t11}, l_{t12}, l_{t13}), (l_{t21}, l_{t22}, l_{t23}) \ldots\ldots (l_{tm1}, l_{tm2}, l_{tm3})]$ and $FA_t = [(\alpha_{t11}, \alpha_{t12}, \alpha_{t13}), (\alpha_{t21}, \alpha_{t22}, \alpha_{t23}) \ldots\ldots (\alpha_{tm1}, \alpha_{tm2}, \alpha_{tm3})]$ be m × 3-dimensional vectors representing a test sign. Let $\mathrm{RF_{ce}}$; c = 1, 2, 3,..N,: e = 1, 2, 3,… n be the representative symbolic feature vectors saved on to knowledge base. During the process of sign classification, kth side/angle's (length and angle orientation of qth triangle, q = 1, 2, 3………,m) where k = 1, 2, 3, and eth sample in cth class, to decide whether the feature value of the test sign lies within in that range and such features are taken as acceptance count. The test sign sample is categorized to class Cj with which it has maximum acceptance count ACj.

Acceptance count ACj for jth class where j = 1, 2, 3…. is given by

$$AC_j = ACL_j + ACA_j \tag{6}$$

$$ACL_j = \sum_{k=1}^{m} \sum_{p=1}^{3} \sum_{i=1}^{n} C\left(l_{tkp}, \left[l_{ip}^{-}, l_{ip}^{+}\right]\right) \tag{7}$$

$$ACA_j = \sum_{k=1}^{m} \sum_{p=1}^{3} \sum_{i=1}^{n} C\left(\alpha_{tkp}, \left[\alpha_{ip}^{-}, \alpha_{ip}^{+}\right]\right) \tag{8}$$

$$C\left(l_{tkp}, \left[l_{ip}^{-}, l_{ip}^{+}\right]\right) = \begin{cases} 1 & if\left(l_{tkp} \geq l_{ip}^{-} \text{ and } l_{tkp} \leq l_{ip}^{+}\right) \\ 0 & otherwise \end{cases}$$

$$C\left(\alpha_{tkp}, \left[\alpha^{-}, \alpha_{ip}^{+}\right]\right) = \begin{cases} 1 & if\left(\alpha_{tkp} \geq \alpha_{ip}^{-} \text{ and } \alpha_{tkp} \leq \alpha_{ip}^{+}\right) \\ 0 & otherwise \end{cases}$$

From the above discussion, the classification is done based on the maximum of number triangles of a test sample matching against the respective class, and hence we name the classifier as a voting-based classifier.

4 Database

The images in the database determine how efficiently the analysis is carried out to recognize the sign languages. Here we created our own database despite of existence of databases. We capture the images by ourselves using camera in an unconstrained environment. Later these images are arranged according to respective classes (A to Z, 0–9). The database consists of American Sign Language Alphabets and Numbers so totally about 360 images divided into 36 classes. Here each class consists of 10 images per class. Figure 7 shows the samples of sign language alphabets and numbers. The variations in different view point, shape, hand orientation makes the database more challenging.

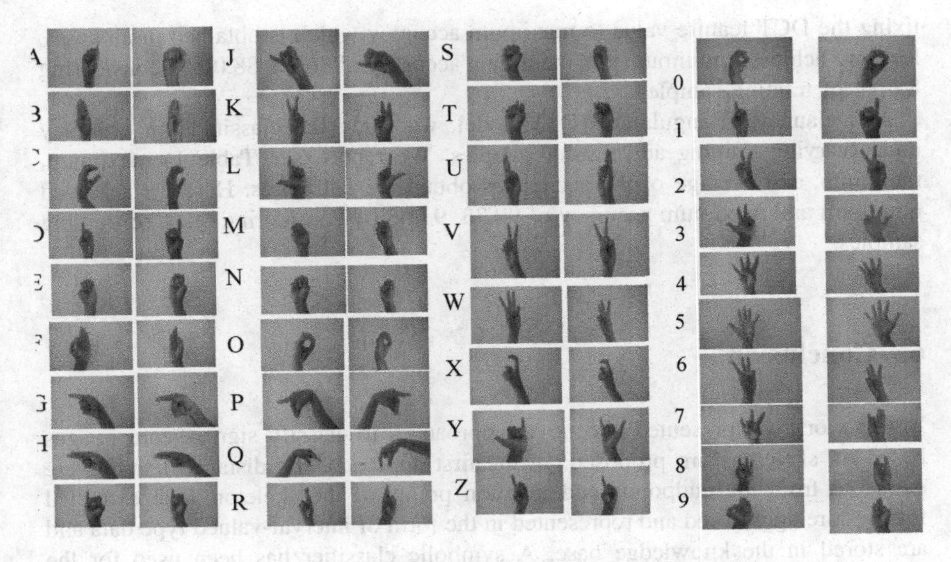

Fig. 7 Examples of sign gesture belonging to different classes of alphabets and numbers

5 Experimentation

During experimentation, we conducted three different sets of experiments. For the first set of experiments, we used 30 % of signs for training and remaining 70 % of signs for testing. In the second set of experiments, the number of training and testing signs is in ratio 50:50. In the third set of experiments, we used 70 % for training and 30 % for testing. In each of these experiments, we considered images randomly from the database and experimentation is repeated at least five times.

In symbolic classification model, we study the classification accuracy under varying features of DCT from 1 to 10. In our experimentation, we have finally fixed the number of DCT feature value to the value that achieves the maximum accuracy. We report (see Table 1) maximum, minimum, and average of the accuracies by

Table 1 Maximum, minimum, and average accuracies of the proposed symbolic and Delaunay triangulation sign gesture classification model

Training set	Symbolic classifier	Triangle based classifiers
80	88.16	91.66
	86.86	90.33
	87.41	90.99
60	86	88.58
	85.16	86.33
	85.58	87.45
40	82.25	83.44
	82.08	82.33
	82.16	82.88

fixing the DCT feature value to maximum accuracy which is obtained in all cases. Here we achieved minimum and maximum accuracy of 86.16, 88.06 by considering 80 per of training samples.

In Delaunay Triangulation (DT) model, we study the classification accuracy under varying training and testing samples. We report (see Table 1) maximum, minimum, and average of the accuracies obtained in all cases. Here we achieved minimum and maximum accuracy of 90.33, 91.66 by considering 80 % of training samples.

6 Conclusion

In this work, we presented two novel approaches to classify sign gesture images based on skeletons are proposed. In the first approach, the distance features are extracted from the endpoints and junction points of the skeleton. The extracted features are aggregated and represented in the form of interval-valued type data and are stored in the knowledge base. A symbolic classifier has been used for the purpose of classification. In the second approach, the concept of spatial topology is combined with symbolic approach for classifying the flower skeletons. From the endpoints and junction points of the skeletons, the triangles are generated using Delaunay Triangulation and for each triangle features like lengths of each side and angles are extracted. The features of each skeleton flower are aggregated and represented in the form of interval-valued type data. A suitable symbolic classifier is designed for the purpose of classification. Experiments are conducted on our own real dataset to evaluate the performance.

References

1. Sawant, SN.: Sign Language recognition system to aid deaf-dumb people using PCA. Int. J. Comput. Sci. Eng. Technol. (IJCSET) vol. 5, No. 05 (2014)
2. Sharma, R., Goyal, S., Sharma, I., Sharma, S., Kane, L., Khanna, P.: Recognition of single handed sign language using contour descriptor. In Proceedings of World Congress on Engineering (WCE) vol. 2, (2013)
3. Goyal, S., Sharma, I., Sharma, S.: Sign language recognition system for deaf and dumb people. In: Proceedings of International Journal of Engineering Research & Technology (IJERT) vol. 2, Issue 4, (2013)
4. Singha, J., Das, K.: Indian Sign language recognition using eigen value weighted euclidean distance based classification technique. Int. J. Adv. Comput. Sci. Appl. **4**(2) (2013)
5. Singha, J., Das, K.: Recognition of Indian sign language in live video. Int. J. Comput. Appl. **70** (19) (2013)
6. Dixit, K., Jalal, A.S.: Automatic Indian sign language recognition system. In: 3rd IEEE International Advance Computing Conference (IACC), (2013)
7. Bhuyan, M.K.: FSM-based recognition of dynamic hand gestures via gesture summarization using key video object planes. Int. Sci. Index **6**(8) (2012)

8. Subha Rajam, P., Balakrishnan, G., Dr..: Real time Indian sign language recognition system to aid deaf-dumb people. In: proceedings of IEEE International Conference on Communication Technology (ICCT), (2011)
9. Shukla, J., Dwivedi, A.: A Method for hand gesture recognition. In: Proceedings of IEEE International Conference on Computer system and Network Technologies (CSNT), (2014)
10. Billard, L., Diday, E.: Symbolic data analysis—conceptual statistics and data mining. Wiley Series In Computational Statistics (Wiley, Chichester 2006)
11. Gonzalez, R.C., Woods, R.E. Eddins, S.L.: Digital image processing using MATLAB. (Prentice-Hall, Inc., Upper Saddle River 2003)
12. Bai, L., Latecki, J., Liu, W.Y.: Skeleton pruning by contour partitioning with discrete curve evolution, IEEE Trans. Patt. Anal. Mach. Intell. **29**(3) 449–462 (2007)
13. Guru, D.S., Prakash, H.N.: Online signature verification and recognition an approach based on symbolic representation. IEEE Trans. Pattern Anal. Mach. Intell. **31,** 1059–1073 (2009)

Modified Cuckoo Search-Based Image Enhancement

Lalit Maurya, Prasant Kumar Mahapatra and Garima Saini

Abstract Swarm intelligence-based evolutionary algorithms have a prominent role in the field of automatic image enhancement that is considered as the optimization problem. In this paper, a modified cuckoo search-based image enhancement is proposed. The modified cuckoo search is used to find the optimal solution of the problem automatically. The proposed method is applied to the lathe tools and some other standard images for improvement of contrast and brightness. The results of the proposed image enhancement method are compared with the other conventional image enhancement techniques like histogram equalization (HE), linear contrast stretching (LCS) and particle swarm optimization (PSO).

Keywords Image enhancement · Modified cuckoo search · Swarm intelligence · Histogram equalization · Linear contrast stretching · Particle swarm optimization

1 Introduction

Image enhancement is the process of manipulating an image so that it is more suitable for a specific application than the original and the information from the enhanced image can be easily collected. According to [1], image enhancement

L. Maurya · P.K. Mahapatra (✉)
V-2(Biomedical Instrumentation Division), CSIR-Central Scientific Instruments Organisation, Sector-30, Chandigarh 160030, India
e-mail: prasant22@csio.res.in; prasant22@gmail.com

L. Maurya
e-mail: lalitmaurya47@gmail.com

L. Maurya · G. Saini
Department of Electronics and Communication Engineering, National Institute of Technical Teachers' Training and Research (NITTTR), Sector-26, Chandigarh 160019, India
e-mail: garimasaini_18@rediffmail.com

S. Das et al. (eds.), *Proceedings of the 4th International Conference on Frontiers in Intelligent Computing: Theory and Applications (FICTA) 2015*, Advances in Intelligent Systems and Computing 404, DOI 10.1007/978-81-322-2695-6_53

techniques can be divided into four main categories: point operation, spatial operation, transformation, and pseudocoloring. In this paper, image enhancement is done on the basis of spatial operation. Spatial operation deals with image in spatial domain based on direct manipulation of pixels in the image. Many images generally suffer from the poor contrast and that is why contrast improvement has been the focus of the researchers. There are several techniques such as histogram equalization (HE) [2] and linear contrast stretching (LCS) [3] which is used to enhance the contrast of images. HE enhances the contrast of images by transforming the intensity values of the image so that the histogram of the improved image approximately matches the histogram of ideal image. Contrast stretching is a process that expands the range of intensity levels in an image to make full use of possible values.

A major problem with image enhancement is that a human interpreter is needed to judge whether an image is suitable for specific application or not. There was no specific benchmark against which the measurement of enhancement could be drawn. Automatic enhancement, which is a method to yield enhanced image without a human interpreter finally came with the introduction of swarm intelligence-based evolutionary algorithms and genetic programs. Several swarm intelligence-based methods have been applied for image-processing application [4–8] including image enhancement problem. In [4], the author applied a contrast-enhancement technique using genetic algorithm. Similarly, several computational techniques like PSO [6], Firefly [7], differential evolution (DE) [8], etc., have been applied to enhance images. In this paper, gray-level image contrast enhancement is proposed using a modified cuckoo search algorithm. The algorithm is applied on lathe tool images and some of the standard images. The enhanced images by proposed method are found to be better when compared with other automatic image enhancement techniques. The algorithm has been developed using MATLAB® software (version: 2014b). The detail approach is discussed in the methodology.

2 Methodology

2.1 Overview of Cuckoo Search Algorithm

Cuckoo Search (CS) is one of the latest nature-inspired metaheurastic algorithms, developed in 2009 by Xin-She Yang [9]. CS is based on the interesting breeding behavior such as brood parasitism of certain species of cuckoos. The cuckoo bird depends on other host bird's nest for laying its eggs. The host bird looks after the egg, assuming it as own. If a host bird discovers that the egg is not its own, then the host bird either throws away the alien egg or simply abandon its nest and build a new nest in a new location. Each egg in a nest represents a solution and the best cuckoo egg represents a new solution. In cuckoo search algorithms, each cuckoo species alter their position as time goes, and every egg in the nest stand for only one new solution. The CS method has been developed based on three idealized assumptions as follows:

1. Each cuckoo lays one egg at a time and dumps its egg in a randomly chosen nest.
2. The best nests with high quality of eggs (solutions) will be kept up to the next generation.
3. The number of available host nests is fixed and the egg laid by a cuckoo is discovered by the host bird with a probability $p_a \in [0, 1]$. In this case, the host bird can either throw the egg away or simply abandon the nest and build a completely new nest in a new location.

Number of studies have indicated that flight behavior of many animals and insects have the typical characteristics of the Levy flights [10]. When generating new solutions X(t + 1) for a cuckoo i, a Levy flight is performed using the following equation:

$$X_i^{(t+1)} = X_i^{(t)} + s \otimes Levy(\lambda) \quad (1)$$

Here s (s > 0) represents a step size. This step size should be related to the scales of the problem, the algorithm is trying to solve. The product symbol ⊗ means an entry-wise multiplication. Levy flights essentially provide a random walk while their random steps are drawn from a Levy distribution for large steps.

$$Levy \sim u = t^{-\lambda}, \quad (1 < \lambda \leq 3), \quad (2)$$

which has an infinite variance with an infinite mean and the consecutive steps/jumps have a power-law step-length distribution with a heavy tail. This algorithm use a balance approach for a local search and far field search, and this is controlled by the switching/discovery probability p_a. The local search is carried out by the Levy walk around the best solution so far. On the other hand, far field search is obtained by far field randomization, far enough from the best solution [10]. There are a few ways exist for generating Levy flights in the literature; however, one of the most efficient and yet straightforward ways is to use the so-called Mantegna algorithm for a symmetric Levy stable distribution. Here 'symmetric' means that the steps can be positive and negative [9, 10].

2.2 Functions Used

Image enhancement based on spatial domain uses a transform function that takes the intensity value of each pixel from the input image and generates a new intensity value for the corresponding pixel to produce the enhanced image. According to literature [5, 6], the image enhancement is done using a transformation function as in Eq. (3) which considers both global and local information and is similar to statistical scaling given in [1].

$$g(i,j) = \frac{k.D}{\sigma(i,j)+b}[f(i,j) - c \times m(i,j)] + m(i,j)^a \tag{3}$$

where, $f(i,j)$ = gray value of the $(i,j)th$ pixel of the input image.

$g(i,j)$ = gray value of the $(i,j)^{th}$ pixel of the enhanced image.

$m(i,j)$ = mean of the $(i,j)^{th}$ pixel over an n × n window.

D = global mean of the pixel of the input image.

M and N = size of original image.

$\sigma(i,j)$ = local standard deviation of the $(i,j)^{th}$ pixel over an n × n window.

a, b, c, k = parameters to produce large variations in the processed image.

To check the quality of an enhanced image without human interaction, an objective function is required which measures the fitness. A good contrast-enhanced image must have high intensity of edges, high number of edgels (edge pixels) and high entropy value [5].

$$F(I_e) = \log(\log(E(I_s))) \times \frac{edgels(I_s)}{M \times N} \times H(I_e) \tag{4}$$

In the above equation, I_e is the enhanced image and I_s is the edge image which is produced by an edge detector. Here, the Sobel edge detector algorithm is used as an automatic threshold detector. $E(I_s)$ is the sum of M × N pixel intensities of the edge image, $edgels(I_s)$ is the number of edge pixels, whose intensity value is above a threshold value in the edge image. $H(I_e)$ is the entropy of the enhanced image.

2.3 Proposed Method

The randomization has an important role in population-based algorithms. The simplified cuckoo search given by Yang, in which a substantial fraction of new solution generated by far field randomization using large step size and uniform random number [9]. In the proposed modified CS algorithm, new nest is generated using levy flight as described in Eq. (1) while a fraction (pa) of worse nest is abandoned and the new solution is created by the Cauchy mutation [11] as follows:

$$X_i^{(t+1)} = W^t \cdot X_i^{(t)} + C \tag{5}$$

where C is Cauchy random number, W^t is the inertia weight. In this process, the inertia weight is gradually decreased from maximum inertia value at each iteration. Therefore, initial inertia component is high and explore the global area in the solution space, but gradually becomes less and exploit better solutions. Inertia value is calculated as

$$W^t = W_{max} - \frac{(W_{max} - W_{min})}{iter} \cdot iter_{max} \tag{6}$$

The Cauchy distribution is a continuous probability distribution. The probability density function (PDF) is given by following equation:

$$f(x; x_0, \gamma) = \frac{1}{\pi}\left[\frac{\gamma}{(x - x_0)^2 - \gamma^2}\right] \tag{7}$$

x_0 is the location parameter, specifying the location of the peak of the distribution, and γ is the scale parameter which specifies the half-width at half-maximum (HWHM). For large values of γ, a fat tail curve is obtained, whereas for smaller values of γ, the shape of the curve changes towards a sharper peak. Since the variance of Cauchy distribution is infinite, so that Cauchy mutation could make a cuckoo to have a long jump. For an optimization algorithm, it should be needed to accomplish a proper balance of exploration and exploitation. In general, exploration means searching search space as much as possible, while exploitation means concentrating on one point. In this paper, the long jump increased the probability of large mutation step sizes, i.e., better exploration. The Levy walk converges the swarm to the best solution so far. Therefore, long jumps in random walk due to the Cauchy mutation as in Eq. (5) increase the probability of escaping from the local optimum.

The transformation function defined in Eq. (3) contains both global and local information to determine a good contrast image. The function has four parameters, namely, a, b, c, and k which are used to produce diverse result and help to find the optimal-enhanced image according to the fitness function. The ranges of these parameters are a ϵ [0, 1.5], b ϵ [0, 0.5], c ϵ [0, 1], and k ϵ [0.5, 1.5] according to [5, 6]. In this proposed method, modified CS is used to find the best set of values for these parameters. The flowchart of the proposed method is shown in the Fig. 1.

Each nest in the proposed technique is initialized with four parameters a, b, c, and k. The stopping criterion of this algorithm is the max number of iterations, when the max number of iterations is achieved, the algorithm is stopped. The input image is taken and the number of nests, probability of detection of alien egg, number of iterations is initialized. Each nest generates an enhanced image with defined parameters. The quality of the image is evaluated by the objective function as defined by Eq. (4). At the end of the iterations, the enhanced image is achieved corresponding to the set of parameters obtained from the best nest. The parameter settings used in modified cuckoo search are set as: the number of nests = 25, the discovery rate of alien egg, p_a = 0.25, location parameter x_0 = 0 and scale parameter γ = 1. The max number of iterations is 40.

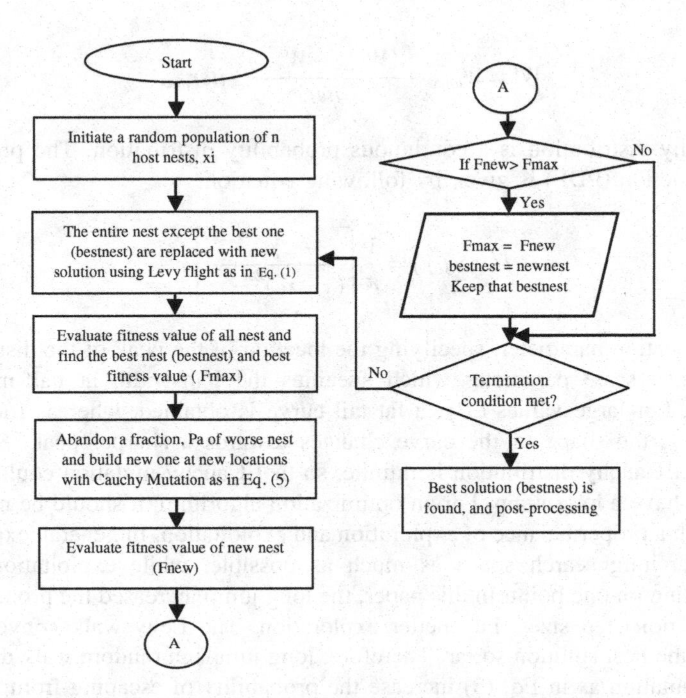

Fig. 1 Flowchart of the proposed modified cuckoo search method for image enhancement

3 Results and Discussions

The proposed algorithm was applied on lathe tool and some of the natural digital images included in the standard image database. The lathe tool images were used because the tool is being used by the authors to carry out different experiments on micro and nanoscale movements for desktop machining. The main objective is to maximize the number of pixels at the edges, increase the overall intensity of the edges, and increase the measure of the entropy so that the histogram of the enhanced image approaches towards the required uniform distribution. The results of the proposed method were compared with HE, LCS, and standard PSO technique. Here, the parameter setting for PSO, [c1 = c2 = 1.80 and w = 0.60, number of swarm particle = 30] have been chosen as recommended in [12]. The gray scale images of lathe tool captured using single AVT Stingray F125B monochrome camera mounted with Navitar lens, and their enhanced images obtained from different enhancement techniques are shown in Figs. 2, 3, 4, 5. Here, also the maximum number of iterations is 40.

In order to evaluate, the qualitative performance of the proposed method and other techniques are illustrated in Figs. 2, 3, 4, 5. The quantitative analyses are done by fitness function and number of edge pixels. The fitness value is calculated by the Eq. (4) which basically measures the contrast of edges and sharpness. The number

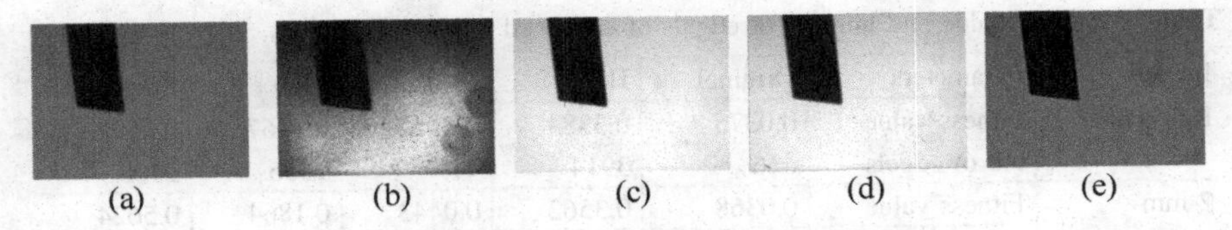

Fig. 2 Lathe tool's reference image **a** original, **b** HE, **c** LCS, **d** PSO, and **e** proposed enhancement

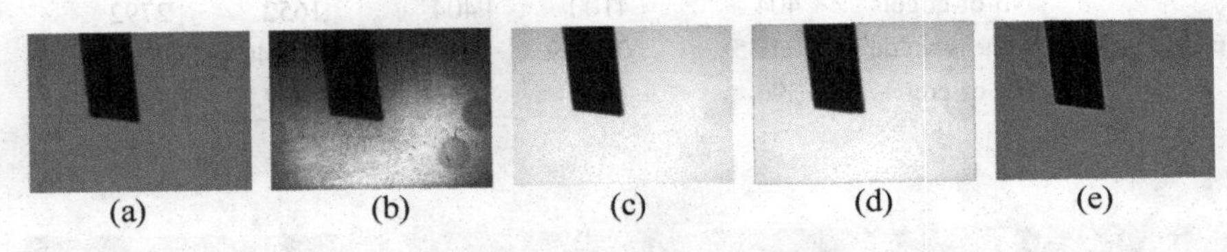

Fig. 3 Lathe tool's 2 mm image **a** original, **b** HE, **c** LCS, **d** PSO, and **e** proposed enhancement

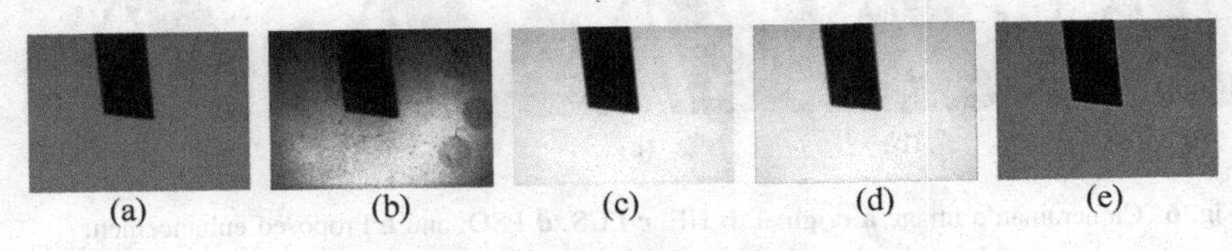

Fig. 4 Lathe tool's 4 mm image **a** original, **b** HE, **c** LCS, **d** PSO, and **e** proposed enhancement

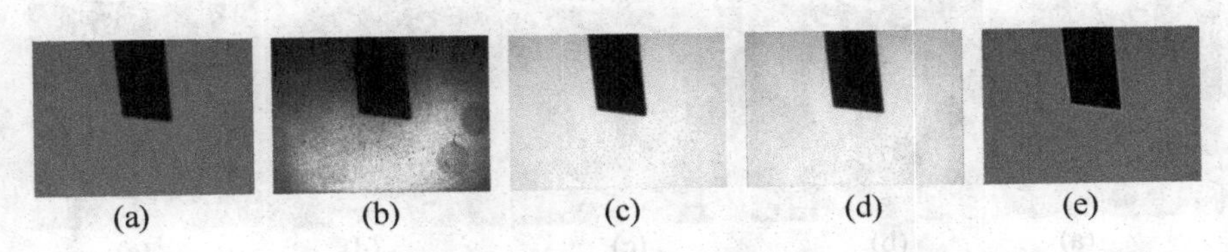

Fig. 5 Lathe tool's 6 mm image **a** original, **b** HE, **c** LCS, **d** PSO and **e** proposed enhancement

of edgels of enhanced image is obtained by the Sobel edge detector method. The value of fitness and number of edgels is shown in Table 1.

To check the effectiveness of the proposed algorithm, it has also been applied to some of the standard images viz. cameraman, mandrill, and the house as provided in [13] that gives similar results. The enhanced image obtained from different enhancement techniques is shown in Figs. 6, 7, 8. The maximum number of iterations for cameraman, mandril, and house image are set as 45, 50, and 40, respectively.

It is observed from the Tables 1, 2 that the fitness value and the number of edge pixels are higher for the proposed method. The comparison of fitness value of proposed method and PSO-based method for standard images in each iteration is shown in Fig. 9. To evaluate the robustness (repeatability of the results) of the

Table 1 Fitness value and number of edgels of enhanced lathe tools images

Images	Parameters	Original	HE	LCS	PSO	Proposed
Reference	Fitness value	0.0375	0.3384	0.0452	0.1867	0.5591
	No of edgels	416	3914	419	1666	2408
2 mm	Fitness value	0.0368	0.3562	0.0443	0.1864	0.5624
	No of edgels	411	3874	411	1659	2419
4 mm	Fitness value	0.0360	0.3760	0.0435	0.1854	0.5053
	No of edgels	404	4100	404	1652	2792
6 mm	Fitness value	0.0353	0.3519	0.0426	0.1840	0.5606
	No of edgels	396	3846	396	1647	2234

Fig. 6 Cameraman's image **a** original, **b** HE, **c** LCS, **d** PSO, and **e** Proposed enhancement

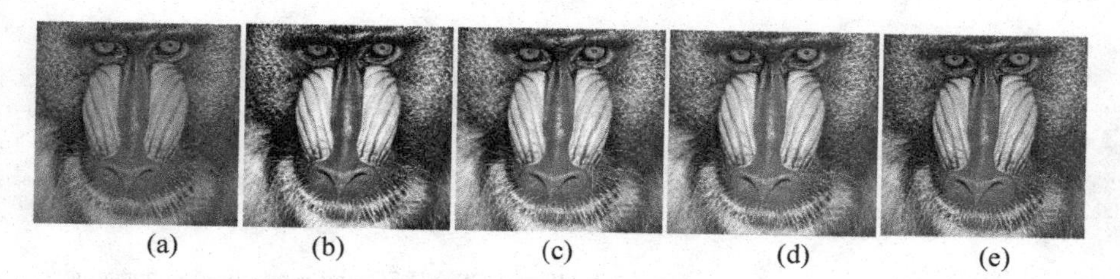

Fig. 7 Mandrill's image **a** original, **b** HE, **c** LCS, **d** PSO, and **e** Proposed enhancement

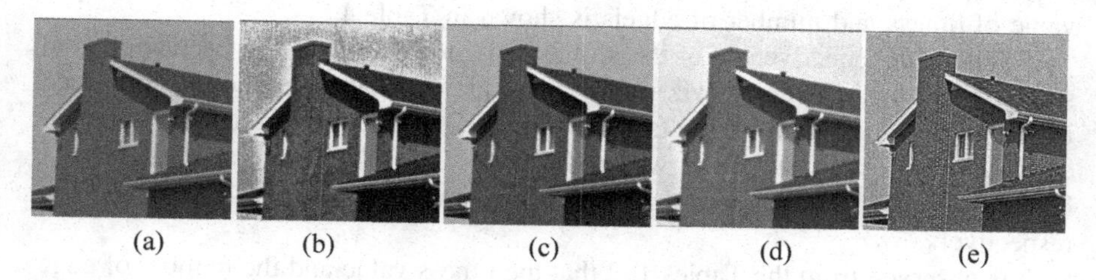

Fig. 8 House's image **a** original, **b** HE, **c** LCS, **d** PSO, and **e** Proposed enhancement

Table 2 Fitness value and number of edgels of enhanced standard images

Images	Parameters	Original	HE	LCS	PSO	Proposed
Cameraman	Fitness value	0.6139	0.5089	0.6133	0.7823	1.0418
	No of edgels	2503	2430	2503	2913	3741
Mandrill	Fitness value	0.7250	0.6793	0.7393	0.7541	0.8795
	No of edgels	2820	3123	2859	2723	2901
House	Fitness value	0.5181	0.4842	0.5203	0.6627	0.8837
	No of edgels	2308	2487	2316	2777	3164

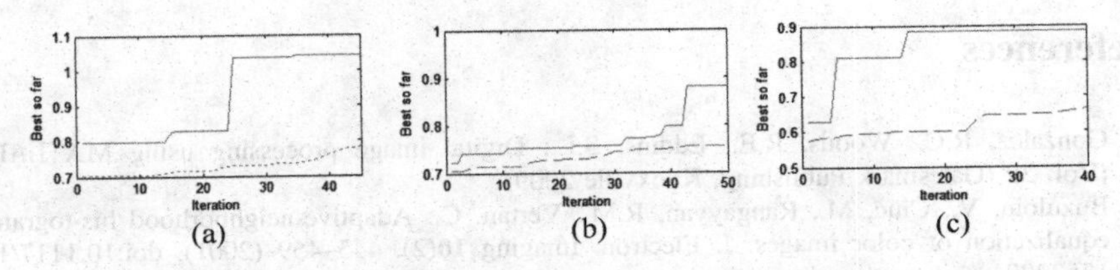

Fig. 9 Graph showing the comparison of the proposed method (*blue line*) and PSO method (*red line*) **a.** cameraman. **b.** Mandrill. **c.** House

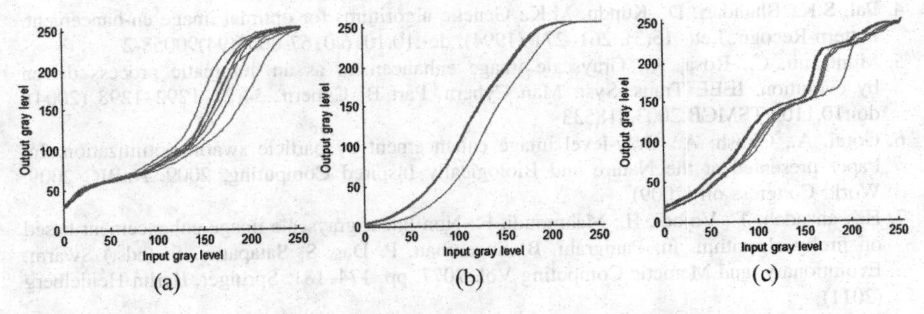

Fig. 10 Transformation curve for standard images of the proposed algorithm (10 times run) **a.** cameraman. **b.** Mandrill. **c.** House

proposed algorithm, 10 independent runs of proposed method were performed and obtained the transformation curve for each image. As illustrated in Fig. 10, the resulted transformation curve has the same shape and are clustered together, i.e., they are similar.

4 Conclusions

In this paper, a modified cuckoo search-based image enhancement has been proposed to enhance the contrast of images. The main aim of the algorithm is to enhance the gray vale of images by increasing the number of edge pixel, edge intensity, and entropy of the images. The proposed method is applied on lathe tool as well as standard images and gives better performance. The results are compared with HE, LCS, and PSO-based image enhancement techniques. In each and every case, the result of the proposed modified cuckoo search algorithm is found better.

Acknowledgments This work is supported by the Council of Scientific and Industrial Research (CSIR, India), New Delhi under the Network program (ESC-112) in collaboration with CSIR-CMERI, Durgapur. Authors would like to thank Director, CSIR-CSIO for his guidance during investigation.

References

1. Gonzalez, R.C., Woods, R.E., Eddins, S.L.: Digital image processing using MATLAB (Vol. 2): (Gatesmark Publishing, Knoxville 2009)
2. Buzuloiu, V., Ciuc, M., Rangayyan, R.M., Vertan, C.: Adaptive-neighborhood his-togram equalization of color images. J. Electron. Imaging **10**(2), 445–459 (2001). doi:10.1117/1.1353200
3. Chi-Chia, S., Shanq-Jang, R., Mon-Chau, S., Tun-Wen, P.: Dynamic contrast en-hancement based on histogram specification. IEEE Trans. Consum. Electron. **51**(4), 1300–1305 (2005). doi:10.1109/TCE.2005.1561859
4. Pal, S.K., Bhandari, D., Kundu, M.K.: Genetic algorithms for optimal image en-hancement. Pattern Recogn. Lett. **15**(3), 261–271 (1994). doi:10.1016/0167-8655(94)90058-2
5. Munteanu, C., Rosa, A.: Grayscale image enhancement as an automatic process driven by evolution. IEEE Trans. Syst. Man Cybern. Part B: Cybern. **34**(2), 1292–1298 (2004). doi:10.1109/TSMCB.2003.818533
6. Gorai, A., Ghosh, A.: Gray-level image enhancement by particle swarm optimization. In: Paper presented at the Nature and Biologically Inspired Computing, 2009. NaBIC 2009. World Congress on (2009)
7. Hassanzadeh, T., Vojodi, H., Mahmoudi, F.: Non-linear grayscale image enhancement based on firefly algorithm. In: Panigrahi, B. Suganthan, P. Das, S. Satapathy, S. (eds.) Swarm, Evolutionary, and Memetic Computing Vol. 7077, pp. 174–181. Springer, Berlin Heidelberg (2011)
8. Sarangi, P.P., Mishra, B.S.P., Majhi, B., Dehuri, S.: Gray-level image enhancement using differential evolution optimization algorithm. In: Paper presented at the Signal Processing and Integrated Networks (SPIN), 2014 International Conference on 20–21 Feb 2014
9. Xin-She, Y., & Deb, S.: Cuckoo Search via Levy flights. In: Paper Presented at the Nature and Biologically Inspired Computing, 2009. NaBIC 2009. World Congress on 9–11 Dec 2009
10. Yang, X.-S. (2008). Nature-Inspired Metaheuristic Algorithms: Luniver Press
11. Stacey, A., Jancic, M., Grundy, I.: Particle swarm optimization with mutation. In: Paper presented at the Evolutionary Computation, 2003, CEC '03, The 2003 Congress on 8–12 Dec 2003
12. Karaboga, D., Akay, B.: A comparative study of Artificial Bee Colony algorithm. Appl. Math. Comput. **214**(1), 108–132 (2009). doi:10.1016/j.amc.2009.03.090
13. http://sipi.usc.edu/database/

A Novel Iterative Salt-and-Pepper Noise Removal Algorithm

Amiya Halder, Sayan Halder and Samrat Chakraborty

Abstract This paper proposes a novel iterative algorithm for removal of salt-and-pepper impulse noises. Even if the noise level is as high as 90 %, this proposed methodology ensures elimination of all the impulse noises while restoring the fine details of gray-scale images. Most of the salt-and-pepper noise removal techniques try to search for impulse noise from the images, but this proposed algorithm searches for the non-noisy points and removes the impulse points found in vicinity. The algorithm does the process iteratively but with low time complexity compared to most of the modern powerful algorithms. This simple straight forward algorithm produces the denoising image that gives better peak signal-to-noise ratio (PSNR) than other existing algorithms.

Keywords Salt-and-pepper noise · Median filtering · PSNR

1 Introduction

Noise is introduced in digital images if the intensity of any point of the original image changes due to some error in storage, during transmission or due to some electronic problem. One of the major kinds of noise is salt-and-pepper impulse noise. In this type of noise, the intensity of some pixels of the images either becomes maximum or minimum (i.e. 255 or 0 for the gray-scale image) [1, 2]. During a long period, there are many non-linear filters [3–8], adaptive filters [9–11] and other algorithms

A. Halder (✉) · S. Halder · S. Chakraborty
St. Thomas College of Engineering and Technology, 4, D.H. Road Kidderpore, Kolkata 700023, West Bengal, India
e-mail: amiya.halder77@gmail.com

S. Halder
e-mail: sayanhaldervms@gmail.com

S. Chakraborty
e-mail: i.m.samrat75@gmail.com

S. Das et al. (eds.), *Proceedings of the 4th International Conference on Frontiers in Intelligent Computing: Theory and Applications (FICTA) 2015*, Advances in Intelligent Systems and Computing 404, DOI 10.1007/978-81-322-2695-6_54

[12–15] published which gives remedy to this type of noise. In images with noises more than 50 %, high blurring takes place due to these types of impulse filtering which causes loss of fine details which is not desirable.

One of the most commonly used non-linear methods of salt-and-pepper noise removal is standard median filter (SMF) in which a local window is taken and the median of the intensities in the local window is used for replacing the centre pixel [1, 2]. This is highly preferred for its simplicity of algorithm. The main drawback of the algorithm is that for noise intensities greater than 50 %, this method fails to remove the noise and causes high amount of blurring. So the fine details are lost. In the modified approach, modified standard median filter (MSMF) [16], a threshold is used and if other values in the local window have the difference from median more than threshold then they are also replaced by median. This algorithm works better than SMF upto 30 % of noise but for noise more than 70 % the image is completely lost and the threshold value is to be defined, which varies from image to image.

Another such filtering technique is decision median filter (DMF) which acts just like SMF but the difference lies in the fact that DMF takes the decision, which of the pixels are corrupted, and removes most of them [17]. DMF acts well but the edge details are lost for images with noise percentage of 60 % or more. As we all know that the impulse noises take the value of maximum or minimum intensity, depending on which logic unsymmetrically trimmed median filter (UTMF) is proposed [18], here, the impulse intensities are trimmed from the local window before taking the decision of what intensity will remove the corrupted pixel. This acts better than DMF but for a continuous train of impulse noises in an image this method fails to give satisfactory results. To overcome the drawbacks of DMF, a better method improvement of decision median filtering (IDMF) is proposed [19]. In this method, change the window size as per requirement to take decision of the pixel to be used to replace the corrupted pixel. This improves the results of DMF a bit more. It gives satisfactory results for images up to 70 % noise density, but the main drawback of this algorithm is that it has a high time complexity compared to others. In another approach to reduce noisy image to non-noisy one, the two methods of DMF and UTMF are cascaded in modified cascaded filter (MCF) [17] which gives satisfactory output up to 70 % of noise but it takes even more time than that of IDMF [19].

To remove the drawback of the above existing filtering algorithms, we introduce a straight forward iterative and simple approach to remove the salt-and-pepper impulse noises even at noises as high as 90 %. This proposed algorithm produces the better visual quality images and in addition our proposed method uses only four neighbourhood pixels for processing and this algorithm takes lower processing time. Also, this proposed straight forward algorithm gives better PSNR than the above existing algorithm.

The rest of the paper is organized as follows. Section 2 describes about the proposed salt-and-pepper noise removal straight forward algorithm. Section 3 contains the experimental results for different images, and finally conclusions are drawn in Sect. 4.

2 Proposed Method

The proposed algorithm is used to remove impulses in an iterative fashion such that it guarantees the removal of all noisy impulses in the image, preserving the fine details and the edge details in images with even 90 % of noise.

Most of the other existing algorithms for removal of salt-and-pepper noise search for the impulse points and remove it depending on the intensities of the points near the impulse. This algorithm searches for non-noisy points present in the image and tries to remove the impulses in its vicinity with its intensity. This algorithm is an iterative one which is a completely new step towards removal of the noise with low time complexity. We consider a gray-scale image F and size of the image is $M \times N$. The steps of the proposed algorithm (also block diagram is shown in Fig. 1) are given below:

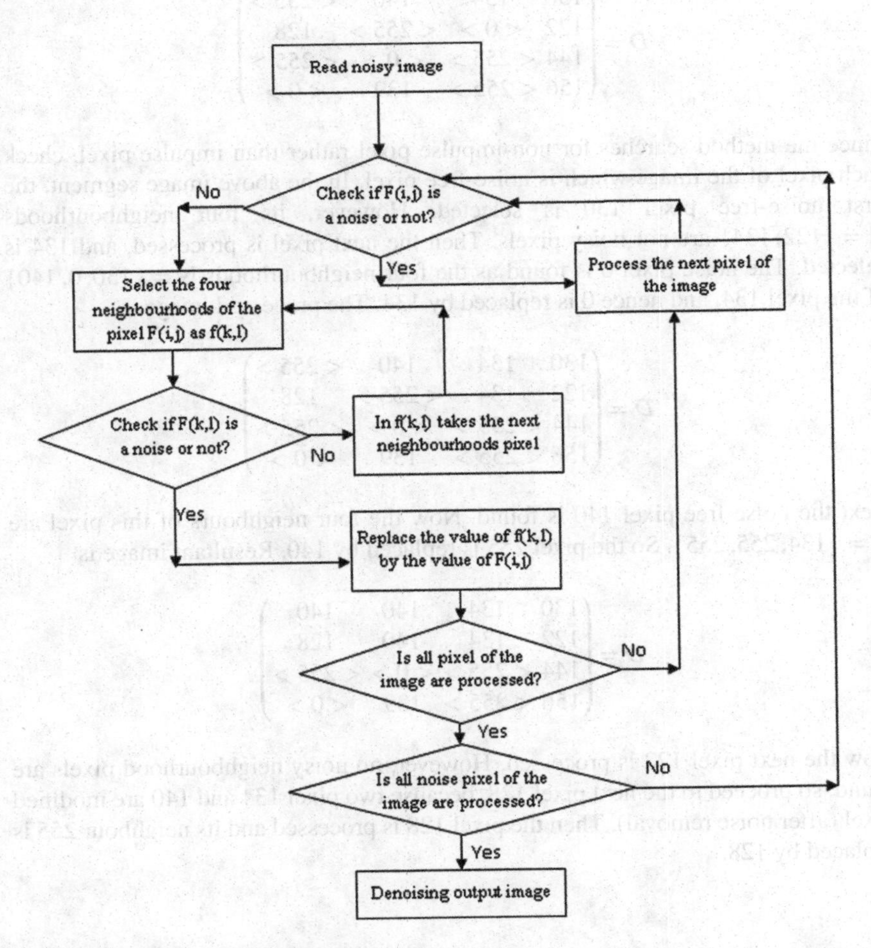

Fig. 1 Block diagram of the proposed method

Algorithm 1 Algorithm for proposed method

Step1: Process each pixel $F(i,j)$. If $0 < F(i,j) < 255$ goto Step2 otherwise process the next pixel of the image
Step2: Select the horizontal and vertical four neighbours of $F(i,j)$ { i.e. $F(i-1,j), F(i+1,j), F(i,j-1), F(i,j+1)$}
Step3: If any value of these four neighbours is impulse (0 or 255) then replace it with the intensity of $F(i,j)$
Step4: Repeat Step2 and Step3 for all pixel of the image
Step5: If no more impulses are present then stop otherwise repeat Step1 to Step4

For the understanding of the above algorithm, a 4×4 subimage is used as an example for illustrating the proposed method as follows:

$$D = \begin{pmatrix} 130 & 134 & 140 & <255> \\ 122 & <0> & <255> & 128 \\ 144 & <255> & <0> & <255> \\ 156 & <255> & 139 & <0> \end{pmatrix}$$

Since the method searches for non-impulse pixel rather than impulse pixel, check each pixel of the image which is noise-free pixel. In the above image segment, the first noise-free pixel 130 is selected. However, its four neighbourhoods $N = \{122, 134\}$ are not noisy pixels. Then the next pixel is processed, and 134 is selected. The noise pixel 0 is found as the four neighbourhoods $N = \{130, 0, 140\}$ of the pixel 134, and hence 0 is replaced by 134. The processed image is

$$D = \begin{pmatrix} 130 & 134 & 140 & <255> \\ 122 & 134 & <255> & 128 \\ 144 & <255> & <0> & <255> \\ 156 & <255> & 139 & <0> \end{pmatrix}$$

Next the noise-free pixel 140 is found. Now the four neighbours of this pixel are $N = \{134, 255, 255\}$. So the pixel 255 is replaced by 140. Resultant image is

$$D = \begin{pmatrix} 130 & 134 & 140 & 140 \\ 122 & 134 & 140 & 128 \\ 144 & <255> & <0> & <255> \\ 156 & <255> & 139 & <0> \end{pmatrix}$$

Now the next pixel 122 is processed. However, no noisy neighbourhood pixels are found, so proceed to the next pixel 128, because two pixel 134 and 140 are modified pixel (after noise removal). Then the pixel 128 is processed and its neighbour 255 is replaced by 128.

$$D = \begin{pmatrix} 130 & 134 & 140 & 140 \\ 122 & 134 & 140 & 128 \\ 144 & < 255 > & < 0 > & 128 \\ 156 & < 255 > & 139 & < 0 > \end{pmatrix}$$

Next the pixel 144 is checked and its neighbour 255 is replaced by 144. The image becomes

$$D = \begin{pmatrix} 130 & 134 & 140 & 140 \\ 122 & 134 & 140 & 128 \\ 144 & 144 & < 0 > & 128 \\ 156 & < 255 > & 139 & < 0 > \end{pmatrix}$$

The next pixel is processed 128 but it was previously noisy pixel 255 and it is replaced by 128. Thus, the pixel is skipped and next pixel 156 is checked and its horizontal neighbour 255 is replaced by 156.

$$D = \begin{pmatrix} 130 & 134 & 140 & 140 \\ 122 & 134 & 140 & 128 \\ 144 & 144 & < 0 > & 128 \\ 156 & 156 & 139 & < 0 > \end{pmatrix}$$

After that the next pixel 139 is processed and its horizontal and vertical neighbour $N = \{0, 0\}$ is replaced by 139. Finally, removing all noisy pixels, the resultant image is shown below:

$$D = \begin{pmatrix} 130 & 134 & 140 & 140 \\ 122 & 134 & 140 & 128 \\ 144 & 144 & 139 & 128 \\ 156 & 156 & 139 & 139 \end{pmatrix}$$

3 Experimental Results

This proposed method has been implemented on Dev C++ platform. This method is tested on more than 100 noisy gray-scale images of different sizes. Experimental results conducted on gray-scale images under a wide range (from 10 % to 90 %) of noise and clearly show that our proposed algorithm performs better results than median filter, DMF, MCF and IDMF visually as well as theoretically. The result shows that the PSNR (quality of the image) value for this proposed algorithm exceeds all the other methods. For typical illustration, the denoising output (using the proposed algorithm and with the four different algorithms, used for comparison) is shown (in Fig. 2) for Lena image. Figures 3, 4 and 5 depict the variation of PSNR value using the proposed algorithm and other four techniques for three images Lena, Parrot, Zelda, Mandrill and Cameraman, respectively. These five figures clearly show

Fig. 2 Denoising output images using different algorithms (**a**) Salt-an-pepper noise of the lena image (10–90 % noise) (**b**) Median filter (**c**) DMF (**d**) MCF (**e**) IDMF and (**f**) proposed method

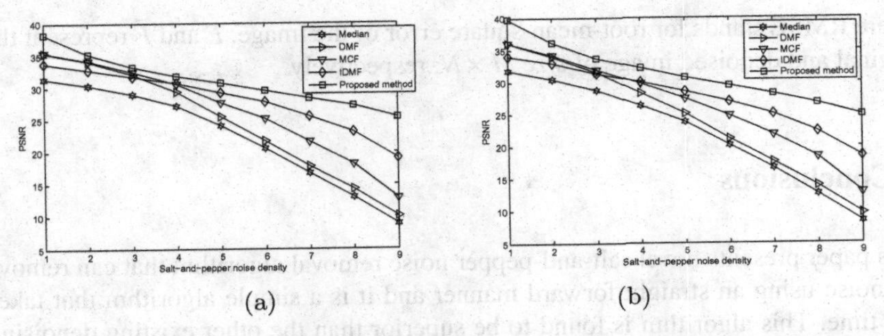

Fig. 3 PSNR value (for 10–90 % noise) of the (**a**) Lena and (**b**) Parrot images using Median filter, DMF, MCF, IDMF and Proposed algorithm

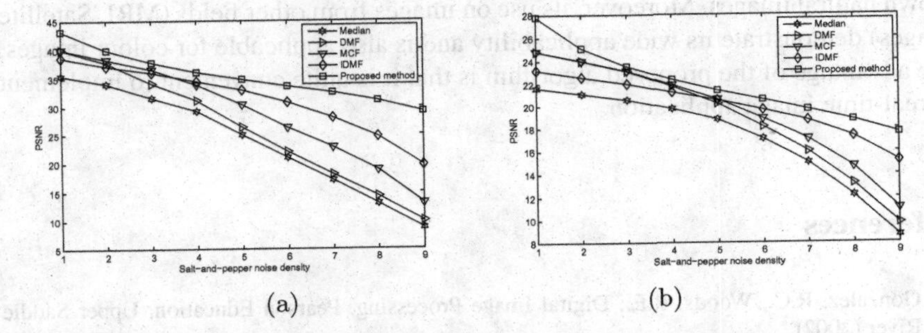

Fig. 4 PSNR value (for 10–90 % noise) of the (**a**) Zelda and (**b**) Mandril images using Median filter, DMF, MCF, IDMF and Proposed algorithm

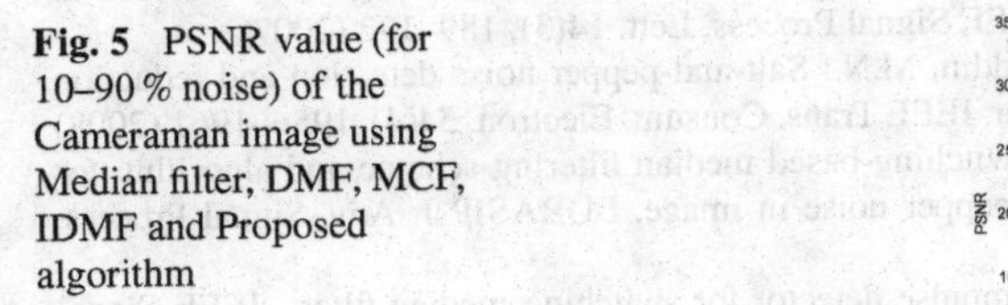

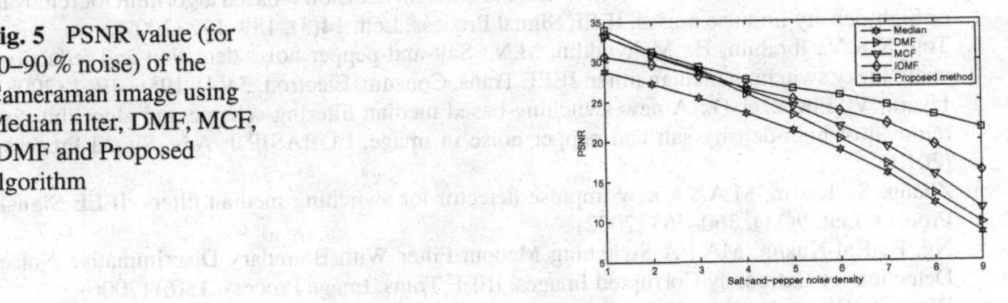

Fig. 5 PSNR value (for 10–90 % noise) of the Cameraman image using Median filter, DMF, MCF, IDMF and Proposed algorithm

that the proposed algorithm gives better PSNR value than the other denoising algorithms. PSNR is calculated from the given equation:

$$PSNR(dB) = 20\log_{10}[\frac{255}{RMSE}] \quad (1)$$

$$RMSE = \sqrt{\frac{\sum_x \sum_y [(F(x,y) - \vec{F}(x,y)]}{M \times N}} \quad (2)$$

where RMSE stands for root-mean-square error of the image. F and $\vec{F}$ represent the original and denoised image of size $M \times N$, respectively.

4 Conclusions

This paper presents novel salt-and-pepper noise removal algorithm that can remove the noise using an straight forward manner and it is a simple algorithm that takes less time. This algorithm is found to be superior than the other existing denoising filtering algorithms. Comparison of the results with that of other filtering methods shows that the proposed technique gives satisfactory results when applied on well-known natural images. Moreover, its use on images from other fields (MRI, Satellite Images) demonstrate its wide applicability and is also applicable for colour images. The advantage of the proposed algorithm is that it is quite convenient to implement on real-time image application.

References

1. Gonzalez, R.C., Woods, R.E.: Digital Image Processing. Pearson Education, Upper Saddle River (2002)
2. Sridhar, S.: Digital Image Processing. Oxford Higher Education. Oxford University Press, New Delhi (2011)
3. Srinivasan, K.S., Ebenezer, D.: A new fast and efficient decision—based algorithm for removal of high-density impulse noises. IEEE Signal Process. Lett. **14**(3), 189–192 (2007)
4. Toh, K.K.V., Ibrahim, H., Mahyuddin, M.N.: Salt-and-pepper noise detection and reduction using fuzzy switching median fillter. IEEE Trans. Consum. Electron. **54**(4), 1956–1961 (2008)
5. Jayaraj V., Ebenezer D.: A new switching-based median filtering scheme and algorithm for removal of high-density salt and pepper noise in image. EURASIP J. Adv. Signal Process. (2010)
6. Zhang, S., Karim, M.A.: A new impulse detector for switching median filters. IEEE Signal Process. Lett. **9**(11), 360–363 (2002)
7. Ng, P., Kai-Kuang, MA.: A Switching Median Filter With Boundary Discriminative Noise Detection for Extremely Corrupted Images. IEEE Trans. Image Process. **15**(6) (2006)
8. Wang Z., Zhang D.: Progressive Switching Median Filter for the Removal of Impulse Noise from Highly Corrupted Images. IEEE Trans. Circuits Syst-II: Analog Digit. Signal Process. **46**(1) (1999)
9. Ibrahim, H., Kong, N.S.P., Foo, T.F.: Simple adaptive median filter for the removal of impulse noise from highly corrupted images. IEEE Trans. Consum. Electron. **54**(4), 1920–1927 (2008)
10. KalVinToh, K., Isa, N.A.M.: Noise adaptive fuzzy switching median filter for salt-and-pepper noise reduction. IEEE Signal Process. Lett. **17**(3) (2010)
11. Halder, A., Shekhar, S., Kant, S., Mubarki, M.A., Pandey, A.: A new efficient adaptive spatial filter for image enhancement. IEEE ICCEA, pp. 244–246 (2010)
12. Chen, P.Y., Lien, C.Y.: An efficient edge-preserving algorithm for removal of salt-and-pepper noise. IEEE Signal Process. Lett. **15**, 833–836 (2008)
13. Aiswarya, K., Jayaraj, V., Ebenezer, D.: A new and efficient algorithm for the removal of high density salt and pepper noise in images and videos. In: Second International Conference Computer Modeling and Simulation, pp. 409–413 (2010)

14. Chan, R.H., Chung-WaHo, Nikolova, M.: Salt-and-pepper noise removal by median-type noise detectors and detail-preserving regularization. IEEE Trans. Image Process. **14**(10) (2005)
15. Zhang, D., Wang, Z.: Impulse noise detection and removal using fuzzy techniques. Electron. Lett. **33**, 378–379 (1997)
16. Pandey, R.: An improved switching median filter for uniformly distributed impulse noise removal. World Academy of Science, Engineering and Technology **38**, (2008)
17. Karthik, B., Kiran Kumar, T.V.U., Dorairangaswamy, M.A., Logashanmugam, E.: Removal of high density salt and pepper noise through modified cascaded filter. Middle-East J. Sci. Res. **20**(10), 1222–1228 (2014)
18. Esakkirajan, S., Veerakumar, T., Subramanyam, A.N., PremChand, C.H.: Removal of high density salt and pepper noise through modified decision based unsymmetric trimmed median filter. IEEE Signal Process. Lett. **18**(5) (2011)
19. Bhateja, V., Verma, A., Rastogi, K., Malhotra, C., Satapathy, S.C.: Performance improvement of decision median filter for suppression of salt and pepper noise. Adv. Intell. Syst. Comput. **264**, 287–297 (2014)

Part XIV
Intelligent System Planning

Queuing Model for Improving QoS in Cloud Service Discovery

V. Viji Rajendran and S. Swamynathan

Abstract One of the fundamental issues in cloud computing is the discovery of cloud services. Given the plethora and range of service offerings in the cloud, a cloud broker is vital in finding the services that meet the functional and nonfunctional requirements of the user. Too much request to the broker increases the waiting time and hence the response time, resulting in a degraded performance. Mathematical formulation using the queuing model is used to show how the response time varies between a single broker system and a multiple broker system in a cloud-computing environment.

Keywords Queuing model · Cloud computing · Broker

1 Introduction

Cloud computing refers to the on-demand provisioning of resources as a service over the Internet. Cloud computing has gained widespread popularity due to the advantages like incremental scalability, elasticity of resources, better capital utilization, and new business opportunities. The increase in the type and number of services offered is massively rising day by day. Most public cloud services do not provide a means to describe and publish their service offerings for discovery and selection. Hence, it has been challenging for the user to choose a service that suits his requirements by comparing various services. User onus can be brought down by introducing cloud brokers that act as an intermediary serving users and cloud

V. Viji Rajendran (✉) · S. Swamynathan
Department of Information Science and Technology,
CEG Campus, Anna University, Chennai, India
e-mail: vijirajv@gmail.com

S. Swamynathan
e-mail: swamyns@annauniv.edu

S. Das et al. (eds.), *Proceedings of the 4th International Conference on Frontiers in Intelligent Computing: Theory and Applications (FICTA) 2015*, Advances in Intelligent Systems and Computing 404, DOI 10.1007/978-81-322-2695-6_55

providers. The key functionality of cloud broker is to find the best cloud service providers based on the user requirement.

Cloud brokers aggregate services from multiple providers to develop a customized solution that meets user requirements. Hence, in the perspective of a user, rendering a broker to gain intelligence for service discovery is a great value addition. The user needs to enter into a waiting line if the broker is busy with processing the request of the previous user. Response time in service discovery is defined as the time for a requested service to be discovered, including the waiting time and the processing time in the broker. Long lines of waiting users lead to lower response time and hence, the cloud broker is modeled by the open Jackson network [8]. Arrivals of cloud service request follow a Poisson distribution. The service discovery request can be simple query with one or two constraints, or complex query with two or simpler ones concatenated together. Moreover, the rate of simple queries is significantly more than complex ones. Since the time taken for discovering the services is proportional to the complexity of the query, the time taken for matching the services follows an exponential distribution [7]. Hence, the service discovery system can be modeled using queuing models.

As the user request arrival time and the time taken by the broker to process the request are random, it is important to trim down the waiting time to improve the QoS in service discovery. Hence, multiple brokers are employed to reduce the waiting time of each user. This paper proposes M/M/c broker queuing model to reduce the waiting time and thus improving the response time to the user. The rest of this paper is structured as follows. The following section examines the related work and need for queuing model in broker-based cloud architecture is described in Sect. 3. Section 4 briefly portrays background information on the concepts of queuing models. Section 5 discusses the impact of various queuing models in minimizing the waiting time to discover services in a cloud environment. Section 6 concludes the paper.

2 Related Work

The cloud broker effectively intervenes and coordinates heterogeneous cloud data and resources from distinctive cloud providers. Consequently the appeal given to broker is high. Various queuing models are discussed to address mean waiting time taken by the broker to process the user request. The cloud architecture is represented by an open Jackson network of M/M/1 and M/M/m interconnected servers [1]. The open Jackson network is employed to model a service discovery system by building a distributed suffix tree index to speed up the discovery procedure [7]. The queuing model has been used for the performance analysis of various research initiatives in cloud computing [1–3]. M/G/m/m+r queuing model is considered to obtain performance evaluation of cloud server centers [2]. The deployment of multiple servers increases the performance of the system by decreasing the overload of the system. Optimal multi-server configuration for profit maximization in a cloud computing has been set up by modeling multi-server system as an M/M/m queuing model [3]. Homogeneous continuous time Markov chain is used to describe a finite-capacity multi-server model with non-preemptive priorities [4]. Multi-server queuing model is considered with Markovian arrival of regular customers and phase-type renewal process arrivals of special customers [5]. The main objective of this paper is to analyze the impact of queuing models in improving the response time of cloud service discovery.

3 Need of Queuing Model in Broker-Based Cloud Architecture

Medium- to large-sized enterprises exploit various heterogeneous platforms to realize their business prerequisites. Hence, an enterprise consuming cloud service might essentially need to interoperate with different cloud service providers and vendors to exploit the best service offerings. The broker is the user's single point of contact for multiple cloud services and permits the user to leverage specialized expertise to provide sophisticated services. The demand of the brokers for provisioning the optimal cloud offerings in the enterprise's business and technical requirements is rapidly rising as many organizations turn to cloud services.

Queuing model in cloud computing differs from traditional queuing systems in various imperative perspectives. The number of services in the cloud is typically of the order of thousands; traditional queuing models rarely consider systems of this size. Due to the dynamic nature of cloud and the diversity of the user's requests, brokers must give expected QoS at broadly differing burdens. The distribution of service times is obscure and does not follow any of the "well-behaved" probability distributions; finally, the traffic intensity can vary to a great degree of range. The service request to a broker may go exponential in certain situations like event of natural calamities, elections, and cyber terrorism. At the point when a disaster like the Tsunami or earthquake happens, telephone lines in debacle regions are completely ruined or overloaded with calls. In this scenario, the demand for cloud-based social networking services are very high as it provides a platform for organizations and individuals to interact seamlessly with each other to render images, access information, impart and realize massive collaboration in real-time from all types of computing devices including mobile handheld gadgets. Similarly, the request for backup and recovery services is unusually high during mass virus attacks. Queuing model is necessary to handle the massive service request in these cases.

4 Queuing Model Notation and Definitions

The processing of user request to discover cloud services have the following features.

- The inter-arrival times between any two consecutive user requests are independent of each other and the arrival of requests follows a Poisson process with an arrival rate λ.
- If the broker is able to find suitable services that meet the user requirements, the discovered services are suggested to the user. The broker may send an exception if it takes too much time to find a suitable service and can ask the user to rephrase their query or to lessen the constraints.
- The system can have single or more than one broker with differing capability. Jockeying (move request from one broker to another) occurs whenever a broker gets overloaded. User can withdraw the request if the waiting time is too long, a phenomenon called reneging.
- When multiple requests are given, the broker may process the user request at first come first out, random order, or based on the user priority. First come first out is the most commonly used pattern.
- Since there exist multiple offerings in the same category of service, the time taken to find the relevant services depends on the service category and the user requirements. Assume that all the service times have identical probability distribution and are independent of inter-arrival times.

Thus the queuing theory can be used to model the processing of user requests, as it captures the features of a typical queuing model. As the population size of cloud services is reasonably high while the probability that a given user will request service is relatively small, arrival process can be modeled as a Markovian process [9]. Exponentially distributed random variables are notated by M, meaning Markovian or memoryless. Queuing model can be described using Kendall Notation (A/B/C/D/E) where

A: Distribution function of arrival rate of user request.
B: Distribution function of service rate of broker.
C: No of brokers. System may contain single or multiple.
D: Maximum number of user requests.
E: Service discipline

4.1 *(M/M/1/∞/FIFO) Queuing Model*

In a cloud environment, the service required by the cloud user and the service times of service provider are random variables. Service request demands arrive from an infinite source with mean arrival rate λ of the Poisson distribution. Service times are

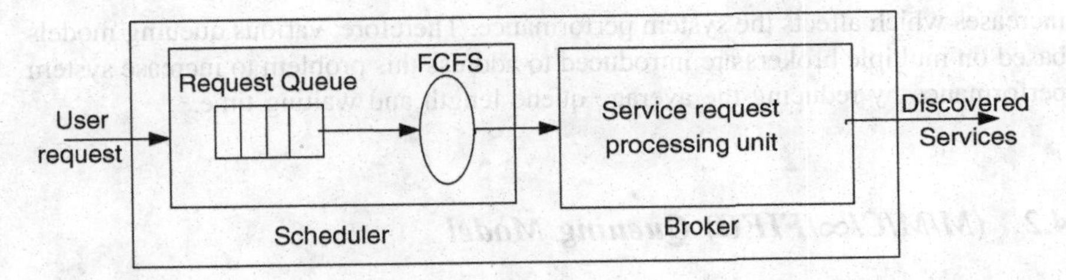

Fig. 1 M/M/1 model for service discovery

assumed to be independently and identically distributed and have an exponential distribution with parameter μ. The user request is stored in a queue in the scheduler. The service discipline is first-come-first-served (FCFS). The scheduler monitors the status of the broker and the request from the queue is given to the service request processing unit of the broker when the broker becomes idle.

The queuing model, as shown in Fig. 1, has the following aspects.

- User request arrivals occur at rate λ according to a Poisson process.
- Service time depends on the constraints in the user request and varies from one request to another, but are independent of one another. Service times have an exponential distribution with parameter $1/\mu$ in the M/M/1 queue, where μ is the average service rate.
- A single broker serves user request one at a time from the front of the queue, according to a first-come-first-served discipline. When the service is complete, the user request leaves the queue and the number of requests in the system reduces by one.
- The capacity of the system is assumed to be infinity and hence there is no restriction on the number of user requests given.
- The average service rate is greater than the average arrival rate. The model is considered stable only if $\lambda < \mu$.

Response time is a measure of the amount of time the broker consumes while processing a user request. The average time spent waiting can be computed using Little's law as

$$W_q = \frac{1}{\mu - \lambda} - \frac{1}{\mu} = \frac{\lambda}{\mu(\mu - \lambda)} \tag{1}$$

The utilization factor for the system is

$$\rho = \frac{\lambda}{\mu} \tag{2}$$

In these traditional approaches of cloud computing, only a single server, called the broker, serves the entire end users and so the overload on that single broker increases which affects the system performance. Therefore, various queuing models based on multiple brokers are introduced to address this problem to increase system performance by reducing the average queue length and waiting time.

4.2 (M/M/C/∞/FIFO) Queuing Model

With a single broker, queue length and hence the response time taken for the user request increases if mean user request rate λ is greater than mean service rate μ. Hence multiple brokers are employed to get (M/M/C/∞/FIFO) model as depicted in Fig. 2. Each broker has an independent identical exponential service time distribution.

The steady state probability is given as

$$P_n = \begin{cases} \frac{1}{n!}\left(\frac{\lambda}{\mu}\right)^n P_0, & 1 \leq n \leq c \\ \frac{1}{c^{n-c}c!}\left(\frac{\lambda}{\mu}\right)^n P_0, & c \leq n \end{cases} \quad (3)$$

where

$$P_0 = \left[\sum_{n=0}^{c-1}\left(\frac{\left(\frac{\lambda}{\mu}\right)^n}{n!} + \frac{\left(\frac{\lambda}{\mu}\right)^c}{c!}\frac{1}{\left(1-\frac{\lambda}{\mu c}\right)}\right)\right]^{-1} \quad (4)$$

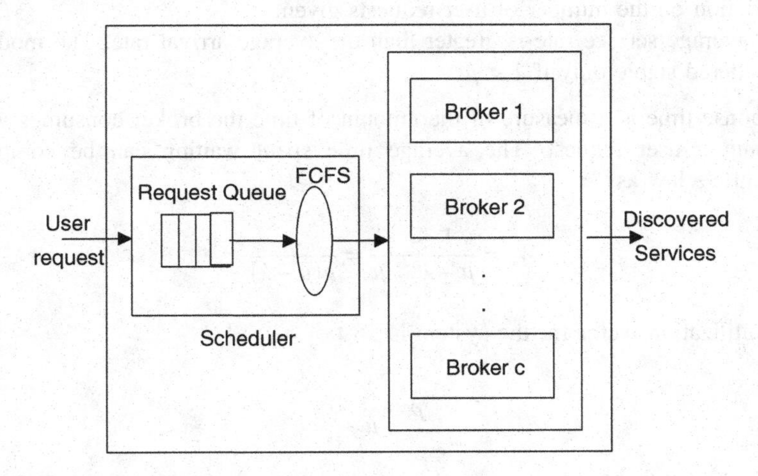

Fig. 2 M/M/c model

The expected number of user requests waiting in the queue is,

$$L_q = \frac{P_0\left(\frac{\lambda}{\mu}\right)^{c+1}}{c!\,c\left(1-\frac{\lambda}{\mu c}\right)^2} = \frac{P_0\left(\frac{\lambda}{\mu}\right)^{c}}{c!}\frac{\rho}{(1-\rho)^2} \tag{5}$$

where utilization factor is

$$\rho = \frac{\lambda}{\mu c} \tag{6}$$

The average waiting time of an arrival is

$$W_q = \frac{L_q}{\lambda} \tag{7}$$

P(arrival has to wait for service is

$$P(n \geq c) = \frac{P_0\left(\frac{\lambda}{\mu}\right)^{c}}{c!}\frac{1}{\left(1-\frac{\lambda}{\mu c}\right)} \tag{8}$$

Thus, compared to the conventional approach of having only single broker, system performance enhances efficiently by reducing the mean waiting time.

4.3 *M/M/n Non-preemptive Queuing Model*

Users who make use of brokers to locate cloud services may be a normal regular user or a premium registered user. A regular user exploits free brokerage services and hence the discovered services may not cover assorted services/providers. Registered users pay for the brokerage and hence it is worth emphasizing that premium users have higher priority than the normal user. However, the service of a normal user is not interrupted by the premium user. This scenario can be modeled using M/M/n non-preemptive queuing model as illustrated in Fig. 3. There are 'n' brokers in the system offering services to regular and premium users. The inter-arrival times of regular, premium user request, and their service times are assumed to be independent of each other. The scheduler maintains separate queues for both the category of users, and services within each queue follow FCFS. Premium and regular user requests arrive according to a Poisson process with rate $\lambda 1$ and $\lambda 2$ respectively. Processing times for all are exponentially distributed with parameter μ.

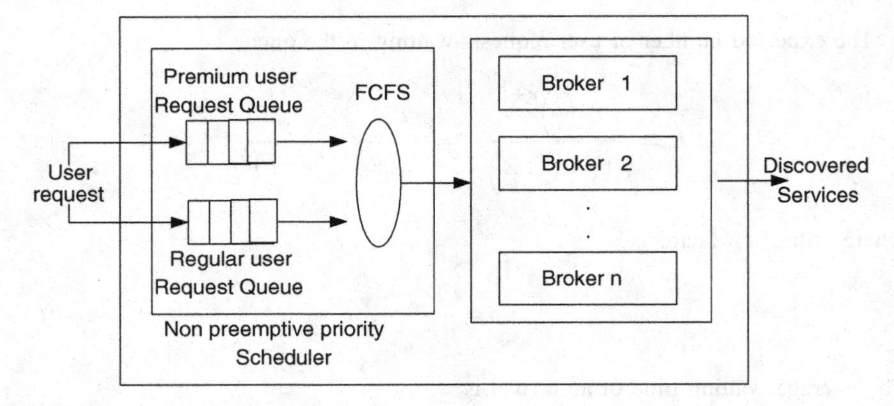

Fig. 3 M/M/n non-preemptive queuing model

Utilization factor due to premium user request

$$\rho_1 = \frac{\lambda_1}{\mu c} \tag{9}$$

and that due to regular user request

$$\rho_2 = \frac{\lambda_2}{\mu c} \tag{10}$$

Here

$$\rho = \rho 1 + \rho 2 < 1. \tag{11}$$

The average waiting times [6] of premium user request and regular user request are given as

$$E[W_1] = \frac{\pi}{(\mu n - \lambda_1)} = \frac{\pi}{\mu n(1-\rho_1)} \tag{12}$$

$$E[W_2] = \frac{\pi}{\mu n(1-\rho_1)(1-\rho)} \tag{13}$$

where π is the probability of waiting in the M/M/c with no priorities

$$\pi = \frac{\left(\frac{\lambda}{\mu}\right)^n}{n\,!(1-\rho)} \left[\sum_{i=0}^{n-1} \frac{\left(\frac{\lambda}{\mu}\right)^i}{i\,!} + \frac{\left(\frac{\lambda}{\mu}\right)^n}{n\,!(1-\rho)} \right]^{-1} \tag{14}$$

and

$$\lambda = \lambda_1 + \lambda_2 \tag{15}$$

5 Discussion

The primary concerns of the user in service discovery are the response time and the accuracy of the discovery results. In general, it may take quite a while to discover all the services required by the user. From the chi-square test applied to conduct the hypothesis test for the service time, it is evident that the time of each service discovery follows an exponential distribution. Service cost and the waiting time cost are considered to find the optimal number of brokers needed in the system. Total cost anticipated is the sum of anticipated service cost and anticipated waiting cost.

$$\text{Total\_cost}_{\text{anticipated}} = N \times C_b + L_s \times C_w \tag{16}$$

where N is the number of brokers, C_b is the service cost of each broker, L_s is the expected number of service request in the system, and C_w is the waiting cost of the user.

In the case of M/M/n non-preemptive, i.e., multi queue-multi broker model,

$$\text{Total\_cost}_{\text{anticipated}} = N \times C_b + N \times L_s \times C_w \tag{17}$$

The values used are randomly picked so as to conform with the pattern in which request for services enters a given system. Assume that C_b and C_s are 1.00$ each. It is observed from Table 1 that as the number of brokers increases, waiting time and the total cost incurred get reduced. Waiting time is almost nil with more than three brokers. However, as the number of broker further increases, there will be no more decrease in the anticipated total cost; instead a slight increase in the cost occurs and this is due to the base power consumption of the brokers.

Table 1 Mean waiting time in the queue and cost with varying number of brokers

λ	μ	M/M/1		M/M/2		M/M/3	
		W_q	C($)	W_q	C($)	W_q	C($)
40	43	0.3101	14.0	0.0064	3.19	0.0009	3.96
150	154	0.2435	38.5	0.0020	2.38	0.0003	4.01
200	205	0.1951	41.0	0.0015	3.28	0.0002	4.02
250	252	0.4960	126	0.0013	3.32	0.0002	4.04

6 Conclusions

With the increase in the number and category of services, the aid of brokers is essential to find most appropriate cloud services. Service request queue is formed if a new request is given to the broker before it completes the previous one. The waiting time for a service request before it is being processed is greater in a system with single broker. Consequently, it is evident that the throughput is higher in a system with multiple brokers. Hence, to have an efficient system in handling service discovery requests in cloud computing it is necessary to have a multi-broker system. Thus queuing hypothesis model can be seamlessly consolidated into system design to enhance system performance.

References

1. Jordi, V., Solsona, F., Teixidó, I., Mateo, J., Abella, F., Rius, J.: A queuing theory model for cloud computing. J. Supercomput. **69**(1), 492–507 (2014)
2. Khazaei, H., Misic, J., Misic, V.B.: Performance analysis of cloud computing centers using M/G/m/m+r queuing systems. IEEE Trans. Parallel Distrib. Syst. **23**(5), 936–943 (2012)
3. Cao, J., Hwang, K., Li, K., Zomaya, A.Y.: Optimal multi server configuration for profit maximization in cloud computing. IEEE Trans. Parallel Distrib. Syst. **24**(6), 1087–1096 (2013)
4. Wagner, D.: Waiting times of a finite-capacity multi-server model with non-preemptive priorities. Eur. J. Oper. Res. **102**(1), 227–241 (1997)
5. Chakravarthy, S.R., Marcel, F.N.: Analysis of a multi-server queuing model with MAP arrivals of regular customers and phase type arrivals of special customers. Simul. Model. Pract. Theory **43**, 79–95 (2014)
6. Kella, O., Yechiali, U.: Waiting times in the non-preemptive priority M/M/c queue. Stoch. Models **1**(2), 257–262 (1985)
7. Jin, B., Weng, H., Wen, Y., Zhang, F.: A service discovery system analyzed with a queueing theory model. In: 33rd Annual IEEE International Computer Software and Applications Conference, COMPSAC'09, pp. 190–198 (2009)
8. Jackson, J.R.: Networks of waiting lines. Oper. Res. **5**(4), 518–521 (1957)
9. Grimmett, G., Stirzaker, D.: Probability and random processes, 3rd edn. Oxford University Press, Oxford (2001)

Adaptive Multilayer Routing for Incremental Design of an SoC

Debasri Saha

Abstract Intellectual properties (IPs) in design form are integrated for incremental design of system-on-chip (SoC). In order to realize proper functionality of an SoC, some global signal nets are often needed to be connected between the design IPs. If the component IPs are in-house, the global nets may be routed through the routing region of the design IPs and this routing may be formulated as an adaptive incremental routing problem where all the routing issues need to be tackled directly in the detailed routing phase only. In this work, we propose a technique for adaptive incremental routing of global nets with the objectives of minimal increase in wire length, congestion, and number of vias. The first objective maintains the desired frequency, the second and the third ones reduce the power overhead, and specifically the third one is effective in keeping the layout manufacture-friendly. The proposed technique is applied on some ISCAS'85 benchmarks and finally on a crypto SoC design which integrates several component designs for crypto-cores. The results on CPU time for routing and the overhead of routing are encouraging.

Keywords Multilayer routing · SoC design · RSMT · Interval cover

1 Introduction

A large number of applications nowadays are performed in hardware because the applications in hardware are faster and the application-specific chips can easily be embedded in a larger system. Circuit or design components for smaller applications are reused and integrated to realize a large application on a single chip (SoC) and these reusable components are known as intellectual property (IP). Some control and clock signal nets may be needed to be connected between the IPs to enforce the sequence and timing of the tasks performed by the IPs. Often, global nets may be inserted for user authentication purpose.

D. Saha (✉)
A.K. Choudhury School of IT, University of Calcutta, Kolkata, India
e-mail: sahadebasri@gmail.com

S. Das et al. (eds.), *Proceedings of the 4th International Conference on Frontiers in Intelligent Computing: Theory and Applications (FICTA) 2015*, Advances in Intelligent Systems and Computing 404, DOI 10.1007/978-81-322-2695-6_56

Integration of IPs may be done in two ways. Hardware IPs may directly be integrated in plug-and-play SoCs, where integration is easy but not optimized. The global nets are routed taking the component IPs as obstrucle. In large SoCs, often interconnects follow the network-on-chip structure. In the second type, merging of IPs is performed in design level and in general, is applied for moderate-sized chips where some or all IPs are in-house and hence transparent. Here, the global nets of SoC design are routed through the routing regions of the IPs in an incremental design stage so that the resultant design is optimized in terms of area, performance, etc.

In general multilayer routing problem, approximate routes for the nets are found by identifying the routing layers and the regions in global routing phase, where routing region is represented as a grid graph of moderate size and further the multilayer grid graph is compressed to a single-layer grid. Next, in detailed routing, each routing region is handled separately to find the exact routes of the nets. The target routing problem is different from the general one in the following aspects:

D1. It is a kind of adapted routing as the new nets to be routed are not known beforehand and are generated according to the requirements in SoC. Here, the routes for the global nets are searched through the detailed routed region of the IPs, and therefore all their routing issues are to be directly handled in detailed routing, where the corresponding grid graph is very large in size.

D2. As the routing is of adaptive type, the advantage of net ordering is not available. A new net may be a critical net, i.e., would be of higher priority in net ordering, but it is to be routed in later stage.

D3. The new net connections are to be routed through routing region nearly congested with large number of routed nets. Therefore, these need considerable detour in intra-layer or inter-layer or in both.

In this paper, the adaptive routing problem for incremental, i.e., IP-based SoC design, is targeted. The proposed technique finds routes for the global nets with minimal wire length and vias, and through minimally congested region. The objective of minimal wire length helps to keep the operating frequency unchanged, and that of minimal congestion minimizes crosstalk and possibility of hotspot. Minimal vias ensures the layout to be fabrication-friendly.

The rest of the paper is organized as follows. Section 2 focuses on prior approaches. The proposed technique for incremental routing in SoCs is presented in Sect. 3. Analysis of the proposed method appears in Sect. 4. Section 5 presents the simulation results, and Sect. 6 presents the concluding remarks.

2 Previous Work

In the context of VLSI routing problem, several algorithms are proposed for generating rectilinear Steiner tree for routing nets in a two-layer reserved Manhattan model. Kahng et. al. in [1] proposed an widely used technique for generating a rectilinear Steiner minimal tree (RSMT) from a minimum Steiner tree (MST). Sarafjadeh et. al.

in [2] proposed an approach to generate more flexible RSMTs from one RSMT to facilitate routability of all nets. The work in [3] proposed a lookup table-based efficient method for generating RSMT. The authors in [4] developed an algorithm for generating all the rectilinear Steiner trees for a given set of terminals or pins.

Among the multilayer routing algorithms, authors in [5] proposed a convex hull-based layer assignment of nets for global routing where each net is entirely routed in a particular layer. Layer assignment based on this rigid constraint may not be close to optimum. In [6], the same group of authors proposed an routing algorithm for 3D ICs consisting of number of device layers. This algorithm is obstacle aware and can handle routing through through-silicon-vias (TSVs).

In multilayer global routing, where the routing region is represented as a grid graph, algorithms are classified into two broad categories. One set of work, e.g., [7], routes all nets directly on the multilayer grid graph representing global routing region. Another set of works, e.g., [8], compresses a multilayer grid graph into a one-layer grid graph, then use a one-layer router to solve the one-layer global routing problem, and finally perform layer assignment to assign each wire in the multilayer grid graph. The work in [8] proposes a congestion-constrained layer assignment technique which comprises net ordering followed by dynamic programming-based single-net layer assignment considering accurate and predictable examination for congestion constraints (APEC). In case of adaptive routing of global nets as additional nets in later stage, the routing technique in [9] considers the IPs as black box obstacles and performs obstacle aware routing for global nets.

3 Proposed Technique

3.1 An Overview of Proposed Routing for the Above Problem

In order to address the adaptive incremental routing problem described in the Introduction with the objectives of minimal increase in wire length, congestion, and vias, the proposed technique *ARoute_SoC* for incremental routing adopts some specific *route searching strategies* through its steps. The major steps of the proposed routing technique are described below.

1. For each new (multiterminal) global net, a rectilinear Steiner minimal tree (RSMT) is generated [Fig. 1a]. The branches (including trunk) of the Steiner tree are identified and the coordinates of Steiner points are obtained.

For each branch, the steps 2 and 3 are performed to generate route for it.

2. a. A bounding cuboid [defined later] enclosing each branch is determined. The reserved multilayer detailed routing region corresponding to the bounding cuboid of a branch is represented as a grid graph [Fig. 1b].
   b. The congestion index of the bounding cuboid is computed. If it exceeds some threshold, an alternative branch is generated and step 2.a is repeated.

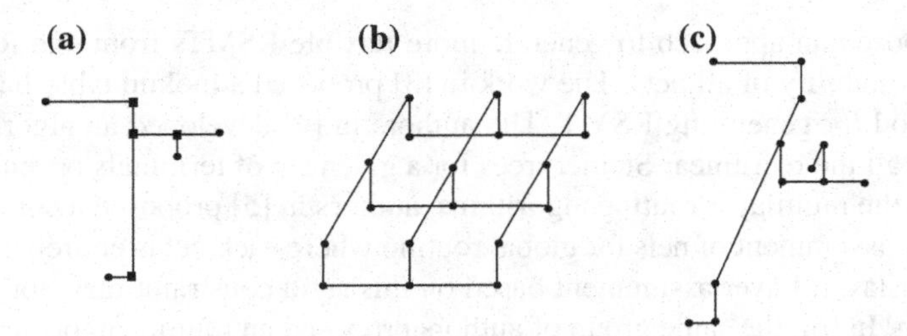

Fig. 1 **a** RSMT for a net, **b** grid structure reprending multiple layers, **c** cuboidal Steiner minimal tree corresponding to the RSMT in (**a**)

3. The route for the branch is searched in the corresponding grid graph. The intervals (with grid points not preoccupied) in tracks parallel to the branch and within the bounding cuboid are stitched with minimal overhead in wire length and vias to generate the route for the branch.

Use of Steiner tree generated in step 1 performs guided search to reduce time for searching route through a congested region. In order to complete the routing of all new nets in spite of large number of occupied grid points, route is separately searched for each branch and finally these routes are stitched. Considering the grid graph only for the bounding cuboid of each branch in step 2.a reduces the searching time. A route through a region with congestion index not exceeding the threshold is searched in step 2.b, and thus a route in low congestion region is preferred. Interval stitching with miminal overhead in wire length and vias in step 3 takes care of the objectives of minimizing wire length and vias. While an alternative branch is generated to avoid an over-congested region, partial flipping is not permitted to avoid additional bends, i.e., vias.

3.2 Generation of Rectilinear Steiner Tree

A rectilinear Steiner minimal tree (RSMT), which connects the terminals and some additional grid points (Steiner points) with edges of minimal length and parallel to x- or y-direction, is generated for each net [Fig. 1a]. A RSMT can be generated either using the look-up table-based method given in [8].

3.3 Defining Bounding Cuboid of a Branch in Low Congestion Region

Bounding cuboid of a branch:

A cuboid enclosing a branch is considered, where its length is equal to length of the branch, its width is δ with the branch in the middle, and its height is equal to some or all metal layers. The routing region corresponding to the bounding cuboid of a branch is extracted and represented as grid graph.

Representing reserved multilayer detailed routing region within a bounding cuboid as a grid graph:

In reserved multilayer model, alternate layers are reserved for horizontal and vertical net segments. The detailed routing region enclosed in the bounding cuboid of a branch is represented by a 3-D grid graph, where grid separation equals to track separation. In odd layers, grid points are connected by horizontal gridlines (x-direction) and in even layers, grid points are connected by vertical gridlines (y-direction). Each grid point is connected to its next layer corresponding grid point a via (z-direction). The size of the grid graph within a bounding cuboid is quite small [Fig. 1b].

The grid graph is not uniform as the separation of tracks in the upper layer is more. That is, the number of tracks in the upper layers is less than that in the lower layer. So, in upper layers, some grid points have no vertical edge.

If a junction is preoccupied in the routed design by a local net, the corresponding grid point in the grid graph is specified as *OFF* grid point. Otherwise, the grid point is specified as *ON* grid point.

Defining congestion index of a bounding cuboid:

The ratio of the number of *OFF* (preoccupied) grid points to the total number of grid points within a bounding cuboid is termed as congestion index of that bounding cuboid.

Determining alternative branch to avoid high congestion region:

If congestion index of a bounding cuboid exceeds a threshold, an alternative to the branch is generated in the following ways [2]:

(i) If the Steiner tree has a L-shaped branch, it is reflected.
(ii) If a Steiner edge of the branch has Steiner points at both its end, the edge may be moved parallel to it.
(iii) If intra-layer alternative to the branch is not available, the branch is moved to upper layer.

3.4 Incremental Routing: Searching Route for Each Branch

A branch having terminals at both end points is defined as rigid branch, a branch having terminal at one end point is moderately rigid, and a branch with Steiner points at both end points is a flexible one. The spans with all *ON* grid points on the tracks parallel to a branch and within its bounding cuboid are taken as (available) intervals [Fig. 2a]. The cost of stitching a pair of intervals with overlapped spans is computed as the distance between the intervals in terms of intra-layer detour (the number of grids) and the number of vias. In order to determine the intervals which can be

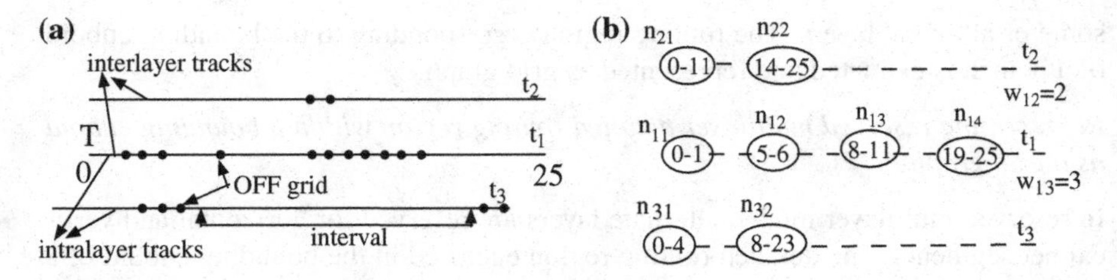

Fig. 2 **a** Intervals and OFF nodes on parallel tracks, **b** corresponding interval graph

stitched with minimal cost to cover the span of the branch, an interval graph is constituted.

Defining interval graph:

Each interval forms a node in an interval graph and an edge between each pair of overlapped intervals is drawn with edge weight equals to cost of stitching those two intervals [Fig. 2b]. Two intra-layer overlapped intervals at track distance x may be stitched using a 'C'-shaped connector with detour x and minimum via count 2. Two overlapped inter-layer intervals one above another at distance z may be stitched using a 'I'-shaped connector with via count z. Any other two intervals at track distance x and layer distance z may be stitched using a 'Z'-shaped connector with detour x and via count z.

Finding interval cover with minimal overhead of stitching:

For rigid branch, the two intervals containing the two terminals are taken as start and end nodes in the graph and the interval with the terminal of moderately rigid branch is considered as start node. The problem of interval covering is solved using the graph [10] and then the farthest interval is repeatedly replaced by a number of intervals to reduce the path cost, i.e., the cost of stitching the intervals. Thus, the path with minimum cost in the graph, i.e., the intervals with minimum cost of stitching and covering the span of the branch, is determined.

Among multiple intervals at same distance, for rigid branch, the intervals above the branch are considered with higher priorities, and for moderately rigid branch with more branches on one side, the intervals on that side are considered with higher priorities. No such priority is given for moderately rigid branches with equal branches on both sides or for flexible branch.

The route for each branch is obtained in terms of the intervals and the stitching connectors. Trunk route construction is done in the similar way. Branch ends which may be in different layers are stitched together to generate the trunk route.

Defining cost of cuboidal Steiner tree obtained:

The route generated for a net is a 3D generalization of rectilinear Steiner tree, may be termed as cuboidal Steiner tree, where different Steiner branches or portions of Steiner branches are in different layers. Therefore, the Steiner tree edges may be horizontal (parallel to x-direction) in even layers, or vertical (parallel to y-direction) in odd layers, or along a via (parallel to z-direction) between two consecutive layers [Fig. 1c]. The cost associated with a cuboidal Steiner tree is expressed as the number of excess grid sides and vias.

4 Time Complexity of the Proposed Technique

In step 1, generation of RMST takes $O(tlogt)$ time, where t is the number of terminals of a new net. For each branch of $O(t)$ branches of RSMT, step 2 and 3 are performed. In step 2, checking of congestion and subsequent reformation each takes time in the order of size of the bounding cuboid. Taking width δ of the cuboid and the number of metal layers constant, this sub-step takes time $O(l)$, where l is the average length of the branch. In step 3, in order to find the intervals whose stitching with minimal overhead covers the length l, we first find the minimum number of intervals whose stitching covers the length l (solution of interval covering problem) in $O(p^2)$ time, where p is the number of intervals in the bounding cuboid of the branch. By solving the interval covering problem $log\delta$ times, solution of interval stitching with minimal overhead can be solved in $O(p^2 log\delta)$, i.e., in $O(p^2)$ time. In worst case, p is $O(l)$ and routing a new net takes time $O(l^2.t)$.

5 Simulation Results

From the verilog files for ISCAS'85 benchmarks, layout is generated using Mentor Graphics Pyxis tool and from the .DEF file of the layout, the description of the routing region is taken in form of the coordinates and the status (whether *ON* or *OFF*) of the grid points. For each benchmark in Table 1, sets of terminal coordinates are randomly generated where each set represents terminals of a net. The terminal coordinates are on the boundary of the circuit layout, which specifies that the nets connect the pins of several neighboring chips. Fast lookup table-based method FLUTE [3] is aplied for generating RSMT structure connecting the terminals of each net.

Table 1 CPU time, wire length and via overhead, congestion index for incremental routing using *ARoute_SoC*

Benchmark circuits	Size of routing region	# nets and # new nets	CPU time (sec) for incr. routing	Values of g and v	Congestion index
c1355	20250x20250x5	587, 8	0.099	1.768, 0.0790	3.623
c3540	32670x32670x5	1719, 15	0.308	2.674, 0.1122	6.577
c5315	38430x38430x5	2485, 21	0.509	2.878, 0.1290	8.083
c7552	45090x45090x5	3720, 28	0.807	3.288, 0.1494	10.312

Table 2 Description of the component designs of a crypto SoC

Component circuits	Size of routing region	# nets of component circuits
DES	60615x9620x6	1400
AES	62640x62640x6	7481
Decoder	610x229x6	9
Sequence recog.	690x641x6	20

Table 3 Results for an incremental design of the crypto SoC from its components

Circuit	# new nets	CPU time (sec) for incr. routing	Values of g and v	Congestion index
Crypto SoC	16	0.861	3.895, 0.0923	16.623

The routing subregions with congestion index less than 20 % are chosen for routing the branches of each RSMT. *ARoute_SoC* is implemented in C on 3GHz Linux machine to obtain the routing results. The values of g are computed as (wire length-HPWL)*100/HPWL and those of v are computed as (#vias*100/HPWL). Congestion indices are given in percentage.

Table 2 contains the description of component designs of a crypto SoC—two crypto-cores DES and AES designs, a decoder as a selection component and a sequence recognizer for user authentication.

Table 3 shows the results of applying *ARoute_SoC* for incremental routing of global nets on a crypto SoC designed from its components. CPU time for incremental routing is negligible. The average values of g and those of v are quite low. The % values for the congestion index after routing remain very low.

6 Conclusion

The performance and the solution quality of *ARoute_SoC* show that this algorithm works efficiently and effectively for incremental adaptive routing of a medium-sized SoC designed from its in-house components.

References

1. Kahng, A.B., Robins, G.: A new class of Steiner tree heuristics with good performance: the iterated 1-Steiner approach. In: Proceedings of the International Conference on Computer-Aided Design (1990)
2. Bozorgzadeh, E., Kastner, R., Sarrafzadeh, M.: Creating and exploiting flexibility in rectilinear Steiner trees. IEEE Trans. Comput. Aided Des. Integr. Circuits Syst. **22**(5), 605–615 (2003)

3. Chu, C., Wong, Y.-C.: FLUTE: fast lookup table based rectilinear Steiner minimal tree algorithm for VLSI design. IEEE Trans. Comput. Aided Des. Integr. Circuits Syst. **27**(1), 70–83 (2008)
4. Dourado, M.C., De Oliveira, R.A., Protti, F.: Generating all the Steiner trees and computing Steiner intervals for a fixed number of terminals. Electron. Notes Discret. Math. **35**, 323–328 (2009)
5. Samanta, T., Ghosal, P., Rahaman, H., Dasgupta, P.S.: A method for the multi-net multi-pin routing problem with layer assignment. In: Proceedings of the International Conference on VLSI Design, pp. 387–392 (2009)
6. Ghosal, P., Rahaman, H., Das, S., Das, A., Dasgupta, P.S.: Obstacle aware routing in 3D integrated circuits. In: Proceedings of the Advanced Computing, Networking and Security, LNCS, vol. 7135, pp. 451–460 (2012)
7. Roy, J.A., Markov, I.L.: High performance routing at the nanometer scale. In: Proceedings of the International Conference on Computer-Aided Design, pp. 496–502 (2007)
8. Lee, T.-H., Wang, T.-C.: Congestion-constrained layer assignment for via minimization in global routing. IEEE Trans. Comput. Aided Des. Integr. Circuits Syst. **27**(9), 1643–1656 (2008)
9. Xu, J., Hong, X., Jing, T., Yang, Y.: Obstacle-avoiding rectilinear minimum-delay Steiner tree construction toward IP-block-based SOC design. IEEE Trans. Circuits Syst. II: Express Briefs **53**(4), 309–313 (2006)
10. Edwards, K., Griffiths, S., Kennedy, W.S.: Partial Interval Set Cover—Trade-Offs Between Scalability and Optimality, Approximation, Randomization and Combinatorial Optimization Algorithms and Techniques, LNCS, vol. 8096, pp. 110–125 (2013)

Analyzing and Modeling Spatial Factors for Pre-decided Route Selection Behavior: A Case Study of Fire Emergency Vehicles of Allahabad City

Mainak Bandyopadhyay and Varun Singh

Abstract Developing countries like India seriously lack computing infrastructure for evaluating the performance of emergency response services. Most of the time it is assumed that drivers take the shortest distance route to reach the incident location. In this paper a model is proposed for the pre-decided route selection behavior of Fire Emergency Vehicle drivers. The characteristics of routes selected by Fire Emergency Vehicle drivers of Allahabad City are analyzed spatially to comprehend the influence of proximity area of road segments on pre-decided route selection. A model using shortest path algorithm and spatial factors, i.e., length, width, land use of the proximity, and population density of the area are developed and compared with Shortest Distance assumption.

Keywords Fire emergency vehicle · Emergency response · GPS trajectory · Shortest path algorithm · Land-use type · Population density · GIS

1 Introduction

For real-time traffic monitoring and related information dissemination, cities in developing countries like India lack advanced transportation infrastructure in comparison to developed countries. It is also quite difficult to estimate the emergency response time and conduct any study related to the performance assessment of emergency response. Mostly due to the non-availability of real-time directional information system to drivers, the route followed by the driver of emergency fire vehicle depends on his experience and knowledge of the road network and

M. Bandyopadhyay (✉)
GIS Cell, MNNIT-Allahabad, Allahabad, India
e-mail: Ermainak@gmail.com

V. Singh
Department of Civil Engineering, MNNIT-Allahabad, Allahabad, India
e-mail: Vsingh.ce@gmail.com

S. Das et al. (eds.), *Proceedings of the 4th International Conference on Frontiers in Intelligent Computing: Theory and Applications (FICTA) 2015*, Advances in Intelligent Systems and Computing 404, DOI 10.1007/978-81-322-2695-6_57

vehicular traffic. The driver always intends to take routes that can minimize the response time. Generally, the routes followed by drivers are pre-decided before dispatching of emergency vehicle. In order to study the selection of routes, it is necessary to capture the routes taken by the drivers in real time.

In this study, due to lack of standard transportation infrastructure to collect real-time driving data of Fire Emergency Vehicle (FEV) in Allahabad City, the real-time route selections from fire station to incident location by the drivers are captured using low cost GPS logger HOLUX M1000C. The GPS trajectories collected are analyzed from a spatial point of view to explore the influence of spatial factors in the route selection behavior. In the present work the main objective is to design a computational model to represent the predetermined route selection behavior of FEV driver of the Allahabad Fire Department. A model using shortest path algorithm, Dijkstra's Algorithm, and spatial factors, i.e., length, width, land use of the proximity, and population density of the area is proposed with a matching percentage of more than 80 %. This model in the future will help in evaluating the performance of emergency response by the Allahabad Fire Department.

2 Spatial Context in Route Selection and Dataset

In the present work emphasis is given to the FEV driver's route selection behavior. Given the type of activity carried out and size of vehicle (heavy vehicles) driven by the drivers, delay in reaching the destination specifically due to congestion is one of the key factors. The spatial cognition developed through experience while carrying out long years of service and interaction with other drivers, helps in selecting routes through less endogenously congested areas and minimizing delays [1, 2]. Therefore, a mechanism is needed to incorporate spatial cognitive results into a computational model. For the same it is necessary to construct the required *Spatial Environment* to simplify the incorporation of cognitive results. Here the *Spatial Environment* is constructed around the road network with the following additional spatially related information:-

(a) Characteristics of road segment, i.e., length and width.
(b) Proximity of road segment, i.e., population density of the area and land use type.

The endogenous congestion on a road segment in an urban city depends on factors like width, length of the road segment and population density, land use of the proximity of the road segment [3, 4]. The land use of an area provides an indication of the daily activities occurring in that area [5]. The daily activity of an area certainly affects the driver's intention to route through the area [6]. In Fig. 1 difference can be seen clearly between the path selected by the FEV and the shortest distance path.

As per the data provided by *Allahabad Nagar Nigam* population density of an area is categorized into less than 200, between 200–400, and more than

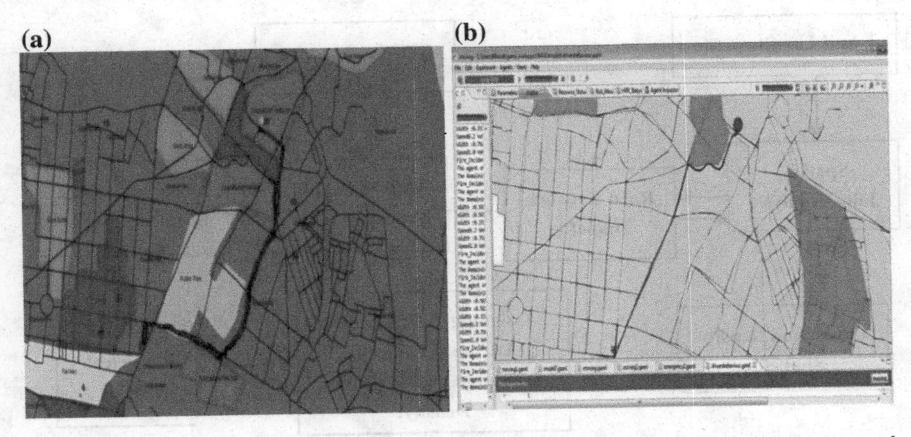

Fig. 1 **a** GPS trajectory of FEV from fire station to incident location. **b** Shortest distance path from fire station to incident location

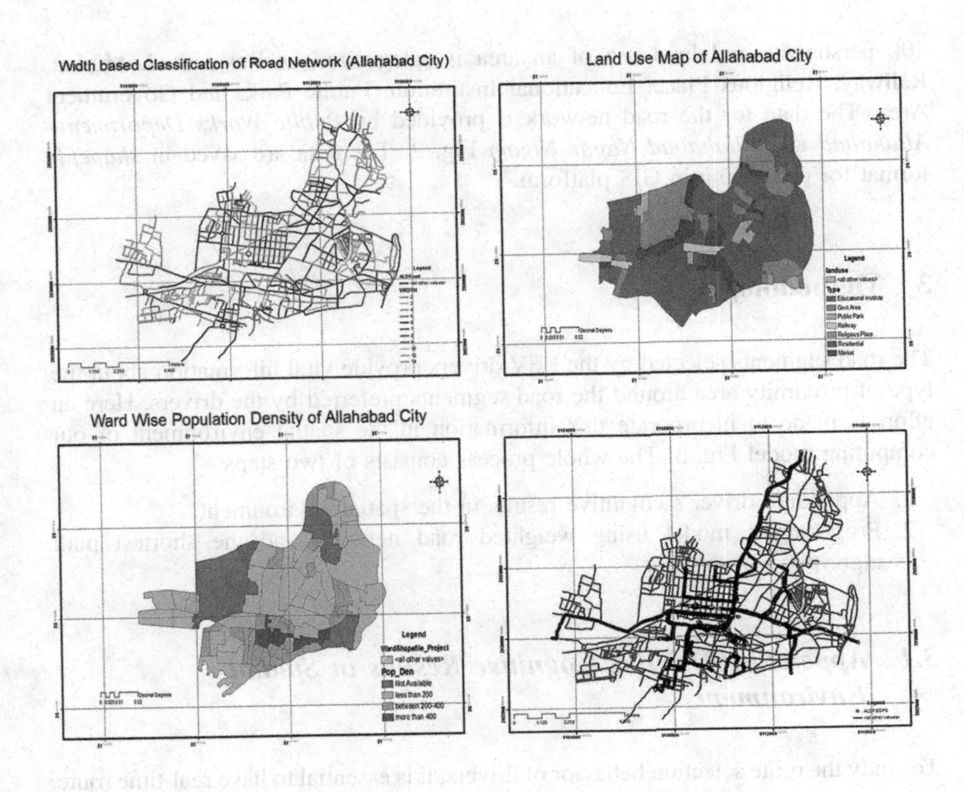

Fig. 2 Components of spatial environment and collected GPS trajectories

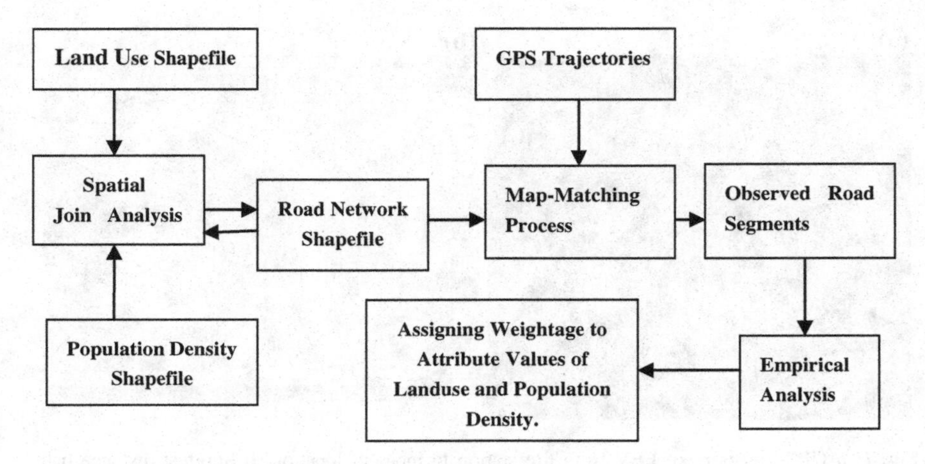

Fig. 3 Incorporation of driver's spatial proximity preference

400 persons/ha and land use of an area is categorized as Residential, Market, Railway, Religious Place, Educational Institution, Public Park, and Government Area. The data for the road network is provided by *Public Works Department–Allahabad* and *Allahabad Nagar Nigam* Fig. 2. The data are saved in *shapefile* format for processing in GIS platform.

3 Methodology

The road segments selected by the FEV drivers provide vital information about the type of proximity area around the road segments preferred by the drivers. Here an effort is made to incorporate this information in the spatial environment of our computing model Fig. 3. The whole process consists of two steps:-

1. Appending driver's cognitive results in the spatial environment.
2. Proposing a model using weighted road network and the shortest path algorithm.

3.1 Appending Driver's Cognitive Results in Spatial Environment

To study the route selection behavior of drivers, it is essential to have real-time route selection data. In the current work for tracking the FEV routes, low cost GPS Logger HOLUX M1000C is used. The GPS logger stores the track Id, latitude, longitude, timestamp, and speed of each GPS point. The GPS trajectories collected have a sampling interval of 1 s. The GPS trajectories are preprocessed and map matching is done using GIS functionalities [7]. Map matching is the process of matching GPS points to the correct location in the digitized road network. The map matching process determines the road segments selected by the driver while routing. The road network Shapefile contains data related to the spatial characteristics of road segments, i.e., length and width. The spatial join operation of road network *Shapefile* with land use Shapefile and population density *Shapefile* results in road network *Shapefile* having spatial proximity details like land use type and population density.

Weightage for different land use and population density values. The values of attribute land use and population density are categorical and interval data respectively. For computational purposes numerical values are assigned to categorical and interval data values based on the real dataset collected. Various attribute values are given weightage in the range of [0–1] with 0 max weightage and 1 min weightage. The road segments selected by the drivers are collectively analyzed to determine two parameters, x_{ijT} and S_{ij}.

$$x_{ijT} = \frac{x_{ij}}{x_T} \times 100 \tag{1}$$

x_{ij} = The total number of segments with value i of attribute j in the collected trajectories.
x_T = Total number of segments in the collected trajectories.
x_{ijT} = Percentage of segments with value i of attribute j in the collected trajectories.

$$r_{ijT} = \frac{r_{ij}}{r_T} \times 100 \tag{2}$$

r_{ij} = The total number of segments with value i of attribute j in the road network.
r_T = Total number of segments in the road network.
r_{ijT} = Percentage of segments with value i of attribute j in the road network.

$$S_{ij} = \frac{x_{ijT}}{r_{ijT}} \tag{3}$$

S_{ij} = provides relevance of value i of attribute j in path selection by drivers.

The value of S_{ij} helps in the identification of least significant attribute values; a threshold value of 0.5 is taken here.

Weightage for width and length of road segment. The road segments with large width and lengths are generally preferred by the drivers. The values of attribute width and length of road segment are numerical values. The values are normalized using feature scaling to bring within [0–1] range.

$$X' = 1 - \frac{X - X_{\min}}{X_{\max} - X_{\min}}. \tag{4}$$

3.2 Generation of Weighted Road Network

The most simplified representation of road network topology is through Graph Theory. The road network is abstracted as a weighted graph $G = (V, E, w)$ where V is set of vertices, E is set of edges, and $w: E \rightarrow \mathbb{R}_0^+$. The weighted edges in the graph represent the spatial weights of road segments and vertices represent intersections. Here we propose a model utilizing single source shortest path algorithm to mimic the pre-decided path selection behavior of FEV drivers. In the present work Dijkstra's Algorithm is used for shortest path determination in a graph.

Function for representing road segment weights. A weighted summation function calculates the weightage of each road segment and represents the local spatial preference of drivers in the road network graph data structure.

$$Sw_i = \sum_{j=1}^{4} \alpha_j \times w_{ij} \tag{5}$$

$$Sw_i = \alpha_1 w_{i\text{Length}} + \alpha_2 \times w_{i\text{Width}} + \alpha_3 \times w_{i\text{PopulationDensity}} + \alpha_4 \times w_{i\text{LandUse}} \tag{6}$$

w_{ij} = Weight of attribute j of i road segment.

In the present model the value of $\alpha_{1\ldots4}$ is limited within the range of [0, 2] with an interval of 0.5. The weighted road network is used in Dijkstra's Algorithm with Fire Emergency Station as source and Fire Incident locations as destinations. A parameter called *matching percentage* is used for comparing the road segments resulted from the proposed model with the road segments actually selected by the drivers. The value of $\alpha_{1\ldots4}$ is updated until a satisfactory *matching percentage* is achieved.

$$m = \frac{\sum_{i=1}^{T} n_i}{\sum_{i=1}^{T} t_i} \times 100 \tag{7}$$

n_i = Number of road segments matched with actual road segments selected during routing in ith trajectory.
t_i = Total number of road segments selected during routing in ith trajectory.
m = Matching percentage.
T = Number of trajectories.

4 Implementation Details and Results

The implementation is done in two folds, for digitization, map matching and spatial operations ArcGIS 10 is used and for determining the values of $\alpha_{1...4}$ GAMA 1.5.1 a GIS-based agent modeling and simulation platform is used. In the present study a total of 35 GPS trajectory data of FEVs of Allahabad Fire Station while attending Fire turnouts has been collected during March–May 2014 (Fire Season) from 9:00 a.m. to 9:00 p.m. After preprocessing and map matching GPS trajectories, the weightage of various land use and population density values are calculated as shown in Tables 1 and 2. The value of S_{ij} helps in the identification of the least significant attribute values and a weightage of 1 is assigned to attribute values; a threshold value of 0.5 is taken here.

The weightage of other attribute values are assigned based on the value of x_{ijT}. The attribute values having greater values are assigned greater weightage. For population density with *Not Available* value, weightage of 0.5 is assigned.

The proposed model has a matching percentage of more than 80 % with $\alpha_1 = 1.5$, $\alpha_2 = 1$, $\alpha_3 = 0.5$ and $\alpha_4 = 2$. The proposed model is compared with the shortest distance approach; results reported in Fig. 4 show that in the former model 612 segments matched with a matching percentage of more than 80 %, whereas in the later 342 segments matched with a matching percentage of less than 50 % out of total 727 segments. As shown in Fig. 5 the road segments selected by the proposed model

Table 1 Weightage for population density values

Population density	Number of segments	x_{ijT}	r_{ijT}	S_{ij}	Weightage
Not available	41	–	–	–	**0.5**
<200	459	63.14	56	1.13	**0.2**
200–400	202	27.79	24.59	1.13	**0.6**
>400	25	3.44	5.39	0.64	**0.8**

Table 2 Weightage for various land use values

LandUse type	Number of segments	x_{ijT}	r_{ijT}	S_{ij}	Weightage
Residential	491	67.54	65.74	1.03	**0.2**
Market	137	18.84	12.12	1.55	**0.6**
Railway	1	0.14	1.49	0.0923	**1.0**
Religious place	3	0.413	1.03	0.04	**1.0**
Educational institution	20	2.751	5.046	0.544	**0.8**
Public park	18	2.48	2.132	1.16	**0.8**
Government area	55	7.57	12.153	0.622	**0.7**

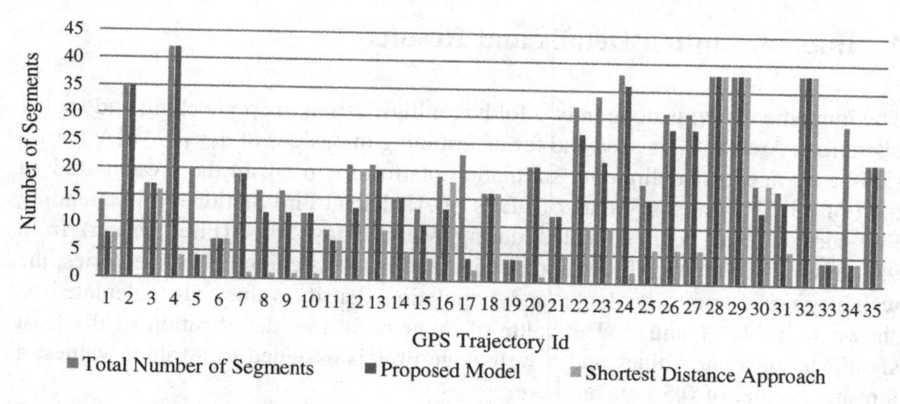

Fig. 4 Comparison of proposed model with shortest distance approach

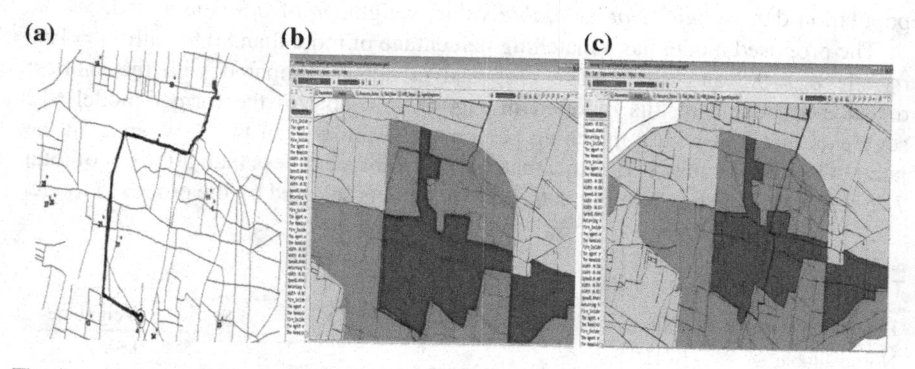

Fig. 5 **a** Collected GPS trajectory of fire emergency vehicle from Allahabad fire station to fire incident with Id 4. **b** Path selected according to proposed model. **c** Path selected according to shortest distance

provide exact resemblance with the collected GPS trajectory than the shortest distance approach. In some of the observations like trajectory 12, 16, and 30 matching of segments is better in shortest distance approach than in the proposed model. In case of trajectories 6, 18, 19, 28, 29, 32, 33, and 35 both the models provide exact matching of road segments; one such result is shown in Fig. 6. Overall the results show that the pre-decided route selection by FEV drivers is not totally based on length but also on factors like width, population density, and land use type.

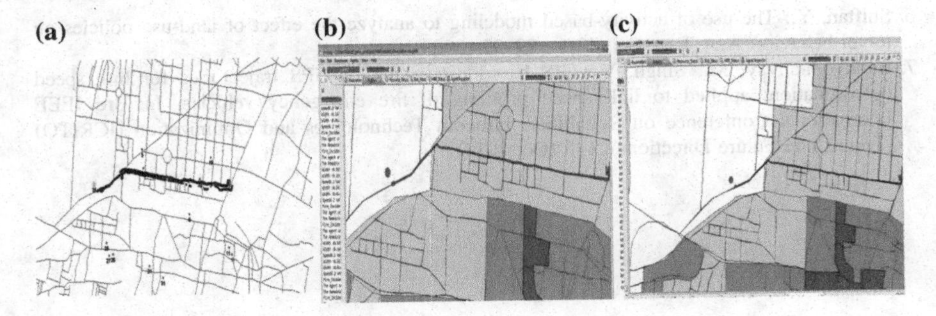

Fig. 6 **a** Collected GPS trajectory of fire emergency vehicle from Allahabad fire station to fire incident with Id 35. **b** Path selected according to proposed model. **c** Path selected according to shortest distance

5 Conclusion

In this paper, a model is proposed using weighted road network and shortest path algorithm (Dijkstra's Algorithm) for representing the pre-decided route selection by FEVs of Allahabad Fire Station. Here the GPS trajectories of FEV have been collected to determine the preference of characteristics of road segment and spatial proximity of the drivers. A weighted road network is developed using the road segment characteristics (i.e., length and width) and spatial proximity characteristics, i.e., land use type and population density. The results show that the proposed model has a matching percentage of more than 80 %. The paper also negates the assumption that Fire Emergency services always consider paths with the shortest distance. The present work will certainly help in evaluating the emergency response capability of FEV of urban cities in India.

References

1. Denis, M., Mores, C., Gras, D., Gyselinck, V., Daniel, M.P.: Is memory for routes enhanced by an environment's richness in visual landmarks? Spat. Cogn. Comput. **14**(4), 284–305 (2014)
2. Manley, E.J., Addison, J.D., Cheng, T.: Shortest path or anchor-based route choice: a large-scale empirical analysis of minicab routing in London. J. Transp. Geogr. **43**, 123–139 (2015)
3. Litman, T.: Evaluating Transportation Land Use Impacts: Considering the Impacts, Benefits and Costs of Different Land Use Development Patterns. Victoria Transport Policy Institute (2011)
4. Bok, K.S., Li, H., Lim, J.T., Yoo, J.S.: Discovering congested routes using vehicle trajectories in road networks. Adv. Multimedia (2014)
5. Sun, L., Yao, L., Wang, S., Qiao, J., Rong, J.: Properties analysis on travel intensity of land use patterns. Math. Prob. Eng. (2014)

6. Shiftan, Y.: The use of activity-based modeling to analyze the effect of land-use policies on travel behavior. Ann. Reg. Sci. **42**(1), 79–97 (2008)
7. Bandyopadhyay, M., Singh, V.: GIS based processing of GPS trajectories for link speed determination: applied to link speed profiling of fire emergency vehicles. In: 3rd IEEE International Conference on Reliability, Infocom Technologies and Optimization (ICRITO) (Trends and Future Directions), 1–5 (Oct 2014)

Fault Tolerant Scheduling with Enhanced Performance for Onboard Computers: Evaluation

Archana Sreekumar and V. Radhamani Pillay

Abstract Dependability as a factor in safety critical systems like satellite launch vehicles avoids catastrophic effects. Onboard computers with traditional dual redundancy mechanism for fault tolerance performs acceptably on fault conditions but underutilization of computational resources are prevalent during fault-free operation. Augmentation of computational resources in such systems with a task allocation paradigm dealt with in earlier studies, helps in achieving better performance and improved safety margins keeping the constraints like size, weight, and power. Testing of a fault tolerant algorithm for a flexible augmented model for hardware fault has been implemented with simulation using TORSCHE real-time scheduling tool and a user interface using Matlab GUI. Experimental hardware implementation on ARM Cortex-M has also been developed. Using performance metrics and verification for faults tolerance, an evaluation of the system performance projects the benefits of implementing such models for weight critical space applications.

Keywords Safety critical system · Onboard computer · Permanent hardware fault · Dependability · Resource augmentation

1 Introduction

Safety critical systems like avionics system, patient monitoring system have to ensure stringent reliability and dependability demands. Onboard computers in satellite launch vehicle perform navigation, guidance and control tasks, and other complicated computations needed for the reliable and correct performance of the

A. Sreekumar (✉) · V. Radhamani Pillay
Amrita Vishwa Vidyapeetham, Coimbatore, India
e-mail: archanasreekumar2007@gmail.com

V. Radhamani Pillay
e-mail: radhamanipillayv@yahoo.co.in

S. Das et al. (eds.), *Proceedings of the 4th International Conference on Frontiers in Intelligent Computing: Theory and Applications (FICTA) 2015*, Advances in Intelligent Systems and Computing 404, DOI 10.1007/978-81-322-2695-6_58

system. In space applications like satellite launch vehicle permanent faults may occur; junction shortage, thermal aging, improper manufacture, etc., are the reasons for such faults. Dependable performance of onboard computers requires incorporation of fault tolerance strategies like hardware redundancy to provide redundant spare capacity to tolerate faults.

Design and development phase and component selection phase cannot overcome the chances of hardware failure completely. The dual redundancy with a hot standby approach is used for fault tolerance in some systems. Resources normally available during a fault-free operation are underutilized, while sufficient dependability is made available during occurrence of fault.

The power, cost, and size constraints directs for an effective utilization of computational resources available in the system. Augmented frame work of onboard computers based on task criticality achieves improved performance of the system by efficient utilization of resources and by addition of extra task into the system. This paper discusses about a user friendly interface developed for testing the hardware fault tolerance on an augmented framework onboard computer system. Different modes of operation for adding flexibility to the augmented model is also put forth in the paper. Validation of the resource augmented system has been done with respect to dual redundant system and a hardware fault injection prototype has been being developed.

2 Literature Survey

Real-time systems need to meet critical timing constraints when operated in a variety of conditions [1]. In hard real-time systems deadlines have to be met, missing of deadlines in a mission critical system like satellite launch vehicle can cause huge financial loss [2]. The reliable working of onboard computers which perform complicated computations in launch vehicle is important [3]. Evolution of onboard computers aims at increasing the performance with a decrease in size, power, and weight [4]. Fault tolerance options for a fly by wire system with design analysis and validation are explained by Hitt [2]. Dependability impairments, means and measures to tolerate faults are discussed by Laprie [5]. The basic concepts, motivation, and techniques for fault tolerance are elucidated in [6].

Static table-driven scheduling ensures that the deadlines are met during the execution of safety critical systems [7]. Method to exploit the spare capacity and improving the system utility by scheduling optional tasks has been explained in [8]. An algorithm for improving the performance of onboard computers using augmentation based on task criticality is introduced in [9]. A fast hardware prototyping of multiprocessor fault tolerant algorithm for cruise control system using resource management is put forward in [10].

Functionalities of navigation, guidance, and control modules in space vehicle along with architectures, algorithms, and strategies are discussed in [11]. An introduction to TORSCHE scheduling tool box for Matlab for offline and online scheduling is given in [12]. For validating the effectiveness of the resource augmentation different performance metrics like average utilization, speed-up, process execution time, and time to recovery are used [13–15].

3 Background Study

Onboard computers control the functionality of satellite launch vehicle during launch and flight phase [3], increased dependence of launch vehicle on onboard computers and other flight critical functions executed necessitate the dependable performance of such a system [2, 4]. An efficient safety critical system should produce correct and timely output in spite of the faults occurred.

System fails when the delivered service deviates from the specified service [5]. Dependability can be achieved using different methods like fault avoidance, fault tolerance, error removal, and error forecasting [5]; fault tolerance is achieved through some kind of redundancy or some other approach where alternative resources are invocated [2]. Hardware redundancy technique for permanent fault tolerance consists of extra components in the system. Hardware redundancy approaches can be divided into static or dynamic redundancies, static redundancy uses detection mechanism to detect the faults and recovery approaches are activated on the detection of faults [6].

Satellite launch vehicle relies on hot standby dual redundant system [3] than triple modular redundancy because the use of triple modular redundancy increases the weight, power, and synchronization difficulties. The thorough testing of the components, quality assurance conducted before the implementation of the system and the use of dual redundancy approach for fault tolerance make onboard computers dependable. Most of the critical systems like flight control of fly-by-wire system are double fail operational [2].

Cross-strapping connection of onboard computer modules helps in detection of failures and also for synchronization [3]. Self-checking softwares which runs periodically in the modules send regular health signal to the succeeding modules. Fault detection can be done using watchdog timers, comparators, timing checks, replication checks, etc. [2]. Allocation of tasks in safety critical systems is done offline, schedule is calculated before the runtime and tasks both critical and non-critical are dispatched in accordance to generated schedule. Static table-driven scheduling is applicable to periodic tasks; this method ensures whether the task meet deadlines a prior [7].

The hot standby dual redundancy approach helps the system to carry forward with the operation when confronted with any permanent fault in any of processing unit. The extra processing unit performs tasks for the successful completion of the flight operations. The system utility and performance can be improved by scheduling optional tasks in the spare capacity obtained by augmentation [8, 9]. The hardware prototyping and evaluation of a proposed resource management model for cruise control system indicates the tremendous improvement compared to traditional system [10].

Incorporation of fault tolerance causes the increase in complexity, overhead, and validation challenges [2]. Validation of the system ensures that the system will achieve desired functionality when confronted with fault permanent faults. The performance metrics like average utilization [13], process execution time [14], slack time [15], speed-up [14], and mean time to recovery [15] can be used to analyze the algorithm performance. Average utilization gives how effectively the resources in a processor are utilized. Slack time in a processor is the time available in the processor where the resources are not utilized by any task. Speed-up gives the relative performance improvement in the system and the ratio of execution time taken by the old algorithm to that of the new algorithm. Execution time indicates the total time a task uses the resources of a processor. Mean time to recovery gives the recovery latency time that the system takes to detect the fault; it is the difference between the time of occurrence and time of detection of faults.

4 Objectives

Validation using user friendly interface and prototype testing of fault tolerance to permanent hardware faults on a resource augmented dual redundant system frame work of onboard computers.

- Algorithm for permanent hardware fault tolerance with task precedence.
- Performance evaluation with simulation on dual and augmented systems.
- Evaluation with user-friendly interface of augmented system using fault injection model—hardware and simulation.

5 Assumptions and System Model

The assumptions considered for development and evaluation of augmented system model for onboard computers to tolerating permanent hardware faults are:

- A minimum time is available for the fault to recover.
- The communication costs are considered to be negligible.
- The noncritical tasks are preemptable.
- Synchronization of the NGC units with the bidirectional and cross-strapping buses.
- Watch dog timer detects fault.
- A global memory holds task set for the system.
- Negligible time overhead for fault injection.
- Minimum time gap between fault occurrences.

5.1 *Resource Augmented Onboard Computer Model*

Onboard computers in satellite launch vehicle have a cross-strapped dual redundant model, where primary chain and secondary chain in the system has navigation, guidance, and control modules with local memories. This cross-strapped model helps in achieving tight synchronization and also tolerating two independent unit failures in the system. The primary chain modules and secondary chain modules works synchronously, navigation modules performs task related to calculation of position, velocity, etc. Guidance unit performs the corrections required to bring the vehicle into predetermined correct orbit while the commands to actuators have been provided from control unit [11].

Figure 1 shows dual redundant connection between the primary and secondary chain modules in onboard computers. The tasks allocated in the primary and secondary chain modules have periodic tasks which include both critical and noncritical tasks. Optional tasks have been included in the augmented framework, optional tasks like iterative refinement, multiple versions, and approximate processing help in utility enhancement which improves the precision, reliability, or confidence level [8]. Critical tasks have been indicated using $\mathfrak{C}$, noncritical tasks by $\mathfrak{N}$, and optional tasks by $\mathfrak{O}$. Navigation, guidance, and control units in primary chain have been represented using N1, G1, and C1, respectively, while corresponding modules in secondary chain have been indicated using N2, G2, and C2. Health checking of the system has been performed during every minor cycle and tasks of navigation and guidance unit is having a hyper period equal to the major cycle.

Here we consider navigation, guidance, and control modules of onboard computers as a single module. A synthesized task set has been framed as well as precedence constraints exist among navigation, guidance, and control unit (Table 1).

ζ_1, ζ_2, ζ_3 indicates the critical tasks and noncritical tasks of the navigation module. ζ_{No1} denotes the optional task to be scheduled in primary chain while ζ_{No2} indicates the optional task of navigation unit to be scheduled in secondary chain. ζ_4, ζ_5 implies the critical tasks of the guidance unit and ζ_6 and ζ_7 denotes the noncritical tasks of the guidance module. ζ_{Go1} and ζ_{Go2} indicates the optional tasks of the guidance module to be scheduled in primary and secondary chains. ζ_8, ζ_9 and ζ_{10} forms the critical and noncritical tasks of control unit.

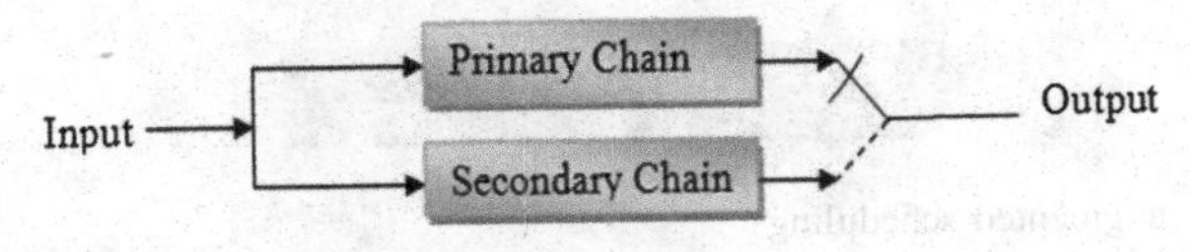

Fig. 1 Dual redundant hot standby connection

Table 1 Task table

Task	Criticality	Execution time (units)	Time period (units)
ζ_1	$\mathfrak{C}$	15	50
ζ_2	$\mathfrak{N}$	98	500
ζ_3	$\mathfrak{N}$	99	500
ζ_4	$\mathfrak{C}$	15	50
ζ_5	$\mathfrak{C}$	30	500
ζ_6	$\mathfrak{N}$	120	500
ζ_7	$\mathfrak{N}$	120	500
ζ_8	$\mathfrak{C}$	5	20
ζ_9	$\mathfrak{N}$	7	20
ζ_{10}	$\mathfrak{N}$	5	20

6 Approach

In a traditional dual redundant system, the tasks have been duplicated in both the processing units. Output has been taken from primary processing unit during fault-free operation while a failure of the primary unit causes the internal logic to switch and output to be taken from redundant processing unit. Augmented system duplicates the entire critical task in both processing units while shares the independent noncritical tasks between the processing units.

Figure 2 indicates task allocation on both primary and secondary processing units in the traditional dual redundant system; Fig. 3 shows the task allocation in primary and secondary chains of resource augmented system. No optional tasks have been considered for control unit as the commands to the actuator have been generated by control unit.

Fig. 2 Task allocation in dual redundant system

Fig. 3 Resource augmented scheduling

6.1 Synchronization and Fault Detection

In the framework of augmented onboard computer critical tasks in system have been executed synchronously and health checking of the system has been performed in every minor cycle. A periodic health signal has been send by secondary unit to all the succeeding units and the redundant unit; similarly a health signal has been send by primary unit to succeeding units and secondary unit. When health signal from any of the unit has not arrived in time the system switches to a different mode of operation, where the extra computational processor available in the system other than the failed processor has to take care of all activities.

6.2 Modes of Operation

In fault-free operation, only one optional task has been available for the whole system. Secondary chain has no optional task to be performed, while the noncritical task in primary chain has been further shared among the primary and secondary chains. The global memory holds all tasks of the system, while task to be scheduled by each processor during fault-free condition has been present in local memory. Occurrence of faults causes the healthy unit to load tasks of faulty unit from global memory and execute. On fault occurrence, the system switches two different mode of operation:

Mode 1.

In mode 1 operation, the healthy processor schedules critical task and all other noncritical tasks for the system without missing the deadlines. When a module has been affected by fault after the completion of noncritical task, the healthy module schedules optional tasks and noncritical tasks allotted for healthy unit which has been similar to mode 2.

Mode 2.

In mode 2 operation, the healthy processor schedules noncritical tasks and optional task allocated for the functioning processor. The noncritical task in the failed processor has been discarded assuming that the optional task allocated to healthy processor to be more efficient than noncritical task in failed processor.

6.3 Pseudocode

Permanent fault tolerance and fault recovery

Input: Task set for Navigation, guidance, and control modules of onboard computers

Output: Fault tolerant scheduling

1. **Start**
2. Schedule tasks in the system in mode3
3. Check for the health signal
4. **If** health signal is available
5. Continue scheduling tasks in the system in mode 3
6. **Else**
7. Select the mode of operation—mode1/mode2
8. **If** mode is 1
9. Schedule critical and noncritical tasks in the healthy processor
10. **Else if** mode is 2
11. Schedule critical, noncritical and optional task allocated for the healthy processor.
12. **End**
13. **End**

6.4 User Friendly Interface

A user friendly interface for the evaluation of the performance and verification of the algorithm has been implemented using Matlab graphical user interface (GUI) given in Fig. 5. Selection mechanisms for selection of components and mode of operation have been added into the GUI, the user interface also displays result of scheduling after the fault tolerance. The instants of permanent hardware fault follows a Poisson's distribution and the instant has been displayed along with run time calculated slack margin and effective utilization in GUI. The module encounter with the permanent hardware fault has been considered failed while the other units in primary chain go on executing the tasks. The secondary chain unit corresponding to the failed unit takes care of the operations to be performed by the failed unit.

6.5 Performance Metrics

The instants of fault occurrences have been considered to be same in both traditional dual redundant system and augmented dual redundant system. Average utilization has been obtained by finding the average of the utilizations of both processors present in a system. Slack time attained by locating the total free slots available in the processors present in the system. Process execution time takes in amount the time for which a process uses the resources available in a system. Speed-up factor considers ratio of total execution of the traditional dual redundant system to the total execution time of the augmented system. Recovery latency has been obtained by the difference of time instants of fault occurrence in the system and detection of fault in the system.

7 Simulation

Matlab-based scheduling toolbox TORSCHE (time optimization, resources, scheduling) used for the simulation and testing of the algorithm. The tool provides a collection of different data types that can be used to formalize different offline and online real-time scheduling problems [12].

Figure 4 displays the task scheduling during the simulation of the algorithm. The instant of fault occurrence has been generated using a poisons distribution, the program generates random number indicating time of occurrence of the fault and user selects the chain affected by fault. Fault has been detected during minor cycle, and recovery is performed instantly. Fault recovery latency finds the difference between the time of recovery and time of occurrence of fault.

The user interface in Fig. 5 shows the scheduling trace and performance metrics. It displays occurrence of single fault in navigation system, the mode 1 operation has been selected. The fault has affected the N1 module after scheduling of noncritical

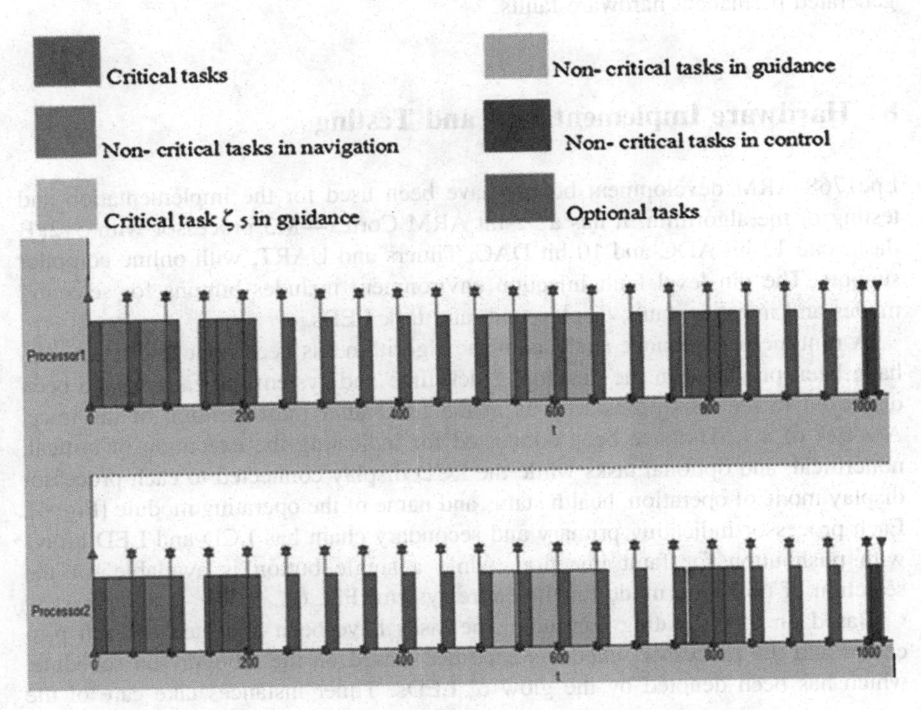

Fig. 4 Scheduling trace under resource augmented algorithm (fault-free condition)

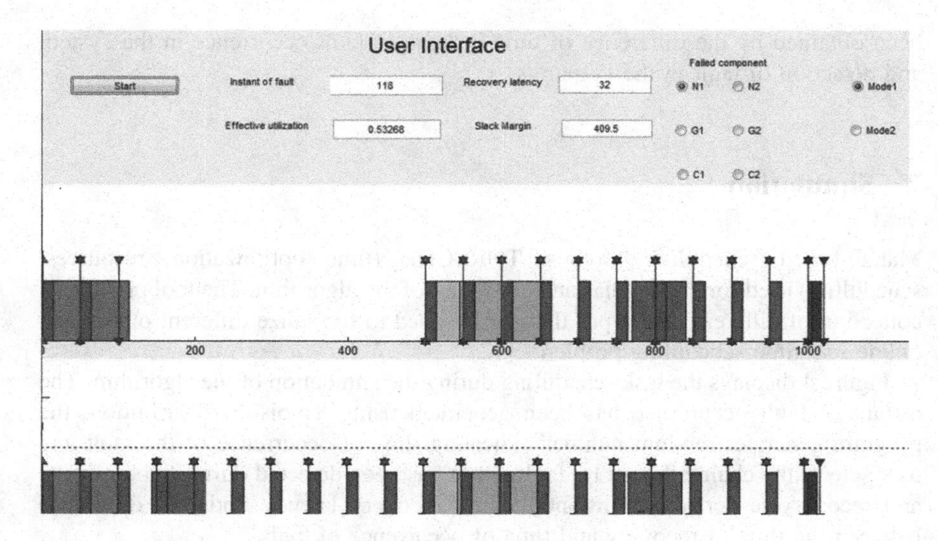

Fig. 5 User interface for fault at 118th time instant (Mode 1)

task, so the N2 module schedules noncritical task and optional task allotted to the N2. The window displays recovery latency, effective utilization, and instant of generated permanent hardware faults.

8 Hardware Implementation and Testing

Lpc1768 ARM development boards have been used for the implementation and testing of the algorithm. It has a 32 bit ARM Cortex—M3 processor with 512kB flash, one 12 bit ADC and 10 bit DAC, Timers and UART, with online compiler support. The pin level fault injection environment includes buttons for selecting modes and injecting fault, display unit, and task LEDs.

A runtime performance analysis of the algorithm has been done and the results have been projected in the desktop. Slack time and system utilization have been displayed in the desktop as well as in the LCD after the execution of the trace. A series of 4 LEDs have been connected for indicating the execution of critical, noncritical, and optional tasks while the LCD display connected to each processor display mode of operation, health status and name of the operating module [Fig. 6]. Each processor indicating primary and secondary chain has LCD and LED arrays with pushbutton for fault injection, while a single button is available for the selection of operating mode for the entire system [Fig. 6].

Based on the operating condition, the tasks have been allocated to each processor and the processor schedules each task based on the table-driven schedule, which has been denoted by the glow of LEDs. Timer instances take care of the

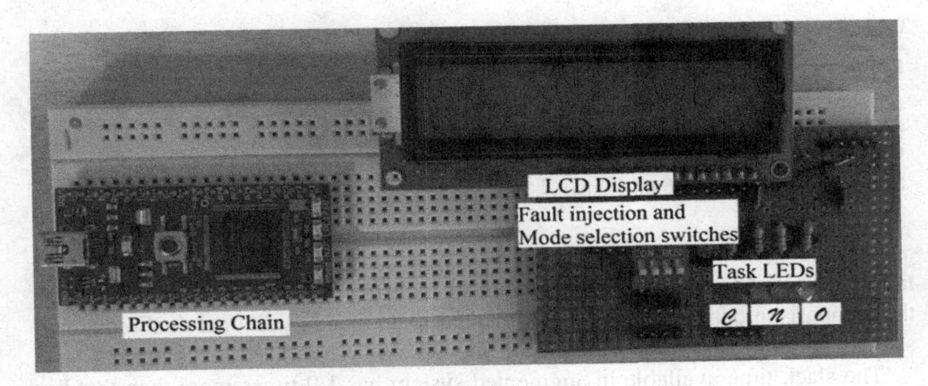

Fig. 6 Single chain along with the evaluation board

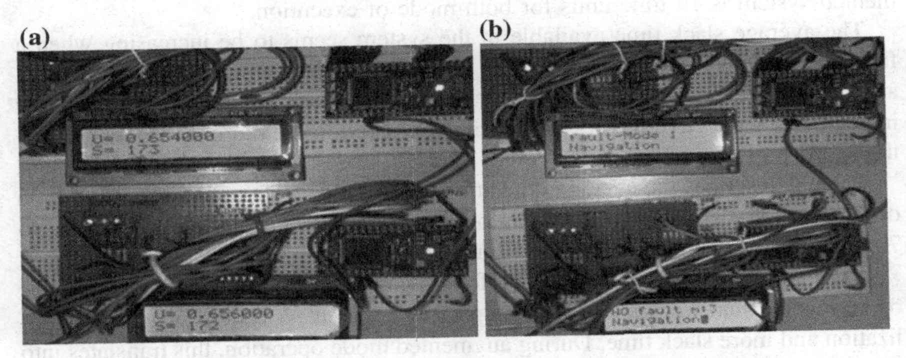

Fig. 7 Hardware fault injection model **a** fault-free operation **b** permanent hardware fault in secondary chain

execution time of each tasks and bidirectional communication for synchronization has been done during the execution of critical tasks.

Figure 7 shows occurrence of permanent hardware fault in secondary processing chain. Permanent fault has been injected into the chain by cutting off the supply to the processor using a switch. In accordance to the input given by mode selection switch, the healthy processor continues scheduling the tasks according to mode 1 or mode 2.

9 Results

A simulation run of 50 is conducted for obtaining the results of both augmented and traditional dual redundant system models. The average utilization of augmented system reduces by 21.47 %, which indicates that the system can schedule more

Table 2 Performance metrics during fault-free execution

Parameter	Traditional dual redundant system	Augmented dual redundant system
Average utilization	77.45 %	55.98 %
Slack time	236 time units	454.5 time units

tasks compared to the traditional system in fault-free condition. Table 2 indicates the comparison between resource-augmented system and traditional dual redundant system, no optional are considered for augmented system.

The slack time available in augmented system has 1.9 times more than that has been available in traditional dual redundant system. Recovery latency indicates the time required for the detection of fault, the average recovery latency for the augmented system is 19 time units for both mode of execution.

The average slack time available in the system seems to be increasing when a fault occurs in any one of the processor but there has been an increase in utilization and decrease in available slack time for the healthy processor. The speed-up factor indicates how effective the new algorithm compared to the old one, Table 3 shows the speed-up factor:

Table 4 describes the average utilization, slack time, and average execution time during fault mode execution. Here an optional task with processing times equal to 70 and 120 ms has been considered for mode 2 performance calculations.

The metrics indicate the system performs effectively and efficiently in presence of permanent hardware faults with respect to augmented system in terms of utilization and more slack time. During augmented mode operation, this translates into reduced power and fuel consumption. These available resources can be better utilized for fault tolerance strategies and enhancing performance.

Table 3 Speed-up factor

Fault-free execution	Fault tolerant execution	
1.1869	Mode	
	1	1
	2	1.356

Table 4 Performance metrics for fault condition operation

Parameters	Augmented system		Traditional dual redundant system
	Mode 1	Mode 2	
Average utilization	0.65	0.64	0.774
Slack time (ms)	451.5	441.833	366.9
Average execution time (ms)	568.5	578	653.1667

10 Conclusion

This paper discuses a resource augmented algorithm along with fault tolerance implementation for onboard computers of satellite launch vehicle. A fault injection model based on LPC1768 is implemented for validating the algorithm and a user friendly interface for simulation based testing is developed in MATLAB. Performance metrics like average utilization, slack time, average execution time, recovery latency, and speed-up factor indicate the resource augmented algorithm performs better compared to the traditional dual redundancy. Improved slack margin and computational resources available helps in scheduling tasks that enhance the functionality and working of the system. The future work includes software and transient fault tolerance implementation in the resource augmented system, validating the system using fault injection models.

References

1. Liu, Z., Joseph, M.: Verification of fault tolerance and real time. In: Proceedings of Annual Symposium on Fault Tolerant Computing, Sendai, pp. 220–229 (1996)
2. Hitt, E.F., Mulcare, D.: The fault tolerant avionics. In: Spitzer, C.R. (ed.) Digital Avionics Handbook, 2nd edn, chapter 28(CRC press LLC, Boca Raton 2007)
3. Basu, D., Prasad Rao, K.V.S.S., Varaprasad, S.V.L.A., Kurien, T., Jayasri, R., Bharathi, M.: A fault tolerant computer systems for Indian satellite launch vehicle programme. Sadhana, **11**, 221–231 (1987)
4. Cooper, A.E., Chow, W.T.: Development of on-board space computer systems. IBM J. Res. Develop **20**, 5–19 (1976)
5. Laprie, J.C.: Dependable computing and fault tolerance: concepts and terminology. In: Proceedings of Fault Tolerant Computing FTCS-15, 2–11 (1985)
6. Avizienis, A.: Fault-tolerant systems. IEEE Trans. Comput. **c-25**(12), 1304–1312 (1976)
7. Ramamritham, K., Stankovic, J.: Scheduling algorithms and operating systems support for real-time systems. Proc. IEEE. **8**(I), 55–67 (1994)
8. Davis, R.I.: On exploiting spare capacity in hard real time systems. Ph.D. thesis, University of York, England (1995)
9. Pillay, R.V., Punnekkat, S., Dasgupta, S.: An improved redundancy scheme for optimal utilization of onboard computers. In: Proceedings of IEEE INDICON pp. 1–4 (2009)
10. Swetha, A., Pillay, R.V., Chandran, S.K., Dasgupta, S.: A real-time performance monitoring tool for dual redundant and resource augmented framework of cruise control system. Int. J. Comput. Appl. **92**(14), 44–49 (2014)
11. Stesina, F.: Design and Verification of Guidance. Navigation and Control systems for Space application. Ph.D. thesis, Porto Institutional repository, Polytechnic university of Turtin Italy (2014)
12. Sucha, P., Kutil, M., Sojka, M., Hanzalek, Z.: TORSCHE scheduling toolbox for matlab. In: Proceedings of 2006 IEEE International Conference on Control Applications, Computer Aided Control System Design, pp. 1181–1186 Munich, October (2006)
13. Davis, R.I., Burns, A.: A survey of hard real-time scheduling for multiprocessor systems. ACM Comput. surveys **43**(4), number 35(2010)
14. Wikipedia, Speedup. http://en.wikipedia.org/wiki/Speedup
15. Krishna, C.M., Shin, K.G.: Real Time Systems (Tata McGraw Hill edition, 2010)

A Multiagent Planning Algorithm with Joint Actions

Satyendra Singh Chouhan and Rajdeep Niyogi

Abstract In this paper, we consider multiagent planning with joint actions—that refer to the same action being performed concurrently by a group of agents. There are few works that study specification of joint actions by extending PDDL. Since there are no multiagent planners that can handle joint actions, we propose a multiagent planning algorithm, which is capable of handling joint actions. In a multiagent setting, each agent has a different capability. The proposed algorithm obtains the number of agents involved in a joint action based on the capability of the individual agents. We have implemented our algorithm and compared its efficiency with some state-of-the-art classical planners. The results show that when the problem size increases, our algorithm can solve such problems whereas it cannot be solved by the classical planners.

Keywords Multiagent planning · Joint action · PDDL

1 Introduction

Multiagent system refers to a system of agents working together to reason, plan, solve problems, and make decision Here an agent can be an intelligent system, process, robot, sensor [1]. Multiagent system is applicable to various domains such as supply chain management, robotics, games, military operation, logistics, and security [2–4].

S.S. Chouhan (✉) · R. Niyogi
Department of Computer Science and Engineering,
Indian Institute of Technology Roorkee, Roorkee 247667, India
e-mail: satycdec@iitr.ernet.in

R. Niyogi
e-mail: rajdpfec@iitr.ernet.in

S. Das et al. (eds.), *Proceedings of the 4th International Conference on Frontiers in Intelligent Computing: Theory and Applications (FICTA) 2015*, Advances in Intelligent Systems and Computing 404, DOI 10.1007/978-81-322-2695-6_59

In recent years, several models and algorithms have been developed in multiagent planning (MAP). A multiagent planning problem is given a set of initial states, goal states, and actions of the agents, to find plans that transform the initial state to goal state of each agent. Typically, a multiagent plan consists of concurrent actions [5, 6]. In some domains, the problem may be decomposed into subproblems and each subproblem is assigned to a different agent [7, 8]. Unlike classical planning (single agent), very few state-of-the-art multiagent planning systems exists. A recent development is FMAP [9].

Concurrent actions come naturally in multiagent planning problems. For example, consider a two-agent domain [6] where both agents are in a room R1 and want to move into the room R2 through a single doorway. The agents have the operators 'G' to go through the door and 'W' to wait. Then possible concurrent actions would be (G, G), (G, W), (W, G), and (W, W). However, a successful plan would be obtained when one agent waits and allows the other agent to pass through the gate. Therefore, coordination is needed between the actions of different agents [10]. Concurrent interacting actions are said to consistent if there is no adverse effect of actions on to each other. Concurrent actions may have negative interaction also.

Joint actions can be considered as a special case of concurrent actions. In some planning domains, an action should be performed jointly by more than one agent to get a desired effect. For example, in order to lift a heavy table, two agents have to lift the opposite sides of the table simultaneously, i.e., both agents have to perform the same action on a particular object at the same time [11]. In this paper, we focus on multiagent planning problems involving joint actions.

Joint actions in multiagent planning raise two main issues: specification and planning. In [11], PDDL [12] is extended for specifying joint actions by including concurrent action constraints in action schema. Another approach to handle joint action is proposed in [13]. In this work, agents are encoded as parameters in action schema and necessary preconditions are added in the precondition list. However, this representation can only specify positive interaction. However, there is no multiagent planning system that can handle joint actions.

In this paper, we make an attempt in this direction to come up with a multiagent planning system that handles joint actions. For this, we propose a multiagent planning algorithm in Sect. 3. The rest of the paper is structured as follows. Section 2 presents specification of joint actions. Section 4 presents the implementation results. Conclusions are given in Sect. 5.

2 Specification of Joint Action

We discuss some approaches that extend the planning domain definition language (PDDL) to represent joint actions. PDDL is widely used to formally describe planning problems. A basic syntax of action in PDDL is as follows:

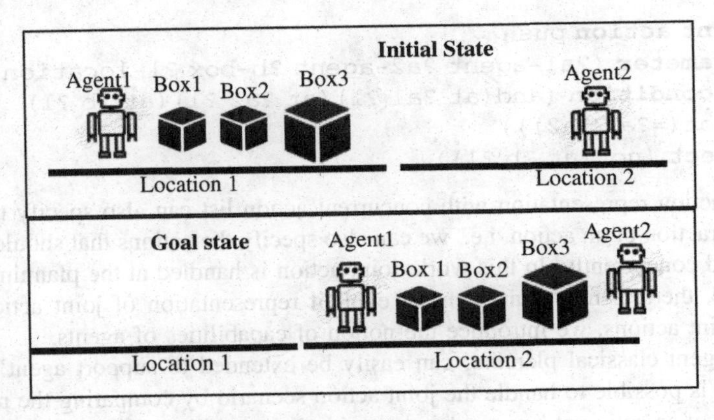

Fig. 1 A Box-pushing domain with possible configuration

```
:action <action_name>
:parameter <parameter_list>
:precondition <precondition list>
:effect <effect list>
```

The above syntax is extended in [11] to introduce concurrent interacting actions. Two main issues that are handled include: (i) 'who' is performing the action and (ii) 'what other actions' should (not) be performed concurrently.

The first issue is handled by including an agent as a parameter in the parameter list of an action. The second issue is dealt by adding a 'concurrent' action list in the action schema. In the concurrent action list, all the actions that are to be performed concurrently are mentioned.

Consider a box-pushing domain with varying size of boxes. Some boxes are heavy; a single agent cannot move it. Hence, more than one agent has to perform the action 'push' simultaneously to move heavy boxes, i.e., a joint action should be performed to achieve the desired effect. Figure 1 shows a simple box-pushing domain with two agent and three boxes. PDDL syntax of the action 'push' is given below:

```
:action push
:parameter (?a1-agent ?b -box ?l- location)
:precondition (and(at ?a1 ?l) (at ?b ?l))
:concurrent (and (push ?a2 ?b ?l)(not(= ?a1 ?a2))
:effect (not (at ?b ?l))
```

Other representation of joint action is proposed in [14]. In this representation, agents are specified in the parameter list and joint action preconditions are mentioned in the precondition list. There is no such additional syntax or special tag to represent concurrency.

```
:joint action push
:parameter (?a1-agent ?a2-agent ?b-box ?l-location)
:precondition (and(at ?a1 ?l)(at ?a2 ?l)(at ?b ?l)
  (not(=?a1 ?a2)))
:effect (not(at ?b ?l))
```

Joint action representation with concurrent action list can also specify the negative interaction of an action, i.e., we can also specify the actions that should not be performed concurrently. In this work, joint action is handled at the planning level. Therefore, there is no requirement of explicit representation of joint actions. To handle joint actions, we introduce the notion of capabilities of agents.

Multiagent classical planning can easily be extended to support agent's capabilities. It is possible to handle the joint action scenario by comparing the required capability of an action at a particular state with the capability of agent performing at that time. Consider the example, Fig. 1, to move a heavy box that may involve at least two agents. The number of agents involved in the joint actions depends on the capability of the individual agents. So the number of agents need not be decided *a priori*.

3 Algorithm for Multiagent Planning with Joint Actions

In this section, we present the brief overview of the proposed multiagent planning algorithm. Input to the planner is given by a PDDL input problem file. Input file contains the domain description that includes the number of agents, set of actions agent can perform, and other objects in the domain. It also specifies the capabilities of each agent. Each action is associated with the required capability to perform the action. Planner agent uses a heuristic forward search to search in state space.

3.1 Description of the Algorithm

The algorithm maintains an individual state of each agent and a global state. For each agent 'i' it maintains an open list of states that needed to be explored and closed list of states that are already explored. For each agent it selects an action from the applicable action list of current state. Applicable action list is ordered based on heuristic values. Here the heuristic value is taken as number of different propositions in the current state and goal state. After selecting actions for all agents, it merges the effects of all agents' selected actions to the current global state. The algorithm eventually stops when the current global state is same as the goal state. The algorithm is given below.

Pseudocode of the algorithm

Algorithm: MA-planner

```
1.  current state = initial state
2.  While (goal is not achieved)
3.      for each agent i
4.          If (! Agent_busy)
5.              App_act_list_i = Applicable_Action(current state, Actionlist_i)
6.              Opt_Act_i = get_optimal_action(App_act_list_i, current state, goal
                state)
7.          else
8.              Opt_Act_i = No_op
9.              Add Opt_Act_i into the MA-plan
10.             If (Check status(agent_i) then set agent_i = free;
11.         end if
12.     end for
13.     current_state = merge_effect (Current_state, Opt_Act_1, Opt_Act_2 ....., Opt_Act_k)
14. end while
```

In the algorithm, planner extracts the applicable action list for agent i using **Applicable_Action**(.) function. Applicable action list contains all the actions that can agent apply at current state (step 5). Form the Applicable action list, optimal action is selected based on heuristic value using **get_optimal_action**(.) (step 6). Pseudocode for **get_optimal_action**(.) is given below.

Algorithm: get_optimal_action (*App_act_list, current state, goal state*)

```
1.  for each action act in App_act_list,
2.      calculate heuristic values
3.  end for
4.  sort App_act_list according to heuristic values
5.  for each act in sorted_App_act_list
6.  If (!conflict(act))
7.          If (coordinated_capbility_act == 0)
8.                  opt_act = act;
9.                  coordinated_capbility_act = act_req_cap - agent_Cap
10.             If (coordinated_capbility_act > 0)
11.                 set agent = busy;
12.             end if
13.             return opt_act
14.         else
15.             opt_act = act
16.             coordinated_capbility_act =agent_Cap – act_req_cap
17.             return opt_act
18.         end if
19. end if
20. end for
21. return no -op
```

In the above algorithm, for each action in applicable action list of agent i heuristic value is calculated (steps 1–3). Applicable action list is sorted according to the heuristic values step (4). For each action from the sorted applicable action list, algorithm checks whether this action can be added in MA-plan or not. If so then optimal action is return otherwise no-op action is return steps (5–21).

3.2 Specification of Multiagent Plan

Planner agent generates the parallel sequence of actions for multiple agents, i.e., at particular state a set of action $A = \{a_1, a_2, a_3... a_n\}$ is selected. Where a_1 is an action of agent 1, a_2 is an action of agent 2, and so on (possibly included no-op action). Sequence of the action sets $A_1;...; A_k$ is called a multiagent plan. This MA-plan is consistent and coordinated with respect to all agents. There can be more than one possible MA-plan for same problem instance. The planner outputs one possible MA-plan for the problem and it may be the case that it is not optimal.

For example, consider the initial state of and goal state of the box-pushing domain as shown in Fig. 1. Objective is to move all the boxes to the location2. Here box3 is heavy therefore can be move if both agents jointly perform the push action at same time. Figure 2 shows one of the possible MA-plan structures for the configuration shown in Fig. 1.

In the MA-plan (Fig. 2), agent1 is waiting at time T3 for agent2 to perform the joint action. At time T4, agentl and agent2 performs the joint action ‘push’. The plan length of the MA-plan is four. For various problem instances summarized results shown in Table 1.

Here the multiagent plan would be the sequence of these concurrent actions set of agentl and agent2, i.e., {Push(B1, location2), Move (location2, location1)}; {Move (location2, location1), Push (B2, location2)};{No-op, Move (location2, location1)}; {Push(B3, location2), Push (B3, location2)}, where the elements in a set are executed concurrently or jointly.

Time	T0	T1	T3	T4
Agent 1	Push(B1, location2)	Move(location2, location1)	No-op	**Push(B3, location2)**
Agent 2	Move(location2, location1)	Push(B2, location2)	Move(location2, location1)	**Push(B3, location2)**

Fig. 2 A possible MA-plan structure for the configuration shown in Fig. 1

Table 1 Experimental results in Box-pushing domain and extended block world domain

	No. of agents	Problem size (s, m, l)	Solved?			MA-plan length		
			BB	PP	FF	BB	PP	FF
Box-pushing domain	2	(4, 2, 0)	Y	Y	Y	7	8	11
	2	(5, 3, 0)	Y	Y	N	7	12	–
	2	(6, 2, 0)	Y	Y	N	9	12	–
	3	(6, 2, 1)	Y	Y	N	9	9	–
	3	(8, 4, 2)	N	Y	N	–	17	–
Extended-block world domain	2	(3, 0, 0)	Y	Y	Y	4	4	4
	2	(2, 1, 0)	Y	Y	Y	6	6	6
	2	(2, 2, 0)	N	Y	N	–	6	–
	3	(1, 2, 1)	N	Y	N	–	8	–
	3	(1, 1, 2)	N	Y	N	–	7	–

[a]*BB* Blackbox planner, *PP* Proposed planning system, *FF* Fast Forward planner *N* No, *Y* Yes

4 Implementation Results

We have implemented the forward search algorithm with joint action using Java. Experiments were run on the Intel core i5 with 2.53 GHz machine with 4 GB of RAM. We ran our algorithm on two domains: box-pushing domain and extended block world domain. Box-pushing domain is explained in Sect. 2.

In the block-world domain, the goal is to find a plan to rearrange blocks from the initial configuration to the goal configuration. Blocks can be put on the table or on top of each other. A block can be moved only if it is the top block; additionally, only one block can be moved at a time. It is a classical planning problem. In our experiment, we considered block world domain with multiple agents and different sizes of blocks. In this experiment, light box can be moved or pickup by single agent. However to move heavy boxes more than agent should pick up the box jointly.

We have compared the performance of our algorithm with some classical planners like black box (BB) [15] and fast forward (FF) planner [14]. Black box and fast forward takes planning problems defined in STRIPS or PDDL. We have described the domains using the joint action specification technique suggested in [14]. With some modifications in PDDL, these planners (FF and BB) can support joint action up to a certain level. Table 1 shows the implementation results.

In the table, column1 shows the number of agents in the system. Column2 shows the problem size. The notation (s, m, l) is the number of small, medium, large boxes present in the problem, respectively. A small box needs one agent to push it. A medium box requires two agents simultaneously (jointly). A large box needs more than two agents simultaneously. Column3 shows whether the centralized planner was able to generate a solution plan. Column4 shows the MA-plan length.

Experimental results show that problems having more than two agents and large objects cannot be solved by classical planners, as in the (8, 4, 2) case. It is because these classical planner search in the joint state space of all agents, which becomes exponentially large with the problem size.

5 Conclusion and Future Work

In this paper, we considered a multiagent planning problem where multiple agents with different capabilities may be involved in performing some actions. We have developed a planner that can generate such plans. We have compared the performance of our planner with some existing state-of-the-art planners on some benchmark domains. Experimental results show that some problem instances cannot be solved by the classical planners. However, the proposed planner can easily handle such problem instances. There are few multiagent planners available; none of these planners can handle joint actions. Thus, our work is a first attempt toward the development of such planners. As part of future work, we would like to develop a distributed planner capable of handling joint actions.

References

1. Parker, L.E.: Distributed intelligence: overview of the field and its application to multi-robot systems. J. Phys. Agents **2**(1), 5–14 (2008)
2. Marcello C., Pecora F., Andreasson, H., Uras, T., Koenig, S.: Integrated motion planning and coordination for industrial vehicles. In: Proceedings of International Conference on Automated Planning and Scheduling (ICAPS) (2014)
3. Sara B., Fox, M., Long, D.: Planning the behaviour of low-cost quadcopters for surveillance missions. In: Proceedings of International Conference on Automated Planning and Scheduling (ICAPS) (2014)
4. Wu, F., Zilberstein, S., Chen, X.: Online planning for multi-agent systems with bounded communication. Artif. Intell. **175**, 487–511 (2011)
5. Brenner, M., Nebel, B.: Continual planning and acting in dynamic multi-agent environments. In: Proceedings of Autonomous Agent Multi-agent System, pp. 297–331 (2009)
6. Bowling, M., Jensen, R., Veloso, M.: A formalization of equilibria for multi agent planning, In: Proceedings of International Joint Conference on Artificial Intelligence (IJCAI), pp. 1460–1462 (2003)
7. Brafman, R.I., Domshlak, C.: From one to many: planning for loosely coupled multi-agent systems, In: Proceedings of International Conference on Automated Planning and Scheduling (ICAPS), pp. 28–35 (2008)
8. Ephrati, E., Rrosenschein, J.S.: Divide and conquer in multi-agent planning. In: Proceedings of the Twelfth National Conference on Artificial Intelligence, pp. 375–380 (1994)
9. Torreño, A., Onaindia E., ÓscarSapena: FMAP: distributed cooperative multi-agent planning. Appl. Intell. **41**, 606–626 (2014)
10. Larbi, R.B., Konieczny, S., Marquis, P.: Extending classical planning to the multi-agent case: a game-theoretic approach. In: ECSQARU. Lecture Notes in Computer Science, vol. 4724, pp. 731–742. Springer (2007)

11. Boutilier, C., Brafman, R.I.: Planning with concurrent interacting actions. J. Artif. Intell. Res. (JAIR), **14**, 105–136 (2001)
12. Fox, M., Long, D.: PDDL2.1: an extension to PDDL for expressing temporal planning domains. J. Artif. Intell. Res. (JAIR) **20**, 61–124 (2003)
13. Brafman, R.I., Zoran, U.: Distributed heuristic forward search for multi-agent system. In: Proceedings of 2nd Distributed and Multi-agent Planning Workshop (ICAPS DMAP), pp. 1–7 (2014)
14. Hoffmann, J., Nebel, B.: The FF planning system: fast plan generation through heuristic search. J.Artif. Intell. Res. (JAIR) **14**, 253–302 (2001)
15. Kautz, H., Selman, B.: BLACKBOX: a new approach to the application of theorem proving to problem solving. In: AIPS98 Workshop on Planning as Combinatorial Search (1998)

4-Directional Combinatorial Motion Planning Via Labeled Isotonic Array P System

Williams Sureshkumar and Raghavan Rama

Abstract In this paper we propose a method to find collision free path from the starting position to the target in the 2D grid via a labeled isotonic array P system which enables only 4-directional movements of the robot by excluding diagonal moves. The proposed grammatical model finds a collision free path in polynomial time.

Keywords Isotonic array grammar · Labeled isotonic array P system · Combinatorial motion planning · Priority

1 Introduction

A P system is a computational model which works in a parallel distributed manner. The basic cell structure is abstracted as membranes holding the computations at different levels. In 1998 the concept was first introduced by Gh. Paun in [1] whose last name is the origin of the letter P in P systems. Several variants of P systems led to the formation of a new branch of research known as *Membrane Computing* [2].

Motion planning is frequently referred to as motion of a robot in 2D or 3D space that may or may not contain obstacles. Planning a motion of a robot means to determine appropriate motions for the robot so that it reaches the target position without colliding into obstacles. In order to find paths via continuous configuration space, one can use combinatorial approaches. Such approaches will be called as 'Combinatorial path/motion planning approaches'. In such approaches it is important to carefully define the input and obstacles. The set of configurations that avoid collisions with

This work was funded by the Project No: MAT/15-16/046/DSTX/KALP, Department of Science and Technology, Government of India.

W. Sureshkumar (✉) · R. Rama
Department of Mathematics, Indian Institute of Technology Madras,
Chennai 600036, India
e-mail: wisureshkumariit@gmail.com

R. Rama
e-mail: ramar@iitm.ac.in

S. Das et al. (eds.), *Proceedings of the 4th International Conference on Frontiers in Intelligent Computing: Theory and Applications (FICTA) 2015*, Advances in Intelligent Systems and Computing 404, DOI 10.1007/978-81-322-2695-6_60

obstacles is called the free space and it is difficult to explicitly compute this space. In [3] the authors have given a method of obtaining a path in the free space by moving the robot in 8-directions. This method is different from the traditional decomposition method [4]. The work space of the robot is in 2D plane having only rectangular obstacles.

We propose in this paper a labeled P system which uses isotonic array rules for the movement of a robot in a rectangular grid in only 4-directions unlike the 8-directional movement given in [3]. We there by avoid the diagonal movements in our model.

In this paper define the input as a rectangular grid where one of the position is indicated by symbol *A* say, and this symbol corresponds to the starting position of a robot (which is resembling a continuous configuration space). The obstacles are suitably coded by special symbols. The set of transformations that will be applied on the robot will be a collection of isotonic rules.

This paper is divided into four sections. In Sect. 2 we give the basic definitions. Section 3 we present the procedure to model the combinatorial motion planning via *LIAPS*. The future direction research is presented in the concluding section.

2 Preliminary

We refer the reader to [5] for basic notions, notations and results about formal language theory. For definition of Isotonic array grammar (*IAG*) the reader is referred to [6].

We define the isotonic *restricted monotonic* array grammar G. An isotonic grammar G is said to be *restricted monotonic* if for each isotonic array rule $\alpha \to \beta$, α contains exactly one non-terminal symbol and α may contain #'s and terminal symbols. Image of each non-# symbol of α is a non-# symbol whereas image of each # symbol is either a # symbol or a non-# symbol. Hence, each symbol of β is in $N \bigcup T \bigcup \{\#\}$. Note that #'s cannot be created. We call this new restricted grammar by Isotonic Restricted Monotonic Array Grammar (*IRMAG*). For example rules look like

$$\begin{matrix} 0\,0 \\ \#A \end{matrix} \to \begin{matrix} 0\,0 \\ A\,1 \end{matrix}, \quad \begin{matrix} \#\,0 \\ \#\,B \end{matrix} \to \begin{matrix} \#\,0 \\ \#\,1 \end{matrix}, \ldots \text{etc.}$$

In this paper, we use only the rules of type isotonic regular, isotonic context-free and isotonic restricted monotonic. Hereafter if we mention isotonic rules, it means only these three types of rules.

2.1 Isotonic Array P Systems

An Isotonic Array P Systems (*IAPS*) of degree m ($\geq$1) is a construct $\Pi = (V, T, \#, \mu, I_1, \ldots, I_m, (R_1, \rho_1), \ldots, (R_m, \rho_m), i_o)$, where V is the alphabet of terminals and non-terminals, $T \subseteq V$ is the terminal alphabet, $\# \notin V$ is the blank or background symbol. Here μ represents the membrane structure and the number of membranes are numbered as $1, 2, \ldots, m$. Membrane 1 is the outermost membrane or skin membrane. The set of initial arrays are I_i associated with region i, $1 \leq i \leq m$. For definition of membrane region the reader is referred to [1]. R_i is the set of evolution isotonic array rules associated with region i, $1 \leq i \leq m$. ρ_i describes the priority among the rules in R_i, $1 \leq i \leq m$. The rules are written as $\mathcal{A} \rightarrow \mathcal{B}\,(tar)$, where $\mathcal{A}, \mathcal{B}$ are isotonic arrays and $tar \in \{here, out, in\}$ which means that the array $\mathcal{A}$ is replaced by an identical (in shape) array $\mathcal{B}$ and the resultant modified array is sent to a membrane indicated by *tar*. Here, i_o is the output membrane.

The evolution in the above defined system happens in a parallel distributed manner. However, each array will be computed by only one rule at a time. The evolution of the arrays may also be dominated by priorities. An evolution of the array may or may not always halt. Such an output condition is in general ignored or avoided by suitable array rewriting rules. Hence, the output of this system will be only the collection of arrays that appear in a designated output membrane for which the system halts. Moreover, the output is a collection of terminal arrays.

2.2 *Labeled Isotonic Array P Systems*

A Labeled Isotonic Array P System (*LIAPS*) Π of degree m ($\geq$1) is a construct $\Pi = (V, T, \#, \mu, I_1, \ldots, I_m, (R_1, \rho_1), \ldots, (R_m, \rho_m), i_o, lab)$, where: $V, T, \#, \mu, I_1, \ldots, I_m$, $R_1, \ldots, R_m, \rho_1, \ldots, \rho_m, i_o$ are same as in Sect. 2.1. Let $R = \bigcup_{i=1}^{m} R_i$. Here we assign a label to every rule in R where the labels are chosen from a finite alphabet *lab* or the labels can be λ (empty label). Define a labeling function $f : R \rightarrow lab \cup \{\lambda\}$ that assign a label to each rule in R. Note that more than one rule may have the same label, but the same rule in different membranes cannot be assigned different labels. For label sequence $S = l_1\, l_2\, \ldots\, l_k \in R^*$, we extend the labeling as follows : $f(\lambda) = \lambda$ and $f(l_1\, l_2\, \ldots\, l_k) = f(l_1)f(l_2 \ldots l_k)$. A transition $C \overset{e}{\Rightarrow} C'$ between two successive configurations uses only rules with the same label e and rules labeled with λ. If at least one rule has a label $e \in lab$ then the transition is called λ-restricted transition. If we allow all rules with λ label in an evolution then the transition is called λ-unrestricted transition (or λ-transition).

The *LIAPS* Π generates a set of strings α over *lab* where all symbols in α are consumed and the system halts.

3 *LIAPS* for Combinatorial Motion Planning

In this section, we can show how the *LIAPS* can be used to simulate the movement of robot from its initial position to target position in the 2D square grid. The problem will be to find a collision free (in the case of obstacle present in the grid) path from the initial position to target position. In the simulation we assume the following:

(i) The 4 movements of the robot can happen only in upward, downward, leftward and rightward directions.

(ii) The 2D square grid is an $n \times n$ array, where n is even.
(iii) The simulation is done on a grid with obstacles.
(iv) The obstacles are rectangular in shape.

In the subsequent sections we discuss the two different cases of robot movements. That is,

1. Robot Movement with obstacles adjacent to the boundary of the grid.
2. Robot Movement with obstacles scattered everywhere in the grid.

These are shown in Figs. 2a and 4a respectively.

The 2D grid of the robot will be represented as an array in the initial configuration of the *LIAPS* as follows:

- The boundary of the square grid is filled with 1's.
- The initial and target position are denoted by a non-terminal A and a $ symbol.
- + denotes the obstacles.
- # which is defined as in the isotonic array grammar.

3.1 Robot Movement with Obstacles Adjacent to the Boundary of the Grid

In this case a rectangular obstacle present in the grid, one of whose side is connected with the boundary of the grid and is shown in Fig. 1a. The connected isotonic array representation of Fig. 1a is shown as I_1. I_1 will be the initial connected input array of Π_1, the P system. The *LIAPS* Π_1 with one membrane is given as, $\Pi_1 = (\{A, \star, +, 0, 1, \$\}, \{\star, +, 0, 1, \$\}, \#, [_1]_1, I_1, R_1, 1, \{a, b\})$, where

$$I_1 = \left\{ \begin{matrix} 1 & 1 & 1 & 1 & 1 & 1 & 1 & 1 & 1 & 1 & 1 & 1 \\ 1 & \# & \# & \# & \# & \# & \# & \# & \# & \# & \$ & 1 \\ 1 & \# & \# & \# & \# & \# & \# & \# & \# & \# & \# & 1 \\ 1 & \# & \# & \# & \# & \# & \# & \# & \# & \# & \# & 1 \\ 1 & \# & \# & \# & \# & \# & \# & \# & \# & \# & \# & 1 \\ 1 & + & + & + & \# & \# & \# & \# & \# & \# & \# & 1 \\ 1 & + & + & + & \# & \# & \# & \# & \# & \# & \# & 1 \\ 1 & + & + & + & \# & \# & \# & \# & \# & \# & \# & 1 \\ 1 & \# & \# & \# & \# & \# & \# & \# & \# & \# & \# & 1 \\ 1 & \# & \# & \# & \# & \# & \# & \# & \# & \# & \# & 1 \\ 1 & A & \# & \# & \# & \# & \# & \# & \# & \# & \# & 1 \\ 1 & 1 & 1 & 1 & 1 & 1 & 1 & 1 & 1 & 1 & 1 & 1 \end{matrix} \right\}$$

and $R_1 = \left\{ a : \begin{matrix} \# \\ A \end{matrix} \rightarrow \begin{matrix} A \\ \star \end{matrix} > b : A\# \rightarrow \star A \ , \ a : \begin{matrix} \$ \\ A \end{matrix} \rightarrow \begin{matrix} \star \\ \star \end{matrix} \ , \ b : A\$ \rightarrow \star\star \right\}.$

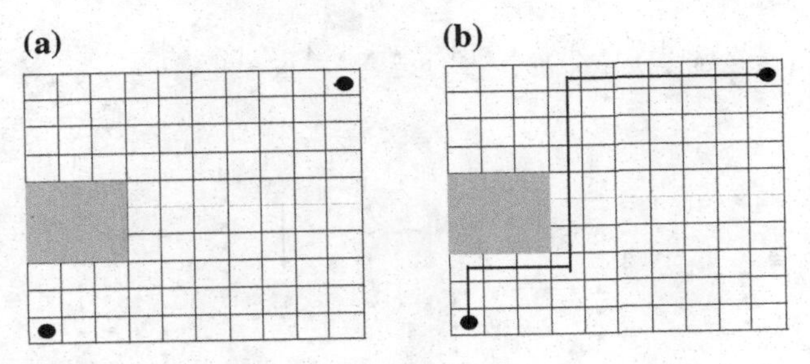

Fig. 1 Obstacles adjacent to the boundary of the grid

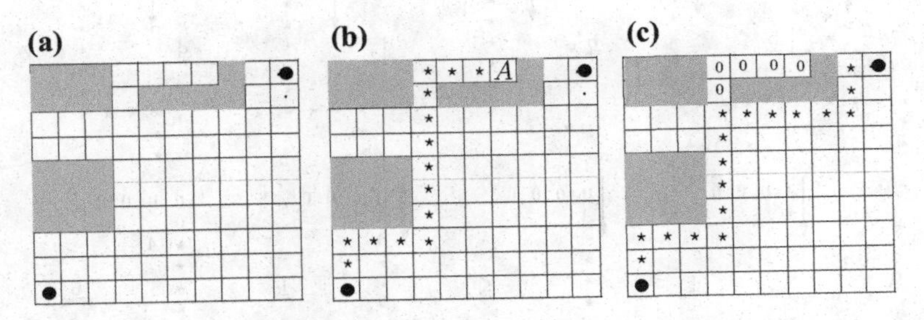

Fig. 2 Obstacles adjacent to the boundary of the grid with deadlock

The computation of *LIAPS* starts from its initial configuration (the source), and stops at its halting configuration (the target). In the initial configuration it generates the starting position of the path of the robot and in the halting configuration it generates the target position of the path of the robot. The string over *lab* corresponding to Fig. 1a will be $w = a^2b^3a^7b^6$ and the path shown in Fig. 1b. Always this will not happen. In some occasions the system may halt but it will not generate the target position. Then we say that the path reaches *deadlock* as shown in Fig. 2b.

To recover from *deadlock* and generate the collision free path, we need to redefine the rules of Π_1. The modified rules are as follows:

$$R_1 = \left\{ a : \begin{matrix} \# \\ A \end{matrix} \rightarrow \begin{matrix} A \\ \star \end{matrix} \;>\; b : A\# \rightarrow \star A \;>\; c : \begin{matrix} A \\ \star \end{matrix} \rightarrow \begin{matrix} 0 \\ A \end{matrix} \;>\; d : \star A \rightarrow A\,0, \right.$$

$$\left. a : \begin{matrix} \$ \\ A \end{matrix} \rightarrow \begin{matrix} \star \\ \star \end{matrix}\,,\; b : A\$ \rightarrow \star\star \right\}.$$

The successful collision free path from the starting to target position is given in Fig. 2c. It contains not only $\star$'s but also contains 0's. Now we have to prune these 0's from the path. The entire computation (we are not showing the whole array instead we show only the path) will be as in Fig. 3.

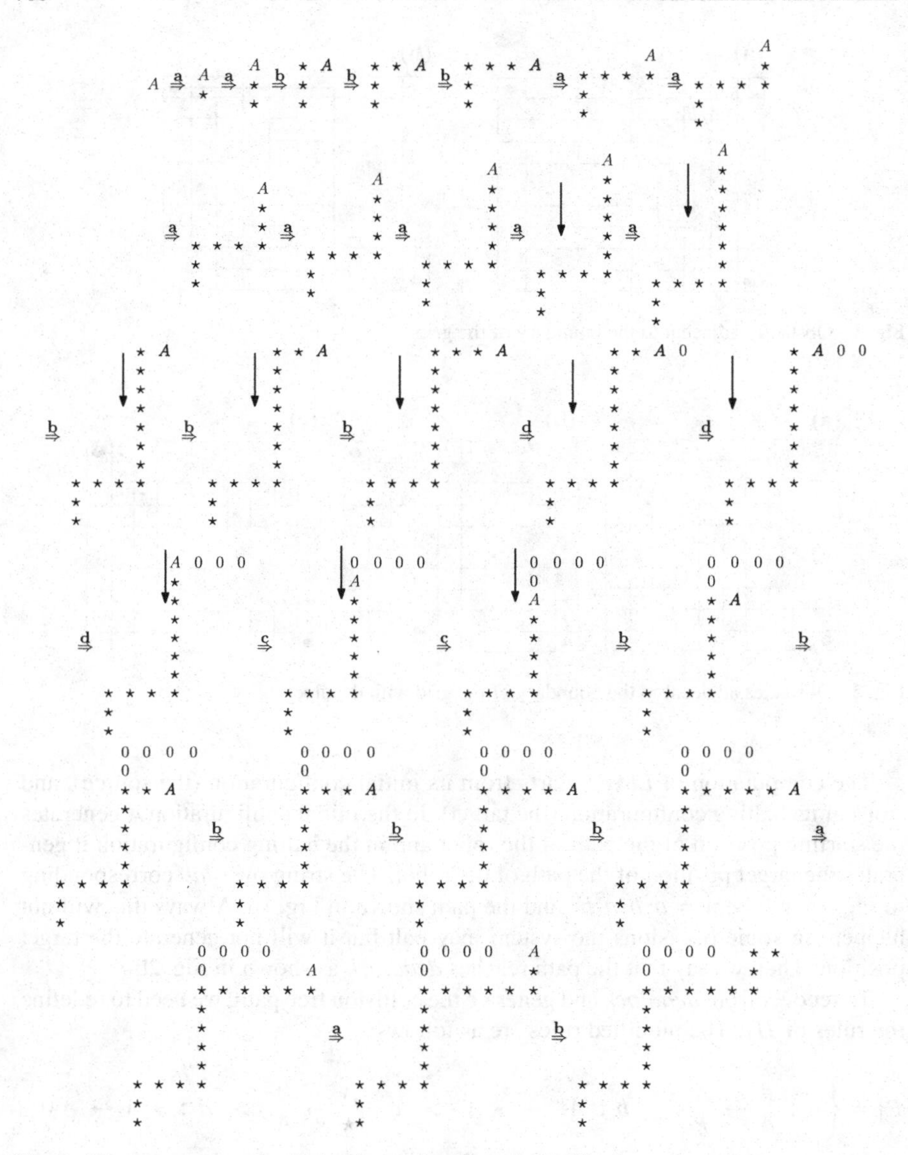

Fig. 3 The computation of collision free path of the robot in Fig. 2c

The regulated string generating the collision free path in the above computation is $a^2b^3a^7b^3d^3c^2b^5a^2b = a^2b^3a^5(\mathbf{a^2b^3d^3c^2})\ b^5a^2b$. Now we have to prune the useless part of the path in the following manner: Consider the computation of the collision free path (Fig. 3), if the configuration generates the last 0 in the computation, mark it with ↓ (right most) symbol (Fig. 3). The configuration next to this is overwriting the cell which is already filled by $\star$. But this $\star$ filled by the earlier configuration is indicated by the leftmost $\downarrow$. Now prune the configurations corresponding to the substring $\mathbf{a^2b^3d^3c^2}$ from the leftmost $\downarrow$ to rightmost $\downarrow$ in the computation. The resulting sting becomes $a^2b^3a^5b^5a^2b$. Since the grid contains n^2 cells, the number of moves to generate this path require less than n^2 moves. Hence the collision free path can be generated in polynomial time.

3.2 *Robot Movement with Obstacles Scattered Everywhere in the Grid*

In this case the rectangular obstacles are scattered everywhere in the grid as shown in Fig. 4a. If the obstacles are placed in the grid, the situation is to model the grid as a connected array which requires an additional terminal symbol, say c in place of #'s. The P system which works on prioritized rules. The *LIAPS* Π_2 with one membrane is given as, $\Pi_2 = (\{A, c, \star, +, 0, 1, \$\}, \{c, \star, +, 0, 1, \$\}, \#, [_1]_1, I_1, R_1, 1, \{a, b, d, e\})$, where

$$I_1 = \left\{\begin{matrix} 1&1&1&1&1&1&1&1&1&1&1&1\\ 1&c&c&c&c&c&c&c&c&c&\$&1\\ 1&c&+&+&c&c&c&c&c&+&c&1\\ 1&c&+&+&c&c&c&c&c&c&c&1\\ 1&c&c&c&c&+&+&+&c&c&c&1\\ 1&c&c&c&c&c&c&c&c&c&c&1\\ 1&c&c&c&+&c&c&c&c&+&c&1\\ 1&+&c&c&+&c&c&c&+&+&c&1\\ 1&c&c&c&c&c&c&c&c&+&c&1\\ 1&c&c&c&c&+&+&c&c&c&c&1\\ 1&A&c&+&c&c&c&c&c&c&c&1\\ 1&1&1&1&1&1&1&1&1&1&1&1 \end{matrix}\right\}$$

and $R_1 = \left\{ a : \begin{matrix} c \\ A \end{matrix} \rightarrow \begin{matrix} A \\ \star \end{matrix} \;>\; b : A\,c \rightarrow \star A \;>\; d : \begin{matrix} A \\ \star \end{matrix} \rightarrow \begin{matrix} 0 \\ A \end{matrix} \;>\; e : \star A \rightarrow A\,0\,,\; a : \begin{matrix} \$ \\ A \end{matrix} \rightarrow \begin{matrix} \star \\ \star \end{matrix}\,,\; b : A\$ \rightarrow \star\star \right\}$. In the initial configuration, each # symbol is replaced with a terminal symbol c. This is because throughout the computation the connected nature of the array is to be preserved.

The string over *lab* corresponding to Fig. 4a will be $w = a^2ba^4b^2a^3b^6$ and the path shown in Fig. 4b. The *LIAPS* Π_2 generates the collision free path from starting to target position. The resulting array in the halting configuration of Π_2 is given in Fig. 4b. Since the grid has n^2 cells and the collision free path (irrespective of target position) requires less than n^2 moves, it can be generated in polynomial time.

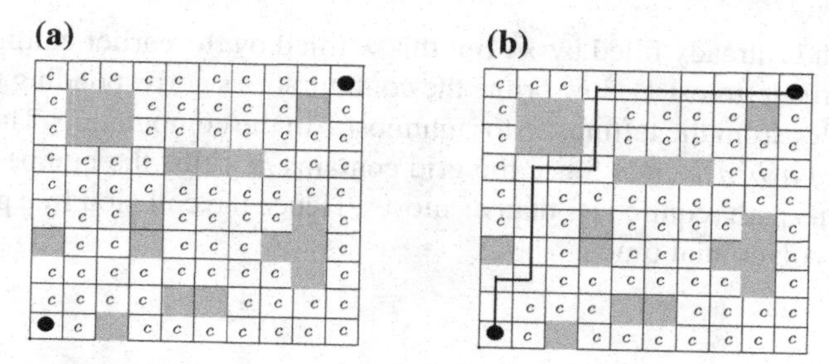

Fig. 4 Obstacles scattered everywhere in the grid

3.3 Multiple Robot Movement-Model

Suppose there are n robot $R_1, R_2, \ldots, R_n$ moving in a grid with obstacles with different start and target positions. One can construct a *LIAPS* with $(n+1)$ membranes where robot R_i is set to evolve in the grid which is placed in the ith membrane. The ith membrane has no other robot in it. The evolution rules are defined uniformly in all the n membranes as defined in Sect. 3.2. The membrane structure will be $[_{n+1}[_1]_1[_2]_2[_3]_3 \cdots [_n]_n]_{n+1}$. The output string of each membrane i will be ejected to membrane $(n+1)$ which completely describes the path of the robot R_i. The n strings ejected to membrane $(n+1)$ can be suitably compared by a table filling algorithm to find the overlapping position in a linear time and then a delay unit can be added to robot which are having overlapping positions. That is the robot movements are in the same configuration in their respective membranes which is to be determined by means of the string over *lab*. Definition of *LIAP* can be defined so that all the membranes contain the same set of rules. Each membrane evolves parallely at the same time independently. That is there will not be any restrictions while applying the labeled rules.

4 Concluding Remarks

In this paper we propose a grammatical method via labeled *IAPS* to obtain a collision free path from starting position to target of a robot moving in 2D workspace. This is the first model in the design of combinatorial motion planning. The search for such a path is an output by our *LIAPS*. The input to this system will be 2D isotonic array suitably modeled. The output of the system is a word w which describes one collision free path. It will be interesting to study the extension of the motion planning in a 2D space which contains obstacles of any shape i.e. polygonal obstacles regions.

References

1. Paun, Gh.: Computing with membranrs. J. Comput. Syst. Sci. **61**(1), 108–143 (2000)
2. Paun, Gh., Rozenberg, G., Salomaa. A. (eds.): The Oxford Handbook of Membrane Computing. Oxford University Press, New York (2010)
3. Sedă, M., Březina, T.: Robot motion planning in fight directions, In: Proceedings of the World Congress on Engineering and Computer Science (WCECS 2009), vol. II, pp. 691–695. San Francisco, USA, 20–22 Oct 2009
4. Russell, S., Norvig, P.: Artificial Intelligence: A Modern Approach, 2nd edn. Prentice Hall, Englewood Cliffs (2003)
5. Rozenberg, G., Salomaa, A. (eds.): Handbook of Formal Languages, vol. I–III. Berlin, word, language, grammar (1997)
6. Rosenfeld, A.: Isotonic grammars, parallel grammars, and picture grammars. In: Machine Intelligence VI, pp. 281–294. Elsevier, New York (1971)

Intelligent Control of DFIG Using Sensorless Speed Estimation and Lookup Table-Based MPPT Algorithm to Overcome Wind and Grid Disturbances

D.V.N. Ananth and G.V. Nagesh Kumar

Abstract Doubly fed induction generator (DFIG) needs to get adapted to situations like rapid change in wind speed and sudden requirement for grid disturbances to meet modern grid rules. The paper explains the design of controllers for MPPT algorithm for the turbine, DFIG converters, and sensorless rotor speed estimation. These intelligent controllers are used to maintain equilibrium in rotor speed, generator torque, and stator and rotor voltages. They are also designed to meet desired reference real and reactive power during the turbulences like sudden change in reactive power or voltage with concurrently changing wind speed. The turbine blade angle changes with variations in wind speed and direction of wind flow and improves the coefficient of power extracted from turbine using MPPT algorithm. Rotor side converter (RSC) helps to achieve optimal real and reactive power from generator, which keeps rotor to rotate at optimal speed and to vary current flow from rotor and stator terminals. For tracking reactive power demand from grid and in maintaining synchronism when grid voltage changes due to fault at a very faster and stable way Grid side converter (GSC) is helpful. Rotor speed is estimated using stator and rotor flux estimation algorithm. Parameters like stator and rotor voltage, current, real, reactive power, rotor speed, and electromagnetic torque are studied using MATLAB simulation. The performance of DFIG is compared when there is in wind speed change only; alter in reactive power and variation in grid voltage individually along with variation in wind speed.

Keywords Doubly fed induction generator · MPPT algorithm · Reactive power control · Rotor speed estimation

---

D.V.N. Ananth
VITAM College of Engineering, Visakhapatnam, India

G.V. Nagesh Kumar (✉)
GITAM University, Visakhapatnam, India
e-mail: gundavarapu_kumar@yahoo.com

S. Das et al. (eds.), *Proceedings of the 4th International Conference on Frontiers in Intelligent Computing: Theory and Applications (FICTA) 2015*, Advances in Intelligent Systems and Computing 404, DOI 10.1007/978-81-322-2695-6_61

1 Introduction

Renewable energy resources like wind and solar electric power generation systems are getting significance due to retreating of primary fuels and eco-friendly nature and are available from few kilo-watt powers to megawatt rating and the back-to-back converters are rated from 25 to 35 % of generator stator rating [1]. Capability to extract more or maximum power point tracking theorem (MPPT) [2], The DFIG is getting importance compared to permanent magnet synchronous generator (PMSG) or asynchronous generator because of operation under variable speed conditions [3–5], effective performance during unbalanced and flickering loads [6].

Maximum power extraction from DFIG using pitch angle controller and optimal power coefficient at low and high speed is analyzed in [7]. Direct and indirect control of reactive power control with an aim to meet active and reactive power equal to the reference values as achieved in [8]. MPPT-based WECs design facilitates the wind turbine has to operate in variable speed as per the ideal cube law power curve [9–13]. The constant power mode of operation can be achieved by (i) including energy storage devices [13–17] and (ii) employing pitch control [9, 18, 19].

In this paper, performance of DFIG was compared and analyzed under situations like, with variation in wind speed alone, with reactive power variation and with grid voltage variation for same variation in wind speed. In these cases, variation in tip speed ratio and coefficient of turbine power, effect on real and reactive power flows, voltages and current from stator, rotor, rotor speed and electromagnetic torque are examined. The paper was organized with overview of WECS with wind turbine modeling and pitch angle controller in Sect. 2, study of mathematical modeling of DFIG in Sect. 3, the Sect. 4 describe RSC architecture and design; Sect. 5 analyses the performance of DFIG for three cases like effect of variation on wind speed variation, reactive power demand along with variation in wind speed and grid voltage variation with wind speed. Conclusion was given in Sect. 6. System parameters are given in “Appendix”.

2 Wind Energy Conversion Systems (WECS)

The wind turbine is the prime mover which facilitates in converting kinetic energy of wind into mechanical energy which further converted into electrical energy.

2.1 Pitch Angle Controller

The wind turbine blade angles are controlled by using servo mechanism to maximize turbine output mechanical power during steady state and to protect the turbine during high wind speeds. This control mechanism is known as pitch angle

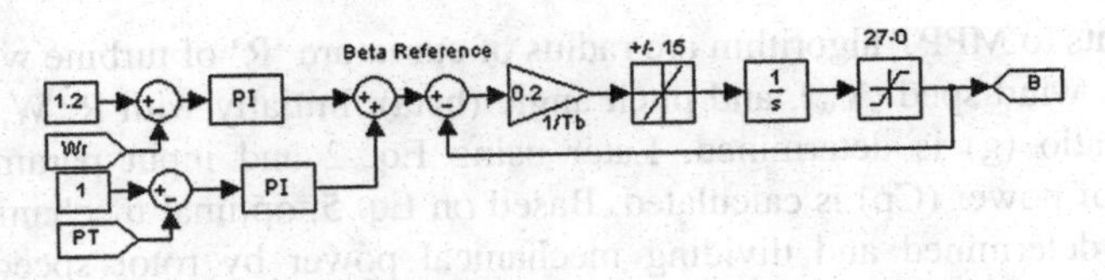

Fig. 1 Pitch angle controller design for wind turbine

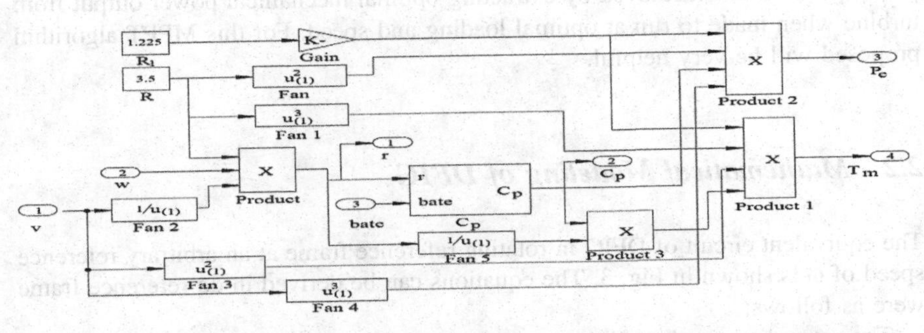

Fig. 2 Reference electrical power generation control circuit and mechanical torque output from turbine with MPPT algorithm

controller. When wind speed is at cut-in speed, the blade pitch angle is set to produce optimal power, at rated wind speed; it is set to produce rated output power from generator. At higher wind speeds, this angle increases and makes the turbine to protect from over-speeding.

In this system reference generator speed is $W_r^{ref} = 1.2$ pu or is obtained from MPPT algorithm and actual speed of the generator is Wr. The actual speed can be estimated using an encoder or using sensorless estimation strategy. The error between reference and actual values is controlled using PI controller. In the similar way, the difference in reference (P* = 1) and actual power outputs from turbine (PT) is controlled by PI controller. Both the outputs from PI controller are designed to get reference pitch angle controller (βref). The closed loop control of pitch angle is obtained as shown in Fig. 1.

The reference real power (Pe) or actual mechanical output from turbine and mechanical output torque (Tm) is shown in Fig. 2. Using Tip Speed Ratio (TSR) and Coefficient of Power (C_p) are used to generate reference power and the control scheme is useful to extract maximum mechanical power, thereby more mechanical tore Tm by using the MPPT algorithm. The aim of MPPT algorithm is to generate optimal mechanical power output from turbine and mechanical input torque to give to doubly fed induction generator. The mechanical power is given as reference to grid side converter (GSC) to make the rotor to rotate at optimal speed. The optimal input torque to DFIG is so as to operate for extracting maximum power from the generator.

The inputs to MPPT algorithm are radius of curvature 'R' of turbine wings, rotor speed (Wr), wind sped (V_w), and pitch angle (beta). Initially with R, W_r, and Vw, tip speed ratio (g) is determined. Later using Eq. 2 and input parameter beta, coefficient of power (Cp) is calculated. Based on Eq. 5, optimal mechanical power (P_{m_opt}) is determined and dividing mechanical power by rotor speed, optimal torque (T_{m_opt}) is determined. The pitch angle (beta) is determined as shown in Fig. 2. The application of P_{m_opt} and T_{m_opt} is shown in Figs. 4 and 5. Optimal real power generation is achieved by extracting optimal mechanical power output from turbine when made to run at optimal loading and speed. For this MPPT algorithm proposed will be very helpful.

2.2 *Mathematical Modeling of DFIG*

The equivalent circuit of DFIG in rotating reference frame at an arbitrary reference speed of ω is shown in Fig. 3. The equations can be derived in dq reference frame were as follows:

The rotor direct and quadrature axis are derives as

$$V_{rd} = R_r I_{rd} - (\omega_s - \omega_r)\psi_{rq} + \frac{d\psi_{rd}}{dt} \tag{1}$$

$$V_{rq} = R_r I_{rq} + (\omega_s - \omega_r)\psi_{rd} + \frac{d\psi_{rq}}{dt} \tag{2}$$

The difference between stator speed (ω_s) and rotor speed (ω_r) is known as slip speed ($s\omega_s$). For motoring action, this difference is less than zero and for generating, the slip speed is negative.

The stator real power in terms of two axis voltage and current is

$$P_s = \frac{3}{2}(V_{sd}I_{sd} + V_{sq}I_{sq}) \tag{3}$$

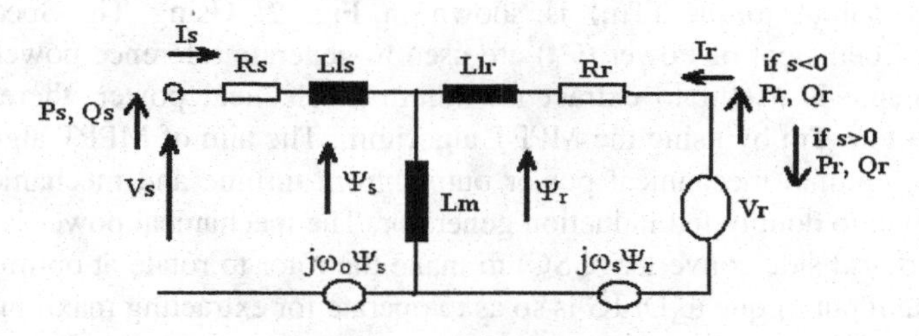

Fig. 3 Equivalent circuit of DFIG in rotating reference frame at speed ω

The rotor real power in terms of two axis voltage and current is

$$P_r = \frac{3}{2}\,(V_{rd}I_{rd} + V_{rq}I_{rq}) \tag{4}$$

The stator reactive power in terms of two axis voltage and current is

$$Q_s = \frac{3}{2}\,(V_{sq}I_{sd} - V_{sd}I_{sq}) \tag{5}$$

The rotor reactive power in terms of two axis voltage and current is

$$Q_r = \frac{3}{2}\,(V_{rq}I_{rd} - V_{rd}I_{rq}) \tag{6}$$

The output electromagnetic torque is given by the equation

$$T_e = \frac{3}{2}pL_m(I_{sq}I_{rd} - I_{sd}I_{rq}) \tag{7}$$

3 Rotor Side Controller (RSC) and Grid Side Controller (GSC) Architecture and Design

3.1 Operation of GSC and RSC Controllers

The rotor side converter (RSC) is used to control the speed of rotor and also helps in maintaining desired grid voltage as demanded. The control circuit for grid side controller (GSC) is shown in Fig. 4 and rotor side controller (RSC) is shown in Fig. 5 with internal circuits for deriving RSC PLL for 2–3 phases inverse Parks transformation is shown in Fig. 6. This Fig. 6 helps to inject current in rotor winding at slip frequency. The GSC and RSC have four control loops each, later has one speed control loop, other is reactive power and last two are direct and quadrature axis current control loops. The speed and reactive power control loops are called outer control loop and direct and quadrature axis control loops are called inner control loops.

The reference rotor speed is derived from the wind turbine optimal power output P_{mOpt} as shown in Fig. 4 and grid power demand. In total, the reference power input to the lookup table as shown in Fig. 5 is $P_{m,gOpt}$. Based on the value of $P_{m,gOpt}$, the rotor is made to rotate at optimal speed so as to extract maximum power from DFIG set. The difference between reference speed of generator and actual speed of generator is said to be rotor speed error. Speed error is minimized and maintained nearly at zero value by using speed controller loop which is a PI controller with K_{pn} and K_{in} as proportional and integral gain parameters. The output from speed controller is

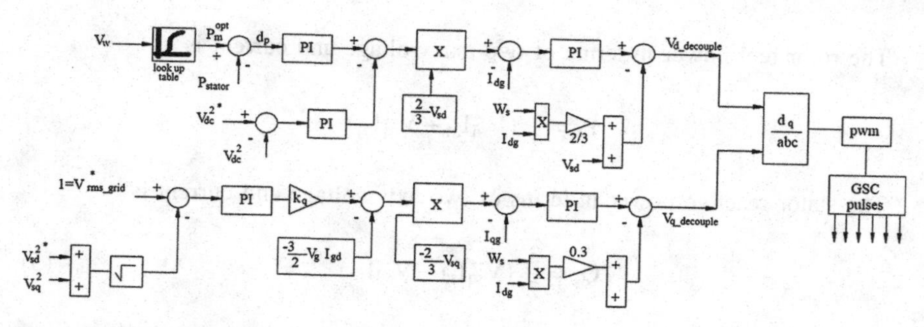

Fig. 4 Grid side controller for DFIG

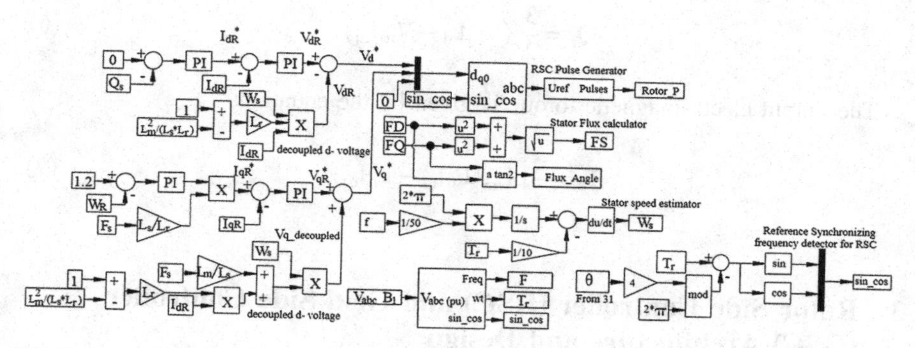

Fig. 5 Rotor side controller for DFIG

Fig. 6 Design of overall DFIG system with RSC and GSC

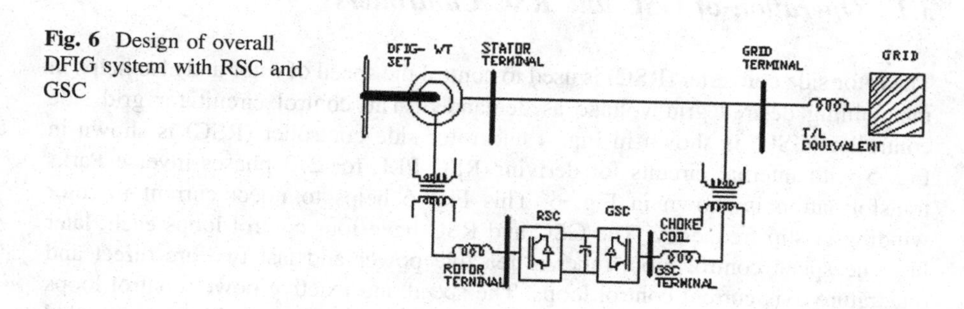

multiplied with stator flux (F_s) and ratio of stator and rotor (L_s and L_r) inductances to get reference quadrature current (I_{qr}) for rotor. The error in reference and actual reactive power give reference direct axis current (I_{qr}). The difference between these reference and actual two axis currents is controlled by tuned PI controller to get respective direct and quadrature axis voltages. The output from each PI controller is manipulated with disturbance voltages to get reference voltage for pulse generation as shown in Fig. 5. It must be noted that the pulses are regulated at slip frequency for RSC rather than at fundamental frequency and slip frequency synchronizing for inverse Park's transformation can also be seen in the figure.

The block diagram of GSC is shown in Fig. 4. For a given wind speed, reference or control power from turbine is estimated using lookup table. From Eq. (3), stator real power (P_{stator}) is calculated and the error in powers is difference between these two powers (dP) which is to be maintained near zero by PI controller. The output from PI controller is multiplied with real power constant (K_p) gives actual controllable power after disturbance. The difference in square of reference voltage across capacitor dc link (V^*_{dc}) and square of actual dc link voltage (V_{dc}) is controlled using PI controller to get reference controllable real power. The error in the reference and actual controllable power is divided by using $2/3V_{sd}$ to get direct axis (d-axis) reference current near grid terminal (Igdref). Difference in Igdref and actual d-axis grid current is controlled by PI controller to get d-axis voltage. But to achieve better response during transient conditions, decoupling d-axis voltage is added as in case of separately excited DC motor. This decoupling term helps in controlling steady state error and fastens transient response of DFIG during low voltage ride through (LVRT) or during sudden changes in real or reactive powers from/to the system.

Similarly from stator RMS voltage (V_s) or reference reactive power, actual stator voltage or reactive power is subtracted by PI controller and multiplied with appropriate reactive power constant (K_q) to get actual reference reactive power compensating parameter. From Eq. (5), actual reactive power is calculated and the difference in this and actual compensating reactive power and when divided by 2/3Vsq, we get quadrature axis (q-axis) reference current (I_{qref}). When the difference in I_{qref} and stator actual q-axis current (I_q) is controlled by PI controller, reference q-axis voltage is obtained. As said earlier, to improve transient response and to control steady state error, decoupled q-axis voltage has to be added as shown in Fig. 4. Both d- and q- axis voltage parameters so obtained are converted to three axis abc parameters by using inverse Park's transformation and reference voltage is given to scalar PWM controller to get pulses for grid side controller. The DFIG grid connected system general layout is shown in Fig. 6.

The rotating direct and quadrature reference voltages of rotor are converted into stationary abc frame parameters by using inverse parks transformation. Slip frequency is used to generate sinusoidal and cosine parameters for inverse parks transformation.

3.2 *Rotor Speed Sensing by Using Sensorless Control Technique*

The sensorless speed control for DFIG system with stator and rotor flux observers are shown in Fig. 7. The internal sub-systems to produce two axis qauadrature and direct axis flux is given in Fig. 8a, b. The three phase stator voltage and currents are converted into two phase dq voltages and current by using Park's transformation. The dq axis stator voltage and current are transformed into dq axis stator flux based

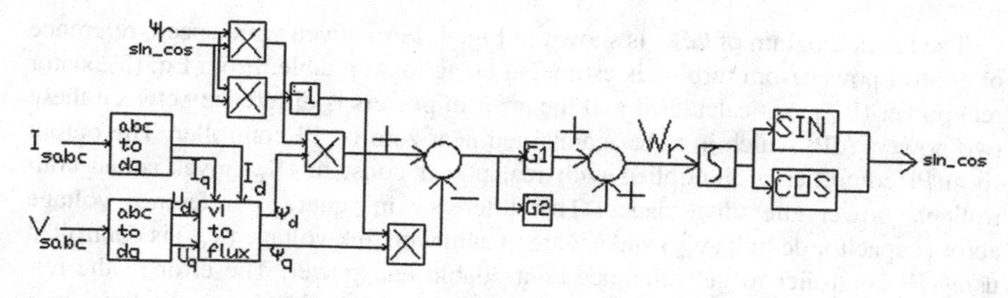

Fig. 7 Estimation of rotor speed with stator voltage and current and rotor flux as inputs

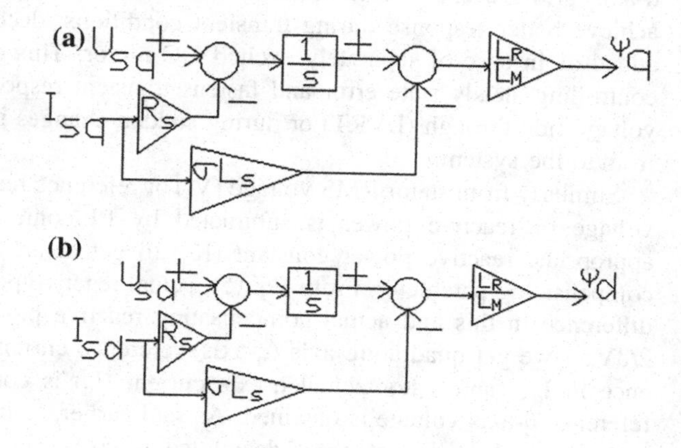

Fig. 8 **a** Derivation of stator q-axis flux from q-axis stator voltage and current Eq. 2. **b** Derivation of stator d-axis flux from d-axis stator voltage and current from Eq. 1

on Eqs. 1 and 2. The internal structure for dq axis flux derivations are shown in Fig. 8a, b.

The derived rotor and stator flux are compared and is controlled to estimate rotor speed by using PI controller. The blocks G1 and G2 are PI controller functional blocks. The speed is estimated and is termed as W_r and is integrated to get rotor angle. The angle is multiplied with trigonometric SIN and COS terms and is given to mux to get sin_cos parameters and the total setup can be used as phase locked loop (PLL). This sin_cos helps in estimating exact phase sequence and for locking the new system to reference grid. The estimated speed W_r is given as input for RSC controller as shown in Fig. 5. From the lookup table, reference rotor speed is estimated from optimal power block, which is obtained from MPPT algorithm.

4 Results and Discussion

The dynamic performance of the DFIG system as shown in Fig. 6 is investigated with three different cases and rating specifications for DFIG and wind turbine parameters are given in "Appendix". The wind speed change in all cases in meters per seconds as 8, 15, 20, and 10 at 15, 25, and 35 s. The reactive power and voltage value change in individual two cases with change in time is from −0.6 pu at 12 s to 0 pu change at 20 s. It was further changed from 0 to +0.6 pu magnitude at 30 s. DFIG will become better generator source if immediately it can supply any desired reactive power effectively. The change in grid terminal voltage takes place when suddenly switching on or off large loads or due to small faults near point of common coupling (PCC). The effect of change in wind speed, change in wind speed with reactive power, and change in wind speed with grid terminal voltages on generator and turbine parameters are studied.

Case A: Change in TSR and C_p with wind speed, reactive power, and grid voltage

The changes in tip-speed ration and power coefficient C_p with change in wind speed alone is shown in Fig. 9(i), with both reactive power and wind speed variation in Fig. 9 (ii) and variation with grid terminal voltage and wind speed both is shown in Fig. 9(iii). It can be observed that when wind speed is at 8 m/s, tip speed ratio (TSR) is high near 4.8° and slowly decreases to 2.6° at 15 s when speed increases to 15 m/s, further increased to 1.9° at 25 s when speed of wind is 20 m/s and decreased to 3.9° when wind speed decreased to 10 m/s at 35 s. In the similar way, C_p is also changing from 3.25 to 1.7 at 15 s, and further decreased to 1.25 at 25 s, and then increased to 2.55 at 35 s with wind speed variation from 8 to 15 and then to 20, and 10 m/s. The variation in TSR and C_p with change in reactive power is independent and has no effect as shown in Fig. 9(i), (ii). However, with change in grid terminal voltage, a very small change in TSR and C_p can be observed. It is shown in Fig. 9(iii). It is due to the fact that the TSR and C_p depends on parameters as described by Eqs. 1–5 and is independent on voltage and reactive power. The TSR and C_p are blade size and shape with change in ambient temperature and wind speed dependant natural parameters.

To meet the desired grid reactive power, both stator and rotor has to supply for faster dynamics with an aid to RSC and GSC control schemes and is achieving as shown in Fig.9(ii). With the change in wind speed and reactive power, real power from generator is matching its reference value for case 2, but small deviation can be observed from time 30 to 35 s is due to sudden change in reactive power demand from grid and the deviation in real power is from 0.8 to 0.7 pu which is small. However, reactive power is following its trajectory within 1 s. In case 3, both voltage and reactive power changing with time, the deviation in real power and is following the trajectory nearly accurate with maximum deviation of 5 % in real power. The reactive power change with grid voltage is high when voltage decreased from 1 to 0.8 pu volts. When wind speed is increasing, mechanical and electrical torques are increasing without any change in stator reactive power.

Variation in grid reactive power causes quadrature currents on both stator and rotor to change but torque, speed or real powers from stator or rotor remains unaltered. The change in three cases is tabulated below. The variation in reactive power and grid voltage variations during the respective time period is shown in Tables 2 and 3. In Table 1, due to change in wind speed input to turbine alone, generator and wind turbine parameters change are summarized.

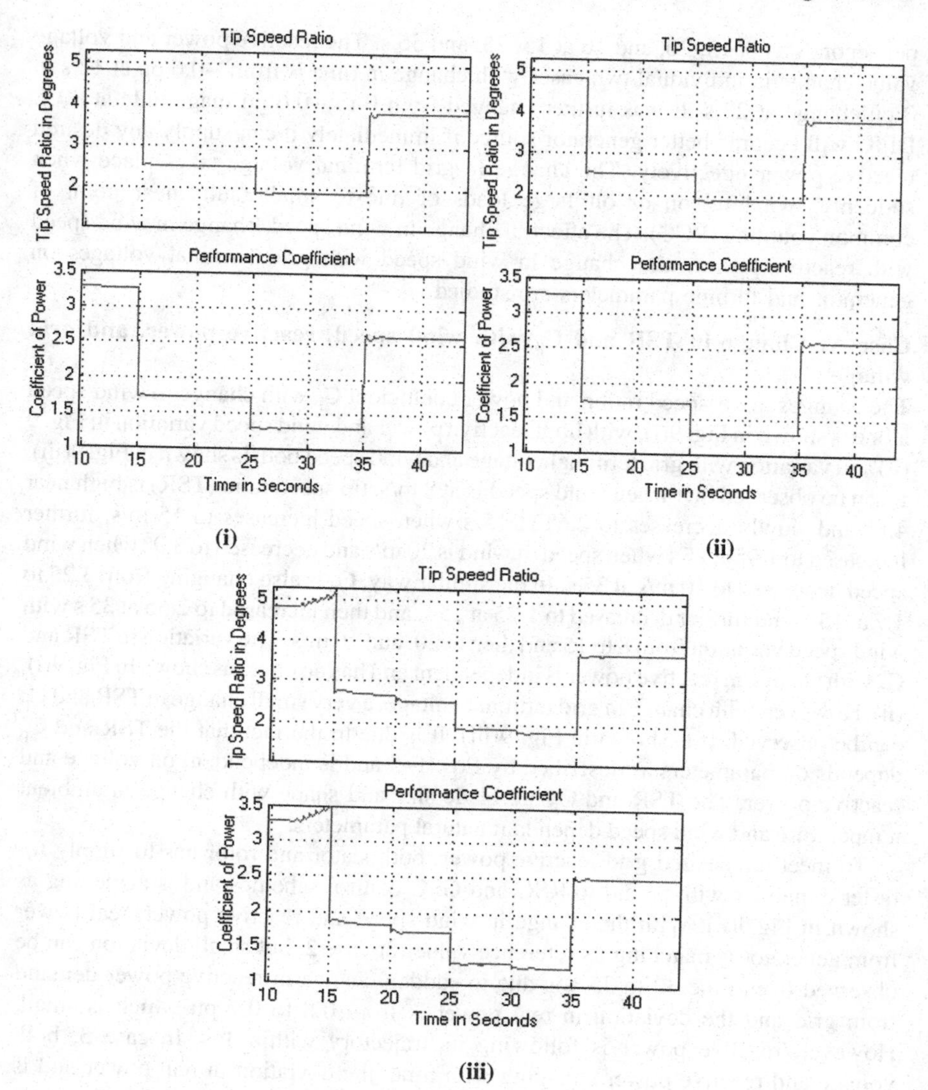

Fig. 9 Tip speed ratio and coefficient of power C_p for **i** change in wind speed alone, **ii** reactive power change and wind speed variation, **iii** both grid voltage and wind speed changes

Case B: Change in electromagnetic torque and rotor speed with wind speed, reactive power, and grid voltage

The reference mechanical turbine torque and generator torque with magnitudes overlapping and variation of rotor speed for all three cases comparison is shown in Fig. 10. In this the reference and actual torque waveform with blue color is turbine reference torque and pink color lines are for generator torque. It can be observed in

Table 1 Change in reactive power during the time period along with change in wind speed

Time range (s)	0–20	20–30	30–50
Reactive power (pu)	−0.6	0	0.6

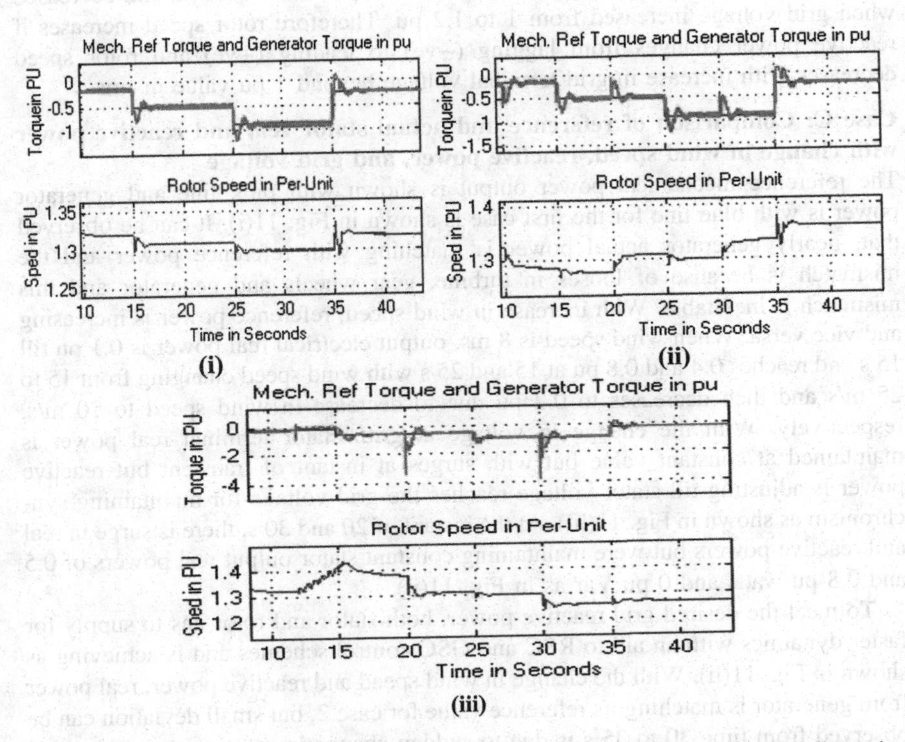

Fig. 10 Reference and actual generator torque and rotor for **i** change in wind speed, **ii** reactive power and wind speed variation, **iii** both grid voltage and wind speed changes

Fig. 10(i), (ii), and (iii), with increase in wind speed, torque is increasing and vice versa. Till time up to 15 s, wind speed is at low value of 8 m/s, so torque is at −0.2 pu and increased to −0.5 pu at 15 s with increase in wind speed to 15 m/s. The torque further increased to −0.9 pu when wind speed is 20 m/s and decreased to −0.28 pu when speed decreased to 10 m/s. There are small surges in torque waveform because of sudden change in wind speed.

In the first case, reactive power was at 0 pu and grid terminal voltage is 1 pu. The changes in torque have effect with change in reactive power as in Fig. 10(ii) and further more surges been observed when grid voltage disturbance occurred as in Fig. 10(iii) is taking place. When reactive power is lagging at −0.6 pu, there is a small surge in torque at 20 s. Generator speed is also low at 1.27 pu at −0.6 pu reactive power, while at 0 pu reactive power, it is 1.32 pu speed. But rotor speed increased to 1.4 pu speed at low terminal grid voltage of 0.8 pu. When, reactive power changes to 0 pu from −0.6 pu, rotor speed increased to 1.3 pu from 1.27 pu and grid terminal voltage changes to 1 pu from 0.8 pu between 20 to 30 s. Speed further increased to 1.35 pu with leading reactive power of +0.6 pu and decreased when grid voltage increased from 1 to 1.2 pu. Therefore rotor speed increases if reactive power changes from lagging (−ve) to leading (+ve) and rotor speed decreases with increase in grid terminal voltage beyond 1 pu value in rms.

Case C: Comparison of reference and actual stator real and reactive power with change in wind speed, reactive power, and grid voltage
The reference mechanical power output is shown with pink line and generator power is with blue line for the first case is shown in Fig. 11(i). It can be observed that, nearly generator actual power is matching with reference power and the mismatch is because of looses in turbine, gear wheels and generator and this mismatch is inevitable. With increase in wind speed, reference power is increasing and vice versa. When wind speed is 8 ms, output electrical real power is 0.1 pu till 15 s and reaches 0.4 and 0.8 pu at 15 and 25 s with wind speed changing from 15 to 25 m/s and then decreases to 0.2 pu due to decrease in wind speed to 10 m/s, respectively. With the change in voltage at grid, stator terminal real power is maintained at constant value but with surges at instant of transient but reactive power is adjusting till stator voltage reaches the grid voltage for maintaining synchronism as shown in Fig. 11(iii). At the instant of 20 and 30 s, there is surge in real and reactive powers but were maintaining constant stator output real powers of 0.5 and 0.8 pu watts and 0 pu Var as in Fig. 11(ii).

To meet the desired grid reactive power, both stator and rotor has to supply for faster dynamics with an aid to RSC and GSC control schemes and is achieving as shown in Fig. 11(ii). With the change in wind speed and reactive power, real power from generator is matching its reference value for case 2, but small deviation can be observed from time 30 to 35 s is due to sudden change in reactive power demand from grid and the deviation in real power is from 0.8 to 0.7 pu which is small. However, reactive power is following its trajectory within 1 s. In the case 3, both voltage and reactive power changing with time, the deviation in real power and is following the trajectory nearly accurate with maximum deviation of 5 % in real power. The reactive power change with grid voltage is high when voltage decreased from 1 to 0.8 pu volts. When wind speed is increasing, mechanical and electrical torques are increasing without any change in stator reactive power.

Variation in grid reactive power causes quadrature currents on both stator and rotor to change but torque, speed or real powers from stator or rotor remains unaltered. The change in three cases is tabulated below. The variation in reactive power and grid voltage variations during the respective time period is shown in Tables 2 and 3. In Table 1, due to change in wind speed input to turbine alone, generator and wind turbine parameters change are summarized. There are surges

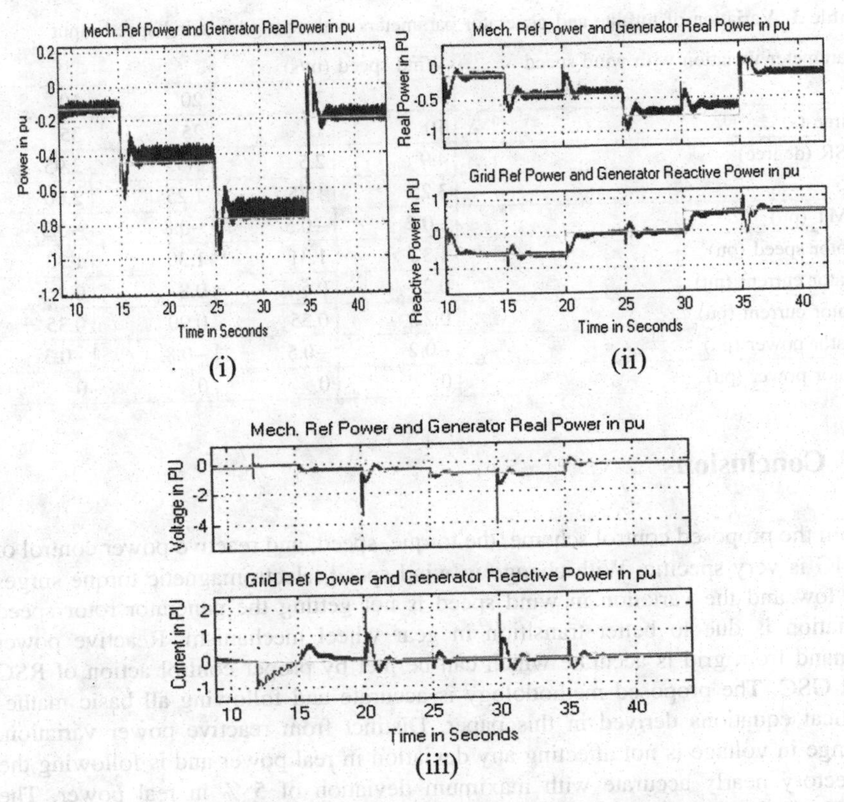

Fig. 11 Generator reference and actual real power waveform with time for **i** change in wind speed alone, generator real and reactive power flows for actual and reference for change in **ii** reactive power change and wind speed variation, **iii** both grid voltage and wind speed changes

Table 2 Change in grid voltage during the time period along with change in wind speed

Time range (s)	0–12	12–20	20–30	30–50
Reactive power (pu)	1	0.8	1	1.2

produced in the electromagnetic torque (EMT) due to variations in reactive power and grid voltage. Large spikes in stator current and rotor current are produced due to sudden increase or decrease in grid voltage. Certain deviations in rotor speed can be observed due to change in reactive power or grid voltage. It is due to variation in current flow in the rotor circuit, thereby variation in rotor flux and hence rotor speed.

Table 3 Variation of turbine and generator parameters with change in wind speed input

Parameter variation with wind speed	Wind speed (m/s)			
	8	15	20	10
Time (s)	10	15	25	35
TSR (degree)	4.9	2.5	1.95	3.95
C_p	3.25	1.75	1.25	2.60
EMT (pu)	−0.2	−0.5	−0.8	−0.3
Rotor speed (pu)	1.32	1.31	1.30	1.33
Stator current (pu)	0.2	0.5	0.8	0.3
Rotor current (pu)	0.25	0.55	0.90	0.35
Stator power (pu)	−0.2	−0.5	−0.8	−0.3
Rotor power (pu)	0	0	0	0

5 Conclusions

From the proposed control scheme, the torque, speed, and reactive power control of DFIG is very specific. With change in wind speed, electromagnetic torque surges are low and the variation in wind speed is not getting the generator rotor speed variation is due to better transition in gear wheel mechanism. Reactive power demand from grid is accurate which can be met by proper control action of RSC and GSC. The proposed methodology is accurate and following all basic mathematical equations derived in this paper. Distinct from reactive power variation, change in voltage is not affecting any deviation in real power and is following the trajectory nearly accurate with maximum deviation of 5 % in real power. The reactive power change with grid voltage is high when voltage decreased from 1 to 0.8 pu volts. Hence the proposed control scheme can be applied with ever changing transients like large variation in wind speed, reactive power and grid voltage. The system is very stable without losing synchronism when grid voltage is increasing or decreasing to a ±0.2 pu change from nominal voltage value.

Appendix

The parameters of DFIG used in simulation are: Rated Power = 1.5 MW, Rated Voltage = 690 V, Stator Resistance Rs = 0.0049 pu, rotor Resistance Rr = 0.0049 pu, Stator Leakage Inductance Lls = 0.093 pu, Rotor Leakage inductance Llr1 = 0.1 pu, Inertia constant = 4.54 pu, Number of poles = 4, Mutual Inductance Lm = 3.39 pu, DC link Voltage = 415 V, Dc link capacitance = 0.2 F, Wind speed = 14 m/s. Grid Voltage = 25 kV, Grid frequency = 60 Hz. Grid side Filter: Rfg = 0.3 Ω, Lfg = 0.6 nH. Rotor side filter: Rfr = 0.3 mΩ, Lfr = 0.6 nH. Wind speed variations: 8, 15, 20 and 10 at 15, 25 and 35 s. Reactive power change: −0.6 to 0 and +0.6 pu at 20 and 30 s. Grid voltage change: 0.8–1 and to 1.2 pu at 20 and 30 s.

References

1. Aghanoori, N., Mohseni, M., Masoum, M.A.S.: Fuzzy approach for reactive power control of DFIG-based wind turbines. In: IEEE PES Innovative Smart Grid Technologies Asia (ISGT), pp: 1–6 (2011)
2. Kazmi, S.M., Goto, H., Guo, H.-J., Ichinokura, O.: A novel algorithm for fast and efficient speed-sensor less maximum power point tracking in wind energy conversion systems. IEEE Trans. Ind. Electron. **58**(1), 29–36 (2011)
3. Iwanski, G., Koczara, W.: DFIG-based power generation system with UPS function for variable-speed applications. IEEE Trans. Ind. Electron. **55**(8), 3047–3054 (2008)
4. Aktarujjaman, M., Haque, M.E., Muttaqi, K.M., Negnevitsky, M., Ledwich, G.: Control dynamics of a doubly fed induction generator under sub- and super-synchronous modes of operation. In: IEEE Power and Energy Society General Meeting—Conversion, and Delivery of Electrical Energy in the 21st Century, pp. 1–9 (2008)
5. Iwanski, G., Koczara, W.: DFIG-based power generation system with UPS function for variable-speed applications. IEEE Trans. Ind. Electron. **55**(8), 3047–3054 (2008)
6. Shukla, R.D., Tripathi, R.K.: Low voltage ride through (LVRT) ability of DFIG based wind energy conversion system II. In: Students Conference on Engineering and Systems (SCES), pp. 1–6 (2012)
7. Díaz, G.: Optimal primary reserve in DFIGs for frequency support. Int. Electr. Power Energy Syst. **43**, 1193–1195 (2012)
8. Poitiers, F., Bouaouiche, T., Machmoum, M.: Advanced control of a doubly-fed induction generator for wind energy conversion. Electr. Power Syst. Res. **79**, 1085–1096 (2009)
9. Cardenas, R., Pena, R., Perez, M., Clare, J., Asher, G., Wheeler, P.: Power smoothing using a flywheel driven by a switched reluctance machine. IEEE Trans. Ind. Electron. **53**(4), 1086–1093 (2006)
10. Muljadi, E., Butterfield, C.P.: Pitch-controlled variable-speed wind turbine generation. IEEE Trans. Ind. Appl. **37**(1), 240–246 (2001)
11. Wei, Q., Wei, Z., Aller, J.M., Harley, R.G.: Wind speed estimation based sensorless output maximization control for a wind turbine driving a DFIG. IEEE Trans. Power Electron. **23**(3), 1156–1169 (2008)
12. Sharma, S., Singh, B.: Control of permanent magnet synchronous generator-based stand-alone wind energy conversion system. IET Power Electron. **5**(8), 1519–1526 (2012)
13. Kazmi, S.M.R., Goto, H., Hai-Jiao, G., Ichinokura, O.: A novel algorithm for fast and efficient speed-sensorless maximum power point tracking in wind energy conversion systems. IEEE Trans. Ind. Electron. **58**(1), 29–36 (2011)
14. Mathiesen, B.V., Lund, H.: Comparative analyses of seven technologies to facilitate the integration of fluctuating renewable energy sources. IET Renew. Power Gener. **3**(2), 190–204 (2009)
15. Takahashi, R., Kinoshita, H., Murata, T., et al.: Output power smoothing and hydrogen production by using variable speed wind generators'. IEEE Trans. Ind. Electron. **57**(2), 485–493 (2010)
16. Bragard, M., Soltau, N., Thomas, S., De Doncker, R.W.: The balance of renewable sources and user demands in grids: power electronics for modular battery energy storage systems. IEEE Trans. Power Electron. **25**(12), 3049–3056 (2010)
17. Bhuiyan, F.A., Yazdani, A.: Reliability assessment of a wind-power system with integrated energy storage. IET Renew. Power Gener. **4**(3), 211–220 (2010)
18. Sen, P.C., Ma, K.H.J.: Constant torque operation of induction motor using chopper in rotor circuit. IEEE Trans. Ind. Appl. **14**(5), 1226–1229 (1978)
19. Geng, H., Yang, G.: Robust pitch controller for output power leveling of variable-speed variable-pitch wind turbine generator systems. IET Renew. Power Gener. **3**(2), 168–179 (2009)

Erratum to: Query-Based Extractive Text Summarization for Sanskrit

Siddhi Barve, Shaba Desai and Razia Sardinha

Erratum to: S. Das et al. (eds.), *Proceedings of the 4th International Conference on Frontiers in Intelligent Computing: Theory and Applications (FICTA) 2015*, Advances in Intelligent Systems and Computing 404, DOI 10.1007/978-81-322-2695-6_47

Erratum DOI: 10.1007/978-81-322-2695-6_62

In original version of this chapter Tables 1–4 consist of scrambled texts that have been corrected now.

2. The preprocessing steps are performed on the selected document. The steps include sentence length feature, tokenization, stop word removal, morphological analysis, and sandhi and compound resolution.
   Below is an illustration of morphological analysis and sandhi/compound resolution for the document "कदली."

The online version of the original chapter can be found under
10.1007/978-81-322-2695-6_47

S. Barve (✉) · S. Desai · R. Sardinha
Department of Information Technology, Padre Conceicao College of Engineering,
Goa University, Goa, India
e-mail: barvesiddhi@gmail.com

S. Desai
e-mail: shaba.desai@gmail.com

R. Sardinha
e-mail: razia.sardinha@gmail.com

S. Das et al. (eds.), *Proceedings of the 4th International Conference on Frontiers in Intelligent Computing: Theory and Applications (FICTA) 2015*, Advances in Intelligent Systems and Computing 404, DOI 10.1007/978-81-322-2695-6_62

- Morphological Analysis
  Analysis for the word “आम्रं” follows:
  आम्रं = आम्र{पुं}{2;एक}
  आम्रं{नपुं}{1;एक}
  आम्रं{नपुं}{2;एक}
- Sandhi/Compound Resolution
  Suppose if the encountered word, “रक्तवर्णीयम्” is compound, then the split can be obtained as रक्तवर्णीयम्= रक्त-वर्णि-इयम्.
  The word obtained could also be a Sandhi; if it is Sandhi, then the splits are obtained as चेदपि = चेत्+अपि.
  After preprocessing is carried out, the user is prompted to enter the query corresponding to the document. The query is “वृक्षाः कदली शलाटुः फलं.”

Table 1 Raw Sanskrit document

आम्रं पनसं, कदलीफलं च प्रमुखफलत्रयत्वेन परिगण्यन्ते । एवम् एव कदली-आम्र-उदुम्बर तिन्त्रिणीवृक्षाः प्रमुखाः वृक्षाः इति उच्यन्ते ।एतान् अधिकृत्य काचित विचित्रा कथा केषुचित प्रदेशेषुश् श्रूयते ।एते पञ्च वृक्षाः अपि कस्मिंश्चित् काले मनुष्यरुपेण सहोदर्यः आसन् ।बहुकालं यावत् तासां विवाहः न जातः ।कदाचित् देवः प्रत्यक्षीभूय ‘विवाहम् इच्छन्ति वा ?’ इति ताः पृष्टवान् ।तदा चतस्रः सहोदर्यः विवाहम् अङ्घीकृतवत्यः । किन्तु अन्तिमा सहोदरी न अङ्गीकृतवती । सा सन्तानमात्रं प्रार्थितवती । अनन्तरं देवः ताः सर्वाः वृक्षरुपेण परिवर्त्य - "यः एतान् वृक्षान् आरोहति सः एव पतिः ’ इति उक्तवान् । अतः एव् विवाहम् अङ्गीकृतवतः वृक्षान सर्वे आरोढुं शक्नुअन्ति चेदपि कदलीवृक्षं कोऽपि आरोढुं न शक्नोति । कदली इति संस्कृतभाषया, हिन्दिभाषया केला इति, आंगलभाषया बनाना प्लाण्टन्, आडम्स् आपिल इत्यपि उच्यते । काभिश्चित् भारतीयभाषाभिः कदलीइत्येव उच्यते ।तमिळु मलयाळभाषया च वाळ्:ऐ इति, तेलुगुभाषया अरटि इति च उच्यते । बौद्धाः कदलीवृक्षं पवित्रं मन्यन्ते । बौद्धसाहित्ये उक्तं मोचापानं कदलीफलेन एव सज्जीक्रियते । अस्माकं पुराणे श्रूयते यत् -हनुमान् हिमालयप्रान्ते कदलीवने आसीत् इति । कदलीवृक्षस्य पुष्पं, शलाटुः, फलं च आहररुपेण उपयुज्यते । कदलीवृक्षस्य शलाटुः हरितवर्णीयः भवति। तस्य फलं हरितं ,पीतं, पीतहरितमिश्रितं, रक्तवर्णीयं वा भाति । तदा आकारेण लघु बृहत् चापि भवति । कदल्यां ५,००० प्रभेदाः सन्ति इति श्रूयते । पच्चेकदली, बूदिकदली, वृक्षकदली, रसकदली, एलाकदली, अरण्यकदली इति अष्ट, दश वा विधाः प्रसिद्धाः सन्ति । कदलीपर्णं बृहदाकारकं पञ्चषपादमितदीर्धम्, अधिकविशालं च भवति । दक्षिणाभारते केषुचित स्थलेषु भोजनार्यम् एतस्य उपयोगः क्रियते । जनाः विवाहादिशुभकार्येषु वितानं मण्डपं द्वारं च पुष्पगुच्छसहितेन कदलीवृक्षेण अलङ्कुर्वन्ति । एतं शुभसङ्केतं मन्येन्ते जनाः । पक्वानि कदलीफलानि कदलीपुष्पम् यदा शलाटुः पक्वं भवति तदा वृक्षं कर्तयन्ति । कर्तितवृक्षस्य प्रकाण्डं परितः स्थितेभ्यः कन्देभ्यः नूतनसस्यानि उत्पद्यन्ते । एवं किदलीसन्तानः वर्धते ।

Table 2 Summary generated by average tf.isf approach

आम्रं पनसं, कदलीफलं च प्रमुख फल त्रयत्वेन परिगण्यन्ते । एवम् एव कदली-आम्र-उदुम्बर तिन्त्रिणीवृक्षाः प्रमुखाः वृक्षाः इति उच्यन्ते । अतः एव् विवाहम् अङ्गीकृतवतः वृक्षान सर्वे आरोढुं श क्नुअन्ति चेत् अपि कदलीवृक्षं कः अपि आरोढुं न शक्नोति । कदली इति संस्कृतभाषया, हिन्दिभाषया केला इति, आंगलभाषया बनाना प्लाण्टन्, आडम्स् आपिल इति अपि उच्यते । अस्माकं पुराणे श्रूयते यत् हनुमान् हिमालयप्रान्ते कदलीवने आसीत् इति । प-च्चेकदली, बूदिकदली, वृक्ष-कदली, रस-कदली, एला-कदली, अरण्य-कदली इति अष्ट, दश वा विधाः प्रसिद्धाः सन्ति ।

Table 3 Summary generated by graph-based approach

कदलीवृक्षस्य शलाटुः हरितवर्णीयः भवति | कदलीपर्णं बृहदाकारकं पञ्चषप अदमितदीर्धम्, अधिक विशालम् च भवति| बौद्धाः कदली-वृक्षं पवित्रं मन्यन्ते | कदलीवृक्षस्य पुष्पं, शलाटुः, फलं च आहररुपेण उपयुज्यते | काभिः चित् भारतीयभाषाभिः कदली इति एव उच्यते | पक्वानि कदलीफलानि कदली-पुष्पम् यदा शलाटुः पक्वं भवति तदा वृक्षं कर्तयन्ति |

Table 4 Summary generated by VSM approach

कदली इति संस्कृतभाषया, हिन्दिभाषया केला इति, आंगलभाषया बनाना प्लाण्टन्, आडम्स् आपिल इति अपि उच्यते| काभिःचित् भारतीयभाषाभिः कदली इति एव उच्यते | कदलीवृक्षस्य पुष्पं, शलाटुः, फलं च आहररुपेण उपयुज्यते | पक्वानि कदलीफलानि कदली-पुष्पम् यदा शलाटुः पक्वं भवति तदा वृक्षं कर्तयन्ति | एवम् एव कदली-आम्र-उदुम्बर तिन्त्रिणीवृक्षाः प्रमुखाः वृक्षाः इति उच्यन्ते | कदलीवृक्षस्य शलाटुः हरितवर्णीयः भवति |

Author Index

S. Das et al. (eds.), *Proceedings of the 4th International Conference on Frontiers in Intelligent Computing: Theory and Applications (FICTA) 2015*, Advances in Intelligent Systems and Computing 404, DOI 10.1007/978-81-322-2695-6

Printed in the United States
By Bookmasters